Feature Papers in Electronic Materials Section

Feature Papers in Electronic Materials Section

Editor

Fabrizio Roccaforte

MDPI • Basel • Beijing • Wuhan • Barcelona • Belgrade • Manchester • Tokyo • Cluj • Tianjin

Editor
Fabrizio Roccaforte
Istituto per la Microelettronica e
Microsistemi (IMM)
Consiglio Nazionale delle
Ricerche (CNR)
Catania
Italy

Editorial Office
MDPI
St. Alban-Anlage 66
4052 Basel, Switzerland

This is a reprint of articles from the Special Issue published online in the open access journal *Materials* (ISSN 1996-1944) (available at: www.mdpi.com/journal/materials/special_issues/feature_electronic_materials).

For citation purposes, cite each article independently as indicated on the article page online and as indicated below:

LastName, A.A.; LastName, B.B.; LastName, C.C. Article Title. *Journal Name* **Year**, *Volume Number*, Page Range.

ISBN 978-3-0365-3227-1 (Hbk)
ISBN 978-3-0365-3226-4 (PDF)

Contents

About the Editor

Fabrizio Roccaforte

He received the M.Sc. Degree in Physics from the University of Catania (Italy) in 1996, and the Ph.D. from the University of Göttingen (Germany) in 1999. Then, he was visiting scientist at the University of Göttingen, and scientific consultant at STMicroelectronics (Italy). In December 2001, he joined the permanent staff of CNR-IMM in Catania as a Researcher, and he became Senior Researcher in 2007 and Research Director in 2020. At CNR-IMM is team leader of the "Power- and High-frequency devices group".

His research interests are mainly focused in the field of wide band gap (WBG) semiconductors, (e.g., SiC, GaN, Ga_2O_3) materials and devices processing for power electronics devices. In particular, he has a recognized experience on metal/semiconductor and dielectric/semiconductor interfaces on WBG semiconductors.

He is co-author of about 350 papers in international journals and conference proceedings (scopus h-index=43), several review articles, 8 book chapters, 5 patents, and he has given several invited talks and lectures on SiC and GaN at international conferences.

He has been the chairperson of the conferences HeteroSiC-WASMPE2009, WOCSDICE2011, and ICSCRM2015. Currently, he is member of the Steering Committee of the conferences ECSCRM and EXMATEC and Section Editor-in-Chief of the journal *Materials* (MDPI).

He is or has been responsible for the CNR-IMM research unit in several European and National projects, bilateral collaborations with other European institutions, and industrial research contracts.

Preface to "Feature Papers in Electronic Materials Section"

Today, electronic materials are widespread in our society. In fact, cellular phones and wireless communication systems, computers, energy conversion devices, power modules, solid-state lighting devices, and sensors and detectors, are typically fabricated using a variety of electronic materials (e.g., semiconductors, metals, insulators, etc.) with specific functionalities. Hence, the deep comprehension of the fundamental properties of these materials, the development of their processing technology, and their integration in real electronic devices require continuous efforts from the scientific community.

This book entitled *Feature Papers in Electronic Materials Section*is a collection of selected regular and review papers recently published on the journal Materials, focusing on the latest advances in electronic materials and devices in different fields (e.g., power- and high-frequency electronics, optoelectronic devices, detectors, etc.).

In particular, in the first part of the volume, many papers are dedicated to wide band gap semiconductors (WBG), a class of materials that are pervasively entering the power- and high-frequency electronic devices market, replacing the traditional silicon devices. Among them, silicon carbide (SiC) and gallium nitride (GaN) are the most advanced ones in terms of crystalline quality and technology maturity. In fact, a variety of devices, e.g., Schottky diodes, MOSFETs, HEMTs, MISHEMT, etc., are already commercialized worldwide. In this context, the key technological issues related to contacts, doping, and dielectrics for SiC and GaN devices, together with important reliability issues associated to GaN-on-Si devices, are specifically addressed in some papers. Moreover, the radiation hardness of SiC, useful for detectors applications, and the perspectives of cubic silicon carbide (3C-SiC) materials and devices technology are also discussed. Finally, the current status of some emerging, but still less mature, WBG materials, such as gallium oxide (Ga_2O_3) and diamond, is also presented. On the other hand, the second part of the volume is a miscellaneous of other electronics materials for various applications, e.g., going from gallium arsenide (GaAs) nanomembranes for HEMT devices, to iron selenide (FeSe) for high-temperature superconductivity and silica for optical communications, up to the popular two-dimensional (2D) materials, such as graphene and molybdenum disulfide (MoS_2), which can be applied in both optoelectronic and high-frequency devices. Furthermore, some recent advances in materials and flexible sensors for bioelectronics and medical applications are also reviewed at the end of the book.

Fabrizio Roccaforte
Editor

materials

MDPI

Review

Materials and Processes for Schottky Contacts on Silicon Carbide

Marilena Vivona *, Filippo Giannazzo and Fabrizio Roccaforte

Consiglio Nazionale delle Ricerche, Istituto per la Microelettronica e Microsistemi (CNR-IMM), Strada VIII, n. 5—Zona Industriale, I-95121 Catania, Italy; filippo.giannazzo@imm.cnr.it (F.G.); fabrizio.roccaforte@imm.cnr.it (F.R.)

* Correspondence: marilena.vivona@imm.cnr.it; Tel.: +39-095-5968-243

Abstract: Silicon carbide (4H-SiC) Schottky diodes have reached a mature level of technology and are today essential elements in many applications of power electronics. In this context, the study of Schottky barriers on 4H-SiC is of primary importance, since a deeper understanding of the metal/4H-SiC interface is the prerequisite to improving the electrical properties of these devices. To this aim, over the last three decades, many efforts have been devoted to developing the technology for 4H-SiC-based Schottky diodes. In this review paper, after a brief introduction to the fundamental properties and electrical characterization of metal/4H-SiC Schottky barriers, an overview of the best-established materials and processing for the fabrication of Schottky contacts to 4H-SiC is given. Afterwards, besides the consolidated approaches, a variety of nonconventional methods proposed in literature to control the Schottky barrier properties for specific applications is presented. Besides the possibility of gaining insight into the physical characteristics of the Schottky contact, this subject is of particular interest for the device makers, in order to develop a new class of Schottky diodes with superior characteristics.

Keywords: silicon carbide; 4H-SiC; Schottky barrier; Schottky diodes; electrical characterization

check for updates

Citation: Vivona, M.; Giannazzo, F.; Roccaforte, F. Materials and Processes for Schottky Contacts on Silicon Carbide. *Materials* **2022**, *15*, 298. https://doi.org/10.3390/ma15010298

Academic Editor: Alexander A. Lebedev

Received: 12 November 2021
Accepted: 22 December 2021
Published: 31 December 2021

1. Introduction

Nowadays, the wide bandgap semiconductors SiC and GaN are considered as the basis of a huge advancement in power electronics, enabling the definition of a game-changing generation of devices with superior performance if compared with that currently achieved by traditional Si-based devices [1]. This is due to the outstanding physical properties of this class of materials, such as wide bandgap, high critical electrical field and high saturation velocity, that push forward the limits reached by Si-based power electronics [2]. In addition to a more efficient performance, the superior properties of wide-band gap semiconductors also translate into devices able to operate in high temperature, high power and high frequency regimes, with the possibility of extending the field of applications for power electronics [3].

Among the wide bandgap semiconductors, one of the hexagonal polytypes of silicon carbide, i.e., 4H-SiC, plays a pivotal role in power electronics, owing to the excellent quality achieved from the commercially available substrates and epilayers and the high level of compatibility with the existing Si-based device manufacturing facilities and technology. Quantitatively, 4H-SiC features a wide bandgap of 3.26 eV, high critical electric field >2 MV/cm, high thermal conductivity of 4.9 $WK^{-1}cm^{-1}$ and saturated drift velocity higher than 2×10^7 cm s^{-1} [4], making it the material of choice for a variety of power devices operating in the medium/high voltage range (600–3000 V) [3].

Along the lines of Si-based technology, several 4H-SiC power devices have been developed, with a mature technology level in terms of processing implementation and real-world applications [5]. Among them, n-type 4H-SiC-based Schottky barrier diodes (SBDs) are currently commercialized on a large scale in the power electronics market and have penetrated our daily lives in various fields, e.g., consumer electronics, electric and hybrid vehicles, industrial processes, energy conversion from renewable sources, sensors, photodetectors and so on [6–8].

A schematic cross-section view of a typical SBD is depicted in Figure 1. In this device, the drift layer is given by a lightly doped n-type 4H-SiC epitaxial layer (doping density N_D ~ 10^{16} cm^{-3}) grown onto a heavily doped n-type 4H-SiC substrate ($N_D > 5 \times 10^{18}$ cm^{-3}). The connection with external components is given by metal electrodes consisting of a front-side Schottky contact to the 4H-SiC epitaxial layer and a back-side Ohmic contact to the 4H-SiC substrate. In addition, the Figure schematically illustrates a p-type implanted edge termination needed to optimize the breakdown behavior. For the sake of completeness (not reported in the scheme), a few μm thick front-side metallization (typically Al based) acting as device bonding pad and a polymeric passivation finalize the top part of the device, while a thick metal layer, soldered to the back-side and consisting of a gold-plated metal frame wrapped in ceramic, operates as connection to the package. A schematic fabrication process flow, with the sequential steps typically adopted for the Schottky diode fabrication, can be found in [9].

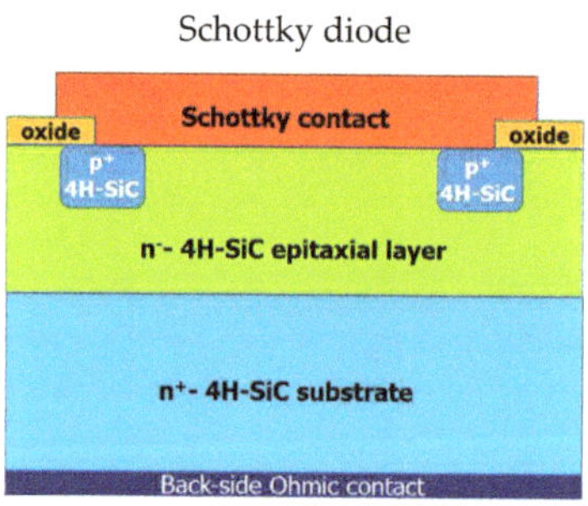

Figure 1. Schematic cross-section view of a 4H-SiC Schottky barrier diode (SBD).

The core of an SBD is the metal/semiconductor junction (the so-called Schottky contact) and the properties of this system must be carefully investigated to assess and optimize the electrical performance of the whole device [9].

Over the years, different approaches, ranging from the choice of materials for the Schottky barrier formation to semiconductor treatments or even considering the device layout, have been developed to improve and gain control on the Schottky contact properties. Moreover, unconventional methods were also explored as alternative solutions in improving and controlling the Schottky barrier systems.

In this paper, after a brief discussion on the fundamentals of the metal/4H-SiC Schottky barrier formation and its electrical characterization, we will give an overview on the current materials and processing solutions for the fabrication of Schottky contacts to 4H-SiC. Afterwards, besides the consolidated approaches, we will present a variety of the nonconventional methods proposed in literature to control the Schottky barrier properties for specific applications.

2. Schottky Contacts to n-Type 4H-SiC

In the last three decades, many efforts have been devoted to the study and characterization of the metal/4H-SiC interfaces. Despite the large amount of literature data published on this system, the physics at the base of the Schottky properties and carrier transport through the contact is not yet fully understood, and the research on Schottky contacts to 4H-SiC remains a scientifically open topic. Besides its fundamental scientific character, this research is also strongly pushed by industrial requirements for achieving a more efficient performance of SBDs. In this Section, we briefly report on the Schottky barrier fundamentals, the well-established and the most promising metallization schemes, discussing also some aspects related to the device layouts proposed to optimize the electrical properties of the diodes.

2.1. Fundamentals on Schottky Barriers

When metal and semiconductor are put in intimate contact, a Schottky barrier forms, whose height (ϕ_B) is the most significant electrical parameter associated to this system. According to the well-known Schottky–Mott rule [10,11], the Schottky barrier height ideally depends only on the metal work function (W_M) and semiconductor electron affinity (χ_S) [12], as expressed by:

$$\phi_B = W_M - \chi_S. \tag{1}$$

Hence, considering the work function of the typical metals used for SBDs (in the range 4–5.5 eV [13]) and the electron affinity of 4H-SiC (χ_{SiC} = 3.2 eV [14]), Schottky barrier height values between 1.0 and 2.3 eV are expected, i.e., much higher than in silicon. However, in real cases, the Schottky barrier height does not simply obey to the Schottky–Mott rule, but may depend on other factors, like the surface preparation, metal deposition techniques and/or post-metallization annealing treatments [15,16]. In particular, the electronic properties of the metal/semiconductor contact are affected by the presence of surface states, which can be related to roughness, surface contaminants, residual thin interfacial oxide layers and so on, and are responsible for a deviation from the Schottky–Mott prediction [17,18].

The experimentally measured value of the Schottky barrier height ϕ_B also depends on the methods used for its determination [19]. Generally, the barrier height in a metal/semiconductor system can be determined by means of different techniques, such as electrical characterizations (current–voltage (I–V) and capacitance–voltage (C–V) measurements) or internal photoemission (IPE) measurements that are based on the photo-generated current detection [19,20]. Obviously, each method shows advantages and drawbacks. The electrical characterization methods require the fabrication of appropriate test patterns, namely Schottky diodes, and provide interesting insights of the Schottky barrier nature. For example, the I–V characterization is very sensitive to the presence of inhomogeneity of the barrier, with some well-established models, such as the Tung's model [17]) or the Werner's and Güttler's model [21], developed to take this aspect into account. On the other hand, the C–V characterization supplies information about the space-charge region width [22]. As a consequence, for a given Schottky contact, the barrier height value extrapolated by I–V analysis is typically lower than that derived from the C–V characteristics. This aspect can be explained by the fact that lower barrier height regions are preferential paths for the current, while C–V characteristics account for an overall electrical behavior with the largest regions (usually with highest barrier) dominating in the barrier extraction [22]. On the other hand, IPE measurements are independent of the local barrier inhomogeneity and supply a reliable value for the Schottky barrier. However, the photocurrent detection requires special equipment and semi-transparent front or back contacts, thus making these kind of measurements less common with respect to the electrical characterizations.

The Schottky barrier height determines the electrical behavior of a metal/semiconductor contact by governing the current flow through the metal/semiconductor interface. Generally, for doping density N_D in the range $1 \times 10^{15} < N_D < 1 \times 10^{17}$ cm^{-3}, under forward voltage the current transport mechanism through the metal/4H-SiC interface is ruled by the thermionic emission (TE) theory. In this model the current I_{TE} can be expressed by [12]

$$I_{TE} = AA^*T^2 \exp\left[\left(-\frac{q\phi_B}{k_BT}\right)\right] \exp\left[\left(\frac{qV_F}{nk_BT}\right) - 1\right] \tag{2}$$

where A is the device area, A^* is the effective Richardson's constant (146 A·cm^{-2}·K^{-2} for 4H-SiC [23]), k_B is the Boltzmann's constant, q is the elementary charge, V_F is the applied forward voltage and T is the absolute temperature.

As can be seen in Equation (2), besides the Schottky barrier height ϕ_B, another important electrical parameter that characterizes the Schottky contact is the ideality factor n.

These parameters can be determined by the intercept and the slope of a linear fit in semilog scale of the forward current–voltage characteristic, using Equation (2) for $V_F > 3\ k_BT/q$, where the term -1 can be neglected.

Basically, the TE theory assumes a temperature-independent ideality factor and barrier height. However, in order to justify the experimentally observed temperature dependence of these parameters in real 4H-SiC Schottky contacts, the TE theory was modified, introducing models taking into account local fluctuations (inhomogeneity) of the Schottky barrier over the contact interface, as discussed in the Tung's [17] model or in the Werner's and Güttler's [21] model. Specifically, the Tung's model [17] assumes a local lateral inhomogeneity at nanometric scale by considering the presence of low barrier regions (patches) embedded in a high-barrier background, while the Werner's and Güttler's [21] model considers a Gaussian distribution of barrier heights around an apparent temperature-dependent barrier height.

On the other hand, when the doping concentration of the semiconductor exceeds 10^{17} cm^{-3}, the high electric field at the interface and the thin barrier width make dominant for the current transport a thermionic filed emission (*TFE*) mechanism, which involves a tunneling component for thermally excited electrons [19,24]. Specifically, for doping in the range $1 \times 10^{17} < N_D < 10^{19}$ cm^{-3}, the *TFE* describes the electrical behavior of the system [24], according to the following current–voltage relationship [25]:

$$I_{TFE} = I_{0,TFE}(V_F) \times \exp\left(q\frac{V_F}{E_0}\right) \tag{3}$$

The term $I_{0,TFE}\ (V_F)$ corresponds to the saturation current and is given by

$$I_{0,TFE}(V_F) = \frac{AA^*T}{k_B cosh(qE_{00}/k_BT)} \times \sqrt{\pi E_{00}(\phi_B - \Delta E_F - V_F)} \times \exp\left(-\frac{q\Delta E_F}{k_BT} - \frac{\phi_B - \Delta E_F}{E_0}\right) \tag{4}$$

with $E_0 = E_{00} \times coth\left(\frac{qE_{00}}{k_BT}\right)$ dependent on the doping concentration N_D through the parameter $E_{00} = \frac{h}{4\pi} \times \sqrt{\frac{N_D}{m^*\varepsilon_{SiC}}}$. The other symbols are the Planck's constant h, the effective mass $m^* = 0.38\ m_0$ (with m_0 the electron mass) [26] and the dielectric constant of the semiconductor $\varepsilon_{SiC} = 9.76\ \varepsilon_0$ (with ε_0 the vacuum permittivity) [4]. ΔE_F is the difference between the bottom of the conduction band and the semiconductor Fermi level. In this case, the barrier ϕ_B and the doping concentration N_D can be considered the parameters to be determined from a best-fit procedure of the experimental detected current–voltage characteristics.

Figure 2 depicts a schematic energy band diagram for a metal/semiconductor junction under forward bias V_F, when the predominant current transport mechanism is ruled by the TE (Figure 2a) or by the *TFE* (Figure 2b) model, according to the doping of the semiconductor layer at the interface. Noteworthy, as shown in Equations (2)–(4), in both models, the current has an exponential dependence on the Schottky barrier height. Thus, a good control of the ϕ_B must be guaranteed, as it significantly affects the current level through the contact.

2.2. Survey of Literature Data on Schottky Contacts to n-Type 4H-SiC

In literature, many studies on the metal/n-type 4H-SiC systems have been reported and focused on the choice of the metal and its evolution in the Schottky contact formation [27–33]. A collection of literature results related to some of the most diffused metal/n-type 4H-SiC contacts is reported in Table 1, including as deposited (unannealed) Schottky contacts or contacts subjected to thermal annealing treatments. The reported barrier height values were determined by I–V measurements on Schottky diodes.

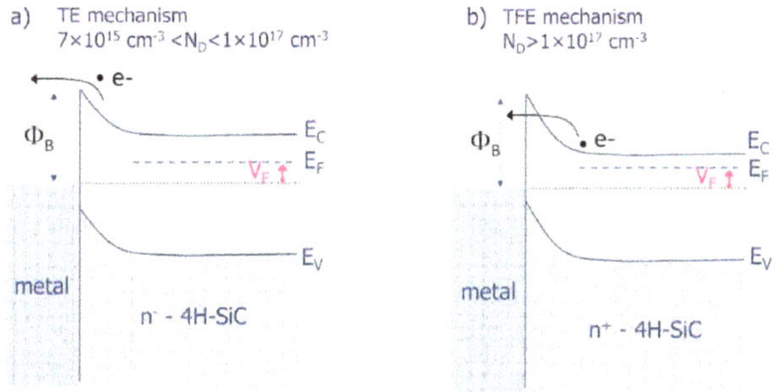

Figure 2. Schematic energy band diagrams for the metal/4H-SiC contact under forward bias V_F, according to the predominance of the (**a**) thermionic emission (TE) or (**b**) thermionic field emission (*TFE*) current transport mechanism.

Table 1. Schottky barrier height for metal/n-type 4H-SiC system for different metals. The values were determined by I–V measurements on Schottky diodes.

Metal	ϕ_B (eV)	Thermal Treatment	Reference
Ta	1.10	none	[34]
Ti	0.95	none	[35,36]
Ti	0.78	none	[37]
Ti	0.96	none	[38]
Ti	1.15	600 °C for 10 min in Ar	[37]
Ti	0.95	500 °C for 60 h in vacuum	[35]
W	1.11	none	[39]
W	1.14–1.25	475–700 °C for 10 min in N_2	[39]
W	1.17	none	[37]
W	1.09	600 °C for 10 min in Ar	[37]
W	1.11	500 °C in N2	[40]
Mo	1.04	none	[29]
Mo	1.11	none	[37]
Mo	1.21	600 °C for 10 min in Ar	[37]
Mo	1.17	none	[41]
Mo	1.01	400 °C for 30 min in Ar	[41]
Ni	1.45	none	[29]
Ni	1.62	none	[27]
Ni	1.52	none	[38]
Ni	1.52	400 °C, RTA	[29]
Au	1.73	none	[27]
Ir	1.31	none	[42]
Pt	1.39	none	[43]
Pt	1.72	200 °C	[44]

As can be seen, a large variety of barrier height values is found, depending on the metals and post-metallization thermal treatments. Especially, by reporting the barrier height values ϕ_B versus the metal work function W_M, it is possible to determine the correlation between ϕ_B and W_M, which represents the so-called "interface index" S = $d\phi_B/d\phi_m$ [45].

Typically, for real 4H-SiC-based Schottky contacts, a linear correlation is found, with S values between the Bardeen limit (i.e., S = 0, indicating interface properties independent of the metal) and the ideal Schottky–Mott behavior (S = 1) [4,46], suggesting the occurrence of a partial Fermi level pinning at the interface. Figure 3 displays this kind of plot for the unannealed and low-temperature annealed metal/4H-SiC contacts, with the values taken from Table 1. The slope values confirm an intermediate behavior, with a slight improvement towards the Schottky–Mott behavior from 0.46 to 0.54 for the low-temperature annealed contacts.

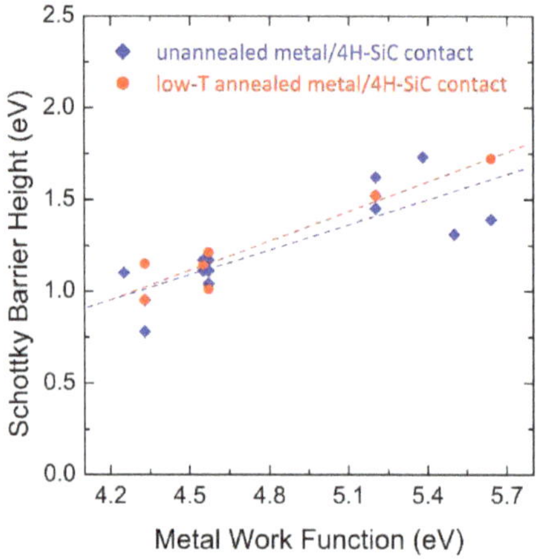

Figure 3. Experimental dependence of the barrier height ϕ_B on the metal work function W_M in unannealed and low-temperature annealed metal/n-type 4H-SiC systems. All the reported barrier values were determined by I–V characterization of Schottky diodes. Data are taken from Refs. [21–44].

Throughout the years, titanium (Ti) and nickel silicide (Ni_2Si) have come out as widely diffused barrier metals for 4H-SiC Schottky diodes in different applications. Ti- and Ni_2Si-based metallization schemes are currently well-established technology, offering a high level of reproducibility for the Schottky barrier height values. The representative forward I–V characteristics of the 4H-SiC Schottky diodes, employing Ti and Ni_2Si barrier metals and acquired at three different temperatures (173, 298 and 373 K), are reported in Figure 4a,b, respectively [23,47]. All the curves were analyzed according to the TE model, by fitting the linear region in a semilog plot of the forward I–V curve according to Equation (2) approximated for the linear region. From this analysis, the extrapolation of an ideality factor n very close to 1 for both contacts confirms the predominance of the TE mechanism in the current transport through the metal/semiconductor interface. Specifically, the values of the Schottky barrier height typically obtained at room temperature were ϕ_B = 1.27 eV for the Ti/4H-SiC and 1.60 eV for the Ni_2Si/4H-SiC contacts [23,47]. Consequently, Ti is used as the Schottky barrier material in power electronics applications, for which a low barrier height is desired [48], whereas Ni_2Si is preferred for sensing or detection applications [49,50], where the higher barrier permits a low leakage current to be obtained. Moreover, the possibility of a "self-aligned" process given by the Ni_2Si formation is used for the fabrication of semi-transparent interdigit electrodes in UV-detectors [51].

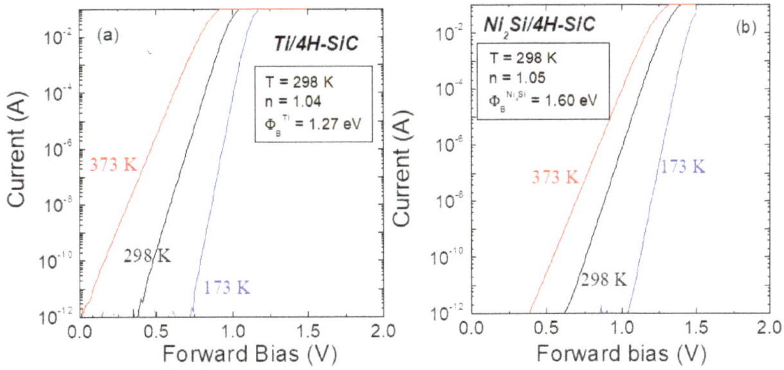

Figure 4. Semilog plot of the temperature-dependent forward-voltage characteristics of 4H-SiC Schottky diodes based on (**a**) Ti and (**b**) Ni_2Si. Adapted with permission from Ref. [23]. Copyright 2021 AIP Publishing.

2.3. Diode Layout

Managing the power dissipation in Schottky diodes is becoming an aspect of increasing interest, due to the need to reduce electricity consumption in modern electronic systems. In particular, a conventional Schottky diode (as that schematically depicted in Figure 1) is a majority carrier device, in which the dynamic power dissipation is negligible with respect to the conduction static power losses [52]. In such a device, the barrier height, ideality factor and reverse leakage current are important parameters affecting the static power dissipation. In particular, the conduction static power dissipation P_D can be expressed as [53]:

$$P_D = \%_{ON} \times (V_F \times J_F) + (1 - \%_{ON}) \times (V_R \times J_R) \quad (5)$$

where $\%_{ON}$ is the ON duty cycle, V_F and J_F are the voltage and current density under forward bias, while V_R and J_R are the voltage and current density under reverse bias. Hence, considering the case of the conventional 4H-SiC Schottky diodes, the forward current-voltage behavior is described by the TE model, with the V_F expressed as function of the forward current density J_{TE} as follows, accordingly to the linear region approximation of Equation (2):

$$V_{F,TE}(J_{TE}) = n\phi_B + \frac{nkT}{q} ln\left(\frac{J_F}{A^* T^2}\right) \quad (6)$$

Moreover, due to the high electric field in the space-charge region that entails a sharp band bending and thus a thin barrier, the reverse current is typically described according to the *TFE* model, with the relationship between current density J_{TFE} and reverse voltage V_R given by

$$J_{R,TFE}(V_R) == A^* T^2 \sqrt{\frac{q^2 \pi E_{00}}{(kT)^2}} \sqrt{V_R + \frac{\phi_B}{cosh^2\left(\frac{qE_{00}}{kT}\right)}} \exp\left(-\frac{\phi_B}{E_0}\right) \exp\left(\frac{V_R}{E_1}\right) \quad (7)$$

where $E_1 = E_{00} \times (qE_{00}/kT - tanh(qE_{00}/kT))^{-1}$ and the other parameters of Equation (7) as described before. Hence, accordingly to Equation (5), the static power dissipation depends on the Schottky barrier height ϕ_B.

Figure 5a shows the calculated conduction power loss for conventional 4H-SiC Schottky diode as a function of the barrier height, in the temperature range 25–150 °C. The calculation has been performed using Equations (5)–(7), assuming a duty cycle %ON = 50%, a forward current density J_F = 100 A/cm^2 and reverse voltage V_R = 650 V. As highlighted in Figure 5a, for a given temperature, a reduction of the Schottky barrier entails a reduction of the power losses, up to a certain lower limit, where the losses show a sudden increase, due to the significant increase of the reverse leakage current with the barrier reduction, especially at higher temperatures. Evidently, based on these considerations, in power electronics applications, Schottky contacts with low barrier height are sought after, as lowering the Schottky barrier height leads to a reduction of the power consumption [48]. However, it must be considered that a lowering of the barrier ϕ_B could lead to an increase of the leakage current density J_R and thus, a good compromise between the diode forward and reverse behavior must be found for the minimization of the power dissipation P_D.

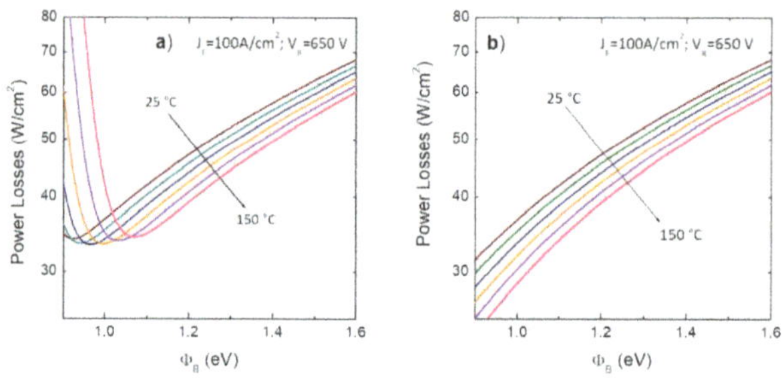

Figure 5. Schottky barrier height dependence of the static power losses for (**a**) a conventional Schottky diode and (**b**) a modern JBS diode, in the temperature range 25–150 °C. The curves were simulated by considering the forward electrical behavior ruled by thermionic emission model and the reverse characteristics ruled by thermionic field emission model. In the case of the JBS diode (**b**), the leakage current contribution has been neglected.

For that reason, as discussed later in this paragraph, in modern 4H-SiC-device technology, the high-reverse *TFE* leakage current, which typically characterized the conventional Schottky diodes, has been strongly reduced by acting on the device layout. In this case, the contribution of the leakage current can be neglected, and the static conduction losses decrease with a reduction of the barrier height, as shown in Figure 5b.

Since the middle of the 1990s, new diode designs have been proposed to achieve improved rectifier characteristics of semiconductor-based Schottky diodes. Mehrotra et al. [54] demonstrated that a design involving metal-oxide-semiconductor (MOS) regions, built into a trench region of the device front, was successful in pushing higher the limit given by the reverse blocking voltage, allowing the device to support larger doping of the semiconductor epitaxial layer, and thus an improvement of the on-state characteristics. A schematic view of this device is reported in Figure 6a: as can be seen, the MOS structure is formed on the bottom and sidewalls of a trench, while the Schottky contact is on the top surface. With this layout, called trench MOS barrier Schottky rectifier (TMBS), a reduced level of electrical field is achieved on the Schottky interface, producing a smaller Schottky barrier height lowering and thus a reduced leakage current level, if compared with the standard Schottky structure.

Afterward, another approach of "Schottky-pinch rectifier" was proposed by Zhang et al. [55] for 4H-SiC Schottky rectifiers. Differently from the previously discussed solution, this diode layout consisted of integrating MOS-structures (with a thermally grown oxide) together with Ni Schottky contacts on the same plane, as shown in Figure 6b. This layout, called the planar MOS Schottky diode structure (MOSSD), was able to maintain an acceptable level of forward current, up to 90% with respect to a conventional Schottky diode of the same footprint area, while reducing the leakage current by one order of magnitude.

However, these designs including MOS structures, could give some limitations, with the possible occurrence of the oxide breakdown before the 4H-SiC critical electrical field was reached. To overcome this limitation, a dual-metal-trench (DMT) device structure, implementing low and high Schottky barrier height materials (i.e., Ti and Ni metals, respectively) was suggested by Schoen et al. [56]. The device scheme is given in Figure 6c:

under forward bias, the mesa was not pinched off and the electrical characteristics are given by the low barrier contact (Ti/4H-SiC). In contrast, under reverse bias, the mesa structure became fully pinched-off and the high barrier height of the Ni Schottky contact prevails, limiting the electric field. A further evolution of the DMT has been later proposed by Roccaforte et al. [57], who combined the advantages of Ni_2Si and Ti in a dual-metal-planar (DMP) Schottky diode (schematically depicted in Figure 6d) which exhibited a forward voltage drop close to that of a Ti/4H-SiC diode (lower barrier) and a reverse current comparable to that of a Ni_2Si/4H-SiC (higher barrier).

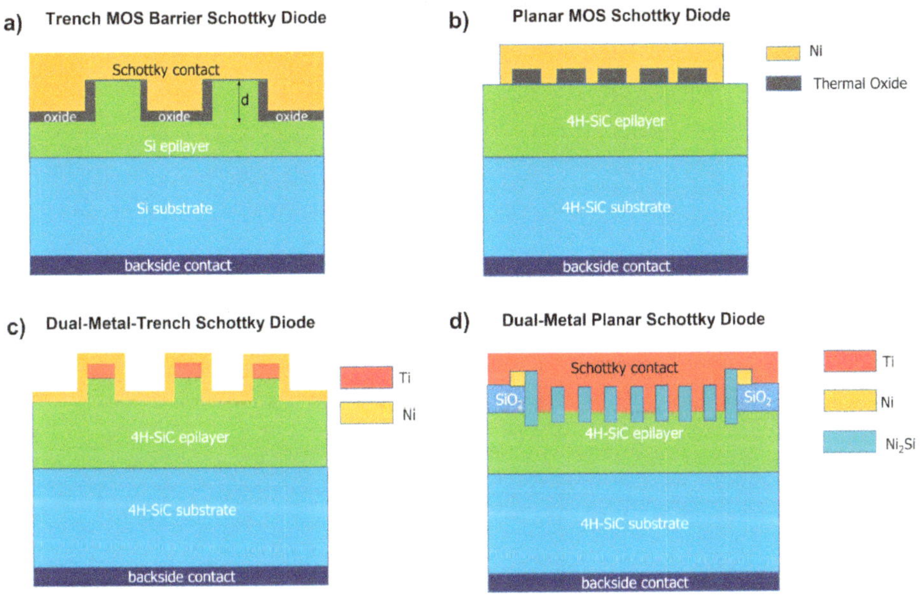

Figure 6. Different Schottky diode layouts proposed to achieve an optimal trade-off between the forward and reverse characteristics. (**a**) Trench MOS barrier Schottky (TMBS) diode. (**b**) Planar MOS Schottky diode (MOSSD). (**c**) Dual-metal-trench (DMT) Schottky diode with Ti and Ni Schottky contact. (**d**) Dual-metal-planar (DMP) Ti-Ni_2Si/4H-SiC Schottky diode.

The electrical behavior of the DMP structure can be explained by an equivalent system with two parallel diodes which have two different barrier heights, specifically the low barrier of the Ti layer determines the current flow under forward bias, and the high Ni_2Si barrier dominates the reverse conduction by the pinch-off of the low barrier Ti regions. The DMP diode presented an ideality factor $n = 1.25$, a barrier height $\phi_B = 1.23$ eV (close to that of Ti contact) and leakage current at -100 V of 5.8×10^{-4} A/cm^2, 30 times lower than the leakage current observed in the Ti/4H-SiC diode.

All the aforementioned diode layouts showed an improvement of the trade-off between the forward and reverse characteristics of the diodes.

Nowadays, the so-called junction barrier Schottky (JBS) diode is the most widely used Schottky-like architecture in SiC technology with significant improvement with respect to the standard Schottky diode [58]. This device consists in p^+-type regions (usually achieved by p-type ion implantation and electrical activation) embedded within an n-type Schottky epitaxial area, as schematically shown in Figure 7 [59,60]. This layout mitigates the reverse leakage current of the Schottky diode and achieves a hard breakdown, as typical of a p-n junction [58]. Specifically, under low forward bias, the current flows in the regions between the p^+-wells, exploiting the Schottky barrier characteristics given by the top metal, while under reverse bias, these regions are pinched-off and the electrical characteristics are given by the p-n junction. The distance d between two adjacent p^+-wells and the size s of p^+-well are important parameters that must be carefully designed for optimizing the trade-off between forward and reverse characteristics. These parameters, together with the depletion width W_D, defined the cell pitch p ($p = d + s + W_D$). For instance, the on-state voltage drop decreases as the cell pitch is reduced, while the leakage current decreases as the p^+-well distance is reduced for a constant value of p^+-well size [61]. The schematic process flow, with the sequential steps typically adopted for JBS diode fabrication, can be found in Ref. [9]. As occurs in SBDs, in these devices the Schottky contact between the metal and the n-type epitaxial 4H-SiC is also a key part for optimizing the overall electrical performance of the device.

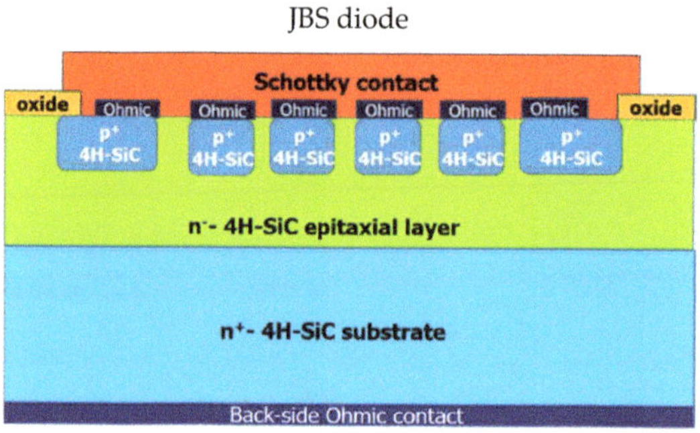

Figure 7. Schematic cross-section view of a 4H-SiC junction barrier Schottky (JBS) diode.

Beyond the diode layout with the well-established JBS design for 4H-SiC-based Schottky rectifiers, the employment of materials with low work function is of particular interest for minimizing the power dissipation of Schottky diodes and they are currently explored with promising results. This aspect will be discussed in detail in the next subsection.

2.4. Low Work Function and Refractory Metals

In the last years, regarding the choice of the Schottky metal for 4H-SiC-JBS, increasing interest has been devoted to metallization schemes containing metals (and their compounds) with low work function and a certain degree of stability with 4H-SiC and the environment. In the facts, these low work function materials (such as the refractory metals Mo, W, Nb, etc.) can guarantee a minimization of the on-state conduction losses, making this configuration highly aimed at industrial 4H-SiC-based Schottky device development [48]. Moreover, since these metals exhibit a high melting point, they could be indicated for harsh environment applications, requiring temperature-resistant materials [62].

In recent literature, many papers have dealt with the electrical characterization of Mo/4H-SiC Schottky contacts for power electronics [63–67], highlighting the possibility of achieving a barrier height value as low as 1.010 eV and an ideality factor of 1.045 [65]. As reported in those studies, the Mo/4H-SiC behaved as an inhomogeneous contact, with the current conduction dominated by a TE mechanism and a slight discrepancy from the ideal behavior explained either according to the Werner and Güttler [21] or the Tung model [17].

Very recently, Renz et al. [63] studied a series of surface passivation treatments to achieve an improvement of the Mo/4H-SiC Schottky diode electrical properties. In particular, after the deposition or thermal growth of an oxide layer, annealing processes, similar to those employed in metal-oxide-semiconductor field-effect transistors (4H-SiC MOSFETs) technology, were considered [68–70]. These treatments included thermal oxidation in O_2 or N_2O environments at temperatures of 1400 and 1300 °C, respectively, or the deposition of a phosphorus pentoxide (P_2O_5) layer at 1000 °C for 2 h. The first two processes consume the SiC surface while the third one does not. In all the samples, the oxide on the surface was removed by cleaning in dilute HF (10%) solution prior to Mo Schottky metal deposition. Figure 8a shows this approach schematically, with the treated area of the semiconductor depicted as a patterned blue layer. An electrical I–V_F characterization at room temperature was performed on a set of equivalent Mo/4H-SiC diodes fabricated under different conditions and, for comparison, on an untreated Mo/4H-SiC contact (labelled as "control"). The I–V_F curves were analyzed according to TE model, obtaining an almost ideal behavior after the treatments. The lowest barrier height value (ϕ_B = 1.27 ± 0.03 eV) was observed for the contact subjected to a prior deposition of P_2O_5 (left scale of Figure 8b). Surprisingly, although a reduced value of barrier height was obtained in the P_2O_5-treated contacts with respect to the control sample, this process enabled the lowest value of the leakage current (right scale of Figure 8b) to be obtained. The authors explained this effect with the capability of the oxide to homogenize the interface by filling the nanopits, as witnessed by means of morphological (AFM) and microstructural (TEM) analyses. The high density (5×10^9 cm^{-2}) of these nanopits allows one to believe that they are different from those typically related to the threading dislocations arriving on the surface of 4H-SiC and observed after removal of surface electrodes [71]: if the threading dislocation-related nanopits were demonstrated to be potential leakage paths for the current, plausibly, in the case of treated P_2O_5 deposited on 4H-SiC surface, the nanopits were oxide-filled, with a barrier lowering due to two contributions, one associated with the phosphorous-rich region below the contact, which increases the n-type doping, and the other related to a homogenization of the barrier height after oxide termination of the surface defects. This could explain the reduction of the barrier, with a simultaneous decreasing of the leakage current. It is worth noting that the absence of silicide reaction at the Mo/SiC interface, which would otherwise consume the top few nanometers of 4H-SiC, enabled these beneficial changes in the contact subsurface.

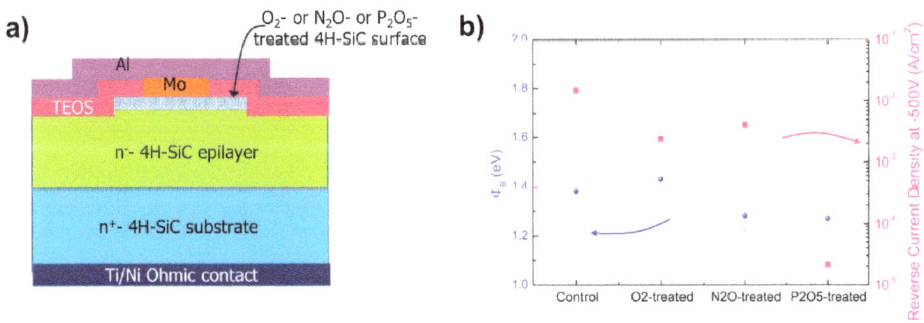

Figure 8. (**a**) Schematic view of Mo/4H-SiC Schottky diodes with premetallization treatments of the semiconductor surface, consisting of thermally grown oxidation in O_2 and N_2O or oxide deposition of P_2O_5, followed by oxide removal prior to Mo deposition in all cases. (**b**) Schottky barrier height and reverse leakage current density at −500V values averaged over a set of I–V_F curves of equivalent Mo/4H-SiC Schottky diodes with 4H-SiC surface pretreated under different conditions. Panel (**b**) is adapted with permission from Ref. [63]). Copyright 2021 AIP Publishing.

Furthermore, Mo-based Schottky contacts were also investigated as a possible route for an improved control of the Schottky contact properties. For instance, Stöber et al. [72] proposed the use of molybdenum nitride (MoNx) thin film metallization for adjusting the barrier height within a large range by varying the nitrogen N_2 fraction in the reactive sputtering metal deposition step in the fabrication process. The total gas flow, i.e., sum of argon and nitrogen, was kept constant at 80 sccm while the nitrogen fraction $\chi = N_2/(Ar + N_2)$ in the gas composition varied from 0 to 80%, by increasing the content of N_2 in the chamber. For a pure Mo contact (with χ = 0%), the Schottky barrier was ϕ_B = 0.68 eV at room-temperature, increasing up to 1.03 eV for χ = 80%. Due to the polycrystalline nature of the Mo_2N thin films, the barriers showed an inhomogeneous behavior, probably arising from different microstructures with regard to the nitrogen fraction used in the processing.

In parallel to Mo-based 4H-SiC Schottky contacts, Schottky contacts based on W have also been largely studied, obtaining a Schottky barrier ranging between 0.94 to 1.29 eV [62,73,74]. According to these papers, a certain degree of inhomogeneity was observed in the W/4H-SiC Schottky contact, successfully explained by means of the Tung's model [73] or the Werner and Güttler's model [74].

Noteworthy, for both Mo and W metals, also the carbide compounds were considered as possible electrode material [75–78]. As an example, Knoll et al. [76] investigated a Schottky barrier based on tungsten carbide, fabricated by depositing a thin layer of W (2 nm) followed by a rapid thermal annealing in vacuum for 5 min at temperatures ranging from 600 to 1200 °C. Then, a 500 nm thick Al layer was deposited on the top of the structure to define the diode structures. At temperatures > 1000 °C, a W_2C hexagonal structure layer in epitaxial relation with 4H-SiC was produced, stable up to the highest tested temperature of 1200 °C. In this system, they observed a barrier height of 0.94 eV extrapolated under forward I–V characterization.

Recently, we investigated 4H-SiC Schottky diodes with an 80 nm thick layer of tungsten carbide (WC) barrier metal, deposited by magnetron sputtering and defined by optical lithography and lift-off [39,78] process. The Schottky diodes were characterized both before (as-deposited) and after some annealing treatments with temperatures varying form 475 °C to 700 °C for 10 min in N_2 atmosphere by I–V measurements and applying the thermionic emission (TE) model to the analysis of the electrical characteristics. The Schottky barrier height ϕ_B, derived by fitting the linear region of the semilog forward J–V_F curves reported in Figure 9, had an average value of 1.12 eV in the as-deposited contact and decreased down to about 1.05 eV after annealing at 700 °C. In our experimental conditions, the ideality factor was only slightly affected by the annealing treatment, with a value decreasing from 1.08 to 1.03. For sake of comparison, we also reported in the same Figure 9, a representative J–V_F curve of a similar Ti/4H-SiC Schottky contact, for which a barrier height of 1.21 eV was observed [79].

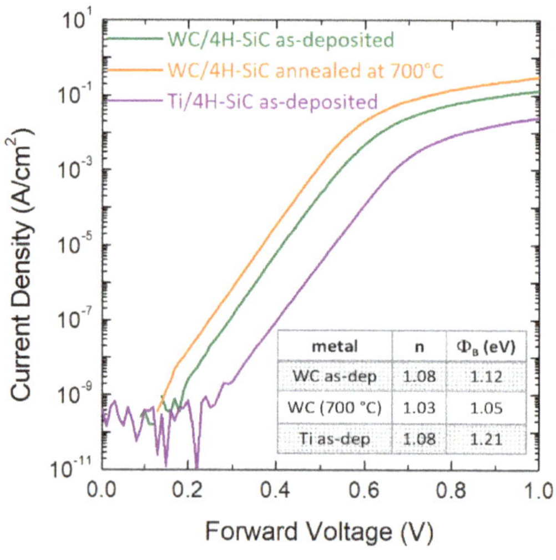

Figure 9. Representative forward I–V characteristics of WC/4H-SiC Schottky diodes for the as-deposited contact and after thermal annealing at 700 °C. The I–V characteristic of a reference Ti/4H-SiC is also reported for comparison. The data are taken from Refs. [78,79].

In those studies, we observed a temperature-dependence of the ϕ_B and n, extrapolated by means of I–V–T characterization, for the annealed-WC/4H-SiC contact [78], as well as for the W/4H-SiC contact fabricated and annealed under similar conditions [39]. This indicated the presence of a nanoscale lateral inhomogeneity for both Schottky contacts, that was fully described by means of the Tung's model, with an effective barrier ϕ_{Beff} of 1.15 and 0.96 eV and a homogeneous barrier ϕ_{B0} of 1.28 and 1.11 eV for the W/4H-SiC and WC/4H-SiC contact, respectively.

Essentially, the promising results obtained for the 4H-SiC Schottky diodes based on these low-work function refractory materials (mainly W and Mo) enable the study of this kind of contacts to be pushed forward towards a better understanding of the Schottky properties and inhomogeneity, and a suggestion of possible solutions for a better barrier uniformity and interface quality.

3. Unconventional Approaches for the Control of 4H-SiC Schottky Interfaces

Parallel to the standard metallization stacks and layouts presented in the previous Section, a variety of innovative contacts, chemical compounds or alternative metal stacks have been proposed as new routes to control the Schottky barrier height values on 4H-SiC. In the next subsections, we will discuss some of the representative papers on these unconventional methods.

3.1. Manipulation of the Schottky Interface

Since the early 2000s, some studies demonstrated the possibility of lowering the barrier height in the Schottky diode by the incorporation of nanostructures in the metal layer of the Schottky contact [80–83]. One of the first attempts in SiC was reported by Lee et al. [80], who studied the effect of Au-nanoparticle embedding in Ti/n-type 4H-SiC contact. The diodes were fabricated by first depositing Au-aerosol nanoparticles (diameter of 20 nm, density of 90 μm^{-2}) and then depositing 200 nm thick Ti layer in an evaporation chamber. The schematic view of the final Schottky diode structure is depicted Figure 10a. In Figure 10b, the I–V curves for these Schottky-diode embedding nanoparticles are compared to those of a control Ti/4HSiC Schottky diode.

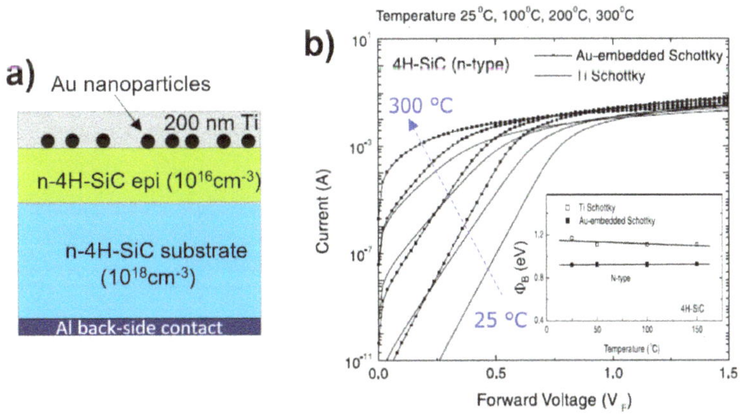

Figure 10. (**a**) Schematic view of Ti/4H-SiC Schottky contact with embedded Au nanoparticles on 4H-SiC. (**b**) Forward current–voltage characteristics of Au-nanoparticle embedded Ti/4H-SiC contact and particle-free Ti/4H-SiC control contact for different measurement temperatures (25, 100, 200 and 300 °C). Inset: comparison between the Schottky barrier height value of Au-nanoparticle-embedded Ti/4H-SiC contact and Ti/4H-SiC control contact, as a function of the measurement temperature. Figures adapted with permission from Ref. [80]. Copyright 2021 Elsevier Ltd.

From the comparison, carried out at four different measurement temperatures (25, 100, 200 and 300 °C), it was possible to point out that for each testing temperature, the I–V curve related to the Au-nanoparticle-embedded Ti/4H-SiC contact was shifted towards a lower voltage than the control sample (Figure 10b). The Schottky barrier height, derived by a fit in the linear region according to TE theory and reported in the inset of Figure 10b, was lowered of 0.19 eV in the sample with nanoparticles. To explain this result, the authors invoked the enhancement of the electric field under the interface in the depletion region, due to the small size of the embedded particles and the large Schottky barrier height difference obtained by using two metals as Ti and Au. This is, in part, confirmed by theoretical calculation according to the Tung's dipole-layer approach [18].

Later on, other studies reported on Schottky contacts with Au- or Ag-nanoparticles embedded in a Ni-metal [82] or Al-metal [83] layer, obtaining a similar reduction of the Schottky barrier height, which was associated with a reduction of the metal work function induced by the presence of the interfacial nanoparticles.

Besides working on the barrier material, many efforts have been dedicated also to the preparation and treatments of the semiconductor surface, to obtain a higher degree of homogeneity of the contact.

As an example, the inhomogeneity observed in Schottky contacts to 4H-SiC could be reduced by suitable treatments of the semiconductor surface, such as passivation with the insertion of an insulating thin film between the semiconductor and the metal [84–86].

For example, Shi et al. [86] demonstrated that the presence of an ultrathin Al_2O_3 layer, deposited on the semiconductor surface by atomic-layer deposition before the metal stack (Al 300 nm/Ti 100 nm) annealed at in Ar at 300 °C for 5 min (Figure 11a), enabled a reduction of the barrier height. Three different oxide-layer thicknesses were investigated in that work (0.8, 1.2 and 2 nm). The forward I–V characteristics, shown in Figure 11b for all the tested Al_2O_3 thicknesses, indicated a reduction of the Schottky barrier height with the increase of the Al_2O_3 thickness, down to a value lower than 1 eV (inset of Figure 11b). In particular, the cross-section TEM analyses showed that the insertion of Al_2O_3 reduces the diffusion of Ti into 4H-SiC and, hence, the possible occurrence of solid-state reactions between metal and semiconductor. In this way, the formation of new titanium silicide and carbide phases is prevented, thus resulting in an improvement of the interface homogeneity.

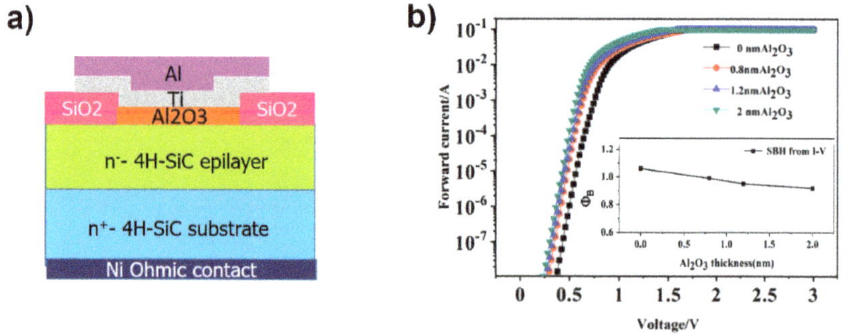

Figure 11. (**a**) Schematic view of a Ti/4H-Si Schottky diode with the insertion of an ultrathin Al_2O_3 layer between metal and semiconductor surface. (**b**) Forward I–V characteristics of the Ti/4H-SiC Schottky contact with increasing thickness of the inserted Al_2O_3-layer (0, 0.8, 1.2 and 2 nm). The trend of the Schottky barrier height as function of Al_2O_3 thickness is reported in the inset. Panel (**b**) is adapted with permission from Ref. [86]. Copyright 2021 Elsevier Ltd.

In another case, the insertion of an ultrathin amorphous-hydrogenated SiC layer (a-SiC:H) in the Ti/4H-SiC contact has been assessed with promising results (see schematic view in Figure 12a) [87]. The amorphous layer, with a thickness between 0.7 and 4 nm, was grown on the 4H-SiC surface by means of plasma-enhanced chemical vapor deposition prior to Ti deposition. Thermal annealing in a vacuum at 600 °C was also performed. The value of the Schottky barrier height varied between 0.78 and 1.16 V. These values, derived from room temperature I–V measurements, are reported in Figure 12b. As one can see, the Schottky barrier height depends on the amorphous layer thickness and thermal annealing duration. Specifically, while a slight influence of the amorphous layer thickness was observed on the Schottky barrier value, the duration of the 600 °C annealing, supposed to result in the formation of the Ti_5Si_3 phase [88], had a more significant impact on the barrier height. In particular, the lowest barrier value was obtained after the longest annealing treatment.

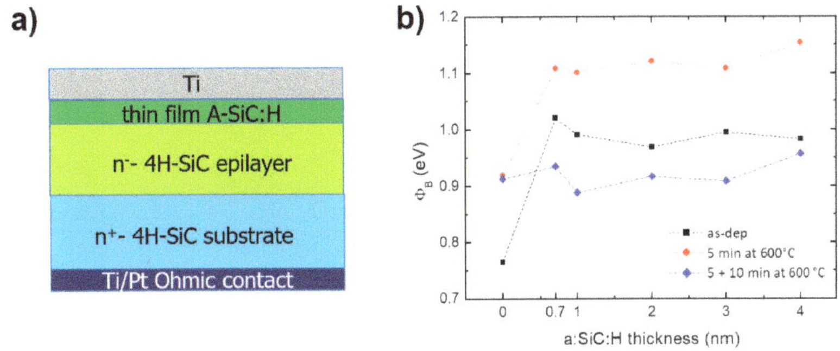

Figure 12. (**a**) Schematic view of a Ti/4H-SiC contact with an ultrathin amorphous SiC:H layer inserted between Ti and 4H-SiC. (**b**) Barrier height ϕ_B for different thicknesses of the amorphous layer and duration of the annealing treatment. Panel (**b**) adapted with permission from Ref. [87]). Copyright 2021 Elsevier Ltd.

Another method to modify, at atomic level, the surface where the Schottky contact is formed, consisted of the graphitization of the 4H-SiC surface [89,90]. It is known that as-grown monolayer graphene (MLG) on hexagonal SiC consists of a buffer layer (BL), similar to graphene but still covalently bond to SiC, plus a graphene overlayer [91]. The as-grown MLG contact exhibits ohmic characteristics, which have been explained by a low Schottky barrier height (ϕ_B = 0.36 ± 0.1 eV) with SiC [92]. Such a low ϕ_B value has been ascribed to the positively charged dangling Si bonds at the BL/SiC interface, which cause a Fermi level pinning of graphene close to the SiC conduction band, as well as a high n-type doping of graphene itself ($n \approx 1 \times 10^{13}$ cm^{-2}) [92]. Differently, after an annealing treatment of the MLG in H_2-atmosphere, the covalent bonds between the BL and SiC break up and H_2 saturates the dangling bonds, converting the electrically inactive BL into an additional real graphene layer [93]. This quasi-freestanding bilayer graphene (QFBLG) is moderately hole-doped ($p \approx 8 \times 10^{12}$ cm^{-2}) and provides a Schottky contact to 4H-SiC [94]. As discussed by Hertel et al. [94], these two different graphene/4H-SiC interfaces can be used side-by-side on the same chip in a real 4H-SiC-based MESFET device, as illustrated in Figure 13. Specifically, the MLG on 4H-SiC is used as ohmic contact for the source and drain electrodes (marked as "contact graphene"), while the QFBLG/4H-SiC–Schottky interface serves as a gate electrode (marked as "gate graphene") [94].

In this system, nanoscale conductive atomic force microscopy (C-AFM) on QFBLG showed a dependence of the Schottky barrier height on the diode area, from values in the range (0.9–1) eV obtained for large contacts, up to values approaching ~1.5 eV for the smallest contacts. The behavior of this kind of contact was explained by considering that SiC step edges and facets are preferential current paths causing the effective lowering of the barrier. The reduced barrier height in these regions can be explained in terms of a reduced doping of QFBLG from SiC substrate at (11–20) step edges with respect to the p-type doping on the (0001) terraces [93].

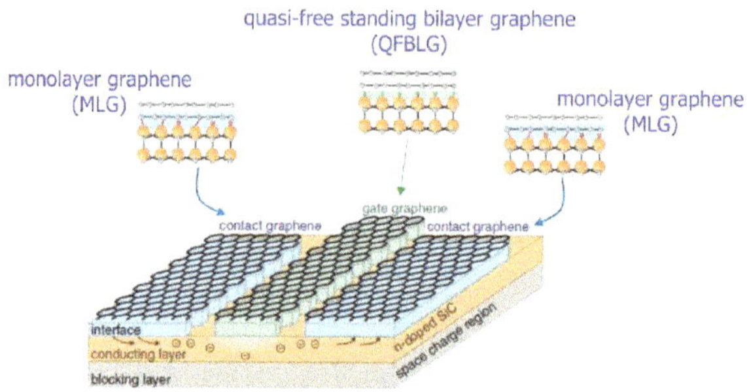

Figure 13. Schematic view of graphene/4H-SiC-based transistor, with two different interfaces, i.e., MLG and QFBLG contact acting as ohmic and Schottky gate contacts, respectively. Figure adapted with permission from Ref. [94]). Copyright 2012 Nature Portfolio.

A final example regards the work of Lin et al. [95], who explored a new way to fabricate tunable Schottky diodes with ns-pulsed excimer laser (193 nm)-modified n-type single-crystal 4H-SiC. The diodes were analyzed both by macroscopic I–V measurements by using Au-layer as electrode and by nanoscale characterization by means of atomic force microscopy in PeakForce TUNA configuration, this latter schematized in Figure 14a. Particularly, as noticed from the macroscopic I–V characterization on pristine and irradiated contacts, the most notable change in the I–V behavior was observed for the contact exposed to 2 J/cm^2 (not shown here). For the contact irradiated at such fluence, the nanoscale I–V characterization for different numbers of pulses (1–20 pulses), directly on the bare laser-exposed surface of the semiconductor, showed a rectifying electrical behavior of the contact, with the Schottky barrier increasing from 0.38 up to 1.82 V in the range 3–20 pulses (reported in Figure 14b). A combined analysis with Raman spectroscopy for the sample irradiated at 2 J/cm^2 demonstrated a graphitization of the 4H-SiC surface after laser irradiation, which is probably at the base of the barrier height increase in contact to the laser-modified 4H-SiC surface. For fluence as high as 5 J/cm^2, the appearance of the peak corresponding to monocrystalline silicon (~520 cm^{-1}) was observed.

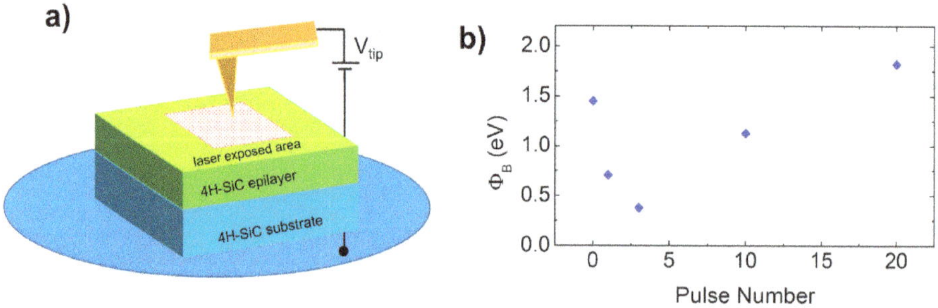

Figure 14. Nanoscale current–voltage characterization (by PeakForce TUNA mode of AFM) of Au/4H-SiC contact to laser-irradiated semiconductor surface: (**a**) scheme of the nanoscale current–voltage measurement set-up; (**b**) Schottky barrier height values extrapolated by I–V analysis for 4H-SiC surface irradiated with different pulse numbers. Figure adapted with permission from Ref. [95]. Copyright 2021 Elsevier Ltd.

3.2. N-Type Doping of the Interface

The capability of 4H-SiC to sustain a high electric field (if compared to conventional semiconductors, such as Si) enables the possibility of tailoring the Schottky barrier height by varying the doping concentration (and hence the electric field) below the contact. If, under reverse bias, the effect of a larger electric field has been widely investigated with an experimentally observed larger leakage current explained by the *TFE* model [96,97] and mitigated by the use of the JBS layout, under forward bias, the effect of a modification of the electric field requires deeper understanding. For instance, ion-irradiation-induced damage below the interface in Ti/4H-SiC Schottky diodes showed the possibility of increasing the barrier height by a deactivation of the dopant and a reduction of the electric field at the interface following a re-ordering of the crystal structure [98].

On the other hand, as mentioned above, a way to increase the electric field consists of increasing the doping concentration in the semiconductor below the contact layer. In this context, it is interesting to study the effects on the barrier height and carrier transport mechanisms in a heavily-doped 4H-SiC layer [19,99]. Hara et al. [19] studied the dependence of the barrier height and forward carrier transport mechanism on the doping concentration N_D in Ni/SiC Schottky barrier diodes with 4H-SiC epitaxial layer. In particular, they investigated a range of doping concentrations, varying from 6.8×10^{15} up to 1.8×10^{19} cm^{-3}. The increase in the doping concentration entailed a shift of the forward I–V characteristics (that means larger current observed for higher doping concentration for a given voltage value). This shift corresponded to a lower turn-on voltage, increasingly stronger from the lightly-doped sample ($N_D = 6.8 \times 10^{15}$ cm^{-3}) to the heavily-doped sample ($N_D = 1.8 \times 10^{19}$ cm^{-3}). On the other hand, a modification was observed for the predominant current transport mechanism, sweeping from TE to a *TFE* mechanism for a higher doping concentration ($N_D > 2.6 \times 10^{17}$ cm^{-3}).

The predominance of the *TFE* mechanism for Schottky contacts on a heavily-doped 4H-SiC layer was also demonstrated for Ni/4H-SiC with a n+-type implanted layer of 4H-SiC ($N_D = 1.97 \times 10^{19}$ cm^{-3}) [100], whose forward J-V_F characteristics are shown in Figure 15a. The inset reports the schematic energy band diagram for the metal/4H-SiC interface when a *TFE* current transport mechanism is predominant. This contact exhibited a lower value of turn-on voltage if compared to a reference Ni/4H-SiC contact formed on the 4H-SiC epilayer without implanted layer and standard epitaxial layer doping concentration, as clearly highlighted by the graph in Figure 15b. The possible increase of the leakage current could be mitigated by an appropriate choice of the device layout, as in the JBS diode. This last point was theoretically investigated in Ref. [100].

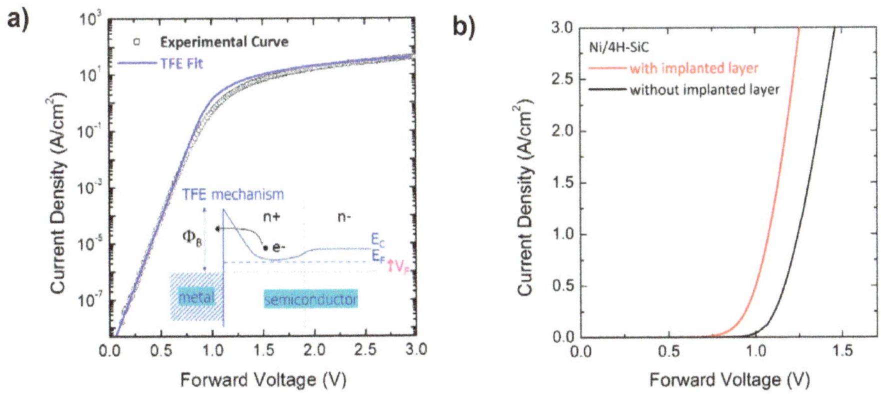

Figure 15. (**a**) Forward current density–voltage characteristics (open symbols) for Ni/n-type implanted-4H-SiC Schottky diode and fitting curve according to the *TFE* model (continuous line). In the inset, schematic energy band diagram for the metal/4H-SiC contact under forward bias, according to the *TFE* current transport mechanism. (**b**) Forward current density–voltage characteristics of the Ni Schottky contacts to n-type 4H-SiCwith or without a heavily doped n-type implanted layer. (Figure extracted from Ref. [100]).

4. Conclusions

In this paper, we overviewed some approaches applied in the 4H-SiC Schottky contact development in order to improve the performance of the Schottky devices.

After a short discussion on the fundamentals of the metal/4H-SiC Schottky contact formation and the typical electrical characterization by I–V measurements, we pointed out the well-established technology of Schottky diodes, using Ti or Ni-based Schottky barriers and discussed the current solutions, including the most promising low work function and highly chemically stable metallization schemes and appropriate diode layouts. Then, we presented some unconventional methods based on the manipulation of the metal/semiconductor interface and aimed at an improved control of the Schottky properties of the contact. As a matter of fact, although the metal/4H-SiC system has been studied for a long time, many aspects in the contact formation are still unclear and require a deeper understanding, both from a fundamental and a technological standpoint, in order to obtain superior control of the Schottky contact electrical properties.

Nevertheless, some solutions have shown interesting outcomes. For instance, the introduction of metal nanoparticles in the metal layer has been considered for the advantages given in terms of barrier reduction. Other solutions act on the semiconductor side, for example, with treatments before metal deposition, in order to homogenize the surface and narrow the barrier heights and ideality factor distribution. The effects on the Schottky barrier related to an increase of the doping density of the semiconductor layer have also been investigated. Although these are early studies, they are very promising for the practical implications in Schottky diode technology. In fact, in addition to an improvement of the electrical properties in terms of uniformity, these solutions addressed the superior control of the Schottky barrier height, with the ultimate capability to tailor and tune its value. Besides the possibility of obtaining insight into the physical characteristics of the Schottky contact, this aspect is of particular interest for the device makers, for the development of a new class of Schottky diodes with tailored characteristics.

Author Contributions: Writing—original draft preparation, experimental investigation and data analyses M.V.; writing—review and editing, M.V., F.R. and F.G.; conceptualization, supervision, project management and funding acquisition, F.R. All authors have read and agreed to the published version of the manuscript.

Funding: This research was in part funded by the ECSEL-JU project REACTION (first and euRopEAn siC eigTh Inches pilOt liNe), Grant Agreement No. 783158.

Institutional Review Board Statement: Not applicable.

Informed Consent Statement: Not applicable.

Data Availability Statement: The data that support the findings of this study are available from the corresponding author upon reasonable request.

Acknowledgments: The authors would like to acknowledge their colleagues G. Greco, P. Fiorenza and R. Lo Nigro from CNR-IMM and G. Bellocchi and S. Rascunà from STMicroelectronics for fruitful discussions. S. Di Franco (CNR-IMM) is acknowledged for his technical support during device fabrication.

Conflicts of Interest: The authors declare no conflict of interest.

References

1. Ren, F.; Zopler, J.C. *Wide Band Gap Electronic Devices*; World Scientific Publishing: Singapore, 2003.
2. Roccaforte, F.; Fiorenza, P.; Greco, G.; Nigro, R.L.; Giannazzo, F.; Iucolano, F.; Saggio, M. Emerging trends in wide band gap semiconductors (SiC and GaN) technology for power devices. *Microelectron. Eng.* **2018**, *187–188*, 66–77. [CrossRef]
3. She, X.; Huang, A.Q.; Lucia, O.; Ozpineci, B. Review of Silicon Carbide Power Devices and Their Applications. *IEEE Trans. Ind. Electron.* **2017**, *64*, 8193–8205. [CrossRef]
4. Kimoto, T.; Cooper, J.A. *Fundamentals of Silicon Carbide Technology*; John Wiley & Sons: Singapore, 2014; p. 20.
5. Kimoto, T. Material science and device physics in SiC technology for high-voltage power devices. *Jpn. J. Appl. Phys.* **2015**, *54*, 040103. [CrossRef]

6. STMicroelectronics. Silicon-Carbide—STPower SiC MOSFETs and SiC Diodes, Industrial and Automotive-STMicroelectronics. Available online: https://www.st.com/en/sic-devices.html (accessed on 9 April 2021).
7. Wolfspeed. Power Products. Available online: https://www.wolfspeed.com/power/products (accessed on 9 April 2021).
8. ROHM Co., Ltd. Silicon-Carbide (SiC) Power Devices. Discrete Semiconductors. ROHM Semiconductor-ROHM Co., Ltd. Available online: https://www.rohm.com/products/sic-power-devices (accessed on 9 April 2021).
9. Roccaforte, F.; Brezeanu, G.; Gammon, P.M.; Giannazzo, F.; Rascunà, S.; Saggio, M. Schottky contacts to silicon carbide: Physics, technology and applications. In *Advancing Silicon Carbide Electronics Technology I*; Zekentes, K., Vasilevskiy, K., Eds.; Materials Research Foundations LLC: Millersville, PA, USA, 2018; Volume 37, pp. 127–190.
10. Schottky, W. Zur Halbleitertheorie der Sperrschicht- und Spitzengleichrichter. *Eur. Phys. J. A* **1939**, *113*, 367–414. [CrossRef]
11. Mott, N.F. The theory of crystal rectifiers. *Proc. R. Soc. A* **1939**, *171*, 27–38. [CrossRef]
12. Sze, S.M.; Kwok, K.N. *Physics of Semiconductor Devices*; John Wiley & Sons: Hoboken, NJ, USA, 2007.
13. Lide, D.R. (Ed.) Physical Constants of Organic Compounds. In *CRC Handbook of Chemistry and Physics*; Internet Version 2005; CRC Press: Boca Raton, FL, USA, 2005.
14. Davydov, S.Y. On the electron affinity of silicon carbide polytypes. *Semiconductors* **2007**, *41*, 696–698. [CrossRef]
15. Roccaforte, F.; Giannazzo, F.; Iucolano, F.; Eriksson, J.; Weng, M.H.; Raineri, V. Surface and interface issues in wide band gap semiconductor electronics. *Appl. Surf. Sci.* **2010**, *256*, 5727–5735. [CrossRef]
16. Badila, M.; Brezeanu, G.; Millan, J.; Godignon, P.; Banu, V. Silicon carbide Schottky and ohmic contact process dependence. *Diam. Relat. Mater.* **2002**, *11*, 1258–1262. [CrossRef]
17. Tung, R.T. Electron transport at metal-semiconductor interfaces: General theory. *Phys. Rev. B* **1992**, *45*, 13509–13523. [CrossRef]
18. Tung, R.T. The physics and chemistry of the Schottky barrier height. *Appl. Phys. Rev.* **2014**, *1*, 011304. [CrossRef]
19. Hara, M.; Asada, S.; Maeda, T.; Kimoto, T. Forward thermionic field emission transport and significant image force lowering caused by high electric field at metal/heavily-doped SiC Schottky interfaces. *Appl. Phys. Express* **2020**, *13*, 041001. [CrossRef]
20. Yakimova, R.; Hemmingsson, C.; Macmillan, M.F.; Yakimov, T.; Janzén, E. Barrier height determination for n-type 4H-SiC schottky contacts made using various metals. *J. Electron. Mater.* **1998**, *27*, 871–875. [CrossRef]
21. Werner, J.H.; Güttler, H.H. Barrier inhomogeneities at Schottky contacts. *J. Appl. Phys.* **1991**, *69*, 1522–1533. [CrossRef]
22. Schroder, D.K. *Semiconductor Material and Device Characterization*; John Wiley & Sons: Hoboken, NJ, USA, 2006.
23. Roccaforte, F.; La Via, F.; Raineri, V.; Picrobon, R.; Zanoni, E. Richardson's constant in inhomogeneous silicon carbide Schottky contacts. *J. Appl. Phys.* **2003**, *93*, 9137–9144. [CrossRef]
24. Roccaforte, F.; La Via, F.; Raineri, V. Ohmic contacts to sic. *Int. J. High Speed Electron. Syst.* **2005**, *15*, 781–820. [CrossRef]
25. Padovani, F.A.; Stratton, R. Field and thermionic-field emission in Schottky barriers. *Solid-State Electron.* **1966**, *9*, 695–707. [CrossRef]
26. Pensl, G.; Ciobanu, F.; Frank, M.; Krieger, M.; Reshanov, S.; Schmid, F.; Weidner, M. SiC-Materials and Devices. In *Selected Topics in Electronics and Systems*; Shur, M., Rumyantsev, S., Levinshtein, M., Eds.; World Scientific Publishing: Singapore, 2006; Volume 40, pp. 1–41.
27. Itoh, A.; Kimoto, T.; Matsunami, H. High performance of high-voltage 4H-SiC Schottky barrier diodes. *IEEE Electron. Device Lett.* **1995**, *16*, 280–282. [CrossRef]
28. Park, J.S.; Landry, K.; Perepezko, J. Kinetic control of silicon carbide/metal reactions. *Mater. Sci. Eng. A Struct.* **1999**, *259*, 279–286. [CrossRef]
29. Perrone, D.; Naretto, M.; Ferrero, S.; Scaltrito, L.; Pirri, C.F. 4H-SiC Schottky Barrier Diodes Using Mo-, Ti- and Ni-Based Contacts. *Mater. Sci. Forum* **2009**, *615–617*, 647–650. [CrossRef]
30. Lee, S.-K.; Zetterling, C.-M.; Ostling, M. Schottky diode formation and characterization of titanium tungsten to n- and p-type 4H silicon carbide. *J. Appl. Phys.* **2000**, *87*, 8039–8044. [CrossRef]
31. Ramesha, C.K.; Reddy, V.R. Influence of annealing temperature on the electrical and structural properties of palladium Schottky contacts on n-type 4H–SiC. *Superlattices Microstruct.* **2014**, *76*, 55–65. [CrossRef]
32. Huang, L.; Wang, D. Temperature dependent electrical characteristics of Pt Schottky barriers fabricated on lightly and highly doped n-type 4H-SiC. *Jpn. J. Appl. Phys.* **2015**, *54*, 114101. [CrossRef]
33. Gora, V.E.; Chawanda, A.; Nyamhere, C.; Auret, F.D.; Mazunga, F.; Jaure, T.; Chibaya, B.; Omotoso, E.; Danga, H.T.; Tunhuma, S.M. Comparison of nickel, cobalt, palladium, and tungsten Schottky contacts on n-4 H -silicon carbide. *Phys. B Condens. Matter* **2018**, *535*, 333–337. [CrossRef]
34. Choi, K.J.; Han, S.Y.; Lee, J.-L. Degradation mechanism of Schottky diodes on inductively coupled plasma-etched n-type 4H-SiC. *J. Appl. Phys.* **2003**, *94*, 1765–1768. [CrossRef]
35. Oder, T.N.; Kundeti, K.C.; Borucki, N.; Isukapati, S.B. Effects of deposition temperature on the electrical properties of Ti/SiC Schottky barrier diodes. *AIP Adv.* **2017**, *7*, 125311. [CrossRef]
36. Itoh, A.; Matsunami, H. Analysis of Schottky Barrier Heights of Metal/SiC Contacts and Its Possible Application to High-Voltage Rectifying Devices. *Phys. Status Solidi* **1997**, *162*, 389–408. [CrossRef]
37. Nakamura, T.; Miyanagi, T.; Kamata, I.; Jikimoto, T.; Tsuchida, H. A 4.15 kV 9.07-Ω cm^2 4H-SiC Schottky-barrier diode using Mo contact annealed at high temperature. *IEEE Electron. Device Lett.* **2005**, *26*, 99–101. [CrossRef]
38. Kang, I.-H.; Kim, S.-C.; Moon, J.-H.; Bahng, W.; Kim, N.-K. Fabrication of a 600-V/20-A 4H-SiC schottky barrier diode. *J. Korean Phys. Soc.* **2014**, *64*, 1886–1891. [CrossRef]

39. Vivona, M.; Bellocchi, G.; Nigro, R.L.; Rascuna, S.; Roccaforte, F. Electrical evolution of W and WC Schottky contacts on 4H-SiC at different annealing temperatures. *Semicond. Sci. Technol.* **2022**, *37*, 015012. [CrossRef]
40. Toumi, S.; Ouennoughi, Z.; Weiss, R. Temperature analysis of the Gaussian distribution modeling the barrier height inhomogeneity in the Tungsten/4H-SiC Schottky diode. *Appl. Phys. A* **2021**, *127*, 1–8. [CrossRef]
41. Hong, J.; Kim, K.H.; Kim, K.H. Rectifying Characteristics of Thermally Treated Mo/SiC Schottky Contact. *Coatings* **2019**, *9*, 388. [CrossRef]
42. Kwietniewski, N.; Sochacki, M.; Szmidt, J.; Guziewicz, M.; Kaminska, E.; Piotrowska, A. Influence of surface cleaning effects on properties of Schottky diodes on 4H–SiC. *Appl. Surf. Sci.* **2008**, *254*, 8106–8110. [CrossRef]
43. Saxena, V.; Su, J.N.; Steckl, A.J. High-voltage Ni- and Pt-SiC Schottky diodes utilizing metal field plate termination. *IEEE Trans. Electron. Devices* **1999**, *46*, 456–464. [CrossRef]
44. Shalish, I.; De Oliveira, C.E.M.; Shapira, Y.; Burstein, L.; Eizenberg, M. Thermal stability of Pt Schottky contacts to 4H–SiC. *J. Appl. Phys.* **2000**, *88*, 5724–5728. [CrossRef]
45. Schlüter, M. Chemical trends in metal-semiconductor barrier heights. *Phys. Rev. B* **1978**, *17*, 5044–5047. [CrossRef]
46. Mönch, W. *Silicon Carbide: Recent Major Advances*; Choyke, W.J., Matsunami, H., Pensl, G., Eds.; Springer: Berlin/Heidelberg, Germany, 2004.
47. Roccaforte, F.; Vivona, M.; Greco, G.; Nigro, R.L.; Giannazzo, F.; Rascunà, S.; Saggio, M. Metal/Semiconductor Contacts to Silicon Carbide: Physics and Technology. *Mater. Sci. Forum* **2018**, *924*, 339–344. [CrossRef]
48. Rupp, R.; Elpelt, R.; Gerlach, R.; Schomer, R.; Draghici, M. A new SiC diode with significantly reduced threshold voltage. In Proceedings of the 2017 29th International Symposium on Power Semiconductor Devices and IC's (ISPSD), Sapporo, Japan, 28 May–1 June 2017; pp. 355–358. [CrossRef]
49. Pristavu, G.; Brezeanu, G.; Pascu, R.; Drăghici, F.; Bădilă, M. Characterization of non-uniform Ni/4H-SiC Schottky diodes for improved responsivity in high-temperature sensing. *Mater. Sci. Semicond. Process.* **2019**, *94*, 64–69. [CrossRef]
50. Sciuto, A.; Roccaforte, F.; Raineri, V. Electro-optical response of ion-irradiated 4H-SiC Schottky ultraviolet photodetectors. *Appl. Phys. Lett.* **2008**, *92*, 093505. [CrossRef]
51. Sciuto, A.; Roccaforte, F.; Di Franco, S.; Raineri, V.; Billotta, S.; Bonanno, G. Photocurrent gain in 4H-SiC interdigit Schottky UV detectors with a thermally grown oxide layer. *Appl. Phys. Lett.* **2007**, *90*, 223507. [CrossRef]
52. Morisette, D.T.; Cooper, J.A., Jr. Theoretical comparison of SiC PiN and Schottky diodes based on power dissipation considerations. *IEEE Trans. Electron. Devices* **2002**, *49*, 1657–1664. [CrossRef]
53. Baliga, J.B. *Modern Power Devices*, 1st ed.; Wiley-Interscience: New York, NY, USA, 1987.
54. Mehrotra, M.; Baliga, B.J. Trench MOS Barrier Schottky (TMBS) rectifier: A Schottky rectifier with higher than parallel plane breakdown voltage. *Solid-State Electron.* **1995**, *38*, 801–806. [CrossRef]
55. Zhang, Q.; Madangarli, V.; Sudarshan, T. SiC planar MOS-Schottky diode: A high voltage Schottky diode with low leakage current. *Solid-State Electron.* **2001**, *45*, 1085–1089. [CrossRef]
56. Schoen, K.J.; Henning, J.P.; Woodall, J.M.; Coopern, J.A., Jr.; Melloch, M.R. A Dual-Metal-Trench Schottky Pinch-Rectifier in 4H-SiC. *IEEE Electron. Dev. Lett.* **1998**, *19*, 97. [CrossRef]
57. Roccaforte, F.; La Via, F.; La Magna, A.; Di Franco, S.; Raineri, V. Silicon carbide pinch rectifiers using a dual-metal Ti-Ni/sub 2/Si Schottky barrier. *IEEE Trans. Electron. Devices* **2003**, *50*, 1741–1747. [CrossRef]
58. Perez, R.; Mestres, N.; Vellvehi, M.; Godignon, P.; Millan, J. Analysis of 1.2 kV JBS rectifiers fabricated in 4H-SiC. *Semicond. Sci. Technol.* **2006**, *21*, 670–676. [CrossRef]
59. Singh, R.; Capell, D.C.; Hefner, A.R.; Lai, J.; Palmour, J.W. High-power 4H-SiC JBS rectifiers. *IEEE Trans. Electron. Devices* **2002**, *49*, 2054–2063. [CrossRef]
60. Dahlquist, F.; Lendenmann, H.; Östling, M. A High Performance JBS Rectifier—Design Considerations. *Mater. Sci. Forum* **2001**, *353–356*, 683–686. [CrossRef]
61. Baliga, B.J. *Silicon Carbide Power Devices*; World Scientific Publishing: Singapore, 2005.
62. Berthou, M.; Godignon, P.; Montserrat, J.M.; Millan, J.D.R.; Planson, D. Study of 4H-SiC JBS Diodes Fabricated with Tungsten Schottky Barrier. *J. Electron. Mater.* **2011**, *40*, 2355–2362. [CrossRef]
63. Renz, A.B.; Shah, V.A.; Vavasour, O.J.; Bonyadi, Y.; Li, F.; Dai, T.; Baker, G.W.C.; Hindmarsh, S.; Han, Y.; Walker, M.; et al. The improvement of Mo/4H-SiC Schottky diodes via a P2O5 surface passivation treatment. *J. Appl. Phys.* **2020**, *127*, 025704. [CrossRef]
64. Zaman, M.Y.; Perrone, D.; Ferrero, S.; Scaltrito, L.; Naretto, M. Barrier Inhomogeneities of a Medium Size Mo/4H-SiC Schottky Diode. *Mater. Sci. Forum* **2012**, *711*, 188–192. [CrossRef]
65. Ouennoughi, Z.; Toumi, S.; Weiss, R. Study of barrier inhomogeneities using I–V–T characteristics of Mo/4H–SiC Schottky diode. *Phys. B Condens. Matter* **2015**, *456*, 176–181. [CrossRef]
66. Zhang, T.; Raynaud, C.; Planson, D. Measure and analysis of 4H-SiC Schottky barrier height with Mo contacts. *Eur. Phys. J. Appl. Phys.* **2019**, *85*, 10102. [CrossRef]
67. Zeghdar, K.; Dehimi, L.; Pezzimenti, F.; Megherbi, M.L.; Della Corte, F.G. Analysis of the Electrical Characteristics of Mo/4H-SiC Schottky Barrier Diodes for Temperature-Sensing Applications. *J. Electron. Mater.* **2020**, *49*, 1322–1329. [CrossRef]
68. Okamoto, D.; Yano, H.; Hirata, K.; Hatayama, T.; Fuyuki, T. Improved Inversion Channel Mobility in 4H-SiC MOSFETs on Si Face Utilizing Phosphorus-Doped Gate Oxide. *IEEE Electron Device Lett.* **2010**, *31*, 710–712. [CrossRef]

69. Fiorenza, P.; Swanson, L.K.; Vivona, M.; Giannazzo, F.; Bongiorno, C.; Frazzetto, A.; Roccaforte, F. Comparative study of gate oxide in 4H-SiC lateral MOSFETs subjected to post-deposition-annealing in N2O and POCl3. *Appl. Phys. A* **2014**, *115*, 333–339. [CrossRef]
70. Sharma, Y.K.; Ahyi, A.C.; Issacs-Smith, T.; Shen, X.; Pantelides, S.B.; Zhu, X.; Feldman, L.C.; Rozen, J.; Williams, J.R. Phosphorous passivation of the SiO2/4H–SiC interface. *Solid-State Electron.* **2012**, *68*, 103–107. [CrossRef]
71. Katsuno, T.; Watanabe, Y.; Fujiwara, H.; Konishi, M.; Naruoka, H.; Morimoto, J.; Morino, T.; Endo, T. Analysis of surface morphology at leakage current sources of 4H–SiC Schottky barrier diodes. *Appl. Phys. Lett.* **2011**, *98*, 222111. [CrossRef]
72. Stober, L.; Konrath, J.P.; Patocka, F.; Schneider, M.; Schmid, U. Controlling 4H–SiC Schottky Barriers by Molybdenum and Molybdenum Nitride as Contact Materials. *IEEE Trans. Electron. Devices* **2016**, *63*, 578–583. [CrossRef]
73. Hamida, A.F.; Ouennoughi, Z.; Sellai, A.; Weiss, R.; Ryssel, H. Barrier inhomogeneities of tungsten Schottky diodes on 4H-SiC. *Semicond. Sci. Technol.* **2008**, *23*, 045005. [CrossRef]
74. Toumi, S.; Ferhat-Hamida, A.; Boussouar, L.; Sellai, A.; Ouennoughi, Z.; Ryssel, H. Gaussian distribution of inhomogeneous barrier height in tungsten/4H-SiC (000-1) Schottky diodes. *Microelectron. Eng.* **2009**, *86*, 303–309. [CrossRef]
75. Suzuki, T.; Wakabayashi, H.; Tsutsui, K.; Iwai, H.; Kakushima, K. Laminated Mo/C Electrodes for 4H-SiC Schottky Barrier Diodes with Ideal Interface Characteristics. *IEEE Electron. Device Lett.* **2016**, *37*, 618–620. [CrossRef]
76. Knoll, L.; Teodorescu, V.; Minamisawa, R.A. Ultra-Thin Epitaxial Tungsten Carbide Schottky Contacts in 4H-SiC. *IEEE Electron. Device Lett.* **2016**, *37*, 1318–1320. [CrossRef]
77. Chen, K.-H.; Cao, F.; Yang, Z.-Y.; Li, X.-J.; Yang, J.-Q.; Shi, D.-K.; Wang, Y. Improved interface characteristics of Mo/4H-SiC schottky contact. *Solid-State Electron.* **2021**, *185*, 108152. [CrossRef]
78. Vivona, M.; Greco, G.; Bellocchi, G.; Zumbo, L.; Di Franco, S.; Saggio, M.; Rascunà, S.; Roccaforte, F. Electrical properties of inhomogeneous tungsten carbide Schottky barrier on 4H-SiC. *J. Phys. D Appl. Phys.* **2021**, *54*, 055101. [CrossRef]
79. Bellocchi, G.; Vivona, M.; Bongiorno, C.; Badalà, P.; Bassi, A.; Rascuna, S.; Roccaforte, F. Barrier height tuning in Ti/4H-SiC Schottky diodes. *Solid-State Electron.* **2021**, *186*, 108042. [CrossRef]
80. Lee, S.-K.; Zetterling, C.-M.; Östling, M.; Åberg, I.; Magnusson, M.H.; Deppert, K.; Wernersson, L.-E.; Samuelson, L.; Litwin, A. Reduction of the Schottky barrier height on silicon carbide using Au nano-particles. *Solid-State Electron.* **2002**, *46*, 1433–1440. [CrossRef]
81. Ruffino, F.; Crupi, I.; Irrera, A.; Grimaldi, M.G. Pd/Au/SiC Nanostructured Diodes for Nanoelectronics: Room Temperature Electrical Properties. *IEEE Trans. Nanotechnol.* **2010**, *9*, 414–421. [CrossRef]
82. Kang, M.-S.; Ahn, J.-J.; Moon, K.-S.; Koo, S.-M. Metal work-function-dependent barrier height of Ni contacts with metal-embedded nanoparticles to 4H-SiC. *Nanoscale Res. Lett.* **2012**, *7*, 75. [CrossRef] [PubMed]
83. Gorji, M.S.; Cheong, K.Y. Au nanoparticles embedded at the interface of Al/4H-SiC Schottky contacts for current density enhancement. *Appl. Phys. A* **2015**, *118*, 315–325. [CrossRef]
84. Kumta, A.; Rusli; Xia, J.H. Passivation of 4H-SiC Schottky barrier diodes using aluminum based dielectrics. *Solid-State Electron.* **2009**, *53*, 204–210. [CrossRef]
85. Tsui, B.-Y.; Cheng, J.-C.; Lee, L.-S.; Lee, C.-Y.; Tsai, M.-J. Schottky barrier height modification of metal/4H-SiC contact using ultrathin TiO_2 insertion method. *Jpn. J. Appl. Phys.* **2014**, *53*, 04EP10. [CrossRef]
86. Shi, D.-K.; Wang, Y.; Wu, X.; Yang, Z.-Y.; Li, X.-J.; Yang, J.-Q.; Cao, F. Improving the barrier inhomogeneity of 4H-SiC Schottky diodes by inserting Al2O3 interface layer. *Solid-State Electron.* **2021**, *180*, 107992. [CrossRef]
87. Triendl, F.; Pfusterschmied, G.; Berger, C.; Schwarz, S.; Artner, W.; Schmid, U. Ti/4H-SiC schottky barrier modulation by ultrathin a-SiC:H interface layer. *Thin Solid Films* **2021**, *721*, 138539. [CrossRef]
88. Defives, D.; Durand, O.; Wyczisk, F.; Olivier, J.; Noblanc, O.; Brylinski, C. Thermal Annealing Effect on TiN/Ti Layers on 4H-SiC: Metal-Semiconductor Interface Characterization. *Mater. Sci. Forum* **2000**, *338–342*, 411–414. [CrossRef]
89. Giannazzo, F.; Deretzis, I.; Nicotra, G.; Fisichella, G.; Ramasse, Q.M.; Spinella, C.; Roccaforte, F.; La Magna, A. High resolution study of structural and electronic properties of epitaxial graphene grown on off-axis 4H–SiC (0001). *J. Cryst. Growth* **2014**, *393*, 150–155. [CrossRef]
90. Giannazzo, F.; Hertel, S.; Albert, A.; Fisichella, G.; La Magna, A.; Roccaforte, F.; Krieger, M.; Weber, H.B. Electrical Properties of Hydrogen Intercalated Epitaxial Graphene/SiC Interface Investigated by Nanoscale Current Mapping. *Mater. Sci. Forum* **2015**, *821–823*, 929–932. [CrossRef]
91. Varchon, F.; Feng, R.; Hass, J.; Li, X.; Nguyen, B.N.; Naud, C.; Mallet, P.; Veuillen, J.-Y.; Berger, C.; Conrad, E.H.; et al. Electronic Structure of Epitaxial Graphene Layers on SiC: Effect of the Substrate. *Phys. Rev. Lett.* **2007**, *99*, 126805. [CrossRef] [PubMed]
92. Sonde, S.; Giannazzo, F.; Raineri, V.; Yakimova, R.; Huntzinger, J.-R.; Tiberj, A.; Camassel, J. Electrical properties of the graphene/4H-SiC(0001) interface probed by scanning current spectroscopy. *Phys. Rev. B* **2009**, *80*, 241406. [CrossRef]
93. Riedl, C.; Coletti, C.; Iwasaki, T.; Zakharov, A.A.; Starke, U. Quasi-Free-Standing Epitaxial Graphene on SiC Obtained by Hydrogen Intercalation. *Phys. Rev. Lett.* **2009**, *103*, 246804. [CrossRef] [PubMed]
94. Hertel, S.; Waldmann, D.; Jobst, J.; Albert, A.; Albrecht, M.; Reshanov, S.; Schöner, A.; Krieger, M.; Weber, H.B. Tailoring the graphene/silicon carbide interface for monolithic wafer-scale electronics. *Nat. Commun.* **2012**, *3*, 957. [CrossRef] [PubMed]
95. Lin, Z.; Ji, L.; Wu, Y.; Hu, L.; Yan, T.; Sun, Z. Laser-induced interfacial state changes enable tuning of the Schottky-barrier height in SiC. *Appl. Surf. Sci.* **2019**, *469*, 68–75. [CrossRef]

96. Treu, M.; Rupp, R.; Kapels, H.; Bartsch, W. Temperature Dependence of Forward and Reverse Characteristics of Ti, W, Ta and Ni Schottky Diodes on 4H-SiC. *Mater. Sci. Forum* **2001**, *353–356*, 679–682. [CrossRef]
97. Hatakeyama, T.; Shinohe, T. Reverse Characteristics of a 4H-SiC Schottky Barrier Diode. *Mater. Sci. Forum* **2002**, *389–393*, 1169–1172. [CrossRef]
98. Roccaforte, F.; Libertino, S.; Giannazzo, F.; Bongiorno, C.; La Via, F.; Raineri, V. Ion irradiation of inhomogeneous Schottky barriers on silicon carbide. *J. Appl. Phys.* **2005**, *97*, 123502. [CrossRef]
99. Hara, M.; Kaneko, M.; Kimoto, T. Nearly Fermi-level-pinning-free interface in metal/heavily-doped SiC Schottky structures. *Jpn. J. Appl. Phys.* **2021**, *60*, SBBD14. [CrossRef]
100. Vivona, M.; Greco, G.; Spera, M.; Fiorenza, P.; Giannazzo, F.; La Magna, A.; Roccaforte, F. Ni Schottky barrier on heavily doped phosphorous implanted 4H-SiC. *J. Phys. D Appl. Phys.* **2021**, *54*, 445107. [CrossRef]

materials

MDPI

Article

Mechanisms of Ohmic Contact Formation of Ti/Al-Based Metal Stacks on p-Doped 4H-SiC

Matthias Kocher [1,*], Mathias Rommel [1], Pawel Michalowski [2] and Tobias Erlbacher [1,3]

[1] Fraunhofer Institute for Integrated Systems and Device Technology (IISB), Schottkystrasse 10, 91058 Erlangen, Germany; mathias.rommel@iisb.fraunhofer.de (M.R.); tobias.erlbacher@iisb.fraunhofer.de (T.E.)

[2] Łukasiewicz Research Network—Institute of Microelectronics and Photonics, Aleja Lotników 32/46, 02-668 Warsaw, Poland; pawel.michalowski@imif.lukasiewicz.gov.pl

[3] Chair of Electron Devices, Friedrich-Alexander-University Erlangen-Nuremberg, Cauerstraße 6, 91058 Erlangen, Germany

* Correspondence: matthias.kocher@iisb.fraunhofer.de; Tel.: +49-9131-761-392

Abstract: Ohmic contacts on p-doped 4H-SiC are essential for the fabrication of a wide range of power electron devices. Despite the fact that Ti/Al based ohmic contacts are routinely used for ohmic contacts on p-doped 4H-SiC, the underlying contact formation mechanisms are still not fully understood. TLM structures were fabricated, measured and analyzed to get a better understanding of the formation mechanism. SIMS analyses at the Ti_3SiC_2-SiC interface have shown a significant increase of the surface near Al concentration. By using numerical simulation it is shown that this additional surface near Al concentration is essential for the ohmic contact formation.

Keywords: 4H-SiC; ohmic contact; SIMS; Ti_3SiC_2; simulation

check for updates

Citation: Kocher, M.; Rommel, M.; Michalowski, P.; Erlbacher, T. Mechanisms of Ohmic Contact Formation of Ti/Al-Based Metal Stacks on p-Doped 4H-SiC. *Materials* **2022**, *15*, 50. https://doi.org/10.3390/ma15010050

Academic Editor: Fabrizio Roccaforte

Received: 22 October 2021
Accepted: 16 December 2021
Published: 22 December 2021

1. Introduction

Silicon carbide (SiC) is predestined for the fabrication of high power, high temperature and high frequency semiconductor devices, due to its outstanding properties. Despite the commercial availability of SiC power devices, like vertical MOSFETs or Schottky barrier diodes, some topics are not fully understood yet and need further investigations. One major topic is the understanding of ohmic contact formation mechanism on p-doped 4H-SiC [1], due to its importance in the fabrication of bipolar power devices, like pin-diodes or Insulated Gate Bipolar Transistors (IGBTs).

In order to fabricate reliable and low ohmic contacts on p-doped 4H-SiC, various metals and metal stacks have been investigated [1,2]. Due to rather low obtained contact resistivities Ti/Al based metal stacks have become a quasi-standard for ohmic contacts on p-type SiC [1–3]. Numerous studies with different Ti/Al ratios and stacking sequences as well as varying Al surface concentrations fabricated by epitaxial growth or implantation were done in order to fabricate low-ohmic contacts on p-doped 4H SiC [3–11]. These studies have shown the possibility of fabricating ohmic contacts with a specific contact resistance ρ_C down to 10^{-6} Ωcm^2 [2,6].

It was also shown that the formation of Ti/Al based low ohmic contacts correlates with the appearance of Ti_3SiC_2 on the 4H-SiC surface. Several investigations in literature have revealed an epitaxial growth of Ti_3SiC_2 during contact formation [3,8,11–16]. Maeda et al. have described that the epitaxial growth of Ti_3SiC_2 consists of two separate reactions (see Equations (1) and (2)) [14].

$$Ti + 3Al = TiAl_3 \text{ at } 959 \text{ K} \quad (1)$$

$$2SiC + 3TiAl_3 = Ti_3SiC_2 + 9Al + Si \text{ at } 1270 \text{ K} \quad (2)$$

Ti_3SiC_2 formation is key for ohmic contact formation mechanism [1–3,6,8,12,17]. Notwithstanding these results, the formation mechanism of Ti/Al based ohmic contacts on p-doped 4H-SiC is not fully understood [1]. Therefore, this work investigates the underlying formation mechanism and sets up a theory for the contact formation mechanism by analyzing and simulating the Ti/Al based ohmic contacts interface.

2. Materials and Methods

Transfer length method (TLM) structures with increasing pad distances (30 µm to 480 µm) were fabricated on n^- doped 4H-SiC epitaxial layers in order to investigate the contact formation mechanism of Ti/Al based ohmic contact. Al implantation and subsequent high temperature annealing was used to create samples with p^+ doped regions and Al surface concentrations between 3.3×10^{18} cm^{-3} and 5.0×10^{19} cm^{-3}. After depositing and structuring a SiO_2 passivation layer, a metal stack consisting of 80 nm Ti and 300 nm Al was deposited and patterned using a lift-off process. Subsequently ohmic contact formation was done by rapid thermal annealing (RTA) at 980 °C in Ar atmosphere. Finally, metal pads for electrical measurements were deposited and structured. A more detailed description of the fabrication process of sample C can be seen elsewhere [18]. Table 1 gives an overview of the fabricated sets of samples and their fabrication parameters.

Table 1. Parameters of the fabricated sets of samples.

	A	B	C
Implanted Al surface concentration [10^{19} cm^{-3}]	5.0	5.0	0.33 to 5.0 (14 different Al surface conc.)
Implanted Al dose [10^{14} cm^{-2}]	9.0	6.0	0.34 to 5.1 (14 different Al doses)
High temperature implantation annealing	1700 °C, 30 min, Ar atmosphere		

By using 4-point I-V-measurements (Keithley 4200 Parameter Analyzer, Keithley Instruments, USA) at different temperatures (300 K to 450 K), the fabricated TLM structures were electrically characterized and the sheet resistance R_{sh}, the contact resistance R_C as well as the specific contact resistance ρ_C were determined in the given temperature range.

Focused ion beam (FIB) (Helios Nanolab 600, FEI, USA) as well as transmission electron microscopy (TEM) analysis were done to determine the thickness of the Ti_3SiC_2 layer (approx. 100 nm) as well as its stoichiometry.

3. Results

3.1. Sheet Resistance and Determination of the Acceptor Ionization Energy

All fabricated samples show ohmic behavior across all measurement temperatures. The sheet resistance R_{sh} was used to determine the acceptor ionization energies ΔE_A of the fabricated samples. Equation (3) (together with Equations (4)–(8)) can be used to describe the sheet resistance R_{sh} of a semiconductor, where q indicates the elementary charge, t the thickness of the semiconductor layer, p and n the hole and electron concentrations, μ_p and μ_n the hole and electron mobilities, respectively.

$$R_{sh} = \left(q \int_0^t (\mu_n(x)n(x) + \mu_p(x)p(x))dx \right)^{-1} \tag{3}$$

Equations (4)–(6) were used to calculate the hole and electron mobility and their respective temperature dependence, where μ_{const} describes the mobility due to phonon scattering, μ_{dop} the doping dependent mobility degradation, T the temperature, N_D the donor concentration and N_A the acceptor concentration (all other parameters and their values are shown in Table A1 [19].

$$\mu = \left(\mu_{const}^{-1} + \mu_{dop}^{-1} \right)^{-1} \tag{4}$$

$$\mu_{const} = \mu_L \left(\frac{T}{300\text{K}} \right)^{-\xi} \tag{5}$$

$$\mu_{dop} = A_{min} \left(\frac{T}{300\text{K}} \right)^{\alpha_m} + \frac{A_d \left(\frac{T}{300\text{K}} \right)^{\alpha_d}}{1 + \left(\frac{N_A + N_D}{A_N \left(\frac{T}{300\text{K}} \right)^{\alpha_N}} \right)^{A_a \left(\frac{T}{300\text{K}} \right)^{\alpha_a}}} \tag{6}$$

Equations (7) and (8) were used to describe the carrier ionization, where N_D^+ describes the ionized donor concentration and N_A^- describes the ionized acceptor concentration (all other parameters and their values are shown in Table A2 [1]. Here, a negligible carrier compensation ($p \approx N_A^-$) was assumed at first.

$$n \approx N_D^+ = \frac{\eta_n}{2} \left(\sqrt{1 + \frac{4N_D}{\eta_n}} - 1 \right) \text{ with } \begin{array}{l} \eta_n = \frac{N_C}{g_D} exp\left(-\frac{\Delta E_D}{kT} \right) \\ N_C = N_{C,300\text{K}} \left(\frac{T}{300\text{K}} \right)^{1.5} \end{array} \tag{7}$$

$$p \approx N_A^- = \frac{\eta_p}{2} \left(\sqrt{1 + \frac{4N_A}{\eta_p}} - 1 \right) \text{ with } \begin{array}{l} \eta_p = \frac{N_V}{g_A} exp\left(-\frac{\Delta E_A}{kT} \right) \\ N_V = N_{V,300\text{K}} \left(\frac{T}{300\text{K}} \right)^{1.5} \end{array} \tag{8}$$

In Figure 1a the normalized average measured sheet resistances and the associated standard error of all sets of samples with an implanted Al surface concentration of 5×10^{19} cm^{-3} are shown. It can be seen that all sets of samples show similar temperature dependent behavior despite differences in sheet resistance values (see inset in Figure 1a)

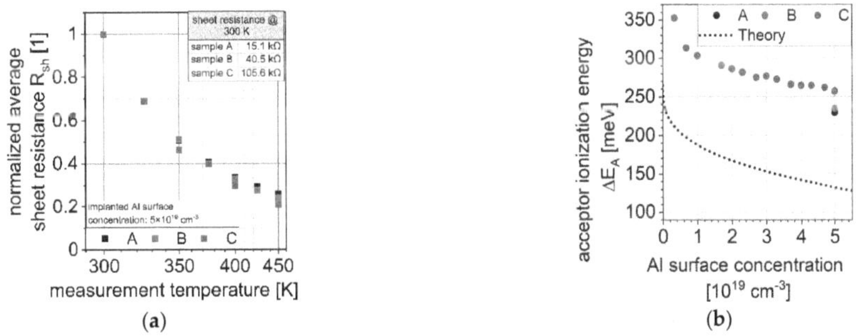

Figure 1. (**a**) Measurement temperature dependence of normalized average sheet resistance R_{sh}; (**b**) Determined effective acceptor ionization energies ΔE_A.

The associated acceptor ionization energies ΔE_A were determined by fitting the theoretical sheet resistance to the measured ones. The theoretical sheet resistance was determined by using the simulated implantation profile and assuming 100% activation of the dopants.

The determined effective acceptor ionization energies ΔE_A as well as the theoretical acceptor ionization energies (see Equation (9) with $\Delta E_{A,0}$ = 0.265 eV [19,20]) are shown in Figure 1b. It can be seen that the determined acceptor ionization energies differ significantly from the theoretical ones, which can be explained by a significant amount of carrier compensation centers (see Equation (8)). As discussed in Section 3.3, these compensation centers might be modelled by donor-like defects that trap free holes.

$$\Delta E_A = \Delta E_{A,0} - \sqrt[3]{N_D + N_A} \tag{9}$$

3.2. Determination of Schottky Barrier Height

Based on the Thermionic Field Emission (TFE) model [21], Equation (10) can be used to determine the Schottky barrier height ϕ_B, where k describes the Boltzmann constant, h the Planck constant, m^* the effective tunneling mass (here 0.91 electron masses [22–24]), ϵ_0 the vacuum permittivity and ϵ_S the relative permittivity of 4H-SiC (here 9.7 [4,22,23]).

$$\rho_{C,TFE} \propto \frac{q\phi_B}{E_{00}\coth\left(\frac{E_{00}}{kT}\right)} \; with \; E_{00} = \frac{qh}{2\pi}\sqrt{\frac{p}{m^*\epsilon_0\epsilon_S}} \quad (10)$$

The Schottky barrier height ϕ_B itself can be calculated by Equation (11), where E_g describes the bandgap of the semiconductor, ϕ_M the metal workfunction of the ohmic contact material, χ_S the electron affinity of the semiconductor and V_i the built-in voltage [21,25]. It can be seen that the Schottky barrier height decreases slightly with increasing N_A^-.

$$\phi_B = \frac{E_g}{q} - (\phi_M - \chi_S) - \sqrt[4]{\frac{q^3 N_A^- V_i}{8\pi^2\epsilon_0^3\epsilon_S^3}} \quad (11)$$

Figure 2a shows the normalized average specific contact resistances of all sets of samples with an implanted Al surface concentration of 5×10^{19} cm^{-3}. It can be observed that all sets of samples show quite similar temperature dependent behavior despite different absolute values of the specific contact resistances (see inset of Figure 2a) and despite the deviation of sample B at temperatures higher than 375 K (indicated by the open red squares). The origin of these deviations is not fully understood. Therefore, these values are not used further.

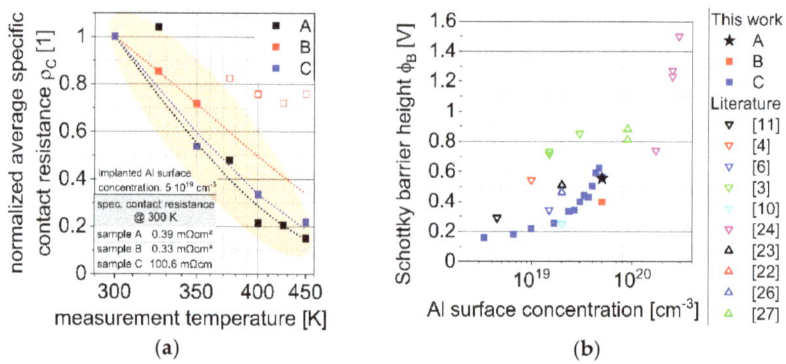

Figure 2. (**a**) Normalized average specific contact resistance (**b**) Determined Schottky barrier heights form this work and literature. Ohmic contacts from literature fabricated on epitaxial regions are indicated with ∇ ([3,4,6,10,11,24]), ohmic contacts from literature fabricated on implanted regions are indicated with Δ ([22,23,26,27]).

By fitting the theoretical specific contact resistance to the measured ones, the Schottky barrier heights ϕ_B were determined (see Equation (10)). Figure 2b shows the determined Schottky barrier heights from this work and compares them with Schottky barrier heights known from literature. It can be seen that the determined Schottky barrier heights increase with increasing Al surface concentration, which is in contrast to the theoretically predicted decreasing of the Schottky barrier height with increasing Al surface concentration (see Equation (11)).

In order to investigate this contradiction Secondary Ion Mass Spectrometry (SIMS) analyses were carefully done at the 4H-SiC/Ti_3SiC_2-interface of sample A by using a CAMECA IMS SC Ultra SIMS tool which allows a sub-nm resolution [28,29]. The sub-nm depth resolution was achieved for O^{2+} primary ions with an impact energy of 250 eV. The

Al concentration was calibrated using a reference sample consisting of a SiC substrate implanted with Al ions with an energy of 100 keV and a dose of 10^{14} cm^{-2}.

Figure 3a shows the measured Al concentration, the measured Ti and Si counts per second (CPS) as well as the implanted Al profile. While no Al could be detected in the Ti_3SiC_2 layer, the Al concentration at the 4H-SiC-Ti_3SiC_2-interface is significantly increased. This additional Al concentration decreases within approx. 3 nm from a peak concentration of approx. 10^{21} cm^{-3} to the implanted Al concentration (5×10^{19} cm^{-3}). Furthermore no significant amount of Ti could be detected in the 4H-SiC layer.

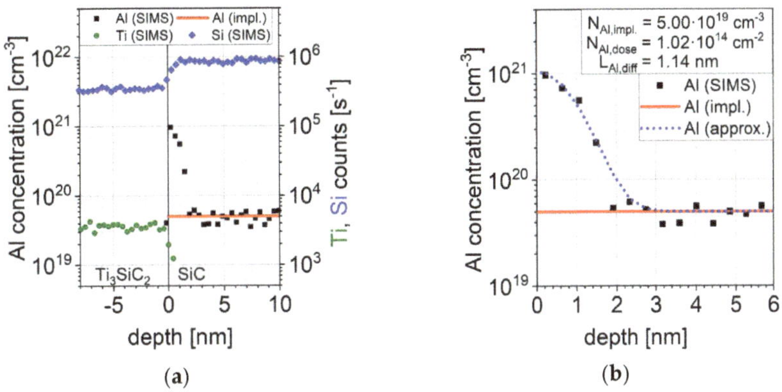

Figure 3. (**a**) Measured SIMS profiles of Al, Ti and Si on the Ti_3SiC_2-SiC interface (**b**) Approximation of the measured Al profile.

This increase of the Al concentration can be explained by a diffusion of Al during Ti_3SiC_2 formation. The total resulting Al profile can be approximated by the superposition of the implanted Al profile $N_{Al,impl.}$ and the increase of the Al concentration at the 4H-SiC/Ti_3SiC_2-interface. Equation (12) describes this superposition by using the implanted Al profile $N_{Al,impl.}$, the diffused Al dose during high temperature annealing $N_{Al,dose}$ and the associated diffusion length $L_{Al,diff}$ [25]. Figure 3b shows the approximation as well as the associated parameters. It can be seen, that the approximation fits very well with the measured data.

$$N_{Al}(x) = N_{Al,impl.}(x) + \frac{2\,N_{Al,dose}}{L_{Al,diff}\sqrt{\pi}} \exp\left(-\frac{x^2}{L^2_{Al,diff}}\right) \quad (12)$$

3.3. Numerical Simulation

To investigate the influence of the surface-near increased Al concentration as well as the temperature dependent behavior of sample A a numerical simulation model was developed by using Sentaurus TCAD (Version O_2018.06). Figure 4 shows the scheme of the used simulation model. This model includes Ti_3SiC_2 based ohmic contact pads with a height of 100 nm and a pad distance d between the ohmic contacts. The model includes further a homogeneous n^--doped 4H-SiC epitaxial layer, a p^+-doped region and a p^{++}-doped region beneath the ohmic contacts. The p^+-doped region was created by using a Monte Carlo simulation of the implanted Al profiles. The surface near p^{++}-doped region was created by adding a Gaussian distributed Al profile with diffused Al dose $N_{Al,dose}$ and the associated diffusion length $L_{Al,diff}$. Furthermore the model assumes complete activation of the Al atoms and takes account of incomplete ionization as well as doping and temperature dependent carrier mobility.

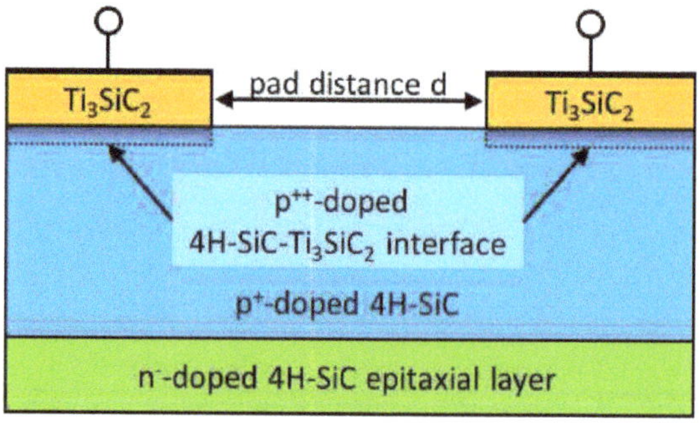

Figure 4. Scheme of the simulation model.

An additional virtual N profile was added in order to model the concentration of carrier compensation centers. The distribution of this additional virtual N profile is identical with the distribution of the implanted Al profile and can be scaled by using the compensation ratio f_{comp}. Due to these additional donor atoms the concentration of free holes can be reduced similarly to compensation by donor-like traps which increases the associated sheet resistance and allows to fit R_{sh}.

Using this numerical simulation model, I-V characteristics depending on the diffused Al dose $N_{Al,dose}$, the associated diffusion length $L_{Al,diff}$ and the compensation ratio f_{comp} for each temperature and each pad distance d can be obtained. This allows to simulate I-V characteristics for TLM structures at different temperatures.

Based on these I–V characteristics the sheet resistance R_{sh} and the contact resistance R_C of the modelled TLM structures were determined. Figure 5a compares the sheet resistance R_{sh}, Figure 5b compares the contact resistance R_C determined from the electrical measurement data with simulated ones of sample A. It can be seen that both fits are in decent agreement with the measurements. It should be mentioned here that no adjustments on the parameters $N_{Al,dose}$ and $L_{Al,diff}$ determined by the SIMS analysis were necessary.

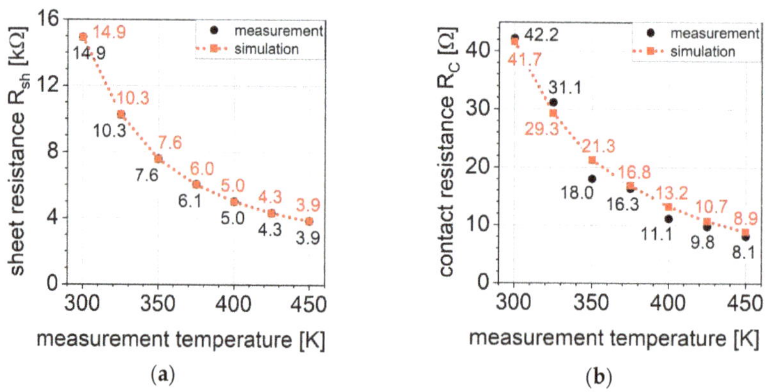

Figure 5. (**a**) Measured and simulated sheet resistance (**b**) Measured and simulated contact resistance.

The determined compensation ratio f_{comp} is dependent on the temperature as shown in Figure 6. It can be seen that f_{comp} increases slightly with increasing temperature from 8.3% at 300 K to 10.5% at 450 K which fits to temperature independent compensation ratios known from literature (10% to 27%) [30–32]. This temperature dependence might be explained by the fact that the ionization energy from nitrogen differs from the ionization energy of the actual compensation centers.

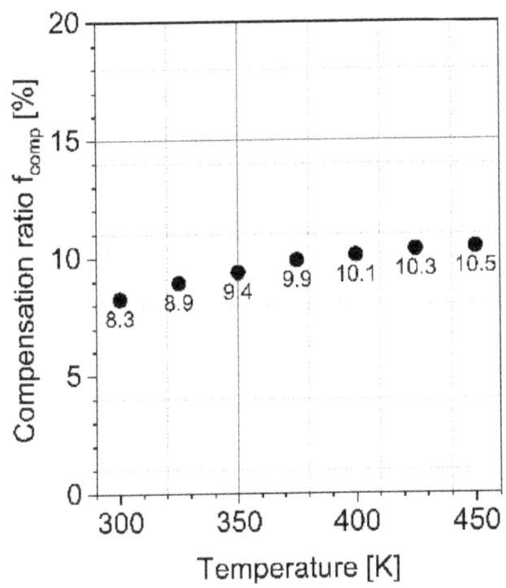

Figure 6. Temperature dependence of determined compensation ratio.

4. Discussion

Due to the well-fitted simulation results, it can be concluded that the numerical simulation model is suitable to describe Ti_3SiC_2 based ohmic contacts on p-doped 4H-SiC temperature dependent. Considering the fact that the simulation model does not show ohmic behavior when not using the surface near Al profile it can be further concluded that the surface near Al profile is essential for the ohmic contact formation. Based on these results it is possible to propose a theory regarding the formation mechanism of Ti/Al based ohmic contact on p-doped 4H-SiC and the role of Ti_3SiC_2 during contact formation.

During the Ti_3SiC_2 formation a certain amount of Al diffuses in the SiC surface via lattice places and increases the surface near Al concentration significantly. This increase of the surface near Al concentration can significantly decrease the specific resistance ρ_C (see Equation (10)) and is therefore the key in the ohmic contact formation.

Further investigations are necessary to verify this model and to obtain a better understanding of the conditions leading to the ohmic contact formation under various process conditions. Nevertheless, the fundamental effects are becoming accessible for process integration and process modelling.

Author Contributions: Conceptualization, M.K.; methodology, M.K.; formal analysis, P.M.; software, M.K.; writing—original draft preparation, M.K.; writing—review and editing, M.R. and T.E. All authors have read and agreed to the published version of the manuscript.

Funding: This research received no external funding.

Institutional Review Board Statement: Not applicable.

Informed Consent Statement: Not applicable.

Data Availability Statement: The datasets generated and analysed during the current study are available from the corresponding author on reasonable request.

Conflicts of Interest: The authors declare no conflict of interest.

Appendix A

Table A1. Mobility parameters [19].

Equation	Parameter	Unit	Electrons	Holes
(4)	μ_L	cm^2/Vs	950	125
(4)	ξ	1	2.4	2.15
(5)	A_{min}	cm^2/Vs	40	0
(5)	α_m	1	−1.536	−0.57
(5)	A_d	cm^2/Vs	910	113.5
(5)	α_d	1	−2.397	−2.6
(5)	A_N	cm^{-3}	2.0×10^{17}	2.4×10^{18}
(5)	α_N	1	0.75	2.9
(5)	A_a	1	0.76	0.69
(5)	a_a	1	0.722	−0.2

Table A2. Incomplete ionization parameters [19].

Equation	Parameter	Unit	Nitrogen	Aluminum
(7)	g_D	1	2	—
(7)	ΔE_D	eV	0.0709	—
(7)	$N_{C,300K}$	cm^{-3}	1.7193×10^{19}	—
(8)	g_A	1	—	4
(8)	ΔE_A	eV	—	0.265
(8)	$N_{V,300K}$	cm^{-3}	—	1.0752×10^{20}

References

1. Kimoto, T.; Cooper, J.A. *Fundamentals of Silicon Carbide Technology: Growth, Characterization, Devices and Applications*; IEEE Xplore: Singapore; Piscataway, NJ, USA, 2014; ISBN 978-1-118-31352-7.
2. Roccaforte, F.; Frazzetto, A.; Greco, G.; Giannazzo, F.; Fiorenza, P.; Lo Nigro, R.; Saggio, M.; Leszczyński, M.; Pristawko, P.; Raineri, V. Critical issues for interfaces to p-type SiC and GaN in power devices. *Appl. Surf. Sci.* **2012**, *258*, 8324–8333. [CrossRef]
3. Abi-Tannous, T.; Soueidan, M.; Ferro, G.; Lazar, M.; Raynaud, C.; Toury, B.; Beaufort, M.-F.; Barbot, J.-F.; Dezellus, O.; Planson, D. A Study on the Temperature of Ohmic Contact to p-Type SiC Based on Ti_3SiC_2 Phase. *IEEE Trans. Electron. Devices* **2016**, *63*, 2462–2468. [CrossRef]
4. Fisher, C.A. On the Schottky Barrier Height Lowering Effect of Ti_3SiC_2 in Ohmic Contacts to P-Type 4H-SiC. *IJFPS* **2014**, *4*, 95–100. [CrossRef]
5. Gao, M.; Tsukimoto, S.; Goss, S.H.; Tumakha, S.P.; Onishi, T.; Murakami, M.; Brillson, L.J. Role of Interface Layers and Localized States in TiAl-Based Ohmic Contacts to p-Type 4H-SiC. *J. Electron. Mater.* **2007**, *36*, 277–284. [CrossRef]
6. Jennings, M.R.; Fisher, C.A.; Walker, D.; Sanchez, A.; Pérez-Tomás, A.; Hamilton, D.P.; Gammon, P.M.; Burrows, S.E.; Thomas, S.M.; Sharma, Y.K.; et al. On the Ti_3SiC_2 Metallic Phase Formation for Robust p-Type 4H-SiC Ohmic Contacts. *Mater. Sci. Forum* **2014**, *778-780*, 693–696. [CrossRef]
7. Johnson, B.J.; Capano, M.A. The effect of titanium on Al–Ti contacts to p-type 4H-SiC. *Solid-State Electron.* **2003**, *47*, 1437–1441. [CrossRef]
8. Johnson, B.J.; Capano, M.A. Mechanism of ohmic behavior of Al/Ti contacts to p-type 4H-SiC after annealing. *Solid-State Electron.* **2004**, *95*, 5616–5620. [CrossRef]
9. Nakatsuka, O.; Takei, T.; Koide, Y.; Murakami, M. Low Resistance TiAl Ohmic Contacts with Multi-Layered Structure for p-Type 4H-SiC. *Mater. Trans.* **2002**, *43*, 1684–1688. [CrossRef]
10. Tang, Y.D.; Shen, H.J.; Zhang, X.F.; Guo, F.; Bai, Y.; Peng, Z.Y.; Liu, X.Y. Effect of Annealing on the Characteristics of Ti/Al Ohmic Contacts to p-Type 4H-SiC. *Mater. Sci. Forum* **2017**, *897*, 395–398. [CrossRef]
11. Tsukimoto, S.; Nitta, K.; Sakai, T.; Moriyama, M.; Murakami, M. Correlation between the electrical properties and the interfacial microstructures of TiAl-based ohmic contacts to p-type 4H-SiC. *J. Electron. Mater.* **2004**, *33*, 460–466. [CrossRef]
12. Tsukimoto, S.; Ito, K.; Wang, Z.; Saito, M.; Ikuhara, Y.; Murakami, M. Growth and Microstructure of Epitaxial Ti_3SiC_2 Contact Layers on SiC. *Mater. Trans.* **2009**, *50*, 1071–1075. [CrossRef]
13. Ideguchi, K.; Maeda, M.; Takahashi, Y. Nucleation and growth of Ti_3SiC_2 on SiC by interfacial reaction. *IOP Conf. Ser. Mater. Sci. Eng.* **2014**, *61*, 12032. [CrossRef]
14. Maeda, M.; Takahashi, Y. Formation of Ti3SiC2 on SiC by Control of Interfacial Reaction between SiC and Ti/Al Multilayer. *Prepr. Natl. Meet. JWS* **2008**, *2008*, 105. [CrossRef]

15. Wang, Z.; Tsukimoto, S.; Saito, M.; Ito, K.; Murakami, M.; Ikuhara, Y. Ohmic contacts on silicon carbide: The first monolayer and its electronic effect. *Phys. Rev. B* **2009**, *80*, 245303. [CrossRef]
16. Vivona, M.; Greco, G.; Bongiorno, C.; Lo Nigro, R.; Scalese, S.; Roccaforte, F. Electrical and structural properties of surfaces and interfaces in Ti/Al/Ni Ohmic contacts to p-type implanted 4H-SiC. *Appl. Surf. Sci.* **2017**, *420*, 331–335. [CrossRef]
17. Buchholt, K.; Ghandi, R.; Domeij, M.; Zetterling, C.-M.; Lu, J.; Eklund, P.; Hultman, L.; Spetz, A.L. Ohmic contact properties of magnetron sputtered Ti_3SiC_2 on n-and p-type 4H-silicon carbide. *Appl. Phys. Lett.* **2011**, *98*, 042108. [CrossRef]
18. Kocher, M.; Rommel, M.; Erlbacher, T.; Bauer, A.J. Influence of Al Doping Concentration and Annealing Parameters on TiAl Based Ohmic Contacts on 4H-SiC. *Mater. Sci. Forum* **2018**, *924*, 393–396. [CrossRef]
19. Synopsys, Inc. *Sentaurus™ Device User Guide: Version O-2018.06*; Synopsys, Inc.: Mountain View, CA, USA, 2018.
20. Koizumi, A.; Suda, J.; Kimoto, T. Temperature and doping dependencies of electrical properties in Al-doped 4H-SiC epitaxial layers. *J. Appl. Phys.* **2009**, *106*, 13716. [CrossRef]
21. Sze, S.M.; Ng, K.K.; Ng, K.K. *Physics of Semiconductor Devices/Physics of Semiconductor Devices*, 3rd ed.; Wiley-Interscience: Hoboken, NJ, USA, 2007; ISBN 0-471-14323-5.
22. Roccaforte, F.; Vivona, M.; Greco, G.; Lo Nigro, R.; Giannazzo, F.; Di Franco, S.; Bongiorno, C.; Iucolano, F.; Frazzetto, A.; Rascunà, S.; et al. Ti/Al-based contacts to p-type SiC and GaN for power device applications. *Phys. Status Solidi A* **2017**, *214*, 1600357. [CrossRef]
23. Frazzetto, A.; Giannazzo, F.; Lo Nigro, R.; Raineri, V.; Roccaforte, F. Structural and transport properties in alloyed Ti/Al Ohmic contacts formed on p-type Al-implanted 4H-SiC annealed at high temperature. *J. Phys. D Appl. Phys.* **2011**, *44*, 255302. [CrossRef]
24. Thierry-Jebali, N.; Vo-Ha, A.; Carole, D.; Lazar, M.; Ferro, G.; Planson, D.; Henry, A.; Brosselard, P. Very low specific contact resistance measurements made on a highly p-type doped 4H-SiC layer selectively grown by vapor-liquid-solid transport. *Appl. Phys. Lett.* **2013**, *102*, 212108. [CrossRef]
25. Colinge, J.P.; Colinge, C.A. *Physics of Semiconductor Devices*; Kluwer Academic Publishers: Boston, MA, USA, 2002; ISBN 978-0-306-47622-8.
26. Roccaforte, F.; Frazzetto, A.; Greco, G.; Lo Nigro, R.; Giannazzo, F.; Leszczyński, M.; Prystawko, P.; Zanetti, E.; Saggio, M.; Raineri, V. Microstructure and Transport Properties in Alloyed Ohmic Contacts to P-Type SiC and GaN for Power Devices Applications. *Mater. Sci. Forum* **2012**, *711*, 203–207. [CrossRef]
27. Zhang, Y.; Guo, T.; Tang, X.; Yang, J.; He, Y.; Zhang, Y. Thermal stability study of n-type and p-type ohmic contacts simultaneously formed on 4H-SiC. *J. Alloys Compd.* **2018**, *731*, 1267–1274. [CrossRef]
28. Michałowski, P.P.; Kozdra, S.; Pasternak, I.; Sitek, J.; Wójcik, A.; Strupiński, W. Precise localization of contaminants in graphene with secondary ion mass spectrometry. *Measurement* **2021**, 110308. [CrossRef]
29. Michałowski, P.P.; Maciążek, D.; Postawa, Z.; Caban, P.A.; Kozdra, S.; Wójcik, A.; Baranowski, J.M. Defect-mediated sputtering process of boron nitride during high incident angle low-energy ion bombardment. *Measurement* **2021**, *179*, 109487. [CrossRef]
30. Parisini, A.; Nipoti, R. Analysis of the hole transport through valence band states in heavy Al doped 4H-SiC by ion implantation. *J. Appl. Phys.* **2013**, *114*, 243703. [CrossRef]
31. Kocher, M.; Yao, B.T.; Weisse, J.; Rommel, M.; Xu, Z.W.; Erlbacher, T.; Bauer, A. Determination of Compensation Ratios of Al-Implanted 4H-SiC by TCAD Modelling of TLM Measurements. *Mater. Sci. Forum* **2019**, *963*, 445–448. [CrossRef]
32. Weiße, J.; Hauck, M.; Sledziewski, T.; Tschiesche, M.; Krieger, M.; Bauer, A.J.; Mitlehner, H.; Frey, L.; Erlbacher, T. Analysis of Compensation Effects in Aluminum-Implanted 4H-SiC Devices. *Mater. Sci. Forum* **2018**, *924*, 184–187. [CrossRef]

materials

Review

Radiation Hardness of Silicon Carbide upon High-Temperature Electron and Proton Irradiation

Alexander A. Lebedev [1,*], Vitali V. Kozlovski [2], Klavdia S. Davydovskaya [1] and Mikhail E. Levinshtein [1]

1 Solid State Electronic Department, Ioffe Institute, Politekhnicheskaya Street 26, 194021 St. Petersburg, Russia; Davidovskaya.Klava@mail.ioffe.ru (K.S.D.); melev@nimis.ioffe.ru (M.E.L.)
2 Department of Experimental Physics, St. Petersburg State Polytechnic University, Polytekhnicheskaya 29, 195251 St. Petersburg, Russia; kozlovski@physics.spbstu.ru
* Correspondence: shura.lebe@mail.ioffe.ru; Tel.: +7-812-292-71-25

Abstract: The radiation hardness of silicon carbide with respect to electron and proton irradiation and its dependence on the irradiation temperature are analyzed. It is shown that the main mechanism of SiC compensation is the formation of deep acceptor levels. With increasing the irradiation temperature, the probability of the formation of these centers decreases, and they are partly annealed out. As a result, the carrier removal rate in SiC becomes ~6 orders of magnitude lower in the case of irradiation at 500 °C. Once again, this proves that silicon carbide is promising as a material for high-temperature electronics devices.

Keywords: silicon carbide; radiation hardness; proton and electron irradiation; charge removal rate; compensation; irradiation temperature

Citation: Lebedev, A.A.; Kozlovski, V.V.; Davydovskaya, K.S.; Levinshtein, M.E. Radiation Hardness of Silicon Carbide upon High-Temperature Electron and Proton Irradiation. *Materials* **2021**, *14*, 4976. https://doi.org/10.3390/ma14174976

Academic Editor: Fabrizio Roccaforte

Received: 30 July 2021
Accepted: 26 August 2021
Published: 31 August 2021

1. Introduction

One of the stimuli in development of the technology of wide-bandgap semiconductors and creating devices on their basis is the high presumed radiation hardness of these materials. Indeed, making higher binding energy of atoms in the lattice of a semiconductor requires a higher energy of particles needed to disintegrate this lattice. Studies carried out in the 1960s demonstrated that silicon carbide substantially surpasses silicon in the radiation hardness [1]. Later, with increasing structural perfection of SiC and decreasing level of background doping, the difference in radiation hardness between SiC and Si decreased. It is noteworthy that the decrease in radiation hardness with the increasing quality of material is also characteristic for other semiconductors. Various structural defects and uncontrollable impurities could serve as drains for radiation defects and, thereby, slow the degradation of material parameters.

However, statements started to appear in the literature suggesting that the radiation hardness of silicon carbide does not surpass, and is even inferior to that of silicon in certain conditions [2–5]. This conclusion seems to be surprising because the energy gap of 4H-SiC (3.2 eV) is nearly three times that of silicon. We found it interesting to consider the situation by using both our results and those of other researchers. Thus, the goal of the present study is to consider the issue of the radiation hardness of silicon carbide and compare it with the similar characteristic for Si.

Our work is focused on the results of high-temperature irradiation, due to the fact that a lot of works have been devoted to the study and analysis of the results of irradiation at room temperature. The great number of published studies are concerned with the radiation hardness of SiC MOSFETs against γirradiation [6–12]. The effect of room temperature electron irradiation on the properties of high-voltage 4H-SiC Schottky diodes also has been studied in many works [13–20]. The effect of room-temperature proton irradiation on the properties of 4H-SiC JBS has been extensively studied [21–28]. Consequently, in this work, we considered it expedient to focus on the results of our work in the field of high-temperature irradiation.

2. Mechanism of Radiation-Induced Degradation of SiC

The radiation-induced degradation of a semiconductor device is commonly understood as the deterioration of its performance under irradiation with high-energy particles. The higher the irradiation dose required for the degradation of a semiconductor, the more radiation-hard it is believed to be.

First, consider the possible mechanisms of the radiation-induced conductivity compensation [29,30].

Assuming, for example, that the main defects generated by fast electrons are vacancies in a SiC sublattice; if the formation of multivacancy complexes is considered unlikely, we have

$$\frac{dV}{dt} = \eta_{FP}\cdot G - \frac{V}{\tau} - \beta\cdot V\cdot N \tag{1}$$

Here V is the concentration of vacancies, G the flux of charged particles, η_{FT} the probability of vacancy formation by a single particle, τ the lifetime of a vacancy, determined by drains; β the probability of vacancy capture by a free (having no captured vacancy) atom of nitrogen impurity, and N the concentration of free nitrogen atoms. The initial conditions are $t = 0$, $V = 0$, $N = N_0$.

The concentration of complexes of secondary defects N_c (vacancy and impurity atom) can be calculated by the formula

$$N_c = N_0 - N \tag{2}$$

where N_c being zero at the initial moment of time. The concentration of carriers, electrons (n), is the difference between the concentrations of impurities (shallow donors) and complexes (in the case of deep acceptors).

$$n = N - N_c = 2N - N_0 \tag{3}$$

Assuming that the lifetime of a vacancy, determined both by drains and by the impurity capture, substantially exceeds the irradiation time, then, the term V/τ in Equation (1) can be neglected.

A semiconductor can be compensated by two mechanisms. First, the radiation-induced defects create deep acceptor levels, to which electrons from shallow donor levels pass. In this case, no vacancy donor level complexes are formed.

Then, the concentration of vacancies linearly grows with increasing irradiation dose

$$V = \eta_{FP}\cdot G \tag{4}$$

and the carrier concentration linearly falls:

$$N = N_0 - \eta_{FP}\cdot G \tag{5}$$

Thus, with this mechanism being operative, the carrier concentration will linearly decrease with increasing irradiation dose.

In the framework of the second mechanism, the radiation defect (vacancy) interacts with a shallow-impurity atom to give an electrically neutral (or acceptor) center. This occurs when the lifetime of a vacancy is substantially shorter than the irradiation duration, being determined by drains. In this case, the vacancy concentration can be considered stationary and be determined from Equation (1) as

$$V = \eta_{FP}\cdot G\cdot\tau \tag{6}$$

In this case, the dependence of the carrier concentration on the irradiation dose is determined by the interaction of a vacancy with an impurity. Thus, the contribution of the secondary radiation defects, vacancy + impurity atom complexes, dominates. In this case, the carrier concentration is equal to the concentration of free impurity atoms (N), $n = N$.

The kinetics of N with second mechanism is described by the equation

$$-\frac{dN}{dt} = \eta_{FP}\cdot\beta\cdot G\cdot\tau\cdot N \tag{7}$$

possessing the following analytical solution

$$N = N_0 \exp(-\eta_{FP}\ \beta\cdot\tau\cdot G\cdot t) \tag{8}$$

In this case, the concentration of the electrically active impurity will exponentially decrease with increasing irradiation dose.

Figure 1 shows experimental data for N_d–N_a = F(ΔD) for SiC and Si, where ΔD is the irradiation dose.

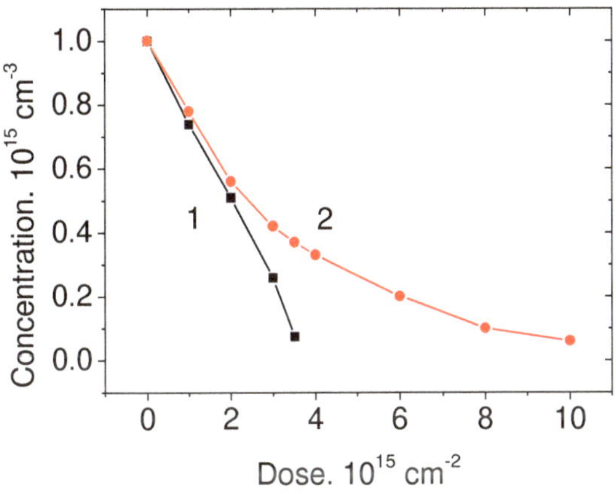

Figure 1. Conductivity compensation in (1) n-4H-SiC and (2) n-Si under irradiation with 0.9 MeV electrons. Points represent experimental data. The straight line 1 represents a calculation according to Equation (9) at a parameter η_{FP} of 0.25 cm^{-1}. Curve 2 represents a calculation by Equations (5) and (8) at a factor (η_{FP} β τ) in the exponent equal to 1.2×10^{-16} cm^2.

As can be seen from Figure 1, for silicon carbide, in the similarity to GaAs, the carrier concentration linearly decreases with increasing irradiation dose. This means that the first compensation mechanism is operative in SiC, this mechanism being associated with the formation of deep acceptor levels and transition to these levels of electrons from shallow donors. The linear dependence of the carrier concentration on the irradiation dose has also been observed in studies by other researchers, see, e.g., [31,32].

3. Experiments on Determining the Carrier Removal Rate

Frequently, the radiation hardness of a semiconductor is evaluated by the parameter "carrier removal rate" V_d, defined by

$$V_d = \frac{N_0 - N_1}{\Delta D} \tag{9}$$

where N_0 is the concentration N_a–N_d in the epitaxial layer prior to irradiation; N_1 the N_a–N_d concentration in the epitaxial layer after the irradiation; and ΔD the irradiation dose.

The value of V_d for Schottky diodes (SBDs) and junction-barrier Schottky (JBS) diodes under irradiation with electrons and protons are listed in Table 1, which also presents the carrier removal rates for silicon under the same irradiation conditions. It can be seen that V_d for SiC is approximately twice as small as that for Si.

Table 1. Comparison of carrier removal rates in devices based on SiC and Si.

SiC Device Type	SBD 600 V	SBD 1200 V	JBS 1700 V	Si
N_d–N_a in base, cm^{-3}	6.6×10^{15}	4.5×10^{15}	3.5×10^{15}	~10^{15}
V_d for electrons (0.9 MeV), cm^{-1}	0.095 [32]	0.073 [32]	0.12 [33]	0.232 [34]
V_d for protons (15 MeV) cm^{-1}	63 [33]	50 [33]	54 [33]	110 [35]

Figures 2 and 3 present the dependences N_d–N_a = F(ΔD) under irradiation with electrons and protons of SiC Schottky diodes manufactured by CREE company. It can be seen that this dependence is linear, which confirms the conclusion made in Section 2 about the mechanism of the radiation compensation of SiC via formation of deep acceptor levels.

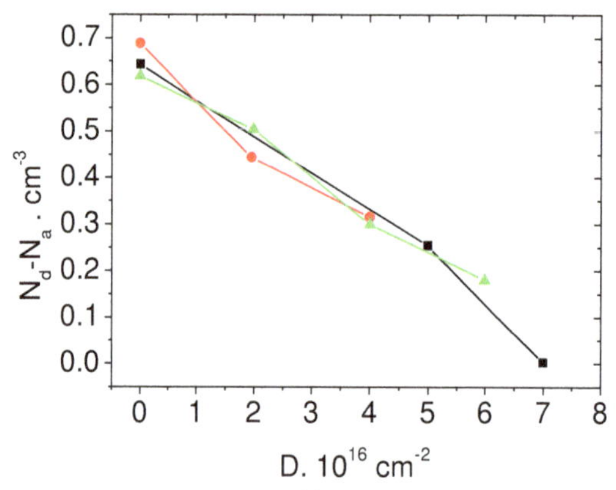

(**a**)

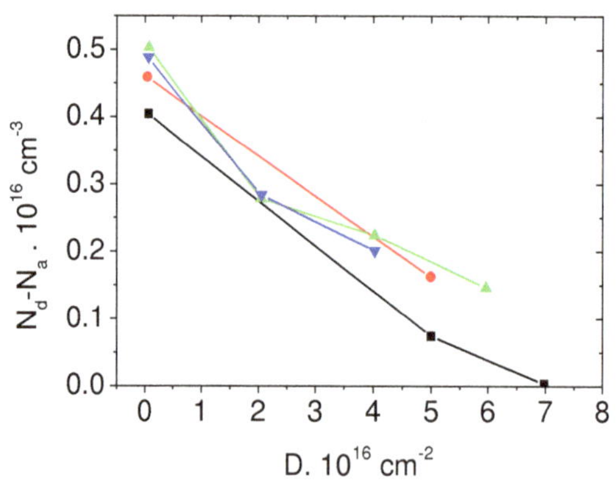

(**b**)

Figure 2. Dependence of the concentration N_d–N_a in Schottky diodes (CREE) with the blocking voltage of (**a**) 600 V and (**b**) 1200 V on the electron irradiation dose at room temperature. Different symbols correspond to different diodes from the same batch.

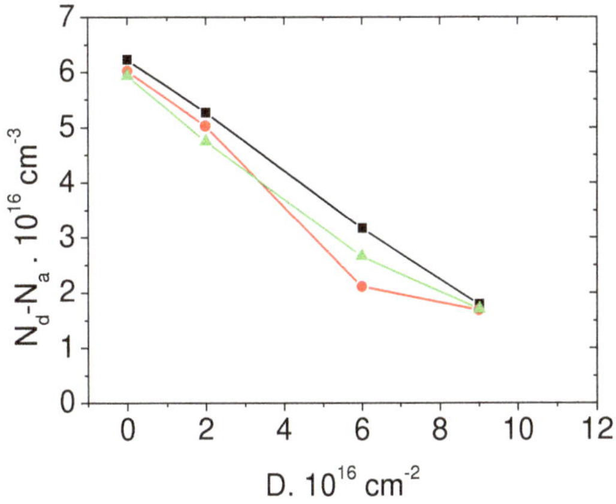

Figure 3. Dependence of the concentration N_d–N_a in Schottky diodes (CREE) with the blocking voltage of 600 V. Irradiation with protons at room temperature. Different symbols correspond to different diodes from the same batch.

4. Degradation of SiC Performance under the Action of Radiation

When a charged particle is decelerated in the semiconductor matrix, the released energy can shift the lattice atoms away from the equilibrium position. This yields the so-called primary radiation defects (Frenkel pairs), vacancies in the lattice and interstitial atoms. Most of formed defects recombine, and the rest of them create levels (deep centers) in the band energy gap of a semiconductor. Also possible is the interaction of primary defects with each other and with impurity atoms to give secondary radiation defects. As a rule, the secondary radiation defects are formed upon an increase in temperature, which is accompanied by the annealing out of the remaining primary defects.

As the irradiation dose increases, radiation defects gradually accumulate, which causes degradation of a semiconductor device, for example, see [36–38]. In SiC pn structures, as well as in pn structures based on other semiconductor materials, irradiation leads to the following effects.

1. The free carrier concentration decreases and, accordingly, the ohmic resistance of the base region grows. This is due to the formation of compensating radiation defects to which free carriers go.
2. The carrier lifetime and their diffusion length become shorter. This is due to the increase in the concentration of recombination centers in the semiconductor.
3. The leakage currents under a reverse bias increase. This may be due to the formation of defect clusters shunting the pn junction.

Figure 4 shows dependences of the quantum efficiency of a SiC UV photodetector before and after the irradiation with heavy ions. The decrease in the quantum efficiency is due to that in the carrier diffusion length.

Figure 5 shows how the conductivity of Schottky diode bases decreases upon irradiation with protons and electrons. The formation of compensating acceptor levels leads to a 6–7 orders of magnitude decrease in the carrier concentration in the base region. In this case, the carrier mobility decreases only slightly.

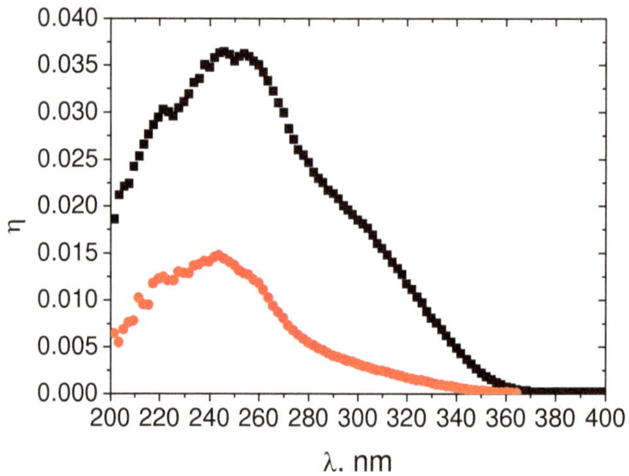

Figure 4. Spectral dependences of the quantum efficiency of a 4H-SiC photodetector with Schottky barriers: (black) initial sample and (red) sample irradiated with 167 MeV Xe ions at a fluence of 6×10^9 cm^{-2}. Room temperature.

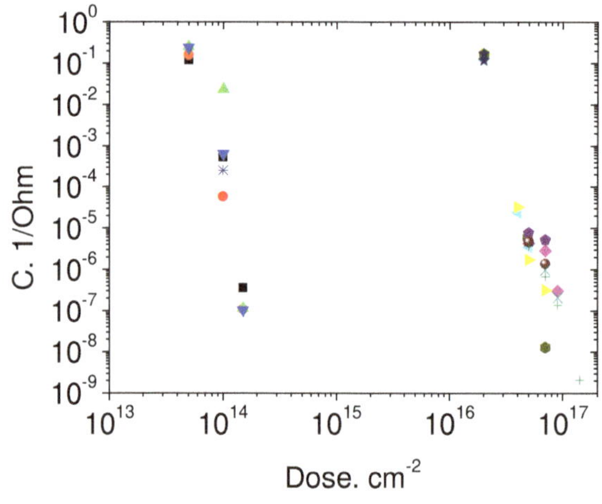

Figure 5. Base conductivity of a Schottky diode (600 V) after (1) protons and (2) electron irradiation. Different symbols correspond to different diodes from the same batch.

Figure 6 shows current-voltage characteristics of a Schottky diode with breakdown voltage of 1200 V after the proton irradiation. The irradiation affects only slightly the voltage dependence of the forward current in the exponential area of the current-voltage characteristic. The irradiation in the pre-exponential (currents of 10^{-12}–10^{-14} A) and post-exponential (high currents) areas drastically affects the current-voltage characteristic. At high currents, the base resistance increases due to the decrease in the free-carrier concentration. Leakage currents grow at low currents.

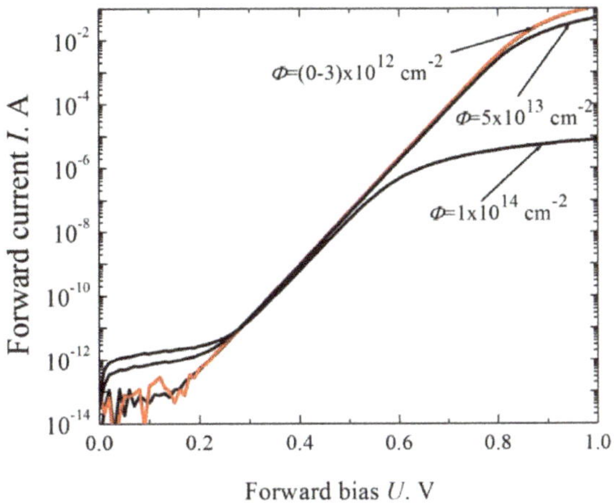

Figure 6. Forward current-voltage characteristic of a Schottky diode (1200 V class) at various doses of irradiation with 15 MeV protons [36].

The reverse current-voltage characteristics of Schottky diodes before and after the proton irradiation were investigated in [33]. It was shown that the leakage currents decrease at low reverse voltages, which is apparently due to the total increase in the resistance of the structure. At high reverse voltages, the reverse currents do increase, which can be attributed to the appearance in the space-charge layer of deep centers (radiation defects) via which carriers are generated.

Somewhat more complicated is the result of irradiation of SiC MOSFETs (a typical structure of such a device is shown in Figure 7). First, as in the case of a pn structure and Schottky diode, the free-carrier concentration decreases and the resistance of the drift region grows. Second, the devices have a subgate insulator layer (SiO_2) in which the charge state of traps changes under irradiation. This may lead to an increase in the output current of a transistor at small irradiation doses. Both of these effects are well represented in the current-voltage characteristics presented in Figure 8.

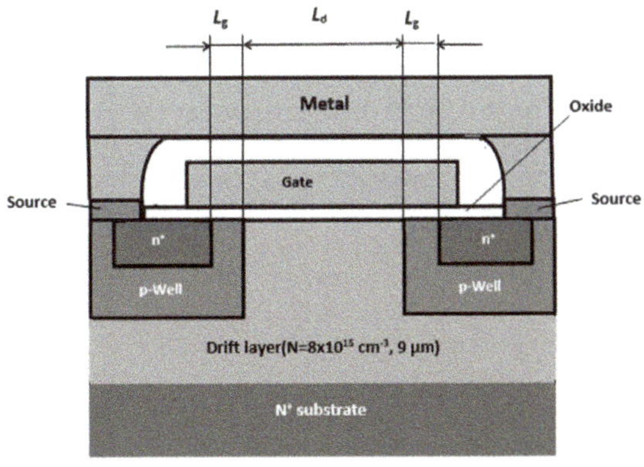

Figure 7. Cross-section of an elementary cell of a 4H-SiC MOSFET. The gate length Lg is 0.5 μm, the oxide thickness d is 60 nm, and the drift (blocking) layer thickness Hd is 9 μm [37].

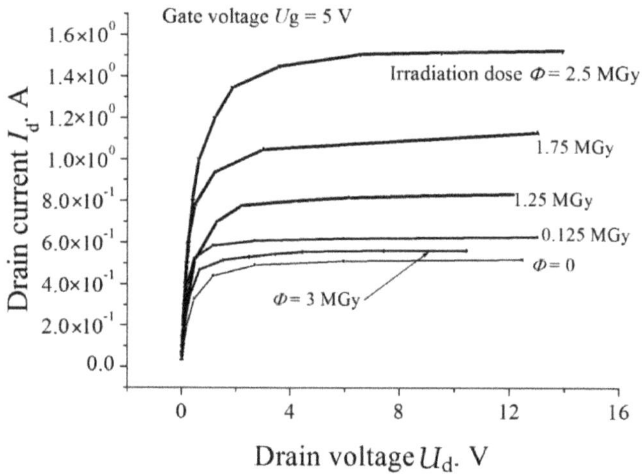

Figure 8. Output characteristics $I_d(V_d)$ of a SiC MOSFETs (1.2 kV class) under study at various irradiation doses. The gate voltage $V_g = 5$ V [37].

5. Comparison of the Radiation Hardnesses of Si and SiC

Since silicon carbide is often viewed as a possible replacement for silicon in power devices, it is interesting and useful to compare the radiation hardness of the two materials. In our opinion, two approaches to such a comparison are possible.

First, two SiC- and Si-based diodes with the same breakdown voltage can be compared

$$U_{brSi} = U_{brSiC} \Rightarrow (E_{crSi} \cdot W_{Si})/2 = (E_{crSiC} \cdot W_{SiC})/2 \Rightarrow W_{cr} = W_{Si} \cdot E_{crSiC}/E_{crSi} \quad (10)$$

Here, U_{br} is the breakdown voltage, E_{cr} the critical electric field, and W the space-charge layer thickness at U_{br}.

Because the critical field in silicon carbide exceeds by an order of magnitude that in silicon, $E_{crSiC}/E_{crSi} \approx 10$, we obtain, with consideration for the fact that $W \approx \sqrt{N_d - N_a}$, $N_{(d-a)Si} = 100\ N_{(d-a)SiC}$, where N_{d-a} is the concentration of the uncompensated impurity in the base.

Thus, at the same breakdown voltage, the SiC diode is doped to a level exceeding by two orders of magnitude that for the Si diode. Consequently, even at equal values of V_d, the compensation of SiC diodes requires a 100 times higher irradiation doses, compared with Si diodes.

Second, the radiation hardness of SiC and Si diodes with the same base thickness can be compared [39]. This is important for fabrication of charged-particle detectors, in which the applied reverse voltage is limited and the maximum thickness of the space-charge layer should be obtained. In this case, the carrier-removal rates are directly compared.

It can be seen in Table 1 that the value of V_d for SiC is only two times smaller than for Si. Because the energy gap E_g of SiC is nearly three times that for silicon, a question arises why the difference between the values of V_d for these two materials is so insignificant.

Table 2 presents the results of an analysis of the annealing-out of radiation defects in 4H and 6H silicon carbide irradiated with various kinds of ions. It can be seen that there are two characteristic temperature ranges in which this annealing occurs: 200–800 °C and ≥1200 °C. Such a situation is also characteristic of other semiconducting materials. In the first stage of annealing, most of the primary radiation defects recombine, with the remaining forming substantially more temperature-resistant complexes, which are annealed out at significantly higher temperatures. However, the position of these annealing stages along the temperature scale depends on the properties of a semiconductor, including its energy gap.

Table 2. Annealing temperatures of radiation defects in SiC after various kinds of irradiation [38].

Refs.	SiC	Kind of Irradiation	Onset Temperature of the Defect Rearrangement, °C	Temperature at Which the RDs are Finally Annealed-Out, °C
[40]	4H-p	e-2.5 MeV	200–400	950–1400
[41]	6H-n	e-0.3–0.4 MeV	400–900	1600
[42]	4H-n	e-15 MeV; p-1.2 MeV	200–800	>1200
[43]	6H-n 4H-n	e-2.5 MeV; p-1 MeV, He	-	1200–1700
[44]	4H-n	e-15 MeV	400–800	1200–2000
[45]	6H,4H,15R-p	p-8 MeV	-	>1200

e stands for irradiation with electrons; p, for irradiation with protons; and He, for irradiation with helium nuclei.

Figure 9 schematically demonstrates how the concentration of radiation defects (R_d) in silicon and silicon carbide varies with temperature.

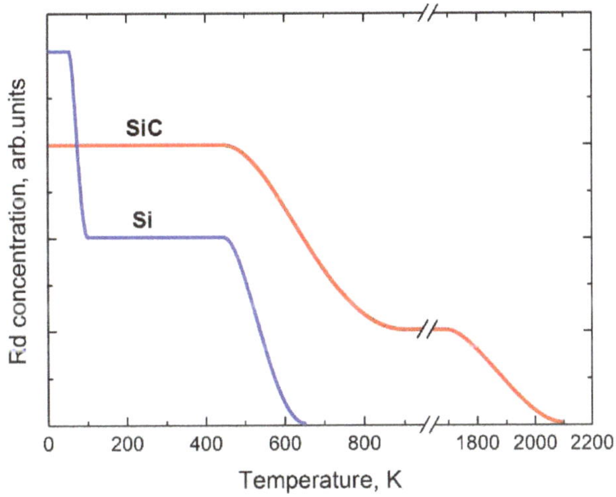

Figure 9. Schematic comparison of how the radiation defects are annealed-out in SiC and Si [38].

The figure shows that, at room temperature, these two materials are in different physical states with respect to the annealing-out of radiation defects. For silicon, the primary annealing stage has already been completed, whereas for SiC it has not yet begun. Thus, even if the concentration of introduced radiation defects was lower immediately after the irradiation (the irradiation temperature is conditionally 0 K), the concentration of defects in silicon when heated to room temperature became lower than that in SiC.

This can explain such a small difference between V_d in SiC and Si and gives impetus to a desire to verify this assumption and irradiate silicon carbide at elevated temperatures.

6. Irradiation of SiC at Elevated Temperatures

Previous experiments with III–V materials have demonstrated that the irradiation temperature can cardinally change the radiation hardness of materials and devices [46]. We examined, for the first time, the influence exerted by the electron irradiation temperature on high-power (blocking voltage 1700 V, working current 10 A) 4H-SiC Schottky diodes within the range 23–500 °C [47–49].

To perform high-temperature irradiations, we designed and constructed a special target chamber. This chamber enabled us to work with irradiations in air at temperatures ranging from room temperature to 600 °C. The accuracy of maintaining the sample temperature during irradiation was ±5 °C. The heating rate was maintained at 0.5 deg/s and the cooling rate was about 0.25 deg/s.

It was found that, at comparatively small values of $\Phi \approx 10^{16}$ cm^{-2}, raising the irradiation temperature from room temperature to 300 °C affects, comparatively slightly, the electron removal rate. With increasing dose, the difference between the base resistivities upon irradiation at room and elevated temperatures monotonically grows and exceeds three orders of magnitude at $\Phi \approx 6 \times 10^{16}$ cm^{-2} (Figure 10).

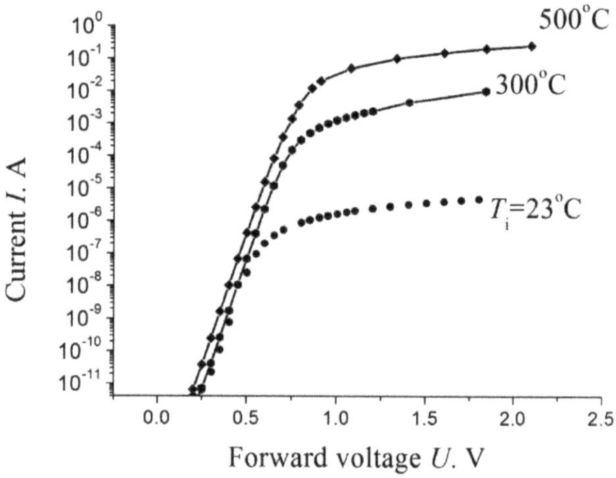

Figure 10. Forward current–voltage characteristics of diodes upon their irradiation with 0.9 eV electrons at three different irradiation temperatures T_i. the dose $\Phi \approx 6 \times 10^{16}$ cm^{-2}. The inset shows how the base resistivity ρ depends on the inverse irradiation temperature [47].

We also examined the effect of a high-temperature irradiation with 15 MeV protons on parameters of high-voltage 4H-SiC Schottky diodes at doses in the range from 7×10^{13} to 2×10^{14} cm^{-2}.

After the irradiation with a dose of 10^{14} cm^{-2} at room temperature, the forward current at a forward voltage U = 2 V decreases by ~10 orders of magnitude (Figure 11). In this case, the cutoff voltage U_c, equal to ~0.6 V in unirradiated devices, decreases to $U_c \approx 0.35$ V. By contrast, irradiation with the same dose at a temperature of 500 °C results in that U_c increases to $U_c \approx 0.8$ V. At the same reference forward voltage U = 2 V, the decrease in current, as compared with the value for unirradiated devices, was ~4 orders of magnitude. In the whole range of doses and irradiation temperatures under study, the forward current-voltage characteristic of the diodes is linear at $U > U_c$ up to $U \geq 2$ V. In unirradiated diodes, the forward current I at the reference forward voltage U = 2 V is $I \approx 12$ A (see details, datasheet, quote on part number: CPW3-1700-S010B-WP).

The resulting set of experimental data indicates an increase in the radiation resistance of diodes with an increase in the temperature of irradiation. The physical reason for this temperature dependence is a decrease in the stationary concentration of radiation defects (RD), which are responsible for compensation of the base conductivity of the Schottky diodes under study, with an increase in the irradiation temperature.

As is known, the main RDs that create deep acceptor levels in n-SiC are mainly carbon vacancies [37,50]. The rate of generation of primary RDs (which are vacancies and interstitial atoms in both silicon carbide sublattices) in the temperature range under study is practically independent of the irradiation temperature [51,52].

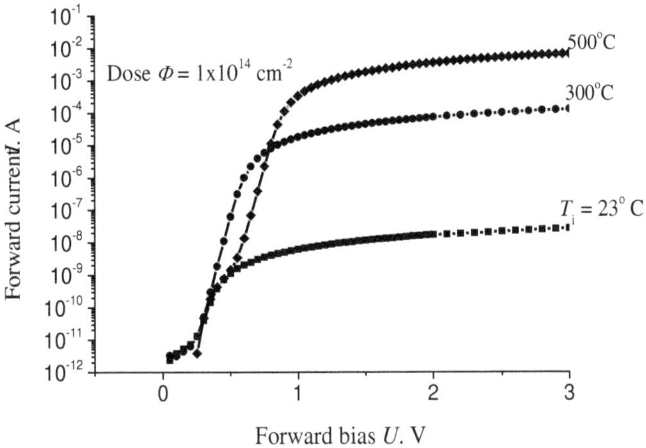

Figure 11. Forward current–voltage characteristics of the diodes upon their irradiation with 15 MeV protons at three different irradiation temperatures T_i, the dose $\Phi = 1 \times 10^{14}$ cm^{-2} [48].

However, the further fate of the generated vacancies (secondary defect formation) can significantly depend on temperature. As the temperature rises, the vacancy mobility increases and the recombination radius with a genetically related interstitial atom increases. Therefore, the fraction of vacancies that have escaped recombination and created deep acceptor levels is greatly reduced. According to our picture of irradiation, this proportion is about 25%, then at 300–400 °C it decreases by a factor of 2–3.

In principle, a second possible reason cannot be ruled out, which is a change in the spectrum of secondary radiation defects created during hot irradiation. A change in the X-ray diffraction spectrum was previously observed under hot electron irradiation of silicon and A_3B_5 materials [47,52].

Thus, we can conclude that the previously made assumption that the radiation hardness of silicon carbide will grow with an increase in the irradiation temperature is valid. In our opinion, this is an important conclusion because SiC is considered to be a promising material, especially for development of high-power and high-voltage devices.

7. Conclusions

- Silicon carbide of *n*-type is compensated under irradiation due to the transition of carriers to the acceptor-type radiation defects being formed. As a result, the concentration difference N_d–N_a (N_a–N_d) linearly decreases with increasing irradiation dose.
- It was shown that the dose Φ_{cr} corresponding to the total degradation of a device satisfies the condition $\Phi_{cr} \approx V_d/n_0$, where V_d is the removal rate of electrons from the blocking layer of the device and n_0 the initial electron concentration in the blocking (drift) layer. In the VMOSFET devises under study (1.2 kV class), $\Phi_{cr} \approx 10^{14}$ cm^{-2}, from a physics viewpoint, the condition $\Phi_{cr} \approx V_d/n_0$ reflects the situation in which the concentration of a deep levels created by irradiation becomes equal to the initial concentration of electrons in the drift region.
- The radiation hardness of SiC devices exceeds by approximately two orders of magnitude that of silicon-based devices with the same breakdown voltage.
- It was shown for the first time that when operating under conditions of increased radiation and elevated temperature, the service life of carbide devices is longer than that for the same devices working at room temperature. Judging by the carrier removal rate or the increase in the base resistance of Schottky diodes under irradiation, it can be said that the service life is at least doubled.

Author Contributions: A.A.L.: Conceptualization, Project administration, Writing—original draft, V.V.K.: Investigation, Data curation. M.E.L.: Investigation, formal analysis, K.S.D.: Investigation, software. All authors have read and agreed to the published version of the manuscript.

Funding: This research received no external funding.

Institutional Review Board Statement: Not applicable.

Informed Consent Statement: Not applicable.

Data Availability Statement: The data underlying this article will be shared on reasonable request from the corresponding author.

Conflicts of Interest: The authors declare no conflict of interest.

References

1. Choyke, W.J. A review of radiation damage in SiC. *Inst. Phys. Conf. Ser.* **1977**, *31*, 58–67.
2. Metcalfe, J. Silicon Detectors for the sLHC. *Nucl. Phys. B Proc. Suppl.* **2011**, *215*, 151–157. [CrossRef]
3. Hallen, A.; Henry, A.; Pelligrino, P.; Swensson, B.G.; Aberg, D. Ion Implantation induced defects in epitaxial 4H-SiC. *Mater. Sci. Eng.* **1999**, *61–62*, 378–381. [CrossRef]
4. Swensson, B.G.; Hallen, A.; Linnarson, M.K.; Kuznetsov, A.Y.; Janson, M.S.; Aberg, D.; Osterman, J.; Persson, P.O.A.; Hultman, L.; Storasta, L.; et al. Doping of Silicon Carbide by ion Implantation. *Mater. Sci. Forum* **2001**, *353–356*, 349–354.
5. Casse, G. Overview of the recent activities of the RD50 collaboration on radiation hardening of semiconductor detectors for the sLHC. *Nucl. Instrum. Methods Phys. Res. A* **2009**, *598*, 54–60. [CrossRef]
6. Murata, K.; Mitomo, S.; Matsuda, T.; Yokoseki, T.; Makino, T.; Onoda, S.; Takeyama, A.; Ohshima, T.; Okubo, S.; Tanaka, Y.; et al. Impacts of gate bias and its variation on gamma-ray irradiation resistance of SiC MOSFETs. *Phys. Status Solidi A* **2017**, *214*, 1600446. [CrossRef]
7. Ashrafi, S.; Eslami, B. Investigation of sensitivity and threshold voltage shift of commercial MOSFETs in gamma irradiation. *Nucl. Sci. Tech.* **2016**, *27*, 144–146. [CrossRef]
8. Akturk, A.; McGarrity, J.M.; Potbhare, S.; Goldsman, N. Radiation effects in commercial 1200 V 24 A silicon carbide power MOSFETs. *IEEE Trans. Nucl. Sci.* **2012**, *59*, 3258–3264. [CrossRef]
9. Tadjer, M.J.; Stahlbush, R.E.; Hobart, K.D.; McMarr, P.J.; Hughes, H.L.; Imhoff, E.A.; Kub, F.J.; Haney, S.K.; Agarwal, A. Spatial localization of carrier traps in 4H-SiC MOSFET devices using thermally stimulated current. *J. Electron. Mater.* **2010**, *39*, 517–525. [CrossRef]
10. Ahyi, A.C.; Wang, S.R.; Williams, J. Gamma irradiation effects on 4H-SiC MOS capacitors and MOSFETs. *Mater. Sci. Forum* **2006**, *527–529 Pt 2*, 1063–1066. [CrossRef]
11. Matsuda, T.; Yokoseki, T.; Mitomo, S.; Murata, K.; Makino, T.; Abe, H.; Takeyama, A.; Onoda, S.; Tanaka, Y.; Kandori, M.; et al. Change in characteristics of SiC MOSFETs by gamma-ray irradiation at high temperature. *Mater. Sci. Forum* **2016**, *858*, 860–863. [CrossRef]
12. Castaldini, A.; Cavallini, A.; Rigutti, L.; Nava, F. Low temperature annealing of electron irradiation induced defects in 4H–SiC. *Appl. Phys. Lett.* **2004**, *85*, 3780–3782. [CrossRef]
13. Danno, K.; Kimoto, T. Investigation of deep levels in n -type 4H-SiC epilayers irradiated with low-energy electrons. *J. Appl. Phys.* **2006**, *100*, 113728. [CrossRef]
14. Kozlovskii, V.V.; Bogdanova, E.V.; Emtsev, V.V.; Emtsev, K.V.; Lebedev, A.A.; Lomasov, V.N. Direct experimental comparisson of the effects of Electron Irradiation on the Charge Carrier Removal rate in n-type Silicon and Silicon carbide. *Mater. Sci. Forum* **2004**, *483–485*, 385–388.
15. Kozlovski, V.; Lebedev, A.; Lomasov, V.; Bogdanova, E.; Seredova, N. Conductivity compensation in n-4H-SiC (CVD) under irradiation with 0.9-MeV electrons. *Semiconductors* **2014**, *48*, 1006–1009. [CrossRef]
16. Strel'chuk, A.M.; Kozlovskii, V.V.; Lebedev, A.A.; Smirnova, N.Y. Influence of Irradiation on Exess Currents in SiC pn structures. *Mater. Sci. Forum* **2004**, *483–485*, 1001–1004.
17. Omotoso, E.; Meyer, W.E.; Auret, F.D.; Paradzah, A.T.; Diale, M.; Coelho, S.M.M.; Janse van Rensburg, P.J. The influence of high energy electron irradiation on the Schottky barrier height and the Richardson constant of Ni/4H-SiC Schottky diodes. *Mater. Sci. Semicond. Process.* **2015**, *39*, 112–118. [CrossRef]
18. Kozlovski, V.V.; Lebedev, A.; Levinshtein, M.E.; Rumyantsev, S.L.; Palmour, J.W. Impact of high energy electron irradiation on high voltage Ni/4H-SiC Schottky diodes. *Appl. Phys. Lett.* **2017**, *110*, 083503. [CrossRef]
19. Mikelsen, M.; Grossner, U.; Bleka, J.H.; Monakhov, E.V.; Swensson, B.G.; Yakimova, R.; Henry, A.; Jansen, E.; Lebedev, A.A. Carrier Removal rate in electron Irradiated 4H and 6H SiC. *Mater. Sci. Forum* **2009**, *600–603*, 425–428.
20. Lebedev, A.; Veinger, A.; Davydov, D.; Kozlovski, V.; Savkina, N.; Strel'chuk, A.M. Doping of n-type 6H–SiC and 4H–SiC with defects created with a proton beam. *J. Appl. Phys.* **2000**, *88*, 6265–6271. [CrossRef]
21. Lebedev, A.A.; Ber, B.Y.; Seredova, N.V.; Kazantsev, D.Y.; Kozlovski, V.V. Radiation-stimulated photoluminescence in electron irradiated 4H-SiC. *J. Phys. D Appl. Phys.* **2015**, *48*, 485106. [CrossRef]

22. Henry, L.; Barthe, M.-F.; Corbel, C.; Desgardin, P.; Blondiaux, G. Silicon vacancy-type defects in as-received and 12-MeV proton-irradiated 6H-SiC studied by positron annihilation spectroscopy. *Phys. Rev. B* **2003**, *67*, 115210. [CrossRef]
23. Castaldini, A.; Cavallini, A.; Rigutti, L. Assessment of the intrinsic nature of deep level Z1/Z2 by compensation effects in proton-irradiated 4H-SiC. *Semicond. Sci. Technol.* **2006**, *21*, 724–728. [CrossRef]
24. Emtsev, V.; Ivanov, A.; Kozlovski, V.; Lebedev, A.; Oganesyan, G.; Strokan, N.; Wagner, G. Similarities and distinctions of defect production by fast electron and proton irradiation: Moderately doped silicon and silicon carbide of n-type. *Semiconductors* **2012**, *46*, 456–465. [CrossRef]
25. Vobecky, J.; Hazdra, P.; Záhlava, V.; Mihaila, A.; Berthou, M. ON-state characteristics of proton irradiated 4H–SiCSchottky diode: The calibration of model parameters for device simulation. *Solid-State Electron.* **2014**, *94*, 32–38. [CrossRef]
26. Hazdra, P.; Popelka, S.; Záhlava, V.; Vobecký, J. Radiation Damage in 4H-SiC and Its Effect on Power Device Characteristics. *Solid State Phenom* **2016**, *242*, 421–426. [CrossRef]
27. Hazdra, P.; Popelka, S. Displacement damage and total ionization dose effects on 4H-SiC power devices. *IET Power Electron.* **2019**, *12*, 3910–3918. [CrossRef]
28. Kozlovski, V.; Korolkov, O.; Lebedev, A.; Toompuu, J.; Sleptsuk, N. Comparative Results of Low Temperature Annealing of Lightly Doped n-Layers of Silicon Carbide Irradiated by Protons and Electrons. *Mater. Sci. Forum.* **2020**, *1004*, 231–236. [CrossRef]
29. Kozlovski, V.V.; Lebedev, A.A.; Bogdanova, E.V. Model for conductivity compensation of moderately doped n- and p- 4H-SiC by high-energy electron bombardment. *J. Appl. Phys.* **2015**, *117*, 155702. [CrossRef]
30. Hazdra, P.; Vobecký, J. Radiation Defects Created in n-Type 4H-SiC by electron Irradiation in the Energy Range of 1–10 MeV. *Phys. Stat. Solidi A* **2019**, *216*, 1900312. [CrossRef]
31. Lebedev, A.A.; Davydovskaya, K.S.; Strel'chuk, A.M.; Kozlovski, V.V. Radiation Resistance of 4H-SiC Schottky Diode under Irradiation with 0.9 –MeV Electrons. *J. Surf. Investig. X-ray Synchrotron Neutron Tech.* **2017**, *11*, 924–926. [CrossRef]
32. Lebedev, A.A.; Davydovskaya, K.S.; Strel'chuk, A.M.; Yakimenko, A.N.; Kozlovski, V.V. A study of the effect of electron and proton irradiation on 4H-SiC device structures. *Tech. Phys. Lett.* **2017**, *43*, 1027–1029. [CrossRef]
33. Kozlovski, V.V.; Emtsev, V.V.; Emtsev, K.V.; Strokan, N.B.; Ivanov, A.M.; Lomasov, V.N.; Oganesyan, G.A.; Lebedev, A.A. Effect of electron irradiation on carrier removal rate in silicon and silicon carbide withy 4H modification. *Semiconductors* **2008**, *42*, 242–247. [CrossRef]
34. Kozlovski, V.V.; Strokan, N.B., Ivanov, A.M.; Lebedev, A.A.; Emtsev, V.V.; Oganesyan, G.A.; Poloskin, D.S. Charge carrier removal rates in n-type silicon and silicon carbide subjected to electron and proton irradiation. *Phys. B* **2009**, *404*, 4752–4754. [CrossRef]
35. Kalinina, E.V.; Lebedev, A.A.; Bogdanova, E.V.; Lebedev, A.A.; Berenquier, B.; Ottaviani, L.; Violina, G.N.; Skuratov, V.A. Irradiation of 4H-SiC UV detectors with heavy ions. *Semiconductors* **2015**, *4*, 540–546. [CrossRef]
36. Kozlovski, V.V.; Lebedev, A.A.; Levinshtein, M.E.; Rumyantsev, S.L.; Palmour, J.W. Electrical and noise properties of proton irradiated 4H-SiC Schottky diodes. *J. Appl. Phys.* **2018**, *123*, 024502. [CrossRef]
37. Lebedev, A.A.; Kozlovski, V.V.; Levinshtein, M.E.; Ivanov, A.E.; Strel'chuk, A.M.; Zubov, A.V. Fursin Leonid, Impact of 0.9 MeV electron irradiation on main properties of high voltage vertical power 4H-SiC MOSFETs. *Radiat. Phys. Chem.* **2020**, *177*, 109200. [CrossRef]
38. Lebedev, A.A.; Kozlovski, V.V. Comparison of the radiation hardness of silicon and silicon carbide. *Semiconductors* **2014**, *10*, 1293–1295. [CrossRef]
39. Zolnai, Z.; Son, N.T.; Hallin, C.; Janzen, E. Annealing behavior of the carbon vacancy in electron—Irradiated 4H-SiC. *J. Appl. Phys.* **2004**, *96*, 2406–2408. [CrossRef]
40. Luo, J.M.; Zhong, Z.Q.; Gong, M.; Fung, S.; Ling, C.C. Isochronal annealing study of low energy electron irradiated Al-doped p-type 6H silicon carbide with deep levels transient spectroscopy. *J. Appl. Phys.* **2009**, *105*, 063711. [CrossRef]
41. Alferi, G.; Monakhov, E.V.; Svensson, B.G.; Hallen, A. Defect energy levels in hydrogen-implanted and electron-irradiated N-type 4H-SiC silicon carbide. *J. Appl. Phys.* **2005**, *98*, 113524. [CrossRef]
42. Weider, M.; Frank, T.; Pensl, G.; Kawasuso, A.; Itoh, H.; Krause-Rehberg, R. Formation and annihilation of intrinsic-related defects centers in high energy electron irradiated of ion-implanted 4H- and 6H- silicon carbide. *Phys. B* **2001**, *308–310*, 633–636. [CrossRef]
43. Alfieri, G.; Monakhov, E.V.; Svensson, B.G.; Linnarsson, M.K. Annealing behavior between room temperature and 2000 °C of deep level defects in electron-irradiated N-type 4H silicon Carbide. *J. Appl. Phys.* **2005**, *98*, 043518. [CrossRef]
44. Lebedev, A.A.; Bogdanova, E.V.; Grigor'eva, M.V.; Lebedev, S.P.; Kozlovski, V.V. Annealing of radiation-compensated silicon carbide. *Tech. Phys. Lett.* **2012**, *38*, 910–912. [CrossRef]
45. Zarharenkov, L.F.; Kozlovski, V.V.; Pil'kevich, Y.Y. Role of irradiation temperature in the conductivity compensation of n-GaAs by radiation defects. *Phys. Stat. Solidi A* **1989**, *111*, K215–K218. [CrossRef]
46. Kozlovski, V.V.; Kolchenko, T.I.; Lomako, V.M.; Zakharenkov, L.F. The influence of irradiation temperature upon the radiation defect formation and conductivity compensation of n-GaAs. *J. Radiat. Eff. Defects Solids* **1996**, *138*, 63–73. [CrossRef]
47. Lebedev, A.A.; Kozlovski, V.V.; Levinshtein, M.E.; Ivanov, A.E.; Davydovskaya, K.S.; Yuferev, V.S.; Zubov, A.V. Impact of high temperature irradiation on characteristics of power SiC Schottky diode. *Radiat. Phys. Chem.* **2021**, *185*, 109514. [CrossRef]
48. Lebedev, A.A.; Kozlovski, V.V.; Levinshtein, M.E.; Ivanov, A.E.; Davydovskaya, K.S. Effect of high temperature irradiation with 15 MeV protons on characteristics of power SiC Schottky diodes. *Solid-State Electron.* **2021**, *181–182*, 108009. [CrossRef]

49. Kaneko, H.; Kimoto, T. Formation of a semi-insulating layer in n-type 4H-SiC by electron irradiation. *Appl. Phys. Lett.* **2011**, *98*, 262106. [CrossRef]
50. Kozlovski, V.; Abrosimova, V. *Radiation Defect Engineering. Selected Topics in Electronics and Systems—v.37*; World Scientific: Singapore; New York, NY, USA; London, UK; Hong Kong, China, 2005; ISBN 981-256-521-3.
51. Claeys, C.; Simoen, E. *Radiation Effects in Advanced Semiconductor Materials and Devices*; Springer: Berlin, Germany, 2002; p. 401.
52. Lindström, J.; Murin, L.; Hallberg, T.; Markevich, V.; Svensson, B.; Kleverman, M.; Hermansson, J. Defect engineering in Czochralski silicon by electron irradiation at different temperatures. *Nucl. Instrum. Methods Phys. Res. Sect. B Beam Interact. Mater. Atoms* **2002**, *186*, 121–125. [CrossRef]

materials

MDPI

Review

New Approaches and Understandings in the Growth of Cubic Silicon Carbide

Francesco La Via [1,*], Massimo Zimbone [1], Corrado Bongiorno [1], Antonino La Magna [1], Giuseppe Fisicaro [1], Ioannis Deretzis [1], Viviana Scuderi [1], Cristiano Calabretta [1], Filippo Giannazzo [1], Marcin Zielinski [2], Ruggero Anzalone [3], Marco Mauceri [4], Danilo Crippa [4], Emilio Scalise [5], Anna Marzegalli [5], Andrey Sarikov [5], Leo Miglio [5], Valdas Jokubavicius [6], Mikael Syväjärvi [6], Rositsa Yakimova [6], Philipp Schuh [7], Michael Schöler [7], Manuel Kollmuss [7] and Peter Wellmann [7,*]

1 Consiglio Nazionale delle Ricerche—Istituto per la Microelettronice e Microsistemi, Strada VIII 5, 95121 Catania, Italy; massimo.zimbone@imm.cnr.it (M.Z.); corrado.bongiorno@imm.cnr.it (C.B.); antonino.lamagna@imm.cnr.it (A.L.M.); Giuseppe.fisicaro@imm.cnr.it (G.F.); ioannis.deretzis@imm.cnr.it (I.D.); viviana.scuderi@imm.cnr.it (V.S.); cristiano.calabretta@imm.cnr.it (C.C.); filippo.giannazzo@imm.cnr.it (F.G.)

2 NOVASIC, Savoie Technolac—Arche Bat.4, Allée du Lac d'Aiguebelette, BP 267, CEDEX, 73375 Le Bourget du Lac, France; mzielinski@novasic.com

3 STMicroelectronics, Stradale Primosole 50, 95121 Catania, Italy; ruggero.anzalone@st.com

4 LPE, Strada XVI, Pantano d'Arci, 95121 Catania, Italy; Marco.Mauceri@lpe-epi.com (M.M.); danilo.crippa@lpe-epi.com (D.C.)

5 L-NESS and Department of Materials Science, Università di Milano-Bicocca, Via R. Cozzi 55, I-20125 Milano, Italy; emilio.scalise@unimib.it (E.S.); anna.marzegalli@unimib.it (A.M.); andrey.sarikov@unimib.it (A.S.); leo.miglio@unimib.it (L.M.)

6 Department of Physics, Chemistry, and Biology (IFM), Linköping University, 581 83 Linköping, Sweden; valjo@ifm.liu.se (V.J.); mikael@alminica.se (M.S.); rosya@ifm.liu.se (R.Y.)

7 Crystal Growth Lab, Materials Department 6 (i-MEET), Friedrich-Alexander University Erlangen-Nürnberg (FAU), 91058 Erlangen, Germany; Philipp.Schuh@fau.de (P.S.); michael.schoeler@fau.de (M.S.); manuel.kollmuss@fau.de (M.K.)

* Correspondence: francesco.lavia@imm.cnr.it (F.L.V.); peter.wellmann@ww.uni-erlangen.de (P.W.)

Citation: La Via, F.; Zimbone, M.; Bongiorno, C.; La Magna, A.; Fisicaro, G.; Deretzis, I.; Scuderi, V.; Calabretta, C.; Giannazzo, F.; Zielinski, M.; et al. New Approaches and Understandings in the Growth of Cubic Silicon Carbide. *Materials* **2021**, *14*, 5348. https://doi.org/10.3390/ma14185348

Academic Editor: Alexander A. Lebedev

Received: 22 June 2021
Accepted: 7 September 2021
Published: 16 September 2021

Abstract: In this review paper, several new approaches about the 3C-SiC growth are been presented. In fact, despite the long research activity on 3C-SiC, no devices with good electrical characteristics have been obtained due to the high defect density and high level of stress. To overcome these problems, two different approaches have been used in the last years. From one side, several compliance substrates have been used to try to reduce both the defects and stress, while from another side, the first bulk growth has been performed to try to improve the quality of this material with respect to the heteroepitaxial one. From all these studies, a new understanding of the material defects has been obtained, as well as regarding all the interactions between defects and several growth parameters. This new knowledge will be the basis to solve the main issue of the 3C-SiC growth and reach the goal to obtain a material with low defects and low stress that would allow for realizing devices with extremely interesting characteristics.

Keywords: 3C-SiC; heteroepitaxy; bulk growth; compliant substrates; defects; stress

1. Introduction

Wide band-gap (WBG) semiconductor devices based on both silicon carbide (SiC) and gallium nitride (GaN) can lead a revolution in power electronics through its faster switching speeds, lower losses, and higher blocking voltages [1]. Furthermore, their properties enable higher operating temperatures and increased power densities, but until now the benefits shown by WBG power electronics have not been fully realized due to the high costs of the material and reliability challenges.

Silicon carbide is a material presenting different crystalline structures called polytypes [2]. To date, only the two hexagonal structures 4H and 6H-SiC are commercialized, while the cubic form (3C-SiC) is not used until now in power devices despite the large effort of the last years and several hundreds of papers published for each year. All these polytypes have similar benefits over silicon such as higher breakdown fields (2–4 MV/cm) and the larger energy band-gap (2.3–3.2 eV). The cubic silicon carbide is the only polytype that can be grown on silicon wafers. This approach reduces the cost as no SiC substrate is used and only the silicon carbide layer thickness required for the specific application is grown on a cheaper Si substrate. This technology also offers the potential for faster scale-up with wafer size compared with the development of larger diameter hexagonal SiC wafers. In principle, with a large reactor, a 300 mm Si wafer can be obtained using the present process.

Both 3C-SiC and GaN can work in the same breakdown voltage range (200–1200 V). 3C-SiC is more appropriate for high-current applications due to its high thermal conductivity, while GaN better fits in RF applications because of the high saturated electron velocity. The lower band-gap of 3C-SiC (2.3 eV) in comparison to 4H-SiC (3.28 eV) is often viewed as a negative aspect with respect to other polytypes. The lower band-gap brings a positive effect because the lowering of the conduction band minimum causes a reduced density of states at the SiO_2/3C-SiC interface [3]. As a consequence, it has been demonstrated that the metal oxide semiconductor field effect transistor (MOSFET) on 3C-SiC has the highest channel mobility (above 300 $cm^2/V/s$) ever presented on any SiC polytype [4]. This produces a large reduction in the power consumption of power switching devices [5]. A remaining challenge in both 3C and 4H-SiC is the electrical activity of extended defects. It is identified as the major problem for electronic device functionality. The mechanisms of defect formation must be clarified and the methods for their reduction developed to reach full functionality and high yield [6]. So far, the growth of 3C-SiC on silicon has been demonstrated on 150 mm Si wafers [7,8]. The process is feasible with up-scaled reactors on 200 mm or 300 mm wafers.

Another problem to solve concerns the intrinsic stress created during the growth process due to the lattice mismatch between 3C-SiC (4.36 Å) and Si (5.43 Å) [9]. Thermo-elastic stress also appears at the post-deposition cooling due to the 8% difference in the thermal expansion coefficients between these two materials [10]. This results in stress which induces the formation of various planar or extended defects in 3C-SiC. These defects produce a considerable degradation of the crystalline quality of the epitaxial layer [11].

There are two types of defects in the epitaxial 3C-SiC layers. Anti-phase boundaries (APBs) are planar defects that are formed at the geometrical separation of two 3C-SiC grains. The grain differs by a 90° rotation in the Si(100) growth plane [12]. These are described as anti-phased domains (APDs) that are formed by the steps present on the Si surface. A second defect type is the stacking fault along the {111} planes. Most research work highlights the intrinsic nature of these defects and points out that a reduction of their density is possible by increasing the film thickness. The electrical activity of the extended defects in 3C-SiC is a dominant problem for electronic device performance. Clearly, a reduction of these defects is essential to improve the yield of power electronic devices [13].

In the last years during the CHALLENGE project [14], we have been active in developing two different approaches. In the first one, we have used several structured substrates (compliance substrates) [15–17] to reduce the defect density and stress. This approach maintains the attractive low cost of the silicon substrate material. However, it does not take advantage of the high heat dissipation of 3C-SiC because it is limited by the low thermal conductivity of silicon. The second approach of the bulk growth has a higher cost due to the higher cost of the 3C-SiC substrate but allows to take advantage of the high thermal conductivity of 3C-SiC in the devices. Furthermore, there is an advantage in the device processing because of the possibility to use the high-temperature processes already developed in 4H-SiC processing (e.g., epitaxy, ion implantation activation, etc.). The use of Si as a substrate has the main advantage, with respect to 4H-SiC, of reaching large diameters directly without the long and expensive process of the traditional diameter enlargement. In addition, the lower band-gap and higher channel mobility of 3C-SiC can produce MOSFET with a lower R_{on} with respect to both Si and 4H-SiC in the breakdown voltage region between 200 and 1200 V [5]. In previous works [18,19], it was observed that in increasing the grown thickness of the material, the density of SFs can be strongly reduced and this will produce an overall improvement in the electrical performance of the devices.

The bulk growth of hexagonal 6H and 4H silicon carbide polytypes can be considered mature while the same cannot be said about the cubic polytype (3C-SiC) [20,21]. Different approaches such as the modified PVT (Physical Vapour Transport) (M-PVT) or continuous-feed PVT (CF-PVT) method have been presented for bulk growth over the last decade [22–24]. Only the methods based on the sublimation sandwich implemented by Tairov et al. [25] could allow for the formation of 3C-SiC [6,26]. A major drawback for all the mentioned bulk growth processes is the lack of available high-quality seeding material. One widely investigated approach makes use of hexagonal SiC wafer materials and explores switching to 3C-SiC. Another promising approach is based on the heteroepitaxial growth of cubic silicon carbide on silicon (3C on Si) using chemical vapor deposition (CVD). This method faces some challenges caused by the lattice mismatch of approximately 20% between silicon and 3C-SiC, and the difference in thermal expansion [27]. The research in this area has advanced in major ways over the last years and has been revitalized to use the material as seed for bulk growth processes [6].

2. Materials and Methods

Several different compliance substrates have been realized to improve the quality of the heteroepitaxial growth.

Inverted silicon pyramids (ISP) were fabricated on (001) Si wafers by deep UV lithography. The structures consisted of 700 nm-wide square geometries with a 1.4 μm pitch. A thin layer of stoichiometric silicon nitride deposited by LPCVD with a thin buffer layer of thermal silicon dioxide was used as a hard mask. The layer was etched by a fluorine-based plasma and the silicon substrate was etched by a 45 wt.% KOH @ 70 °C solution.

In the second compliance substrate, the samples consist of a silicon (0 0 1) substrate with 2 μm of $Si_{1-x}Ge_x$ grown on top and a cap of 10 nm or 20 nm thick Si. From simulations [28], the ideal value of [Ge] for the lattice match was calculated to be around 12%. In our studies, three germanium concentrations were used at 10%, 12%, and 15% The heteroepitaxial growth of the 3C-SiC films on the ISP and Si-Ge buffer substrates was realized using a chemical vapor deposition process by NOVASiC in a horizontal hot-wall reactor operating with standard silane/propane/hydrogen chemistry [29].

In the third compliance substrate, the Si(111) substrates were patterned into hexagonal arrays of pillars on 100 mm wafers. The standard Bosch process in the shape of hexagonal prisms 8 μm deep and 2 or 5 μm wide that were separated by 2 or 3 μm trenches was applied to fabricate the pillars. The pillars had 100 μm hexagonal patches separated by 5 μm trenches to avoid their contact and reduce substrate bowing. To further improve the pillars' properties, a three-step procedure was used: (i) a vertical Bosch process was first applied to dig the pillars; (ii) one isotropic dry carving was next applied to create the under etching and necking; and (iii) some oxidation-stripping cycles were used to finally smooth the sidewalls and reduce the top layer thickness of the pillars. The heteroepitaxial growth was realized in a LPE M10 reactor using a trichlorosilane–ethylene hydrogen chemistry [6].

For the bulk growth, 3C-SiC seeds were realized on Si (100) substrates using a chemical-vapor deposition process in a horizontal hot-wall reactor (LPE M-10, Catania, Italy). The silicon and carbon precursors were trichlorosilane ($SiHCl_3$ or TCS) and ethylene (C_2H_4) using hydrogen (H_2) as a gas carrier. The process was implemented in a low-pressure regime (100 mbar), wherein the epitaxy started at 3 μm/h, then increased to 6 μm/h, and finally further increased to 30 μm/h. This process resulted in a thick layer of about 70 μm. In the next step, the temperature was increased above the melting point of silicon. This resulted in the Si substrate being fully melted inside the CVD reactor. The remaining freestanding SiC layer was used as a seed layer for the following homoepitaxial growth which used low-pressure regime at different temperatures (between 1600 and 1700 °C). A growth rate of 60 µm/h was used to increase the substrate thickness for two hours [6]. The first 20 microns of the 30 µm layer were highly doped and the last 10 microns were low doped for device realization. Nitrogen was used for n+ and n-type layer formation. The total thickness of the 3C-SiC homoepitaxial samples was about 200 µm (confirmed by SEM analysis).

Another approach for the bulk growth is to use enhanced sublimation epitaxy (ESE), which is a modified physical vapor transport (PVT) growth technique originating from a patent-protected epitaxial SiC growth method [30] developed by researchers at the Linköping University. A typical ESE growth setup is shown in Figure 1. The exclusiveness of the ESE is the distance between the source and the substrate, as well as the character of the source material itself. In a standard PVT setup, the distance between the source and the substrate is more than 10 mm, while in the ESE it typically varies between 0.5 and 2 mm. Such a distance is sufficient to create a large enough temperature gradient, which is the main driving force for the growth. In addition, compared to PVT, such a short distance allows for the more direct transfer of SiC vapor species from the source to the substrate, with a much lower interaction with the graphite walls. In PVT, the source material is a polycrystalline SiC powder, while in the ESE, the powder is replaced by a polycrystalline SiC plate. Furthermore, tantalum (Ta) foil is inserted into the graphite crucible. At growth temperatures, it reacts with carbon-bearing species and forms TaC. In this way, the vapor-phase composition inside the crucible is enriched with Si, which is beneficial for the enhancement of 3C-SiC stability. The combination of the short distance, Si-enriched growth ambience, and stochiometric monolithic source makes ESE an excellent technique for the growth of high-quality SiC layers in a vacuum (1×10^{-4} mbar) at temperatures below 2000 °C. Such growth conditions are favorable to induce SiC conversion from hexagonal to cubic polytypes, which are known to be more stable at temperatures below 2000 °C.

Using a transfer process developed at FAU Erlangen-Nurnberg [31], growth of the bulk 3C-SiC with reasonable dimensions and thicknesses was demonstrated using such seeding material and the approach of closed space PVT (CSPVT), [26,32], in its original concept also known as Sublimation epitaxy (SE), was utilized.

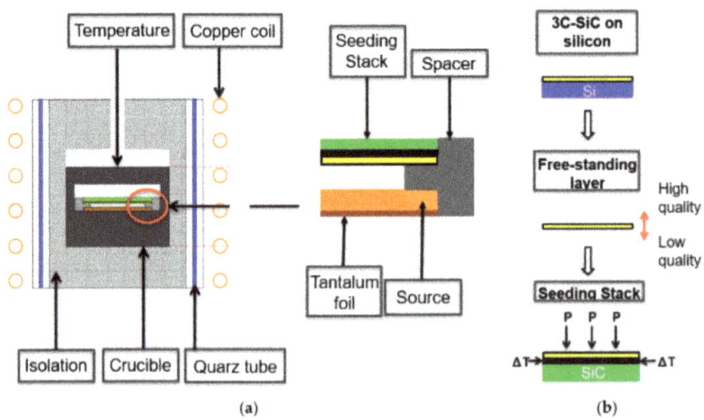

Figure 1. (**a**) PVT reactor used for the sublimation growth and hot zone consisting of a tantalum foil to acquire carbon, the source material, a spacer to separate the source and seed, and the manufactured seeding stack. (**b**) Schematic of the seed manufacturing process. Starting from 3C grown by CVD on a silicon substrate, the silicon is removed by chemical wet etching and subsequently the thin freestanding 3C layer is merged to a polycrystalline SiC carrier for mechanical stabilization and backside protection (see Reference [32]).

Starting from 3C-on-Si material grown by CVD (LPE M10, Catania, Italy), high-temperature stable seeds for bulk growth in a PVT setup can be fabricated according to Figure 1b. First, the CVD seeds are cut to get the desired dimensions. At the start of the project, a diamond wire saw was used. However, the size of the samples that could be prepared was limited to approximately 12.5×12.5 mm^2, as the sawing process induced cracks along the <110> direction in the 3C layer due to the applied mechanical force [33]. Therefore, a change to a multipulse-laser ablation technique [34,35] was made. As the melting point of silicon (1419 °C) is well below the required temperatures for PVT growth (<1800 °C), a removal of the silicon substrate is necessary. Otherwise, the molten silicon would immediately react to silicon carbide with the graphite crucible and graphite isolations used in PVT setups [24]. Therefore, after the cutting of the samples, a wet-chemical etching of the silicon substrate with HNA (HF: HNO_3: H_2O) was performed [36,37], resulting in a thin (typically between 20 and 50 μm) free-standing 3C-SiC layer. The etched layers featured a high-quality growth front as well as a defect-rich backside associated with the former transition area between Si and 3C-SiC. Subsequently, the layers were merged to a polycrystalline SiC-carrier with the high-quality growth front facing up. This step is necessary for the mechanical stabilization of the thin 3C layers and to prevent a backside sublimation during the sublimation growth. For the merging, a carbon glue with the main component of 1-Methoxy-2-propanol acetate [31] was used and both a combined heat and pressure treatment was applied. After the merging, residues of the carbon glue that could remain on top of the growth caused by an overflow of the glue were removed to maintain a high-quality starting point for the sublimation growth [38].

3. Results and Discussions

3.1. Flat Substrates

The 3C-SiC grown on silicon has still high defectivity even though strategies for the elimination of a large plethora of defects are achieved. Common three-dimensional defects of 3C-SiC grown epitaxially on silicon, such as protrusions, twinned regions, antiphase domains, and polytypes inclusion, are eliminated or strongly reduced. On the contrary, the elimination of dislocation, stacking faults, and stress in the 3C-SiC layer is far from being solved. The hetero-interface is the principal "source" of such defects. Indeed, lattice mismatch among Si and SiC, as well as the disparity in thermal expansion coefficients, induced the formation of stress in the film and the formation of both SFs and dislocations.

For example, stacking faults in 3C-SiC can come out from Lomer dislocations' dissociations at the hetero-interface: Lomer dislocation (that forms naturally due to the lattice mismatch at the hetero-interface) with the Burgers vector of $a/2\,[\bar{1}\bar{1}0]$ can dissociate in two partial dislocations, with the Burgers vectors $a/6\,[\bar{2}\bar{1}1]$ and $a/6\,[\bar{1}\bar{2}1]$ where a is the lattice constant. The two stacking faults propagate through the epitaxial layer and can approach the surface. A cross-view image of the Si/SiC hetero-interface is shown in Figure 2. In this figure, SFs are the oblique bright lines. A region near the hetero-interface with a high number of SFs is highlighted and the density of SFs decreases, moving away from the hetero-interface. In the "on-axis" substrate (Figure 2a), the SF can intersect and form another kind of defect such as the "Lomer lock". The formation of this linear extended defect modifies the mechanical properties and more interestingly can avoid the propagation of the SFs in the epilayer. Unlike the case of the "on-axis" substrate on which the SFs can interact with each other in the "off-axis" substrate (Figure 2b), SFs can arrive at the surface because they have the same orientation and are not able to cross each other. Even in this case, a region of high density of SFs is apparent near the hetero-interface.

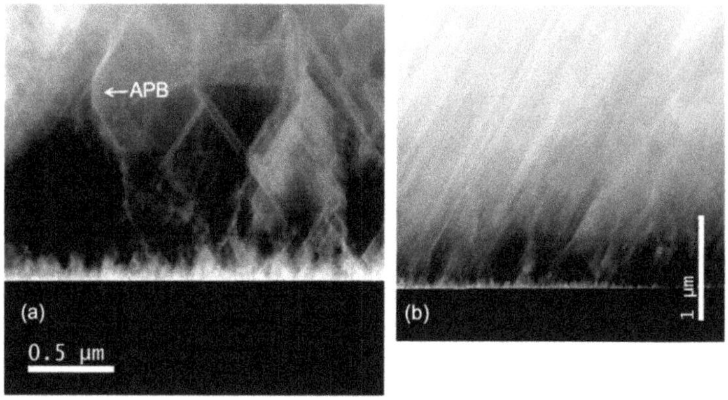

Figure 2. (**a**) TEM cross-section of a 3C-SiC layer grown on on-axis silicon. APBs and SFs in opposite directions are visible. (**b**) TEM cross-section of a 3C-SiC layer grown on off-axis silicon. No APBs can be observed while only SFs in the step direction are visible.

Another defect that is present only in the "on-axis" image is the anti-phase boundary (APB) or also called "inverted domain boundary" (IDB)) (Figure 2a). This defect is a 2D defect and it is the boundary between two 3C crystals rotated by each other by 180° around the [110] axes. It is observed in Figure 2a as a curved bright line. This defect preferentially lies on the (110) or (111) plane and has a particular atomic structure: it is coherent and can couple with SF in a complex way. In the (110) plane, the structure is made up of a bent Si-C bond that generates a square and a semi-octahedral configuration with unaltered bonds, while in the (111) plane it resembles a twin with a Si–Si bond. In the last case, it can be associated with an SF. For more detail, the reader can refer to [39,40]. The propagation of IDBs within the crystal appears to be extremely complex, resulting in "complex IDBs" interacting with SFs. Moreover, we noticed that IDBs can also end and generate SFs. The presence of "disconnection" (which are steps with a Burgers vector associated in the IDB) might cause such behavior.

The different orientations of the substrates ((100), (111), or (110)) produce a different structure of the material on the surface [41] and a difference in the stress after the deposition.

3.2. Compliance Substrates: Pillar Growth

The pillar technology is intended to doctor the thermal strain of the deposited SiC film by growing a suspended (thick) layer on top micrometric Si pillars, which eventually bend to accommodate the larger thermal retraction in the cooling down of the SiC film with respect to the Si substrate. The pillars are patterned in arrays in the Si substrate by a dry etching process [42,43] (Figure 3). The <111> orientation is the most critical one for stress accumulation with film thickness on flat substrates (less than 1 μm without cracks) and also provides a better quality of the deposited material on pillars. Therefore, the shape of the pillars (hexagonal in cross-section) and the arrangement of the arrays (still hexagonal in the pattern), which are suitably rotated with respect to the wafer flat to reduce the slanted 111 facet extension, have been optimized for the <111> growth orientation, wherein the pillar technology could provide the most important contribution.

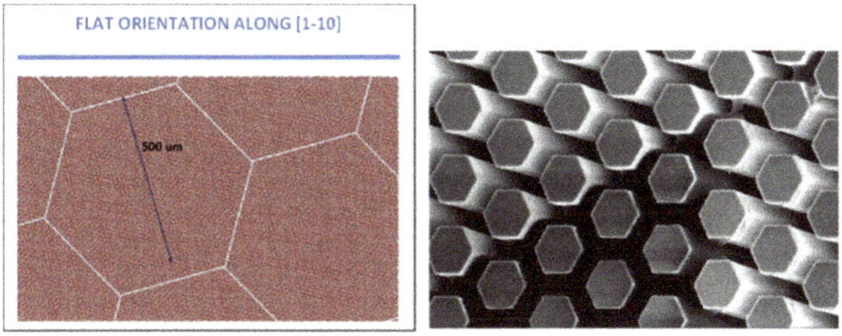

Figure 3. Pattern (left) and shape of the pillars (right). Both pillars and the pattern have a hexagonal structure.

Understanding and controlling the 3D crystal growth and subsequent coalesce dynamics are the keys to optimizing the patterning and obtaining high-quality 3C-SiC suspended layers on the underlying Si pillars. To this goal, an extensive theory-experiment analysis of the evolution of the crystal growth has been performed and detailed in Reference [44]. First, the faceted growth of the individual SiC crystals has been characterized, as illustrated in Figure 4, by comparing the profiles of samples grown at different times with phase-field simulations based on the kinetic growth model of Reference [15]. The unknown facet-dependent growth rates to be set in the model have been extrapolated by fitting the simulation profiles to the experimental ones, resulting in a good match between Figure 4a and b. Once calibrated, the simulations allow us to investigate all intermediate stages of the growth (c), as well as the subsequent dynamics of merging between neighboring crystals.

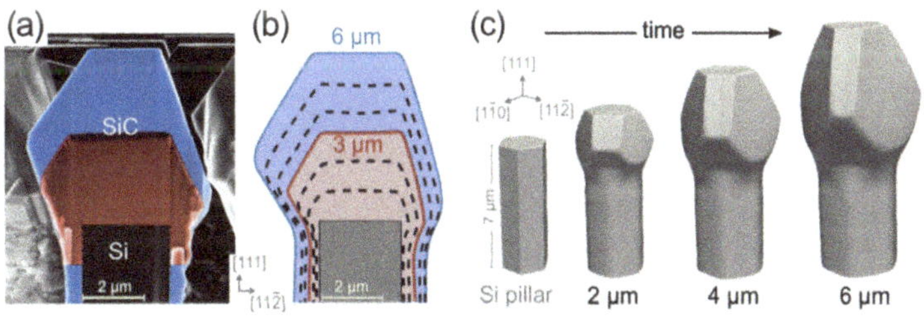

Figure 4. (**a**) SEM (1-10) cross-section view of the upper part of a SiC crystal after 3 µm (red) and 6 µm (blue) deposition on top of a 2 µm-wide Si pillar (gray), which is 8 µm tall. (**b**) Phase-field simulation profiles for the same conditions of (**a**) reproduced every 1 µm deposition. (**c**) 3D view of the evolution sequence obtained from simulations (see Reference [44]).

As illustrated in Figure 5, two limiting cases have been studied, corresponding to a 90° rotation of the hexagonal pillar pattern. In case (a), pillar rows are along the [11-2] directions so that coalescence occurs with a six-fold symmetry by bridging the large {111}-C terminated facets with the smaller {100} ones, leaving six identical holes to fill at the latest stages. In the same way, in case (b), pillar rows are aligned along the [1-10] direction such that coalescence occurs at facet edges, resulting in a three-fold symmetry arrangement with a larger hole in between {111}-C facets and smaller ones at the crossing of {100} facets. As made evident by simulations, this latter arrangement is the most convenient, returning a smoother surface profile after the deposition of about 12 µm.

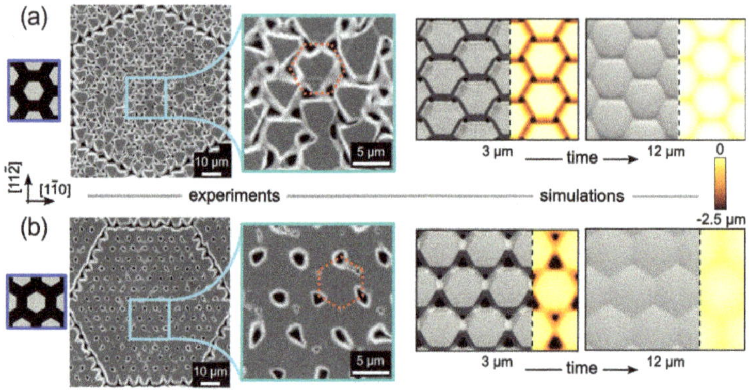

Figure 5. Comparative analysis of the coalescence of SiC crystals grown on Si pillars for the two different patterns with pillar rows along (**a**) the [11-2] and (**b**) [1-10] directions from both experiments and simulations. SEM views are reported for samples obtained after 12 µm SiC deposition on 5 µm large prismatic Si pillars, spaced by 2 µm gaps. The magnified views highlight the different patterns of holes left by partial coalescence. Simulation snapshots are shown for both the 3 and 12 µm deposition. The colored regions show the variations in height by the colormap. A smoother profile is achieved in case (**b**) (see Reference [44]).

The strain relaxation in the 3C-SiC epilayer is enabled by the tilting of the pillars underneath. As reported in Reference [45] for the case of Ge grown on Si pillars, the deformation can be described as a rigid-body rotation of each pillar. It is possible to conclude that the capability to rotate strongly depends on the aspect ratio of prismatic (or paralepidid) pillars. Indeed, as shown in Figure 6a for the pillar at the periphery of the array exhibiting the maximum deformation, the rotation mechanisms and consequently the stress relaxation are larger for a smaller pillar width. Another important parameter that controls the relaxation in the 3C-SiC epilayer is the height of the pillars. Indeed, as observed in Figure 6b for different patch sizes, the higher the pillar, the better is the strain relaxation. The stress (and strain) relaxation at the center of the array decreases when the patch size is increased, at a fixed pillar aspect ratio, asymptotically matching the reference case without any pillar when the patch size tends to be infinite. The relaxation of the elastic energy in the epilayer can be enhanced also by changing the pillar spacing or, more importantly, the pillar shape. Indeed, if compared to the standard parallelepiped pillars, T-shaped ones (Figure 6c) offer a higher capability to rotate, being thinner in the intermediate section of the pillar. This results in a lower residual stress or equivalent strain, as shown in Figure 6a. The T-shape (Figure 6c) case is comparable to the one with parallelepiped pillars, characterized by a base of half-size and a larger pillar spacing, with the advantage that the T-shape ones have a larger top surface for each pillar, above which the SiC can be grown. According to the approach discussed in Reference [46], in Figure 6b, the relation between the width and height of the pillars is plotted to guarantee a curvature radius of the sample that is larger than 10 m. The curvature radius is calculated from the average residual strain in the epilayer according to the Timoshenko formula for planar bilayers [47].

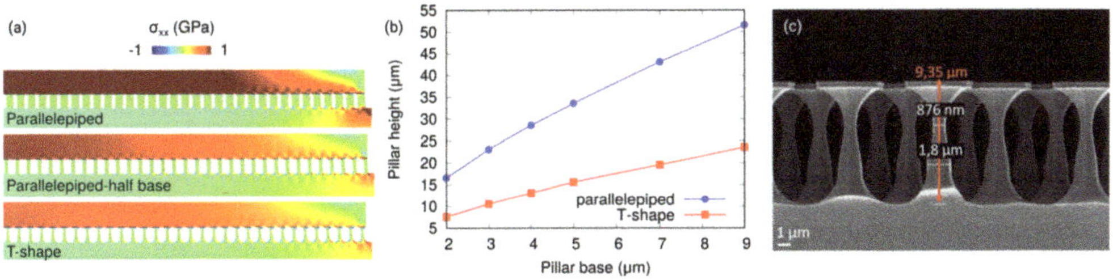

Figure 6. (**a**) Color maps of the xx component of the stress tensor (σ_{xx}) for three 3C-SiC epilayers grown on array of pillars with different geometries. Top: parallelepiped pillars, spaced by 2 μm and with a base width of 5 μm. Center: parallelepiped pillars, spaced by 4.5 μm and with a base width of 2.5 μm. Bottom: T-shape pillars, spaced by 2 μm and with a maximum base width of 5 μm. (**b**) Plot of the height of the pillars as a function of the width of the pillars that is needed to guarantee a curvature radius of the sample that is larger than 10 m (acceptable for post-processing of 4′ wafers). A (111) Si substrate is considered. (**c**) SEM image of the T-shape pillars (adapted from Reference [46]).

3.3. Compliance Substrates: SiGe Buffer Layer

Another attempt to grow a high-quality 3C-SiC epilayer on a silicon substrate was done by introducing a buffer epitaxial layer of $Si_{1-x}Ge_x$ between Si and the SiC. We choose a layer of Si-Ge because Si and Ge have the same FCC structure and are perfectly miscible: the stoichiometry and lattice parameter can be decided a priori. This fact has important implications: fine-tuning the lattice parameter in such a way to minimize the mismatch due to the 4/5 ratio among the Si and SiC atomic layers is possible. Indeed, as already reported, 3C-SiC and Si show roughly a 20% lattice mismatch, implying that four layers of Si almost "equate" to five layers of SiC (the 4/5 rule). The extra plane of SiC creates Lomer and misfit dislocations, thus stacking faults. Nevertheless, the 4/5 rule is not exact and a mismatch (dependent on temperature) also exists between four layers of Si and five of SiC. This mismatch creates stress with the formation of extra dislocations and stacking faults. The adoption of a buffer layer of Si-Ge can also reduce the thermal stress due to the thermal expansion coefficient mismatch (between Si and SiC) caused by the cooling of the sample from the growth temperature (about 1400 °C) to room temperature.

In Figure 7, we show the structure of the sample used in the experiment: 10 nm Si cap on 2 μm $Si_{1-x}Ge_x$ grown over a 300 μm silicon (0 0 1) substrate. The thin Si capping layer thickness is lower than the "critical thickness" to avoid the formation of interfacial defects at the $Si/Si_{1-x}Ge_x$ interface and it was introduced as a seed for the carbonization step. The intensity of the transverse optical (TO) peak of SiC at 796 cm^{-1} is shown in Figure 7b for different carbonization temperatures and buffer layer compositions. The TO Raman peak is forbidden for perfect 3C-SiC grown on (0 0 1) substrates (due to the selection rules) and its presence is associated with twins and poly-crystals. It was observed that greater Ge content and lower temperatures (1000 °C, 15%; 1050 °C, 15%; and 1000 °C, 12%) lead to a high TO mode intensity. In these samples, we find Ge segregation at the interface between the SiC and SiGe layer. Ge segregation implies the formation of poly-crystals, while samples with lower concentrations and higher carbonization temperatures lead to a mirror-like surface morphology and a lower value of TO intensity indicates a higher SiC quality.

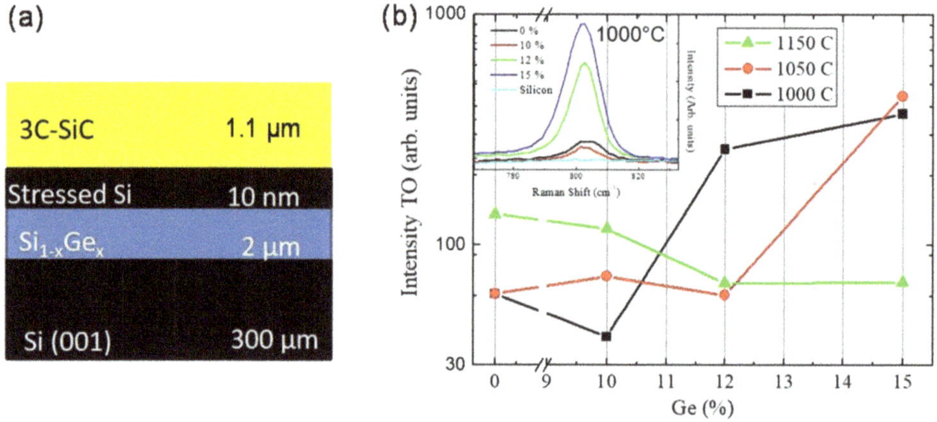

Figure 7. (**a**) Schematic of the sample structure. The image is not to scale. (**b**) 3C-SiC TO peak height with respect to nominal Ge concentration for several carbonization temperatures. The spectra of samples that have undergone 1000 °C carbonization are shown in the inset (adapted from Reference [16]).

We also demonstrated that the carbonization temperature and composition of the layer control the quality of the SiC film. It is also possible to achieve a higher quality with respect to film grown on virgin silicon [16].

3.4. Compliance Substrates: Inverted Silicon Pyramids (ISP)

One of the most interesting attempts to grow a high-quality 3C-SiC epilayer on a silicon substrate was done by creating a structured substrate. The structure came from the following consideration: the SFs lie on {111} planes and can interact with each other, stopping the propagation. Consider two SFs laying, for example, in the (111) and (11-1) planes; they can cross and the structure is able to stop the propagation of one or even both SFs. This clearly improves the crystalline quality of the film surface because the SFs remain buried in the epilayer. The rate of SF annihilation is inversely related to SF density, however, by means of the inverted silicon pyramid (ISP) compliant substrate, allowing for a significant drop in SF concentration just within a few microns.

Its unique shape can concentrate SFs in tiny areas, enhancing the phenomenon of SF annihilation [48].

In Figure 8a, we show a schematic cross-section view of the effect of this compliant substrate. Silicon and silicon carbide are drawn as black and white regions. Blue lines are SFs which either generate an X-shaped defect known as the forest dislocation or self-annihilate, resulting in a system known as the Lomer lock, or end on an existing SF, producing a so-called "λ-shaped" defect. In Figure 8b,c, SEM images of the ISP structure are shown in plane cross-view and plane-view. The four (111) planes of the pyramid are shown, as well as the (001) zone among the two pyramids.

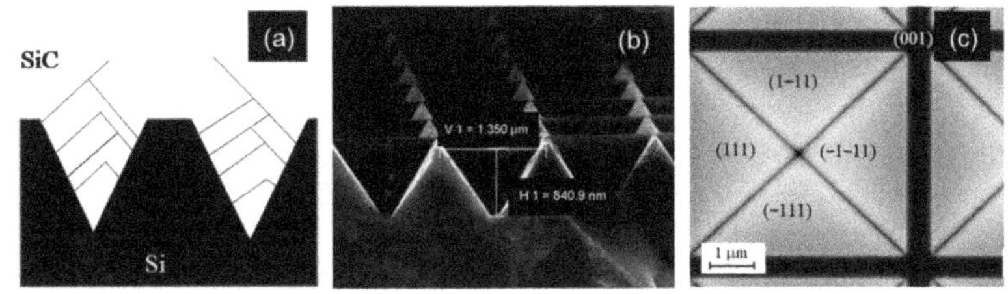

Figure 8. (**a**) Schematic cross-section view of the effect of the ISP compliant substrate on the SFs. Silicon and silicon carbide are drawn as black and white regions. Blue lines are SFs. (**b**) Cross-view SEM image of the ISP structure. (**c**) Plane-view SEM image. The four (111) planes of the pyramid are shown, as well as the (001) region among the two pyramids (adapted from Reference [17]).

The drawback of the use of this substrate is the formation of APB due to the different polarities of the (111) faces of the SiC [17]. Nevertheless, it is well known that the grain boundary density can be greatly decreased through the enhancement of the film thickness.

The APBs coverage with respect to layer thickness is depicted in Figure 9a. Despite the fact that the substrate design yields APBs, their concentration was rapidly reduced. Some tenth of microns of the SiC layer is enough to largely reduce the density of APBs. The ISP morphology also induces the formation of buried voids in the epilayer because the (111) face has a slower growth rate than the (100) face. These voids are observed in Figure 9b in which a cross-view TEM image of the 12 µm-thick epitaxial 3C-SiC layer is shown. The generation of voids may be advantageous in reducing the defectiveness of the epilayer. Voids can annihilate SFs and reduce the residual stress in the layer. SFs that cross the void are not able to propagate into the epilayer, reducing the defectivity. It is also feasible to manage the void height by adjusting the growth rate and conditions. In such a way, it is possible to modulate the concentration of SFs arriving on the surface [17].

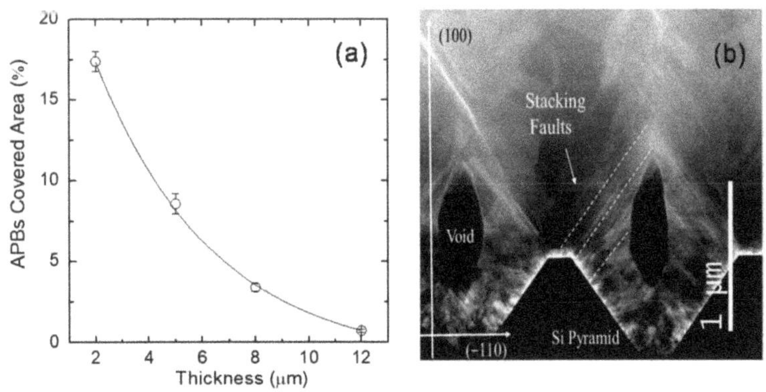

Figure 9. (**a**) Anti-phase boundaries (APBs) covered-area percentages for different thicknesses of the epitaxial growth. (**b**) Cross-section SEM image of 12 um-thick epitaxial 3C-SiC layer grown on ISP (adapted from Reference [17]).

3.5. Compliance Substrates: 4H and 6H-SiC

Hexagonal SiC (4H- or 6H-SiC) is a very promising substrate for the heteroepitaxial growth of 3C-SiC due to the excellent chemical compatibility, thermal expansion, and lattice constant matching. Moreover, contrary to silicon, the hexagonal SiC can be used in high (>1800 °C) temperature processes such as the ESE, the concept of which has been proven to be advantageous for the growth of homo and heteroepitaxial SiC layers at growth rates of up to 1 mm/h [49]. A majority of hexagonal SiC substrates available on the market today can be categorized into off-axis (usually 4 degrees off-oriented towards <11–20> direction) and nominally on-axis substrates. The latter have been commonly used to grow 3C-SiC by sublimation techniques. However, due to difficulties in controlling the spontaneous nucleation of the 3C-SiC island on such substrates and the formation of structural defects called double positioning boundaries (DPBs), it has been challenging to grow 3C-SiC with high crystalline quality. In contrast, the off-axis substrates have been mainly used for homoepitaxy or the bulk growth of hexagonal SiC crystals. The difference between the off-axis and nominally on-axis surfaces is the density of steps. The higher density of steps on off-oriented surfaces enhances the reproducibility of the substrate polytype and significantly reduces the possibility of 2D formation of 3C-SiC on step terraces. Therefore, generally, they have not been considered for the heteroepitaxial growth of 3C-SiC layers. However, it has been demonstrated that under certain growth conditions, excellent polytype stability and a high quality of 3C-SiC can be obtained on research size (7×7 mm^2) 4 degrees off-oriented hexagonal SiC (0001) substrates [22,23].

3.6. Hetero-Epitaxy Process: Carbonization

The heteroepitaxy of 3C-SiC on Si is a complex process that is realized in several steps. After the introduction in the reaction chamber, the first step is the etching of the silicon substrate in a hydrogen flux to remove the native oxide (step 1). Then, the second step is the substrate carbonization wherein a flux of the carbon precursor and carrier gas (hydrogen) is introduced in the chamber at temperatures between 900 and 1200 °C (step 2). Subsequently, the temperature should be increased to grow the 3C-SiC layer at temperatures close to the melting point of silicon (step 3). Finally, the temperature is decreased to room temperature (step 4). All these steps have a considerable impact on the quality of the epitaxial layer.

The most critical in the 3C-SiC/Si growth seems to be the second step: the carbonization of the Si substrate. This process is sometimes referred to as "reactive CVD" (R-CVD) because one of the components of the compound (in this case, silicon) is not supplied from the vapor phase but comes directly from the Si substrate that reacts with the gas ("vapor") species. As a result of carbonization, a thin seed of a few nanometers is formed for the subsequent CVD epitaxy process; it is sometimes denoted as the "carbonization buffer". The characteristics of this seeding layer are fundamental for the crystalline quality and the stress of the film. From previous studies, it has been observed that for given growth conditions, the morphology and thickness of the carbonization buffer depend strongly on the substrate orientation [50]. Consequently, the conditions to obtain an optimal buffer differ between orientations. For any orientation, the maximal thickness is conditioned by the nucleation density (proportional to carbon supply) and the ratio between the vertical and lateral growth rates (controlled by process pressure and temperature).

During carbonization, initial nucleation centers extend progressively, laterally, and vertically into three-dimensional 3C-SiC islands. Their temperature-dependent growth rate is proportional to the carbon flow rate but remains limited by the Si supply from the substrate; high in the initial stage when surface coverage with 3C-SiC is low, it reduces progressively to zero as the 3C-SiC islands extend, coalesce, and block the Si supply. It is important to mention that the R-CVD growth mechanism remains active until a complete coalescence of the 3C-SiC buffer is achieved, which is sometimes a long process. Consequently, in many cases, the CVD mechanism coexists (intentionally or not) with R-CVD during the initial part of step 3 of the heteroepitaxial growth.

The roughness of the carbonization layer has a large effect on the stress of the entire film: high roughness makes the relaxation of intrinsic stress during the growth easier [51]. The main process parameters that influence the roughness of the carbonization buffer are the temperature of the carbon precursor introduction and temperature of the carbonization plateau.

One of the problems related to the carbonization step is the formation of voids (also called "etch pits") in the near-interface region of the silicon substrate. These micrometric cavities form from the coalescence of silicon vacancies created in the Si substrate as a consequence of the R-CVD growth mechanism. The major part of voids does not affect the quality of the 3C-SiC film (their presence is sometimes considered as a stress-relaxing factor), although some of them can be at the origin of surface defects in the epitaxial film. Consequently, the void formation should be reduced. This can be achieved through carbonization under a high C/H_2 ratio, which increases the nucleation density and favors fast-film coalescence that stops the formation of voids. In Figure 10a, the fraction in percentage of the void area with respect to the total observed area as a function of the C/H_2 ratio are reported. In the same figure, the density of the void as a function of the C/H_2 ratio is also reported. The effective void areas decreases from 11% to 5% while increasing the flux of carbon atoms. The reduction can be further enhanced by introducing the silicon precursor during the thermal ramp between the carbonization plateau (step 2) and epitaxy (step 3) to form a transition layer. Such intentional mixing of R-CVD and CVD mechanisms further improves the quality of the interface between the 3C-SiC film and Si substrate [27]. In the same paper, it has been reported that the increase of the C/H_2 ratio also produces an increase in the density of the layer, as well as an increase in the carbonized thickness.

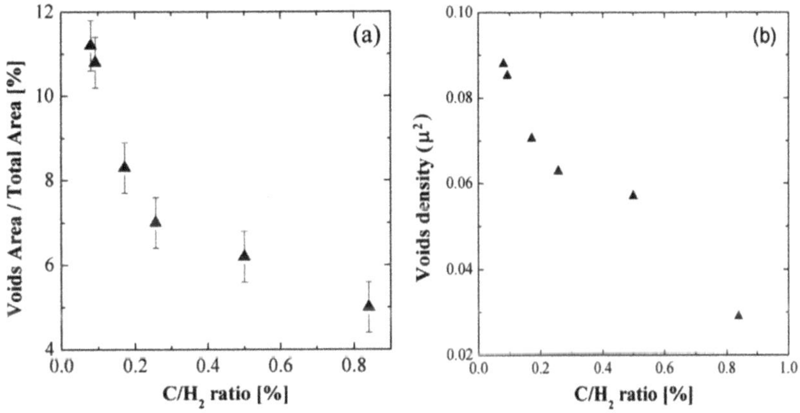

Figure 10. The percentage of void areas occupied with respect to the total observed area (**a**) and voids' density (**b**) are reported as a function of the C/H_2 ratio [%] (see Reference [27]).

The growth on compliance substrates may require a modification of the carbonization step in order to fit particular substrate-related requirements. This is, for instance, the case for substrates with the Si-Ge buffer for which the carbonization temperature was reduced to below 1000 °C and H_2 etching (step 1) was excluded from the process in order to preserve a thin Si cap (10 nm or 20 nm), necessary for correct carbonization. In addition, the temperature of the CVD growth (step 3) was lowered to avoid Si-Ge melting. It is important to underline that such a "low temperature" process resulted in a higher quality of the epilayer on Si-Ge with respect to the film grown on bare silicon.

An alternative approach to the formation of the 3C-SiC seed on the Si substrate was recently proposed. Silicon substrates are "pre-carbonized", meaning that a few nm-thick amorphous carbon (a-C) film is deposited using the plasma immersion ion implantation (PIII) technique. During H_2 annealing (step 1), carbon reacts with silicon to form oriented 3C-SiC seeds through a solid-state epitaxy mechanism. The standard R-CVD carbonization step is no longer necessary. CVD deposition on such seeds gave satisfactory results on all studied orientations: (100), (110), (111), and (112) [52].

3.7. *Effect of Growth Rate: Defects and Stress*

The growth rate has a large effect on the quality of the 3C-SiC both in terms of its structural quality and stress. It has been observed in a previous paper [53] that the growth rate has a large effect on the density of twins. In fact, in decreasing the growth rate from

10 μm/h to 1 μm/h, a decrease of the twin density of almost a factor of 6 can be observed. A similar (but weaker) effect has been observed also on the rocking curve, which is more sensitive to SFs and point defects [6]. The Full Width at Half Maximum (FWHM) of the rocking curve is reduced both by reducing the growth rate and increasing the thickness.

With decreasing growth temperature, the growth rate has to be reduced or otherwise the deposition may become polycrystalline. This is particularly the case for 3C-SiC growth on the Si-Ge buffer and constitutes a potential limitation for further development of this approach.

For the 3C-SiC growth on ISP substrates, we demonstrated that the height of the void created above the vertex of the pyramid (Figure 9b) increased at higher growth rates. Consequently, the initial stage of 3C-SiC growth on ISP substrates, until reaching complete coalescence, should be performed at a low growth rate.

The growth rate has also an influence on the final stress of the 3C-SiC epilayer. Indeed, as demonstrated in [54], during the growth of the 3C-SiC layer, the intrinsic stress in the layer is continuously relaxing. For a given film thickness, by tuning the growth rate, we can adjust the duration of the relaxation (by the duration of the growth), controlling the final stress of the sample.

3.8. Defects in 3C-SiC: SFs and APBs

In 3C-SiC, the most important defects that hinder its use in the microelectronic industry are related to SFs and dislocations. Stacking faults (SFs) are the most important ones dominating over the entire 3C-SiC layer thickness. In literature, three types of SFs are observed depending on the number of atomic planes with the wrong orientation: SFs can have 1, 2, or 3 errors in the stacking sequence and they are called intrinsic (or SF<1>), extrinsic (or SF<2>), or conservative (or SF<3>) [40,55,56].

In 3C-SiC, mechanisms for SFs' self-annihilation exist but there is also the possibility for SFs to be generated [6,40,57,58]. The concomitant presence of these mechanisms leads to the fact that SFs in 3C-SiC can be hardly reduced. The minimum SFs' densities achieved so far amount to about 10^4 cm^{-1} in thin films [58]. In Figure 11a, a TEM image in in-plane view shows four stacking faults that are generated from a grain boundary, while in Figure 11b, a TEM image in cross-view shows the annihilation of the SF. In Figure 11a, a vertical grain boundary generates three clearly visible SFs lying in $(1\bar{1}1)$, $(11\bar{1})$, and $(\bar{1}\bar{1}1)$ planes. Interestingly, two SFs in the $(1\bar{1}1)$ planes limit the SF in the $(\bar{1}\bar{1}1)$ plane; they are limited on the other side by the grain boundary. The intersection between the SFs $(1\bar{1}1)$ and $(\bar{1}\bar{1}1)$ is the Lomer−Cotrell partial dislocation and has a Burgers vector of a/6[011 The place in which the grain boundary intersects the SF $(\bar{1}\bar{1}1)$ is the place in which the SF $(\bar{1}\bar{1}1)$ is generated. In Figure 11b, the crossing of several SFs is shown. This image proves that there are two possible intersections of the SF lying in the (111) and $(11\bar{1})$ planes. The intersection indicated as "1" has an inverted V-shape typical for the formation of a Lomer–Cottrel dislocation. This dislocation, as already reported, has a Burgers vector (a/6[110]) lower than the usual partial dislocation Burger vector (a/6[112]) that borders the stacking fault. The intersection called "2" has a lambda shape; it forms for kinetical reasons. These two configurations can decrease the amount of SF approaching the surface and improve the quality of the epitaxial film.

The annihilation mechanism is considered in more detail in Reference [59]. It is found that the key parameter for the formation of a lambda-shape or an inverted V-shape is the distance between the PDs and the mutual orientation of their Burgers vectors. In the case in which the PDs have Burgers vectors that sum in such a way that the resulting Burger vector is shorter than the initial ones, the partial dislocation attracts to each other and if they are closer by less than about 15 nm, the propagation of both SFs is suppressed with the formation of a Lomer–Cottrell lock. In the case in which the two PDs are far more than 15 nm, they do not interact with each other and can form a Lambda-shape structure. The Lambda-shape can form even if the partial dislocations are close enough, but they repulse. In Figure 12 on the left, a sequence of MD simulation snapshots of the formation of "inverted V"-shaped intersection of stacking faults have been shown. In Figure 12 on the right, MD simulation snapshots of the formation of "λ"-shaped intersections of stacking faults in the case of a large distance between partial dislocations (a–c) and repulsing dislocations (d–f) have been reported. Blue atoms correspond to the Si and C atoms in the cubic diamond lattice, while orange atoms belong to the stacking faults.

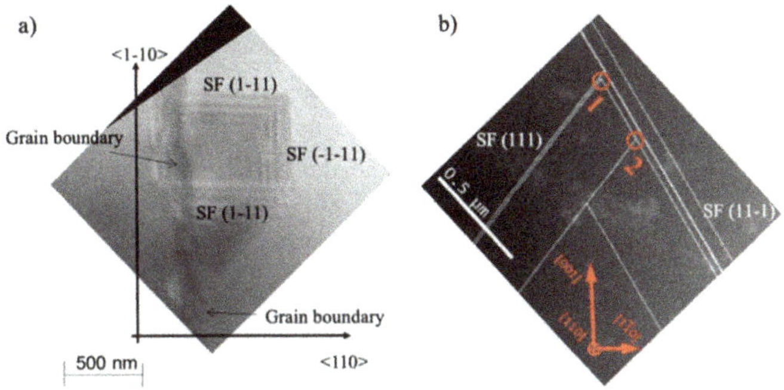

Figure 11. (**a**) TEM image in in-plane view shows four stacking faults that are generated from a grain boundary. (**b**) TEM image in cross-view showing the annihilation of the SF with two different structures (adapted from References [40,60]).

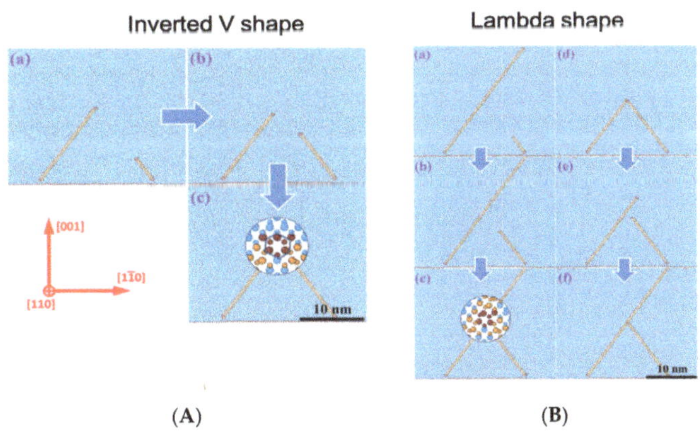

Figure 12. (**A**) Molecular dynamics simulation snapshots of the inverted V-shape configuration. (a–c) The simulation time: (a)—0, (b)—120 ps, and (c)—180 ps. Blue atoms correspond to the Si and C atoms in the cubic diamond lattice, orange atoms belong to the stacking faults. Inset in panel (c) shows the atomic configuration of the formed Lomer–Cottrell lock dislocation. (**B**) Molecular dynamics simulation snapshots of the lambda-shape configuration. in the case of the large distance between the 30° leading dislocations (a–c) and as a result of the interaction of closely spaced 30° dislocations with equal screw components of Burgers vectors (d– f). Simulation time: (a)—0, (b)—360 ps, (c)—540 ps, (d)—0, (e)—60 ps, (f)—200 ps. Inset in panel (c) shows the atomic configuration of the intersection of 30° partial dislocation with crossing stacking fault, also corresponding to the intersection in panel (f). (adapted from Reference [59]).

The SFs can interact also with other extended defects, such as the inverted domain boundary (IDB) (sometimes called the anti-phase boundary, APB). In 3C-SiC grown on (100) "on-axis" silicon, due to the symmetries of the Si lattice, two equivalent dispositions of the SiC crystal are possible. The two possible orientations are rotated 90° around [001] and, due to the SiC symmetries, a rotation of 90° is equivalent to flip the crystal upside-down.

The boundary between two such domains is called IDB (or APB). The SFs can interact strongly with this kind of extended defect of SiC. In Figure 13, a sequence of STEM images showing an IDB interacting with SFs is shown. The image is the projection of the TEM lamellae in the (110) plane. The SFs are observed as straight lines, while the IDB has different lying planes and appears as a ribbon. A close inspection of these images shows that SFs can be generated and annihilated by the IDB: several SFs can be recognized in the figure and some of these are apparent only in the crystal below the boundary, while some others are apparent only in the crystal above the boundary. SFs that are in the lower crystal are not allowed to propagate in the top crystal and in this case, we observe an annihilation of the SF due to the presence of IDB. On the contrary, SFs that belong to the top crystal and are not present in the lower crystal are generated in the IDB. The SFs can be generated during the growth due to interface instability that creates seeds for nucleation; after the nucleation, it expands following the growth of the surface. Eventually, it can collide on an IDB and be annihilated [40]. SFs can be also generated during the cooling down of the temperature after the growth; indeed, temperature gradients can induce stress in the layer. Above critical shear stress, it is known that the formation of dislocations and, in 3C-SiC, the formation of partial dislocation is a thermodynamically favored process.

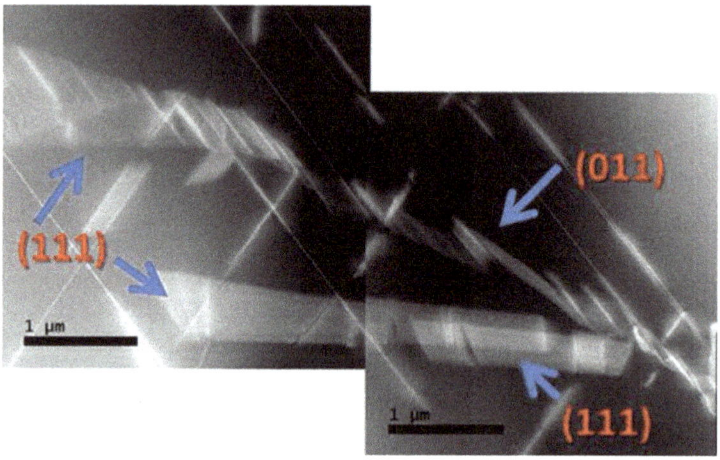

Figure 13. Sequence of STEM (110) cross-view images showing an IDB and its interaction with SFs. The lying planes of IDB and SFs are indicated (adapted from Reference [40]).

As previously discussed at the beginning of Section 3.7, different kinds of SFs are present in 3C-SiC. These different types of SFs can be seen as inclusions of different hexagonal polytypes in the cubic structure. In particular, the intrinsic SF can be called a 2H-like SF, the extrinsic one can be seen as a 4H-like SF, and the conservative one can be seen as a 6H-like SF. These different SFs have different energies [60] and different behaviors of these defects should be expected. The room temperature μ-PL map at 540 nm, taken on a 3C-SiC sample in cross-section, is shown in Figure 14a [61]. Moving from the Si-SiC interface towards the top (from 0 μm to 35 μm), the band-edge peak intensity rises, showing a considerable improvement of the crystalline quality, increasing the growth thickness. Figure 14b,c exhibits μ-Raman maps obtained in the same location and indicate certain areas as well as the 3C-SiC/Si interface with greater signal magnitude at 778 cm^{-1} and 784 cm^{-1}, respectively. Figure 14d–f displays the mean Raman spectra obtained in areas (1), (2), and (3), which reveal the 3C-SiC TO mode centered at about 796.5 ± 0.2 cm^{-1}. Conversely, the mean Raman spectrum obtained in points (2) and (3) reveal an extra peak at 778.3 cm^{-1}, as well as two more peaks correspondingly at 778.0 cm^{-1} and 784.0 cm^{-1}. These additional peaks can be related to the presence of extrinsic (4H-like and 6H-like) stacking faults. From these data, we can observe that, despite the low energy of the 6H-like SF, it appears that this kind of stacking fault can be observed in larger regions and closer to the surface with respect to the 4H-like SF. More investigations should be done concerning this aspect but we suspect that this large presence of 6H-like SFs could be due to kinetic reasons more than energetic ones. In fact, from the energetic point of view, this SF has the lowest formation energy.

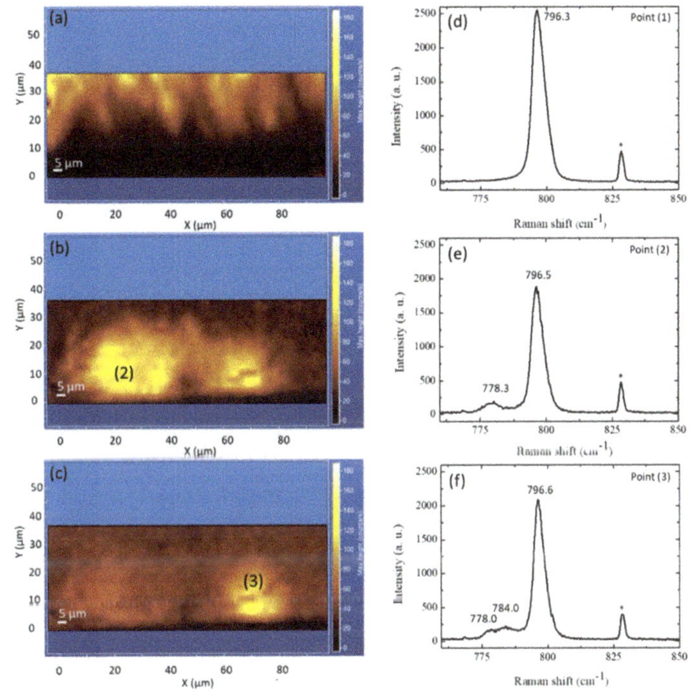

Figure 14. Micro-PL mapping (**a**) at 540 nm and micro-Raman mapping of a 3C-SiC cross-section located at (**b**) 778 cm^{-1} and (**c**) 784 cm^{-1}.The interface with the removed Si substrate is shown by point 0 on the Y-axis. Average Raman spectra achieved in the (**d**) area (1) of the map (**b**), (**e**) zone (2) of the map (**b**), and (**f**) zone (3) of the map (**b**,**c**). The laser probe created the peak located at 828.37 cm^{-1} (*) (see Reference [61]).

3.9. Defects in 3C-SiC: Kinetic Monte Carlo Super Lattice Simulations

The study of the kinetic evolution of a defective system is a difficult task as it requires both atomistic accuracies typical for the molecular dynamics approach and large space-time scales typical for the experimental systems. Within the CHALLENGE project, we developed an ab-initio calibrated kinetic Monte Carlo super lattice (KMCsL) code, [62] offering a good compromise between accuracy and efficiency, which can simulate the results of the growth processes in non-polar SiC as a function of the growth parameters also in terms of defectivity and surface morphologies. Hence, the KMCsL simulations allowed for the investigation of the formation and development of extended as well as point defects over a realistic growth [62]. As an example of a simulation application strongly relevant to the experimental studies, we considered the evolution of anti-phase boundaries (APBs) in 3C-SiC and their interaction with stacking faults (SFs), which is discussed in detail in Reference [57].

Due to the comparable energetics of polytypes, SFs are a frequent and wide-spread extended defect in SiC, with respect to the polytype. They are classified as incorrect atoms sequences in comparison with theoretical polytype stacking arrangements. In the purely hexagonal close-packed (hcp) representation, the polytype sequence is defined as a repeating series of layers made up of Si-C dimers aligned on the hexagonal axis. Dimers in every layer take one of three extremely symmetric locations (often denoted as A, B, and C). If the recurring arrangement is ABC ABC ABC . . . , etc., extending along the crystal <111> axes, the cubic 3C-SiC (zinc-blend) configuration is produced. As a result, the existence of {111} planes on the surface depletion allows for an increase in SF production. The formation of a pair of triple SFs (SF <3>: three bilayers not in the correct crystalline structure) from the surface depletion caused by an APB is shown in Figure 15. A series of pictures of the under-coordinated atoms obtained at various KMC intervals are displayed. The existence of three-fold coordinated Monte Carlo particles at its border leads to the generation of the triple SF in this depiction (i.e., at the corresponding partial dislocation). We note that the stacking sequence of the triple SF (sometimes termed micro-twin, i.e., ABC ABC ACB ABC ABC ABC) divides two crystal areas in pristine epitaxial order. The only atoms out of the right crystal locations are those within the extended defect. The APBs' localized asymmetry as well as the existence of the {111} faceted surface result in the development of a triple SF (due to polytype instability) (see Figure 15a,b). When the APB generates the SF, the two extended defects (SF and APB) maintain separate kinetics: the APB proceeds to move through the [110] plane, while the SF expands on the (111) plane (snapshots b, c, and d of Figure 15). A TEM picture of an SF generated by an APB along the epitaxial growth of a 3C-SiC (001) substrate is shown in Figure 15e. It expands autonomously from the APB kinetics along the {111} planes. Moreover, surface depletion can be seen in the correspondence of the (001) surface. These composed structures of proximal APB and SF-type defects have been also evidenced by the conductance maps in this paper.

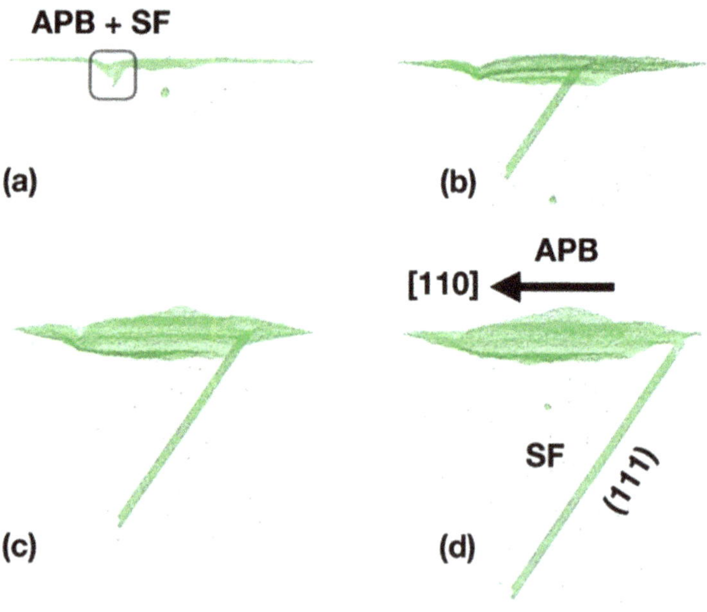

Figure 15. *Cont.*

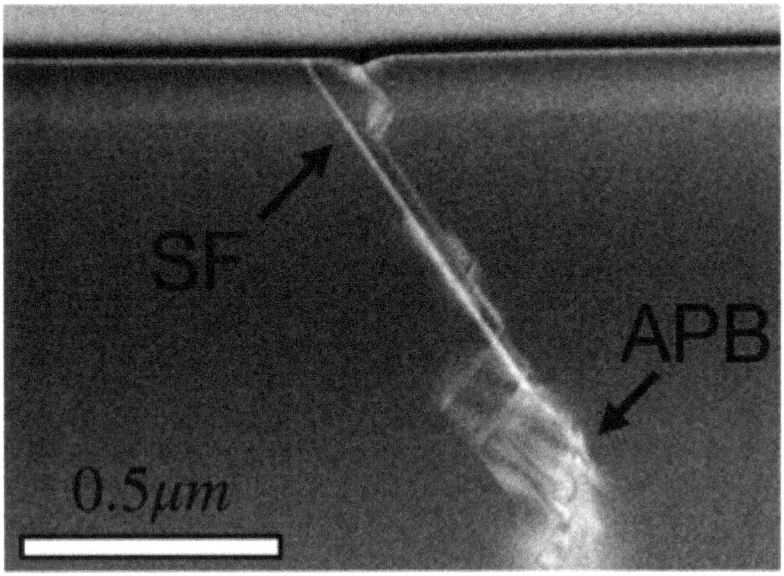

(e)

Figure 15. A pair of triple SFs are generated as a result of the surface depletion caused by an APB during 3C-SiC epitaxy along the [001] z-direction. Under-coordinated atoms from several KMC moments: (**a**) triple SFs created by an APB; (**b–d**) three consecutive images illustrating the autonomous kinetics of the APB traveling towards the [110] axis; and two formed triple SFs expanding along the (111) planes. (**e**) TEM picture of an SF caused by an APB along the epitaxial growth (001) of a 3C-SiC. It expands on the {111} planes autonomously from the APB kinetics. Moreover, the surface depletion is evident at the (001) surface (adapted from Reference [57]).

3.10. Defects in 3C-SiC: Electrical Effects

Nanoscale-resolution current mapping of 3C-SiC by conductive atomic force microscopy (CAFM) provided a direct demonstration of APBs as the main extended defects responsible for the enhanced leakage under reverse-bias, whereas both APBs and SFs were shown to act as current paths under forward polarization. Figure 16a illustrates the experimental configuration used for CAFM measurements on the cubic silicon carbide surface. A typical topographic image collected on a 20 μm × 20 μm scan area is shown in Figure 16b, from which a surface root mean square (RMS) roughness of 3.2 nm was calculated. The nanometer deep "V-shape" depression in the morphology and in the height line-scan (Figure 16b, right panel) were associated to an APB in accordance with the Monte Carlo simulations of Section 3.7. Figure 16c,d report the current maps measured simultaneously to the topography by applying a reverse-bias ($V_{tip} = -0.5$ V) and forward-bias ($V_{tip} = 0.5$ V) polarization to the Pt tip, respectively. This Schottky diode behavior of the Pt/3C-SiC contact was confirmed by the significantly lower current values measured under reverse polarization with respect to those measured under forward-bias. Using the same current range (from 0 to 50 pA) for the two current maps, APBs are the most evident conductive features under reverse-bias, whereas both APBs and SFs (indicated by blue arrows in Figure 16d) contribute to the conduction under forward polarization. Two representative scan lines across the APB for the two opposite tip biases are also shown in the right panels of Figure 16c,d, showing a higher current peak on the APB under forward-bias with respect to the reverse one. This suggests that APBs are mainly responsible for the enhanced reverse leakage current measured in macroscopic Pt/3C-SiC Schottky diodes. In particular, the separation between these extended defects deduced from this microscopic analysis is in the order of tens of micrometers, in very close agreement with the value of L (≈20 μm) deduced from the statistical characterization of Schottky diodes with different areas for thin 3C-SiC layers [63].

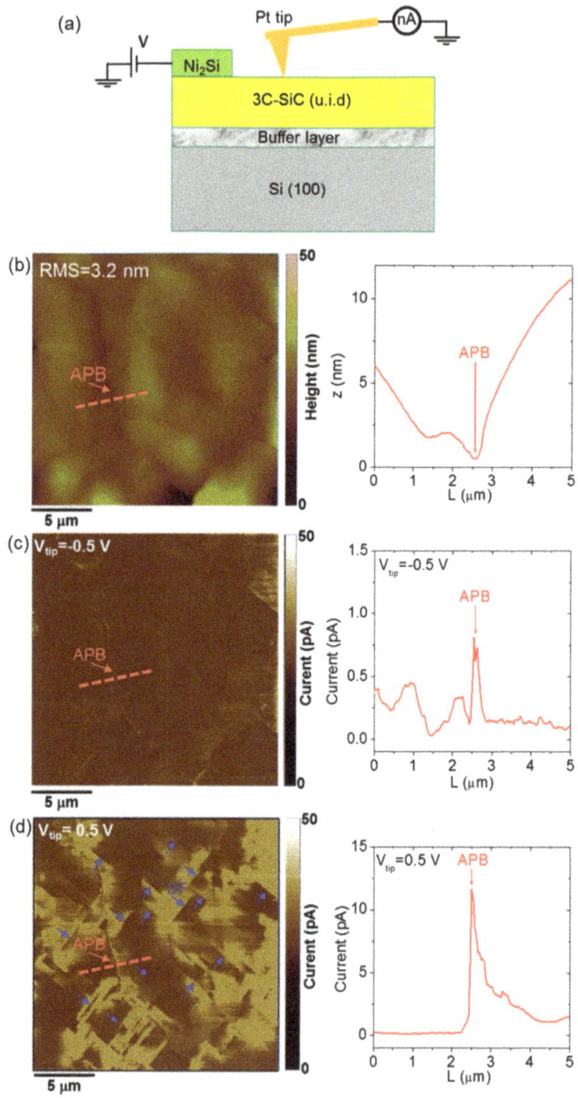

Figure 16. (**a**) Schematic illustration of the CAFM setup. (**b**) Morphology and (**c**) current maps collected under reverse-bias polarization of the tip (Vtip = −0.5 V) and (**d**) forward-bias polarization (Vtip = 0.5 V). An APB is indicated by a red arrow and SFs by blue arrows. Representative line-scans across a grain boundary extracted from the topography ((**b**), right panel), current maps under reverse-bias polarization ((**c**), right panel), and forward-bias polarization ((**d**), right panel) of the tip are shown (see Reference [63]).

3.11. Defects in 3C-SiC: Point Defects

Point defects can be observed in 3C-SiC by PL measurements on different samples grown in different conditions. In particular, the PL spectra in the wavelength range of 1100–1600 nm of different samples grown at different temperatures with the same growth rate are reported in Reference [64]. It was possible to observe that carbon vacancy (V_C), carbon–silicon vacancy (V_CV_{Si}), carbon vacancy–Si antisite (V_CC_{Si}), and Al-related defects are present [65]. The growth rate and growth temperature seem to be the main parameters that influence the point defect formation during the growth, as also reported in a previous simulation paper [66].

3.12. Defects in 3C-SiC: Protrusions

Small defects produced during the carbonization process can have a large influence on the final wafer quality: an example of this effect can be seen in the case of the defect called "protrusions". These defects appear on the surface of 3C-SiC as dark squares, with a peculiar 3D structure similar to inverted pyramids with a vertex close to the SiC/Si interface. In Figure 17a,b, a scanning electron microscope image of protrusion in in-plane (a) and in cross-view (b) are shown. [67] The plan-view image is shown for a 30 µm-thick epilayer to evidence the shape of the structure, while the cross-view is shown for a 3 µm-thick epilayer. Yellow lines are drawn to identify the edge of the defect that is limited by four stacking faults. It was also found that the inner core of the defect consists of nano-crystals twinned with respect to the substrate orientation. The base of the inverted pyramid is a square and the height is the same as the thickness of the epilayer. In Figure 17c, the lateral size of the protrusion is shown as a function of the epilayer thickness. A linear correlation between the size of the protrusion and the thickness of the epilayer is apparent. Again in Figure 17c, two optical images of the protrusion are shown: in the left-upper corner, a defect in the 30 µm-thin film is shown, similar to what was observed in Figure 17a, while in the right-lower corner, the defect in the 150 µm-thick film is shown. Since the quality of the layer depends strongly on the thickness, a higher thickness is preferable. Thus, the presence of even a small density of protrusions must be indeed avoided because it can strongly decrease the quality of the wafer. In stating the importance of reducing the density of such a defect, our group investigated the reason for the formation of protrusions. The seed of this extended 3D defect lies 10 nm above the SiC/Si interface and is probably related to a non-balanced carbon amount during the carbonization step or during the temperature ramp-up after carbonization. Carbonization, as earlier reported, is a process in which the carbon precursor reacts with the bare silicon surface and this is performed to prepare the Si surface for SiC growth. The Si/C ratio during the carbonization and the post-carbonization process is the key parameter to avoid the formation of protrusions. In Figure 17d, the density of protrusions as a function of the C/Si ratio during the rise of the temperature after carbonization is shown. As it is apparent, a ratio lower than 1.2 is able to decrease the density of protrusions by almost two orders of magnitude, leading to a density of 10 cm^{-2}.

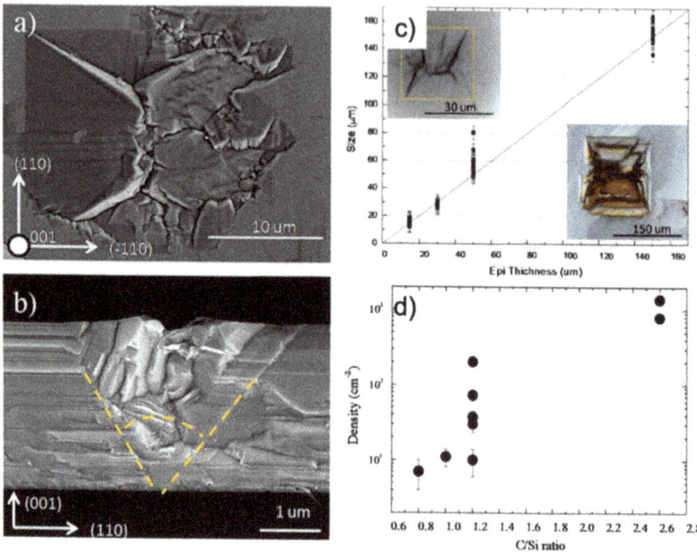

Figure 17. (**a**) Scanning electron microscope plan-view image of a protrusion in a 30 μm-thick epitaxial layer. (**b**) Cross-view obtained after the cleavage of the wafer for a 3 μm-thick epilayer. Yellow lines are drawn in order to identify the edge of the defect. Crystallographic orientations are also drawn. (**c**) The average size of the protrusions vs. the epitaxial layer thickness. (**d**) The density of the protrusion as a function of the C/Si ratio during the buffer layer step (adapted from Reference [67]).

3.13. Stress in 3C-SiC

Another aspect that is crucial in the development of the 3C-SiC material is stress. In this case, we have essentially two different components of the stress:. The first one is called intrinsic stress, related to the different lattice constant between 3C-SiC and the silicon that produces a high concentration of defects at the interface. These defects produce a high level of stress essentially in the first microns of the growth. For thick layers, we observe a reduction of this intrinsic stress and for very thick layers, this component of the stress is close to zero. For the intrinsic stress, it has been observed through using MEMS devices that the reduction of the stress follows an exponential low that is very close to the exponential decrease of the SFs' density vs. thickness [9]. This stress is generally compressive in 3C-SiC (100) while it is tensile in 3C-SiC (111).

The second component of the stress is the "thermal stress", related essentially to the different thermal coefficients between silicon carbide and silicon. In fact, the growth occurs at high temperatures (1350–1390 °C) and the two materials (SiC and Si) decrease their lattice constants in different ways, moving from the growth temperature to room temperature. This component of the stress is always tensile and then reduces the stress or even changes the sign of the stress in the (100) material, while considerably increasing the total stress in the case of the (111) 3C-SiC. For this reason, it is extremely difficult to grow a thick layer of 3C-SiC on the (111) Si without cracking the film or even the substrate during the ramping down of the temperature in the reactor.

In the CHALLENGE project, we used two different approaches to try to solve this problem. The first approach has been described in Section 3.2. In fact, using the pillars' structures, it is possible to considerably reduce the thermal stress with the deflections of the pillars on the edge of different patches (see Figure 7). In this way, it was possible to obtain thick (111) 3C-SiC layers' wafers with a low bow.

The other approach that we used in this project will be described later in Section 3.15. In this approach, after a thick growth (60–90 μm), the silicon substrate was melted inside the reaction chamber and then removed one of the sources of the thermal stress.

Obviously, this process does not remove the intrinsic stress due to the defects at the 3C-SiC/Si interface. Using the SiGe buffer layer approach described in Section 3.3, it is possible to decrease the intrinsic stress but further experiments should be done to completely remove this component.

3.14. Bulk Growth on Hexagonal SiC

As demonstrated [21,22], during the initial stages of the growth, a facet with an on-axis surface is formed at the edge of the grown layer. At specific growth conditions, this facet becomes a preferential 3C-SiC nucleation site. Once 3C-SiC is formed on the facet, it laterally enlarges by covering the entire surface. The lateral enlargement of 3C-SiC from the edge towards the center on the SiC (0001) surface is proportional to ~ tan α, where α is the off-cut angle of the substrate. Therefore, by increasing the off-orientation of the substrate, the total layer thickness needed to cover the entire substrate surface, with the 3C-SiC enlarging from the edge of the sample, also increases. Based on our estimations, to cover a 1-inch 4 degrees off-oriented hexagonal SiC (0001) substrate with 3C-SiC, the layer thickness should be about 4–5 mm. Such a thickness complicates the growth process. Therefore, as a compromise, a hexagonal substrate with a 0.8 degrees off-cut was used to explore the growth on a larger substrate area. As seen on the left side in Figure 18a, a full surface coverage with the 3C-SiC on hexagonal substrates with the size 15 $\times$ 15 mm^2 was obtained. However, when the same growth conditions (T = 1900 °C, average growth rate of ~0.3 mm/h) were applied to the growth on 1-inch substrates, an instability of the 3C-SiC polytype was observed. This was attributed to the dimensional limitations of the graphite container, which does not allow for the obtaining of the uniform supersaturation of SiC vapor species over the entire surface of the substrate. Therefore, a new hot zone for the growth of 3C-SiC on 2-inch substrates was designed and the immediate advantage of it in controlling the stability of 3C-SiC was observed. However, by growing thicker layers on (0001)/Si-face substrates, it was observed that the DPBs tended to branch out into larger structural defects, which deteriorates the quality of the 3C-SiC crystal. To compare the formation of DPBs, the growth of 3C-SiC on the (000-1)/C-face was investigated. It was observed that there was an obvious difference in the DPBs' appearance in the 3C-SiC grown on the (000-1)/C-face. As shown in the scanning electron microscope (SEM) images in Figure 18a, the majority of DPBs on the (000-1)/C-face maintain a line-like propagation path. This means that their propagation is less damaging to the 3C-SiC crystal compared to the ones on the Si-face. In addition, an interesting phenomenon indicating different step dynamics in 3C-SiC layers grown on the Si and C-faces has been observed. Surfaces analysis by atomic force microscope demonstrated that the step-height in both cases is very similar and mostly varies between 0.25 and 0.8 nm, while the terrace width is almost three times larger on the Si-face and varies in a range of ~130–150 nm. The crystalline quality of 3C-SiC layers grown on the Si and C-face on substrates with the same off-cut angle of 0.8 degrees is similar and the full-width at the half-maximum of the XRD ω rocking curve using a footprint of 5 $\times$ 5 mm^2 varies between 200 and 300 arcsec. However, 3C-SiC layers grown on the C-face of a hexagonal substrate with the off-cut angle of 1.5 degrees contain areas with ω rocking curve values of 93 arcsec. This indicates that using hexagonal SiC substrates with even higher off-cut angles could be the right direction for further research, even though the growth on such substrates will require growing much thicker layers to obtain a full coverage with 3C-SiC. Therefore, a comparative study of the 3C-SiC(111) grown on the (0001)/Si-face and (000-1)/C-face on 4 degrees off-oriented 4H-SiC research size (7 $\times$ 7 mm^2) substrates was conducted [68]. Even though the 3C-SiC polytype is more stable on the (0001)/Si-face, it was shown that smoother surfaces of 3C-SiC could be obtained on the C-face. In addition, the transition layer, which is a mixture of various polytypes, between the hexagonal SiC substrate and the 3C-SiC is significantly thinner on the (000-1)/C-face, leading to a direct polytype conversion mechanism.

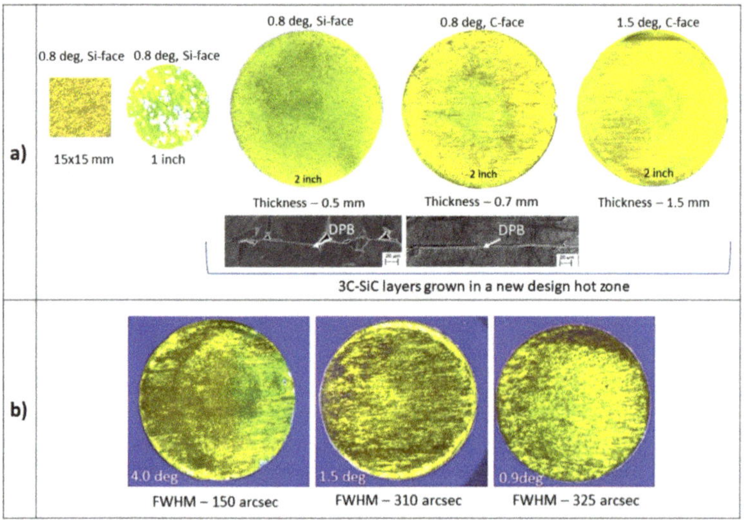

Figure 18. (**a**) 3C-SiC growth on hexagonal SiC substrates using enhanced sublimation epitaxy. (**b**) Polarized light optical micrographs of 2.5 mm-thick 3C-SiC layers grown on 2-inch 4.0, 1.5, and 0.9 degrees off-oriented SiC (000-1) substrates. All samples were grown at 1950 °C in vacuum (5×10^{-4} mbar).

Based on the promising results on the growth of 3C-SiC on the (000-1)/C-face of 4 degrees off-oriented 4H-SiC research size (7×7 mm^2) substrates, a series of experiments were done on a 2-inch area. An example of a 2.5 mm-thick 3C-SiC layer grown on a 4 degrees off-oriented substrate is shown in Figure 18b (picture on the left). The 3C-SiC layer grown on 4 degrees off-oriented substrates still contains double-positioning boundaries that deteriorate the crystalline quality. Despite that, when compared to 2.5 mm-thick 3C-SiC layers grown on substrates with lower off-orientations, the crystalline quality is higher. This was confirmed by the FWHM of HRXRD ω rocking curves, which were measured on three different areas on each layer using a footprint of 2×10 mm^2. The average FWHM values of 3C-SiC layers grown on 4.0, 1.5, and 0.9 degrees off-oriented SiC substrates were 150, 310, and 325 arcseconds, respectively. This indicates that even larger off-orientations could be a potential route for the further improvement of 3C-SiC crystalline quality.

In addition to DPBs, which are dominant defects in 3C-SiC grown on hexagonal SiC substrates, dot-like and arrow-like defects are observed (Figure 19a). These defects usually occur in samples that are thick (>1 mm) and grown by interrupted growth (growth stopped to change polycrystalline SiC plate/source). After selective etching using molten KOH, the dot-like defects appear as triangular etch pits (Figure 19b), which are characteristic features of threading screw dislocations on the (111) crystal surfaces. The selective etching of arrow-like defects revealed elongated groves as shown in the SEM micrograph in Figure 19d. In addition, after KOH etching, the surface around the elongated grooves possessed imprints of stacking faults propagating along (111) planes, which appeared as line-like features rotated to each other by 60 degrees. Based on cross-sectional analysis by optical microscope, the dot-like features on the surface corresponded to the threading defects with a cylindrical path that did not widen while the crystal grew (Figure 19c). They originate in the grown 3C-SiC layer, transition layer, or can be tracked down all the way to the substrate. In the latter case, they can be observed as a continuation of micro-pipes propagating in the substrate material. The arrow-like defects can originate anywhere in the 3C-SiC layer but the most common origin is at the interface between 3C-SiC layers grown by repeated growth runs as shown in Figure 19e. This indicates that thick (>1 mm) 3C-SiC layers should be grown in a single growth run. Otherwise, disturbances in the growth during the cooling down and temperature ramp-up processes may cause the formation of such defects. Moreover, as the thickness of the 3C-SiC layer increases, the arrow-like defects tend to branch out and significantly deteriorate the crystal quality. The formation of such defects can be tackled by using thicker source-material, which would allow for growing thick layers without any interruption.

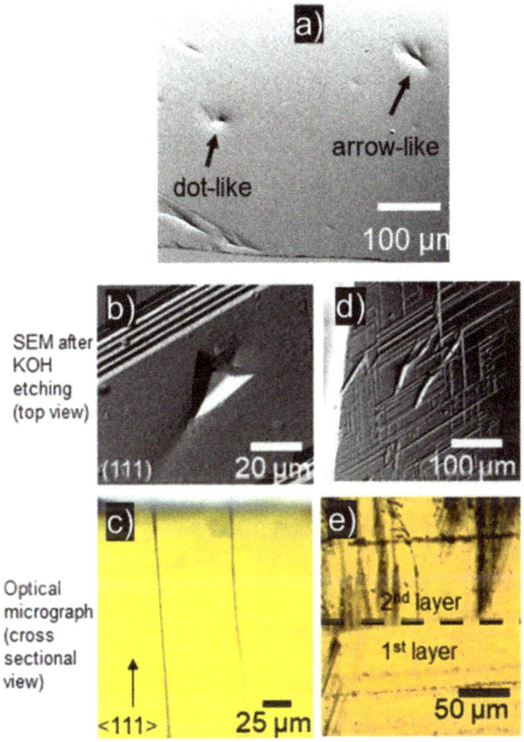

Figure 19. Top view SEM image after KOH etching of (**a**) threading dislocation and (**d**) arrow-like defects. Optical micrograph (cross-sectional view) of (**b**) threading dislocations and (**e**) arrow-like defects. (**c**) SEM image of the surface appearance of defects.

Even though the defect density in the 3C-SiC grown on hexagonal SiC over the 2-inch area is too high for the industrial processing of transistors, it could be used to explore hydrogen generation using solar-driven water splitting [69] or the growth of a large-area monolayer and multilayer graphene [70,71].

3.15. Close-Space PVT Growth of Bulk 3C-SiC on 3C-SiC-on-Si CVD Seeding Layers

Figure 20 gives an overview of the development of the diameter of grown crystals during the project. Starting with a diameter of 0.5 inches at the beginning of the project, the first big milestone was reached in 2018 with the reproducible growth of 2-inch crystals with thicknesses of up to 870 μm. This marked the first time that bulk material, with relevant sizes, was grown regularly using a sublimation method [26]. Such a material could be used as a seed for subsequent growth in other processes such as M-PVT or CF-PVT. With the transfer process described in Section 2, crack-free crystals could only be obtained for up to 2 inches. For larger diameters, cracking of the thin epitaxial layer poses a problem during the preparation of the seeding stack. Nevertheless, the first growth runs on 4-inch materials were performed at the beginning of 2019. Additional improvements of the transfer process were necessary to optimize the results for the large sample size. Although the cracking problem could not be solved at this point, first-cracked but non-broken samples with a thickness of approximately 1 mm could be produced by the end of 2020, consisting of one coherent piece of crystal. Measurements of the XRD rocking curve of the full-width at half-maximum (FWHM) of the (002) reflex resulted in values of 138 and 140 arcsec for 2-inch and 4-inch materials, respectively, confirming that there is no decline in material quality for large diameters.

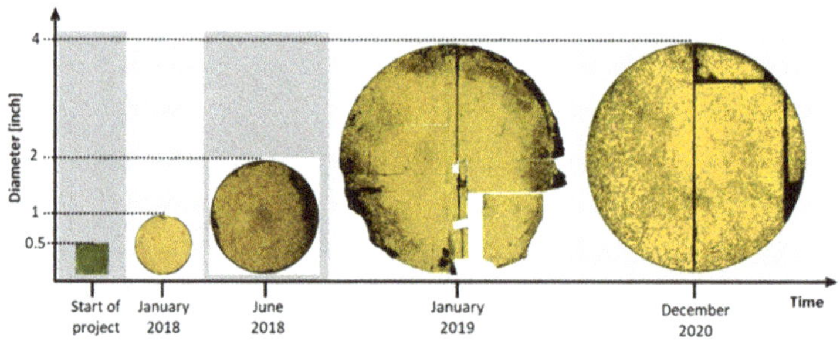

Figure 20. Evolution of diameters for bulk 3C-SiC crystals grown by sublimation growth. The timeline is indicated.

There are two main reasons for cracking during the manufacturing of seeding stacks. The first one is associated with the used starting material grown by CVD. Caused by a lattice mismatch between silicon and 3C-SiC of approximately 20%, a wafer bow occurs during the heteroepitaxial growth [11]. The bow and the accompanying stress in the material will increase the probability of cracking the thin 3C layers during handling. This problem will be even bigger with increasing diameters. The second issue occurs during the etching process. During this step, NO_x species will be created [36]. As the silicon removal starts at the edges and moves towards the center of the samples, the created NO_x species will accumulate in the middle of the sample, leading to the buoyancy of the thin remaining 3C layer. This mechanical stress will lead to cracks. For diameters up to 2 inches, this problem can be neglected but will be present for larger samples. One solution to prevent the uplift concerned the change from a horizontal to a near-vertical etching setup, reducing the effect of sample bending and therefore reducing the cracking probability. Nevertheless, micro defects induced at the edges during the laser ablation process, as well as the wafer bow, still lead to the cracking of the seeding layers.

Due to ongoing research, a new form of seeding material became available. As described by Anzalone et al. [64], the production of freestanding 3C-SiC wafers grown homoepitaxially by CVD at elevated temperatures is possible. The availability of such seeding materials, up to a thickness of approximately 200 µm, offers new possibilities for the continuing growth using CS-PVT. Compared to the thin epitaxial seeds, no transfer process is necessary for such materials. In addition, the material grown by homoepitaxial CVD still has some setbacks concerning the wafer bow and remaining protrusion defects, and first-successful sublimation growth runs on such seeds were carried out. These experiments have proven its suitability as a seed for CS-PVT and therefore represent a promising starting point for the bulk growth of cubic silicon carbide.

Despite the appearance of different defects on the surface, all samples depicted in Figure 20 have a bright yellow appearance typical for the cubic polytype. The results of X-ray diffraction (XRD) and Raman spectroscopy confirm the growth of 3C-SiC. The small black dots visible are associated with protrusion defects that were already present in the CVD seeding layers. These defects increase in size with the increasing layer thickness during the growth process. The darker areas at the edges of the samples were caused by both an overflow of the carbon glue during the merging step and an insufficient cleaning step afterward.

Besides the clearly visible defects, the material quality of the sublimation-grown crystal is quite high. Raman spectrometry is commonly used for the analysis of material quality as it provides a fast and non-destructive method for the evaluation of material quality. The transversal optical (TO) mode should be forbidden for defection-free (100)-oriented on-axis-grown cubic silicon carbide. However, if it is visible, it can be used to determine the stress inside the material, depending on the position of the peak [72,73]. For stress-free materials, the wavenumber of the peak is located at approximately 797.61 cm^{-1} [74]. An increased value is linked to comprehensive stress, whereas a lower value can be related to tensile stress. In [26], the wavenumbers for different types of cubic silicon carbide materials were presented. It could be shown that the layers grown by CVD were tensile-stressed, caused by the lattice mismatch between silicon and 3C-SiC both for on-axis as well as off-axis materials. The value for the crystals grown by the sublimation method, with CVD materials as seeds, displayed similar stress levels. After the sublimation growth, the crystals were usually oxidized at 800 °C to remove the carbon glue and separate the samples from the polycrystalline SiC-carrier. After this treatment, the measured values for the TO mode were near the stress-free value of 797.61 cm^{-1} as reported in the literature, indicating a quasi-stress-free material. In addition, XRD 2θ-ω scans were performed. For a stress-free cubic silicon carbide crystal, the in and out-of-plane lattice constant should be the same based on the cubic lattice. As stress will distort the lattice, a variation of the lattice planes will occur. The data obtained from the measurement confirm the assumption of the stress-free material after the sublimation growth and removal of the carrier. The results can be found in [33].

Typical defects occurring in (100)-orientated 3C-SiC are stacking faults (SF), anti-phase boundaries (APB), and protrusions. To evaluate the evolution of SFs during CSPVT, KOH etching was performed. Depicted in Figure 21 is the SF density for hetero-epitaxially grown CVD material as well as the values for sublimation-grown samples with regard to their thickness. Starting from CVD seeding materials, a defect-rich transition layer will form between the seed and the sublimation-grown material, increasing the SF density. As the CS-PVT growth will continue, this density will decrease with the increasing layer thickness. The density saturates at a level lower than the compared CVD material if the layer thickness reaches a thickness of approximately 200 μm [75].

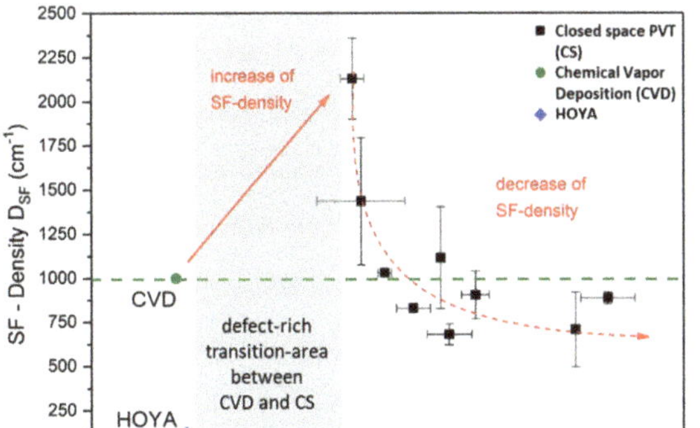

Figure 21. Stacking fault (SF) density of KOH-etched 3C-SiC samples with regard to grown layer thickness using CS-PVT. After an initial rise of SF density due to the defect-rich transition area between CVD and CS, the SF density will decrease with increasing 3C-SiC thickness and even with a value below the value of the used CVD seed. The SF density of the HOYA sample grown by switch-back-epitaxy is presented as a comparison. Adapted from [75].

The characterization of the material using Raman, XRD, and KOH-etching for the evaluation of the SF density shows an improvement in the material quality for CS-PVT compared to the seeding material grown by heteroepitaxial CVD. However, the most important value regarding the real bulk growth is the thickness of the grown crystals. The limitations of the achievable thickness to this point are strongly connected to the type of defect in the material. Protrusion defects were already present in the seeding layers and have their origin in the carbonization step during CVD growth. These three-dimensional defects tended to increase with increasing layer thickness as observed by Zimbone et al. [67] during subsequent CVD growth. In starting with such materials as a seed, a similar trend could be observed for CS-PVT [75]. Optical images of the "as grown" surface for a set of growth runs are depicted in Figure 22. It is clearly visible that the size of protrusions increased with the increasing layer thicknesses. At least for the investigated parameter range, this trend can be described as linear. Consequently, for even large thicknesses (d > 1 mm), the surface became more and more dominated by these defects. For bulk, growth runs with a thickness of approximately 3 mm, with a rough surface completely covered with protrusions, was observed (Figure 22b). In addition to the ragging of the surface, polytype switches towards 6H-SiC can be observed in the crosscut depicted in Figure 22c and can be found near or on top of protrusion defects. The edges of the defects are formed by stacking faults in the (111) planes. As these planes were equivalent to the (0001) faces in the hexagonal system, the probability for the nucleation of 6H-SiC could increase at these sides. Additionally, the growth of 6H-SiC on top of protrusion defects seems to confirm the loss of nucleation information for the continuing growth of the cubic polytype.

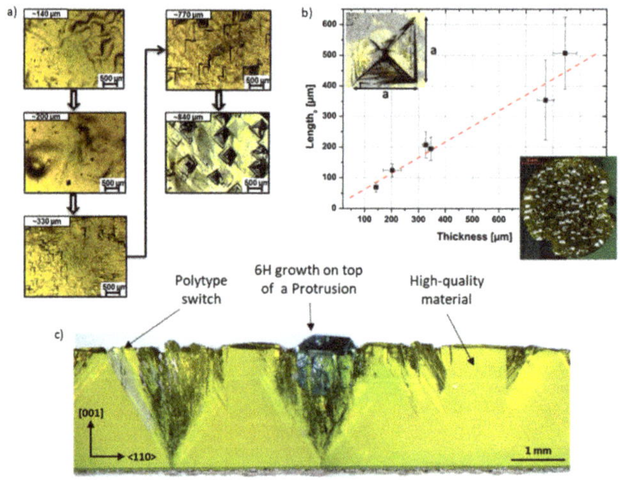

Figure 22. (**a**) Size of protrusion defects for bulk 3C-SiC layers with different 3C-SiC thickness. (**b**) Edge length of protrusions plotted versus 3C-SiC thickness. Additionally, the surface of an approximately 2.7 mm-thick grown crystal with a diameter of 25 mm is visible. The crystal is completely dominated with protrusion defects, leading to a ragged surface. (**c**) Cross-cut of a 3.4 mm-thick crystal revealing polytype switches as well as the inner parts of the protrusion defect. Areas between and underneath the protrusions show high quality material grown by CS-PVT.

The reduction of protrusion density in the CVD seeding layers is an important task towards the real bulk growth using sublimation growth, especially as the material grown by CS-PVT shows a very high quality in the areas in which no protrusion defect is present. Therefore, efforts were made to reduce the protrusion density during the sublimation growth. One approach features the growth on the original transition layer from the CVD growth between the silicon and the cubic silicon carbide. Schuh et al. [76] showed that a slight reduction of the protrusion density could be observed using this transition layer as a starting point for the sublimation growth instead of the original CVD growth front. During regular CS-PVT, a partial overgrowth of protrusion defects could also be observed, as depicted in [77]. This effect could be observed for growth runs on on-axis seeding material as well as for 4° off-orientated-grown seeds. So far, the mechanism behind this overgrowth is not completely understood. It seems that this effect is more pronounced for off-axis-grown samples compared to on-axis grown samples.

3.16. Simulation of PVT Bulk Growth

To determine the growth conditions present inside the growth cell during the sublimation growth of 3C-SiC, numerical modeling of the temperature field and mass transport-related phenomena were performed. The basic aim was to first identify the growth conditions existing in the 50 mm apparatus and then to use such data to ensure stable growth conditions to enlarge wafer sizes of 100 mm and greater. The study examined the effect of the appropriate SiC characteristics and the various carbon materials serving as process values in the computational modeling for both the temperature profile and the associated mass transport. In order to implement the thermal field and mass transport effects, computer simulation was performed through COMSOL Multiphysics. The appropriate selection of the physical parameters related to graphite-based components, as well as for the carbon isolation characteristics over 2000 °C, appeared to not be an easy task. The main issue concerns the ambiguity, if not a complete absence of reliable data, on the temperature behavior of electrical and thermal conductivity at the growth thermal conditions.

Nonetheless, numerical modeling allows for not only the calculation of the small growth cell but also the simulation of the whole growth reactor. Besides the calculation of temperature fields, mass transport, and supersaturation, simulation with COMSOL Multiphysics provided insight into the behavior of magnetic fields as well as into the formation of hot zones.

Worthy to note, despite the fact that there is no unambiguous data on the behavior of thermal conductivity at higher temperatures for graphite crucibles and insulating components, it is possible to assert using calculations and experimental calibrations that the thermal gradient in the gas phase has no effect, unless for the second order of approximation. As a consequence, it is expected that calculating, for example, supersaturation in front of the growth interface would offer accurate findings useful for the design and optimization of growth cells.

The supersaturation of the SiC_2 gas species plays an important role, influencing the growth-limiting parameter at the seed-growth front. This supersaturation can be calculated using the partial pressure of the gas species at the seed and source. Different approaches for the calculation of the supersaturation can be found in literature, for example, by Lilov [78] or Avrov [79]. Each experiment is similar in that for the calculations, the knowledge of the actual temperatures during the growth process is required. A comparison between the simulation data and the measured temperatures for the 50 mm-CS-PVT growth setup at different heating powers, showing a good agreement, is depicted in Figure 23a. A similar trend could be observed for the 100 mm-growth cell. Additionally, an example of the temperature field present in the growth setup can be seen in Figure 23b.

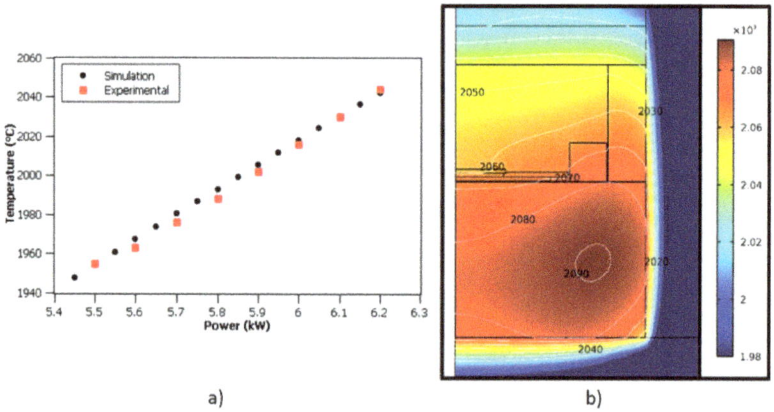

Figure 23. (**a**) Comparisons between simulations using COMSOL Multiphysics and the measured temperatures at the crucible top during growth runs for different heating powers in the 50 mm-CS-PVT setup. (**b**) Typical temperature field for CS-PVT. Some isotherms are indicated with the corresponding temperatures.

Using the temperatures obtained at the seed and source from the simulations, the supersaturations present during the sublimation growth can be calculated using the equation mentioned by Rankl et al. [80]. They found that for the heteroepitaxial growth of 3C-SiC on (0001)-oriented 6H—SiC, a supersaturation as high as s = 0.4 is necessary to achieve a high yield. Based on the development regarding the carbon materials databases, this value was revised to s = 0.24 [32]. In the case of homoepitaxial sublimation growth on seeds already containing the cubic polytype, a supersaturation higher than s = 0.1 was found to be suitable to ensure stable growth. It was also found that for this purpose, a source to seed a distance of 1 mm or smaller is necessary depending on the growth temperature as the supersaturation will decrease with the increasing spacing [31].

A global model for the evaluation of the processes' results in terms of the material growth rate can be obtained from the estimates of the mass transfer rate from the seed to the substrate once the temperature field is evaluated by the chamber simulation, as discussed in the previous section. Assuming that ballistic transport conditions occur for the Si-C molecules, which sublimate at different rates at the two interfaces, approximate estimates of the growth rate for the 3C-SiC in the different positions of the growing substrate (in a fully symmetric configuration) can be obtained from the balance between the atomic species' effective deposition fluxes (j_{dep}), derived from the sources and ruled by the source temperature, and the evaporation flux (j_{ev}), ruled by the substrate local temperature. Due to the composition of the SiC vapor pressure, Si-rich conditions are usually assumed (see Avrov et al. [78]) and the growth rate can be estimated by:

$$Gr = \frac{\rho_{SiC}}{M_{SiC}}\left(j_{dep} - j_{ev}\right)Gr \tag{1}$$

using the atomic carbon effective flux only. In Equation (1), ρ_{SiC} is the SiC density and M_{SiC} is the SiC molar mass. The expression for j_{dep} and j_{ev} are given by

$$\begin{aligned} j_{dep} &= j_{dep}(C) = j_{dep}(Si_2C) + 2j_{dep}(SiC_2) \\ &= \sqrt{\frac{2\pi RT_{Source}}{M_{Si_2C}}}\exp\left(\frac{A_{Si_2C}}{T_{Source}} + B_{Si_2C}\right) + 2\sqrt{\frac{2\pi RT_{Source}}{M_{SiC_2}}}\exp\left(\frac{A_{SiC_2}}{T_{Source}} + B_{SiC_2}\right) \end{aligned} \tag{2}$$

$$\begin{aligned} j_{ev} &= j_{ev}(C) = j_{ev}(Si_2C) + 2j_{ev}(SiC_2) \\ &= \sqrt{\frac{2\pi RT_{Sub}}{M_{Si_2C}}}\exp\left(\frac{A_{Si_2C}}{T_{Sub}} + B_{Si_2C}\right) + 2\sqrt{\frac{2\pi RT_{Sub}}{M_{SiC_2}}}\exp\left(\frac{A_{SiC_2}}{T_{Sub}} + B_{SiC_2}\right) \end{aligned} \tag{3}$$

where T_{Source} and T_{Sub} are the source and the substrate temperatures; M_{Si_2C} and M_{SiC_2} are the molar masses of the Si_2C and SiC_2 molecules in the vapor phase; and A_x and B_x are the experimental parameters that rule the partial pressures of the X species in the vapor mixtures at the thermodynamic equilibrium with the solid counterpart. Different calibrations for the partial pressures-related parameters can be found in the literature and by using the one in Reference [78], a quantitative estimation of the growth rate in accordance with the experimental results can be obtained.

3.17. Bulk Growth by CVD

To obtain a 3C-SiC bulk material of up to 6 inches (or 8 inch in the future), a new epitaxial reactor chamber was designed and tested in several experiments. The major idea was to hetero-epitaxially grow on silicon a 3C-SiC layer and use it as seeds, melt the silicon substrate, and again start the growth at very high temperatures. Thereby, we grew a bulk substrate of 3C-SiC with a low density of SFs and low wafer bow. Most importantly, the bow can be strongly decreased since by removing silicon, the stress due to the different thermal expansion coefficients between the two materials is removed. The silicon melting provides an increase in the growth temperature and in the growth rate. In this way, thicker wafers and better crystal quality of the material can be achieved. For a better understanding of how to melt the silicon substrate, how to drain it, how to etch the remaining silicon, and how to grow the homoepitaxial layer on the 3C-SiC substrate obtained after the silicon melting, many experiments have been performed. In Figure 24 (left), a scheme of the entire process is reported. The first two steps are the standard ones (well-reported and described in the literature).

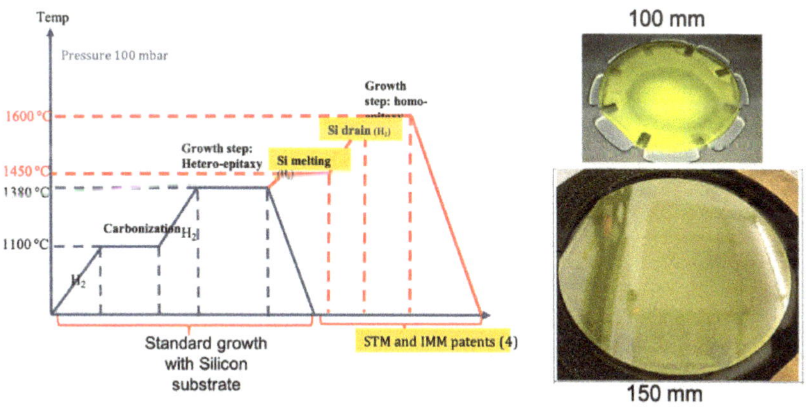

Figure 24. Schematic of the new process for CVD bulk growth (left). 3C-SiC wafers with the dimensions of 100 mm and 150 mm grown with the new process (right, adapted from Reference [64]).

After the melting, the SiC layer was used as a seed layer for subsequent homoepitaxial growth. [64] In this way, growing both 100 mm and 150 mm wafers as reported in Figure 24 (right) was possible.

The effect of temperature on the homo-epitaxial process was observed by X-ray diffraction analysis (Figure 25). The full-width of the half-maximum on the X-ray rocking curve of the 3C-SiC (002) peak was correlated to the crystal structure and defect density (lower FWHM value means better crystal quality). The figure shows the FWHM as a function of the film thickness for several samples. The samples reported in the graph were grown in different conditions (all the samples at a low value of thickness derive from previous experiments [6]). An increase in the film thickness has the effect of the quality of the material increasing [81]. The initial part of the curve (from 0 to about 20 μm of thickness) shows the 3C-SiC sample's growth with the Si substrate. For such samples, the crystal quality was limited by the presence of the silicon substrate that, starting from about 20 microns of thickness, led to cracks and extended defects. These defects are generated during the cooling down process after the growth [6]. The three points between 60 μm and 90 μm are the 3C-SiC samples for which the silicon substrate was melted (Si fusion) and showed good crystal quality (around 200 arcsec), similar to the old 2 inch-3C-SiC wafer provided by the Hoya corporation (dotted line). These wafers were used as a template for the homoepitaxial process. The stared points (three at 200 micron) are the samples growth by using the new melting process explained in the current manuscript. They were grown at three different temperatures (1600 °C, 1640 °C, and 1700 °C). The sample grown at 1600 °C is appreciably better than the other two [64]. The FWHM value at around 100 arcsec is very promising also compared with a thicker sample (about 400 micron) grown at a high temperature by sublimation epitaxy (PVT reactor) [31].

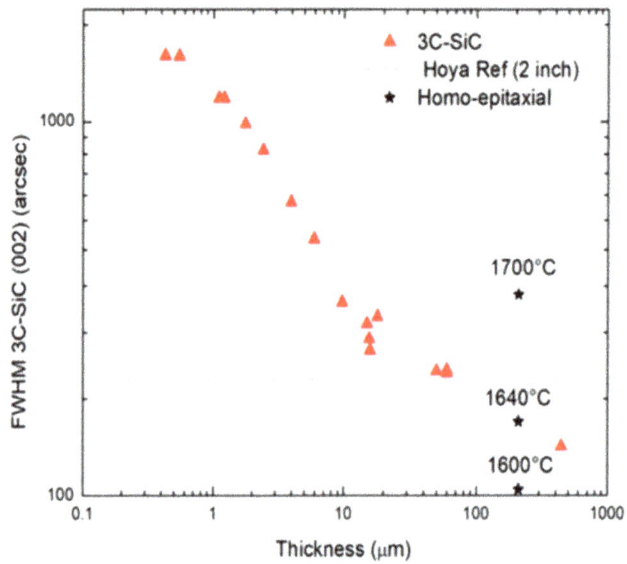

Figure 25. FWHM of the 3C-SiC (002) peak vs. the grown thickness. The decrease of the FWHM at high 3C-SiC thickness can be observed. The low temperature growth shows a much better quality of the material (see Reference [64]).

4. Conclusions

In the CHALLENGE project, two main different approaches were used to grow 3C-SiC. From one hand, several compliance substrates (pillars, SiGe buffer layer, ISP, etc.) were used to reduce both defects and stress. On the other hand, new bulk growth techniques (PVT or CVD) have been developed to improve the quality of this material. During this work, a new understanding of the defects in this material (protrusions, APBs, SFs, and point defects), their interactions, and the effect of the growth process on their formation and reduction have been obtained. This new understanding has been also helped by the simulation codes developed inside the project (KMCsL, MD, phase field, etc.) and by new characterization techniques (C-AFM).

From this large study on the growth of 3C-SiC, several conclusions can be reported:

- The use of the SiGe buffer layer is interesting and in some cases, some good quality samples can be obtained. The main limitation is that a low-temperature growth should be used and the Ge segregation at the 3C-SiC/Si interface can produce the formation of polycrystalline regions. The process window is narrow and thus this process can be difficult to use in a production line.

- The ISP substrate produces a fast decrease of the SFs' density, as reported in the previous papers, but at the same time produces the formation of APBs that can be detrimental for the leakage current of the devices, as observed by C-AFM measurements. To decrease the APBs concentration, a very high thickness can be grown or new ISP structures should be realized.
- The pillar substrates have demonstrated to be extremely interesting in reducing the stress and bow of the wafer, especially for (111) substrates. On this kind of substrate, even 25–30 μm of the 3C-SiC layer could be grown without cracks and with a low bow. The main problem of this process is that the 3D growth of 3C-SiC is not easy to control and thick layers are needed to obtain a continuous substrate. This kind of technology can be interesting in the future when a good 3C-SiC (111) layer should be realized.
- During the last years, we developed a carbonization process that decreases the voids almost to zero at the 3C-SiC/Si interface with a considerable improvement of the material quality and a decrease of the stress. Even the growth during the temperature ramp (buffer layer) between the carbonization and growth steps has a large effect on the stress.
- The main part of the investigation of the last years has been on the study of the evolution of SFs and APBs, as well as their interactions. The different types of SFs (SF<1>, SF<2>, SF<3>), their generation and annihilation, and the influence of the PD on these processes have been observed and studied in great detail. Another aspect that has been studied in detail is the interaction between the APBs and SFs. In fact, APBs can both generate or annihilate SFs and all these processes have been studied both experimentally and by simulations.
- It has been observed that due to this equilibrium between the generation and recombination of SFs, very thick layers are needed to obtain a low value of SFs that can provide the opportunity to realize a good device on 3C-SiC. The growth conditions (growth temperature, growth rate, and doping) can influence this equilibrium between generation and annihilation. It seems also that the different types of SFs have different behaviors during the growth but more investigations on this aspect are needed.
- From these studies, it has been understood that the bulk growth of 3C-SiC is needed to obtain a low value of defects compatible with the realization of power devices. We have developed different kinds of growth techniques (CVD and PVT) to realize bulk wafers with dimensions compatible with the actual standards (100 and 150 mm) but the main problem is related to the formation of the protrusions in the first instant of the growth. This 3D defect can grow and cover the entire surface of the ingot, thus a new process that can eliminate the formation of these defects is needed to obtain a 3C-SiC wafer for the realization of power devices.
- Another aspect of the 3C-SiC bulk growth that needs further investigation is the intrinsic stress of this material. In fact, the high level of stress and bow of the wafers are a problem for the further processing of the 3C-SiC wafers.

Further work should be done to obtain a good material for power devices, but the work performed during this project is a fundamental step for developing new materials for power devices.

Author Contributions: Conceptualization, F.L.V., L.M., A.L.M., R.Y., and P.W.; software, G.F., I.D., A.L.M., E.S., A.M., A.S., and L.M.; validation, M.Z. (Massimo Zimbone), C.B., C.C., F.G., M.Z. (Marcin Zielinski)., R.A., M.M., D.C., V.J., P.S., M.S. (Michael Schöler), V.S., and M.K.; writing—original draft preparation, F.L.V., M.Z. (Massimo Zimbone), A.L.M., R.A., L.M., V.J., V.S., M.S. (Mikael Syväjärvi) and P.W.; writing—review and editing, F.L.V.; supervision, F.L.V.; funding acquisition, F.L.V. All authors have read and agreed to the published version of the manuscript.

Funding: This research study was funded by the European Union within the frame of the project CHALLENGE, grant number 720827.

Data Availability Statement: The data can be made available, upon reasonable request, asking to the corresponding author.

Acknowledgments: P. Fiorenza, G. Greco, and F. Roccaforte (CNR-IMM) are acknowledged for their contribution in the processing and electrical characterization of devices employed to assess 3C-SiC material quality, and S. Di Franco (CNR-IMM) is acknowledged for the skilled technical assistance.

Conflicts of Interest: The authors declare no conflict of interest. The funders had no role in the design of the study; in the collection, analyses, or interpretation of data; in the writing of the manuscript, or in the decision to publish the results.

References

1. Chow, T.P. High-voltage SiC and GaN power devices. *Microelectron. Eng.* **2006**, *83*, 112–122. [CrossRef]
2. Pensl, G.; Ciobanu, F.; Frank, T.; Krieger, M.; Reshanov, S.; Schmid, F.; Weidner, M. SiC material properties. *Int. J. High Speed Electron. Syst.* **2005**, *15*, 705–745. [CrossRef]
3. Schöner, A.; Krieger, M.; Pensl, G.; Abe, M.; Nagasawa, H. Fabrication and Characterization of 3C-SiC-Based MOSFETs. *Chem. Vap. Depos.* **2006**, *12*, 523–530. [CrossRef]
4. Uchida, H.; Minami, A.; Sakata, T.; Nagasawa, H.; Kobayashi, M. High Temperature Performance of 3C-SiC MOSFETs with High Channel Mobility. *Mater. Sci. Forum* **2012**, *717*, 1109–1112. [CrossRef]
5. Van Zeghbroeck, B.; Fardi, H. Comparison of 3C-SiC and 4H-SiC Power MOSFETs. *Mater. Sci. Forum* **2018**, *924*, 774–777. [CrossRef]
6. La Via, F.; Severino, A.; Anzalone, R.; Bongiorno, C.; Litrico, G.; Mauceri, M.; Schoeler, M.; Schuh, P.; Wellmann, P. From thin film to bulk 3C-SiC growth: Understanding the mechanism of defects reduction. *Mater. Sci. Semicond. Process.* **2018**, *78*, 57–68. [CrossRef]
7. Severino, A.; Locke, C.; Anzalone, R.; Camarda, M.; Piluso, N.; la Magna, A.; Saddow, S.E.; Abbondanza, G.; D'Arrigo, G.; La Via, F. 3C-SiC film growth on Si substrates. *ECS Transact.* **2011**, *35*, 99–116. [CrossRef]
8. Wang, L.; Dimitrijev, S.; Han, J.; Iacopi, A.; Hold, L.; Tanner, P.; Harrison, H.B. Growth of 3C–SiC on 150-mm Si(100) substrates by alternating supply epitaxy at 1000 °C. *Thin Solid Film.* **2011**, *519*, 6443–6446. [CrossRef]
9. Anzalone, R.; D'Arrigo, G.; Camarda, M.; Locke, C.; Saddow, S.E.; La Via, F. Advanced Residual Stress Analysis and FEM Simulation on Heteroepitaxial 3C–SiC for MEMS Application. *J. Microelectromech. Syst.* **2011**, *20*, 745–752. [CrossRef]
10. Iacopi, F.; Walker, G.; Wang, L.; Malesys, L.; Ma, S.; Cunning, B.V.; Iacopi, A. Orientation-dependent stress relaxation in hetero-epitaxial 3C-SiC films. *Appl. Phys. Lett.* **2013**, *102*, 011908. [CrossRef]
11. Zielinski, M.; Ndiaye, S.; Chassagne, T.; Juillaguet, S.; Lewandowska, R.; Portail, M.; Leycuras, A.; Camassel, J. Strain and wafer curvature of 3C-SiC films on silicon: Influence of the growth conditions. *Phys. Stat. Sol.* **2007**, *204*, 981–986. [CrossRef]
12. Nagasawa, H.; Yagi, K.; Kawahara, T.; Hatta, N. Reducing Planar Defects in 3C–SiC. *Chem. Vap. Depos.* **2006**, *12*, 502–508. [CrossRef]
13. Nagasawa, H.; Abe, M.; Yagi, K.; Kawahara, T.; Hatta, N. Fabrication of high perfor-mance3C-SiC vertical MOSFETs by reducing planar defects. *Phys. Stat. Sol.* **2008**, *245*, 1272–1280. [CrossRef]
14. La Via, F.; Roccaforte, F.; La Magna, A.; Nipoti, R.; Mancarella, F.; Wellmann, P.J.; Crippa, D.; Mauceri, M.; Ward, P.; Miglio, L.; et al. 3C-SiC Hetero-Epitaxially Grown on Silicon Compliance Substrates and New 3C-SiC Substrates for Sustainable Wide-Band-Gap Power Devices (CHALLENGE). *Mater. Sci. Forum* **2018**, *924*, 913–918. [CrossRef]
15. Albani, M.; Marzegalli, A.; Bergamaschini, R.; Mauceri, M.; Crippa, D.; La Via, F.; von Känel, H.; Miglio, L. Solving the critical thermal bowing in 3C-SiC/Si(111) by a tilting Si pillar architecture. *J. Appl. Phys.* **2018**, *123*, 185703. [CrossRef]
16. Zimbone, M.; Zielinski, M.; Barbagiovanni, E.G.; Calabretta, C.; La Via, F. 3C-SiC grown on Si by using a Si1-xGex buffer layer. *J. Cryst. Growth* **2019**, *519*, 1–6.
17. Zimbone, M.; Zielinski, M.; Bongiorno, C.; Calabretta, C.; Anzalone, R.; Scalese, S.; Fisicaro, G.; La Magna, A.; Mancarella, F.; La Via, F. 3C-SiC Growth on Inverted Silicon Pyramids Patterned Substrate. *Materials* **2019**, *12*, 3407. [CrossRef]
18. Severino, A.R.; Anzalone, M.; Camarda, N.; Piluso, F.; La Via, F. Structural characterization of heteroepitaxial 3C-SiC. *Mater. Sci. Forum* **2012**, *711*, 27–30. [CrossRef]
19. Polychroniadis, E.; Syväjärvi, M.; Yakimova, R.; Stoemenos, J. Microstructural characterization of very thick freestanding 3C-SiC wafers. *J. Cryst. Growth* **2004**, *263*, 68–75. [CrossRef]
20. Wellmann, P.; Neubauer, G.; Fahlbusch, L.; Salamon, M.; Uhlmann, N. Growth of SiC bulk crystals for application in power electronic devices—Process design, 2D and 3D X-ray in situ visualization and advanced doping. *Cryst. Res. Technol.* **2015**, *50*, 2–9. [CrossRef]
21. Wellmann, P.J. Review of SiC crystal growth technology. *Semicond. Sci. Technol.* **2018**, *33*, 103001. [CrossRef]
22. Chaussende, D.; Mercier, F.; Boulle, A.; Conchon, F.; Soueidan, M.; Ferro, G.; Mantzari, A.; Andreadou, A.; Polychroniadis, E.; Balloud, C.; et al. Prospects for 3C-SiC bulk crystal growth. *J. Cryst. Growth* **2008**, *310*, 976–981. [CrossRef]
23. Jokubavicius, V.; Yazdi, G.R.; Liljedahl, R.; Ivanov, I.G.; Sun, J.; Liu, X.; Schuh, P.; Wilhelm, M.; Wellmann, P.; Yakimova, R.; et al. Single Domain 3C-SiC Growth on Off-Oriented 4H-SiC Substrates. *Cryst. Growth Des.* **2015**, *15*, 2940–2947. [CrossRef]
24. Jokubavicius, V.; Yazdi, G.R.; Liljedahl, R.; Ivanov, I.G.; Yakimova, R.; Syväjärvi, M. Lateral enlargement growth mechanism of 3C-SiC on off-oriented 4H-SiC substrates. *Cryst. Growth Des.* **2014**, *14*, 6514–6520. [CrossRef]
25. Tairov, Y.; Tsvetkov, V.; Lilov, S.; Safaraliev, G. Studies of growth kinetics and polytypism of silicon carbide epitaxial layers grown from the vapour phase. *J. Cryst. Growth* **1976**, *36*, 147–151. [CrossRef]

26. Schuh, P.; La Via, F.; Mauceri, M.; Zielinski, M.; Wellmann, P. Growth of Large-Area, Stress-Free, and Bulk-Like 3C-SiC (100) Using 3C-SiC-on-Si in Vapor Phase Growth. *Materials* **2019**, *12*, 2179. [CrossRef] [PubMed]
27. Anzalone, R.; Litrico, G.; Piluso, N.; Reitano, R.; Alberti, A.; Fiorenza, P.; Coffa, S.; La Via, F. Carbonization and transition layer effects on 3C-SiC film residual stress. *J. Cryst. Growth* **2017**, *473*, 11–19. [CrossRef]
28. Camarda, M.; Severino, A.; La Via, F. Manufacture of Wafers of Wide Energy Gap Semiconductor Material for the Integration of Electronic and/or Optical and/or Optoelectronic Devices. US Patent 2014/0264385A1, 18 September 2014.
29. Zielinski, M.; Monnoye, S.; Mank, H.; Moisson, C.; Chassagne, T.; Michon, A.; Portail, M. Structural Quality, Polishing and Thermal Stability of 3C-SiC/Si Templates. *Mater. Sci. Forum* **2018**, *924*, 306–309. [CrossRef]
30. Syväjärvi, M.; Yakimova, R. A Method of Producing Silicon Carbide Epitaxial Layer. U.S. Patent 7,918,937, 5 April 2011.
31. Schuh, P.; Schöler, M.; Wilhelm, M.; Syväjärvi, M.; Litrico, G.; La Via, F.; Mauceri, M.; Wellmann, P.J. Sublimation growth of bulk 3C-SiC using 3C-SiC-on-Si (1 0 0) seeding layers. *J. Cryst. Growth* **2017**, *478*, 159–162. [CrossRef]
32. Schuh, P.; Steiner, J.; La Via, F.; Mauceri, M.; Zielinski, M.; Wellmann, P.J. Limitations during Vapor Phase Growth of Bulk (100) 3C-SiC Using 3C-SiC-on-SiC Seeding Stacks. *Materials* **2019**, *12*, 2353. [CrossRef]
33. Schuh, P. Sublimation Epitaxy of Bulk-Like Cubic Silicon Carbide. Ph.D. Thesis, Friedrich-Alexander-Universitat Erlangen-Nürnberg, Nürnberg, Germany, 2019.
34. Zoppel, S.; Farsari, M.; Merz, R.; Zehetner, J.; Stangl, G.; Reider, G.A.; Fotakis, C. Laser micro machining of 3C–SiC single crystals. *Microelectron. Eng.* **2006**, *83*, 1400–1402. [CrossRef]
35. Fu, C.; Yang, Y.; Huang, Z.; Liu, G.; Zhang, H.; Jiang, F.; Wei, Y.; Jiao, Z. Investigation on the laser ablation of SiC ceramics using micro-Raman mapping technique. *J. Adv. Ceram.* **2016**, *5*, 253–261. [CrossRef]
36. Kulkarni, M.S.; Erk, H.F. Acid-Based Etching of Silicon Wafers: Mass-Transfer and Kinetic Effects. *J. Electrochem. Soc.* **2000**, *147*, 176–188. [CrossRef]
37. Kovacs, G.T.A.; Maluf, N.I.; Petersen, K.E. Bulk micromachining of silicon. *Proc. IEEE* **1998**, *86*, 1536–1551. [CrossRef]
38. Wellmann, P.J.; Schuh, P.; Kollmuss, M.; Schöler, M.; Steiner, J.; Zielinski, M.; Mauceri, M.; La Via, F. Prospects of Bulk Growth of 3C-SiC Using Sublimation Growth. *Mater. Sci. Forum* **2020**, *1004*, 113–119. [CrossRef]
39. Zimbone, M.; Sarikov, A.; Bongiorno, C.; Marzegalli, A.; Scuderi, V.; Calabretta, C.; Miglio, L.; La Via, F. Extended Defects in 3C-SiC: Stacking Faults, Threading Partial Dislocations, and Inverted Domain Boundaries. *Acta Mater.* **2021**, *213*, 116915. [CrossRef]
40. Zimbone, M.; Barbagiovanni, E.G.; Bongiorno, C.; Calabretta, C.; Calcagno, L.; Fisicaro, G.; La Magna, A.; La Via, F. Generation and Termination of Stacking Faults by Inverted Domain Boundaries in 3C-SiC. *Cryst. Growth Des.* **2020**, *20*, 3104–3111. [CrossRef]
41. Anzalone, R.; Severino, A.; D'Arrigo, G.; Bongiorno, C.; Abbondanza, G.; Foti, G.; Saddow, S.; la Via, F. Heteroepitaxy of 3C-SiC on different on-axis oriented silicon substrates. *J. Appl. Phys.* **2009**, *105*, 084910. [CrossRef]
42. Falub, C.V.; von Känel, H.; Isa, F.; Bergamaschini, R.; Marzegalli, A.; Chrastina, D.; Isella, G.; Müller, E.; Niedermann, P.; Miglio, L. Scaling Hetero-Epitaxy from Layers to Three-Dimensional Crystals. *Science* **2012**, *335*, 1330–1334. [CrossRef]
43. Kreiliger, T.; Mauceri, M.; Puglisi, M.; Mancarella, F.; La Via, F.; Crippa, D.; Kaplan, W.; Schöner, A.; Marzegalli, A.; Miglio, L.; et al. 3C-SiC Epitaxy on Deeply Patterned Si(111) substrates. *Mater. Sci. Forum* **2016**, *858*, 151–154. [CrossRef]
44. Masullo, M.; Bergamaschini, R.; Albani, M.; Kreiliger, T.; Mauceri, M.; Crippa, D.; La Via, F.; Montalenti, F.; von Känel, H.; Miglio, L. Growth and Coalescence of 3C-SiC on Si(111) Micro-Pillars by a Phase-Field Approach. *Materials* **2019**, *12*, 3223. [CrossRef]
45. Marzegalli, A.; Cortinovis, A.; Basset, F.B.; Bonera, E.; Pezzoli, F.; Scaccabarozzi, A.; Isa, F.; Isella, G.; Zaumseil, P.; Capellini, G.; et al. Exceptional thermal strain reduction by a tilting pillar architecture: Suspended Ge layers on Si (001). *Mater. Des.* **2016**, *116*, 144–151. [CrossRef]
46. Agati, M.; Boninelli, S.; Calabretta, C.; Mancarella, F.; Mauceri, M.; Crippa, D.; Albani, M.; Bergamaschini, R.; Miglio, L.; la Via, F. Growth of thick [1 1 1]-oriented 3C-SiC films on T-shaped Si micropillars. *Mater. Des.* **2021**, *208*, 109833. [CrossRef]
47. Timoshenko, S. Analysis of Bi-Metal Thermostats. *J. Opt. Soc. Am.* **1925**, *11*, 233–255. [CrossRef]
48. La Via, F.; D'Arrigo, G.; Severino, A.; Piluso, N.; Mauceri, M.; Locke, C.; Saddow, S.E. Patterned sub-strate with inverted silicon pyramids for 3C-SiC epitaxial growth: A comparison with conventional (001) Si substrate. *J. Mater. Res.* **2013**, *28*, 94. [CrossRef]
49. Syväjärvi, M.; Yakimova, R.; Tuominen, M.; Kakanakova-Georgieva, A.; MacMillan, M.; Henry, A.; Wahab, Q.; Janzén, E. Growth of 6H and 4H–SiC by sublimation epitaxy. *J. Cryst. Growth* **1999**, *197*, 155–162. [CrossRef]
50. Severino, A.; D'arrigo, G.; Bongiorno, C.; Scalese, S.; La Via, F.; Foti, G. Thin crystalline 3C-SiC layer growth through carbonization of differently oriented Si substrates. *J. Appl. Phys.* **2007**, *102*, 023518. [CrossRef]
51. Gabriel, F. 3C-SiC hetero-epitaxial growth on silicon: The quest for the Holy Grail. *Crit. Rev. Solid State Mater. Sci.* **2015**, *40*, 56–76.
52. Zielinski, M.; Monnoye, S.; Mank, H.; Torregrosa, F.; Grosset, G.; Spiegel, Y.; Portail, M.; Michon, A. Exploration of solid phase epitaxy of 3C-SiC on Silicon. *Mater. Sci. Forum* **2020**, *1004*, 132–138. [CrossRef]
53. Severino, A.; La Via, F. Microtwin reduction in 3C–SiC heteroepitaxy. *Appl. Phys. Lett.* **2010**, *97*, 181916. [CrossRef]
54. Zielinski, M.; Michaud, J.F.; Jiao, S.; Chassagne, T.; Bazin, A.E.; Michon, A.; Portail, M.; Alquier, D. Experimental observation and analytical model of the stress gradient inversion in 3C-SiC layers on siliconJ. *Appl. Phys.* **2012**, *111*, 053507. [CrossRef]
55. Scalise, E.; Barbisan, L.; Sarikov, A.; Montalenti, F.; Miglio, L.; Marzegalli, A. The origin and nature of killer defects in 3C-SiC for power electronic applications by a multiscale atomistic approach. *J. Mater. Chem. C* **2020**, *8*, 8380–8392. [CrossRef]

56. Yamasaki, J.; Inamoto, S.; Nomura, Y.; Tamaki, H.; Tanaka, N. Atomic structure analysis of stacking faults and misfit dislocations at 3C-SiC/Si(0 0 1) interfaces by aberration-corrected transmission electron microscopy. *J. Phys. D Appl. Phys.* **2012**, *45*, 49. [CrossRef]
57. Fisicaro, G.; Bongiorno, C.; Deretzis, I.; Giannazzo, F.; La Via, F.; Roccaforte, F.; Zielinski, M.; Zimbone, M.; La Magna, A. Genesis and evolution of extended defects: The role of evolving interface instabilities in cubic SiC. *Appl. Phys. Rev.* **2020**, *7*, 021402. [CrossRef]
58. Nagasawa, H.; Gurunathan, R.; Suemitsu, M. Controlling planar defects in 3C-SiC: Ways to wake it up as a practical semiconductor. *Mater. Sci. Forum* **2015**, *821*, 108. [CrossRef]
59. Sarikov, A.; Marzegalli, A.; Barbisan, L.; Zimbone, M.; Bongiorno, C.; Mauceri, M.; Crippa, D.; la Via, F.; Miglio, L. Mechanism of stacking fault annihilation in 3C-SiC epitaxially grown on Si(001) by molecular dynamics simulations. *CrystEngComm* **2021**, *23*, 1566–1571. [CrossRef]
60. Scalise, E.; Marzegalli, A.; Montalenti, F.; Miglio, L. Temperature-Dependent Stability of Polytypes and Stacking Faults in SiC: Reconciling Theory and Experiments. *Phys. Rev. Appl.* **2019**, *12*, 021002. [CrossRef]
61. Scuderi, V.; Calabretta, C.; Anzalone, R.; Mauceri, M.; La Via, F. Characterization of 4H-and 6H-like stacking faults in cross section of 3C-SiC epitaxial layer by room-temperature μ-photoluminescence and μ-raman analysis. *Materials* **2020**, *13*, 1837. [CrossRef]
62. La Magna, A.; Alberti, A.; Barbagiovanni, E.; Bongiorno, C.; Cascio, M.; Deretzis, I.; La Via, F.; Smecca, E. Simulation of the Growth Kinetics in Group IV Compound Semiconductors. *Phys. Status Sol.* **2019**, *216*, 1800597. [CrossRef]
63. Giannazzo, F.; Greco, G.; Di Franco, S.; Fiorenza, P.; Deretzis, I.; La Magna, A.; Bongiorno, C.; Zimbone, M.; La Via, F.; Zielinski, M.; et al. Impact of Stacking Faults and Domain Boundaries on the Electronic Transport in Cubic Silicon Carbide Probed by Conductive Atomic Force Microscopy. *Adv. Electron. Mater.* **2020**, *6*, 1901171. [CrossRef]
64. Anzalone, R.; Zimbone, M.; Calabretta, C.; Mauceri, M.; Alberti, A.; Reitano, R.; La Via, F. Temperature Investigation on 3C-SiC Homo-Epitaxy on Four-Inch Wafers. *Materials* **2019**, *12*, 3293. [CrossRef]
65. Schöler, M.; Lederer, M.W.; Schuh, P.; Wellmann, P.J. Intentional Incorporation and Tailoring of Point Defects during Sublimation Growth of Cubic Silicon Carbide by Variation of Process Parameters. *Phys. Status Sol.* **2019**, *257*, 1900286. [CrossRef]
66. La Via, F.; Camarda, M.; La Magna, A. Mechanisms of growth and defect properties of epitaxial SiC. *Appl. Phys. Rev.* **2014**, *1*, 031301. [CrossRef]
67. Zimbone, M.; Mauceri, M.; Litrico, G.; Barbagiovanni, E.G.; Bongiorno, C.; La Via, F. Protrusions reduction in 3C-SiC thin film on Si. *J. Cryst. Growth* **2018**, *498*, 248–257. [CrossRef]
68. Shi, Y.; Jokubavicius, V.; Höjer, P.; Ivanov, I.G.; Yazdi, G.R.; Yakimova, R.; Syväjärvi, M.; Sun, J. A comparative study of high-quality C-face and Si-face 3C-SiC(1 1 1) grown on off-oriented 4H-SiC substrates. *J. Phys. D Appl. Phys.* **2019**, *52*, 345103. [CrossRef]
69. Jian, J.; Shi, Y.; Ekeroth, S.; Keraudy, J.; Syväjärvi, M.; Yakimova, R.; Helmersson, U.; Sun, J. A nanostructured NiO/cubic SiC p–n heterojunction photoanode for enhanced solar water splitting. *J. Mater. Chem. A* **2019**, *7*, 4721–4728. [CrossRef]
70. Shi, Y.; Zakharov, A.A.; Ivanov, I.G.; Yazdi, G.; Syväjärvi, M.; Yakimova, R.; Sun, J. Epitaxial Graphene Growth on the Step-Structured Surface of Off-Axis C-Face 3C-SiC($1^{-}1^{-}1^{-}$). *Phys. Stat. Sol.* **2020**, *257*, 1900718. [CrossRef]
71. Shi, Y.; Zakharov, A.A.; Ivanov, I.G.; Yazdi, G.R.; Jokubavicius, V.; Syväjärvi, M.; Yakimova, R.; Sun, J. Elimination of step bunching in the growth of large-area monolayer and multilayer graphene on off-axis 3CSiC (111). *Carbon* **2018**, *140*, 533–542. [CrossRef]
72. Capano, M.; Kim, B.; Smith, A.; Kvam, E.; Tsoi, S.; Ramdas, A. Residual strains in cubic silicon carbide measured by Raman spec-troscopy correlated with x-ray diffraction and transmission electron microscopy. *J. Appl. Phys.* **2006**, *100*, 083514. [CrossRef]
73. Feng, Z.C.; Mascarenhas, A.J.; Choyke, W.J.; Powell, J.A. Raman scattering studies of chemical-vapor-deposited cubic SiC films of (100) Si. *J. Appl. Phys.* **1988**, *64*, 3176–3186. [CrossRef]
74. Anzalone, R.; D'Arrigo, G.; Camarda, M.; Piluso, N.; La Via, F. Evaluation of Mechanical and Optical Properties of Hetero-Epitaxial Single Crystal 3C-SiC Squared-Membrane. *Mater. Sci. Forum* **2014**, *778*, 457–460. [CrossRef]
75. Schoeler, M. Characterization of protrusions and stacking faults in 3C-SiC grown by sublimation epitaxy using 3C-SiC-on-Si seeding layers. *Adv. Mater. Proc.* **2017**, *2*, 774–778. [CrossRef]
76. Schuh, P.; Künecke, U.; Litrico, G.; Mauceri, M.; La Via, F.; Monnoye, S.; Zielinski, M.; Wellmann, P.J. Vapor Growth of 3C-SiC Using the Transition Layer of 3C-SiC on Si CVD Templates. *Mater. Sci. Forum* **2019**, *963*, 149–152. [CrossRef]
77. Schöler, M.; la Via, F.; Mauceri, M.; Wellmann, P. Overgrowth of Protrusion Defects during Sublimation Growth of Cubic Silicon Carbide Using Free-Standing Cubic Silicon Carbide Substrates. *Cryst. Growth Des.* **2021**, *21*, 4046–4054. [CrossRef]
78. Lilov, S. Thermodynamic analysis of the gas phase at the dissociative evaporation of silicon carbide. *Comput. Mater. Sci.* **1993**, *1*, 363–368. [CrossRef]
79. Avrov, D.; Bakin, A.; Dorozhkin, S.; Rastegaev, V.; Tairov, Y.M. The analysis of mass transfer in system β-SiC–α-SiC under silicon carbide sublimation growth. *J. Cryst. Growth* **1999**, *198*, 1011–1014. [CrossRef]
80. Rankl, D.; Jokubavicius, V.; Syväjärvi, M.; Wellmann, P.J. Quantitative Study of the Role of Supersaturation during Sublimation Growth on the Yield of 50 mm 3C-SiC. *Mater. Sci. Forum* **2015**, *821–823*, 77–80. [CrossRef]
81. La Via, F. *Silicon Carbide Epitaxy*; T.C. 37/661 (2), Fort P.O., Trivandrum-695 023; Research Signpost: Kerala, India, 2012.

materials

MDPI

Article

Effect of Nitrogen and Aluminum Doping on 3C-SiC Heteroepitaxial Layers Grown on 4° Off-Axis Si (100)

Cristiano Calabretta [1], Viviana Scuderi [1,*], Ruggero Anzalone [2], Marco Mauceri [3], Danilo Crippa [4], Annalisa Cannizzaro [1,5], Simona Boninelli [1] and Francesco La Via [1]

1 Istituto per la Microelettronica e Microsistemi-Consiglio Nazionale delle Ricerche (IMM-CNR), VIII Strada, 5, 95121 Catania, Italy; cristiano.calabretta@imm.cnr.it (C.C.); annalisa.cannizzaro@imm.cnr.it (A.C.); simona.boninelli@imm.cnr.it (S.B.); francesco.lavia@imm.cnr.it (F.L.V.)

2 STMicroelectronics, Stradale Primosole, 50, 95121 Catania, Italy; ruggero.anzalone@st.com

3 LPE, XVI Strada, 95121 Catania, Italy; Marco.Mauceri@lpe-epi.com

4 LPE, Via Falzarego, 8, 20021 Baranzate, Italy; Danilo.Crippa@lpe-epi.com

5 Dipartimento di Fisica e Astronomia, Università di Catania, Via S. Sofia, 64, 95123 Catania, Italy

* Correspondence: viviana.scuderi@imm.cnr.it

Abstract: This work provides a comprehensive investigation of nitrogen and aluminum doping and its consequences for the physical properties of 3C-SiC. Free-standing 3C-SiC heteroepitaxial layers, intentionally doped with nitrogen or aluminum, were grown on Si (100) substrate with different 4° off-axis in a horizontal hot-wall chemical vapor deposition (CVD) reactor. The Si substrate was melted inside the CVD chamber, followed by the growth process. Micro-Raman, photoluminescence (PL) and stacking fault evaluation through molten KOH etching were performed on different doped samples. Then, the role of the doping and of the cut angle on the quality, density and length distribution of the stacking faults was studied, in order to estimate the influence of N and Al incorporation on the morphological and optical properties of the material. In particular, for both types of doping, it was observed that as the dopant concentration increased, the average length of the stacking faults (SFs) increased and their density decreased.

Keywords: 3C-SiC; stacking faults; doping; KOH etching

Citation: Calabretta, C.; Scuderi, V.; Anzalone, R.; Mauceri, M.; Crippa, D.; Cannizzaro, A.; Boninelli, S.; La Via, F. Effect of Nitrogen and Aluminum Doping on 3C-SiC Heteroepitaxial Layers Grown on 4° Off-Axis Si (100). *Materials* **2021**, *14*, 4400. https://doi.org/10.3390/ma14164400

Academic Editor: Ettore Vittone

Received: 6 July 2021
Accepted: 2 August 2021
Published: 6 August 2021

1. Introduction

Silicon carbide is a wide-bandgap semiconductor that shows high mechanical strength, chemical inertness and thermal conductivity. In particular, 3C-SiC is competitive among the SiC polytypes as it is characterized by high mobility and lower density of states at the 3C-SiC/SiO_2 interface with respect to 4H and 6H-SiC. These properties, arising from the higher symmetry related to lower phonon scattering and lower bandgap (2.5 eV), make 3C-SiC ideal for applications in the field of power electronics [1,2] as they lead to many advantages in metal oxide semiconductor (MOS) devices as well as microelectromechanical systems (MEMS) in harsh environments [3]. Indeed, higher channel mobility and high carrier mobility imply low on-state resistance (R_{ON}) for medium-voltage applications working under 1200 V, consequently lowering power dissipation in forward bias [1]. Compared to 4H and 6H hexagonal SiC, 3C-SiC films have the advantage that they can be hetero-epitaxially grown through low-temperature CVD as 3C-SiC is the most thermodynamically stable polytype. Growing a high-quality 3C-SiC epilayer on a large-area substrate would be a significant technical and scientific advancement. As a result of these features, silicon is regarded as the most intriguing substrate and crystal seed for thin epitaxial films and/or following bulk 3C-SiC growth [4]. Current technology involves the use of hetero-epitaxial growth on silicon, which inherently entails a 20% lattice parameter mismatch between Si and SiC and contributes to the generation of compressive intrinsic stress, whereas 8% at different degrees of thermal expansion provides a tensile contribution during the cooling period from the growth temperature to room temperature (RT). Such mismatches give rise to misfit dislocations and stacking faults (SFs) growing along the (111) planes of the face-centered cubic (FCC) lattice. Such defectiveness hinders the realization of devices and constitutes considerable leakage sources not compatible with the development of very large-scale integration (VLSI) technology [5].

Both nitrogen (N_2) and ammonia (NH_3) have been successfully employed as dopant precursors for the n-doping of SiC at temperatures above 1000 °C [6–8]. However, nowadays, N_2 is the principal n-type dopant in all SiC polytypes [9]. Highly p-doped 3C-SiC is interesting for emitters in p–n-junction-based devices, because such emitters will have low resistance and can produce a high concentration of injected holes in the base region [10]. Aluminum is a preferred acceptor in SiC due to its lower activation energy (0.24 eV) compared to other acceptors (0.7 eV for B and 0.33 eV for Ga) [11]. Zielinski et al. observed that nitrogen and aluminum atoms are incorporated, respectively, on carbon and silicon sites [12].

The dopant incorporation into the SiC lattice is known to affect the crystallinity, morphology and mechanical properties of the films [13,14]. Crystallinity is very important when applying doped 3C-SiC in electronic devices because the crystal defects directly influence their leakage current and breakdown voltage.

Although 3C-SiC material for semiconductor applications has been investigated for 30 years, the problem of defect formation at the 3C-SiC/Si contact is still far from being solved. Stacking faults (SFs), partial dislocations (PDs) and anti-phase boundaries (APB) or inverted domain boundaries (IDB) are the most important defects [15]. In particular, IDBs are the main defects responsible for the electrical failure of 3C-SiC/Si-based devices [16,17]. Instead, SFs can be considered highly conducting 2D defects, in the energy range where also the bulk material is conductive. Indeed, under forward polarization, SFs were demonstrated to operate as preferred current pathways, causing a decrease in turn-on voltage [17]. In 1987, Shibahara et al. [18] observed the beneficial effect of the growth of 3C-SiC on an off-axis Si substrate with an off-axis angle from 2° to 5°. In particular, off-axis Si substrates exhibit significant improvements in the anti-phase disorder compared to on-axis ones. Moreover, off-axis growth promotes SF surface propagation lying along the (111) plane, while (−1−11) SFs opposite to the growth step are forbidden.

In this work, we present a comprehensive study of the effects of n-type and p-type doping on hetero-epitaxial 3C growth on different 4° off-axis silicon substrates (100). Micro-Raman, photoluminescence (PL) and stacking fault evaluation through molten KOH etching were performed on different doped samples in order to estimate the influence of N and Al incorporation on the morphological and optical properties of the material. The concentrations investigated in this work, 10^{18} at/cm^3 and 10^{19} at/cm^3 for aluminum and nitrogen, respectively, are compatible with the concentrations necessary for the development of electronic devices on 3C-SiC. In particular, for vertical metal-oxide-semiconductor field-effect transistor (MOSFET) devices, the substrate should have a concentration of approximately 10^{19} at/cm^3 to decrease as much as possible the R_{on} of the device. Instead, for insulated gate bipolar transistors (IGBTs), the p-collector has a doping of 10^{18} at/cm^3 [19]. Today, IGBT production is particularly expensive due to the inability to grow 4H-SiC p-doped wafers, while in the case of 3C-SiC bulk grown by CVD, it is possible to grow such types of wafers.

2. Materials and Methods

In this study, 3C-SiC growth was performed in a horizontal hot-wall chemical vapor deposition (CVD) reactor (ACIS M10 supplied by LPE, Catania, Italy) using (100)-oriented Si substrates. The reaction system used was tri-chloro-silane (TCS), ethylene (C_2H_4) and hydrogen (H_2) as the silicon precursor, carbon precursor and gas carrier, respectively. After the growth of a 75 μm thick layer, at a temperature of 1400 °C and pressure of 100 mbar, the temperature was increased to 1650 °C and the silicon substrate was completely melted inside the CVD chamber, resulting in free-standing samples. For more details, see the references [20,21]. In particular, two different sets of samples were grown: 3C-SiC on Si (100) 4° off-axis along the [110], adding constant nitrogen flux, with a value of 0 (intrinsic), 300, 800 and 1600 sccm; 3C-SiC on Si (100) 4° off-axis along the [100], adding constant trimethylaluminum (TMA) flux, with a value of 0, 1, 2 and 4 sccm. The round brackets "()" indicate a crystallographic plane; square brackets "[]" indicate a direction. Atomic concentration values, for both types of doping, were obtained from calibration curves acquired through secondary-ion mass spectrometry (SIMS) analysis, whose values were determined to be: 1.2×10^{19}, 2.9×10^{19} and 5.8×10^{19} at/cm^3, for nitrogen doping; 1.7×10^{18} at/cm^3, 3.4×10^{18} at/cm^3 and 6.8×10^{18} at/cm^3, for aluminum. In this work, two different doping scales were used, 10^{18} at/cm^3 and 10^{19} at/cm^3 for aluminum and nitrogen, respectively.

Micro-Raman and micro-photoluminescence maps were acquired at room temperature using an HR800 spectrometer integrated system Horiba Jobin Yvon (Horiba, Lille, France) in a backscattering configuration. For both the analyses, the excitation wavelength was supplied by a 325 nm He-Cd continuous-wave laser that was focalized on the sample by a ×40 objective, with numerical aperture (NA) of 0.5. The scattered light was dispersed by an 1800 grooves/mm kinematic grating, for Raman analysis, and by a 300 grooves/mm kinematic grating, for PL analysis.

The etching in potassium hydroxide (KOH) was adopted for the evaluation of SFs. KOH etching was performed at 500 °C for 3 min. The densities were calculated based on the observation of optical and scanning electron microscopy (SEM) images, (FE-SEM ZeissSupra35, Carl Zeiss NTS GmbH, Oberkochen, Germany) operated at an acceleration voltage of 5 kV. An accurate count and identification of SFs were achieved using powerful image analysis software.

3. Results and Discussion

The Raman technique is sensitive to the average quality of the material. In particular, taking into account the wavelength of the laser used (325 nm), the technique is more sensitive to the surface (approximately 2–3 microns). The intensity of the transversal optical (TO) Raman mode and its full width at half maximum (FWHM) are the two parameters most closely related to the defectiveness of the material. Figure 1 shows the FWHM of the TO peak vs. doping concentration for nitrogen (black squares) and aluminum (red points). For the P-type samples, the FWHM of the TO shows a constant trend independent of the doping concentration. The FWHM value is approximately 11 cm^{-1}. For the N-type samples, the FWHM of the TO decreases with increasing nitrogen concentration, reaching values lower than those of the intrinsic sample (FWHM = 10 cm^{-1}) for a nitrogen concentration of 5.8×10^{19} at/cm^3 (FWHM = 7 cm^{-1}). However, the N-type doping is a factor of 10 greater than that of the P-type.

Figure 2 shows the room-temperature PL spectra, centered at 540 nm, acquired on the surfaces of the N-type (left panel) and P-type (right panel) samples.

After increasing the nitrogen doping, we can observe that in the presence of N-type samples, the intensity of the band-edge peak (centered at 540 nm) grows. In N-doped 3C-SiC, there are more electrons near the Fermi level that increase the band-to-band transitions [22] and lead to the enhancement of the detected signals. The shoulder at approximately 560 nm is present in both N-type and P-type samples and it is due to the presence of levels near the conduction band introduced by nitrogen, which cannot be completely eliminated from the chamber during the growth process.

Instead, increasing the aluminum doping, in P-type samples, the intensity of the band-edge peak (centered at 540 nm) decreases. At the same time, a new band is visible in the red spectral region (approximately 600 nm), which rises in intensity when increasing the aluminum doping. In agreement with the literature [22], this band is related to the presence of intra-gap defects, which are responsible for reducing the emission intensity of the band-edge.

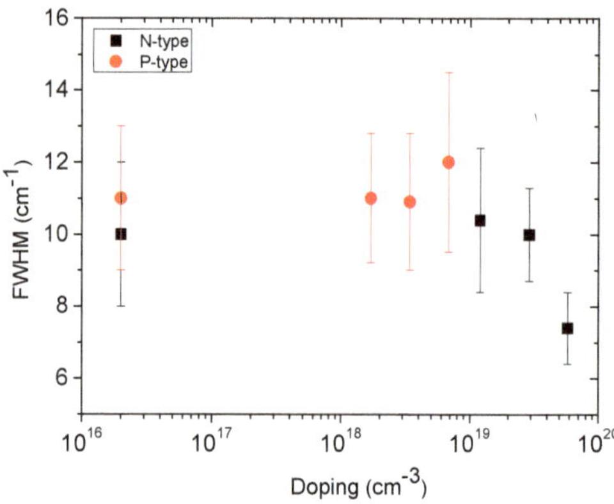

Figure 1. FWHM of the TO peak vs. doping concentration for nitrogen (black squares) and aluminum (red points). N-type and P-type samples were grown on Si (100) off-axis [110] and off-axis [100], respectively.

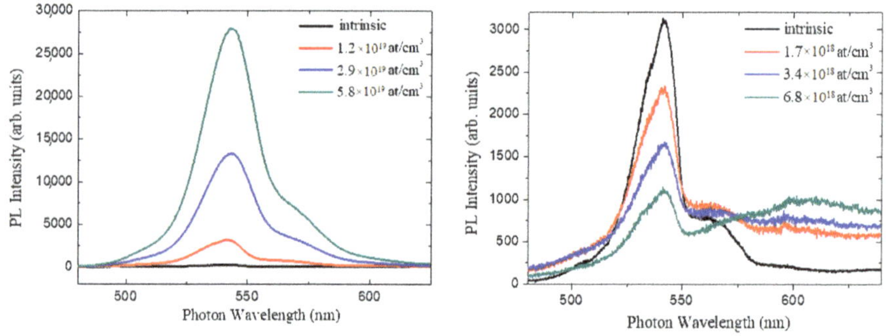

Figure 2. PL spectra, centered at 540 nm, acquired on the surfaces of the N-type (**left panel**) and P-type (**right panel**) samples.

The presence of dopant elements not only influences the crystalline and optical properties of the material but could also influence the presence and development of SFs within 3C-SiC. Due to their high electrical activity, the concentration of SFs in the material is a fundamental parameter. Even today, the concentration of stacking faults is not compatible with the development of VLSI technology.

Therefore, all samples were treated by molten KOH etching to study the effect of doping concentration on the density and mean length of SFs. As an example, Figure 3 shows the optical and SEM images of the intrinsic (Figure 3a,c) and N-type doped (5.8×10^{19} at/cm^3) samples (Figure 3b,d) subjected to selective KOH etching. Here, wet etching highlights the presence of SFs following the variation of the etching rate, which, according to the variation of binding energy in the presence of such defects, promotes groove formation along the extension of the extra planes [23].

In all images, the etched SFs are oriented along a preferential direction. Off-axis growth promotes propagation up to the surface of the SFs lying along the plane (111), while the SFs (−1−11) opposite to the growth step are forbidden. In the absence of the annihilation mechanism, the propagation of the Si-face SFs from the 3C-SiC/Si interface towards the surface is favored. In addition, the etching rate of the C-face SFs, lying on the (−111) and (1−11) crystallographic planes, is uniform to the (001) plane, while the etching rate of the Si-phase SFs is lower, displaying their distribution [24].

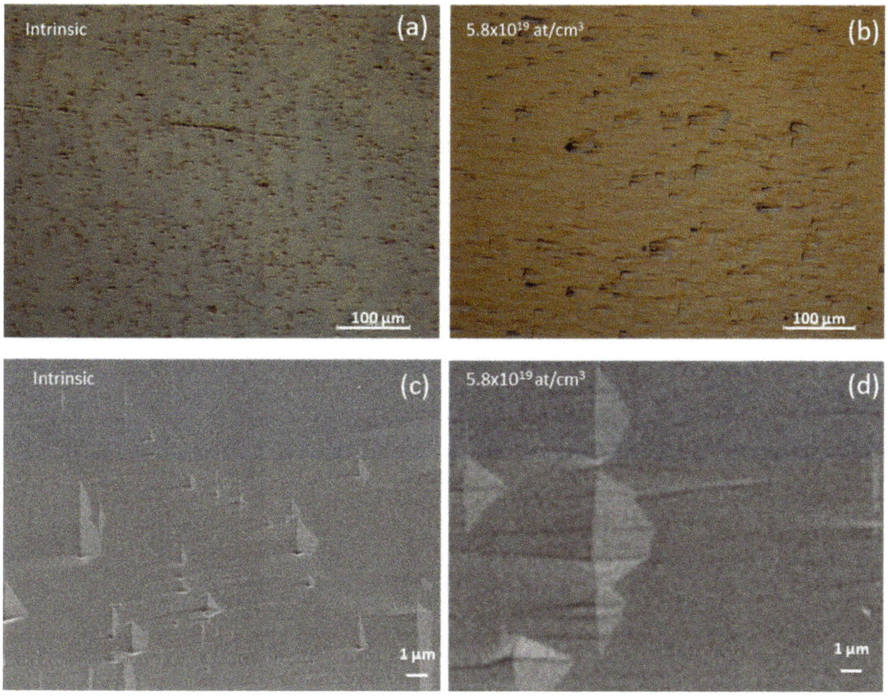

Figure 3. Optical and SEM images of the intrinsic (**a**,**c**) and N-type doped (**b**,**d**) samples subjected to selective KOH etching.

For the observation of the samples attached in KOH, the optical images have proved useful for the detection of large SFs (>1 µm). For SFs smaller than 1 µm, SEM images allowed us to obtain a detection limit of 100–150 nm.

In Figures 4 and 5, we report the distribution of the lengths of the SFs for different N-type and P-type doping, respectively. For N-type samples (Figure 4a–d), we observe that as the nitrogen concentration increases (from intrinsic sample to 5.8×10^{19} at/cm^3), the length distribution of the SFs widens and flattens. Consequently, the maximum shifts to greater lengths. For P-type samples (Figure 5a–d), we observe that as the aluminum concentration increases (from intrinsic sample to 6.8×10^{18} at/cm^3), the maximum of the length distribution is quite constant.

The distributions of the lengths of the SFs were fitted with log-normal type curves, which are better suited to fit the experimental values and to reduce the fit error. From the fits, the mean values of length of the SFs were obtained for all the samples. The values are reported in Table 1 and they are plotted in Figure 6.

We observe that by increasing the dopant concentration for P-type samples (red points, Figure 6, left panel), the mean density values of SFs are closer to the density of the intrinsic counterpart. Indeed, the Al-doped samples exhibit an SF concentration ranging from 5.74×10^3 cm^{-1} for 1.7×10^{18} at/cm^3 to 4.08×10^3 cm^{-1} for the sample with 6.8×10^{18} at/cm^3 doping. For the N-type samples (black squares, Figure 6, left panel), on the other hand, the trend of the density of the SFs tends to move away from the intrinsic value. In fact, if for the intrinsic sample, the average SF density is 2.05×10^3 cm^{-1}, by increasing doping to 2.9×10^{19} at/cm^3 and 5.8×10^{19} at/cm^3, such a value is notably reduced by almost an order of magnitude to 4.7×10^2 cm^{-1} and 2.4×10^2 cm^{-1}. Ab initio DFT calculation shows that when nitrogen is very close to the SFs, their formation energy increases, resulting in a reduction in the density of the SFs inside the material [25]. Comparing the density of the SFs between the two intrinsic samples grown on Si (100) 4° off-axis along the [110] (dashed green line), and on Si (100) 4° off-axis along the [100] (dashed magenta line), two different densities were obtained: 2050 cm^{-1} and 4920 cm^{-1}, respectively. The dependence of density of the SFs on the step evolution and surface morphology on the shear angle was observed for 3C-SiC samples grown on (111) Si substrates with off-cut axes along [110] and [112] [26]. As mentioned above, off-axis growth promotes propagation up to the surfaces of the SFs lying along the plane (111), while the SFs (−1−11) opposite to the growth step are forbidden. However, while, for the N-type samples, the off-axis angle is along the [110] direction and therefore favors the suppression of the SFs along the plane (−1−11), for the P-type samples, the off-axis angle is along the direction [100]. In this way, not all the SFs along the plane (−1−11) are suppressed. At the same time, we observed that this cutting direction does not lead to a complete suppression of the APBs (SEM images not reported here), which act as a source and annihilation point for the SFs [27]. This leads to an increase in the density of the SFs that reach the surface.

Increasing the dopant concentration for the N-type samples (black squares, Figure 6, right panel), the average length of the SFs increases, starting from a value of approximately 2 μm for the intrinsic sample to 5 μm for the sample with a nitrogen concentration of 5.8×10^{19} at/cm^3. Instead, by increasing the dopant concentration for the P-type samples (red points, Figure 6, right panel), the length of the SFs remains constant, less than 1 μm. In the presence of N-type doping, the average values are greater than the value of the intrinsic sample grown off-axis [110] (dashed green line). The reduced density of SFs, increasing the nitrogen concentration (Figure 6 left panel), is susceptible to a reduced mutual annihilation mechanism. This contributes to the increase in the average size of SFs reaching the surface.

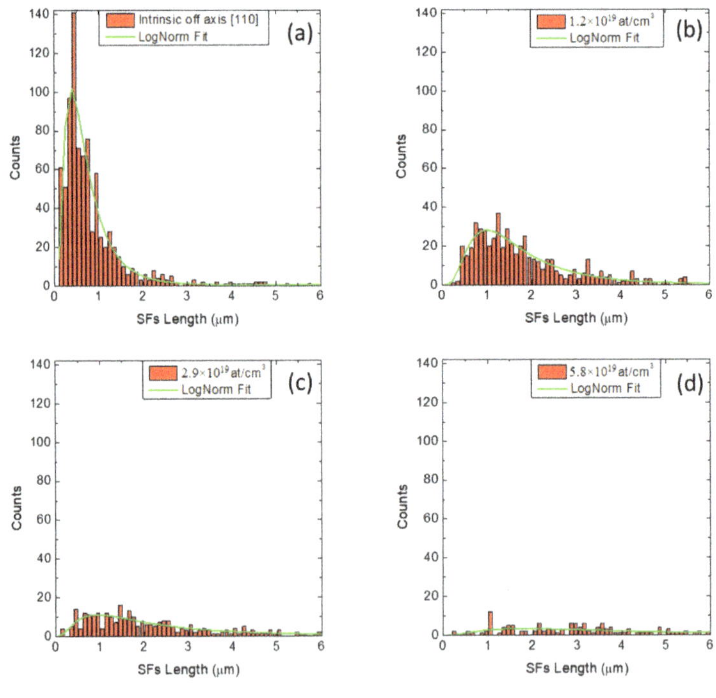

Figure 4. Length distribution of the SFs for different N-type doping. (**a**) Intrinsic sample, (**b**) 1.2×10^{19} at/cm^3, (**c**) 2.9×10^{19} at/cm^3 and (**d**) 5.8×10^{19} at/cm^3. For all samples, there are a few SFs whose lengths are greater than the range reported in the X-axis range. However, all the lengths of the SFs were considered in the calculation of the mean lengths and densities.

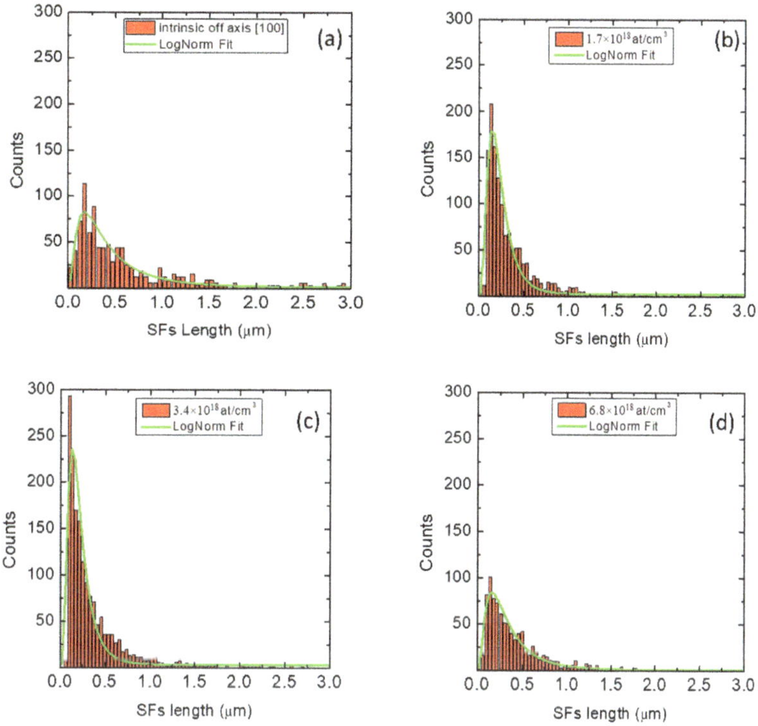

Figure 5. Length distribution of the SFs for different P-type doping. (**a**) Intrinsic sample, (**b**) 1.7×10^{18} at/cm^3, (**c**) 3.4×10^{18} at/cm^3 and (**d**) 6.8×10^{18} at/cm^3. For all samples, there are a few SFs whose lengths are greater than the range reported in the X-axis range. However, all the lengths of the SFs were considered in the calculation of the mean lengths and densities.

Table 1. Values of average lengths and densities for different doping concentrations.

N-Type Concentration (cm^{-3})	N-Type SFs Mean Length (μm)	N-Type SFs Density (cm^{-1})	P-Type Concentration (cm^{-3})	P-Type SFs Mean Length (μm)	P-Type SFs Density (cm^{-1})
Intrinsic [110]	1.87 ± 1.12	2050 ± 19.52	Intrinsic [100]	0.71 ± 0.51	4920 ± 470
1.2×10^{19}	1.79 ± 1.38	690.67 ± 11.40	1.7×10^{18}	0.24 ± 0.15	5740 ± 140
2.9×10^{19}	2.64 ± 1.97	473.78 ± 8.03	3.4×10^{18}	0.22 ± 0.14	6360 ± 150
5.8×10^{19}	5.24 ± 2.60	244.48 ± 10.73	6.8×10^{18}	0.37 ± 0.33	4080 ± 100

In the presence of P-type doping, instead, the average values are smaller than both values of the intrinsic samples grown off-axis (green and magenta line). As mentioned above, for the P-type samples, the off-axis angle is along the direction [100]. In this way, not all the SFs along the plane (−1−11) are suppressed. This leads to an increase in the mutual annihilation mechanism, responsible for reducing the average lengths of the SFs on the surface.

Both values of density and mean length of the SFs depend on the direction of the cutting angle and, in particular, on the doping concentration. It is interesting to note that the trend of the density of the SFs, when increasing the doping concentration, is in agreement with the trend of the crystalline quality (FWHM of the TO) reported in Figure 1. For P-type samples, as the dopant concentration increases, a constant SF density and crystalline quality were observed. Instead, for N-type samples, as the dopant concentration increases, a clear reduction in the density of SFs and a constant increase in crystalline quality were obtained. In Figure 7, the FWHM of the TO peak vs. the 3C-SiC thickness (in cross-section) is reported, for the intrinsic (black line) and 5.8×10^{19}at/cm^3 (red line) samples. The zero value on the X-axis corresponds to the removed interface with the silicon.

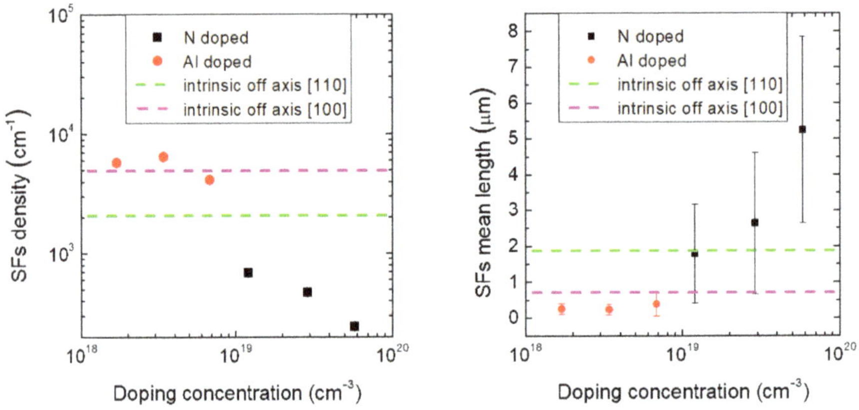

Figure 6. Distribution of SF density (**left panel**) and SF mean length (**right panel**) vs. doping concentration. The graphs also report the average density and average length of the intrinsic samples grown on Si (100) off-axis [110] (dashed green line) and off-axis [100] (dashed magenta line).

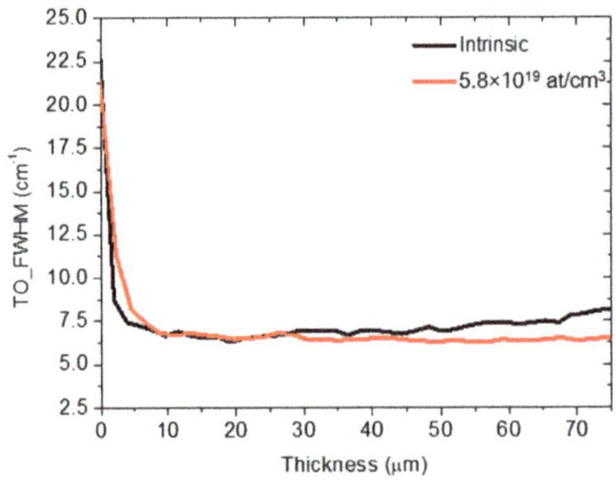

Figure 7. FWHM of the TO peak vs. 3C-SiC thickness in cross-section for intrinsic and 5.8×10^{19} at/cm^3 (red line) samples. The zero value on the X-axis corresponds to the removed interface with the silicon.

In the first 10 μm of the films, a difference in crystalline quality was observed. In particular, lower crystalline quality was observed for the doped sample (red curve). The red curve shows a higher average value in this area. For thicknesses between 10 and 30 μm, the crystalline qualities are constant and show the same value for both samples. For thicknesses greater than 30 μm, a divergence was observed between the two curves. In particular, the doped sample continues to maintain the same crystalline quality, while for the intrinsic sample, the quality tends to worsen, reaching the value reported in the in-plan measurements (Figure 1).

The difference in thermal expansion coefficient and lattice parameters between Si and SiC leads to the formation of dislocations at the hetero-interface and the presence of a local strain field within the epilayer [15]. The presence of residual stress (depending on doping concentration [21]) can drive the formation of perfect and partial dislocations

(PDs) in the epilayer. Every SF is bordered by two PDs, which limit the wrong sequence plane from the perfect 3C crystal. More investigations should be done, particularly in cross-sections, to determine the role of doping concentration in the density, shape and stability of PDs, because the expansion or shrinkage of SFs is driven by the energetics and kinetics of the PDs.

The association of current doping systems alongside compliant substrate adoption, such as inverted silicon pyramids [28] or a $Si_{1-x}Ge_x$ buffer layer [29], could lead to the further implementation of growth strategies capable of reducing the concentration of SFs and leading to the development of efficient 3C-SiC power devices.

4. Conclusions

The 3C-SiC hetero-epitaxial layers, doped with nitrogen or aluminum, were grown in a horizontal hot-wall chemical vapor deposition (CVD) reactor on Si (100) substrate with different 4° off-axis. Stacking fault evaluation through molten KOH etching was performed on different doped samples: 1.2 × 1019, 2.9 × 1019 and 5.8 × 1019 at/cm3, for nitrogen doping; 1.7 × 1018, 3.4 × 1018 and 6.8 × 1018 at/cm3, for aluminum. Samples were grown on Si (100) 4° off-axis along the [110] and on Si (100) 4° off-axis along the [100], for nitrogen and aluminum, respectively.

In particular, it was observed that as the dopant concentration increases, the average length of the SFs increases, from a value of 2 μm to 5 μm, and the density decreases, for N-type doping. Instead, for P-type doping, when increasing the doping concentration, the mean length and density of the SFs remain constant.

Both values (density and mean length) depend on the direction of the cutting angle. In fact, the off-axis angle along the [110] favors the suppression of the SFs along the plane (−1−11), decreasing the density of the SFs on the surface and increasing their mean length. Instead, for an off-axis along the [100], not all the SFs along the plane (−1−11) were suppressed. This leads to an increase in the mutual annihilation mechanism, responsible for reducing the average lengths of the SFs on the surface.

Author Contributions: Conceptualization, V.S., C.C., F.L.V.; methodology, V.S., C.C., synthesis, R.A., M.M., D.C.; investigation, V.S., C.C., A.C., S.B.; data curation, C.C., V.S.; writing—original draft preparation, V.S.; writing—review and editing, V.S., C.C., F.L.V.; supervision of the whole project, F.L.V.; funding acquisition, F.L.V. All authors read and approved the final manuscript.

Funding: This research was funded by the European Union within the framework of the project CHALLENGE, grant number 720827.

Institutional Review Board Statement: Not applicable.

Informed Consent Statement: Not applicable.

Data Availability Statement: The data presented in this study are available on request from the corresponding author. The data are not publicly available due to industrial non-disclosure agreement.

Conflicts of Interest: The authors declare no conflict of interest.

References

1. Chelnokov, V.E.; Syrkin, A.L.; Dmitriev, V.A. Overview of SiC power electronics. *Diam. Relat. Mater.* **1997**, *6*, 1480–1484. [CrossRef]
2. Willander, M.; Friesel, M.; Wahab, Q.U.; Straumal, B. Silicon carbide and diamond for high temperature device applications. *J. Mater. Sci. Mater. Electron.* **2006**, *17*, 1–25. [CrossRef]
3. Azevedo, R.G.; Jones, D.G.; Jog, A.V.; Jamshidi, B.; Myers, D.; Chen, L.; Fu, X.A.; Mehregany, M.; Wijesundara, M.B.; Pisano, A.P. A SiC MEMS resonant strain sensor for harsh environment applications. *IEEE Sens. J.* **2007**, *7*, 568–576. [CrossRef]
4. Stoldt, C.R.; Carraro, C.; Ashurst, W.R.; Gao, D.; Howe, R.T.; Maboudian, R. A low-temperature CVD process for silicon carbide MEMS. *Sens. Actuators A* **2002**, *97*, 410–415. [CrossRef]
5. La Via, F.; Severino, A.; Anzalone, R.; Bongiorno, C.; Litrico, G.; Mauceri, M.; Schoeler, M.; Schuh, P.; Wellmann, P. From thin film to bulk 3C-SiC growth: Understanding the mechanism of defects reduction. *Mater. Sci. Sem. Proc.* **2018**, *78*, 57–68. [CrossRef]
6. Chen, J.; Steckel, A.J.; Loboda, M.J. In Situ N2-Doping of SiC Films Grown on Si (111) by chemical vapor deposition from Organosilanes. *J. Electrochem. Soc.* **2000**, *147*, 2324–2327. [CrossRef]

7. Kern, R.S.; Davis, R.F. Deposition and doping of silicon carbide by gas-source molecular beam epitaxy. *Appl. Phys. Lett.* **1997**, *71*, 1356–1358. [CrossRef]
8. Krotz, G.; Moller, H.; Eickhoof, M.; Zappe, S.; Ziermann, R.; Obermeier, E.; Stomenos, J. Heteroepitaxial growth of 3C-SiC on SOI for sensor applications. *Mater. Sci. Eng. B* **1999**, *61*, 516–521. [CrossRef]
9. Gotz, W.; Schoner, A.; Pensl, G.; Suttrop, W.; Choyke, W.J.; Stein, R.; Leinbenzeder, S. Nitrogen donors in 4*H*-silicon carbide. *J. Appl. Phys.* **1993**, *73*, 3332–3338. [CrossRef]
10. Lebedev, A.A.; Abramov, P.L.; Bogdanova, E.V.; Lebedev, S.P.; Nel'son, D.K.; Oganesyan, G.A.; Tregubova, A.S.; Yakimova, R. Highly doped p-type 3C–SiC on 6H–SiC substrates. *Semicond. Sci. Technol.* **2008**, *23*, 075004. [CrossRef]
11. Ra, M.V.; Griffiths, P.; Kelner, G.; Freitas, J.A., Jr.; Simons, D.S.; Chi, P.H.; Ghezzo, M. Al and B ion-implantations in 6H- and 3C-SiC. *J. Appl. Phys.* **1995**, *77*, 2479–2485. [CrossRef]
12. Zielinski, M.; Arvinte, R.; Chassagne, T.; Michon, A.; Portail, M.; Kwasnicki, P.; Konczewicz, L.; Contreras, S.; Juillaguet, S.; Peyre, H. p-Type Doping of 4H- and 3C-SiC epitaxial layers with aluminum. *Mater. Sci. Forum* **2016**, *858*, 137–142. [CrossRef]
13. Hasegawa, S.; Furuta, N.; Takeshita, T.; Inokuma, T.; Kurata, Y. Phosphorus and nitrogen doping into polycrystalline SiC films prepared by plasma-enhanced chemical vapour deposition at 700 C. *J. Appl. Phys.* **1992**, *72*, 1374–1377. [CrossRef]
14. Zielinski, M.; Portail, M.; Chassagne, T.; Juillaguet, S.; Peyre, H. Nitrogen doping of 3C-SiC thin films grown by CVD in a resistively heated horizontal hot-wall reactor. *J. Cryst. Growth.* **2008**, *310*, 3174–3182. [CrossRef]
15. Zimbone, M.; Sarikov, A.; Bongiorno, C.; Marzegalli, A.; Scuderi, V.; Calabretta, C.; Miglio, L.; La Via, F. Extended defects in 3C-SiC: Stacking faults, threading partial dislocations, and inverted domain boundaries. *Acta Mater.* **2021**, *213*, 116915–116925. [CrossRef]
16. Song, X.; Michaud, J.F.; Cayrel, F.; Zielinski, M.; Portail, M.; Chassagne, T.; Collard, E.; Alquier, D. Evidence of electrical activity of extended defects in 3C-SiC grown on Si. *Appl. Phys. Lett.* **2010**, *96*, 142104. [CrossRef]
17. Giannazzo, F.; Greco, G.; Di Franco, S.; Fiorenza, P.; Deretzis, I.; La Magna, A.; Bongiorno, C.; Zimbone, M.; La Via, F.; Zielinski, M.; et al. Impact of stacking faults and domain boundaries on the electronic transport in cubic silicon carbide probed by conductive atomic force microscopy. *Adv. Electron. Mater.* **2020**, *6*, 1901171. [CrossRef]
18. Shibahara, K.; Nishino, S.; Matsunami, M. Antiphase-domain-free growth of cubic SiC on Si (100). *Appl. Phys. Lett.* **1987**, *50*, 1888–1890. [CrossRef]
19. Li, R.; Huang, M.; Yang, Z.; Ma, Y.; Gong, M. Carrier-storage-enhanced superjunction IGBT with n-Si and p-3C-SiC pillars. *Electron. Lett.* **2019**, *55*, 1353–1355. [CrossRef]
20. Scuderi, V.; Calabretta, C.; Anzalone, R.; Mauceri, M.; La Via, F. Characterization of 4H- and 6H-like stacking faults in cross section of 3C-SiC epitaxial layer by room-temperature μ-photoluminescence and μ-raman analysis. *Materials* **2020**, *13*, 1837. [CrossRef] [PubMed]
21. La Via, F.; Mauceri, M.; Scuderi, V.; Calabretta, C.; Zimbone, M.; Anzalone, R. 3C-SiC bulk growth: Effect of growth rate and doping on defects and stress. *Mater. Sci. Forum* **2020**, *1004*, 120–125. [CrossRef]
22. Wang, K.; Fang, X.Y.; Li, Y.Q.; Yin, A.C.; Jin, H.B.; Yuan, J.; Cao, M.S. Radiative recombination model of degenerate semiconductor and photoluminescence properties of 3C-SiC by P and N doping. *J. Appl. Phys.* **2012**, *112*, 033508. [CrossRef]
23. Van der Berg, N.G.; Malherbe, J.B.; Botha, A.J.; Friedland, E. Thermal etching of SiC. *Appl. Surf. Sci.* **2012**, *258*, 5561–5566. [CrossRef]
24. Katsuno, M.; Ohtani, N.; Takahashi, J.; Yashiro, H.; Kanaya, M. Mechanism of molten KOH etching of SiC single crystals: Comparative study with thermal oxidation. *Jpn. J. Appl. Phys.* **1999**, *38*, 4661–4665. [CrossRef]
25. Umeno, Y.; Yagi, K.; Nagasawa, H. Ab initio density functional theory calculation of stacking fault energy and stress in 3C-SiC. *Phys. Status Solidi B* **2012**, *249*, 1229–1234. [CrossRef]
26. Severino, A.; Camarda, M.; Condorelli, G.; Perdicaro, L.M.S.; Anzalone, R.; Mauceri, M.; La Magna, A.; La Via, F. Effect of the miscut direction in (111) 3C-SiC film growth on off-axis (111) Si. *Appl. Phys. Lett.* **2009**, *94*, 101907. [CrossRef]
27. Jiao, S.; Zielinski, M.; Roy, S.; Chassagne, T.; Michaud, J.F.; Portail, M.; Alquier, D. Detailed study of the influence of surface misorientation on the density of Anti-Phase Boundaries in 3C-SiC layers grown on (001) silicon. *AIP Conf. Proc.* **2010**, *1292*, 15–18.
28. Zimbone, M.; Zielinski, M.; Bongiorno, C.; Calabretta, C.; Anzalone, R.; Scalese, S.; Fisicaro, G.; La Mahna, A.; Mancarella, F.; La Via, F. 3C-SiC growth on inverted silicon pyramids patterned substrate. *Materials* **2019**, *12*, 3407. [CrossRef]
29. Zimbone, M.; Zielinski, M.; Barbagiovanni, E.G.; Calabretta, C.; La Via, F. 3C-SiC grown on Si by using a Si1-xGex buffer layer. *J. Cryst. Growth* **2019**, *519*, 1–6. [CrossRef]

materials

MDPI

Review

Status and Prospects of Cubic Silicon Carbide Power Electronics Device Technology

Fan Li [1], Fabrizio Roccaforte [2], Giuseppe Greco [2], Patrick Fiorenza [2], Francesco La Via [2], Amador Pérez-Tomas [3], Jonathan Edward Evans [4], Craig Arthur Fisher [4], Finn Alec Monaghan [4], Philip Andrew Mawby [5] and Mike Jennings [4,*]

1 Newport Wafer Fab, Cardiff Rd, Duffryn, Newport NP10 8YJ, UK; Fan.Li@nptwf.com
2 Consiglio Nazionale delle Ricerche—Istituto per la Microelettronica e Microsistemi (CNR-IMM), Strada VIII n. 5-Zona Industriale, 95121 Catania, Italy; fabrizio.roccaforte@imm.cnr.it (F.R.); giuseppe.greco@imm.cnr.it (G.G.); patrick.fiorenza@imm.cnr.it (P.F.); francesco.lavia@imm.cnr.it (F.L.V.)
3 Institut Català de Nanociència i Nanotecnologia (ICN2), Universitat Autònoma de Barcelona, 08193 Barcelona, Spain; amador.perez@icn2.cat
4 Faculty of Science, Bay Campus, College of Engineering, Swansea University, Fabian Way, Crymlyn Burrows, Skewen, Swansea SA1 8EN, UK; j.e.evans@swansea.ac.uk (J.E.E.); c.a.fisher@swansea.ac.uk (C.A.F.); f.a.j.monaghan.909302@swansea.ac.uk (F.A.M.)
5 School of Engineering, The University of Warwick, Gibbet Hill Rd, Coventry CV4 7AL, UK; p.a.mawby@swansea.ac.uk
* Correspondence: m.r.jennings@swansea.ac.uk

Citation: Li, F.; Roccaforte, F.; Greco, G.; Fiorenza, P.; La Via, F.; Pérez-Tomas, A.; Evans, J.E.; Fisher, C.A.; Monaghan, F.A.; Mawby, P.A.; et al. Status and Prospects of Cubic Silicon Carbide Power Electronics Device Technology. *Materials* **2021**, *14*, 5831. https://doi.org/10.3390/ma14195831

Academic Editor: Alexander A. Lebedev

Received: 20 July 2021
Accepted: 25 September 2021
Published: 5 October 2021

Abstract: Wide bandgap (WBG) semiconductors are becoming more widely accepted for use in power electronics due to their superior electrical energy efficiencies and improved power densities. Although WBG cubic silicon carbide (3C-SiC) displays a modest bandgap compared to its commercial counterparts (4H-silicon carbide and gallium nitride), this material has excellent attributes as the WBG semiconductor of choice for low-resistance, reliable diode and MOS devices. At present the material remains firmly in the research domain due to numerous technological impediments that hamper its widespread adoption. The most obvious obstacle is defect-free 3C-SiC; presently, 3C-SiC bulk and heteroepitaxial (on-silicon) display high defect densities such as stacking faults and antiphase boundaries. Moreover, heteroepitaxy 3C-SiC-on-silicon means low temperature processing budgets are imposed upon the system (max. temperature limited to ~1400 °C) limiting selective doping realisation. This paper will give a brief overview of some of the scientific aspects associated with 3C-SiC processing technology in addition to focussing on the latest state of the art results. A particular focus will be placed upon key process steps such as Schottky and ohmic contacts, ion implantation and MOS processing including reliability. Finally, the paper will discuss some device prototypes (diodes and MOSFET) and draw conclusions around the prospects for 3C-SiC devices based upon the processing technology presented.

Keywords: 3C-SiC; cubic silicon carbide; power electronics

1. Introduction

Power electronics is a key enabling technology for energy generation, transmission, distribution and motion. The importance of this technology is emphasised by the fact that a 40% increase in energy consumption within 20 years is expected [1]. Moreover, 80% of electrical energy will be processed by a power electronic converter by 2030 [2]. Recently, power electronic converter and device technology has been driven by the huge demand seen within the electric vehicle (EV) sector. EV sales are set to reach 18 million by 2023, representing 16.2% of total global vehicle sales [3]. Together, these circumstances project the ever-increasing demand for power electronics on a global scale. In order to meet this required capacity and while still safeguarding our environment, power converters with near-100% energy-efficiency that are lightweight and compact need to be delivered.

Furthermore, attention must be paid to the lifetime (or longevity) of these systems, meaning increased reliability within the field.

Such a step-change intervention within the world of power electronics requires advancements within the fundamental semiconductor materials that serve to underpin our energy landscape. The underpinning technologies with respect to power electronics are its constituent high-voltage semiconductor devices. Consequently, these devices represent the largest cost associated with the overall power converter (40% of the total bill of materials for a typical 50 kW EV inverter). Traditionally for the last 50 years, silicon (Si) has dominated the power electronics industry as the semiconductor material of choice. However, the demand for increased energy-efficiency and power density together with higher voltage and current operation mean that a new era in semiconductor materials has dawned. Wide bandgap (WBG) semiconductor materials come with the promise to overcome the inherent material limits imposed by Si. 4H-silicon carbide (4H-SiC) and gallium nitride (2H-GaN or GaN) have emerged as the WBG materials of choice that have replaced Si in many power electronic applications.

For the moment, GaN devices that are based mainly on the high electron mobility transistor (HEMT) architecture are limited commercially to a maximum of 650 V. From the reliability perspective, GaN HEMTs have traditionally suffered from a poor thermal conductivity and the "current collapse" phenomenon, degrading their ability to function within harsh environments and high reliability electronics [4]. 4H-SiC, on the other hand, suffers from numerous reliability issues that are hampering its widespread uptake within the automotive sector. In particular, although SiC Trench MOSFETs exhibit superior on-state resistance compared to both GaN and silicon, the ruggedness of the gate oxide is the limiting factor. Gonzalez et al. [5] note that the competing WBG material technologies centre around the 650 V mark.

Early stage research devices are based on so-called ultrawide bandgap oxide materials such as gallium oxide (Ga_2O_3, with β-Ga_2O_3 being the most stable). Thus far, β-Ga_2O_3 suffers from a poor thermal conductivity, a modest bulk mobility and lack of p-type conductivity. Other ultrawide bandgap materials, including diamond and aluminium nitride (AlN), suffer from a lack of n-type conductivity and a poor bulk electron mobility, respectively. It should be noted that GaN, β-Ga_2O_3, and AlN are direct bandgap materials, which severely limits bipolar operation, which is required for higher voltages [6].

This review will place the cubic SiC (3C-SiC) material into the context of power electronic devices; however, it should be noted that other application areas such as biomedical sensors and micro-electromechanical systems (MEMS) are also appropriate and more popular for this SiC polytype. The authors will endeavour to provide a brief insight into some of the advantages of 3C-SiC from the scientific materials perspective in addition to some of the technological issues that must be overcome to realise competitive power MOSFETs and diodes. In particular, the focus will be placed on fundamental semiconductor fabrication technologies; the 3C-SiC/SiO_2 metal-oxide-semiconductor (MOS) interface, ion implantation, ohmic and Schottky contacts.

2. Cubic Silicon Carbide (3C-SiC): Structure and Material Properties for Power Electronic Application

The cubic form of SiC, coined '3C-SiC', is one of many stable polytypes characterised by its wide bandgap and bilayer stacking sequence of ABCABC ... [7]. The resulting structure is a pure zinc-blende exhibiting an energy band gap of 2.3–2.4 eV [8], lower compared to other major SiC polytypes, but with a higher electron mobility and saturation velocity owing to its higher degree of symmetry. Although 3C-SiC has a smaller energy bandgap compared to its wide bandgap counterparts such as 4H-SiC and GaN, this material displays isotropy for many of the desired power device material characteristics such as avalanche coefficients and high electron mobility [9,10]. Another advantage of 3C-SiC is its relatively large thermodynamic stability meaning that bulk material can be grown at reduced thermal budgets (below 1500 °C). Table 1 shows the important physical and electrical properties of 3C-SiC compared to other commercial power device materials such as Si, GaN and 4H-SiC. Likewise included are promising oxide and nitride ultra-WBG materials. The 3C-SiC intrinsic carrier concentration (~10^{-1} cm^{-3}) is several orders of magnitude lower than in Si, but not as low as 4H-SiC or GaN. Moreover, 3C-SiC has a thermal conductivity three times that of Si. Consequently, 3C-SiC devices should have lower leakage currents with the ability to operate at moderately higher temperatures when compared to Si and GaN. Other key aspects are the reasonable critical electric field value resulting in a higher breakdown of the material. On analysis of these material properties, 3C-SiC is a promising semiconductor for power semiconductor devices in the region of 600–1000 V. On reflection, there exists the possibility to obtain a targeted breakdown voltage (V_B) with thinner, more highly doped drift layers, which results in a significant reduction of the specific on-resistance (R_{ON}) compared to Si devices. Therefore, devices that are smaller and more efficient can be fabricated, minimizing both the static and dynamic losses.

Table 1. Appropriate physical and electrical properties of cubic silicon carbide (3C-SiC) compared to other wide bandgap materials (data taken at 300 K).

Material	Band Gap, (eV)	Intrinsic Carrier Conc., (cm^{-3})	Dielectric Constant	Electron Mobility (cm^2/Vs)	Critical Electric Field (MV/cm)	Saturation Velocity (10^7 cm/s)	Thermal Conductivity (W/cmK)	Baliga Figure of Merit
Si	1.12	1.5×10^{10}	11.8	1350	0.2	1.0	1.5	1
GaAs	1.42	1.8×10^{6}	13.1	8500	0.4	1.2	0.55	29
3C-SiC	2.36	1.5×10^{-1}	9.7	800	1.4	2.5	3.2	86
4H-SiC	3.26	8.2×10^{-9}	10	720 [a] 650 [c]	2.8	2.0	4.5	556
2H-GaN	3.39	1.9×10^{-10}	9.9	1000 [a] 2000 **	3.75 [a] 3.3 *	2.5	1.3	3175
Ga_2O_3	4.85	2.6×10^{-9} -1.0×10^{-22}	10	300	8	1.8–2.0	0.1–0.3	6171
Diamond	5.45	1.6×10^{-27}	5.5	3800	10	2.7	22	8.4×10^4
2H-AlN	6.2	10^{-34}	8.5	300	12 *	1.7	2.85	1.8×10^4

Note: [a] is mobility along a-axis, [c] is mobility along c-axis, * refers to an estimated value and ** refers to the 2DEG mobility.

The 3C-SiC Baliga figure of merit (BFOM) and BFOM for high-frequency, high-power unipolar switches (BHFFOM) [11] are 140 and 25, respectively. These values seem very modest compared to the equivalent values for more advanced WBG power semiconductors such as 4H-SiC and GaN. These key performance indicators for power semiconductors quantify the minimum conduction loss during DC operation (BFOM) and the minimum conduction loss at high frequencies (BHFFOM). Indeed, examination of these values suggests that lower resistance devices are possible based on 4H-SiC and GaN when compared to 3C-SiC. However, this advantage must be weighed against power device reliability and field lifetime within a converter application. In this regard, 3C-SiC is the clear winner, benefitting from a favourable metal-oxide-semiconductor (MOS) interface when compared to its 4H-SiC counterpart. The bandgap value (E_g) for 3C-SiC was reported by Bimberg et al. [12] and later by Goldberg et al. [8] (see Table 1). Figure 1 shows the conduction band offsets of the major power semiconductors with silicon dioxide (SiO_2). From the figure it is seen that the band offset (Φ_B) between 3C-SiC and SiO_2 is 3.7 eV. This is significantly larger when compared to the other power semiconductors with their values ranging between 2.7 eV–3.2 eV.

The ramifications of this important property are realised in terms of reduced gate leakage current for a given oxide electric field. The important current transport mechanism which relates to this physical parameter is the Fowler-Nordheim (F-N) tunnelling mechanism. The F-N tunnelling current is given by:

$$J_{FN} = \frac{A}{\Phi_B} E_{ox}{}^2 \exp\left(-\frac{B\Phi_B{}^{3/2}}{E_{ox}}\right) \quad (1)$$

where E_{ox} is the oxide electric field, Φ_B is the barrier height and A, B are constant values. It can be seen that due to F-N tunnelling the oxide electric field value must be reduced by 2–3 times in 4H-SiC compared to the 3C-SiC system.

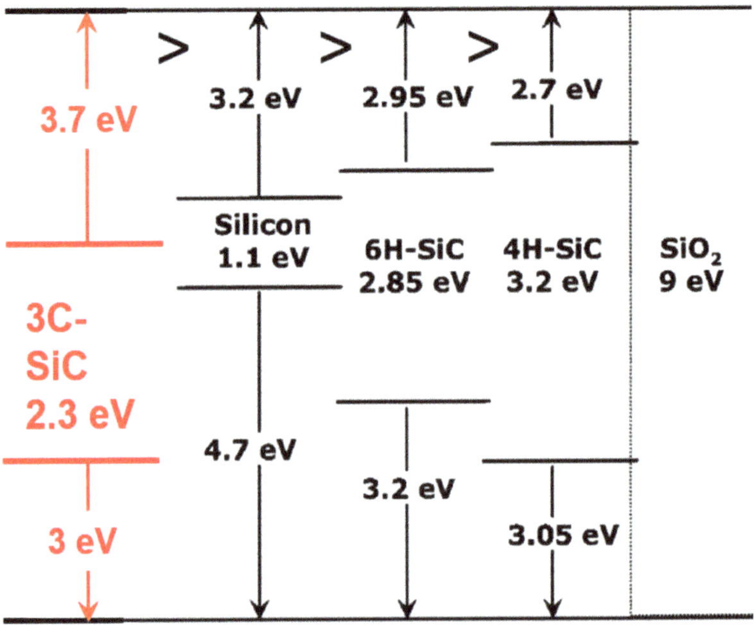

Figure 1. Major power semiconductors' band structure for 3C-SiC, 4H-SiC, 6H-SiC and silicon, illustrating band offsets with silicon dioxide (SiO_2).

Fardi and Van Zeghbroeck [13] developed an empirical breakdown field model based on the breakdown voltage and field values that were obtained from published experimental data [14,15]. This proved to be more than adequate for 3C-SiC device design, having matched electrical breakdown characteristics to many published reports. Moreover, the model has been utilised in commercial 2-dimensional device design suites [16–18]. Fitting these impact ionisation coefficients to the electric field and substituting into the impact ionisation integral leads to closed-form solutions of the breakdown voltage and depletion layer width. These material parameters allow for the initial stages of power device design. The closed-form solutions for the breakdown voltage and parallel-plane depletion region width are given as:

$$BV_{PP} = 7.88 \times 10^{14} N_D{}^{-3/4} \quad (2)$$

$$W_{PP} = 9.12 \times 10^{10} N_D{}^{-7/8} \quad (3)$$

where BV_{PP} is the breakdown voltage, N_D is the doping concentration and W_{PP} is the parallel-plane depletion region width. The breakdown voltage and depletion region widths predicted by Equations (2) and (3), respectively, are shown in Figure 2.

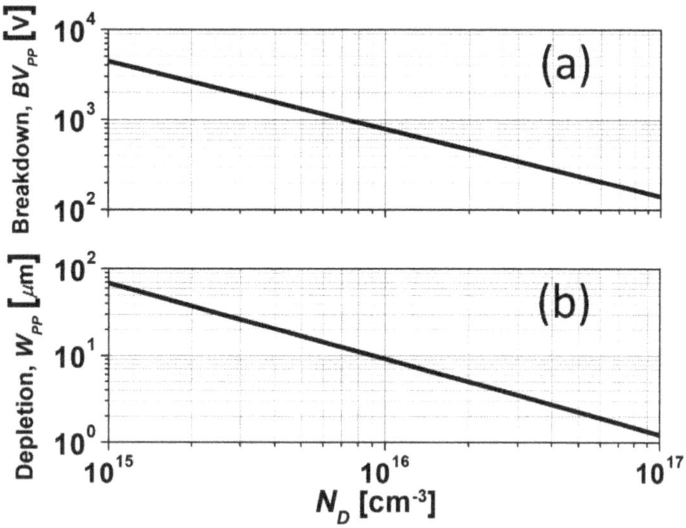

Figure 2. (**a**) Parallel plane breakdown voltage (BV_{PP}) and (**b**) depletion width (W_{PP}) as a function of doping (N_D) for 3C-SiC.

3. Processing Technology for 3C-SiC

3.1. Schottky Contact

One of the main challenges in the processing of electronic devices based on 3C-SiC is the achievement of good quality rectifying contacts, i.e., with almost ideal characteristics and reasonably low leakage current. Several works investigated the properties of Schottky contacts on n-type 3C-SiC over the last three decades. In particular, as summarized in Table 2, most of these works have been performed on 3C-SiC layers grown on Si substrates, using high work-function Schottky contact metals (e.g., Au or Pt). However, the experimental values of the Schottky Barrier Height (SBH), as determined by I-V or C-V measurements, typically lie below 1 eV, i.e., which are much lower than the theoretical predictions of the Schottky–Mott theory.

Eriksson et al. [19] demonstrated the key role of the material quality on the properties of the metal/3C-SiC contacts, showing that double position boundaries (DPB) in 3C-SiC layers grown onto on-axis 4H-SiC can be "killer defects" in large area devices that compromise the functionality of the rectifying barrier [20]. In this work, a novel approach based on Conductive Atomic Force Microscopy (C-AFM) was proposed to characterize Schottky barriers on 3C-SiC in small area devices, establishing a direct relation between the electrical properties of the barrier and the contact area. In particular, reducing the size of the contact resulted in a drastic increase in the measured Au/3C-SiC barrier height, until reaching a value of 1.39 eV for a diode radius of 5 µm, thus demonstrating that the poor rectifying behaviour was due to the high defects density in the material [19].

More recently, using a similar nanoscale approach on 3C-SiC layers grown on Si, Giannazzo et al. [21] confirmed that the device yield, defined as the fraction of diodes with a leakage current lower than 10 μA/cm^2 (see Figure 3a,b) increases with decreasing the device area. Moreover, this work better clarified the role of specific defects by direct probing of the 3C-SiC surface by C-AFM (see Figure 3c–e). In particular, these measurements showed that antiphase boundaries (APBs) are the main defects responsible for reverse leakage current, while both APBs and stacking faults (SFs) worked as preferential current paths under forward bias of the contact.

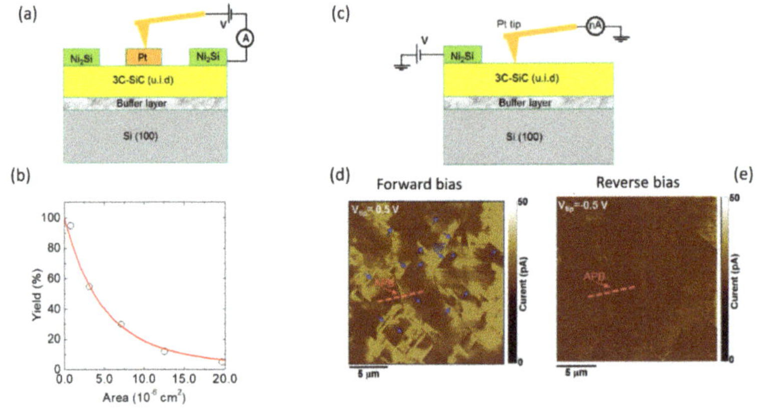

Figure 3. (**a**) Schematic of the C-AFM set-up to probe Pt/3C-SiC Schottky diodes of different areas. (**b**) Percentage of the diodes (yield) with a reverse leakage lower than 10 µA cm^{-2}, as a function of diode area. (**c**) Schematic of the C-AFM set-up to probe the 3C-SiC surface and current maps acquired under forward bias (**d**) and reverse bias (**e**). Adapted with permission from Ref. [21]. Copyright © 2021 Wiley VCH.

Table 2. Collection of literature results on Schottky contacts on 3C-SiC materials.

Metal	3C-SiC Orientation	Growing Substrate	Schottky Barrier Height (eV)	Ideality Factor	Extraction Method	Ref.
Au	100	Si	1.15	N.A.	C-V	[22]
Au	100	Si	1.2	1.5	C-V	[23]
Au	111, 100	Si	1.0–1.6	N.A.	C-V	[24]
Pt			1.3–1.8			
Pt	100	Si	0.95 (as dep) −1.35 (800 °C)	N.A.	C-V	[25]
Pd	100	Si	0.92, 0.95	N.A.	C-V, XPS	[26]
Au			0.87, 0.78			
Co			0.73, 0.69			
Au	100	Si	0.47–0.69	1.58–2.30	I-V	[27]
Pd	100	Si	0.42–0.60	3.02–5.28	I-V	[28]
Ti	100	3C-SiC	0.4, N.A.	N.A.	I-V, C-V	[29]
Au			0.67, 0.65			
Ni			0.56, 0.54			
Au	111	4H-SiC	0.7, 1.39	>2	I-V, I-V by C-AFM	[19]
Pt	100	3C-SiC	0.77 (as dep) −1.12 (500 °C)	N.A.		[30]
Au	111	4H-SiC	0.73–0.76	N.A.	I-V by C-AFM	[31]

Clearly, all these results indicate that a significant improvement of the material quality (namely, a reduction of specific defects' density) remains the only possible route for the achievement of operational Schottky contacts on 3C-SiC materials suitable for power electronics applications.

3.2. Ion Implantation and Activation

High impurity doping is necessary for low ohmic contact and sheet resistance in 3C-SiC power devices. The most commonly used dopants for 3C-SiC are nitrogen or phosphorus for n-type, and mainly aluminium for p-type.

The low diffusivity of typical dopants in SiC below 1800 °C [32] means that highly doped selective regions of SiC power devices are often achieved by ion implantation. As implanted dopant species are nearly always interstitial (not chemically bonded), they are therefore electrically inert. Therefore, an extra post implant annealing (PIA) step is typically deployed to repair the lattice damage and place the implanted dopants into their correct substitutional positions. This is referred to as 'activation'. Extremely high temperatures are required for the SiC PIA; above 1400 °C [33,34] is common for n-type SiC and higher still (>1600 °C) for p-type [35–37]. The higher p-type PIA temperature is required because acceptors sit deeper in the band gap than donors, and are consequently more challenging to activate. Regarding 3C-SiC, the most common form is grown heteroepitaxially on Si. As a consequence, these activation annealing temperatures are often limited to 1412 °C (Si melting point). Performing the ion implantation at a higher temperature helps to reduce the induced lattice damage; thus, it is often applied for high dose implantations. Since the ion implantation induced lattice damage increases with the number of dopants per unit volume (namely the dose), hot implants are almost mandatory when the implant concentration goes above 10^{19} cm^{-3} [38].

High temperature PIA also causes a rough semiconductor surface, which is enhanced within implanted regions and can degrade the performance of critical interfaces such as Schottky contacts and MOSFET channels [39–41]. A graphite capping layer, demonstrated to be effective up to 1800 °C [35], is often utilised to protect the SiC surface during the PIA and reduce the resulting roughness. Comparing the few examples in the literature, n-type implanted 3C-SiC have been extensively studied for varying annealing conditions (1150 °C to 1400 °C) both with a graphite capping layer [42] and without [43,44]. It was shown that there was little advantage demonstrated when using a graphite cap, likely due to the annealing temperature (below 1400 °C due to the Si substrate) not being high enough to roughen the surface. In [45], it was shown that by combing the use of hot implant and pulsed excimer laser processing, which only anneals the surface region, 3C-SiC crystal damage due to implantation can be effectively repaired without degrading the surface morphology (energy density 0.2444 J/cm^2 at 10 Hz), thus providing an alternative solution that allows high temperature PIA to be conducted on Si substrates.

Despite resulting in a rougher surface, a higher temperature is preferred in favour of a higher dopant activation rate. Attributed to a smaller band gap, thus a shallower donor level (55 meV), the activation of n-type dopants in 3C-SiC is easier than in 4H-SiC (80–130 meV) [46]. Studies on n-type 3C-SiC suggest that nitrogen has advantages over phosphorous for use as an n-type dopant, with both fewer defects and lower resistivity achieved [42]. Compared with the N saturation density in 4H-SiC (around 5×10^{19} cm^{-3}) [47], the level in 3C-SiC turns out to be similar at around 7×10^{19} cm^{-3} [48]. With the valence band aligned to other polytypes, the deep acceptor level issue still exists for 3C-SiC. Adding to the limited processing temperature, p-type implant and activation has long been an issue for 3C-SiC-on-Si [38,49]. In recent years, the developments on free standing 3C-SiC materials [50,51] make PIA temperatures above 1400 °C possible, thus facilitating a significant step forward in 3C-SiC power device fabrication. However, the knowledge of p-type 3C-SiC ion implantation and activation is very limited and requires further investigation. Table 3 summarises some past results published on the ion implantation and activation of dopants in 3C-SiC.

Table 3. A summary of literature data on the ion implantation and activation of 3C-SiC.

Material	Implantation	PIA	Activation Rate	Ref.
		N-Type		
2×10^{17} cm^{-3} p-type 3C-SiC(100)/Si	RT [1], N, peak $5 \times 10^{19}/5 \times 10^{20}$ cm^{-3}	None	0.44%/0.55%	[52]
	400 °C, N, peak 5×10^{19} cm^{-3}		1.35%	
	800 °C, N, peak $5 \times 10^{19}/5 \times 10^{20}$cm^{-3}		15%/50.8%	
1×10^{18} cm^{-3} p-type 3C-SiC(100)/Si	800 °C, N, peak 5×10^{19} cm^{-3}	None	12.4%	[53]
	900 °C, N, peak 5×10^{19} cm^{-3}		14.8%	
	1000 °C, N, peak 5×10^{19} cm^{-3}		18.4%	
	1100 °C, N, peak 5×10^{19} cm^{-3}		36.0%	
	1200 °C, N, peak 5×10^{19} cm^{-3}		52.2%	
1×10^{16} cm^{-3} p-type 3C-SiC(100)	RT, N, peak 1×10^{20} cm^{-3}	10 min in Ar at 1500 °C	68%	[49]
1×10^{16} cm^{-3} p-type 3C-SiC(100)	RT, N, peak 6×10^{19} cm^{-3}	10 min in Ar at 1400 °C	80%	[54]
$<1 \times 10^{16}$ cm^{-3} n-type 3C-SiC(100)/Si	RT, N, peak 5×10^{20} cm^{-3}	1 h in Ar at 1150 °C	6.5%	[46]
		1 h in Ar at 1350 °C	13%	
$<1 \times 10^{16}$ cm^{-3} n-type 3C-SiC(100)/Si	RT, N, peak 5×10^{19} cm^{-3}	1 h in Ar at 1150 °C	40%	[44]
		1 h in Ar at 1350 °C	57%	
		1 h in Ar at 1400 °C	100%	
$<1 \times 10^{16}$ cm^{-3} n-type 3C-SiC(100/Si	RT, N, peak $5 \times 10^{19}/5 \times 10^{20}$ cm^{-3}	1 h in Ar at 1350 °C	60%/17%	[55]
$<1 \times 10^{16}$ cm^{-3} n-type 3C-SiC(100/Si	RT, N, peak $1.5 \times 10^{19}/6 \times 10^{20}$ cm^{-3}	1 h in Ar at 1375 °C	100%/12%	[49]
		P-type		
2.8×10^{16} cm^{-3} n-type 3C-SiC(100)/Si	RT and 850 °C, Al and B, peak 5×10^{19}–1×10^{20} cm^{-3}	10 min in N_2 at 1200 °C	Too low, n-type behaviour	[50]
$<1 \times 10^{16}$ cm^{-3} n-type 3C-SiC(100)/Si	500 °C, Al, peak 1×10^{20} cm^{-3}	317–546 h in Ar at 1300 °C	Weak p-type behaviour	[38]

[1] Room temperature.

3.3. Ohmic Contact

Due to the requirement of an extra PIA process, achieving ohmic contacts on implanted regions is more difficult than on epilayers. As is the case in 4H-SiC [56], this is particularly true for p-type 3C-SiC because the acceptor levels are deeper, as previously mentioned. Attributed to a lower conduction band edge (3.8 eV from vacuum level), the theoretical SBH between 3C-SiC and commonly used metals is 0.9 eV lower than for 4H-SiC. This is convenient for n-type ohmic contact fabrication, while p-type remains as challenging as in other polytypes. Most work on SiC ohmic contacts is divided into three topics, namely surface preparation, contact metal, and post metallisation annealing (PMA).

The 3C-SiC epilayer surface roughness can vary significantly, from as low as 1 nm depending on the growth technique [43] to high values reaching tens of nm [57]. To achieve a relatively smooth semiconductor surface for ohmic contact fabrication, chemical mechanical polishing (CMP) is often used prior to any further processing. Noh et al. [58] show that the RMS surface roughness reduced from ≈20 nm to ≈7.5 nm. Consequently, the ohmic contact resistivity ρ_c was reduced by an order of magnitude, from 8.6×10^{-1} Ωcm^2 to 2.8×10^{-2} Ωcm^2. As alluded to previously, practical device fabrication requires a high temperature (above 1400 °C) PIA treatment, which has been shown to degrade the surface following initial CMP. In [43], a detailed discussion was reported around the PIA effects on 3C-SiC surface morphology and its correlation to the resulting ρ_c values. It was communicated that although severe damage to the surface can limit performance, the ρ_c value will not be seriously affected given that the surface roughness value remains below 10 nm.

Many metals or metal stacks, including Al [54,59–61], Ti [54,59–61], Ni [37,54,57,58,60–63], Ni/Ti [43,55,61], Au/Ti [61], Pt [63], W [37], and TiW [64], have been analysed for 3C-SiC n-type ohmic contact fabrication. It was observed that Al contacts typically display the lowest ρ_c, which was explained by the near-zero SBH between Al and 3C-SiC (~0 eV) compared to Ti (0.4 eV) and Ni (0.55 eV) [54]. Nonetheless, both Ti and Al are readily oxidised in air, with Al characterised by a melting point below 600 °C. Conversely, Ni demonstrates a slow rate of oxidation at room temperature combined with a very high melting point. Although Ni reacts with SiC at temperatures higher than 500 °C, the Ni silicide microstructure helps to reduce the SBH. This in turn leads to a lower ρ_c. Consequently, Ni is the most commonly utilised metal contact to n-type SiC.

The effects of PMA on ohmic contacts fabricated on n-type implanted 3C-SiC (Figure 4a) shows a continuous reduction of contact resistivity with increasing annealing temperature up to 1000 °C, above which the resistance increases. Details of the silicide formation are shown by XRD analysis in Figure 4b. It can be inferred that between 500 °C and 600 °C, a coexistence of Ni_2Si (121) and $Ni_{31}Si_{12}$ (300) is present. The $Ni_{31}Si_{12}$ (300) peak gradually diminishes at higher temperature, while the Ni_2Si (002) becomes prominent and enhances continuously to temperatures reaching 1100 °C. Noting that Ni_2Si (121) is readily formed at 600 °C, with no other noticeable phases above that temperature, the Ni_2Si (002) enhanced phase could explain the contact resistance reduction from 800 °C to 1000 °C. It is worth mentioning that, due to the very low SBH of highly doped n-type 3C-SiC/metal interface, as-deposited ohmic contacts can be obtained without PMA processing [59,65]. This makes it possible to integrate SiC transistor technologies with other low temperature technologies, such as atomic layer deposited high k dielectrics (e.g., HfO_2 or Al_2O_3) with relatively low growth temperatures and classic wafer bonded or heterojunction devices.

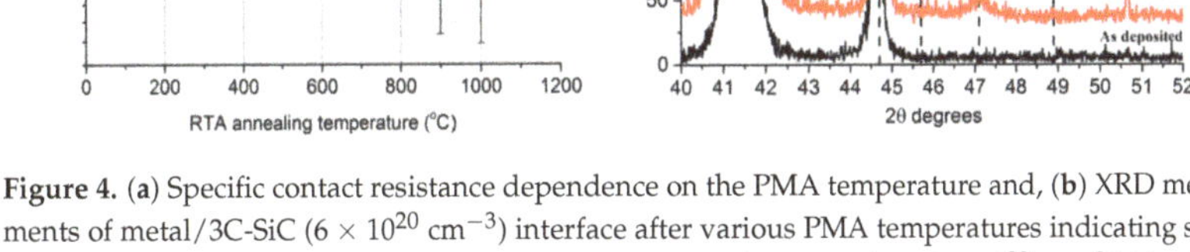

Figure 4. (**a**) Specific contact resistance dependence on the PMA temperature and, (**b**) XRD measurements of metal/3C-SiC (6×10^{20} cm^{-3}) interface after various PMA temperatures indicating silicide formation. Contact was fabricated by depositing (Ti30 nm/Ni100 nm) on 5×10^{20} cm^{-3} N implanted 3C-SiC.

Compared to n-type 3C-SiC, even less is known about p-type 3C-SiC ohmic contacts. As with 4H-SiC, Al based alloys are most commonly used for p-type ohmic contact since very often Al is also the doping species. A Ti interlayer is often applied not only to improve the adhesion, but the TiC product after PMA also helps to reduce the contact resistance [36,66].

Among the very limited data, the lowest specific contact resistances (10^{-5}–10^{-4} Ωcm^2) are obtained from trials made on p-type epilayers [66,67], which eliminates the issue of acceptor activation. However, when fabricating power devices such as MOSFETs, it is crucial to obtain ohmic contacts on selective highly doped, mostly implanted p+ regions. It is reported in [38] that, on 3C-SiC with a Si substrate, a hot implant (500 °C) with high Al concentration (1×10^{20} cm^{-3}) together with very long duration (>300 h) PIA at 1300 °C had to be performed to achieve ohmic contacts, and even so, the resultant ρ_c was still high, around the 10^{-2} Ωcm^2 level. More recently, p-type ohmic contacts (Figure 5a) on Al hot implanted (600 °C, 1×10^{20} cm^{-3}) free standing 3C-SiC have been reported. By increasing the PIA temperature treatment to 1700 °C, a dramatic reduction in annealing time was required—down to 2 h. Even though the ρ_c value is still relatively high ~10^{-3} Ωcm^2, it is promising since the contacts were fabricated on a very rough surface (Figure 5b), which can be further improved either by optimising the 3C-SiC growth process or additional polishing treatments. Table 4 provides a survey of the literature results for ohmic contact processing on 3C-SiC, mostly n-type.

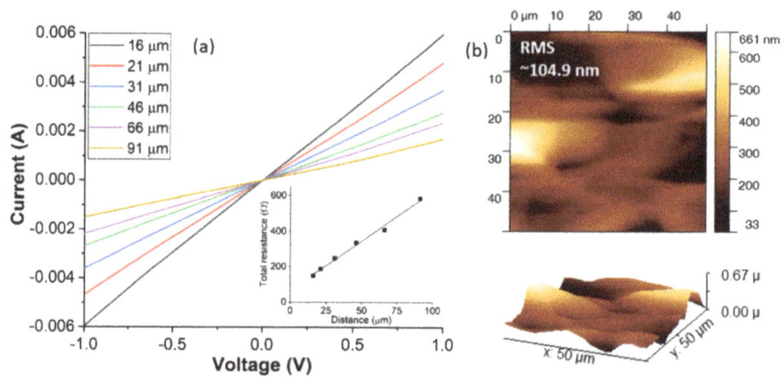

Figure 5. (**a**) I-V characteristic ohmic contact (Ni/Al/Ti) fabricated on Al implanted free standing 3C-Si and, (**b**) surface morphology of the free standing 3C-SiC by AFM.

Table 4. A summary of literature data on the fabrication of 3C-SiC ohmic contact.

Contact	Doping (cm^{-3})	PMA Conditions	ρ_c (Ωcm^2)	Ref.
		N-Type		
Al	5×10^{18} N implanted	As-deposited	1×10^{-4}	[59]
	3×10^{19} N implanted		6×10^{-5}	
	1×10^{20} N implanted		5×10^{-5}	
	3×10^{20} N implanted		1.3×10^{-5}	
	6×10^{19} N implanted	As-deposited/500 °C	$5 \times 10^{-7}/6 \times 10^{-5}$	[54]
	6×10^{18} N implanted	300 °C	5×10^{-7}	[61]
	1×10^{17} N doped epi	As deposited/500 °C	$2 \times 10^{-4}/1 \times 10^{-4}$	[37]
Ti	5×10^{18} N implanted	As-deposited	7×10^{-5}	[59]
	3×10^{19} N implanted		4×10^{-5}	
	1×10^{20} N implanted		2×10^{-5}	
	3×10^{20} N implanted		1.5×10^{-5}	
	6×10^{19} N implanted	As-deposited/500 °C	$5 \times 10^{-6}/6 \times 10^{-5}$	[54]
Ni	Not known, N doped epi	1000°C	3.7×10^{-4}	[58]
	6×10^{19} N implanted	As-deposited/500 °C	$2 \times 10^{-5}/5 \times 10^{-6}$	[54]
	3×10^{19} N doped epi	950 °C	1.2×10^{-5}	[61]
	1×10^{17} N doped epi	As-deposited/500 °C	$5 \times 10^{-4}/5 \times 10^{-5}$	[37]

Table 4. *Cont.*

Contact	Doping (cm^{-3})	PMA Conditions	ρ_c (Ωcm^2)	Ref.
	1×10^{20} P implanted	1000 °C	1.4×10^{-5}	[61]
	Not known, poly crystal epi	As-deposited	1.6×10^{-6}	[65]
	5×10^{17} N doped epi	950 °C	1.5×10^{-5}	[57]
	1×10^{17} N doped epi	950 °C	3.7×10^{-3}	[66]
	5×10^{19} N implanted	As-deposited	7×10^{-4}	[65]
	5×10^{20} N implanted		3×10^{-5}	
	1×10^{19} N implanted	1000 °C	2×10^{-4}	
	5×10^{19} N implanted		4×10^{-5}	
Ni/Ti	5×10^{20} N implanted		9×10^{-6}	
	5×10^{20} N implanted	1000 °C	8×10^{-6}	[43]
	5×10^{20} P implanted		2×10^{-5}	
	$>10^{20}$ N implanted	1050 °C	2×10^{-5}	[61]
	5×10^{19} N implanted	1000 °C	3.2×10^{-6}	[55]
Au/Ti	3×10^{20} N implanted	600 °C	1.2×10^{-5}	[61]
Pt	Not know, poly crystal N doped epi	As-deposited	1.2×10^{-5}	[63]
W	1×10^{17} N doped epi	As-deposited/500 °C	$2 \times 10^{-3}/2 \times 10^{-3}$	[37]
TiW	Not know, N doped epi	1000 °C	4.6×10^{-4}	[64]
	4×10^{19} N/P implanted	As-deposited	ohmic	[68]
		P-type		
Al	1.33×10^{17} Al doped epi	710 °C	1.4×10^{-2}	[67]
Al/Poly			3.5×10^{-4}	
Ni/Al/Ti	5×10^{19} Al doped epi	950 °C	1.8×10^{-5}	[66]
	1×10^{20} Al implanted	1000 °C	10^{-2}	[38]
	1×10^{20} Al implanted	1000 °C	10^{-3}	This work

3.4. MOS Processing

Given the superior electrical performance of SiC and its capacity to be thermally oxidised, it is not surprising that there are copious amounts of SiC MOS devices being demonstrated. The commercialised 4H-SiC polytype is naturally the most frequently reported. Numerous literature reports suggest that MOS interface traps are similar in nature for all SiC polytypes [69]. Therefore, studies relating to the 4H-SiC/SiO_2 interface provide insightful information with respect to the equivalent 3C-SiC system.

In [70], reporting around the possible origins of interface traps identified two primary sources; namely, carbon and oxide defects that accumulate at the MOS interface during the oxidation process. The oxide defect-induced traps (also known as "near-interface traps") have much smaller time constants compared to the carbon-clusters. Therefore, oxide defect-induced traps are also known as "fast traps" while the latter are coined "slow traps". A graphic illustration of the carbon cluster model is shown in Figure 6, including the corresponding specified energy levels of the traps. Figure 6 shows the 4H-SiC conduction band edge is overwhelmingly impacted by π-bonded clusters and carbon near-interface traps, with the latter being most dominant. Both trap forms are acceptor-like, therefore negatively charged when occupied, which explains the positive flat band voltage (V_{fb}) typically detected with respect to 4H-SiC MOS devices. In contrast, the 3C-SiC conduction band is devoid of near-interface traps due to a narrower band gap. However, 3C-SiC is still negatively-impacted by π-bonded carbon clusters. These (carbon clusters near the 3C-SiC conduction band edge) defects are positively charged if occupied as they are donorlike, resulting in a V_{fb} that is negative. Dangling bonds augment the interface defectiveness but are negligible secondary concerns compared to the aforementioned carbon clusters. Consequently, hydrogen annealing is not as effective for SiC when compared to Si. Alternative methods have been demonstrated for high-quality SiC/SiO_2 interface optimisation.

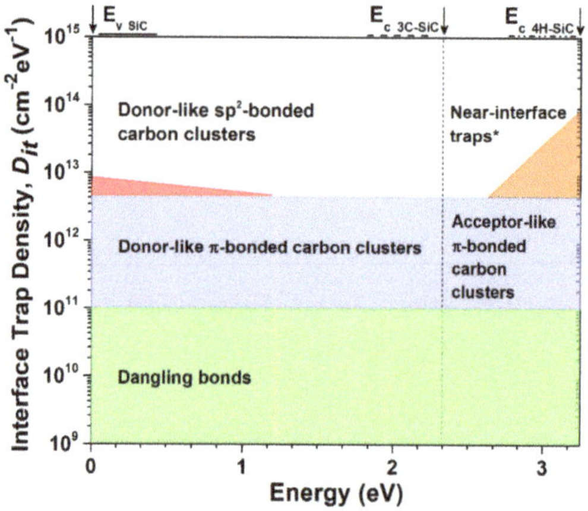

Figure 6. Schematic representation of the "carbon cluster model". Adapted from Ref. [71] with permission from the author (R. Esteve).

Reports focussing on the improvement of the SiC/SiO_2 interface are mainly related to the topic of post oxidation annealing (POA). Former research literature revealed the advantages of including hydrogenation processes either during the (gate) oxidation process or subsequently, via the POA. This has the effect of decreasing the interface trap density (D_{it}) in addition to reducing positive fixed charge (Q_{fc}) [68,72]. Consequently, wet oxidation in conjunction with POA is often utilised for 3C-SiC MOSFET fabrication [73,74]. Regarding the nitridation step, extra deep interface traps revealed by double peak conductance spectra were observed from fabricated MOS capacitors via direct N_2O oxidation and pure O_2 oxidation methods on nitrogen implanted films [72,75].

Figure 7 shows the lateral MOSFET transfer curves on Al implanted 3C-SiC/Si substrates with the gate oxide grown in different atmospheres, but all at the same temperature of 1300 °C [76]. Due to varying oxide thicknesses, for direct comparison the gate field instead of gate voltage is plotted on the x-axis. It can be inferred that the dry oxidized device demonstrates a normally on characteristic with a gate threshold voltage approaching zero. This is in agreement with the previously introduced Carbon Cluster Model, stating that only donor-like states occupy the $3C\text{-}SiC/SiO_2$ interface. Since these states are positively charged when vacant, these donor-like states may be responsible for the inherent negative threshold. The nitrided sample is even further shifted in the negative threshold direction due to the counter doping channel effect [77]. The wet oxidized sample has the most positive gate threshold. A combination of N_2O nitridation and POA (wet) yielded an intermediate threshold field of around −2MV/cm. Clearly, the wet oxidation was successful in shifting the device threshold in a more positive manner, either by forcing a reduction in positive fixed oxide charges or via compensating them with additional negative charge. Both wet POA and oxidized processed devices have a peak field-effect mobility (μ_{FE}) value in the region of 60 cm^2/Vs, which is the lowest compared with the dry oxidized sample (70 cm^2/Vs) and the N_2O nitrided sample (90 cm^2/Vs).

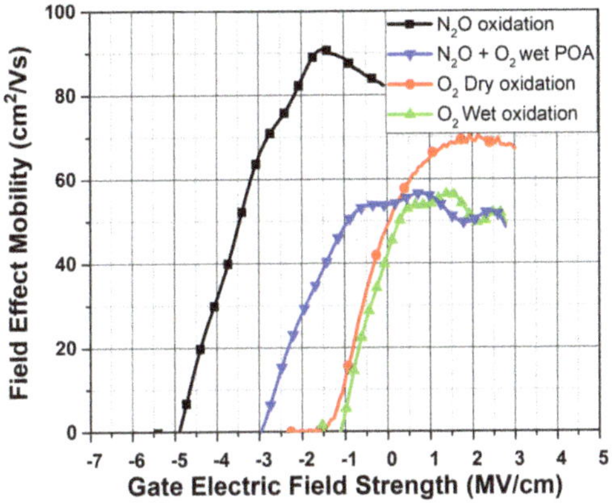

Figure 7. Transfer curves of 1300 °C oxidized lateral MOSFET with various conditions.

As mentioned previously, the reliability of the 3C-SiC MOS system is particularly interesting, yet there has been relatively little study of this topic, mainly due to a shortage of non-defective 3C-SiC material. Figure 8 shows the critical strength (E_c) of SiO_2 layers grown on 3C-SiC/Si substrates in different atmospheres at 1300 °C. As can be seen, by using combined dry O_2 gate oxidation with an N_2O POA process, the noise level was greatly reduced and the critical electric field strength was able to be kept at around 8 MV/cm, the highest value observed.

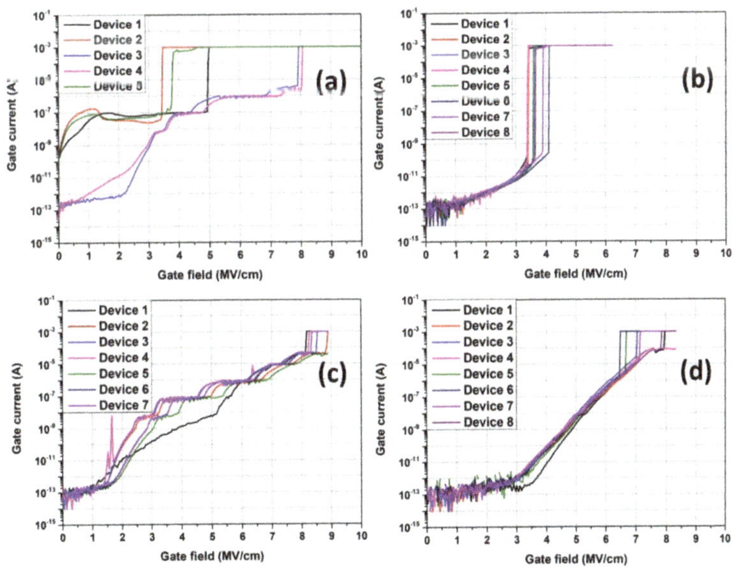

Figure 8. Dielectric breakdown curve of gate oxides fabricated by (**a**) 60 min 1300 °C O_2 dry oxidation, (**b**) 15 min 1300 °C O_2 wet oxidation, (**c**) 120 min 1300 °C N_2O dry oxidation and (**d**) 30 min 1300 °C O_2 dry oxidation + 90 min 1300 °C N_2O POA.

Recently the reliability of 3C-SiC MOS capacitors (dry oxidised and N_2O POA at 1300 °C) has been examined at room temperature by using both v-ramp and time-dependent dielectric breakdown (TDDB) analysis. As can be seen in Figure 9a, the accumulated total failure percentage increases steadily until around 8.5 MV/cm, beyond which the failure number sharply increases to 100%. The failures at lower fields are most likely induced by crystal deficiencies in the 3C-SiC substrate that alter localised material properties. High field (>8.5 MV/cm) failures are characterized by either F-N tunnelling, observed via the increased leakage current, or electron impact ionization energy being reached within the oxide due to elevated electric fields. TDDB analysis is conducted at electric field values of 6, 7.5, 8.5, and 9 MV/cm. The Weibull distributions are displayed in Figure 9b. Even at high fields beyond 8.5 MV/cm, the slope values remain low in the region of ~1, an order of magnitude lower than reported values for 4H-SiC [78], suggesting extrinsic defects are still the dominant failure mechanism.

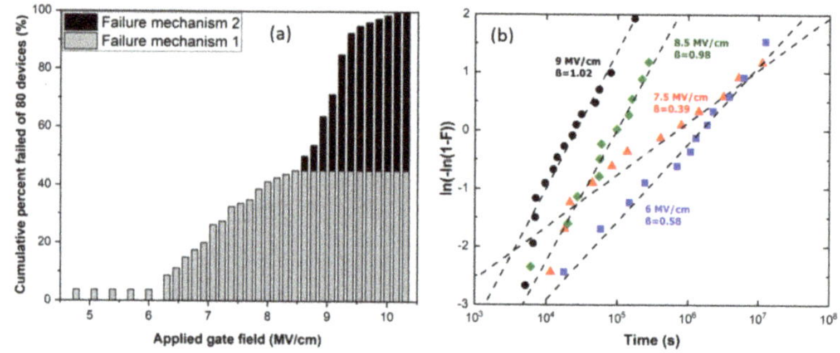

Figure 9. (**a**) Failure distribution of 3C-SiC MOS capacitors in the electric field range of 4.5–10.5 MV/cm, and (**b**) Weibull distributions of device failures at various electric fields.

Besides the application on Schottky contacts described in Section 3.1, nanoscale resolution current mapping by C-AFM can also be a powerful analysis technique for investigation of the dielectric breakdown behaviour of thin insulators. In fact, this method was recently employed by Fiorenza et al. [79] in order to explain the reasons behind the premature breakdown of thermal oxide (SiO_2) grown on 3C-SiC typically observed in MOS capacitors, by stressing the oxide through the application of a bias to the C-AFM tip corresponding to an electric field of 8 MV/cm (see schematic set-up in Figure 10a).

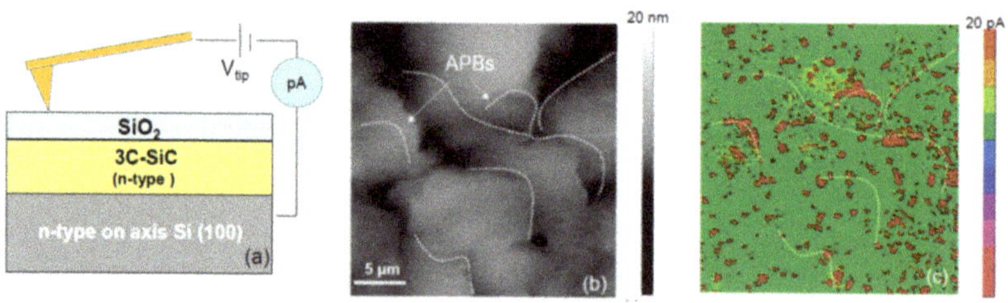

Figure 10. (**a**) C-AFM set-up adopted for the electrical characterization of the SiO_2/3C-SiC system; (**b**) AFM morphology and (**c**) C-AFM current map acquired under the application of an electric field of 8 MV/cm to the tip. Adapted with permission from ref. [79]. Copyright © 2021 Elsevier Ltd.

The C-AFM current map and corresponding AFM surface morphology acquired on the SiO_2/SiC system are reported in Figure 10b,c, respectively. The C-AFM current map in Figure 10c reflects the breakdown distribution of an array of tip/oxide nano-MOS capacitors. Hence, the features on the surface morphology (Figure 10b) could be correlated with the position of the breakdown spots (Figure 10c), which are not randomly distributed, but preferentially appear along the APBs (dashed line in Figure 10b). Here, the straight line conductive aspects associated with SFs on the exposed 3C-SiC surface (see Figure 10c) were not visible in the presence of a thermal oxide. Based on this analysis, the premature dielectric breakdown observed in MOS capacitors could be attributed to the presence of positively charged APBs, causing an electron injection enhancement from the 3C-SiC into the SiO_2.

Table 5 is a list of recent work performed on the study of 3C-SiC MOS interface traps.

Table 5. A summary of literature data on the processing of 3C-SiC MOS interface and relevant information on fixed charges (Q_{fc}), interface trap density (D_{it}), and oxide critical field (E_c), unless specified, the 3C-SiC materials listed are epilayers.

Oxidation Substrate	Oxidation	POA	Q_{fc} (cm^{-2})	D_{it} ($cm^{-2}\ eV^{-1}$)	E_c (MV/cm)	Ref.
n-type 3C-SiC	NO, 1175 °C, 4 h	-	-	~10^{11}	-	[68]
Al implanted 3C-SiC	Dry O_2, 1100 °C, 1.5 h	Wet O_2, 950 °C, 3 h	-	5×10^{12}–1×10^{13}	-	[74]
n-type 3C-SiC	Dry O_2, 1120 °C, 0.5 h	Ar, 1120 °C, 1 h	-	~10^{12}	-	[75]
n-type 3C-SiC	PECVD (SiH_4 + N_2O)	N_2O, 1100 °C, 3 h	2.01×10^{12}	~10^{12}	8.2	[80]
		Wet O_2, 950 °C, 3 h	1.7×10^{11}	~2×10^{12}	9.1	
		Dry O_2, 950 °C, 3 h	1.76×10^{11}	~2×10^{13}	5.9	
		N_2, 1100 °C, 3 h	4.65×10^{12}	~7×10^{12}	6.3	
		N_2, 950 °C, 3 h	2.63×10^{12}	~2×10^{13}	6.2	
n-type 3C-SiC	NO, 1185 °C, 2 h	-	-	~10^{12}	-	[77]
	N_2O, 1185 °C, 1 h	-	-	~8×10^{11}	-	
n-type 3C-SiC	Dry O_2, 1100 °C, 4 h	-	9.3×10^{12}	4.27×10^{13}	-	[73]
	Dry O_2, 1200 °C, 1 h	-	7.1×10^{12}	6.59×10^{13}	-	
	Dry O_2, 1100 °C, 1.5 h	O_2, 950 °C, 3 h	1.3×10^{12}	7.1×10^{12}	-	
	Dry O_2, 1100 °C, 1.5 h	Wet O_2, 950 °C, 3 h	0.9×10^{12}	5.2×10^{12}	-	
	N_2O, 1200 °C, 2 h	-	3.0×10^{12}	1.15×10^{13}	-	
	N_2O, 1250 °C, 2 h		3.1×10^{12}	9.1×10^{12}	-	
	N_2O, 1250 °C, 2 h	Wet O_2, 950 °C, 3 h	1.6×10^{12}	9.4×10^{12}	-	
n-type 3C-SiC	PECVD (SiH_4 + N_2O)	N_2, 950 °C	5.7–7×10^{12}	5×10^{11}–7×10^{12}	-	[81]
n-type 3C-SiC	Dry O_2, 1200 °C	-	1.1×10^{12}	~10^{12}	-	[82]
	Dry O_2, 1300 °C		1.1×10^{12}			
	Dry O_2, 1400 °C		4.1×10^{12}			

4. 3C-SiC Device Prototypes

4.1. Schottky Diode

The study of Schottky and p-n junction diode behaviour on thin film CVD 3C-SiC dates back to the 1980s [83]. Much of this early work was conducted on thin films deposited on silicon and 6H-SiC. The initial studies were concerned with the surface science of fabricating appropriate metal contacts. These diodes demonstrated the first reported 3C-SiC rectification behaviour of up to 200 V, with leakage currents ranging from between 10^{-4}–1 A/cm^2 [84,85]. Vertical heterojunction Schottky diodes based on platinum (Pt) contacts showed a blocking voltage of 85 V with a low forward voltage drop of ~0.85 V [86]. Gold contacts to 3C-SiC for Schottky diode applications displayed a variance of the barrier height with contact area [19]. This can be explained by the defect density inherent within the starting material. More recent Schottky diode reports suggest that the leakage current is not dominated by SF density, as the leakage current had a greater dependency on the barrier height [87]. Barrier height nonuniformities of the Schottky barrier have been observed on lateral 3C-SiC-on-Si diodes, implicating complex trapping/de-trapping phenomena observed within the material [88]. The information acquired has led to validated technology computer aided design (TCAD) models for accurate 3C-SiC device simulation [9].

4.2. PiN Diode

Attributed to its smaller bandgap, 3C-SiC has a lower p-n junction built-in potential (≈1.75 V) than 4H-SiC (≈3 V). In [9] it is shown that, up to 4.5 kV blocking voltage, the forward voltage drop at 250 A/cm^2 remains lower for 3C-SiC than 4H-SiC in PiN diode applications. Until recently, however, fabricating 3C-SiC PiN diodes has been difficult, not only because of the high defect density within 3C-SiC epilayers caused by the lattice mismatch with Si [21], but also due to the lateral nature of structures necessary to avoid the 3C-SiC/Si heterojunction. While there are several reports on achieving n-type conduction in 3C-SiC epi/implanted layers [42,48,66], and p-type conduction in Al doped epilayers [66,89], it remains an obstacle for p-type implanted layers. This is mainly due to the post implantation anneal temperature, which was limited to the Si melting point, 1414 °C, which is not sufficient to activate the deep level Al dopants, even if hot implantation was applied.

Low voltage lateral p-n junction diodes were previously demonstrated via the formation of implanted n+ regions in p-type doped 3C-SiC epilayers grown on Si substrates [90,91]. However, to make the most of its benefits in power applications, a vertical structure is necessary. 3C-SiC growth methods have improved in recent years [6,92,93], and bulk 3C-SiC are now available [51]; thus, a higher annealing temperature can now be applied. Vertical PiN diodes were fabricated on free standing 3C-SiC material by implanting Al in n-type doped epilayer and the forward current density is shown in Figure 11a. The built-in potential of the fabricated PiN diode is around 2 V, slightly higher than the theoretical value 1.75 V [9], but it is still much lower than the typical >3 V for 4H-SiC [94–96]. The forward current density goes above 1000 A/cm^2 at 2.7 V, and the lowest differential resistance is estimated to be 0.5 $m\Omega cm^2$. The device on–off ratio at ±5 V is as high as 10^9, as shown in Figure 11b, and a blocking voltage above 100 V is achieved (Figure 11c). An observation to note with respect to bipolar PiN diode I-V characterisation is that no bipolar degradation has been reported in the literature with respect to 3C-SiC pn diodes. This is most likely due to the fact that attention is being placed upon more fundamental device limiting issues such as SF-induced leakage currents.

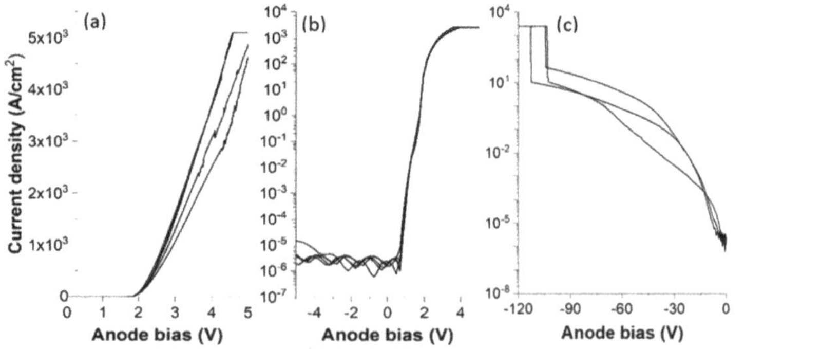

Figure 11. (**a**) Forward J-V characteristics, (**b**) on-off performance at ±5 V, and (**c**) reverse breakdown of bulk 3C-SiC PiN diodes.

4.3. MOSFET

Early 3C-SiC power devices were predominantly demonstrated via heteroepitaxial 3C-SiC grown by chemical vapour deposition (CVD) above silicon substrates in addition to free-standing wafers, provided by HOYA Advanced Semiconductor Technologies Co Ltd. [50,74,97,98]. Power devices were based on diode and MOSFET (lateral and vertical) architectures. Devices demonstrated by 3C-SiC CVD grown on undulant-silicon substrates suffered from premature breakdown voltage and high leakage currents due to APBs and SF inherent within the epitaxial layer of the device [99,100].

Typical characteristics showed that achieving breakdown voltages in excess of 600 V was challenging since the leakage current emanating from the formerly mentioned p-n junction SFs degraded performance in a terminal manner [100]. High current cellular vertical 3C-SiC MOSFETs were demonstrated by Abe et al. [74]. This device achieved an impressive 1220 A/cm^2 current density based on a single cell. This corresponds to a current carrying capability of 41–132 A for a 3 × 3 mm^2, 600 V chip. SF-induced leakage current hampered the off-state performance of this MOSFET. CVD deposited gates produced 600 V-MOSFETs with a high channel mobility of 200 cm^2/Vs [101]. The high channel mobility and low specific on-state resistance of 5–7 mΩcm^2 were brought about by a specific activation anneal of 1600 °C in argon (Ar), in order to realise a smooth 3C-SiC surface prior to deposition of the gate oxide. They used 600 V DMOSFETs to show that material quality has a strong influence on the blocking behaviour. In contrast, the on-state electrical characteristics were unaffected [102]. A 200V reduction in breakdown voltage was observed for DMOSFETs with a high crystal defect density.

Due to the lower interface trap density at the 3C-SiC/SiO$_2$ interface compared with 4H-SiC, MOSFETs are the most studied 3C-SiC devices, targeting for lower on-resistance than 4H-SiC MOSFETs in medium voltage applications (600–1200V). High field-effect mobility values were demonstrated by fabricating 3C-SiC MOSFETs with a high current density of 1220 A/cm^2 and encouraging scaling features were shown in 1 mm × 1 mm and 3 mm × 3 mm devices [74]. In addition, it is shown in [65,68] that by removing the rapid thermal anneal for the ohmic contact, the field-effect mobility can be further improved. Despite the achievements made in forward conditions, reaching blocking ability (BV) close to the theoretical values is still a challenge, mainly because of the high leakage current induced by crystal defects such as SFs [97]. By reducing stacking faults to ~90 cm^{-1}, the device blocking ability (5 × 10^{15} cm^{-3} doped drift region) can be significantly improved to 600 V [50], close to the unipolar limit. Table 6 is a summary of the recent literature results for 3C-SiC MOSFET fabrication.

Table 6. A summary of literature data on the forward and reverse performance of 3C-SiC MOSFETs.

Structure	Channel	Oxidation	POA	μ_{FE} (cm^2/V.s)	B_V (V)	Ref.
Lateral	2 × 10^{17} cm^{-3} p-type epi	Wet O$_2$, 1150 °C, 2.5 h	Ar, 1150 °C, 0.5 h + Wet O$_2$, 950 °C, 2 h	≈165	-	[73]
Lateral	1 × 10^{16} cm^{-3} p-type epi	Wet O$_2$, 1100 °C	Ar, 1150 °C, 0.5 h + Wet O$_2$, 800 °C, 0.5 h	≈229	-	[103]
Lateral	1 × 10^{18} cm^{-3} Al implanted	Dry O$_2$, 1300 °C	-	≈80	-	[65]
Vertical	1 × 10^{18} cm^{-3} Al implanted	Dry O$_2$, 1100 °C, 1.5 h	Wet O$_2$, 950 °C, 3 h	≈28	≈100	[68]
Vertical	1 × 10^{18} cm^{-3} Al implanted	Dry O$_2$, 1100 °C, 1.5 h	Wet O$_2$, 950 °C, 3 h	≈45	550–600	[50]
Vertical	Al implanted	Wet O$_2$, 1150 °C,	-	>100		[104]

5. Conclusions

This paper gave an overview of the processing technology associated with heteroepitaxial 3C-SiC-on-silicon, including the more recently available bulk 3C-SiC studies. This topic is highly relevant today since this material presents some clear advantages over its commercial WBG competitors in terms of MOS channel resistance and reliability. These factors are of the utmost importance when considering that it is the automotive sector that is driving the widespread uptake of WBG technologies. Schottky contact processing on 3C-SiC has mainly been conducted on heteroepitaxy (on-silicon) utilising high work function metals such as Au or Pt. These rectifying contacts are typically characterised by high leakage currents arising from SFs and APBs and it is clear that a step-change in material quality is needed for power device applications. To the best of the authors' knowledge, there remains no semiconductor device grade wafer supplier of bulk 3C-SiC. However, heteroepitaxial 3C-SiC-on-silicon is available up to a wafer diameter of 4 inch. The main obstacle to large diameter 3C-SiC commercialisation remains the SF density that ranges from 200–5000 cm^{-1}. Hence, the future prospects for 3C-SiC are incumbent upon reducing SFs and APBs, which remains key to realising large diameter 3C-SiC bulk wafer production. 3C-SiC-on-silicon demonstrates serious limitations when the ion implantation process is taken into consideration. Therefore, the majority of studies to date have used conventional PIA annealing up to 1400 °C (melting temperature of silicon substrate) and pulsed laser annealing. Generally, dopant activation rates are low in 3C-SiC heteroepitaxy structures, although recently more promising behaviour has been described on free standing (bulk) 3C-SiC. Most recently p-type aluminium doped 3C-SiC has been demonstrated with weak p-type behaviour. N-type ohmic contacts have been consistently achieved using metals such as Ni, Al, Ti, Au and W demonstrating specific contact resistivities as low as 5×10^{-7} Ωcm^2. The success is related to the high n-type ion implantation activation/ionisation rates accompanied by the low donor levels relative to 4H-SiC. P-type ohmicity based on metals including Al, Ni, Ti and poly-silicon have produced resistances in the region of ~10^{-5} Ωcm^2. Compared to n-type donor levels in 3C-SiC, p-type acceptor energy levels are closer to the midgap, resulting in a lower degree of acceptor ionization. Diodes based on Schottky and PiN designs have been demonstrated on 3C-SiC. The state of the art with respect to diodes are bulk PiN structures with a built-in voltage of 2V and current density of 1000 Acm^{-2} observed. The 3C-SiC MOS interface is relatively untroubled by near interface traps when compared to its 4H-SiC counterpart. This can be inferred from experimental results based on nitrogen anneals where channel mobilities approaching 100 cm^2/Vs have been observed. Again nitrogen-based thermal oxidation produced interface trap densities in the region of 10^{11} cm^{-2} eV^{-1}. A reliability analysis of the 3C-SiC MOS interface revealed high breakdown fields in the region of 8MV/cm including cumulative device failure arising primarily from 3C-SiC crystal defects (TDDB). Actual MOSFET demonstrators are plagued by high leakage currents resulting from crystal defects. Thus, 600V 3C-SiC MOSFETs that approach the theoretical unipolar limit have been demonstrated.

Author Contributions: Conceptualization, writing, review and editing, F.L., M.J., F.R.; experimental investigation, F.L., M.J., F.R., G.G., P.F.; data analysis and discussion, J.E.E., F.A.M., F.L., C.A.F., A.P.-T., P.A.M., P.F., M.J., F.R.; funding acquisition, F.L.V. All authors have read and agreed to the published version of the manuscript.

Funding: This research was funded by the European Union within the framework of the project CHALLENGE, grant number 720827.

Institutional Review Board Statement: Not applicable.

Informed Consent Statement: Not applicable.

Data Availability Statement: The data underlying this article will be shared on reasonable request from the corresponding author.

Conflicts of Interest: The authors declare no conflict of interest.

References

1. Roccaforte, F.; Greco, G.; Fiorenza, P.; Iucolano, F. An Overview of Normally-Off GaN-Based High Electron Mobility Transistors. *Materials* **2019**, *12*, 1599. [CrossRef] [PubMed]
2. Tolbert, L.M.; King, T.; Ozpineci, B.; Campbell, J.; Muralidharan, G.; Rizy, D.; Sabau, A.; Zhang, H.; Zhang, W.; Xu, Y.; et al. *Power Electronics for Distributed Energy Systems and Transmission and Distribution Applications: Assessing the Technical Needs for Utility Applications;* U.S. Department of Energy Office of Scientific and Technical Information: Washington, DC, USA, 2005. [CrossRef]
3. Yole Development. *From Technologies to Markets: Compound Semiconductor Service Compound Montior;* Quarterly Update—Q1 2020; Yole Development: Lyon, France, 2020.
4. Ueda, T. Reliability Issues in GaN and SiC Power Devices. *IEEE Int. Reliab. Phys. Symp. Proc.* **2014**, 1–6. [CrossRef]
5. Gonzalez, J.O.; Wu, R.; Jahdi, S.; Alatise, O. Performance and Reliability Review of 650 v and 900 v Silicon and SiC Devices: MOSFETs, Cascode JFETs and IGBTs. *IEEE Trans. Ind. Electron.* **2020**, *67*, 7375–7385. [CrossRef]
6. Chow, T.P.; Omura, I.; Higashiwaki, M.; Kawarada, H.; Pala, V. Smart Power Devices and ICs Using GaAs and Wide and Extreme Bandgap Semiconductors. *IEEE Trans. Electron Devices* **2017**, *64*, 856–873. [CrossRef]
7. La Via, F.; Severino, A.; Anzalone, R.; Bongiorno, C.; Litrico, G.; Mauceri, M.; Schoeler, M.; Schuh, P.; Wellmann, P. From Thin Film to Bulk 3C-SiC Growth: Understanding the Mechanism of Defects Reduction. *Mater. Sci. Semicond. Process.* **2018**, *78*, 57–68. [CrossRef]
8. Levinshein, M.; Sergey, L.; Shur, M. (Eds.) *Properties of Advanced Semiconductor Materials: GaN, AIN, InN, BN, SiC, SiGe*, 1st ed.; John Wiley & Sons, Inc.: New York, NY, USA, 2001.
9. Lebedev, A.; Lebedev, S.; Dayvdov, V.; Novikov, S.; Makarov, Y. Growth and Investigation of SiC Based Herterostructures. In Proceedings of the 2016 15th Biennial Baltic Electronics Conference (BEC), Tallinn, Estonia, 3–5 October 2016; pp. 4–5.
10. Arvanitopoulos, A.E.; Antoniou, M.; Perkins, S.; Jennings, M.; Guadas, M.B.; Gyftakis, K.N.; Lophitis, N. On the Suitability of 3C-Silicon Carbide as an Alternative to 4H-Silicon Carbide for Power Diodes. *IEEE Trans. Ind. Appl.* **2019**, *55*, 4080–4090. [CrossRef]
11. Silicon Carbide Properties. Available online: https://www.ece.rutgers.edu/~{}jzhao/SiC-properties.html (accessed on 20 May 2021).
12. Bimberg, D.; Altarelli, M.; Lipari, N.O. A Calculation of Valence Band Masses, Exciton and Acceptor Energies and the Ground State Properties of the Electron-Hole Liquid in Cubic SiC. *Solid State Commun.* **1981**, *40*, 437–440. [CrossRef]
13. Fardi, H.; Van Zeghbroeck, B. Design and Simulation of 3C-SiC Vertical Power MOSFETs. *Int. J. Electron.* **2021**, *108*, 841–857. [CrossRef]
14. Salupo, C.S.; Larkin, D.J.; Powell, J.A.; Matus, L.G. Electrical Properties of Epitaxial 3C- and 6H-SiC p-n Junction Diodes Produced Side-by-Side on 6H-SiC Substrates. *IEEE Trans. Electron Devices* **1994**, *41*, 826–835. [CrossRef]
15. Spry, D.J.; Trunek, A.J.; Neudeck, P.G. High Breakdown Field P-Type 3C-SiC Schottky Diodes Grown on Step-Free 4H-SiC Mesas. In *Silicon Carbide and Related Materials 2003;* Materials Science Forum; Trans Tech Publications Ltd.: Bäch, Switzerland, 2004; Volume 457, pp. 1061–1064. [CrossRef]
16. Tirino, L.; Weber, M.; Brennan, K.F.; Bellotti, E.; Goano, M. Temperature Dependence of the Impact Ionization Coefficients in GaAs, Cubic SiC, and Zinc-Blende GaN. *J. Appl. Phys.* **2003**, *94*, 423–430. [CrossRef]
17. Bellotti, E.; Nilsson, H.-E.; Brennan, K.F.; Ruden, P.P. Ensemble Monte Carlo Calculation of Hole Transport in Bulk 3C–SiC. *J. Appl. Phys.* **1999**, *85*, 3211–3217. [CrossRef]
18. Fardi, H.; Van Zeghbroeck, B. Breakdown Field Model for 3C-SiC Power Device Simulations. *Mater. Sci. Forum* **2018**, *924*, 617–620. [CrossRef]
19. Eriksson, J.; Weng, M.H.; Roccaforte, F.; Giannazzo, F.; Leone, S.; Raineri, V. Toward an Ideal Schottky Barrier on 3C-SiC. *Appl. Phys. Lett.* **2009**, *95*, 81907. [CrossRef]
20. Lanzia, M. (Ed.) *Conductive Atomic Force Microscopy: Applications in Nanomaterials;* Wiley VCH: Weinheim, Germany, 2017.
21. Giannazzo, F.; Greco, G.; Di Franco, S.; Fiorenza, P.; Deretzis, I.; La Magna, A.; Bongiorno, C.; Zimbone, M.; La Via, F.; Zielinski, M.; et al. Impact of Stacking Faults and Domain Boundaries on the Electronic Transport in Cubic Silicon Carbide Probed by Conductive Atomic Force Microscopy. *Adv. Electron. Mater.* **2020**, *6*, 1–8. [CrossRef]
22. Yoshida, S.; Sasaki, K.; Sakuma, E.; Misawa, S.; Gonda, S. Schottky Barrier Diodes on 3C-SiC. *Appl. Phys. Lett.* **1985**, *46*, 766–768. [CrossRef]
23. Ioannou, D.E.; Papanicolaou, N.A.; Nordquist, P.E. The Effect of Heat Treatment on Au Schottky Contacts on β-SiC. *IEEE Trans. Electron Devices* **1987**, *34*, 1694–1699. [CrossRef]
24. Fujii, Y.; Shigeta, M.; Furukawa, K.; Suzuki, A.; Nakajima, S. Dependence on the Schottky Metal and Crystal Orientation of the Schottky Diode Characteristics of B-SiC Single Crystals Grown by Chemical Vapor Deposition. *J. Appl. Phys.* **1988**, *64*, 5020–5025. [CrossRef]
25. Papanicolaou, N.A.; Christou, A.; Gipe, M.L. Pt and PtSix Schottky Contacts on N-type B-SiC. *J. Appl. Phys.* **1989**, *65*, 3526–3530. [CrossRef]
26. Waldrop, J.R.; Grant, R.W. Formation and Schottky Barrier Height of Metal Contacts to B-SiC. *Appl. Phys. Lett.* **1990**, *56*, 557–559. [CrossRef]
27. Constantinidis, G.; Kuzmic, J.; Michelakis, K.; Tsagaraki, K. Schottky Contacts on CF4/H2 Reactive Ion Etched β-SiC. *Solid State Electron.* **1998**, *42*, 253–256. [CrossRef]

28. Roy, S.; Jacob, C.; Basu, S. Current Transport Properties of Pd/3C–SiC Schottky Junctions with Planar and Vertical Structures. *Solid State Sci.* **2004**, *6*, 377–382. [CrossRef]
29. Satoh, M.; Matsuo, H. Evaluation of Schottky Barrier Height of Al, Ti, Au, and Ni Contacts to 3C-SiC. *Mater. Sci. Forum* **2006**, *527*, 923–926. [CrossRef]
30. Eriksson, J.; Roccaforte, F.; Reshanov, S.; Giannazzo, F.; Lo Nigro, R.; Raineri, V. Evolution of the Electrical Characteristics of Pt/3C-SiC Schottky Contacts upon Thermal Annealing. *AIP Conf. Proc.* **2010**, *1292*, 75–78. [CrossRef]
31. Alassaad, K.; Vivona, M.; Soulière, V.; Doisneau, B.; Cauwet, F.; Chaussende, D.; Giannazzo, F.; Roccaforte, F.; Ferro, G. Ge Mediated Surface Preparation for Twin Free 3C-SiC Nucleation and Growth on Low Off-Axis 4H-SiC Substrate. *ECS J. Solid State Sci. Technol.* **2014**, *3*, P285–P292. [CrossRef]
32. Zhe, F. *Silicon Carbide: Materials, Processing & Devices*, 1st ed.; CRC Press: Boca Raton, FL, USA, 2003.
33. Li, M.; Ahyi, A.C.; Zhu, X.; Chen, Z.; Isaacs-Smith, T.; Williams, J.R.; Crofton, J. Nickel Ohmic Contacts to N-Implanted (0001) 4H-SiC. *J. Electron. Mater.* **2010**, *39*, 540–544. [CrossRef]
34. Vivona, M.; Greco, G.; Giannazzo, F.; Lo Nigro, R.; Rascunà, S.; Saggio, M.; Roccaforte, F. Thermal Stability of the Current Transport Mechanisms in Ni-Based Ohmic Contacts on n- and p-Implanted 4H-SiC. *Semicond. Sci. Technol.* **2014**, *29*, 75018. [CrossRef]
35. Jones, K.A.; Wood, M.C.; Zheleva, T.S.; Kirchner, K.W.; Derenge, M.A.; Bolonikov, A.; Sudarshan, T.S.; Vispute, R.D.; Hullavarad, S.S.; Dhar, S. Structural and Chemical Comparison of Graphite and BN/AlN Caps Used for Annealing Ion Implanted SiC. *J. Electron. Mater.* **2008**, *37*, 917–924. [CrossRef]
36. Frazzetto, A.; Giannazzo, F.; Nigro, R.L.; Raineri, V.; Roccaforte, F. Structural and Transport Properties in Alloyed Ti/Al Ohmic Contacts Formed on p-Type Al-Implanted 4H-SiC Annealed at High Temperature. *J. Phys. D Appl. Phys.* **2011**, *44*, 255302. [CrossRef]
37. Jacob, C.; Pirouz, P.; Kuo, H.-I.; Mehregany, M. High Temperature Ohmic Contacts to 3C–Silicon Carbide Films. *Solid State Electron.* **1998**, *42*, 2329–2334. [CrossRef]
38. Nipoti, R.; Canino, M.; Zielinski, M.; Torregrosa, F.; Carnera, A. 1300 °C Annealing of 1 × 1020 °C m^{-3} Al + Ion Implanted 3C-SiC/Si. *ECS J. Solid State Sci. Technol.* **2019**, *8*, P480–P487. [CrossRef]
39. Capano, M.A.; Ryu, S.; Cooper, J.A.; Melloch, M.R.; Rottner, K.; Karlsson, S.; Nordell, N.; Powell, A.; Walker, D.E. Surface Roughening in Ion Implanted 4H-Silicon Carbide. *J. Electron. Mater.* **1999**, *28*, 214–218. [CrossRef]
40. Negoro, Y.; Katsumoto, K.; Kimoto, T.; Matsunami, H. Electronic Behaviors of High-Dose Phosphorus-Ion Implanted 4H-SiC (0001). *J. Appl. Phys.* **2004**, *96*, 224–228. [CrossRef]
41. Vassilevski, K.V.; Wright, N.G.; Nikitina, I.P.; Horsfall, A.B.; O'Neill, A.G.; Uren, M.J.; Hilton, K.P.; Masterton, A.G.; Hydes, A.J.; Johnson, C.M. Protection of Selectively Implanted and Patterned Silicon Carbide Surfaces with Graphite Capping Layer during Post-Implantation Annealing. *Semicond. Sci. Technol.* **2005**, *20*, 271–278. [CrossRef]
42. Song, X.; Biscarrat, J.; Michaud, J.-F.; Cayrel, F.; Zielinski, M.; Chassagne, T.; Portail, M.; Collard, E.; Alquier, D. Structural and Electrical Characterizations of N-Type Implanted Layers and Ohmic Contacts on 3C-SiC. *Nucl. Instrum. Methods Phys. Res. Sect. B Beam Interact. Mater. Atoms* **2011**, *269*, 2020–2025. [CrossRef]
43. Bazin, A.E.; Michaud, J.F.; Autret-Lambert, C.; Cayrel, F.; Chassagne, T.; Portail, M.; Zielinski, M.; Collard, E.; Alquier, D. Ti–Ni Ohmic Contacts on 3C–SiC Doped by Nitrogen or Phosphorus Implantation. *Mater. Sci. Eng. B* **2010**, *171*, 120–126. [CrossRef]
44. Song, X.; Bazin, A.E.; Michaud, J.F.; Cayrel, F.; Zielinski, M.; Portail, M.; Chassagne, T.; Collard, E.; Alquier, D. Electrical Characterization of Nitrogen Implanted 3C-SiC by SSRM and C TLM Measurements. *Mater. Sci. Forum* **2011**, *679*, 193–196. [CrossRef]
45. Lee, K.Y.; Huang, Y.H.; Huang, C.F.; Chung, C.Y.; Lin, S.C.; Zhao, F. XRD Characterization for Al- and N-Doped 3C-SiC on Si (100) Substrate after Pulsed Excimer Laser Anneal. *Mater. Sci. Forum* **2012**, *717*, 497–500. [CrossRef]
46. Taguchi, E.; Suzuki, Y.; Satoh, M. Electrical Properties of N Ion Implanted Layer in 3C-SiC(100) Grown on Self-Standing 3C-SiC Substrate. *Mater. Sci. Forum* **2007**, *556*, 579–582. [CrossRef]
47. Khemka, V.; Patel, R.; Ramungul, N.; Chow, T.P.; Ghezzo, M.; Kretchmer, J. Characterization of Phosphorus Implantation in 4H-SiC. *J. Electron. Mater.* **1999**, *28*, 167–174. [CrossRef]
48. Li, F.; Sharma, Y.; Shah, V.; Jennings, M.; Pérez-Tomás, A.; Myronov, M.; Fisher, C.; Leadley, D.; Mawby, P. Electrical Activation of Nitrogen Heavily Implanted 3C-SiC(100). *Appl. Surf. Sci.* **2015**, *353*, 958–963. [CrossRef]
49. Rao, M.V.; Griffiths, P.; Holland, O.W.; Kelner, G.; Freitas, J.A.; Simons, D.S.; Chi, P.H.; Ghezzo, M. Al and B Ion-implantations in 6H- and 3C-SiC. *J. Appl. Phys.* **1995**, *77*, 2479–2485. [CrossRef]
50. Nagasawa, H.; Abe, M.; Yagi, K.; Kawahara, T.; Hatta, N. Fabrication of High Performance 3C-SiC Vertical MOSFETs by Reducing Planar Defects. *Phys. Status Solidi Basic Res.* **2008**, *245*, 1272–1280. [CrossRef]
51. La Via, F.; Mauceri, M.; Scuderi, V.; Calabretta, C.; Zimbone, M.; Anzalone, R. 3C-SiC Bulk Growth: Effect of Growth Rate and Doping on Defects and Stress. *Mater. Sci. Forum* **2020**, *1004*, 120–125. [CrossRef]
52. Lossy, R.; Reichert, W.; Obermeier, E. Characterization of 3C-SiC Doped by Nitrogen Implantation. *Mater. Sci. Eng. B* **1997**, *46*, 156–159. [CrossRef]
53. Lossy, R.; Reichert, W.; Obermeier, E.; Skorupa, W. Doping of 3C-SiC by Implantation of Nitrogen at High Temperatures. *J. Electron. Mater.* **1997**, *26*, 123–127. [CrossRef]

54. Suzuki, Y.; Taguchi, E.; Nagata, S.; Satoh, M. Evaluation of Specific Contact Resistance of Al, Ti, and Ni Contacts to N Ion Implanted 3C-SiC(100). *Mater. Sci. Forum* **2007**, *556*, 705–708. [CrossRef]
55. Song, X.; Biscarrat, J.; Bazin, A.E.; Michaud, J.F.; Cayrel, F.; Zielinski, M.; Chassagne, T.; Portail, M.; Collard, E.; Alquier, D. Dose Influence on Physical and Electrical Properties of Nitrogen Implantation in 3C-SiC on Si. *Mater. Sci. Forum* **2012**, *711*, 154–158. [CrossRef]
56. Jennings, M.R.; Fisher, C.A.; Walker, D.; Sanchez, A.; Pérez-Tomás, A.; Hamilton, D.P.; Gammon, P.M.; Burrows, S.E.; Thomas, S.M.; Sharma, Y.K.; et al. On the Ti3SiC2 Metallic Phase Formation for Robust P-Type 4H-SiC Ohmic Contacts. *Mater. Sci. Forum* **2014**, *778*, 693–696. [CrossRef]
57. Eriksson, J.; Roccaforte, F.; Giannazzo, F.; Lo Nigro, R.; Raineri, V.; Lorenzzi, J.; Ferro, G. Improved Ni/3C-SiC Contacts by Effective Contact Area and Conductivity Increases at the Nanoscale. *Appl. Phys. Lett.* **2009**, *94*, 112104. [CrossRef]
58. Noh, J.I.; Nahm, K.S.; Kim, K.C.; Capano, M.A. Effect of Surface Preparation on Ni Ohmic Contact to 3C-SiC. *Solid State Electron.* **2002**, *46*, 2273–2279. [CrossRef]
59. Moki, A.; Shenoy, P.; Alok, D.; Baliga, B.J.; Wongchotigul, K.; Spencer, M.G. Low Resistivity As-Deposited Ohmic Contacts to 3C-SiC. *J. Electron. Mater.* **1995**, *24*, 315–318. [CrossRef]
60. Roy, S.; Jacob, C.; Basu, S. Ohmic Contacts to 3C-SiC for Schottky Diode Gas Sensors. *Solid State Electron.* **2003**, *47*, 2035–2041. [CrossRef]
61. Bazin, A.E.; Michaud, J.F.; Cayrel, F.; Portail, M.; Chassagne, T.; Zielinski, M.; Collard, E.; Alquier, D. High Quality Ohmic Contacts on N-type 3C-SiC Obtained by High and Low Process Temperature. *AIP Conf. Proc.* **2010**, *1292*, 51–54. [CrossRef]
62. Wan, J.; Capano, M.A.; Melloch, M.R. Formation of Low Resistivity Ohmic Contacts to N-Type 3C-SiC. *Solid State Electron.* **2002**, *46*, 1227–1230. [CrossRef]
63. Zhang, J.; Howe, R.T.; Maboudian, R. Nickel and Platinum Ohmic Contacts to Polycrystalline 3C-Silicon Carbide. *Mater. Sci. Eng. B* **2007**, *139*, 235–239. [CrossRef]
64. Chung, G.-S.; Yoon, K.-H. Ohmic Contacts to Single-Crystalline 3C-SiC Films for Extreme-Environment MEMS Applications. *Microelectron. J.* **2008**, *39*, 1408–1412. [CrossRef]
65. Li, F.; Sharma, Y.; Walker, D.; Hindmarsh, S.; Jennings, M.; Martin, D.; Fisher, C.; Gammon, P.; Pérez-Tomás, A.; Mawby, P. 3C-SiC Transistor with Ohmic Contacts Defined at Room Temperature. *IEEE Electron Device Lett.* **2016**, *37*, 1189–1192. [CrossRef]
66. Spera, M.; Greco, G.; Lo Nigro, R.; Bongiorno, C.; Giannazzo, F.; Zielinski, M.; La Via, F.; Roccaforte, F. Ohmic Contacts on N-Type and p-Type Cubic Silicon Carbide (3C-SiC) Grown on Silicon. *Mater. Sci. Semicond. Process.* **2019**, *93*, 295–298. [CrossRef]
67. Jiang, Y.; Zhao, C.; Liu, S.; Huang, Q. Polysilicon-Al Based Ohmic Contact on p-Type 3C-SiC Film Grown on Silicon Substrate. In Proceedings of the 2006 8th Internation Conference on Solid-State and Integrated Circuit Technology Proceedings, Shanghai, China, 23–26 October 2006; pp. 938–940.
68. Schöner, A.; Krieger, M.; Pensl, G.; Abe, M.; Nagasawa, H. Fabrication and Characterization of 3C-SiC-Based MOSFETs. *Chem. Vap. Depos.* **2006**, *12*, 523–530. [CrossRef]
69. Afanas'ev, V.V.; Ciobanu, F.; Pensl, G.; Stesmans, A. Contributions to the Density of Interface States in SiC MOS Structures. In *Silicon Carbide: Recent Major Advances*; Choyke, W.J., Matsunami, H., Pensl, G., Eds.; Springer: Berlin/Heidelberg, Germany, 2004; pp. 343–371. [CrossRef]
70. Afanasev, V.; Bassler, M.; Pensl, G.; Shulz, M. Intrinsic SiC/SiO_2 Interface States. *Phys. Stat. Sol.* **1997**, *162*, 321–337. [CrossRef]
71. Esteve, R. Fabrication and Characterization of 3C- and 4H-SiC MOSFETs. Doctoral Thesis, KTH, School of Information and Communication Technology (ICT), Integrated Devices and Circuits, Stockholm, Sweden, 2011.
72. Esteve, R.; Schöner, A.; Reshanov, S.A.; Zetterling, C.-M.; Nagasawa, H. Comparative Study of Thermally Grown Oxides on N-Type Free Standing 3C-SiC (001). *J. Appl. Phys.* **2009**, *106*, 44513. [CrossRef]
73. Wan, J.; Capano, M.A.; Melloch, M.R.; Cooper, J.A. N-Channel 3C-SiC MOSFETs on Silicon Substrate. *IEEE Electron Device Lett.* **2002**, *23*, 482–484. [CrossRef]
74. Abe, M.; Nagasawa, H.; Ericsson, P.; Strömberg, H.; Bakowski, M.; Schöner, A. High Current Capability of 3C-SiC Vertical DMOSFETs. *Microelectron. Eng.* **2006**, *83*, 24–26. [CrossRef]
75. Krieger, M.; Beljakowa, S.; Trapaidze, L.; Frank, T.; Weber, H.B.; Pensl, G.; Hatta, N.; Abe, M.; Nagasawa, H.; Schöner, A. Analysis of Interface Trap Parameters from Double-Peak Conductance Spectra Taken on N-Implanted 3C-SiC MOS Capacitors. *Phys. Status Solidi Basic Res.* **2008**, *245*, 1390–1395. [CrossRef]
76. Li, F.; Vavasour, O.J.; Walker, M.; Martin, D.M.; Sharma, Y.K.; Russell, S.A.O.; Jennings, M.R.; Pérez-Tomás, A.; Mawby, P.A. Physical Characterisation of 3C-SiC(001)/SiO_2 Interface Using XPS. *Mater. Sci. Forum* **2017**, *897*, 151–154. [CrossRef]
77. Arora, R.; Rozen, J.; Fleetwood, D.M.; Galloway, K.F.; Zhang, C.X.; Han, J.; Dimitrijev, S.; Kong, F.; Feldman, L.C.; Pantelides, S.T.; et al. Charge Trapping Properties of 3C- and 4H-SiC MOS Capacitors With Nitrided Gate Oxides. *IEEE Trans. Nucl. Sci.* **2009**, *56*, 3185–3191. [CrossRef]
78. Matocha, K.; Beaupre, R. Time-Dependent Dielectric Breakdown of Thermal Oxides on 4H-SiC. *Mater. Sci. Forum* **2007**, *556*, 675–678. [CrossRef]
79. Fiorenza, P.; Schilirò, E.; Giannazzo, F.; Bongiorno, C.; Zielinski, M.; La Via, F.; Roccaforte, F. On the Origin of the Premature Breakdown of Thermal Oxide on 3C-SiC Probed by Electrical Scanning Probe Microscopy. *Appl. Surf. Sci.* **2020**, *526*, 146656. [CrossRef]

80. Esteve, R.; Schöner, A.; Reshanov, S.A.; Zetterling, C.-M.; Nagasawa, H. Advanced Oxidation Process Combining Oxide Deposition and Short Postoxidation Step for N-Type 3C- and 4H-SiC. *J. Appl. Phys.* **2009**, *106*, 44514. [CrossRef]
81. Anzalone, R.; Privitera, S.; Camarda, M.; Alberti, A.; Mannino, G.; Fiorenza, P.; Di Franco, S.; La Via, F. Interface State Density Evaluation of High Quality Hetero-Epitaxial 3C–SiC(001) for High-Power MOSFET Applications. *Mater. Sci. Eng. B* **2015**, *198*, 14–19. [CrossRef]
82. Sharma, Y.K.; Li, F.; Jennings, M.R.; Fisher, C.A.; Pérez-Tomás, A.; Thomas, S.; Hamilton, D.P.; Russell, S.A.O.; Mawby, P.A. High-Temperature (1200–1400 °C) Dry Oxidation of 3C-SiC on Silicon. *J. Electron. Mater.* **2015**, *44*, 4167–4174. [CrossRef]
83. Furukawa, K.; Uemoto, A.; Shigeta, M.; Suzuki, A.; Nakajima, S. 3C-SiC P-n Junction Diodes. *Appl. Phys. Lett.* **1986**, *48*, 1536–1537. [CrossRef]
84. Davis, R.F.; Kelner, G.; Shur, M.; Palmour, J.W.; Edmond, J.A. Thin Film Deposition and Microelectronic and Optoelectronic Device Fabrication and Characterization in Monocrystalline Alpha and Beta Silicon Carbide. *Proc. IEEE* **1991**, *79*, 677–701. [CrossRef]
85. Neudeck, P.G.; Larkin, D.J.; Starr, J.E.; Powell, J.A.; Salupo, C.S.; Matus, L.G. Greatly Improved 3C-SiC p-n Junction Diodes Grown by Chemical Vapor Deposition. *IEEE Electron Device Lett.* **1993**, *14*, 136–139. [CrossRef]
86. Shenoy, P.; Moki, A.; Baliga, B.J.; Alok, D.; Wongchotigul, K.; Spencer, M. Vertical Schottky Barrier Diodes on 3C-SiC Grown on Si. In Proceedings of the 1994 IEEE International Electron Devices Meeting, San Francisco, CA, USA, 11–14 December 1994; pp. 411–414. [CrossRef]
87. Cherkaoui, K.; Duane, R.; Ward, P.; Blake, A. Fabrication and Characterisation of Silicide/3C-SiC/Si Contacts for Schottky Barrier Diode Application. *ECS Meet. Abstr.* **2020**, *23*, 1334. [CrossRef]
88. Arvanitopoulos, A.; Li, F.; Jennings, M.R.; Perkins, S.; Gyftakis, K.N.; Antoniou, M.; Mawby, P.; Lophitis, N. Experimental Investigation and Verification of Traps Affecting the Performance of 3C-SiC-on-Si Schottky Barrier Diodes. In Proceedings of the 2019 IEEE Energy Conversion Congress and Exposition (ECCE), Baltimore, MD, USA, 29 September–3 October 2019; pp. 1941–1947. [CrossRef]
89. Wang, L.; Dimitrijev, S.; Han, J.; Tanner, P.; Iacopi, A.; Hold, L. Demonstration of P-Type 3C–SiC Grown on 150 mm Si(100) Substrates by Atomic-Layer Epitaxy at 1000 °C. *J. Cryst. Growth* **2011**, *329*, 67–70. [CrossRef]
90. Shibahara, K.; Takeuchi, T.; Matsunami, H.; Nishino, S. Electrical Properties of Undoped and Ion-Implanted Cubic SiC Grown on Si(100) by Chemical Vapor Deposition. *Jpn. J. Appl. Phys.* **1989**, *28*, 1341–1347. [CrossRef]
91. Tyagi, R.; Chow, T.P. Self-Enclosed vs. LOPOS-Terminated Lateral Planar p/Sup +/n and n/Sup +/p Junctions in 3C-SiC/Si. In Proceedings of the 8th International Symposium on Power Semiconductor Devices and Ics, ISPSD '96 Proceedings, Lahaina, HI, USA, 20–23 May 1996; pp. 115–118. [CrossRef]
92. Hatta, N.; Kawahara, T.; Yagi, K.; Nagasawa, H.; Reshanov, S.A.; Schöner, A. Reliable Method for Eliminating Stacking Fault on 3C-SiC(001). *Mater. Sci. Forum* **2012**, *717*, 173–176. [CrossRef]
93. Fisicaro, G.; Bongiorno, C.; Deretzis, I.; Giannazzo, F.; La Via, F.; Roccaforte, F.; Zielinski, M.; Zimbone, M.; La Magna, A. Genesis and Evolution of Extended Defects: The Role of Evolving Interface Instabilities in Cubic SiC. *Appl. Phys. Rev.* **2020**, *7*, 21402. [CrossRef]
94. Singh, R.; Cooper, J.A.; Melloch, M.R.; Chow, T.P.; Palmour, J.W. SiC Power Schottky and PiN Diodes. *IEEE Trans. Electron Devices* **2002**, *49*, 665–672. [CrossRef]
95. Bu, Y.; Yoshimoto, H.; Watanabe, N.; Shima, A. Fabrication of 4H-SiC PiN Diodes without Bipolar Degradation by Improved Device Processes. *J. Appl. Phys.* **2017**, *122*, 244504. [CrossRef]
96. Fisher, C.A.; Jennings, M.R.; Sharma, Y.K.; Hamilton, D.P.; Gammon, P.M.; Pérez-Tomás, A.; Thomas, S.M.; Burrows, S.E.; Mawby, P.A. Improved Performance of 4H-SiC PiN Diodes Using a Novel Combined High Temperature Oxidation and Annealing Process. *IEEE Trans. Semicond. Manuf.* **2014**, *27*, 443–451. [CrossRef]
97. Bakowski, M.; Schöner, A.; Ericsson, P.; Stromberg, H.; Nagasawa, H.; Abe, M. Development of 3C-SiC MOSFETs. *J. Telecommun. Inf. Tech.* **2007**, *2*, 49–56.
98. Japanese Push SiC Power. *III-Vs Rev.* **2004**, *17*, 35.
99. Nagasawa, H.; Yagi, K.; Kawahara, T.; Hatta, N.; Abe, M.; Schöner, A.; Bakowski, M.; Ericsson, P.; Pensl, G. Challenges for Improving the Crystal Quality of 3C-SiC Verified with MOSFET Performance. *Mater. Sci. Forum* **2009**, *600*, 89–94. [CrossRef]
100. Nagasawa, H.; Yagi, K.; Kawahara, T.; Hatta, N.; Abe, M. Hetero- and Homo-Epitaxial Growth of 3C-SiC for MOS-FETs. *Microelectron. Eng.* **2006**, *83*, 185–188. [CrossRef]
101. Kobayashi, M.; Uchida, H.; Minami, A.; Sakata, T.; Esteve, R.; Schöner, A. 3C-SiC MOSFET with High Channel Mobility and CVD Gate Oxide. *Mater. Sci. Forum* **2011**, *679*, 645–648. [CrossRef]
102. Schoner, A.; Bakowski, M.; Ericsson, P.; Stromberg, H.; Nagasawa, H.; Abe, M. Vertical MOSFET Devices Fabricated on 3C-SiC with High and Low Material Quality. *MRS Online Proc. Libr.* **2011**, *911*, 1303. [CrossRef]
103. Lee, K.K.; Ishida, Y.; Ohshima, T.; Kojima, K.; Tanaka, Y.; Takahashi, T.; Okumura, H.; Arai, K.; Kamiya, T. N-Channel MOSFETs Fabricated on Homoepitaxy-Grown 3C-SiC Films. *IEEE Electron Device Lett.* **2003**, *24*, 466–468. [CrossRef]
104. Uchida, H.; Minami, A.; Sakata, T.; Nagasawa, H.; Kobayashi, M. High Temperature Performance of 3C-SiC MOSFETs with High Channel Mobility. *Mater. Sci. Forum* **2012**, *717*, 1109–1112. [CrossRef]

materials

MDPI

Review

Structural and Insulating Behaviour of High-Permittivity Binary Oxide Thin Films for Silicon Carbide and Gallium Nitride Electronic Devices

Raffaella Lo Nigro *, Patrick Fiorenza, Giuseppe Greco, Emanuela Schilirò and Fabrizio Roccaforte

Consiglio Nazionale delle Ricerche—Istituto per la Microelettronica e Microsistemi (CNR-IMM), 95121 Catania, Italy; patrick.fiorenza@imm.cnr.it (P.F.); giuseppe.greco@imm.cnr.it (G.G.); emanuela.schiliro@imm.cnr.it (E.S.); fabrizio.roccaforte@imm.cnr.it (F.R.)
* Correspondence: raffaella.lonigro@imm.cnr.it; Tel.: +39-095-596-8218

Abstract: High-κ dielectrics are insulating materials with higher permittivity than silicon dioxide. These materials have already found application in microelectronics, mainly as gate insulators or passivating layers for silicon (Si) technology. However, since the last decade, the post-Si era began with the pervasive introduction of wide band gap (WBG) semiconductors, such as silicon carbide (SiC) and gallium nitride (GaN), which opened new perspectives for high-κ materials in these emerging technologies. In this context, aluminium and hafnium oxides (i.e., Al_2O_3, HfO_2) and some rare earth oxides (e.g., CeO_2, Gd_2O_3, Sc_2O_3) are promising high-κ binary oxides that can find application as gate dielectric layers in the next generation of high-power and high-frequency transistors based on SiC and GaN. This review paper gives a general overview of high-permittivity binary oxides thin films for post-Si electronic devices. In particular, focus is placed on high-κ binary oxides grown by atomic layer deposition on WBG semiconductors (silicon carbide and gallium nitride), as either amorphous or crystalline films. The impacts of deposition modes and pre- or postdeposition treatments are both discussed. Moreover, the dielectric behaviour of these films is also presented, and some examples of high-κ binary oxides applied to SiC and GaN transistors are reported. The potential advantages and the current limitations of these technologies are highlighted.

Keywords: insulators; binary oxides; high-κ dielectrics; power electronics; wide band gap semiconductors

check for updates

Citation: Lo Nigro, R.; Fiorenza, P.; Greco, G.; Schilirò, E.; Roccaforte, F. Structural and Insulating Behaviour of High-Permittivity Binary Oxide Thin Films for Silicon Carbide and Gallium Nitride Electronic Devices. *Materials* **2022**, *15*, 830. https://doi.org/10.3390/ma15030830

Academic Editor: Alexander N. Obraztsov

Received: 17 December 2021
Accepted: 19 January 2022
Published: 22 January 2022

Publisher's Note: MDPI stays neutral with regard to jurisdictional claims in published maps and institutional affiliations.

1. Introduction

Today, it is widely recognized that microelectronic devices have improved the quality of our daily lives, strongly contributing to the development of human civilization. In the 1940s–1950s, the first microelectronic devices appeared, and they were based on germanium. However, silicon (Si) gradually began to be the semiconductor of choice, driving the power electronics revolution with the introduction of the first p-n-p-n transistors in 1956 at Bell Laboratories [1,2]. About two decades later, the introduction of metal-oxide-semiconductor field-effect transistors (Si-MOSFETs) set the foundations for the development of the modern CMOS technology [3]. Hence, for about fifty years, microelectronics have been based mainly on Si semiconductors. The great success of digital technology may apparently indicate that Si is still the most suitable material for microelectronic devices. However, in other fields, such as electronic systems for power transmission or distribution (power converters, base stations, wireless connections, etc.) and optoelectronics (light emitting diodes—LEDs, lasers), the achievement of the ultimate silicon performances opened the route for the post-Si era. In this context, wide band gap (WBG) semiconductors emerged as the most suitable materials for this technological revolution, especially in high-power and high-frequency electronics [4–7].

Among the WBG semiconductors, silicon carbide (SiC) and gallium nitride (GaN) are the most attractive candidates because they already provide a good compromise between their theoretical properties (blocking voltage capability, operation temperature, and switching frequency) and commercial availability [4–6]. Their wide band gaps result in higher breakdown voltage and operation temperature with respect to Si, so both are excellent candidates to replace Si in the next generation of high-power and high-frequency electronics. Because of their different physical and electronic properties in terms of carrier mobility and thermal conductivity [8,9], SiC and GaN will cover different market segments in the post-Si technologies [10]. In particular, SiC is more suitable for high-power applications based on vertical devices, while GaN is more efficient for high-frequency applications based on lateral transistors. In any case, both materials can provide superior performances with respect to the existing Si devices [5,6], although the different technological steps for transistor fabrication need to be appropriately integrated.

Gate insulators are certainly the most important brick for transistor operation, even in the post-Si era, since the device performances critically depend on the choice of the insulating material. However, gate insulator technology is rather different in SiC and GaN, thus leading to a variety of issues to be faced when developing devices on these two WBG semiconductors.

Traditional dielectric materials, such as silicon oxide or silicon nitride, have also been widely investigated [11–14] for applications based on WBG semiconductors. However, the performance of the ideal Si/SiO_2 system has been not achieved, and attention has been focused on the so-called "high-κ" oxides [15–20]. Among all the high-κ materials, some binary oxides (such as Al_2O_3 [21,22], HfO_2 [22], NiO [23,24], CeO_2 [25], Sc_2O_3 [26,27], La_2O_3 [28], Gd_2O_3 [28], Y_2O_3 [28,29], ZrO_2, [17,18], Ga_2O_3 [30], etc.) potentially represent a suitable solution for the integration in WBG-based devices because of their higher chemical stability and/or lower fabrication cost. Some other possible materials have been studied, such as ternary oxides and nitrides, but those materials are beyond the topic of this review paper.

Table 1 shows a summary of the possible oxide candidates for the replacement of the SiO_2 dielectric material and their principal physical properties, such as dielectric constant values, band gaps, and crystallization temperatures.

Table 1. Principal physical properties of high-κ gate binary oxides.

Oxide	Dielectric Constant	Band Gap (eV)	Crystallization Temperature	Ref
Al_2O_3	10	9	900 °C	[17,18]
HfO_2	~20	5.6–5.8	500 °C	[17,18,22]
NiO	11.7	4	300 °C	[23,24]
CeO_2	26	6	500 °C	[25]
Sc_2O_3	12–14	6.0	>400 °C	[26–29]
Y_2O_3	10	5.5	>400 °C	[26,28,29]
Gd_2O_3	~20	5.0–5.45	>400 °C	[26,28]
La_2O_3	~20	5.4–5.6	>400 °C	[17,18,27,29]
ZrO_2	25	5.8	>400 °C	[17,18]
Ga_2O_3	~10	5	>500 °C	[30]

Figure 1a reports the values of the band gaps of different insulators as a function of their relative permittivity (in units of the vacuum permittivity ε_0). The general trend (highlighted by the continuous line) is a decrease in the band gap with increasing permittivity. Hence, the reduced band gap of high-permittivity oxides can represent a concern in terms of leakage current. For this reason, insulators with appropriate band alignment with the semiconductor must be preferred. In this context, Figure 1b shows the band alignment of several high-κ oxides with the semiconductor materials under consideration (i.e., Si, 4H-SiC, and GaN). The offset between the conduction bands of the semiconductors and insulators is reported in scale.

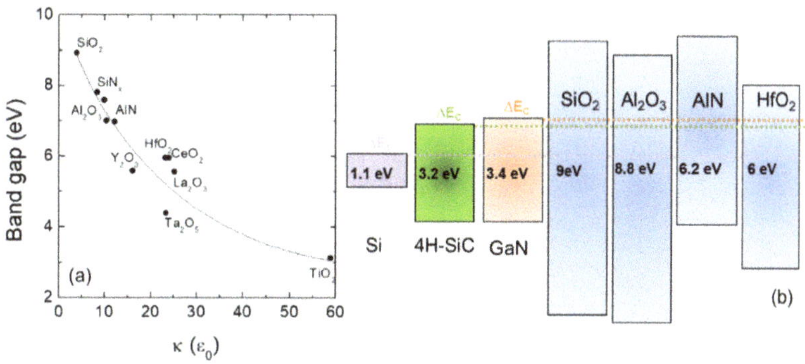

Figure 1. (**a**) Band gap values as a function of relative permittivity (in units of the vacuum permittivity ε_0) for different insulators. The continuous line is a guide; (**b**) schematic illustration (in scale) of the band alignments of some common insulators with the semiconductor materials under consideration (i.e., silicon, 4H-SiC, and GaN). The light purple, green, and orange dotted lines indicate the conduction band edge of the Si, 4H-SiC, and GaN semiconductors, respectively.

Hence, in terms of physical properties, the guidelines for the choice of the ideal gate dielectric material are: (i) high dielectric constant value; (ii) appropriate alignment of the band gap with respect to the substrates (in particular, the band offset should be greater than 1 eV); (iii) thermal stability during the fabrication process (many steps have to be carried out at high temperatures for short periods of time) [17–19].

Moreover, since the gate oxide is directly in contact with the device channel, another important requirement is good quality of the gate oxide/semiconductor interface in terms of low roughness and low density of electronic defects [5].

These requirements could be met throughout two possible approaches, i.e., a crystalline gate oxide epitaxially grown on the semiconducting substrate or an amorphous oxide. Electronic defects can be thus minimized either by exactly or randomly saturating the dangling bonds, respectively. Generally, amorphous oxides are the preferred solution, since they possess isotropic dielectric constants due to the fluctuation of the polarized bonds and do not possess rough edges. By contrast, the advantage of the epitaxial oxides is the abruptness of the interface [17,18].

In general, as schematically illustrated in Figure 2, structural and compositional defects of binary oxides (e.g., oxygen vacancies, impurities, etc.) can generate the presence of energetic levels within the band gap or at the interface, and the trapped charges in these states are undesirable for the following reasons: (i) they are responsible for a shift in the voltage threshold of the transistor; (ii) they may change over time and determine the instability of the transistor output characteristics; (iii) they scatter the carriers in the inversion channel and, consequently, limit the channel mobility; (iv) they compromise the transistor reliability because they are the main cause of the dielectric breakdown [17,18].

Silicon dioxide (SiO_2) [15] was considered an ideal dielectric during the Si era because it possesses a very low electronic defect density. The reason for this is the low coordination number, which guarantees the possibility to "repair" the dangling bonds. On the other hand, alternative high-κ oxides possess chemical bonds that cannot easily relax, thus inevitability leading to a higher electronic defect density. Hence, there is a need to reduce the number of electronic defects in these materials by annealing treatments or by optimizing their deposition processes.

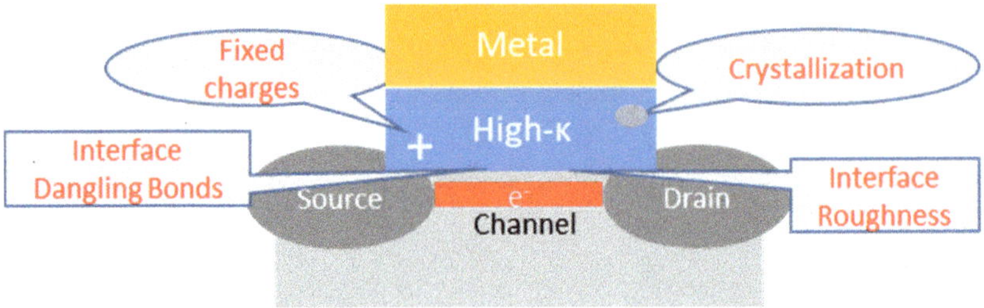

Figure 2. Schematic representation of the main issues affecting the functionality of high-κ binary gate oxides in a transistor.

In this context, the important role of the growth technique for the deposition of the high-κ dielectric layers is clear. Certainly, many deposition techniques based on either physical or chemical principles are available. However, the semiconductor industry currently demands manufacturing techniques able to achieve good surface coverage on large areas, high conformity on three-dimensional structures, high growth rate, reliability, and compatibility with the thermal budget required for the device fabrication [31,32].

Table 2 compares the main features of the common growth techniques used [29] for the deposition of high-κ oxide thin films for microelectronics applications, considering the different deposition parameters. High deposition rates and large varieties of available materials are certainly the main advantages of molecular beam epitaxy (MBE) or chemical vapor deposition (CVD) methods. By contrast, these techniques are characterized by the need for high deposition temperatures. Physical vapor deposition (PVD)-based techniques are generally preferred for metals rather than for insulator deposition and lack uniformity over large areas.

Table 2. Comparison of the main features of the common deposition techniques for high-κ oxides in microelectronics [29,31,32].

	ALD	MBE	CVD	PVD
Thickness range	≤2000 Å	≤2000 Å	≥100 Å	≥100 Å
Deposition rate	Low 1–5 nm/min	High 0.01–0.3 μm/min	High 1–10 μm/h	Medium 0.1–1 μm/h
Step coverage **Aspect ratio**	100% 60:1	25–50% 1:1	70% 1:1	25–50% 1:1
Deposition temperature	25–400 °C	500–1000 °C	300–1100 °C	200–500 °C
Film type availability	High (limited for metals)	High (limited for Metals)	High (limited for metals)	High for metals and conductive materials

However, judging from the latest industrial trends and looking forward at the nanometric-scale miniaturization process of electronic devices, the employment of deposition methods with atomic-level accuracy has become mandatory. From this perspective, atomic layer deposition (ALD) is the most promising deposition technique, and it is gradually replacing CVD and PVD techniques in many applications.

ALD is an innovative thin-film growth method that belongs to the general class of CVD techniques. As in a typical CVD process, films are deposited from gaseous chemical precursors, one for each element of the desired compound. However, unlike the traditional CVD mechanism, the ALD process is characterized by "self-limited" reactions, first between precursor and pristine surface and second on a surface saturated by one "monolayer" of precursor fragments [31]. This deposition mechanism allows subnanometer control of film thickness, conformal coating of nonplanar substrates (step coverage ~100%), and high-quality films deposited at relatively low temperatures [32]. For these reasons, the employment of ALD can give several advantages over that of either CVD or PVD. Finally, the low growth rate of the classical thermal ALD (T-ALD) process has been now significantly improved by the implementation of plasma enhanced ALD (PE-ALD). PE-ALD is an energy-enhanced deposition technique based on plasma ignition to enhance the co-reactants' reactivity. The high reactivity of the plasma species produces a higher density of reactive surface sites. Consequently, higher growth rates and better properties of the resulting films in terms of density, impurity content, and electrical parameters can be obtained. Another advantage of PE-ALD is the possibility to control additional process parameters, such as the operating pressure, plasma power, and plasma exposure time. Varying the plasma parameters enable fine tuning the properties of the deposited films.

A great part of the results presented in the following Sections are related to high-κ oxides grown by ALD techniques.

2. Amorphous High-κ Oxides on WBG Semiconductors

Several amorphous materials have been studied in the last decades as possible high-κ gate oxides for WBG semiconductors. Among them, because of their high crystallization temperature, Al_2O_3 thin films have certainly been the most widely investigated solution as amorphous dielectric layers. Some studies have reported on Al_2O_3 formed by reactive ion sputtering [33–35], oxidation of Al in oxygen ambient at high temperatures [36], and a few others nonconventional techniques [37,38]. The major drawbacks of these solutions are the low breakdown fields (around 5–6 $MVcm^{-1}$) of the deposited films and their poor thickness uniformity on large areas. These limitations have been overcome by the implementation of the ALD technique, which has been the method of choice to study the potentiality of Al_2O_3 thin films [39–44].

However, several issues still remain objects of investigation in order to optimize the quality of deposited materials and their interfaces with the WBG semiconductors. Moreover, though the growth of high-κ oxides amorphous films is generally carried out at low deposition temperatures (in the 200–300 °C range), some interfacial interaction could occur in SiC and GaN substrates, resulting in the presence of unwanted materials or deposition by products.

In this context, the cleaning of the substrate surface before dielectric deposition, as well as the postdeposition annealing treatments, are discussed in the next subsections, illustrating as examples some relevant case studies of amorphous high-κ oxides on SiC and GaN substrates.

2.1. *Growth of Amorphous High-κ Oxides on SiC*

Unlike that of thermal silicon dioxide (SiO_2), the growth of high-κ oxides on silicon carbide is much more affected by the quality of the semiconductor surface. In fact, in order to limit the amount of the interface state density (D_{it}), appropriate cleaning of the SiC surfaces is always required.

A variety of SiC surface-cleaning treatments have been proposed, based either on wet chemical solutions [44–46] or plasma [47–49]. The most used chemical solutions for SiC cleaning are combinations of diluted sulfuric acid, hydrogen peroxide, isopropanol, diluted hydrofluoric acid. Suvanam et al. [46] demonstrated that RCA treatment [45], followed by HF diluted solution and finally isopropanol, was a good route to improve the interfacial electrical characteristics of Al_2O_3 films on SiC, obtaining a density of interface states $D_{it} = 1.5 \times 10^{11}$ eV^{-1} cm^{-2} at $E_C - E_t = 0.2$ eV below the 4H-SiC conduction band edge, which was about two orders of magnitude lower than the values found with thermal SiO_2. In regard to plasma treatment before high-κ deposition, H_2 plasma has been also evaluated in some works [47–49], since it represents an efficient route for the passivation of dangling bonds on SiC surfaces. Heo et al. [49] measured promising values of interface state density ($D_{it} = 6 \times 10^{12}$ eV^{-1} cm^{-2} at $E_C - E_t = 0.2$ eV) when a 15 min long H_2 plasma treatment was performed before deposition and after the post-metallization step.

As a matter of fact, besides surface treatments before the dielectric deposition, post-deposition annealings are of great importance to optimize the dielectric properties. Many parameters can in principle be varied, such as ambient atmosphere, annealing temperature and time, etc. However, these processing steps must be ultimately compatible with complete SiC device fabrication, in which, e.g., the formation of metal contacts is achieved at high temperatures (900–1000 °C) and fixed gas atmospheres (N_2 or Ar). Generally, a large number of high-κ oxides possess crystallization temperatures of about 400–500 °C, with Al_2O_3 being the most thermally stable at up to 800 °C. However, independently of the chemical nature of the high-κ oxide, the annealing process can improve dielectrical properties. For instance, Wang et al. [50] demonstrated the beneficial effects of high-temperature annealings (800–1000 °C) performed in O_2 atmosphere on Al_2O_3 films. In particular, they showed that although Al_2O_3 films started crystallizing at 900 °C, capacitance vs. voltage (C–V) measurements revealed their improved electrical characteristics (i.e., reduced hysteresis phenomena). Hence, the authors concluded that annealing at 900 °C represented the best option in terms of both surface morphology and dielectric quality. On the other hand, many other papers demonstrated that such high annealing temperatures induce the formation of a thin stoichiometric or sub-stoichiometric silicon oxide interfacial layer [33,50–52]. This oxidation phenomenon can have a detrimental impact on the properties of high-κ/SiC interfaces, including in the case of abrupt Al_2O_3/4H-SiC interfaces obtained by ALD growth [40,53–55]. In this context, annealing in N_2 atmosphere can be the preferred solution, although uncontrolled SiO_x formation can occur in N_2 atmosphere for high annealing temperatures. Moreover, Avice et al. [42] and Khosa et al. [36] showed that an additional effect of incomplete SiC oxidation was the formation of C clusters if not enough oxygen was present to enable the out-diffusion of carbon as carbon monoxide. The formation of the SiOx interfacial layers was observed independently of the annealing temperature or ambient. In fact, this phenomenon has been observed even in vacuum or at only 300 °C annealing temperature [55]. Hence, it is expected that the elimination of residual O_2 molecules in the annealing ambient is one the key issues for the limitation of SiOx formation.

In general, most of the reported postdeposition annealing studies were carried out in oxidizing (O_2 or N_2O) or non-oxidizing (Ar, N_2 or forming gas) ambient, in the 500–1100 °C temperature range, and for short (1 min) or long (1–2 h) times. An interaction at the interface has always been observed by the formation of the silicon oxide layers and carbon clusters. The control of the chemical nature of the interface products, which in turn strongly affects the electrical characteristics, is not trivial.

In this context, Schilirò et al. [39,40] reported an interesting comparison between the properties of Al_2O_3 thin films grown by PE-ALD on bare 4H-SiC and on a 5 nm thermal SiO_2/SiC stack. TEM analyses (shown in Figure 3a,b) showed uniform interfaces and well adherent films. The surface morphology of the films (determined by AFM) was very similar, with root-mean-square (RMS) values measured over a 1 μm^2 area of 0.670 nm and 0.561 nm for Al_2O_3/SiC and Al_2O_3/SiO_2/SiC samples, respectively.

Though the interface structural quality appears analogous, quite different electrical properties were measured on MOS capacitors. In fact, current vs. voltage (I–V) measurements (Figure 3d) showed a higher leakage current in the Al_2O_3/SiC than in the Al_2O_3/SiO_2/SiC stack. Furthermore, the breakdown fields, i.e., 5.7 MV/cm for the Al_2O_3/SiC and 7 MV/cm for the Al_2O_3/SiO_2/SiC, demonstrated the better electrical quality obtained by the introduction of the SiO_2 at the interface. Moreover, the relative permittivity values, evaluated from the C–V curves (Figure 3c), were $\varepsilon \approx 6.7$ and $\varepsilon \approx 8.4$ for the Al_2O_3/SiC and the Al_2O_3/SiO_2/SiC samples, respectively.

These results can be explained by considering both the larger conduction band offset between the SiO_2 and the SiC substrate (Figure 1b) and the different chemical impact of the substrate surface on the Al_2O_3 nucleation process. This latter is schematically depicted in Figure 4, showing that the presence of the OH species on the SiO_2 surface favours the nucleation process by increasing the number of nucleation sites and the formation of denser Al_2O_3 films.

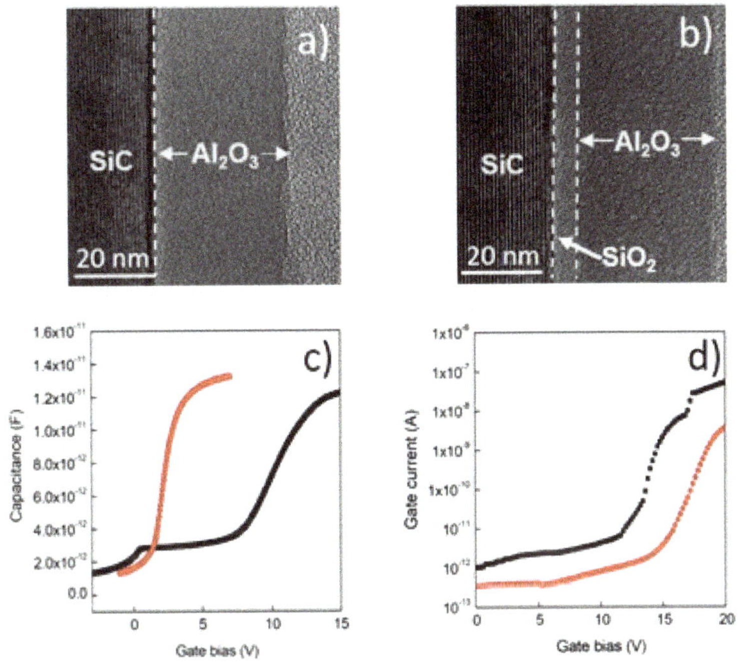

Figure 3. TEM images of Al_2O_3 thin films grown by PE-ALD on 4H-SiC (**a**) and SiO_2/4H-SiC (**b**) substrates and their relative electrical characteristics in terms of C-V curves (**c**) and I-V measurements (**d**) performed on MOS capacitors. Black and red lines are related to Al_2O_3 thin films deposited on SiO_2/4H-SiC and 4H-SiC substrates, respectively. Reproduced with permission from [40]. Copyright © 2016 WILEY-VCH Verlag GmbH & Co. KGaA.

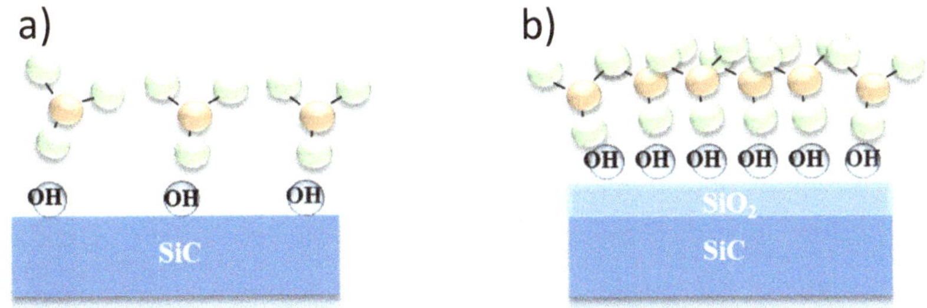

Figure 4. Schematic representation of the chemical impact of the different substrate surfaces on the Al_2O_3 nucleation processes, in the case of a bare SiC substrate (**a**) or a SiC substrate with a thin SiO_2 layer on the top (**b**). Reproduced with permission from [40]. Copyright © 2016 WILEY-VCH Verlag GmbH & Co. KGaA.

Other high-κ oxides have been also grown on SiC substrates as thin amorphous films, such as HfO_2 [56–58], La_2O_3 [59,60], Ta_2O_5 [61], and TiO_2 [62]. Among these materials, HfO_2 thin films have been widely investigated because of their superior theoretical properties, such as much higher permittivity. However, the main drawback for their implementation on SiC-based devices is the imperfect alignment of both conduction and valence band offsets (about 0.7 and 1.74 eV, respectively) with those of SiC. Cheong et al. [56,57] reported on HfO_2 films with a very high dielectric constant value (20), but the interface state density D_{it} was as high as 2×10^{13} eV^{-1} cm^{-2}, which give no advantage with respect to the SiO_2/SiC system. Moreover, very high leakage current densities of 1 mA cm^{-2} were already recorded in an electric field as low as 0.3 $MVcm^{-1}$ by Afanas'ev et al. [58]. While in this case, the high leakage current could in principle be mitigated by the introduction of a SiO_2 layer at the SiC interfaces, a further issue to be considered is the low thermal stability of HfO_2 at temperatures higher than 500 °C, when crystallization starts to occur.

In order to maintain the best features of HfO_2 (i.e., high permittivity) and Al_2O_3 (i.e., high crystallization temperature), these two materials have been evaluated in combined laminated systems.

In this context, some Al_2O_3/HfO_2 bilayer systems deposited on thermally oxidized 4H-SiC substrate have been studied, the most complex stack being an Al_2O_3/HfO_2 multilayer laminated system [63]. The Al_2O_3/HfO_2 nanolaminate shown (Figure 5a) had a total thickness of 38 nm and perfectly distinguishable sublayers, each with thickness of about 1.4–1.8 nm. After annealing treatment at 800 °C in N_2 atmosphere, the interfaces between the sublayers (Figure 5b) became less sharp, and an intermixing process occurred. Notably, both the as-deposited and annealed samples showed amorphous structures. AFM investigation pointed to a smooth surface morphology with a low RMS value of 0.6 nm, which was maintained in the annealed sample. A dielectric constant value of 12.4 was determined by the accumulation capacitance in MOS capacitors, taking into account of the SiO_2 interfacial layer. However, on the as-deposited sample, a high value of oxide trapped charge (N_{ot}) of 2.7×10^{12} cm^{-2} was found. Nevertheless, after the annealing treatment at 800 °C in N_2, the nanolaminated stack showed an improvement of the dielectric properties, since the dielectric constant value increased to 13.4 and the N_{ot} value decreased to 1.15×10^{12} cm^{-2}.

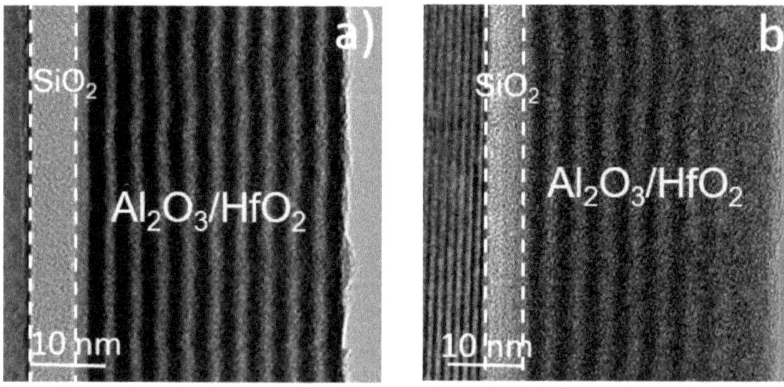

Figure 5. TEM image of (**a**) as deposited and (**b**) 800 °C annealed Al_2O_3/HfO_2 nanolaminate, deposited onto SiO_2/SiC substrate. Reproduced from [63]. Copyright © 2020 Authors.

Few other papers have been dedicated to thin films of simple high-κ oxides such as La_2O_3 [59,60], Ta_2O_5 [61], or TiO_2 [62], which, when directly grown on 4H-SiC, showed analogous results as in the case of simple HfO_2 oxide. Generally, they demonstrated good dielectric constant values, but their high interface state density and low breakdown voltages made them still far from possible implementation in real devices.

In summary, among the pure high-κ oxides, Al_2O_3 thin films represent the best compromise, especially in combination with a very thin SiO_2 interfacial layer. Some possible other high-κ bilayers, such as HfO_2/Al_2O_3 [64], Y_2O_3/Al_2O_3 [65], or ZrO_2/SiO_2 [66], exhibited some potentiality, although not many reports have been made available to date, especially regarding devices.

In regard to dielectric properties, the relevant results on the electrical performances of high-κ oxides integrated in SiC MOSFETs are reported in more detail in Section 4.

2.2. Growth of Amorphous High-κ Oxides on GaN-based Materials

The surfaces of GaN-based materials (GaN, AlGaN, InGaN, etc.) are typically characterized by the presence of large concentrations of defects (e.g., nitrogen vacancies, structural/morphological imperfections, residual contaminations, etc.) that can result in large leakage current and low performance and device reliability. Kerr et al. [67] demonstrated by density functional theory simulations that the defect sites, such as Ga dangling bonds and Ga-Al metal bonds, are responsible for the formation of states in the band gap. These interfacial trap states could be removed by annealing procedures before or after gate dielectric deposition. Moreover, especially from the perspective of high-κ gate oxide deposition, the removal of contaminations is crucial for increasing the density of precursor nucleation sites. Hence, pre-deposition surface treatments are needed to improve high-κ oxide quality. Systematic studies [68–76] have reported on the effect of several pre-treatments, and the principal cleaning/activation methods have been based on the use of wet chemical solutions [68–72,77,78] or plasma/gas actions [73–76]. Generally, the piranha (H_2O_2:H_2SO_4) solution is used for the cleaning of carbon contaminations, but some oxidation of the nitride surface can occur [70,71]. On the other hand, chloride acid (HCl) solution is efficient for the removal of metallic contaminations (eventually present from device processing) or residual oxygen on the surface. However, chlorine itself could be a residual contamination of the system [70]. Finally, hydrofluoric acid (HF) treatment is effective for the elimination of unwanted native oxide formation but is not efficient for carbon contamination [70,71]. Brennan et al. [71] compared the nucleation efficiency of the Al precursor with/without the cleaning of the surface by sequential use of acetone, methanol, isopropanol, and HF 2% solution. It was clear, from the results of an XPS study after each ALD cycle, that the decrease in the Ga-O concentration induced by the HF etch resulted in a stronger interaction between the Al precursor and the Ga surface. Nepal et al. [69] compared the effects of three different chemical solutions (i.e., piranha, diluted HF, and diluted HCl), finding that: (i) the single HCl pre-treatment provides 10–30 nm-sized particles, indicating a three-dimensional nucleation; (ii) the HF-based treatments produced an improvement in the electrical behaviour; (iii) the best dielectric properties, in terms of smaller hysteresis and lower density-trap state values, were obtained on the piranha-treated surface. Finally, Schilirò et al. [72] showed a comparison among several chemical solution combinations (i.e., piranha, HCl/ HF, and piranha/HF). In particular, it was shown that, although the Al_2O_3 thin films treated with each solution possessed identical structural properties, adherent, uniform, and amorphous, there were some intrinsic differences depending on the adopted surface pre-cleaning. In fact, under a TEM electron beam, the films deposited after piranha treatment showed the formation of polycrystalline grains, while epitaxial layers were formed for samples deposited after HF based treatments. This was an indication that in the case of HF-based treatments, the deposition process occurred on a very clean AlGaN surface, which could act as seed layer for the formation of epitaxial films. Moreover, investigation of the initial growth stages by AFM demonstrated that the smallest three-dimensional grain nucleation resulted in deposition on HF-HCl-treated surfaces, which could ensure a cleaner surface in order to allow ideal layer-by-layer ALD growth.

It could be concluded that the pre-deposition treatments of GaN-based surfaces with HF cleaning provided Al_2O_3 films with the best dielectric properties [69,71,72].

An alternative route to cleaning by chemical solution is represented by "in situ" cleaning process based on H_2/N_2 (forming gas) or NH_3 plasma actions [68]. The impact of N_2 and forming gas on the growth and interfacial characteristics of Al_2O_3 on AlGaN/GaN heterostructures was explored by Qin et al. [73], who demonstrated by XPS investigation that C contamination was effectively reduced by both N_2 and forming gas plasma. The latter also decreased the number of Ga-O bonds, improving the Al_2O_3 nucleation. In regard to plasma action effects before high-κ deposition, the same group contributed with a large variety of studies [73,75,76]. In particular, the effects of O_2, N_2, and forming gas plasma annealing were evaluated, comparing the electrical behaviour in terms of interface state density with the results obtained by XPS analyses. The formation of oxynitride bonds (Ga-O-N) increased the number of interface defects and that among all the studied treatments, the forming gas action was the most efficient.

In this context, it has to be emphasized that the semiconductor surface preparation and the deposition conditions may induce different insulting behaviours after the first film growth stages. As an example, Schilirò et al. [79] recently reported different behaviour in the early growth stages of Al_2O_3 thin films deposited on AlGaN/GaN heterostructures by thermal or plasma-enhanced ALD. In particular, they provided evidence that the PE-ALD process occurred under ideal layer-by-layer growth because of the efficiency of the O_2-plasma agent, which acted directly on the Al precursor. On the other hand, the T-ALD approach resulted in a nucleation process of the Al_2O_3 film similar to the island-growth model and a higher susceptibility to charge trapping [79].

Summarizing, surface preparation prior to high-κ oxide deposition is a crucial issue, including in the case of GaN-based materials, and can be carried out by many procedures. The aim is the cleaning of C residues, which are detrimental for the oxides' nucleation, and the elimination of Ga-N-O bonds, which are the main centres of interfacial electronic defects. These two issues are generally addressed by non-oxidizing plasma action or by HF treatments.

3. Epitaxial Growth of High-κ Oxides on WBG Semiconductors

While different oxides have been studied as gate insulators on SiC and GaN [15–19], only some of them can be grown epitaxially on the WBG semiconductor single-crystal surface [23–25,80–84]. The epitaxial growth of high-κ oxides on WBG semiconductor substrates can offer some advantages. Generally, the principal improvement is related to better saturation of interface unbonded atoms. In particular, the most commonly used SiC and GaN polymorphs for microelectronics applications possess the wurtzite structure, with hexagonal surface atomic arrangements. However, though in principle this strategy can be applied to both SiC and GaN technologies, practical studies have been performed mainly on GaN-based substrates. In fact, the few studies of epitaxial high-κ materials on SiC substrates were limited to γ-Al_2O_3 phase films [80] and direct growth of NiO thin films by metal organic chemical vapour deposition (MOCVD) [81]. The γ-Al_2O_3 phase films were initially grown by Tanner et al. [80] by the ALD process as amorphous layers, and the epitaxy on 4H-SiC substrate was obtained under a post-annealing crystallization process at a very high (1100 °C) temperature. The epitaxy was observed for the alignment of the γ-Al_2O_3 (111) planes with the (001) 4H-SiC substrate, having a lattice mismatch of about 8.8%. On the basis of the performed reflection high-energy electron diffraction analysis, the (111) γ-Al_2O_3 oriented films showed quite good structural properties for film thickness up to 20 nm, even though some twinned grains were present. Moreover, upon increasing the film thickness, the crystallization process was no more efficient, and amorphous regions were observed under TEM investigation.

Epitaxial NiO films, by contrast, have been directly grown onto 4H-SiC epilayers at the deposition temperature of 550 °C [81,85]. A high-resolution TEM micrograph of the NiO/4H-SiC interface (Figure 6a) confirmed the presence of an axially-oriented (111) NiO film, but a "non-ideal" interface was observed, because a discontinuous amorphous SiO_2 layer was detected, probably formed during MOCVD growth. Furthermore, the presence of Moiré fringes generated by the superposition of twinned NiO grains was observed. The C–V characteristics of NiO/4H-SiC capacitors (Figure 6b) were used to calculate the dielectric constant, the value of which, at 6.2, was much lower than the theoretical 11.9. This result was justified by the presence of the discontinuous silicon oxide interfacial layer.

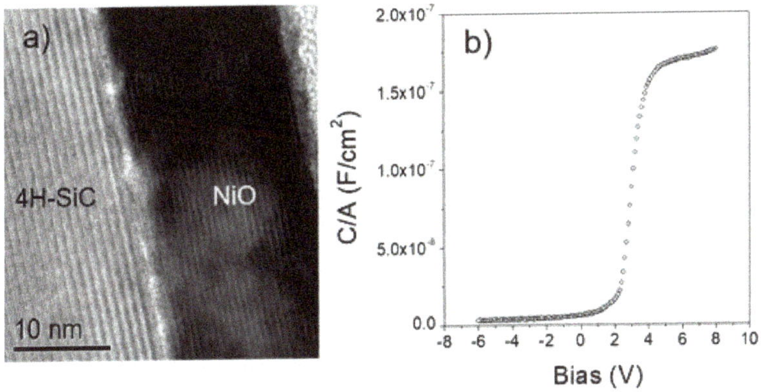

Figure 6. (**a**) High-resolution cross-section TEM image of a NiO film deposited by MOCVD on 4H-SiC at 500 °C; (**b**) C-V curve acquired on a NiO/4H-SiC MOS capacitor. Reproduced with permission from [81]. Copyright © 2013 Elsevier Ltd.

More studies on growing epitaxial oxides have been carried out on GaN-based materials. The materials under investigation comprise some lanthanide oxides, such as gadolinium [82], scandium [83,84], and lanthanum [83] oxides, as well as nickel [23,24,81] and cerium oxides [25,81]. The lanthanides oxides possess bixbyite symmetry, while NiO and CeO_2 are face cubic centred (fcc) oxides. However, the (111) planes of the latter two possess a hexagonal oxygen structure, which is suitable for epitaxy with the (0001) GaN superficial planes. Their structural and physical properties are summarized in Table 3.

Table 3. Physical and structural properties of high-κ oxides epitaxially grown on GaN.

Oxide	Dielectric Constant	Lattice Constant (Å)	Mismatch to (0001) GaN (%)	Deposition Technique	Ref.
Gd_2O_3	9	10.813	20.1	MBE	[82]
Sc_2O_3	13–14	9.845	9.2	PVD and MBE	[83,84]
La_2O_3	18–27	4.211	6.5	MBE	[83]
CeO_2	15–26	5.411	6	MOCVD	[25,81]
NiO	11.9	4.177	5	Thermal oxidation or MOCVD	[23,24,81]

The epitaxial growth of Sc_2O_3 thin films was performed on a GaN substrate at about 700 °C by the pulsed laser deposition (PLD) technique [84]. Herrero et al. [84] demonstrated that the most critical deposition parameter to obtain perfectly stoichiometric and epitaxial Sc_2O_3 thin films was the oxygen partial pressure. In particular, above 50 millitorr oxygen partial pressure, more than one preferential growth direction was observed. The epitaxial growth of Sc_2O_3 was also evaluated by Jur et al. [83] by the MBE technique. Their investigation extended to La_2O_3, which in principle can provide a dielectric constant of 26 in its hexagonal structure. Nevertheless, La_2O_3 growth was demonstrated not to be trivial, since La_2O_3 is a hygroscopic material and tends to form an amorphous layer at the interface with the GaN substrates. Nevertheless, the authors demonstrated that it was possible to inhibit the water diffusion by the introduction of a thin Sc_2O_3 layer between GaN and the growing La_2O_3 films. The MBE technique was also used for the growth of Gd_2O_3 epitaxial gate oxide on an AlGaN/GaN heterostructure [82]. Sakar et al. [82] showed the impact of a Gd_2O_3 epitaxial oxide layer on the electrical performance of an HEMT device. Gd_2O_3 films were deposited at 650 °C. The authors demonstrated that the Gd_2O_3 layer underwent phase transition upon increasing its film thickness. The first layers, up to about 3 nm, possessed hexagonal structure, which changed to monoclinic phase when the thickness of 15 nm was reached. This phase transformation had a great impact on the electrical properties, especially in terms of interface trap density, which showed a minimum value of 2.98×10^{12} cm^{-2} eV^{-1} in Gd_2O_3 film 2.8 nm thick. The authors' conclusion was that the epitaxial lattice strain also positively affected the two-dimensional electron gas density at the AlGaN/GaN interface by about 40%.

Nickel and cerium oxides (NiO and CeO_2) have also been deposited onto AlGaN/GaN systems. The first report on NiO-oriented film as a gate insulating layer in AlGaN/GaN devices was related to thermal oxidation of Ni metal layers [86]. In particular, the fabrication process relied on a heating treatment, in the 300–600 °C temperature range for 5 min in air ambient, of a 10 nm-thick Ni metal layer. Besides the observation of a colour change from the dark Ni metal layer to the transparent NiO film, no details were provided on the structural or compositional characteristics of the formed NiO layers. Generally, the thermal oxidation of Ni metal layers can lead to the formation of voids in the oxide layer and/or of randomly oriented films, since the process initiates at the grain boundaries and then expands in all directions. The growth kinetics of NiO film seem to depend on the texture and crystallite size of the initial Ni metallic layer [87]. It has been shown that the strong (111) texture of the Ni layer results in slow NiO growth. These slow oxidation kinetics are related to the stronger resistance to oxidation of the Ni (111) planes [88]. Therefore, the NiO growth proceeds mainly from other crystallographic planes, mostly located at the grain boundaries. Indeed, most of the Ni grains have a (111) texture. This nonuniform growth results in increased surface roughness after oxidation.

The growth of NiO and CeO_2 thin films on AlGaN/GaN heterostructures was carried out by MOCVD at 500 °C [23–25]. TEM analysis demonstrated the formation of 16 nm-thick NiO (Figure 7a,b) and 20 nm-thick CeO_2 (Figure 7c,d), both compact and uniform films. Since no intermediate layers were visible at the interface, the occurrence of any interaction and/or oxidation of the substrate during the growth process was ruled out. Moreover, the (111) NiO planes were perfectly parallel to the (0001) planes of the AlGaN/GaN substrate.

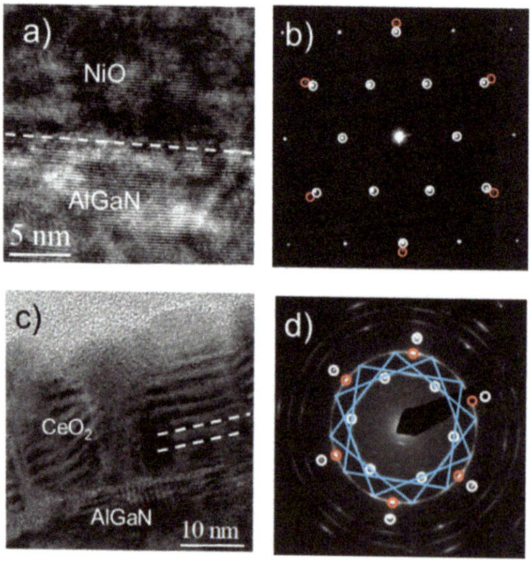

Figure 7. High-magnification cross-section TEM images (**a**) and in-plane SAED patterns (**b**) of NiO thin film deposited by MOCVD on AlGaN/GaN heterostructure at 500 °C. High-magnification cross-section TEM image (**c**) and in-plane SAED patterns (**d**) of CeO_2 thin film deposited by MOCVD on AlGaN/GaN heterostructure at 500 °C. Panel (**a**): reproduced with permission from [23]. Copyright © 2012 AIP Publishing; Panel (**d**): reproduced with permission from [25]. Copyright © 2013 AIP Publishing.

The selected area electron diffraction (SAED) pattern (Figure 7b) indicated that the external spots related to the NiO were perfectly aligned to the internal ones from the AlGaN. In particular, the white spots at 2.77 Å and 1.59 Å plane distances could be related to the (100) and (110) AlGaN/GaN planes and represented the typical 0001 zone axis pattern for a hexagonal single crystal, while the red spots forming the hexagonal pattern at 1.47 Å can be related to the (220) NiO plane; thus, only the 111 NiO zone axis is visible. The NiO spots are perfectly aligned to the AlGaN/GaN spots at 1.59 Å. Hence, it is possible to conclude that an epitaxial growth of the (111) NiO planes on the (0001) substrate plane occurred. The occurrence of the epitaxial growth can be explained by considering the threefold symmetry of the (111) NiO, which makes possible an epitaxial relationship between the hexagonal (0001) planes from the AlGaN substrate and the (111) planes of the NiO film. In particular, the lattice mismatch between the two hexagonal arrangements from the NiO and AlGaN, calculated from the electron diffraction images, was about 5%. Moreover, it is worth noting that the XRD peak position of the NiO (111) reflection was very close to that of bulk NiO, thus indicating that relaxed NiO thin films with strong diffraction intensity could be obtained under the described operating conditions. Hence, it can be concluded that NiO deposited samples were epitaxial and stress-free films and possessed excellent interface quality. TEM analysis also defined the structural relationship between the deposited CeO_2 films and the AlGaN/GaN substrate. A TEM cross-section image showed the formation of 20 nm-thick CeO_2 film and an almost perfect film/substrate interface (Figure 7c). The presence of differently oriented grains is evident, as can be deduced by the appearance of Moiré fringes. In-plane SAED was also recorded, and diffraction patterns of three different zone axes were visible. The 0001 zone axis pattern of the substrate is represented by the white circles in Figure 7d. The CeO_2 SAED pattern demonstrated that the CeO_2 film grew along two different orientations, namely, the (111) and (100) directions. In fact, the 111 zone axis pattern is represented by the red spots lying at 1.93 Å plane distances, and the 100 zone axis pattern is represented by dots lying at the vertex and at the centre of each side of the yellow squares at 1.93 Å and 2.70 Å plane distances, respectively. The 100 CeO_2 zone axis is represented by three equivalent configurations 30° rotated in the plane.

Hence, the NiO films (111) epitaxially grew on (0001) AlGaN/GaN substrate, while the CeO_2 film was not a single crystal epitaxial layer but formed by two sets of differently oriented grains (namely, (111)-oriented and (100)-oriented grains) aligned in the (0001) substrate plane of AlGaN.

The electrical characteristics of the oriented NiO and CeO_2 thin films allowed determining their experimental permittivity values. In fact, from the analysis of the C–V curves, it was possible to estimate permittivity values of 11.7 and 26 for NiO and CeO_2 films, respectively. These values were very close to those of the NiO and CeO_2 bulk permittivity (11.9 and 26) and properly higher than that of AlGaN alloys. These good values were probably due the oriented growth of the two films, which represented almost an "ideal" bulk system, in contrast to amorphous and/or polycrystalline films, which generally show lower values with respect the bulk materials.

Another key parameter to be considered in dielectric material integration onto WBG semiconductors is the effective density of the trapping states. The maximum of the trapping states determined in the AlGaN/GaN metal insulator semiconductor (MIS) diodes were 5×10^{12} $cm^{-2}eV^{-1}$ for the CeO_2 films and 6×10^{11} $cm^{-2}eV^{-1}$ for the NiO films. The trapping states of the CeO_2 were higher than those of the NiO, which could be attributed to the better structural characteristics of the NiO/AlGaN interface. While (111) NiO thin insulating layers seem to be an appealing choice as an epitaxial gate oxide, their integration into a real transistor has not been attempted yet.

4. Application of High-κ Oxides as Gate Dielectrics in SiC and GaN Transistors

As already mentioned in the introduction, most powered electronic devices based on silicon have used silicon dioxide (SiO_2) as a gate dielectric. However, the use of SiO_2 in modern devices based on WBG semiconductors can be a bottleneck for the full exploitation of the intrinsic properties of these materials because of the low value of the dielectric permittivity of SiO_2.

Figure 8 shows the schematics of common insulated gate transistors based on wide band gap semiconductors (SiC and GaN), i.e., a 4H-SiC metal oxide semiconductor field effect transistor (MOSFET) (Figure 8a), an AlGaN/GaN metal insulator semiconductor high electron mobility transistor (MISHEMT) (Figure 8b), and a recessed gate hybrid AlGaN/GaN MISHEMT (Figure 8c).

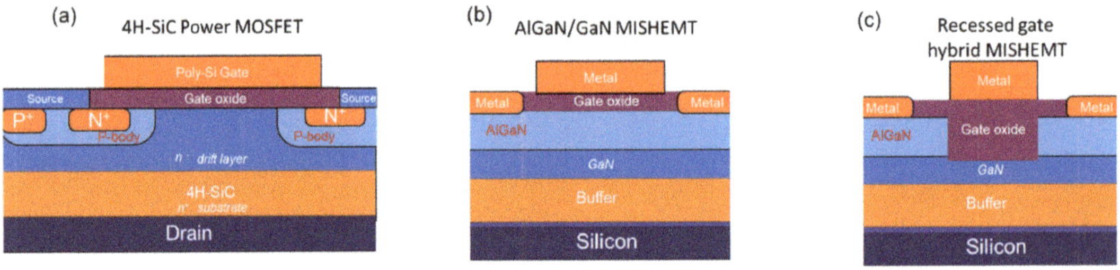

Figure 8. Schematic cross section of (**a**) a 4H-SiC power MOSFET, (**b**) an AlGaN/GaN MISHEMT, and (**c**) a recessed gate hybrid MISHEMT.

A first advantage of using a high-κ dielectric in a power device is related to the distribution of the electric field at the gate dielectric region. In particular, according to Gauss's law, the electric field in a gate dielectric E_{ins} that is placed on a semiconductor substrate, e.g., in the gate of a transistor, is given as:

$$E_{ins} = \frac{\kappa_s}{\kappa_{ins}} E_s \tag{1}$$

where κ_s and κ_{ins} are the relative dielectric permittivity values of the semiconductor and insulator, respectively, and E_s is the electric field in the semiconductor [89].

Considering as an example that the relative dielectric permittivity of 4H-SiC is 9.7 while that for SiO_2 is 3.9, according to Equation (1), the electric field in the gate oxide is about a factor of 2.5 times that in the semiconductor. Hence, when the critical electric field of 4H-SiC is reached, the maximum electric field in the oxide exceeds 9 MV/cm, thus meaning that the insulator is subjected to a significant stress, and the device reliability is penalized. In recognition of this problem, it has been proposed to replace the conventional SiO_2 gate dielectric by a high-κ insulator, with a permittivity comparable to that of SiC, so that the electric field in the gate dielectric would become closer to that in the semiconductor. In this way, the maximum electric field in the gate dielectric could be reduced, which should be satisfactory for reliable device operation. Moreover, the changes in the electric field distribution have a strong impact on the drift layer thickness required to sustain the targeted drain bias. In fact, using a high-permittivity gate dielectric allows using the optimal semiconductor drift region for the targeted breakdown, thus minimizing the specific on-resistance of the device.

Moreover, considering always the case of a SiC MOSFET (Figure 8a), the total specific on-resistance $R_{on,sp}$ of the device is given by the sum of different contributions [89]:

$$R_{on,sp} = R_{ch} + R_a + R_{JFET} + R_{drift} + R_{sub} \tag{2}$$

where R_{ch} is the channel resistance, R_a is the accumulation region resistance, R_{JFET} is the resistance of the JFET region, R_{drift} is the resistance of the drift region after the current spreading from the JFET region, and R_{sub} is the resistance of the n-type doped substrate.

R_a and R_{JFET} can be minimized by appropriately scaling the device layout, and R_{sub} can be reduced by thinning the substrate. Hence, the control of the channel resistance contribution R_{ch} is a critical point in 4H-SiC MOSFET fabrication. In particular, the channel resistance contribution R_{ch} is given by:

$$R_{ch} = \frac{(L_{ch} \cdot p)}{\mu_{inv} C_{ox} (V_G - V_{th})} \quad (3)$$

where p is the pitch of the MOSFET elementary cell, L_{ch} is the channel length, μ_{inv} is the mobility for electrons in the channel (inversion layer), C_{ox} is the specific capacitance of the gate oxide, V_{th} is the threshold voltage, and V_G is the applied gate bias. The gate oxide capacitance term C_{ox} increases with the insulator permittivity. Hence, it has a direct impact on the channel resistance and ultimately on the device's total resistance.

As pointed out by theoretical works [89,90], the use of high-κ is ideally desirable for future application in trench MOSFET technology [91].

One of the interesting features of the GaN semiconductor and its related AlGaN alloys is the possibility of growing AlGaN/GaN heterostructures. AlGaN/GaN heterostructures are characterized by the presence of a two-dimensional electron gas (2DEG) formed at the interface and possessing a high sheet charge density (in the order of 10^{13} cm^{-2}) and a high mobility (above 1000 $cm^2V^{-1}s^{-1}$) [92,93]. Moreover, GaN-based materials have a high critical electric field (above 3 MV/cm). Thanks to these unique properties, high-electron mobility transistors (HEMTs) based on AlGaN/GaN heterostructures with excellent performances have been demonstrated in recent years and are suitable candidates for high-frequency applications [94,95]. These devices are based on a Schottky barrier at the gate electrode to modulate the channel current. However, particularly for high-voltage applications in which the gate electrode is strongly reverse biased with respect to the drain, a high gate leakage current at the Schottky junction can limit the performance of these transistors [96]. Hence, a dielectric must be introduced under the gate in order to reduce the leakage current, creating a metal–insulator–semiconductor high-electron mobility transistor (MISHEMT), as schematically shown in Figure 8b. In this case, however, the choice of the gate dielectric represents a key issue for improving device performance [21,97,98] and optimizing the parasitic capacitance and the gate leakage current [19,99].

Similarly, the benefits of using high-κ materials on the characteristics of insulated gate transistors in SiC and GaN can be understood from the theoretical calculations shown in Figure 9a,b. In particular, Figure 9a shows our calculation of the threshold voltage as a function of the thickness of different high-κ dielectrics for 4H-SiC MOSFETs. As the gate dielectric thickness is increased to reduce the gate leakage current, the threshold voltage of the device (V_{th}) also increases. Hence, while an improvement in the off-state characteristics of the MOSFET is achieved, this is accompanied by a degradation in the on-state performance. However, using high-κ dielectrics as insulating gate materials instead of the conventional SiO_2, the rate of increase in the threshold voltage with the dielectric thickness is reduced. In this way, the leakage can be reduced, with a minor side effect on the output current.

Figure 9b shows the calculation of the threshold voltage of a GaN-based MISHEMT as a function of the gate dielectric layer thickness of different high-κ insulators. In this case, the V_{th} of the device is negative because of the inherent normally-on nature of these devices [92]. The negative value of V_{th} increases with increasing thickness of the gate insulator. However, the rate of this negative shift is reduced with increasing dielectric permittivity [100]. Hence, the use of high-κ gate insulators in GaN-based MISHEMTs is beneficial for reducing the power consumption of the devices.

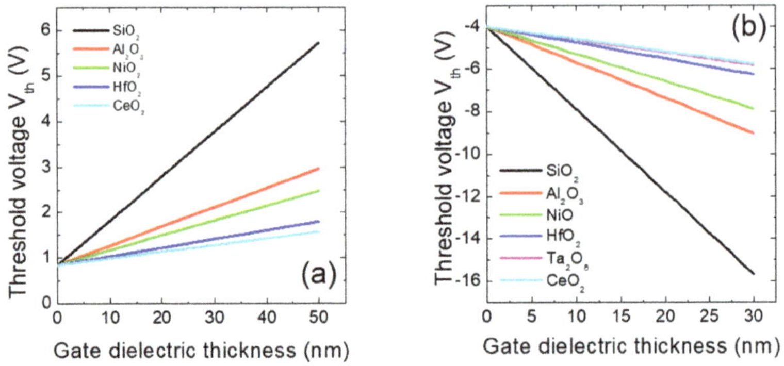

Figure 9. Calculated threshold voltage of SiC MOSFETs (**a**) and AlGaN/GaN MISHEMTs (**b**) as a function of the gate insulator layer thickness for different high-κ materials. Panel (**b**): reproduced with permission from [100]. Copyright © 2014 WILEY-VCH Verlag GmbH & Co. KGaA.

4.1. Binary High-κ Oxides in 4H-SiC MOSFETs

Since the band gap for SiC is three times larger than that for Si, the band offset at the SiO_2/SiC interface is smaller than that in the SiO_2/Si system. Hence, in SiC MOS-based systems, a higher tunnelling current than in Si is expected for a given oxide thickness [8].

Because of its high permittivity (20), hafnium oxide (HfO_2) has been widely used in Si technology. Hence, this material has attracted also the attention of the SiC scientific community. In particular, the investigation started by studying the electronic structure of the HfO_2/SiC interface [101]. However, it was clear that HfO_2 alone is not suitable for SiC because of the low conduction band offset (in the range 0.5–0.7 eV)) at the HfO_2/SiC interface, which may not provide an adequate barrier height for electron injection from the substrate [101,102]. Because of the intrinsic limitation of the band alignment, attention moved to the study of the HfO_2/SiO_2/SiC system [102].

Moreover, other high-κ binary oxides with larger band gaps and more favourable band alignment with SiC, such as Al_2O_3 [101], La_2O_3 [59,103], and ZrO_2 [104,105], have been investigated.

In general, in order to mitigate the fundamental limitations of high-κ binary oxides, the introduction of a SiO_2 interlayer between the high-κ material and SiC is often adopted [58,102].

A good survey of the literature on high-κ dielectrics for SiC was recently reported by Siddiqui et al. [106].

As described before, using high-κ dielectrics in 4H-SiC MOS-based devices can be beneficial to fully exploit the properties of the material and reduce the device's on-resistance. However, combined interaction with the SiOx layer can give further improvements. As an example, high channel mobility in 4H-SiC MOSFETs with Al_2O_3 gate insulators fabricated at low temperatures by MOCVD (64 $cm^2V^{-1}s^{-1}$) can be obtained when the Al_2O_3 gate insulator is deposited at 190 °C. According to Hino et al. [107], this result could be further improved up to an extremely high field-effect mobility of 284 $cm^2V^{-1}s^{-1}$ when the 4H-SiC MOSFET was fabricated with an ultrathin thermally grown SiOx layer inserted between the Al_2O_3 and SiC interface [107].

On this particular aspect, the impact of a thin SiO_2 layer thickness inserted between Al_2O_3 and SiC on the channel mobility in Al_2O_3/SiC MOSFETs was investigated by Hatayama et al. [108]. They demonstrated that the peak value of the field-effect mobility in Al_2O_3/SiO_2/SiC MOSFETs could reach 300 $cm^2V^{-1}s^{-1}$ for an SiO_2 thickness of 1 nm. On the other hand, when the SiO_2 layer increased up to 2 nm, the field-effect mobility drastically reduced to 40 $cm^2V^{-1}s^{-1}$ [108], as illustrated in Figure 10.

Another possible approach is employing a semiconductor surface treatment prior to gate insulator deposition. Lichtenwalner et al. [43] reported the use of a NO annealing at

1175 or 1100 °C for 20 min of a 4H-SiC semiconductor in an attempt to control the interface state density D_{it}. This procedure allowed obtaining a peak field-effect mobility in 4H-SiC MOSFETs of 106 $cm^2V^{-1}s^{-1}$ using an Al_2O_3 film deposited by ALD as gate dielectric with postdeposition annealing at 400 °C for 30 s.

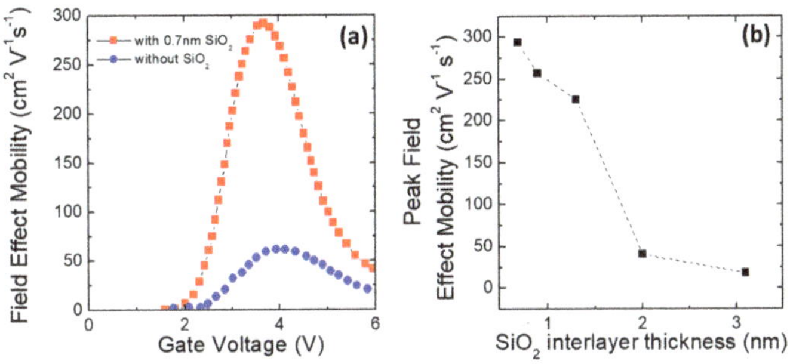

Figure 10. (**a**) Comparison between the field-effect mobility obtained in 4H-MOSFETs fabricated using Al_2O_3 insulators with and without an ultrathin thermally grown SiOx layer inserted between the Al_2O_3 and SiC interface. (**b**) Peak value of the field-effect mobility obtained using SiOx layers with different thicknesses. The data are taken from [107,108].

However, a key aspect is the channel mobility at the operative gate bias. In fact, the remarkable peak values of the field-effect mobility are often accompanied by a rapid decrease due to an increase in the gate bias close to the value at which the device should operate. This particular phenomenon can be understood analysing the single components limiting the channel mobility. As an example, a rapid decrease in the field-effect mobility is associated with a dominant phonon-scattering mechanism, while a smooth decrease with an increase in the gate bias is associated with coulombic scattering [109,110]. In particular, Arith et al. [111] demonstrated a process for forming aluminium oxide (by ALD) as a gate insulator in 4H-SiC MOSFET that did not involve the insertion or formation of SiO_2 at the interface, eliminating traps that may be present in SiO_2. This was achieved with hydrogen plasma pre-treatment followed by annealing in forming gas. Hydrogen treatment was effective at reducing D_{it} at the interface of aluminium oxide and SiC without a SiO_2 interlayer.

Clearly, because of the large differences in the mobility behaviour of the MOSFETs processed under different conditions, this topic has been strongly debated. In particular, Yoshioka et al. [47] demonstrated optimization of the interface of aluminium oxide and SiC without a SiO_2 interlayer, resulting in a low D_{it} for the metal oxide semiconductor (MOS) capacitor of 1.7×10^{12} $cm^{-2}eV^{-1}$ at $E_C - E_t = 0.2$ eV and a peak field-effect mobility of 57 $cm^2V^{-1}s^{-1}$ that was quite constant with the variation of the gate bias. Other works have tried to figure out the right combination of semiconductor surface pre-treatments and postdeposition annealing in order to improve the electrical properties of Al_2O_3/SiC interfaces [41,46].

Other processing steps have been explored to improve the performance of 4H-SiC MOSFETs, e.g., by appropriate manipulation of the SiO_2/SiC interface. In particular, Yang et al. [112] deposited 30 nm of SiO_2 by ALD and subsequently performed a postdeposition annealing (PDA) in a nitrous oxide (N_2O) ambient. The highest electron mobility of 26 $cm^2V^{-1}s^{-1}$ was achieved by performing PDA at 1100 °C for 40 s. The gate oxide could withstand effective fields up to 6 MV/cm within a leakage current range of 1×10^{-7} A/cm^2. This value of maximum electric field was small compared to that of thermally grown SiO_2, which can typically withstand up to 10 MV/cm. In another work, Yang et al. [113] inserted 1 nm of lanthanum silicate (LaSiOx) between ALD-deposited SiO_2 and 4*H*-SiC to form a gate stack. Peak mobility of 132.6 $cm^2V^{-1}s^{-1}$ was found, with three times larger current capability compared to gate oxide without La_2O_3, but no field oxide data were given. Figure 11 shows a summary of the discussed results.

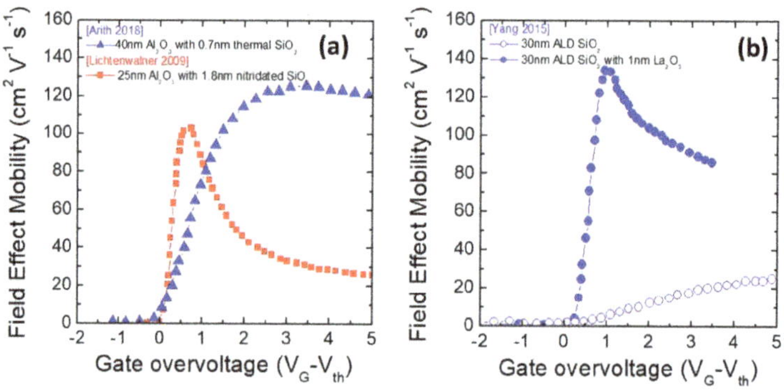

Figure 11. (**a**) Comparison between the field-effect mobility obtained in 4H-MOSFETs fabricated using Al_2O_3 insulators with the insertion of ultrathin thermally grown or nitridated SiO_2 layers. (**b**) Comparison between the field-effect mobility obtained in 4H-SiC MOSFETs fabricated using SiO_2 insulators with the insertion of ultrathin La_2O_3 layer. The data are taken from Refs. [43,111,113].

It has to be mentioned that ternary insulators have also been investigated for MOSFET application in 4H-SiC. In particular, AlON films provided interesting and reliable results both in MOS and MOSFET applications [114–116]. However, ternary elements are not the focus of this review.

Very recently, Jayawardhena et al. [117] pointed out the relevance of the appropriate pre-treatment of the semiconductor to achieve reliable and stable electric characteristics by employing ALD Al_2O_3 films directly in contact to the bare 4H-SiC surface with no interlayers. In particular, their best results were obtained with the preparation of a nitrided surface via NO annealing, i.e., a process known to passivate surface defects, and a hydrogen exposure followed by Al_2O_3 deposition on the bare 4H-SiC surface [117].

A summary of the most relevant 4H-SiC MOSFETs with different high-κ gate dielectrics is reported in Table 4.

Table 4. Survey of literature data on 4H-SiC MOSFETs with different high-κ gate dielectrics.

Gate Insulator	Thickness (nm)	V_{th} (V)	μ_{FE} ($cm^2V^{-1}s^{-1}$)	D_{it} ($cm^{-2}eV$) at $E_C - E_t = 0.2$ eV	Ref.
Al_2O_3	35	2.8	64	8×10^{11}	[108]
Al_2O_3	33	0.5 -3	52	1×10^{11} cm^{-2} (integral)	[117]
Al_2O_3 on SiO_2	35 + 2	2.8	18	8×10^{11}	[108]
	35 + 0.7	2.8	300	5×10^{11}	[108]
	40 + 0.7	2	120	6×10^{11}	[111]
	25 + 1.8	0.8	106	-	[43]
SiO_2 on La_2O_3	30 + 1	3	132	-	[113]
AlON	60 + 10	> 0	26.9	1×10^{11}	[115]

4.2. *Binary High-κ Oxides for GaN-based MISHEMTs*

Standard AlGaN/GaN MISHEMTs (see Figure 8b) are obtained by insertion of the dielectric between the metal gate and the AlGaN layer. The introduction of the gate dielectric, instead of a standard Schottky barrier gate, gives the advantage of reducing the leakage current that could limit the off-state and the gate voltage swing of the device [118]. A typical example of gate current reduction observed in HfO_2 or CeO_2 MISHEMTs is displayed in Figure 12a. Indeed, a gate leakage reduction of several orders of magnitude can be observed in both forward and reverse characteristics. This achievement allows a higher voltage swing in the device, which in turn results in a higher maximum drain current saturation value (I_{DSmax}). Another great advantage is the very high I_{ON}/I_{OFF} current ratio. Indeed, high I_{ON}/I_{OFF} current ratios between 10^6 and 10^8 have been reported in AlGaN/GaN MISHEMTs. In Figure 12b, the I_{ON}/I_{OFF} current ratio was plotted as function of the I_{DSmax}. Interesting, two families of MISHEMTs can be observed depending on the leakage current level. Despite their non-outstanding I_{DSmax}, some devices can exhibit very high I_{ON}/I_{OFF} current ratios because of their very low leakage current. On the other hand, in other cases, despite slightly higher leakage current, extraordinary I_{DSmax} values have been demonstrated. Table 5 summarizes a survey of the most promising results obtained in normally-on AlGaN/GaN MISHEMTs using different high-κ dielectrics. Indeed, not only are Al_2O_3 [119–121] and HfO_2 [122–125] indicated as suitable dielectrics, but many other gate oxide layers (Y_2O_3 [126], HZO [127], Ta_2O_5 [128], La_2O_3 [125], ZrO_2 [129–131], Gd_2O_3 [132]) have shown promising results when integrated into GaN HEMT technology.

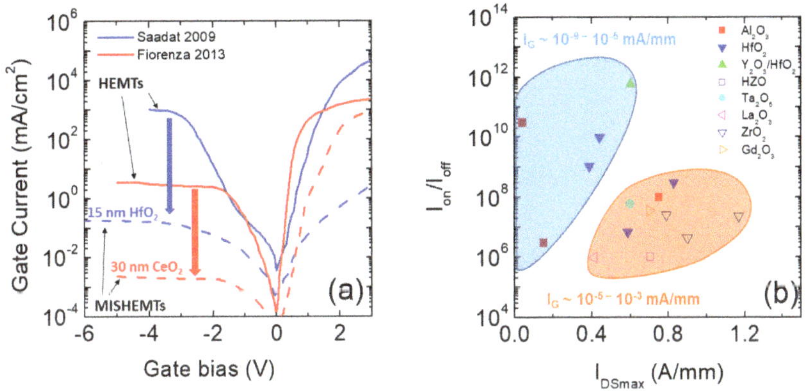

Figure 12. (**a**) Comparison of the gate current–voltage characteristics of AlGaN/GaN HEMTs (Schottky gate) and MISHEMTs employing HfO_2 and CeO_2 gate insulators. The data are taken from [25,118]. (**b**) I_{ON}/I_{OFF} versus I_{DSmax} for MISHEMTs using different gate oxides. The data are taken from Table 6 and references therein.

Table 5. Survey of literature data on normally-on AlGaN/GaN MISHEMTs with different high-κ gate dielectrics.

Dielectric	Thickness (nm)	V_{th} (V)	I_{Dmax} (mA/mm)	$I_{G\text{-}leak}$ (mA/mm)	I_{ON}/I_{OFF}	Ref.
Al_2O_3	25	−7.0	150	5.0×10^{-5}	3.0×10^6	[119]
	15	−7.0	750	8.0×10^{-5}	1.0×10^8	[120]
	30	−8.0	40	1.0×10^{-8}	3.0×10^{10}	[121]
HfO_2	20	−1.1	440	2.2×10^{-7}	1.0×10^{10}	[122]
	12	−8.0	386	1.1×10^{-9}	1.1×10^9	[123]
	23	−6.0	830	3.0×10^{-6}	3.0×10^8	[124]
	8	−3.7	585	6.5×10^{-5}	6.9×10^6	[125]
Y_2O_3/HfO_2	1/12	−5.0	600	3.0×10^{-9}	6.0×10^{11}	[126]
Ta_2O_5	24	−9.7	600	1.0×10^{-5}	6.0×10^7	[128]
La_2O_3	8	−2.9	409	1.0×10^{-4}	9.7×10^5	[125]
ZrO_2	30	−7	1168	5.4×10^{-4}	2.3×10^7	[129]
	10	−4.2	900	2.0×10^{-4}	4.5×10^6	[130]
	10	−3.9	790	3.0×10^{-5}	2.6×10^7	[131]
$HfZrO_x$	20	−12	705	6.0×10^{-4}	1.0×10^7	[127]
Gd_2O_3	4	−6.5	700	1.0×10^{-6}	3.5×10^7	[132]

A relevant concern often characterizing the behaviour of high-κ binary oxides is the occurrence of charge-trapping phenomena upon bias stress [39], which can be the cause of reliability issues in GaN insulated gate transistors. Nevertheless, the electron trapping inside the Al_2O_3 gate insulator in GaN MISHEMTs can be used to intentionally induce a positive shift in the threshold voltage and finally obtain a normally-off operation [121]. In this context, Fiorenza et al. [133] recently studied the temperature stability of these effects, demonstrating the presence of two competitive electron trapping/de-trapping mechanisms in Al_2O_3 films, which were likely related to the presence of oxygen vacancies in the material.

Slightly different is the case of normally-off recessed gate hybrid MISHEMTs (see Figure 8c). In this case, the AlGaN layer below the gate region is removed, interrupting the 2DEG channel and resulting in a positive threshold voltage. The gate region is formed by a metal/oxide/GaN (MOS) interface, which requires a positive gate voltage to accumulate electrons at the oxide/GaN interface to restore the channel device. Though this approach seems to solve the crucial issue of normally-off behaviour, the complexity of these systems generates additional concerns. As an example, the lack of a 2DEG channel in the gate region causes a notable increase the channel resistance, leading to a high final on-resistance (R_{ON}) and a reduced I_{DSmax}. To avoid this problem, it is very important to achieve high electron mobility values [134]. Hence, the oxide/GaN interface quality is clearly a key aspect to ensure a high mobility, as are the morphology of the recessed gate region and the presence of electrically active defects [135]. In this context, the choice of dielectric gate becomes crucial. The use of SiO_2 resulted into a poor interface quality displaying fast (interface) and slow (border) traps [136]. Dielectrics such as AlN [137], SiN [138], and their combination [139] have been also investigated as beneficial solutions to passivate surface N-vacancy, especially after recess etching damage in the gate region [140]. However, despite the good quality of the achieved interface and improved electron mobility, it was very difficult to obtain positive threshold voltages V_{th} well beyond the zero [137]. For these reasons, an increasing number of studies are focused on high-permittivity binary oxide layers for normally-off behaviour of AlGaN/GaN MISHEMTs. Table 6 shows the most promising results obtained in recessed gate hybrid MISHEMTs. ALD-deposited Al_2O_3 is one of the most diffused solutions for normally-off recessed gate hybrid MISHEMTs [141–147]. However, an excessive threshold voltage instability has been observed for Al_2O_3 gate insulators [121,148]. This phenomenon has been attributed to the large number of negative fixed charges incorporated in the gate stack [148,149]. As an alternative solution, ALD gate oxides with even higher dielectric constants, such as HfO_2 [150] or ZrO_2 [151–153], have been investigated for normally-off recessed MISHEMTs. Furthermore, in this case, trapped or fixed charges result in V_{th} instability issues. Other opportunities have been found in ternary oxide layers, such as $HfSiO_x$ [154] or $LaHfO_x$ [155].

Another important challenge in normally-off recessed gate hybrid MISHEMTs is the possibility of obtaining a very high saturation current I_{DSmax} with a well positive V_{th} value. In Figure 13, experimental values of I_{DSmax} are plotted as a function of the threshold voltage V_{th}. However, the values of I_{DSmax} seem to decrease in correspondence with an increase in V_{th}, thus suggesting the existence of a trade-off between a high output current and a more positive threshold voltage. In this context, a partial recession of the AlGaN barrier layer has also been explored to realize normally-off hybrid MISHEMTs. In this way, a higher 2DEG channel density is obtained. On the other hand, a more uniform and accurate AlGaN etching process is required to obtain a positive threshold voltage and normally-off devices.

Finally, to achieve normally-off behaviour in GaN-based HEMTs, the use of appropriate gate oxides with p-type semiconducting behaviour has been proposed. In fact, similarly to the most diffused p-GaN gate approach [156], the use of a p-type semiconducting oxide can lift up the conduction band at the AlGaN/GaN interface, resulting in the depletion of the 2DEG. By applying a positive gate bias V_G, it is possible to realign the conduction band of the structures, restoring the 2DEG and the channel conduction. Among these p-type semiconducting oxides, oxides such as NiO and CuO have been taken in consideration for normally-off HEMT fabrication [157–159]. The origin of the p-type doping of these oxides is still debated. The existence of negatively charged Cu or Ni vacancies and the presence of interstitial oxygen [160,161] have both been considered as possible causes. Moreover, the possibility of epitaxial CVD growth on an AlGaN or GaN template makes this approach for threshold voltage engineering in GaN technology interesting [23].

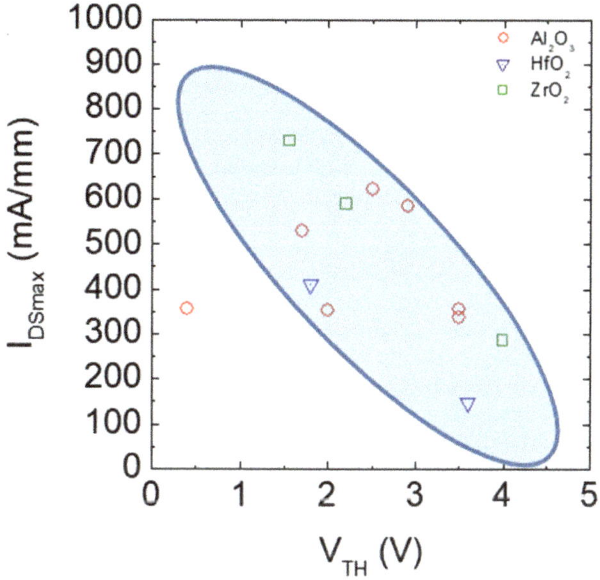

Figure 13. I_{DSmax} versus V_{th} value for recessed hybrid MISHEMTs using different high-κ binary gate oxides. The data are taken from [141–147,150–153]

Table 6. Survey of literature data on normally-off recessed gate hybrid MISHEMTs with different high-κ dielectrics.

Dielectric	Thickness (nm)	V_{TH} (V)	Mobility (cm^2/Vs)	R_{ON} (Ωmm or mΩcm^2)	I_{DSS} (mA/mm)	Ref.
Al_2O_3	30	2	225	7.8 Ωmm	353	[141]
	38	3.5	55	27 Ωmm	336	[142]
	10	1.7	251	9.8 Ωmm	528	[143]
	20	2.9	148	7.2 Ωmm	585	[144]
	30	3.5	170	9.5 Ωmm	355	[145]
	30	2.5	192	9.6 Ωmm	620	[146]
	23	0.4	396	13.3 Ωmm	356	[147]
HfO_2	30	1.8 (partial recessed)	876	5.2 mΩcm^2	411	[150]
		3.6 (total recessed)	118	12.2 mΩcm^2	146	
$HfSiO_x$	15	2.2	520	10.1 Ωmm	519	[154]
$LaHfO_x$	8	0.35		9.4 Ωmm	648	[144]
ZrO_2	20	3.99	210	24 Ωmm	286	[151]
	23	2.2 (partial recessed)	850	9.2 Ωmm	590	[152]
	16	1.55 (partial recessed)	1450	7.1 Ωmm	730	[153]

5. Conclusions

High-permittivity binary oxides for silicon carbide (SiC) and gallium nitride (GaN) electronic devices have attracted significant interest in the last decade because of the potential benefit they can bring in the device performances. In particular, special attention has been placed on the most suitable deposition techniques for their synthesis and on their implementation in real device fabrication, in which all the processes must be compatible with industrial environments and scalable to large areas. Surely the most widely investigated binary oxide is Al_2O_3, as well as its combination with HfO_2 and other materials. In fact, Al_2O_3 provides a good compromise among all the basic physical properties to be fulfilled by the gate dielectric for wide band gap semiconductors, namely, a dielectric constant close to that of the semiconductor, a large band gap, an appropriate band offset, a high critical electrical field, and good thermal stability. On the other hand, HfO_2 and other oxides possess higher dielectric constants than Al_2O_3, but their band alignments and crystallization temperatures represent a concern in application. The most affirmed method for their synthesis has been demonstrated to be the ALD approach, which can be considered the deposition technique of choice for the fabrication of very thin films with high uniformity and conformal growth on large areas. All these capabilities render ALD as very appealing for industrial implementation. In this context, beyond the fundamental study on the impact of the deposition parameters on the films' properties, the pre- and post-deposition conditions are relevant features for the development of a reliable high-κ technology for SiC and GaN. Cleaning treatments before high-κ thin film deposition, e.g., based on wet chemical solutions are the most suitable approach for both SiC and GaN substrates in order to limit the creation of interface defects. In spite of the "gentle" nature of the wet cleaning, interface states, as well as fixed charges within the binary oxides, still represent a great concern in practical applications. Hence, post-deposition and post-metallization annealing treatments need to be optimized in order to achieve the desired device performance. A common problem in SiC technology is the formation of an uncontrolled SiOx layer at the interface as well as residual carbon. Hence, the intentional Al_2O_3/SiO_2 combination has been proposed as a possible solution, although the presence of the SiO_2 interfacial layer partially reduces the advantage offered by the high-κ Al_2O_3. For that reason, the search for other material combinations and/or post-deposition treatments limiting the interfacial interaction has become mandatory.

In regard to GaN-based devices, the implementation of Al_2O_3 thin films is also the most investigated and promising solution. The interaction at the interface is limited to a partial oxidation of the substrate, which in turn might be source of electrically active defects when oxynitride bonds are present. In this case, the epitaxial growth of crystalline oxides has also been widely explored as a possible route to gate insulation in GaN-based devices, considering other oxides, such as lanthanide oxides (Gd_2O_3, Sc_2O_3, and La_2O_3) or NiO and CeO_2. However, the main limitations of the epitaxial oxides' implementation are the number of structural defects occurring after the initial layers and the presence of preferential leakage current paths at the grain boundaries.

In terms of practical device application, high-κ binary oxides have already been implemented in both 4H-SiC MOSFETs and GaN-based MISHEMTs, with Al_2O_3 being the most widely used system. In this case, while promising results in terms of channel mobility and R_{ON} have been reported, charge-trapping effects occurring in these oxides remain a limiting factor that has to be addressed by appropriate surface preparation techniques and post-annealing conditions. In particular, the integration of high-κ oxides as gate insulators in 4H-SiC MOSFETs will require optimization of the process flow, with particular attention to the thermal budget required for ohmic contact formation, which must be compatible with the crystallization temperature of the oxide.

Author Contributions: Conceptualization, R.L.N. and F.R.; methodology, P.F., G.G. and E.S.; formal analysis, P.F., G.G. and E.S.; investigation, E.S., P.F. and G.G.; resources, E.S.; data curation, E.S.; writing—original draft preparation, R.L.N., P.F. and G.G.; writing—review and editing, R.L.N. and F.R.; visualization, P.F. and G.G.; supervision, R.L.N. and F.R.; project administration, F.R.; funding acquisition, F.R. All authors have read and agreed to the published version of the manuscript.

Funding: This work was partially funded by the ECSEL-JU project WInSiC4AP (Wide Band Gap Innovative SiC for Advanced Power)—grant agreement no. 737483 and the national project EleGaNTe (Electronics on GaN-based Technologies)—PON ARS01_01007. Moreover, the authors would like to acknowledge the European project GaN4AP (GaN for Advanced Power Applications)—grant agreement no. 101007310 for funding part of their current GaN activities.

Data Availability Statement: The data that support the findings of this study are available from the corresponding author upon reasonable request.

Acknowledgments: The authors would like to acknowledge their colleagues at CNR-IMM: F. Giannazzo and M. Vivona, for the fruitful discussions and contributions in SiC and GaN experiments, and S. Di Franco and C. Bongiorno, for the precious technical support during device fabrication and TEM analyses. M. Saggio and F. Iucolano from STMicroelectronics are greatly acknowledged for their fruitful collaboration on wide band gap semiconductor research activities. The authors also thank Graziella Malandrino of the Department of Chemistry, University of Catania, for the fruitful collaboration in the realization of NiO and CeO_2 gate dielectrics by MOCVD.

Conflicts of Interest: The authors declare no conflict of interest.

References

1. Bose, K.B. Power Electronics–An Emerging Technology. *IEEE Trans. Ind. Electr.* **1989**, *36*, 403–412. [CrossRef]
2. Brunco, D.P.; De Jaeger, B.; Eneman, G.; Mitard, J.; Hellings, G.; Satta, A.; Terzieva, V.; Souriau, L.; Leys, F.E.; Pourtois, G.; et al. Germanium MOSFET Devices: Advances in Materials Understanding, Process Development, and Electrical Performance. *J. Electrochem. Soc.* **2008**, *155*, H552–H561. [CrossRef]
3. Wong, H.-S.P. Beyond the conventional transistor. *IBM J. Res. Dev.* **2002**, *46*, 133–168. [CrossRef]
4. Ostling, M.; Ghandi, R.; Zetterling, C.-M. SiC power devices Present status, applications and future perspective. In Proceedings of the 2011 IEEE 23rd International Symposium on Power Semiconductor Devices and ICs, San Diego, CA, USA, 23–26 May 2011; IEEE: Piscataway, NJ, USA, 2011; pp. 10–15.
5. Roccaforte, F.; Fiorenza, P.; Greco, G.; Nigro, R.L.; Giannazzo, F.; Iucolano, F.; Saggio, M. Emerging trends in wide band gap semiconductors (SiC and GaN) technology for power devices. *Microelectron. Eng.* **2018**, *187–188*, 66–77. [CrossRef]
6. Chen, K.J.; Haberlen, O.; Lidow, A.; Tsai, C.L.; Ueda, T.; Uemoto, Y.; Wu, Y. GaN-on-Si Power Technology: Devices and Applications. *IEEE Trans. Electron Devices* **2017**, *64*, 779–795. [CrossRef]
7. Roccaforte, F.; Fiorenza, P.; Nigro, R.L.; Giannazzo, F.; Greco, G. Physics and technology of gallium nitride materials for power electronics. *Riv. Nuovo Cim.* **2018**, *41*, 625–681.
8. Kimoto, T.; Cooper, J. *Fundamentals of Silicon Carbide Technology: Growth, Characterization, Devices and Applications*; JohnWiley & Sons: Hoboken, NJ, USA, 2014.
9. Roccaforte, F.; Leszczynski, M. Introduction to Gallium Nitride Properties and Applications. In *Nitride Semiconductor Technology: Power Electronics and Optoelectronic Devices*, 1st ed.; Roccaforte, F., Leszczyński, M., Eds.; Wiley-VCH Verlag: Weinheim, Germany; GmbH & Co. KGaA: Lippstadt, Germany, 2020; pp. 1–39.
10. Saggio, M.; Guarnera, A.; Zanetti, E.; Rascunà, S.; Frazzetto, A.; Salinas, D.; Giannazzo, F.; Fiorenza, P.; Roccaforte, F. Industrial Approach for Next Generation of Power Devices Based on 4H-SiC. *Mater. Sci. Forum* **2015**, *821–823*, 660–666. [CrossRef]
11. Arulkumaran, S.; Egawa, T.; Ishikawa, H.; Jimbo, T.; Sano, Y. Surface passivation effects on AlGaN/GaN high-electron-mobility transistors with SiO_2, Si_3N_4, and silicon oxynitride. *Appl. Phys. Lett.* **2004**, *84*, 613–615. [CrossRef]
12. Nakano, Y.; Jimbo, T. Interface properties of SiO_2/n-GaN metal–insulator–semiconductor structures. *Appl. Phys. Lett.* **2002**, *80*, 4756–4758. [CrossRef]
13. Cook, T.E.; Fulton, C.C.; Mecouch, W.; Davis, R.; Lucovsky, G.; Nemanich, R.J. Band offset measurements of the Si_3N_4/GaN (0001) interface. *J. Appl. Phys.* **2003**, *94*, 3949–3954. [CrossRef]
14. Placidi, M.; Constant, A.; Fontserè, A.; Pausas, E.; Cortes, I.; Cordier, Y.; Mestres, N.; Pérez, R.; Zabala, M.; Millán, J.; et al. Deposited Thin SiO_2 for Gate Oxide on n-Type and p-Type GaN. *J. Electrochem. Soc.* **2010**, *157*, H1008–H1013. [CrossRef]
15. Schlom, D.G.; Guha, S.; Datta, S. Gate Oxides Beyond SiO_2. *MRS Bull.* **2008**, *33*, 1017–1025. [CrossRef]
16. Kingon, A.I.; Maria, J.-P.; Streiffer, S.K. Alternative dielectrics to silicon dioxide for memory and logic devices. *Nature* **2000**, *406*, 1032–1038. [CrossRef] [PubMed]
17. Wallace, R.M.; Wilk, G.D. High-k dielectric material for microelectronics. *Critic. Rev. Solid State Mater. Sci.* **2003**, *28*, 231–285. [CrossRef]
18. Wilk, G.D.; Wallace, R.; Anthony, J.M. High-κ gate dielectrics: Current status and materials properties considerations. *J. Appl. Phys.* **2001**, *89*, 5243–5275. [CrossRef]
19. Roccaforte, F.; Fiorenza, P.; Greco, G.; Vivona, M.; Nigro, R.L.; Giannazzo, F.; Patti, A.; Saggio, M. Recent advances on dielectrics technology for SiC and GaN power devices. *Appl. Surf. Sci.* **2014**, *301*, 9–18. [CrossRef]
20. Gaskins, J.T.; Hopkins, P.E.; Merrill, D.R.; Bauers, S.R.; Hadland, E.; Johnson, D.C.; Koh, D.; Yum, J.H.; Banerjee, S.; Nordell, B.J.; et al. Review—Investigation and Review of the Thermal, Mechanical, Electrical, Optical, and Structural Properties of Atomic Layer Deposited High-kDielectrics: Beryllium Oxide, Aluminum Oxide, Hafnium Oxide, and Aluminum Nitride. *ECS J. Solid State Sci. Technol.* **2017**, *6*, N189–N208. [CrossRef]
21. Ye, P.D.; Yang, B.; Ng, K.K.; Bude, J.; Wilk, G.D.; Halder, S.; Hwang, J.C.M. GaN metal-oxide-semiconductor high-electron-mobility-transistor with atomic layer deposited Al_2O_3 as gate dielectric. *Appl. Phys. Lett.* **2005**, *86*, 63501. [CrossRef]
22. Yue, Y.; Hao, Y.; Zhang, J.; Ni, J.; Mao, W.; Feng, Q.; Liu, L. AlGaN/GaN MOS-HEMT with HfO_2 Dielectric and Al_2O_3 Interfacial Passivation Layer Grown by Atomic Layer Deposition. *IEEE Electron Device Lett.* **2008**, *29*, 838–840. [CrossRef]

23. Roccaforte, F.; Greco, G.; Fiorenza, P.; Raineri, V.; Malandrino, G.; Nigro, R.L. Epitaxial NiO gate dielectric on AlGaN/GaN heterostructures. *Appl. Phys. Lett.* **2012**, *100*, 63511. [CrossRef]
24. Fiorenza, P.; Greco, G.; Giannazzo, F.; Nigro, R.L.; Roccaforte, F. Poole-Frenkel emission in epitaxial nickel oxide on AlGaN/GaN heterostructures. *Appl. Phys. Lett.* **2012**, *101*, 172901. [CrossRef]
25. Fiorenza, P.; Greco, G.; Fisichella, G.; Roccaforte, F.; Malandrino, G.; Nigro, R.L. High permittivity cerium oxide thin films on AlGaN/GaN heterostructures. *Appl. Phys. Lett.* **2013**, *103*, 112905. [CrossRef]
26. Milanov, A.P.; Xu, K.; Cwik, S.; Parala, H.; Arcos, T.D.L.; Becker, H.-W.; Rogalla, D.; Cross, R.; Paul, S.; Devi, A. Sc_2O_3, Er_2O_3, and Y_2O_3 thin films by MOCVD from volatile guanidinate class of rare-earth precursors. *Dalton Trans.* **2012**, *41*, 13936–13947. [CrossRef] [PubMed]
27. Chen, J.-J.; Gila, B.P.; Hlad, M.; Gerger, A.; Ren, F.; Abernathy, C.R.; Pearton, S. Band offsets in the Sc_2O_3/GaN heterojunction system. *Appl. Phys. Lett.* **2006**, *88*, 142115. [CrossRef]
28. Prokofiev, A.; Shelykh, A.; Melekh, B. Periodicity in the band gap variation of Ln_2X_3 (X = O, S, Se) in the lanthanide series. *J. Alloy. Compd.* **1996**, *242*, 41–44. [CrossRef]
29. Fanciulli, M.; Scarel, G. *Rare Earth Oxide Thin Films*; Springer: Berlin/Heidelberg, Germany, 2007.
30. Fiedler, A.; Schewski, R.; Galazka, Z.; Irmscher, K. Static Dielectric Constant of β-Ga_2O_3 Perpendicular to the Principal Planes (100), (010), and (001). *ECS J. Solid State Sci. Technol.* **2019**, *8*, Q3083–Q3085. [CrossRef]
31. Kääriäinen, T.; Cameron, D.; Kääriäinen, M.L.; Sherman, A. *Atomic Layer Deposition, Principles Characteristics and Nanotechnolody Applications*; Scrivener, M., Carmical, P., Eds.; John Wiley & Sons: Hoboken, NJ, USA, 2013.
32. Ahvenniemi, E.; Akbashev, A.R.; Ali, S.; Bechelany, M.; Berdova, M.; Boyadjiev, S.; Cameron, D.C.; Chen, R.; Chubarov, M.; Cremers, V.; et al. Review Article: Recommended reading list of early publications on atomic layer deposition—Outcome of the Virtual Project on the History of ALD. *J. Vac. Sci. Technol. A Vac. Surf. Film.* **2017**, *35*, 10801. [CrossRef]
33. Shukla, M.; Dutta, G.; Mannam, R.; DasGupta, N. Electrical properties of reactive-ion-sputtered Al_2O_3 on 4H-SiC. *Thin Solid Film.* **2016**, *607*, 1–6. [CrossRef]
34. Paskaleva, A.; Ciechonski, R.R.; Syväjärvi, M.; Atanassova, E.; Yakimova, R. Electrical behavior of 4H-SiC metal-oxide-semiconductor structures with Al_2O_3 as gate dielectric. *J. Appl. Phys.* **2005**, *97*, 124507. [CrossRef]
35. Fiorenza, P.; Vivona, M.; Di Franco, S.; Smecca, E.; Sanzaro, S.; Alberti, A.; Saggio, M.; Roccaforte, F. Properties of Al_2O_3 thin films deposited on 4H-SiC by reactive ion sputtering. *Mater. Sci. Semicond. Process.* **2019**, *93*, 290–294. [CrossRef]
36. Khosa, R.Y.; Thorsteinsson, E.B.; Winters, M.; Rorsman, N.; Karhu, R.; Hassan, J.U.; Sveinbjörnsson, E.Ö. Electrical characterization of amorphous Al_2O_3 dielectric films on n-type 4H-SiC. *AIP Adv.* **2018**, *8*, 25304. [CrossRef]
37. Wolborski, M.; Bakowski, M.; Ortiz, A.; Pore, V.; Schöner, A.; Ritala, M.; Leskelä, M.; Hallén, A. Characterisation of the Al_2O_3 films deposited by ultrasonic spray pyrolysis and atomic layer deposition methods for passivation of 4H–SiC devices. *Microelectron. Reliab.* **2006**, *46*, 743–755. [CrossRef]
38. Huang, S.-W.; Hwu, J.-G. Ultrathin Aluminum Oxide Gate Dielectric on N-Type 4H-SiC Prepared by Low Thermal Budget Nitric Acid Oxidation. *IEEE Trans. Electron Devices* **2004**, *51*, 1877–1882. [CrossRef]
39. Schilirò, E.; Nigro, R.L.; Fiorenza, P.; Roccaforte, F. Negative charge trapping effects in Al_2O_3 films grown by atomic layer deposition onto thermally oxidized 4H-SiC. *AIP Adv.* **2016**, *6*, 75021. [CrossRef]
40. Schilirò, E.; Fiorenza, P.; Di Franco, S.; Bongiorno, C.; Saggio, M.; Roccaforte, F.; Nigro, R.L. Effect of SiO_2 interlayer on the properties of Al_2O_3 thin films grown by plasma enhanced atomic layer deposition on 4H-SiC substrates. *Phys. Status Solidi* **2016**, *214*, 1600365. [CrossRef]
41. Tanner, C.M.; Perng, Y.-C.; Frewin, C.L.; Saddow, S.E.; Chang, J.P. Electrical performance of Al_2O_3 gate dielectric films deposited by atomic layer deposition on 4H-SiC. *Appl. Phys. Lett.* **2007**, *91*, 203510. [CrossRef]
42. Avice, M.; Grossner, U.; Pintilie, I.; Svensson, B.G.; Servidori, M.; Nipoti, R.; Nilsen, O.; Fjellvåg, H. Electrical properties of Al_2O_3/4H-SiC structures grown by atomic layer chemical vapor deposition. *J. Appl. Phys.* **2007**, *102*, 54513. [CrossRef]
43. Lichtenwalner, D.J.; Misra, V.; Dhar, S.; Ryu, S.-H.; Agarwal, A. High-mobility enhancement-mode 4H-SiC lateral field-effect transistors utilizing atomic layer deposited Al_2O_3 gate dielectric. *Appl. Phys. Lett.* **2009**, *95*, 152113. [CrossRef]
44. Cheong, K.Y.; Moon, J.H.; Kim, H.J.; Bahng, W.; Kim, N.-K. Analysis of current conduction mechanisms in atomic-layer-deposited Al_2O_3 gate on 4H silicon carbide. *Appl. Phys. Lett.* **2007**, *90*, 162113. [CrossRef]
45. Kern, W.; Puotinen, D.P. Cleaning Solutions Based on Hydrogen Peroxide for use in Silicon Semiconductor Technology. *RCA Rev.* **1970**, *31*, 187.
46. Suvanam, S.S.; Usman, M.; Martin, D.; Yazdi, M.G.; Linnarsson, M.K.; Tempez, A.; Gotelid, M.; Hallen, A. Improved interface and electrical properties of atomic layer deposited Al_2O_3/4H-SiC. *Appl. Surf. Sci.* **2018**, *433*, 108–115. [CrossRef]
47. Yoshioka, H.; Yamazaki, M.; Harada, S. Reduction of interface states by hydrogen treatment at the aluminum oxide/4H-SiC Si-face interface. *AIP Adv.* **2016**, *6*, 105206. [CrossRef]
48. Plummer, J.D.; Deal, M.; Griffin, P.D. *Silicon VLSI Technology: Fundamentals, Practice and Modelling*, 1st ed.; Prentice Hall: Hoboken, NJ, USA, 2000.
49. Heo, S.C.; Lim, D.; Jung, W.S.; Choi, R.; Yu, H.-Y.; Choi, C. Remote plasma atomic layer deposited Al_2O_3 4H SiC MOS capacitor with remote H_2 plasma passivation and post metallization annealing. *Microelectron. Eng.* **2015**, *147*, 239–243. [CrossRef]
50. Wang, Y.-Y.; Shen, H.-J.; Bai, Y.; Tang, Y.-D.; Liu, K.-A.; Li, C.-Z.; Liu, X.-Y. Influences of high-temperature annealing on atomic layer deposited Al_2O_3/4H-SiC. *Chin. Phys. B* **2013**, *22*, 78102. [CrossRef]

51. Idris, M.I.; Wright, N.G.; Horsfall, A.B. Effect of Post Oxide Annealing on the Electrical and Interface 4H-SiC/Al_2O_3 MOS Capacitors. *Mater. Sci. Forum* **2018**, *924*, 486–489. [CrossRef]
52. Linnarsson, M.K.; Hallén, A.; Khartsev, S.; Suvanam, S.S.; Usman, M. Interface between Al_2O_3 and 4H-SiC investigated by time-of-flight medium energy ion scattering. *J. Phys. D: Appl. Phys.* **2017**, *50*, 495111. [CrossRef]
53. Liu, X.-Y.; Wang, Y.-Y.; Peng, Z.-Y.; Li, C.-Z.; Wu, J.; Bai, Y.; Tang, Y.-D.; Liu, K.-A.; Shen, H.-J. Charge trapping behavior and its origin in Al_2O_3/SiC MIS system. *Chin. Phys. B* **2015**, *24*, 87304. [CrossRef]
54. Usman, M.; Suvanam, S.S.; Yazdi, M.G.; Göthelid, M.; Sultan, M.; Hallén, A. Stoichiometry of the ALD-Al_2O_3/4H–SiC interface by synchrotron-based XPS. *J. Phys. D Appl. Phys.* **2016**, *49*, 255308. [CrossRef]
55. Usman, M.; Suvanam, S.S.; Linnarsson, M.; Hallén, A. Improving the quality of Al_2O_3/4H-SiC interface for device applications. *Mater. Sci. Semicond. Process.* **2018**, *81*, 118–121. [CrossRef]
56. Cheong, K.Y.; Moon, J.H.; Kim, H.J.; Bahng, W.; Kim, N.-K. Current conduction mechanisms in atomic-layer-deposited HfO2/nitrided SiO_2 stacked gate on 4H silicon carbide. *J. Appl. Phys.* **2008**, *103*, 84113. [CrossRef]
57. Cheong, K.Y.; Moon, J.H.; Park, T.J.; Kim, J.H.; Hwang, C.S.; Kim, H.J.; Bahng, W.; Kim, N.-K. Improved Electronic Performance of HfO_2/SiO_2 Stacking Gate Dielectric on 4H SiC. *IEEE Trans. Electron Devices* **2007**, *54*, 3409–3413. [CrossRef]
58. Afanas'Ev, V.V.; Stesmans, A.; Chen, F.; Campbell, S.A.; Smith, R. HfO_2-based insulating stacks on 4H–SiC(0001). *Appl. Phys. Lett.* **2003**, *82*, 922. [CrossRef]
59. Moon, J.H.; Eom, D.I.; No, S.Y.; Song, H.K.; Yim, J.H.; Na, H.J.; Lee, J.B.; Kim, H.J. Electrical Properties of the La_2O_3/4H-SiC Interface Prepared by Atomic Layer Deposition Using La(iPrCp)$_3$ and H_2O. *Mater. Sci. Forum* **2006**, *527–529*, 1083. [CrossRef]
60. Wang, Y.; Jia, R.; Zhao, Y.; Li, C.; Zhang, Y. Investigation of Leakage Current Mechanisms in La_2O_3/SiO_2/4H-SiC MOS Capacitors with Varied SiO_2 Thickness. *J. Electron. Mater.* **2016**, *45*, 5600–5605. [CrossRef]
61. Zhao, P.; Rusli; Lok, B.; Lai, F.; Tin, C.; Zhao, J.; Yar, R. Investigation of Ta_2O_5/SiO_2/4H-SiC MIS capacitors. *Microelectron. Eng.* **2005**, *83*, 58–60. [CrossRef]
62. Wolborski, M.; Bakowski, M.; Pore, V.; Ritala, M.; Leskelä, M.; Schöner, A.; Hallén, A. Characterization of Aluminium and Titanium Oxides Deposited on 4H-SiC by Atomic Layer Deposition Technique. *Mater. Sci. Forum* **2005**, *483–485*, 701–704. [CrossRef]
63. Nigro, R.L.; Schilirò, E.; Fiorenza, P.; Roccaforte, F. Nanolaminated Al_2O_3/HfO_2 dielectrics for silicon carbide based devices. *J. Vac. Sci. Technol. A* **2020**, *38*, 32410. [CrossRef]
64. Usman, M.; Henkel, C.; Hallén, A. HfO_2/Al_2O_3 Bilayered High-kDielectric for Passivation and Gate Insulator in 4H-SiC Devices. *ECS J. Solid State Sci. Technol.* **2013**, *2*, N3087–N3091. [CrossRef]
65. Zhao, F.; Amnuayphol, O.; Cheong, K.Y.; Wong, Y.H.; Jiang, J.-Y.; Huang, C.-F. Post deposition annealing effect on properties of Y_2O_3/Al_2O_3 stacking gate dielectric on 4H-SiC. *Mater. Lett.* **2019**, *245*, 174–177. [CrossRef]
66. Król, K.; Sochacki, M.; Taube, A.; Kwietniewski, N.; Gierałtowska, S.; Wachnicki, Ł.; Godlewski, M.; Szmidt, J. Influence of Atomic Layer Deposition Temperature on the Electrical Properties of Al/ZrO_2 /SiO_2 /4H SiC Metal Oxide Semiconductor Structures. *Phys. Status Solidi* **2018**, *215*, 1700882. [CrossRef]
67. Kerr, A.J.; Chagarov, E.; Gu, S.; Kaufman-Osborn, T.; Madisetti, S.; Wu, J.; Asbeck, P.M.; Oktyabrsky, S.; Kummel, A.C. Preparation of gallium nitride surfaces for atomic layer deposition of aluminum oxide. *J. Chem. Phys.* **2014**, *141*, 104702. [CrossRef] [PubMed]
68. Eller, B.S.; Yang, J.; Nemanich, R.J. Electronic surface and dielectric interface states on GaN and AlGaN. *J. Vac. Sci. Technol. A Vac. Surf. Film.* **2013**, *31*, 50807. [CrossRef]
69. Nepal, N.; Garces, N.Y.; Meyer, D.J.; Hite, J.; Mastro, M.A.; Eddy, J.C.R. Assessment of GaN Surface Pretreatment for Atomic Layer Deposited High-kDielectrics. *Appl. Phys. Express* **2011**, *4*, 55802. [CrossRef]
70. Duan, T.L.; Pan, J.S.; Ang, D.S. Interfacial chemistry and valence band offset between GaN and Al_2O_3 studied by X-ray photoelectron spectroscopy. *Appl. Phys. Lett.* **2013**, *102*, 201604. [CrossRef]
71. Brennan, B.; Qin, X.; Dong, H.; Kim, J.; Wallace, R.M. In Situ atomic layer deposition half cycle study of Al_2O_3 growth on AlGaN. *Appl. Phys. Lett.* **2012**, *101*, 211604. [CrossRef]
72. Schilirò, E.; Fiorenza, P.; Greco, G.; Roccaforte, F.; Nigro, R.L. Plasma enhanced atomic layer deposition of Al_2O_3 gate dielectric thin films on AlGaN/GaN substrates: The role of surface predeposition treatments. *J. Vac. Sci. Technol. A* **2017**, *35*, 1B140. [CrossRef]
73. Qin, X.; Dong, H.; Brennan, B.; Azacatl, A.; Kim, J.; Wallace, R.M. Impact of N_2 and forming gas plasma exposure on the growth and interfacial characteristics of Al_2O_3 on AlGaN. *Appl. Phys. Lett.* **2013**, *103*, 221604. [CrossRef]
74. Long, R.D.; McIntyre, P.C. Surface Preparation and Deposited Gate Oxides for Gallium Nitride Based Metal Oxide Semiconductor Devices. *Materials* **2012**, *5*, 1297–1335. [CrossRef]
75. Qin, X.; Dong, H.; Kim, J.; Wallace, R. A crystalline oxide passivation for Al_2O_3/AlGaN/GaN. *Appl. Phys. Lett.* **2014**, *105*, 141604. [CrossRef]
76. Qin, X.; Wallace, R.M. In situ plasma enhanced atomic layer deposition half cycle study of Al_2O_3 on AlGaN/GaN high electron mobility transistors. *Appl. Phys. Lett.* **2015**, *107*, 81608. [CrossRef]
77. Benrabah, S.; Legallais, M.; Besson, P.; Ruel, S.; Vauche, L.; Pelissier, B.; Thieuleux, C.; Salem, B.; Charles, M. H_3PO_4-based wet chemical etching for recovery of dry-etched GaN surfaces. *Appl. Surf. Sci.* **2021**, *28*, 152309. [CrossRef]

78. Vauche, L.; Chanuel, A.; Martinez, E.; Roure, M.-C.; Le Royer, C.; Bécu, S.; Gwoziecki, R.; Plissonnier, M. Study of an Al_2O_3/GaN Interface for Normally Off MOS-Channel High-Electron-Mobility Transistors Using XPS Characterization: The Impact of Wet Surface Treatment on Threshold Voltage VTH. *ACS Appl. Electron. Mater.* **2021**, *3*, 1170–1177. [CrossRef]
79. Schilirò, E.; Fiorenza, P.; Greco, G.; Monforte, F.; Condorelli, G.G.; Roccaforte, F.; Giannazzo, F.; Nigro, R.L. Early Growth Stages of Aluminum Oxide (Al_2O_3) Insulating Layers by Thermal- and Plasma-Enhanced Atomic Layer Deposition on AlGaN/GaN Heterostructures. *ACS Appl. Electron. Mater.* 2021, *in press*. [CrossRef]
80. Tanner, C.M.; Toney, M.F.; Lu, J.; Blom, H.-O.; Sawkar-Mathur, M.; Tafesse, M.A.; Chang, J.P. Engineering epitaxial γ-Al_2O_3 gate dielectric films on 4H-SiC. *J. Appl. Phys.* **2007**, *102*, 104112. [CrossRef]
81. Nigro, R.L.; Fiorenza, P.; Catalano, M.R.; Fisichella, G.; Roccaforte, F.; Malandrino, G. Binary and complex oxide thin films for microelectronic applications: An insight into their growth and advanced nanoscopic investigation. *Surf. Coat. Technol.* **2013**, *230*, 152–162. [CrossRef]
82. Sarkar, R.; Bhunia, S.; Nag, D.; Barik, B.C.; Das Gupta, K.; Saha, D.; Ganguly, S.; Laha, A.; Lemettinen, J.; Kauppinen, C.; et al. Epi-Gd_2O_3/AlGaN/GaN MOS HEMT on 150 mm Si wafer: A fully epitaxial system for high power application. *Appl. Phys. Lett.* **2019**, *115*, 63502. [CrossRef]
83. Jur, J.S.; Wheeler, V.D.; Lichtenwalner, D.J.; Maria, J.-P.; Johnson, M.A.L. Epitaxial growth of lanthanide oxides La_2O_3 and Sc_2O_3 on GaN. *Appl. Phys. Lett.* **2011**, *98*, 42902. [CrossRef]
84. Herrero, A.M.; Gila, B.P.; Abernathy, C.R.; Pearton, S.; Craciun, V.; Siebein, K.; Ren, F. Epitaxial growth of Sc_2O_3 films on GaN. *Appl. Phys. Lett.* **2006**, *89*, 92117. [CrossRef]
85. Nigro, R.L.; Greco, G.; Swanson, L.; Fisichella, G.; Fiorenza, P.; Giannazzo, F.; Di Franco, S.; Bongiorno, C.; Marino, A.; Malandrino, G.; et al. Potentialities of Nickel Oxide as Dielectric for GaN and SiC Devices. *Mater. Sci. Forum* **2013**, *740–742*, 777–780. [CrossRef]
86. Oh, C.S.; Youn, C.J.; Yang, G.M.; Lim, K.Y.; Yang, J.W. AlGaN/GaNAlGaN/GaN metal-oxide-semiconductor heterostructure field-effect transistor with oxidized Ni as a gate insulator. *Appl. Phys. Lett.* **2004**, *85*, 4214–4216. [CrossRef]
87. Courtade, L.; Turquat, C.; Muller, C.; Lisoni, J.; Goux, L.; Wouters, D.; Goguenheim, D.; Roussel, P.; Ortega, L. Oxidation kinetics of Ni metallic films: Formation of NiO-based resistive switching structures. *Thin Solid Film.* **2008**, *516*, 4083–4092. [CrossRef]
88. Graham, M.J.; Hussey, R.J.; Cohen, M. Influence of Oxide Structure on the Oxidation Rate of Nickel Single Crystals. *J. Electrochem. Soc.* **1973**, *120*, 1523–1529. [CrossRef]
89. Baliga, B.J. *Silicon Carbide Power Devices*; World Scientific Co. Pte. Ltd.: Singapore, 2005.
90. Suri, R.; Kirkpatrick, C.J.; Lichtenwalner, D.J.; Misra, V. Energy-band alignment of Al_2O_3 and HfAlO gate dielectrics deposited by atomic layer deposition on 4H–SiC. *Appl. Phys. Lett.* **2010**, *96*, 42903. [CrossRef]
91. Wright, N.G.; Poolamai, N.; Vassilevski, K.; Horsfall, A.B.; Johnson, C.M. Benefits of High-k Dielectrics in 4H-SiC Trench MOSFETs. *Mater. Sci. Forum* **2004**, *457–460*, 1433–1436. [CrossRef]
92. Ambacher, O.; Smart, J.A.; Shealy, J.R.; Weimann, N.; Chu, K.; Murphy, M.J.; Schaff, W.J.; Eastman, L.F.; Dimitrov, R.; Wittmer, L.L.; et al. Two-dimensional electron gases induced by spontaneous and piezoelectric polarization charges in N- and Ga-face AlGaN/GaN heterostructures. *J. Appl. Phys.* **1999**, *85*, 3222–3233. [CrossRef]
93. Ibbetson, J.P.; Fini, P.T.; Ness, K.D.; DenBaars, S.P.; Speck, J.S.; Mishra, U.K. Polarization effects, surface states, and the source of electrons in AlGaN/GaN heterostructure field effect transistors. *Appl. Phys. Lett.* **2000**, *77*, 250–252. [CrossRef]
94. Mishra, U.K.; Likun, S.; Kazior, T.E.; Wu, Y.-F. GaN-Based RF Power Devices and Amplifiers. *Proc. IEEE* **2008**, *96*, 287–305. [CrossRef]
95. Ishida, M.; Uemoto, Y.; Ueda, T.; Tanaka, T.; Ueda, D. GaN power switching devices. In Proceedings of the 2010 International Power Electronics Conference-ECCE ASIA, Sapporo, Japan, 21–24 June 2010; pp. 1014–1017. [CrossRef]
96. Kikkawa, T.; Makiyama, K.; Ohki, T.; Kanamura, M.; Imanishi, K.; Hara, N.; Joshin, K. High performance and high reliability AlGaN/GaN HEMTs. *Phys. Status Solidi* **2009**, *206*, 1135–1144. [CrossRef]
97. Chang, Y.C.; Chang, W.H.; Chiu, H.C.; Tung, L.T.; Lee, C.H.; Shiu, K.H.; Hong, M.; Kwo, J.; Hong, J.M.; Tsai, C.C. Inversion-channel GaN metal-oxide-semiconductor field-effect transistor with atomic-layer-deposited Al_2O_3 as gate dielectric. *Appl. Phys. Lett.* **2008**, *93*, 53504. [CrossRef]
98. Hashizume, T.; Ootomo, S.; Hasegawa, H. Suppression of current collapse in insulated gate AlGaN/GaN heterostructure field-effect transistors using ultrathin Al_2O_3 dielectric. *Appl. Phys. Lett.* **2003**, *83*, 2952–2954. [CrossRef]
99. Chen, K.J.; Zhou, C. Enhancement-mode AlGaN/GaN HEMT and MIS-HEMT technology. *Phys. Status Solidi A* **2011**, *208*, 434. [CrossRef]
100. Roccaforte, F.; Fiorenza, P.; Greco, G.; Nigro, R.L.; Giannazzo, F.; Patti, A.; Saggio, M. Challenges for energy efficient wide band gap semiconductor power devices. *Phys. Status Solidi* **2014**, *211*, 2063–2071. [CrossRef]
101. Tanner, C.M.; Choi, J.; Chang, J.P. Electronic structure and band alignment at the HfO_2/4H-SiC interface. *J. Appl. Phys.* **2007**, *101*, 34108. [CrossRef]
102. Mahapatra, R.; Chakraborty, A.K.; Horsfall, A.B.; Wright, N.; Beamson, G.; Coleman, K. Energy-band alignment of HfO_2/SiO_2/SiC gate dielectric stack. *Appl. Phys. Lett.* **2008**, *92*, 42904. [CrossRef]
103. Wang, Y.; Jia, R.; Li, C.; Zhang, Y. Electric properties of La_2O_3/SiO_2/4H-SiC MOS capacitors with different annealing temperatures. *AIP Adv.* **2015**, *5*, 87166. [CrossRef]
104. Chan, L.S.; Chang, Y.H.; Lee, K.Y. Analysis of C-Face 4H-SiC MOS Capacitors with ZrO_2 Gate Dielectric. *Mater. Sci. Forum* **2014**, *778–780*, 635–638. [CrossRef]

105. Kwietniewski, N.; Masłyk, M.; Werbowy, A.; Taube, A.; Gierałtowska, S.; Wachnicki, Ł.; Sochacki, M. Electrical characterization of ZnO/4H-SiC n-p heterojunction diode. *Phys. Status Solidi* **2016**, *213*, 1120–1124. [CrossRef]
106. Siddiqui, A.; Khosa, R.Y.; Usman, M. High-k dielectrics for 4H-silicon carbide: Present status and future perspectives. *J. Mater. Chem. C* **2021**, *9*, 5055–5081. [CrossRef]
107. Hino, S.; Hatayama, T.; Kato, J.; Tokumitsu, E.; Miura, N.; Oomori, T. High channel mobility 4H-SiC metal-oxide-semiconductor field-effect transistor with low temperature metal-organic chemical-vapor deposition grown Al_2O_3 gate insulator. *Appl. Phys. Lett.* **2008**, *92*, 183503. [CrossRef]
108. Hatayama, T.; Hino, S.; Miura, N.; Oomori, T.; Tokumitsu, E. Remarkable Increase in the Channel Mobility of SiC-MOSFETs by Controlling the Interfacial SiO_2 Layer Between Al_2O_3 and SiC. *IEEE Trans. Electron Devices* **2008**, *55*, 2041–2045. [CrossRef]
109. Fiorenza, P.; Greco, G.; Iucolano, F.; Patti, A.; Roccaforte, F. Channel Mobility in GaN Hybrid MOS-HEMT Using SiO_2 as Gate Insulator. *IEEE Trans. Electron Devices* **2017**, *64*, 2893–2899. [CrossRef]
110. Frazzetto, A.; Giannazzo, F.; Fiorenza, P.; Raineri, V.; Roccaforte, F. Limiting mechanism of inversion channel mobility in Al-implanted lateral 4H-SiC metal-oxide semiconductor field-effect transistors. *Appl. Phys. Lett.* **2011**, *99*, 72117. [CrossRef]
111. Arith, F.; Urresti, J.; Vasilevskiy, K.; Olsen, S.; Wright, N.; O'Neill, A. Increased Mobility in Enhancement Mode 4H-SiC MOSFET Using a Thin SiO_2/Al_2O_3 Gate Stack. *IEEE Electron Device Lett.* **2018**, *39*, 564–567. [CrossRef]
112. Yang, X.; Lee, B.; Misra, V. Electrical Characteristics of SiO_2 Deposited by Atomic Layer Deposition on 4H–SiC After Nitrous Oxide Anneal. *IEEE Trans. Electron Devices* **2016**, *63*, 2826–2830. [CrossRef]
113. Yang, X.; Lee, B.; Misra, V. High Mobility 4H-SiC Lateral MOSFETs Using Lanthanum Silicate and Atomic Layer Deposited SiO_2. *IEEE Electron Device Lett.* **2015**, *36*, 312–314. [CrossRef]
114. Watanabe, H.; Kirino, T.; Uenishi, Y.; Chanthaphan, A.; Yoshigoe, A.; Teraoka, Y.; Mitani, S.; Nakano, Y.; Nakamura, T.; Hosoi, T.; et al. Impact of Stacked $AlON/SiO_2$ Gate Dielectrics for SiC Power Devices. *ECS Trans.* **2011**, *35*, 265–274. [CrossRef]
115. Hosoi, T.; Kagei, Y.; Kirino, T.; Watanabe, Y.; Kozono, K.; Mitani, S.; Nakano, Y.; Nakamura, T.; Watanabe, H. Improved Characteristics of 4H-SiC MISFET with AlON/Nitrided SiO_2 Stacked Gate Dielectrics. *Mater. Sci. Forum* **2010**, *645–648*, 991–994. [CrossRef]
116. Chanthaphan, A.; Hosoi, T.; Nakano, Y.; Nakamura, T.; Shimura, T.; Watanabe, H. Improved bias-temperature instability characteristics in SiC metal-oxide-semiconductor devices with aluminum oxynitride dielectrics. *Appl. Phys. Lett.* **2014**, *104*, 122105. [CrossRef]
117. Jayawardhena, I.U.; Ramamurthy, R.P.; Morisette, D.; Ahyi, A.C.; Thorpe, R.; Kuroda, M.A.; Feldman, L.C.; Dhar, S. Effect of surface treatments on ALD Al_2O_3/4H-SiC metal–oxide–semiconductor field-effect transistors. *J. Appl. Phys.* **2021**, *129*, 75702. [CrossRef]
118. Saadat, O.I.; Chung, J.W.; Piner, E.L.; Palacios, T. Gate-First AlGaN/GaN HEMT Technology for High-Frequency Applications. *IEEE Electron Device Lett.* **2009**, *30*, 1254–1256. [CrossRef]
119. Zhang, Y.; Sun, M.; Joglekar, S.J.; Fujishima, T.; Palacios, T. Threshold voltage control by gate oxide thickness in fluorinated GaN metal-oxide-semiconductor high-electron-mobility transistors. *Appl. Phys. Lett.* **2013**, *103*, 33524. [CrossRef]
120. Hou, B.; Ma, X.; Zhu, J.; Yang, L.; Chen, W.; Mi, M.; Zhu, Q.; Chen, L.; Zhang, R.; Zhang, M.; et al. 0.9-A/mm, 2.6-V Flash-Like Normally-Off Al_2O_3/AlGaN/GaN MIS-HEMTs Using Charge Trapping Technique. *IEEE Electron Device Lett.* **2018**, *39*, 397–400. [CrossRef]
121. Schilirò, E.; Fiorenza, P.; Bongiorno, C.; Spinella, C.; Di Franco, S.; Greco, G.; Nigro, R.L.; Roccaforte, F. Correlating electron trapping and structural defects in Al_2O_3 thin films deposited by plasma enhanced atomic layer deposition. *AIP Adv.* **2020**, *10*, 125017. [CrossRef]
122. Kubo, T.; Egawa, T. Electrical characteristics and interface properties of ALD-HfO_2/AlGaN/GaN MIS-HEMTs fabricated with post-deposition annealing. *Semicond. Sci. Technol.* **2017**, *32*, 125016. [CrossRef]
123. Huang, Z.; Liu, J.; Huang, X.; Yao, J.; Zhang, J.; Guo, Y. Analysis of Interface Properties in AlGaN/GaN MIS-HEMTs with HfO_2 and SiNx Gate Dielectric. In Proceedings of the 2020 10th International Conference on Power and Energy Systems (ICPES), Chengdu, China, 25–27 December 2020; pp. 549–552.
124. Liu, C.; Chor, E.F.; Tan, L.S. Investigations of HfO_2/AlGaN/GaN metal-oxide-semiconductor high electron mobility transistors. *Appl. Phys. Lett.* **2006**, *88*, 173504. [CrossRef]
125. Lin, Y.C.; Lin, T.W.; Wu, C.H.; Yao, J.N.; Hsu, H.T.; Shih, W.C.; Kakushima, K.; Tsutsui, K.; Iwai, H.; Chang, E.Y. Optimization of gate insulator material for GaN MIS-HEMT. In Proceedings of the 2016 28th International Symposium on Power Semiconductor Devices and ICs (ISPSD), Prague, Czech Republic, 12–16 June 2016; pp. 115–118. [CrossRef]
126. Shi, Y.-T.; Xu, W.-Z.; Zeng, C.-K.; Ren, F.-F.; Ye, J.-D.; Zhou, D.; Chen, D.-J.; Zhang, R.; Zheng, Y.; Lu, H. High-k HfO_2-Based AlGaN/GaN MIS-HEMTs With Y_2O_3 Interfacial Layer for High Gate Controllability and Interface Quality. *IEEE J. Electron Devices Soc.* **2020**, *8*, 15–19. [CrossRef]
127. Cui, X.; Cheng, W.; Hua, Q.; Liang, R.; Hu, W.; Wang, Z.L. Enhanced performances of AlGaN/GaN HEMTs with dielectric engineering of HfZrOx. *Nano Energy* **2020**, *68*, 104361. [CrossRef]
128. Kumar, S.; Kumar, H.; Vura, S.; Pratiyush, A.S.; Charan, V.S.; Dolmanan, S.B.; Tripathy, S.; Muralidharan, R.; Nath, D.N. Investigation of Ta_2O_5 as an Alternative High-k Dielectric for InAlN/GaN MOS-HEMT on Si. *IEEE Trans. Electron Devices* **2019**, *66*, 1230–1235. [CrossRef]

129. Balachander, K.; Arulkumaran, S.; Ishikawa, H.; Baskar, K.; Egawa, T. Studies on electron beam evaporated ZrO_2/AlGaN/GaN metal-oxide-semiconductor high-electron-mobility transistors. *Phys. Status Solidi* **2005**, *202*, R16–R18. [CrossRef]
130. Kuzmik, J.; Pozzovivo, G.; Abermann, S.; Gonschorek, M.; Feltin, E.; Grandjean, N.; Bertagnolli, E.; Pogany, D.; Carlin, J.-F.; Strasser, G. Technology and Performance of InAlN/AlN/GaN HEMTs With Gate Insulation and Current Collapse Suppression Using ZrO_2 or HfO_2. *IEEE Trans. Electron Devices* **2008**, *55*, 937–941. [CrossRef]
131. Ye, G.; Wang, H.; Arulkumaran, S.; Ng, G.I.; Hofstetter, R.; Li, Y.; Anand, M.J.; Ang, K.S.; Maung, Y.K.T.; Foo, S.C. Atomic layer deposition of ZrO_2 as gate dielectrics for AlGaN/GaN metal-insulator-semiconductor high electron mobility transistors on silicon. *Appl. Phys. Lett.* **2013**, *103*, 142109. [CrossRef]
132. Gao, Z.; Romero, M.F.; Redondo-Cubero, A.; Pampillon, M.A.; Andres, E.S.; Calle, F. Effects of Gd_2O_3 Gate Dielectric on Proton-Irradiated AlGaN/GaN HEMTs. *IEEE Electron Device Lett.* **2017**, *38*, 611–614. [CrossRef]
133. Fiorenza, P.; Schilirò, E.; Greco, G.; Vivona, M.; Cannas, M.; Giannazzo, F.; Nigro, R.L.; Roccaforte, F. Temperature and time dependent electron trapping in Al_2O_3 thin films onto AlGaN/GaN heterostructures. *Appl. Surf. Sci.* **2021**, *579*, 152136. [CrossRef]
134. Li, Z.; Chow, T.P. Channel scaling of hybrid GaN MOS-HEMTs. *Solid-State Electron.* **2011**, *56*, 111–115. [CrossRef]
135. Roccaforte, F.; Greco, G.; Fiorenza, P.; Iucolano, F. An Overview of Normally-Off GaN-Based High Electron Mobility Transistors. *Materials* **2019**, *12*, 1599. [CrossRef] [PubMed]
136. Fiorenza, P.; Greco, G.; Iucolano, F.; Patti, A.; Roccaforte, F. Slow and fast traps in metal-oxide-semiconductor capacitors fabricated on recessed AlGaN/GaN heterostructures. *Appl. Phys. Lett.* **2015**, *106*, 142903. [CrossRef]
137. Hosoi, T.; Watanabe, K.; Nozaki, M.; Yamada, T.; Shimura, T.; Watanabe, H. Mobility enhancement in recessed-gate AlGaN/GaN MOS-HFETs using an AlON gate insulator. *Jpn. J. Appl. Phys.* **2019**, *58*, SCCD16. [CrossRef]
138. Oka, T.; Nozawa, T. AlGaN/GaN Recessed MIS-Gate HFET With High-Threshold-Voltage Normally-Off Operation for Power Electronics Applications. *IEEE Electron Device Lett.* **2008**, *29*, 668–670. [CrossRef]
139. Greco, G.; Fiorenza, P.; Iucolano, F.; Severino, A.; Giannazzo, F.; Roccaforte, F. Conduction Mechanisms at Interface of AlN/SiN Dielectric Stacks with AlGaN/GaN Heterostructures for Normally-off High Electron Mobility Transistors: Correlating Device Behavior with Nanoscale Interfaces Properties. *ACS Appl. Mater. Interfaces* **2017**, *9*, 35383–35390. [CrossRef]
140. Asahara, R.; Nozaki, M.; Yamada, T.; Ito, J.; Nakazawa, S.; Ishida, M.; Ueda, T.; Yoshigoe, A.; Hosoi, T.; Shimura, T.; et al. Effect of nitrogen incorporation into Al-based gate insulators in AlON/AlGaN/GaN metal–oxide–semiconductor structures. *Appl. Phys. Express* **2016**, *9*, 101002. [CrossRef]
141. Im, K.-S.; Ha, J.-B.; Kim, K.-W.; Lee, J.-S.; Kim, D.-S.; Hahm, S.-H.; Lee, J.-H. Normally off GaN MOSFET Based on AlGaN/GaN Heterostructure with Extremely High 2DEG Density Grown on Silicon Substrate. *IEEE Electron Device Lett.* **2010**, *31*, 192–194. [CrossRef]
142. Kim, K.-W.; Jung, S.-D.; Kim, D.-S.; Kang, H.-S.; Im, K.-S.; Oh, J.-J.; Ha, J.-B.; Shin, J.-K.; Lee, J.-H. Effects of TMAH Treatment on Device Performance of Normally off Al_2O_3/GaN MOSFET. *IEEE Electron Device Lett.* **2011**, *32*, 1376–1378. [CrossRef]
143. Wang, Y.; Wang, M.; Xie, B.; Wen, C.P.; Wang, J.; Hao, Y.; Wu, W.; Chen, K.J.; Shen, B. High-Performance Normally-Off Al_2O_3/GaN MOSFET Using a Wet Etching-Based Gate Recess Technique. *IEEE Electron Device Lett.* **2013**, *34*, 1370–1372. [CrossRef]
144. Wang, M.; Wang, Y.; Zhang, C.; Xie, B.; Wen, C.P.; Wang, J.; Hao, Y.; Wu, W.; Chen, K.J.; Shen, B. 900 V/1.6 $m\Omega{\cdot}cm^2$ Normally Off Al_2O_3/GaN MOSFET on Silicon Substrate. *IEEE Trans. Electron Devices* **2014**, *61*, 2035–2040. [CrossRef]
145. Yao, Y.; He, Z.; Yang, F.; Shen, Z.; Zhang, J.; Ni, Y.; Li, J.; Wang, S.; Zhou, G.; Zhong, J.; et al. Normally-off GaN recessed-gate MOSFET fabricated by selective area growth technique. *Appl. Phys. Express* **2013**, *7*, 16502. [CrossRef]
146. He, L.; Yang, F.; Li, L.; Chen, Z.; Shen, Z.; Zheng, Y.; Yao, Y.; Ni, Y.; Zhou, D.; Zhang, X.; et al. High Threshold Voltage Uniformity and Low Hysteresis Recessed-Gate Al_2O_3/AlN/GaN MISFET by Selective Area Growth. *IEEE Trans. Electron Devices* **2017**, *64*, 1554–1560. [CrossRef]
147. Tao, M.; Liu, S.; Xie, B.; Wen, C.P.; Wang, J.; Hao, Y.; Wu, W.; Cheng, K.; Shen, B.; Wang, M. Characterization of 880 V Normally Off GaN MOSHEMT on Silicon Substrate Fabricated With a Plasma-Free, Self-Terminated Gate Recess Process. *IEEE Trans. Electron Devices* **2018**, *65*, 1453–1457. [CrossRef]
148. Ťapajna, M.; Válik, L.; Gucmann, F.; Gregusova, D.; Fröhlich, K.; Haščík, Š.; Dobročka, E.; Tóth, L.; Pécz, B.; Kuzmík, J. Low-temperature atomic layer deposition-grown Al_2O_3 gate dielectric for GaN/AlGaN/GaN MOS HEMTs: Impact of deposition conditions on interface state density. *J. Vac. Sci. Technol. B* **2017**, *35*, 1A107. [CrossRef]
149. Fiorenza, P.; Greco, G.; Schilirò, E.; Iucolano, F.; Nigro, R.L.; Roccaforte, F. Determining oxide trapped charges in Al_2O_3 insulating films on recessed AlGaN/GaN heterostructures by gate capacitance transients measurements. *Jpn. J. Appl. Phys.* **2018**, *57*, 050307. [CrossRef]
150. Zhao, Y.; Wang, C.; Zheng, X.; Ma, X.; He, Y.; Liu, K.; Li, A.; Peng, Y.; Zhang, C.; Hao, Y. Effects of recess depths on performance of AlGaN/GaN power MIS-HEMTs on the Si substrates and threshold voltage model of different recess depths for the using HfO_2 gate insulator. *Solid-State Electron.* **2020**, *163*, 107649. [CrossRef]
151. Anderson, T.J.; Wheeler, V.D.; Shahin, D.I.; Tadjer, M.J.; Koehler, A.D.; Hobart, K.D.; Christou, A.; Kub, F.J.; Eddy, C.R. Enhancement mode AlGaN/GaN MOS high-electron-mobility transistors with ZrO_2 gate dielectric deposited by atomic layer deposition. *Appl. Phys. Express* **2016**, *9*, 11–14. [CrossRef]
152. Jiang, H.; Tang, C.W.; Lau, K.M. Enhancement-Mode GaN MOS-HEMTs With Recess-Free Barrier Engineering and High- *k* ZrO_2 Gate Dielectric. *IEEE Electron Device Lett.* **2018**, *39*, 405–408. [CrossRef]

153. Cai, Y.; Zhang, Y.; Liang, Y.; Mitrovic, I.Z.; Wen, H.; Liu, W.; Zhao, C. Low on -State Resistance Normally- off AlGaN/GaN MIS-HEMTs with Partially Recessed Gate and ZrOx. *IEEE Trans. Electron Devices* **2021**, *68*, 4310–4316. [CrossRef]
154. Hu, Q.; Li, S.; Li, T.; Wang, X.; Li, X.; Wu, Y. Channel Engineering of Normally-OFF AlGaN/GaN MOS-HEMTs by Atomic Layer Etching and High-κ Dielectric. *IEEE Electron Device Lett.* **2018**, *39*, 1377–1380. [CrossRef]
155. Lin, Y.C.; Huang, Y.X.; Huang, G.N.; Wu, C.H.; Yao, J.N.; Chu, C.M.; Chang, S.; Hsu, C.C.; Lee, J.H.; Kakushima, K.; et al. Enhancement-Mode GaN MIS-HEMTs With LaHfOx Gate Insulator for Power Application. *IEEE Electron Device Lett.* **2017**, *38*, 1101–1104. [CrossRef]
156. Greco, G.; Iucolano, F.; Roccaforte, F. Review of technology for normally-off HEMTs with p-GaN gate. *Mater. Sci. Semicond. Process.* **2018**, *78*, 96–106. [CrossRef]
157. Li, Y.; Guo, Y.; Zhang, K.; Zou, X.; Wang, J.; Kong, Y.; Chen, T.; Jiang, C.; Fang, G.; Liu, C.; et al. Positive Shift in Threshold Voltage Induced by CuO and NiOx Gate in AlGaN/GaN HEMTs. *IEEE Trans. Electron Devices* **2017**, *64*, 3139–3144. [CrossRef]
158. Kaneko, N.; Machida, O.; Yanagihara, M.; Iwakami, S.; Baba, R.; Goto, H.; Iwabuchi, A. Normally-off AlGaN/GaN HFETs using NiOx gate with recess. In Proceedings of the 2009 21st International Symposium on Power Semiconductor Devices & IC's, Barcelona, Spain, 14–18 June 2009; pp. 25–28. [CrossRef]
159. Wang, L.; Li, L.; Xie, T.; Wang, X.; Liu, X.; Ao, J.-P. Threshold voltage tuning in AlGaN/GaN HFETs with p-type Cu_2O gate synthesized by magnetron reactive sputtering. *Appl. Surf. Sci.* **2018**, *437*, 98–102. [CrossRef]
160. Kohmoto, O.; Nakagawa, H.; Isagawa, Y.; Chayahara, A. Effect of heat treatment on the oxygen content and resistivity in sputtered NiO films. *J. Magn. Magn. Mater.* **2001**, *226–230*, 1629–1630. [CrossRef]
161. Raebiger, H.; Lany, S.; Zunger, A. Origins of thep-type nature and cation deficiency in Cu_2O and related materials. *Phys. Rev. B* **2007**, *76*, 45209. [CrossRef]

materials MDPI

Review

Status of Aluminum Oxide Gate Dielectric Technology for Insulated-Gate GaN-Based Devices

Anthony Calzolaro [1,2,*], Thomas Mikolajick [1,2] and Andre Wachowiak [2]

[1] Nanoelectronics, TU Dresden, D-01062 Dresden, Germany; thomas.mikolajick@tu-dresden.de
[2] NaMLab gGmbH, Nöthnitzer Str. 64a, D-01187 Dresden, Germany; andre.wachowiak@namlab.com
* Correspondence: anthony.calzolaro@namlab.com

Abstract: Insulated-gate GaN-based transistors can fulfill the emerging demands for the future generation of highly efficient electronics for high-frequency, high-power and high-temperature applications. However, in contrast to Si-based devices, the introduction of an insulator on (Al)GaN is complicated by the absence of a high-quality native oxide for GaN. Trap states located at the insulator/(Al)GaN interface and within the dielectric can strongly affect the device performance. In particular, although AlGaN/GaN metal–insulator–semiconductor high electron mobility transistors (MIS-HEMTs) provide superior properties in terms of gate leakage currents compared to Schottky-gate HEMTs, the presence of an additional dielectric can induce threshold voltage instabilities. Similarly, the presence of trap states can be detrimental for the operational stability and reliability of other architectures of GaN devices employing a dielectric layer, such as hybrid MIS-FETs, trench MIS-FETs and vertical FinFETs. In this regard, the minimization of trap states is of critical importance to the advent of different insulated-gate GaN-based devices. Among the various dielectrics, aluminum oxide (Al_2O_3) is very attractive as a gate dielectric due to its large bandgap and band offsets to (Al)GaN, relatively high dielectric constant, high breakdown electric field as well as thermal and chemical stability against (Al)GaN. Additionally, although significant amounts of trap states are still present in the bulk Al_2O_3 and at the Al_2O_3/(Al)GaN interface, the current technological progress in the atomic layer deposition (ALD) process has already enabled the deposition of promising high-quality, uniform and conformal Al_2O_3 films to gate structures in GaN transistors. In this context, this paper first reviews the current status of gate dielectric technology using Al_2O_3 for GaN-based devices, focusing on the recent progress in engineering high-quality ALD-Al_2O_3/(Al)GaN interfaces and on the performance of Al_2O_3-gated GaN-based MIS-HEMTs for power switching applications. Afterwards, novel emerging concepts using the Al_2O_3-based gate dielectric technology are introduced. Finally, the recent status of nitride-based materials emerging as other gate dielectrics is briefly reviewed.

Keywords: GaN; gate dielectric; aluminum oxide; interface; traps; instability

Citation: Calzolaro, A.; Mikolajick, T.; Wachowiak, A. Status of Aluminum Oxide Gate Dielectric Technology for Insulated-Gate GaN-Based Devices. *Materials* **2022**, *15*, 791. https://doi.org/10.3390/ma15030791

Academic Editor: Fabrizio Roccaforte

Received: 16 December 2021
Accepted: 16 January 2022
Published: 21 January 2022

1. Introduction

Owing to the large bandgap of 3.43 eV, resulting in a high electric breakdown field of 3.3 MV/cm and in a low intrinsic carrier concentration, and to the large saturation velocity of 2.5×10^7 cm/s, GaN is one of the most promising semiconductors for the future energy-efficient generation of high-power, high-frequency and high-temperature electronics [1–4]. Besides the unique intrinsic material properties, one of the most attractive properties of GaN is the possibility to exploit the polar nature of GaN-based materials to form AlGaN/GaN heterostructures featuring a two-dimensional electron gas (2DEG) at the heterointerface with a high carrier density of over 1×10^{13} cm^{-2} and high mobility values exceeding 2000 cm^2 V^{-1} s^{-1} [5,6]. AlGaN/GaN heterostructures enable the fabrication of high electron mobility transistors (HEMTs) which can significantly outperform the traditional Si power devices in terms of breakdown strength, on-resistance and switching speed, achieving higher power density and higher energy efficiency [7,8].

Nowadays, GaN-on-Si HEMTs qualified for 200 V and 650 V high voltage power switching applications with operating frequency capabilities in the MHz range are commercially available and on the way towards 1.2 kV applications using engineered substrates [1,9,10]. For targeting higher voltage capabilities up to 1.7–1.8 kV, current aperture vertical electron transistors (CAVETs) adopting AlGaN/GaN heterojunctions have also recently attracted significant attention [11–13], where the high conductivity of the 2DEG channel is combined with the better field distribution of the vertical device geometry, and hence with the capability of vertical architectures of achieving an even higher breakdown voltage without enlarging the device area, in contrast to lateral transistors. In addition, GaN-based HEMTs with downscaled gate lengths to the sub-100 nm regime have also been demonstrated to achieve maximum current gain cutoff frequencies over 200 GHz, which are well suited for radio frequency (RF) high power amplifiers for 5G and beyond applications [14–18].

Despite the potentiality of AlGaN/GaN HEMTs, one of the most serious problems degrading the device performance and reliability is represented by the exceedingly high leakage currents through the Schottky-gate contact, especially under forward gate bias, which limits the gate voltage swing and the maximum on-state current of the device, resulting in reduced power efficiency and weak device failure protection [19]. In particular, a small gate swing is a strong limiting factor for power switching applications due to faulty gate voltage overshoots often occurring in circuits, which can eventually lead to early device failures. Moreover, since GaN-based HEMTs are naturally normally on (or depletion-mode) transistors with a negative threshold voltage (V_{th}), normally off (or enhancement-mode) HEMTs with a positive V_{th} are highly preferred to guarantee safe operation and for the reduced power consumption in power switching devices [20,21]. However, since normally off devices require a large positive gate voltage to be turned on, the problem of gate leakage currents becomes even more critical in normally off HEMTs. Similarly, in RF applications, power amplifiers using Schottky-gate HEMTs can suffer from reduced gain and efficiency caused by large gate inputs, which can drive the devices into deep forward bias regimes [22].

The employment of a metal–insulator–semiconductor (MIS) gate is an efficient way to suppress the gate leakage currents of AlGaN/GaN HEMTs, enabling reduced power consumption, a larger gate bias swing and a better immunity to gate breakdown [23–25]. However, in contrast to Si-based devices, the introduction of an insulator in AlGaN/GaN metal–insulator–semiconductor high electron mobility transistors (MIS-HEMTs) is complicated by the absence of a high-quality native oxide for (Al)GaN. Trap states located at the dielectric/(Al)GaN interface or within the dielectric can lead to dynamic charge/discharge processes, which are especially critical in the case of wide bandgap GaN-based materials where the traps can be deeply located in the bandgap and can cause severe operational instability due to their slow detrapping behavior [26–29]. The instability of the threshold voltage in AlGaN/GaN MIS-HEMTs is one of the major challenges [30–33]. In particular, a serious V_{th} shift induced by the "spill-over" of electrons from the 2DEG channel towards the dielectric/(Al)GaN interface in forward gate bias conditions has often been reported [34–37]. Another problem is the degradation of the current linearity in the transfer characteristics of AlGaN/GaN MIS-HEMTs, which can be responsible for gain loss and the degradation of large signal linearity in power amplifiers [38].

Besides insulated-gate GaN-based transistors adopting AlGaN/GaN heterostructures, other emerging GaN-based devices in the form of MIS-FETs use a gate dielectric layer on a GaN channel, such as lateral hybrid GaN MIS-FETs [39], vertical GaN trench MIS-FETs [40] and vertical FinFETs [41]. Figure 1 schematically summarizes the main configurations of lateral and vertical GaN-based transistors employing a gate dielectric layer. A Schottky-gate HEMT is reported for comparison in Figure 1a. Hybrid GaN MIS-FETs obtained by a fully recessed AlGaN barrier layer are especially attractive for normally off operation and large gate voltage operation [42], while GaN trench MIS-FETs have drawn attention among other vertical transistor concepts since they are inherently normally off with a $V_{th} > 3$ V and do not need the regrowth of the AlGaN/GaN channels [43,44]. Similar to MIS-HEMTs, instabilities over the gate dielectric affecting the device performance can arise in GaN devices with MIS-FET configurations due to the trap states at the dielectric/GaN interface influencing the V_{th} and reducing the current drive capability or/and bulk or border traps within the dielectric itself, which might mostly affect the long-term reliability performance of the device [43,45]. Moreover, differently from the MIS-HEMTs, where the 2DEG formed at the AlGaN/GaN interface benefits from the spatial separation from the dielectric/AlGaN interface, minimizing the interface scattering processes [46], interface traps in GaN MIS-FET configurations are located in the proximity of the electron channel and are more prone to act as impurity scattering centers, additionally affecting the carrier mobility [47]. This is particularly challenging for transistor concepts including a recess of the AlGaN barrier layer, as in the case of trench MIS-FETs and vertical FinFETs, since the etching process can critically affect the properties of the dielectric/GaN interface [48].

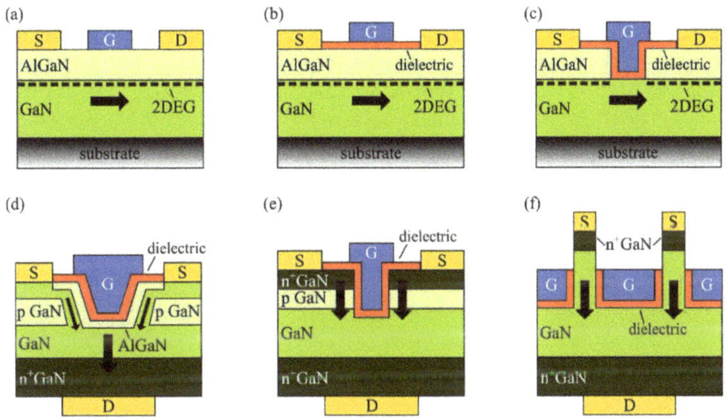

Figure 1. Schematic of representative GaN-based transistors: (**a**) Schottky-gate HEMT; (**b**) MIS-HEMT; (**c**) hybrid MIS-FET with fully recessed AlGaN barrier; (**d**) trench CAVET; (**e**) trench MIS-FET; (**f**) vertical FinFET.

Therefore, regardless of the transistor concept and design, trap states need to be minimized to ensure the safe operation and long-term lifetime of the insulated-gate GaN-based transistors. In particular, a gate dielectric technology aiming to improve the dielectric/(Al)GaN interface and dielectric bulk quality is essential to enhance the performance of the device. In general, various insulator materials have been employed, with SiO_2, SiN and Al_2O_3 as the most commonly used dielectrics [7,29,49,50]. The same dielectric layer deposited both underneath the gate as well as between the gate and the ohmic contacts of the source and drain usually functions both as the gate dielectric and the passivation layer [51]. The latter has been indeed reported to mitigate the effects of drain current collapse and leakage currents at the (Al)GaN surface due to the passivation of trap states at the surface [52–54]. However, even though excellent device characteristics have been obtained, trap states in MIS gate structures still remain one of the biggest challenges for insulated-gate GaN-based transistors, and the practical implementation of these devices has been hindered by the concerns over the gate dielectric stability and reliability [1].

In this paper, the current status of the gate dielectric technology employing Al_2O_3 for insulated-gate GaN-based transistors is reviewed. First, the relevant aspects taken into account for selecting a suitable gate dielectric for GaN-based transistors are highlighted and the influence of this additional layer on the device parameters and performance is discussed. Afterwards, the state of the art of Al_2O_3 as a gate dielectric is presented with a particular attention to the recent progress in engineering high-quality Al_2O_3/(Al)GaN interfaces and to the performance of Al_2O_3-gated GaN-based MIS-HEMTs for power switching applications. Novel emerging concepts using the Al_2O_3-based gate dielectric technology are also introduced. Finally, the recent status of nitride-based materials emerging as other gate dielectrics is briefly reviewed.

2. Gate Dielectrics on (Al)GaN

The design of a MIS gate structure for insulated-gate GaN-based transistors requires consideration of the properties of the bandgap, the band offset to (Al)GaN, the permittivity and the chemical stability of the insulators [7,29,49–51]. For a sufficient suppression of the gate leakage currents, even at forward gate bias operation, a large bandgap material as well as large band offsets to (Al)GaN are necessary, in particular for power switching devices. On the other hand, a high value of permittivity is favorable to obtain high transconductance [55]. In particular, in the case of MIS-HEMTs, since the introduction of a dielectric leads to a reduction of the gate-to-channel capacitance with respect to Schottky-gate HEMTs, a high permittivity dielectric reduces the capacitive contribution of the gate dielectric, enabling it to obtain a stronger coupling between the gate and the 2DEG channel, and hence to maintain a high transconductance, which is especially important for RF devices. At the same time, in normally on MIS-HEMTs, high-permittivity materials can minimize the shift of the threshold voltage towards negative values when compared to Schottky-gate HEMTs, which is beneficial to reduce the static power consumption and to improve the energy efficiency of the device [7].

Various insulator materials have already been considered as gate dielectrics in insulated-gate GaN-based transistors. Figure 2 reports the relationship between the bandgap and permittivity for the relevant insulators and nitride compounds. Figure 3a shows the band offsets of the insulators on the GaN as calculated by Robertson and Falabretti, who first predicted the band alignment of the GaN and the insulators based on the calculation of the charge neutrality levels (E_{CNL}) [56]. The band offsets of the dielectrics on $Al_{0.3}Ga_{0.7}N$, recently determined by Reddy et al. using the same method, are illustrated in Figure 3b [57]. Note that, as shown from the comparison of Figure 3a,b, the different values of the energy bandgap of the same insulators are used in the calculations performed by Robertson and Falabretti [56] and by Reddy et al. [57].

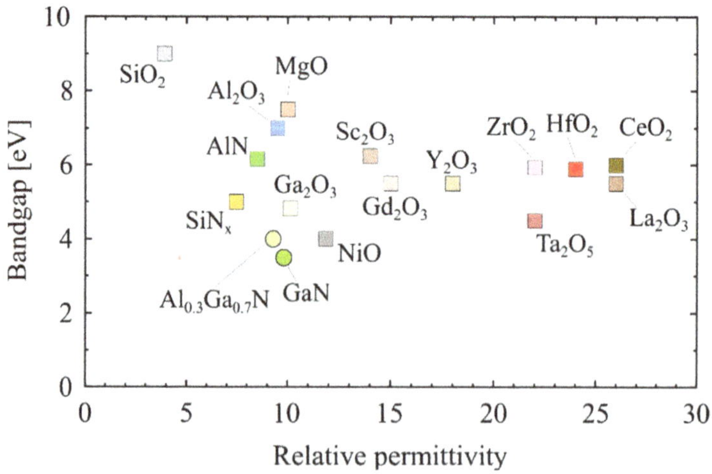

Figure 2. Energy bandgap versus permittivity for major insulators and GaN compounds. Data taken from [7,29,49–51].

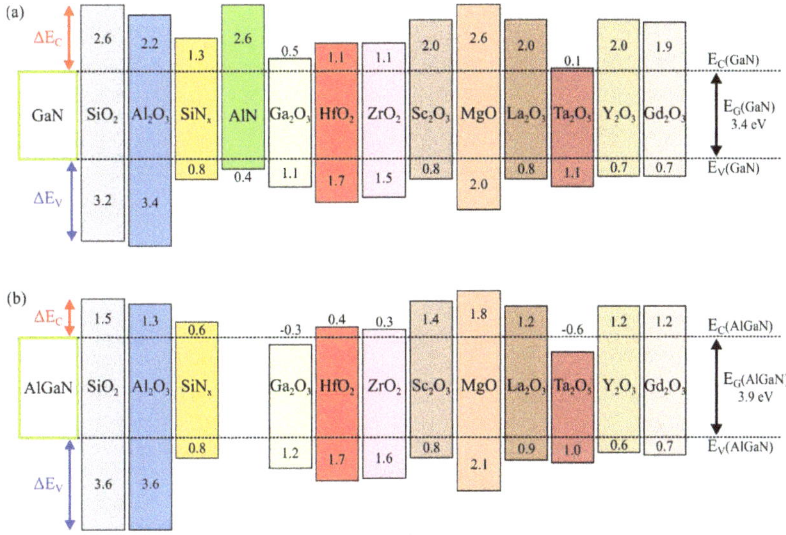

Figure 3. Conduction band offset (ΔE_C) and valence band offset (ΔE_V) of various dielectric materials with respect to (**a**) GaN, calculated by Robertson and Falabretti [56], and to (**b**) $Al_{0.3}Ga_{0.7}N$, calculated by Reddy et al. [57]. Note that, in (**a**,**b**), the different energy bandgaps of the insulators were assumed in the calculations. The conduction band (E_C) and valence band (E_V) of GaN and AlGaN are marked as dashed lines. The energy bandgap (E_G) of GaN and AlGaN is also indicated.

SiO_2 is an attractive insulator due to its large bandgap, large band offset to (Al)GaN and chemical stability. In fact, after Khan and coworkers first applied SiO_2 to AlGaN/GaN MIS-HEMTs to control the gate leakage currents and improve the gate voltage swing capability [24], further high-performance MIS-HEMTs using SiO_2 have been demonstrated [58,59]. Nevertheless, the relatively low dielectric constant of SiO_2 represents a disadvantage compared to other dielectrics. From this perspective, various high-permittivity dielectrics such as HfO_2, ZrO_2, Ta_2O_5, La_2O_3, CeO_2, TiO_2, etc., have been applied to the MIS gate structures of GaN HEMTs [60–72]. Although higher g_m values have been achieved in some cases, most of these insulators have reported to be relatively susceptible to leakage problems due to the relatively small band offsets with respect to (Al)GaN [49,68,73,74]. Similar observations of high gate leakage currents were reported for MIS gate structures employing dielectrics such as SiN_x and Ga_2O_3 due to the small conduction band offsets [51,75–77]. Ga_2O_3 would be appealing as a native oxide grown by thermal or chemical processes. However, in addition to the small band offset to GaN, Ga_2O_3 grown by thermal oxidation at low temperatures has a slow growth rate, while surface damage can be caused at higher growth temperatures [51]. Moreover, the growth of Ga_2O_3 is even more difficult on AlGaN since Al is more easily oxidized than Ga. Differently, SiN_x deposited by in situ metal organic chemical vapor deposition (MOCVD) or by low-pressure chemical vapor deposition (LPCVD) has emerged as a promising candidate as a gate dielectric as well as a passivation layer [42,78]. Similarly, AlN has also been reported in a few studies to be suitable as a gate insulator and passivation layer, especially due to its small lattice mismatch to (Al)GaN [51,79–81]. Other attempts have also used dielectrics like NiO, MgO and Sc_2O_3 [82–87], stacked dielectric layers like SiN_x/SiO_2, SiN_x/Al_2O_3 and HfO_2/Al_2O_3 [88–90] or engineered alloys such as SiON, $HfSiO_x$ and $LaLuO_3$ in order to tune the dielectric constant and band gap of the insulators [90–93]. A comprehensive overview and comparison of the various insulators which have been considered as gate dielectrics for insulated-gate GaN-based devices is given in [7,29,49–51].

Among the insulators, Al_2O_3 remains one of the most attractive insulators as a gate dielectric because of its large bandgap and conduction band offset to (Al)GaN, relatively high permittivity (~9) as well as high breakdown field (~10 MV/cm) and thermal and chemical stability against (Al)GaN [75,94,95]. Additionally, the considerable technological progress in the atomic layer deposition (ALD) process enables the deposition of high-quality Al_2O_3 films to the gate structures in GaN transistors. In the next section, the status of the gate dielectric technology using Al_2O_3 for GaN-based devices is reviewed.

3. Al_2O_3 for Insulated-Gate GaN Devices

Table 1 reports the physical parameters of the energy bandgap (E_G), conduction band offset (ΔE_C) and valence band offset (ΔE_V) obtained experimentally from amorphous Al_2O_3 films deposited on GaN and AlGaN by various deposition methods. Note that the bandgap of the amorphous Al_2O_3 ranges between 6.7 eV and 7.6 eV depending on the method of the oxide film growth, and it is lower than the value for the crystalline bulk α-Al_2O_3 (8.8 eV–9 eV) considered in the theoretical calculations (Figure 3). In fact, it is well known that the E_G of Al_2O_3 compounds strongly depends on its crystallographic phase [96,97]. Momida et al. investigated the structure of amorphous Al_2O_3 by first-principles calculations, concluding that the reduction of the bandgap of amorphous Al_2O_3 compared to crystalline Al_2O_3 could be related to the changes in the density of the Al_2O_3 compounds and the average coordination number of Al atoms [98]. Toyoda et al. showed that annealing at temperatures of 800 °C led to phase transformations of the Al_2O_3 films from amorphous to crystalline, which correlated to a significant increase in the energy bandgap and the modification of the conduction band discontinuity [99]. Afanas'ev et al. pointed out that for Al_2O_3 films treated at temperatures above 800 °C, the widening of the Al_2O_3 bandgap with the phase transformation from amorphous to crystalline mostly occurred at the valence band side [96,97]. Differently, Yang et al. revealed that the annealing processes at a lower temperature of 650 °C can affect the band bending of GaN but has almost no effect on the Al_2O_3/GaN band offset [100]. The decrease of the bandgap of amorphous Al_2O_3 has also been associated with defect-induced states located in the bandgap [101]. This could explain the large discrepancy between the theoretical (Figure 3) and experimental (Table 1) values of ΔE_V. In fact, since in the case of Al_2O_3 the valence band maximum states are associated with the O 2p states, and the conduction band minimum states are associated with the Al 3s, 3p states [102], the rehybridization between Al 3s, 3p and O 2p modifies the charge transfer between Al and O and consequently decreases the bandgap, thus increasing the valence band maximum [51]. In contrast to ΔE_V, the experimental values of ΔE_C obtained for the Al_2O_3/(Al)GaN system are consistent with the theoretical predictions and make Al_2O_3 a suitable dielectric for insulated-gate GaN-based transistors.

In addition to the physical properties of the bandgap of Al_2O_3 and the band offsets in the Al_2O_3/(Al)GaN system, high-quality dielectric layers in terms of defects and bulk traps and an Al_2O_3/(Al)GaN interface with a low interface trap density are required to deliver a high performance and highly efficient MIS gate structure, as discussed above. It is important to mention that these properties strongly depend on the deposition technique and temperature, the crystalline structure of the film and the surface and annealing treatments [51]. Among the techniques explored for the deposition of Al_2O_3 films, such as sputtering [103], the oxidation of a thin Al layer [53] and MOCVD [104–106], the ALD technique is widely used. The main advantages of the ALD method are the low deposition temperature (<350 °C), the excellent film thickness control as well as the high uniformity and conformality, which have enabled the deposition of high-quality Al_2O_3 films and Al_2O_3/(Al)GaN interfaces compared to other methods. Nevertheless, despite substantial progress in the ALD technology, large amounts of defects in the as-deposited Al_2O_3 bulk material and interface traps at the Al_2O_3/(Al)GaN interface are still present and still hinder the success of the insulated-gate GaN devices [1].

Table 1. Energy bandgap (E_G), conduction band offset (ΔE_C) and valence band offset (ΔE_V) measured for Al_2O_3 films on GaN and AlGaN. The deposition method is reported in the second column, where ALD = atomic layer deposition; PEALD = plasma-enhanced atomic layer deposition; CVD = chemical vapor deposition; MBD = molecular beam epitaxy; ECR = electron cyclotron resonance. In addition, the measurement method is noted in column 3, where C–V = capacitance–voltage measurements; F–N = Fowler–Nordheim characteristics; IPE = internal photoemission; XPS = X-ray photoelectron spectroscopy; UPS = ultraviolet photoelectron spectroscopy; XAS = X-ray absorption spectroscopy.

Structure	Deposition Method	Measurement Method	E_G (eV)	ΔE_C (eV)	ΔE_V (eV)	Ref.
Al_2O_3/GaN	ALD	C–V	-	-	1.2	[107]
Al_2O_3/GaN	ALD	XPS and F–N plot	6.7	2.2	-	[108]
Al_2O_3/GaN	ALD	XPS	6.6	2.0	1.2	[109]
Al_2O_3/GaN	ALD	IPE and C–V	-	2.2	-	[110]
Al_2O_3/GaN	PEALD	XPS and UPS	6.7	2.1	1.2	[111]
Al_2O_3/GaN	PEALD	XPS and UPS	-	1.3	1.8	[100]
Al_2O_3/GaN	CVD	XPS and XAS	7.6	2.7	1.5	[99]
$Al_2O_3/Al_{0.3}Ga_{0.7}N$	MBD + ECR plasma oxidation	XPS	7.0	2.1	0.8	[53,75]
$Al_2O_3/Al_{0.25}Ga_{0.75}N$	ALD	XPS	6.9	1.8	1.2	[112]
$Al_2O_3/Al_{0.25}Ga_{0.75}N$	ALD	XPS	6.7	1.8	0.9	[26]

3.1. Al_2O_3/(Al)GaN Structures

The presence of defects acting as traps or fixed charge centers within the Al_2O_3 films and at the Al_2O_3/(Al)GaN interface is of critical importance because of their potential to affect the threshold voltage and the gate leakage currents of the MIS gate structures [51], eventually deteriorating the operational stability and the reliability of the insulated-gate GaN-based devices.

For ALD-Al_2O_3/(Al)GaN structures, a positive fixed charge arising from donor-type interface states and/or defect levels in the bulk Al_2O_3 was often reported [112–115]. In this regard, Esposto et al. [107] and Son et al. [116] pointed out that fixed charges at the Al_2O_3/GaN interface shifted the flat-band voltage (V_{FB}) in the C–V curves of Al_2O_3/GaN capacitors. A shift of the V_{FB} towards the negative bias direction in Al_2O_3/GaN structures was observed by Kaneki et al. [115]. Similar shifts in the C–V characteristics attributed to interface states acting as fixed charges were reported for Al_2O_3/AlGaN/GaN structures by Mizue et al. [26] and Yatabe et al. [73]. Nishiguchi et al. [38] reproduced the observed negative shift in the C–V curve of Al_2O_3/AlGaN structures, assuming an effective fixed positive charge of $+1.2 \times 10^{13}$ cm^{-2} in the Al_2O_3 layer or at the Al_2O_3/AlGaN interface.

In line with this, annealing treatments have been reported to affect the V_{FB} and V_{th} of Al_2O_3/(Al)GaN structures as a result of a change in the defect levels in Al_2O_3 films [117,118]. For example, Hashizume et al. [114] reported a V_{FB} recovery of Al_2O_3/GaN structures after a postmetallization annealing (PMA) in N_2 at 200–400 °C, possibly attributed to the reduction of the donor-type interface states and/or the defect levels in the bulk. Similarly, Hung et al. [119] obtained a V_{FB} recovery by PMA in H_2/N_2 forming gas at 400–550 °C. Zhou et al. [120] showed a permanent positive shift of the V_{th} in ALD-Al_2O_3-gated MIS-HEMTs after a postdeposition annealing (PDA) at 600 °C in N_2, which was also suggested to be caused by a reduction of the deep-level bulk or interface traps. For similar reasons, a recovery of the V_{th} of MIS-HEMTs towards positive bias values was reported by Nishiguchi et al. [38] when using a reverse-bias anneal at 300 °C in air, and by Nakazawa et al. [121] with an anneal process at 750 °C in O_2 atmosphere.

The exact nature of the fixed charges in the bulk of the as-deposited Al_2O_3 or in the vicinity of the Al_2O_3/(Al)GaN interfaces is still under debate, with native defects in the oxide layer or dangling bonds at the interface being the major candidates. Choi et al. [122] investigated the impact of native point defects in Al_2O_3 by first-principle calculations, revealing that oxygen vacancies introduce charge-state transition levels near the GaN conduction band edge, which can act as border traps close to the Al_2O_3/n-GaN interface or as source of leakage current through the dielectric. However, other defects such as aluminum vacancies and interstitials have been identified to act as fixed-charge centers [122]. Weber et al. [123] also suggested that aluminum vacancy and oxygen interstitial defects introduce negatively charged centers while the aluminum interstitials act as positively charged centers, affecting carrier scattering in the channel and the threshold voltage of the device. Moreover, Liu et al. [124] studied the energy levels of the oxygen vacancy in Al_2O_3. Shin et al. [125] and Kim et al. [126] identified oxygen and Al dangling bonds as the origin of the fixed charges in ALD-Al_2O_3. Huang et al. [127] suggested that these defective dangling bonds, which are also associated to fixed positive charges and acceptor-like border traps, can be suppressed by the substitution of H_2O as an oxygen source with O_3 for the ALD deposition of Al_2O_3. Other groups have also demonstrated the influence of using different ALD precursors and different deposition temperatures on oxide charges, as well as the interface traps of Al_2O_3 films [128–132].

Defect states inside Al_2O_3 can affect the leakage current of the MIS gate structures through trap-assisted tunneling mechanisms. For Al_2O_3-gated MIS-HEMTs under forward bias, Liu et al. [133] and Yoshitsugu et al. [131] showed that trap-assisted tunneling (TAT) and Poole–Frenkel emissions (PFE) are dominant at medium electric fields and temperatures above 0 °C, whereas Fowler–Nordheim tunneling (FNT) dominates at high electrical fields and temperatures below 0 °C. In addition, Yoshitsugu et al. [131] estimated a TAT-related trap energy of about 1.0 eV below the conduction band minimum of Al_2O_3. Wu et al. [134] instead suggested that TAT is the dominant transport mechanism in high oxide fields, with trap energies of ~1.1–1.2 eV, while PFE was responsible for medium oxide field gate current transport. Recently, Heuken et al. [135] also suggested that the time-dependent dielectric breakdown (TDDB) of ALD-Al_2O_3 films occurs with the presence of an initial defect density in the film and is then related to the formation of a percolation path by randomly generated defects in the oxide under stress bias. The time to breakdown was found to be thermally activated, with an activation energy of 1.25 eV, similar to the reported values of the activation energy of TAT in Al_2O_3 at a high oxide field [131,134].

While defect states and bulk traps acting as fixed charges mostly affect the absolute value of the threshold voltage, the charging and discharging of bulk traps, especially border traps near the Al_2O_3/(Al)GaN interface, and interface traps deeply located in the bandgap of the (Al)GaN at the Al_2O_3/(Al)GaN interface can induce significant dynamic instabilities of the threshold voltage and of the drain current during device operation due to their slow detrapping behavior. A schematic illustration of the band diagram of Al_2O_3/GaN and Al_2O_3/AlGaN/GaN structures, including border and interface traps, is shown in Figure 4.

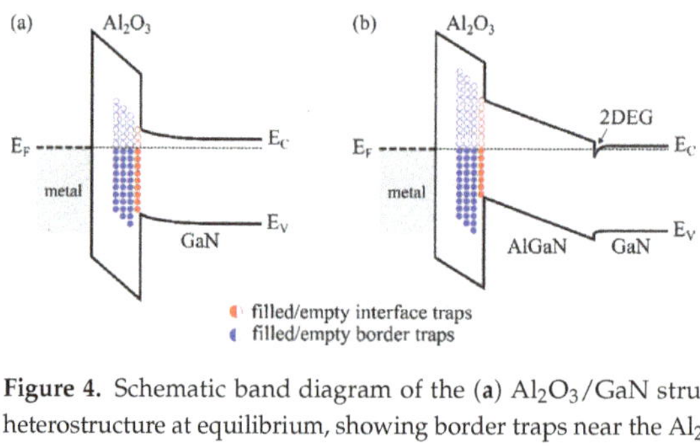

Figure 4. Schematic band diagram of the (**a**) Al_2O_3/GaN structure and (**b**) Al_2O_3/AlGaN/GaN heterostructure at equilibrium, showing border traps near the Al_2O_3/(Al)GaN interface and interface traps at the Al_2O_3/(Al)GaN interface. E_C and E_V are the conduction and valence bands of (**a**) GaN and (**b**) AlGaN, respectively. E_F denotes the Fermi energy.

For these reasons, many groups have focused their efforts on the characterization and minimization of trap states at the dielectric/(Al)GaN interface of the MIS gate structures. Figure 5 illustrates a summary of the interface trap density (D_{it}) distributions reported in the literature for Al_2O_3/(Al)GaN structures. Note that the best results reported in each reference have been illustrated in Figure 5. The Terman method [136] and conductance method [137] are often used to estimate the interface trap state densities of Al_2O_3/GaN structures. Differently, since for Al_2O_3/AlGaN/GaN structures the evaluation of interface trap states is more challenging due to the presence of a double interface (Al_2O_3/AlGaN and AlGaN/GaN) complicating the potential distribution over the structure, more advanced techniques such as conductance dispersion techniques [138–140] and frequency and/or temperature-dependent capacitance voltage measurements [26,73,141] are employed. More detailed overviews on the characterization of the electronic states at the insulator/(Al)GaN interfaces of GaN-based MIS-HEMTs with respect to their applicability and potential limitations are given by Ramanan et al. [142] and Yatabe et al. [49].

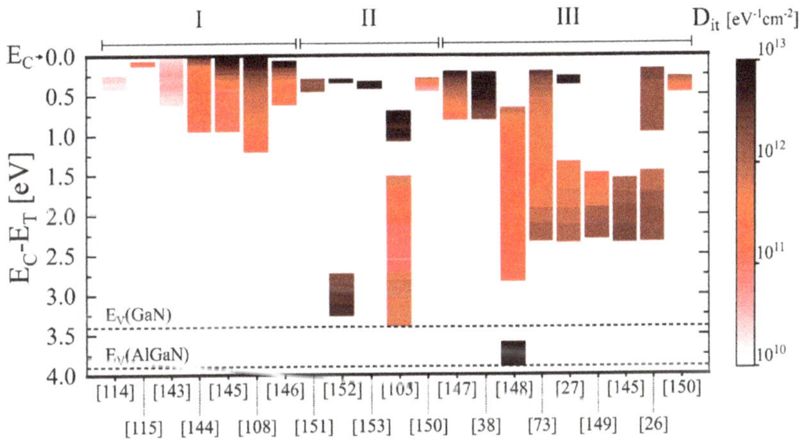

Figure 5. Interface density distributions (column) extracted from literature for (I) Al_2O_3/GaN, (II) Al_2O_3/GaN/$Al_xGa_{1x}N$/GaN and (III) Al_2O_3/$Al_xGa_{1-x}N$/GaN structures. The corresponding reference is indicated at the bottom of the graph for each column. The conduction band minimum E_C of GaN and AlGaN is set at 0 eV as reference. The valence band maximum E_V of GaN and AlGaN, accordingly to the bandgap values of 3.4 eV and 3.9 eV, respectively, are also illustrated as dashed lines.

The results reported in Figure 5 highlight the presence of high-density interface trap states, especially at energies close to the conduction and valence band edges of (Al)GaN. For Al_2O_3/GaN interfaces, minimum values of the interface state densities in the range of 10^{10}–10^{11} cm^{-2} eV^{-1} have been reported [114,115,143–146]. In comparison, Al_2O_3/AlGaN interfaces have shown minimum values of interface state densities that are about one order of magnitude higher [26,27,38,73,145,147–150]. Mizue et al. [26] suggested that this difference can be due to oxygen incorporation into AlGaN or to a higher density of defects in the AlGaN layer. Note also that some groups investigated Al_2O_3/GaN/AlGaN/GaN structures where a thin GaN layer (~1–3 nm) was present on top of the AlGaN layer, possibly affecting the distribution of the interface trap states [105,150–153]. A very thin GaN cap layer is indeed often included in the AlGaN/GaN epitaxial material, as it also helps to protect the AlGaN surface and to reduce leakage currents. Gregušová et al. [150] obtained an interface trap state density that was two to three times lower for the Al_2O_3-gated AlGaN/GaN structures with a GaN cap compared to ones without a GaN cap. On the contrary, Ťapajna et al. [106] reported almost the same C–V characteristics and interface trap state distributions for Al_2O_3/(GaN)/AlGaN/GaN structures with and without a GaN cap layer. For ALD-Al_2O_3/AlGaN/GaN structures, Mizue et al. [26] estimated the trap states density distribution at the ALD-Al_2O_3/AlGaN interface for the first time, showing that trap states with densities higher than 1×10^{12} cm^{-2} eV^{-1} exist at the Al_2O_3/AlGaN interface. To evaluate the near-midgap electronic states at room temperature (RT), a photoassisted C–V method using photon energies less than the AlGaN bandgap was developed [26,73]. For states close to the valence band of (Al)GaN, Matys et al. [154,155] developed a method based on the measurement and simulations of the photo-capacitance of MIS gate heterostructures. Combining this method with the photoassisted capacitance–voltage technique, the interface state density in the entire band gap at the Al_2O_3/AlGaN interface was determined, revealing the presence of a large amount of trap states with D_{it} values higher than 1×10^{13} cm^{-2} eV^{-1} also near the valence band edge [148].

When using Al_2O_3 films on (Al)GaN, particular attention has to be given to the temperature processes applied after the dielectric deposition. Hori et al. [108] showed that the annealing process at 800 °C for the ohmic contact formation applied after the ALD-Al_2O_3 deposition created a large number of microcrystalline regions in the Al_2O_3 layer, causing a pronounced increase of the leakage current of the Al_2O_3/n-GaN structures. To prevent this effect, an "ohmic-first" approach with a SiN protection layer was applied, which maintained the amorphous phase in the atomic configuration of Al_2O_3, leading to a sufficient suppression of the leakage current. In addition, protecting the surface with a SiN layer during annealing resulted in the low interface trap densities of less than 1×10^{12} cm^{-2} eV^{-1} extracted from the C–V characteristics of the Al_2O_3/GaN structures.

Other processing steps for the fabrication of GaN devices are also critical and can affect the interface quality and the electrical properties of the Al_2O_3/(Al)GaN structures. To achieve normally off operation, recessed gates are often employed in MIS-HEMTs or hybrid MIS-FETs. For this reason, the influence of inductively coupled plasma (ICP) etching on the interface properties of Al_2O_3/(Al)GaN structures has also been investigated. Yatabe et al. [73] estimated the state density distribution at the Al_2O_3/AlGaN interface of MIS structures subjected to ICP dry etching of the AlGaN surface, using for the first time the combination of the photoassisted C–V method and the modeling of the C–V curves [26,156]. Trap state densities higher than 2×10^{12} cm^{-2} eV^{-1} were obtained at the Al_2O_3/AlGaN interface of the ICP-etched structures [73]. Without the ICP etching of AlGaN, a near-midgap D_{it} of about 1×10^{12} cm^{-2} eV^{-1} or less was obtained. Similarly, Kim et al. [144] also investigated the effects of a Cl_2-based ICP etching on the interface properties of Al_2O_3/GaN structures. From the X-ray photoelectron spectroscopy (XPS) and transmission electron microscopy (TEM) analyses, it was shown that the ICP etching caused a disorder of the chemical bonds at the GaN surface. This resulted in high-density trap states with a density larger than 1×10^{13} cm^{-2} eV^{-1} near the conduction band edge of the GaN at the Al_2O_3/GaN interface, which was suggested to include defects related to nitrogen vacancy (V_N) levels. A decrease of the interface state density was obtained by applying a PDA process in N_2 at 400 °C, which partially recovered the V_N-related levels, thus increasing the chemical bond order at the GaN surface. Yatabe et al. [149] also reported that the ICP etching of the AlGaN surface introduced a monolayer-level crystalline roughness, the disorder of the chemical bonds and various types of defect complexes including V_N, resulting in high trap state densities of up to 8×10^{12} cm^{-2} eV^{-1} at the Al_2O_3/AlGaN interface. Fang et al. [157] also reported that Cl_2-based ICP etching enhanced the deep centers at the GaN surface originating from V_N and other defect complexes.

Other studies have demonstrated the importance of PDA and PMA treatments to minimize the interface trap states at the Al_2O_3/(Al)GaN interface. From the TEM investigations, Hashizume et al. [114] revealed that PMA in N_2 at 300–400 °C led to a uniform distribution of the lattice constant near the interface of the ALD-Al_2O_3/GaN MIS structures, which resulted in excellent C–V characteristics almost without frequency dispersion and a reduced D_{it} ranging from 1 to 4×10^{10} cm^{-2} eV^{-1} at energies near the conduction band edge. Similar values of D_{it} at the Al_2O_3/GaN interface after PMA in N_2 at 400 °C were also very recently obtained by Ando et al. [158]. Ando et al. [147] also demonstrated that a PMA in N_2 at 300 °C led to a similar reduction of the electronic states at the ALD-Al_2O_3/AlGaN interface. Kaneki et al. [115] pointed out that annealing under reverse bias at 300 °C in air for 3 h is also beneficial to decrease the interface state density of ALD-Al_2O_3/GaN structures, and it is more effective than PDA in N_2 at 400–700 °C, probably due to a relaxation of the dangling bonds and/or the point defects at the GaN surface. Moreover, almost no shift of the V_{FB} with respect to the expected value was observed in the C–V curves due to the reduction of the donor-type interface states and/or defect levels in the bulk Al_2O_3. Similar effects of the reverse-bias annealing were obtained by Nishiguchi et al. [38] for ALD-Al_2O_3/AlGaN structures. Winzer et al. [143] reported that PDA in O_2 or forming gas (H_2/N_2) at 500 °C were more efficient for decreasing the traps at the Al_2O_3/GaN interface than PDA in N_2 at the same temperature. A very low interface trap density of less than 5×10^{11} cm^{-2} eV^{-1} was achieved for Al_2O_3/GaN structures treated by forming gas PDA at 500 °C. However, it was also reported that forming gas PDA resulted in a detrimental increase of the leakage currents of the Al_2O_3 films. Similar results were reported by Long et al. [159], where the effect of trap passivation during the forming gas anneal was correlated to the incorporation of hydrogen at the interface.

Similar to annealing processes, surface treatments are also effective in reducing interface trap states at the Al_2O_3/(Al)GaN interface. Hori et al. [27,145] demonstrated that an N_2O-radical treatment can decrease interface states both at the Al_2O_3/GaN and Al_2O_3/AlGaN interfaces. For Al_2O_3/AlGaN structures, the interface state density was estimated to be 1×10^{12} cm^{-2} eV^{-1} or less around the midgap and 8×10^{12} cm^{-2} eV^{-1} near the conduction band edge [27]. Calzolaro et al. [151] recently reported a significant reduction of frequency dispersion of the C–V characteristics of Al_2O_3/GaN/AlGaN/GaN structures after a remote O_2 plasma-based surface treatment prior to the ALD-Al_2O_3 deposition combined with a PMA in N_2 at 350 °C. The D_{it} was estimated to be reduced to a value in the order of 2×10^{12} cm^{-2} eV^{-1} near the conduction band edge.

Trapping mechanisms at the Al_2O_3/(Al)GaN interface are especially critical for AlGaN/GaN MIS-HEMTs under forward gate bias, where electrons can spill over from the 2DEG channel towards the dielectric by overcoming the AlGaN barrier and become trapped at the Al_2O_3/(Al)GaN interface [34–37]. Similarly, charge trapping in high-density electronic states at the interface has been reported to lead to a significant screening of the gate electric field and the consequent loss of control of the surface potential of the barrier layer, causing the degradation of the current linearity and the saturation of the current at forward bias in AlGaN/GaN MIS-HEMTs [38]. In this regard, the next section focuses on reviewing the recent progress on the performance of Al_2O_3-gated MIS-HEMTs.

3.2. Al_2O_3-Gated MIS-HEMTs

Among the issues facing the MIS gate toward the improvement of the performance of AlGaN/GaN MIS-HEMTs, the dynamic V_{th} instability caused by the trapping mechanisms involving the gate dielectric is the one major concern [1]. The instability of the V_{th} has been reported under various bias conditions [31,32,34,78,160,161]. In particular, the large V_{th} shift induced by forward gate bias stress due to electron trapping at the dielectric/(Al)GaN interface is one of the most serious problems for the operational stability and reliability of the device [34–37]. For this reason, many groups have focused their efforts on studying the origin of the V_{th} instability and various fabrication processing strategies to overcome this issue.

For Al_2O_3-gated MIS-HEMTs, Lu et al. [32] reported that a larger V_{th} shift towards the forward bias direction was induced by increasing the gate positive bias stress in the pulsed current-voltage (I-V) measurements. Similar results were obtained by other groups [28,31,35,151,160,162,163]. Bisi et al. [160] pointed out that the large positive shift of the V_{th} can also promote the current collapse of MIS-HEMTs. Regarding the origin of the V_{th} instability, Ťapajna et al. [105] discussed the effect of interface states and bulk traps on the V_{th} shift in Al_2O_3-gated MIS-HEMTs. Wu et al. [153] and Zhu et al. [33] pointed out that the V_{th} shift during a positive gate bias stress was highly correlated to the trap states at the dielectric/(Al)GaN interface but also to the border traps near the interface. Fixed charges within the dielectric are also involved in the V_{th} shift mechanism [107,116].

A reduction of the interface and/or border traps by means of annealing and surface treatments can lead to an improvement of the dynamic V_{th} instability of MIS-HEMTs. In addition, as mentioned before, the current linearity and the saturation of current at forward bias of MIS-HEMTs can be also affected by a change in the density of the electronic states at the dielectric/(Al)GaN interface [38]. Hori et al. [27] reported that the reduction of the interface states obtained by applying an N_2O-radical treatment on the AlGaN surface prior to the ALD-Al_2O_3 deposition led to a higher maximum drain current of the MIS-HEMTs at the positive gate bias and a suppressed V_{th} instability under the negative gate bias stress even at 150 °C. Nishiguchi et al. [38] showed that the improvement of the Al_2O_3/AlGaN interface by the reverse-bias anneal at 300 °C in air for 3 h of Al_2O_3-gated MIS-HEMTs gave a better gate control of the current even at forward gate bias, effectively enhancing the current linearity, subthreshold behavior and the maximum drain current of the device. Moreover, reduced gate leakage currents and more stable V_{th} under forward bias stress and at higher temperatures were obtained. Similarly, Ando et al. [147] recently reported on the improved gate controllability and current linearity of MIS-HEMTs with the Al_2O_3 gate dielectric as a result of a reduction of the electronic states at the Al_2O_3/AlGaN interface after PMA in N_2 at 300 °C. A subthreshold slope of 68 mV dec^{-1} and excellent V_{th} and operation stability up to 150 °C were also achieved, as shown in Figure 6. Note that in this case Ando et al. [147] pointed out that the improvement of the device performance also benefited from using epitaxial GaN layers grown on free-standing GaN substrates from hydride vapor phase epitaxy (HVPE) with a low dislocation density. Very recently, Calzolaro et al. [151] reported that the reduction of interface trap states by a remote O_2 plasma-based surface treatment before the ALD-Al_2O_3 deposition combined with a PMA in N_2 at 350 °C resulted in a better V_{th} stability in pulsed I-V measurements. It is worth mentioning that, despite the benefits of the PMA treatments, specific attention has to be paid to the employment of higher PMA temperatures, as it can affect the gate leakage currents of the devices using ALD-grown Al_2O_3 films [119,164]. Therefore, a trade-off must be considered when using the PMA treatment between the quality of the Al_2O_3/(Al)GaN interface and the gate leakage currents in a certain voltage range of operation.

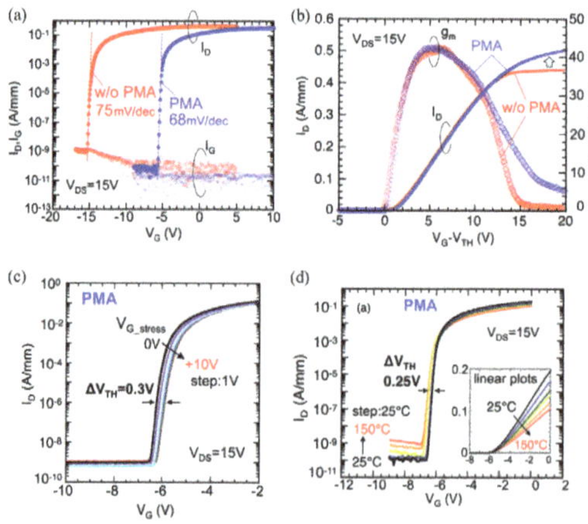

Figure 6. Transfer characteristics of Al_2O_3-gated AlGaN/GaN MIS-HEMTs fabricated on free-standing HVPE GaN substrates and subjected to PMA at 300 °C in N_2 atmosphere, reported by Ando et al. [147]. In (**a**,**b**), the transfer characteristics of MIS-HEMTs with and without PMA are compared in a semi-log scale and as a function of the gate overdrive voltage, respectively. Transfer curves in (**c**,**d**) were obtained after applying an initial gate voltage stress up to 10 V and by increasing the temperature up to 150 °C, respectively.

As in the case of the surface and annealing treatments, various strategies in the fabrication process of the devices can also be adopted to influence the trap states at the interface and, therefore, suppress the V_{th} instability. Szabó et al. [31] reported that for MIS-HEMTs where the deposition of the Al_2O_3 gate dielectric was performed before the ohmic contacts formation and at annealing temperature of 650 °C resulted in an improvement of the V_{th} stability compared to devices where the Al_2O_3 was deposited after the ohmic contacts formation was obtained with a high temperature anneal of 850 °C. It was suggested that this result was a consequence of a better Al_2O_3/(Al)GaN interface quality. Nakazawa et al. [165] applied an interesting approach based on the selective area regrowth of AlGaN to reduce the impact on the ALD-Al_2O_3/AlGaN interface of the dry etching process used for the fabrication of normally off AlGaN/GaN MIS-HEMTs with recessed gate structures. With this approach, they reported a reduced V_{th} instability compared to Al_2O_3-gated MIS-HEMTs with dry-etched recessed gates.

Trapping mechanisms related to the gate dielectric can lead to the failure of the device. For this reason, reliability tests of the gate dielectric are also essential to bring the MIS-HEMT devices to industrial maturity. In this regard, Meneghesso et al. [30] performed an extensive analysis of trapping mechanisms and the reliability issues of AlGaN/GaN MIS-HEMTs using different insulators. They reported a significant correlation between the dynamic V_{th} shift and gate leakage currents under forward gate bias stress and suggested that trapping effects were determined by the electrons trapped in the gate insulator or at the AlGaN/insulator interface. Wu et al. [166] investigated the positive bias temperature instability (PBTI) in hybrid GaN MIS-FETs. Since the defect distribution inside the ALD-Al_2O_3 was found to be centered at about 1.15 eV away from the conduction band of the GaN with a narrow spread in energy, the ALD-Al_2O_3 gate dielectric was suggested to be very promising to improve the PBTI reliability. Meneghesso et al. [30] also measured the TDDB characteristics of MIS-HEMTs with Al_2O_3 as gate dielectrics. Since the time-to-failure of devices indicated a Weibull distribution with slopes larger than 1.0, they demonstrated high robustness for ALD-Al_2O_3. Similarly, a Weibull distribution with a slope of 2.87 was extracted from the TDDB measurements of the Al_2O_3-gated MIS structures by Wu et al. [134]. Huang et al. [127] also achieved good TDDB behavior and a high breakdown electric field of 8.5 MV cm^{-1} in recessed-gate MIS-HEMTs with a gate dielectric stack consisting of 13 nm of ALD-Al_2O_3 deposited using O_3 as an oxygen source and grown on top of 2 nm of ALD-Al_2O_3 deposited using a H_2O oxygen source. For the ALD-Al_2O_3 films on the GaN, Kachi et al. [167] reported a TDDB lifetime at RT and 150 °C of more than 20 years at an electric field of 3 MV cm^{-1}. Kikuta et al. [168] obtained a time-to-breakdown for the ALD-Al_2O_3 on a dry-etched GaN of more than 40,000 years at 3 MV cm^{-1} and RT. In contrast, a time-to-breakdown of only 10^2–10^3 s was obtained at 250 °C, which was suggested to be caused by large TAT leakage currents.

As mentioned before, the dielectric layer employed in MIS-HEMTs can be used both as a gate dielectric and a passivation layer to reduce current collapse. Hashizume et al. [53,75] first demonstrated the use of an Al_2O_3 layer as a gate dielectric and a passivation scheme to control the current collapse in AlGaN/GaN HEMTs. Moreover, comparing the effects of surface passivation on MIS-HEMTs and Schottky-gate HEMTs, Tajima and Hashizume [169] showed a more pronounced reduction of the current collapse in Al_2O_3-gated MIS-HEMTs in contrast to Schottky-gated HEMTs, with Al_2O_3 serving only as a surface passivation. The suppression of the current collapse with a passivation layer, arising from negative surface charges, injected from gate edges to surface states was generally attributed to a reduction of electronic states at the AlGaN surface and of the peak field near the gate edge. Park et al. [94] reported for the first time on the use of Al_2O_3 deposited by ALD as a gate dielectric and passivation layer for AlGaN/GaN MIS-HEMTs. Park et al. [94] and Ye et al. [23] reported on the excellent electrical characteristics of AlGaN/GaN MIS-HEMTs using ALD-Al_2O_3 as a gate dielectric and passivation layer. Despite the improvements obtained by Al_2O_3-based passivation schemes for MIS-HEMT devices, further work is still required to limit and fully understand the current collapse phenomena in GaN transistors [1]. A more detailed overview about surface passivation for GaN-based transistors can be found in [29,49–51].

3.3. Modified Al_2O_3 Gate Dielectrics

Besides the use of pure Al_2O_3 films, other approaches involving the use of Al_2O_3-based bilayer gate stack dielectrics, interface engineering techniques or Al_2O_3-based compound materials have been investigated to combine the properties of Al_2O_3 with the favorable properties of other dielectric materials.

Kambayashi et al. [170] applied a SiO_2/Al_2O_3 gate stack (layers indicated from top to bottom) in hybrid GaN MIS-FETs, thus demonstrating a high-performance device with a channel mobility of 192 cm^2/Vs. Using a SiO_2/Al_2O_3 gate stack, Guo and del Alamo [171,172] studied the origin of PBTI and negative bias temperature instability (NBTI) in hybrid GaN MIS-FETs. It was shown that for a composite SiO_2/Al_2O_3 gate oxide, the resulting V_{th} shifts are due to electron trapping or detrapping in pre-existing oxide traps and the generation of oxide traps near the oxide/GaN interface. Van Hove et al. [173] applied an ALD-Al_2O_3/in situ MOCVD-Si_3N_4 gate bilayer stack in AlGaN/GaN MIS-HEMTs to achieve excellent electrical device characteristics with lower gate leakage currents, more stable threshold voltages and reduced current collapse when compared to Al_2O_3-gated MIS-HEMTs. Capriotti et al. [174] investigated the fixed interface charges between the AlGaN and the Al_2O_3/in situ SiN gate stack of AlGaN/GaN MIS-HEMTs. Colon and Shi [90] fabricated AlGaN/GaN MIS-HEMTs with low gate leakage currents using an ALD-HfO_2/Al_2O_3 bilayer stack as well as an ALD-$HfAlO_x$ ternary compound as gate dielectrics to achieve a higher dielectric constant than Al_2O_3 and a higher conduction band offset, thermal stability and crystallization temperature than HfO_2. However, both the HfO_2/Al_2O_3 and $HfAlO_x$-gated MIS-HEMTs still showed low transconductance, high interface state density and pronounced current collapse. The energy band alignment of MOCVD-HfAlO to GaN was investigated by Liu et al. [175,176], reporting a conduction band offset of 2.2 eV and minimum values of interface trap density in the range of 1–3 $\times$ 10^{11} cm^{-2} eV^{-1} at the HfAlO/GaN interface. Hatano et al. [177] demonstrated reduced gate leakage and the improved operation and thermal stability of AlGaN/GaN MIS-HEMTs using a ZrO_2/Al_2O_3 gate stack dielectric.

Other approaches based on the use of Al_2O_3-based composite materials have also been reported. Partida-Manzanera et al. [178] investigated the potential of a ternary phase of Ta_2O_5 and Al_2O_3 as gate dielectrics to achieve higher permittivity than Al_2O_3, and hence enhance the transconductance of AlGaN/GaN MIS-HEMTs. Although a higher transconductance and reduced gate leakage current were achieved, the C–V curves did not feature the characteristic step at the forward bias in the spill-over regime, indicating a high density of trap states at the dielectric/AlGaN interface. Kikuta et al. [179] applied Al_2O_3/SiO_2 nanolaminate films deposited by ALD on GaN to obtain a gate dielectric material with a larger conduction band offset to GaN and a higher crystallization temperature than pure Al_2O_3 films in order to reduce gate leakage currents. The composition of Al and Si in the oxide and the resulting oxide properties of the permittivity, breakdown field and leakage currents could be controlled and tuned by the numbers of ALD cycles. Compared to pure Al_2O_3 films, a higher breakdown field and better reliability were obtained for the SiO_2 composition, from 0.21 to 0.69. Similarly, Mitrovic et al. [180] suggested that Al_2O_3/TiO_2 nanolaminates can also be favorable as gate dielectrics, and they very recently investigated the band alignment to the GaN and the permittivity of the Al_2O_3 layers doped with Ti, corresponding to $Ti_xAl_{1-x}O_y$. Although the permittivity of $Ti_xAl_{1-x}O_y$ increased significantly with the increasing Ti content, a small conduction band offset for all compositions was obtained. However, Le et al. [181,182] reported excellent characteristics with good insulating properties for MIS-HEMTs using AlTiO deposited by ALD as a gate dielectric.

Current research has also focused on the "doping" by fluorine ions (F^-) of Al_2O_3 gate dielectric films in order to control the threshold voltage of MIS-HEMTs towards normally off operation [183,184]. The latter can be obtained by implanting F^- ions into the AlGaN barrier prior to the dielectric ALD. After the ALD-Al_2O_3 deposition, the incorporated F^- ions can act as a source of negative fixed charges, compensating the intrinsic positive charges in the dielectric and shifting the V_{th} of the device in positive bias direction. It is worth mentioning that a previous physical approach based on the fluorine incorporation via plasma etching under the gate to shift the device threshold voltage was demonstrated by Cai et al. [185]. Using an ALD-Al_2O_3 gate dielectric combined with a fluorine-based plasma treatment, Chu et al. [186] demonstrated normally off Al_2O_3-gated MIS-HEMTs with a breakdown voltage of 1200 V.

An interesting process was used by Liu et al. [79] and Yang et al. [187], who improved the performance and the V_{th} stability of the Al_2O_3-gated hybrid MIS-FETs by inserting a monocrystalline AlN interfacial layer via plasma-enhanced atomic layer deposition (PEALD) at the Al_2O_3/GaN interface to block oxygen from the GaN surface and prevent the formation of oxygen-related interface traps. Al_2O_3/AlN/GaN structures showed a small frequency dispersion in the C–V curves and a D_{it} in the range of 10^{11}–10^{12} cm^{-2} eV^{-1}, determined using the conventional conductance method. Similarly, Yang et al. [188] and Chen et al. [189] used an in situ low-damage plasma treatment based on NH_3 and N_2 prior to the ALD-Al_2O_3 deposition to effectively remove the native oxide while forming an ultrathin monocrystal-like nitridation interlayer (NIL) at the Al_2O_3/GaN interface. The N_2 plasma treatment was also demonstrated to compensate for V_N-related defects at the surface. After a PDA was carried out at 500 °C in O_2 ambient, the Al_2O_3/NIL-gated MIS structures showed a lower interface trap density in the range of 1–6 × 10^{12} cm^{-2} eV^{-1}, resulting in AlGaN/GaN MIS-HEMTs with improved performance [189].

Finally, a very promising approach proposed by Asahara et al. [190] consists in using a sputtered AlON film as a gate dielectric, obtained by introducing nitrogen into Al_2O_3. An atomically abrupt high quality AlON/AlGaN interface with extremely low D_{it} values ranging from 1.2 to 1.4 × 10^{11} cm^{-2} eV^{-1} and improved bulk properties were achieved, resulting in excellent C–V characteristics with negligible frequency dispersions and a markedly suppressed gate leakage current. Similar results were obtained by Wang et al. [191], who deposited AlON films by inserting thin AlN alternating layers into Al_2O_3. As shown in Figure 7, Ueda et al. [192] very recently applied AlON films deposited by ALD combined to a PDA in O_2 for shifting the V_{th} so to realize the normally off operation in the recessed-gate AlGaN/GaN MIS-HEMTs, with a negligible hysteresis in the transfer characteristics, a reduced off-state leakage current, a breakdown voltage of 730 V, an on-state resistance of 270 mΩ for a 10 A drain current rating and impressive switching performance, indicating the great potential of AlON as gate dielectric technology.

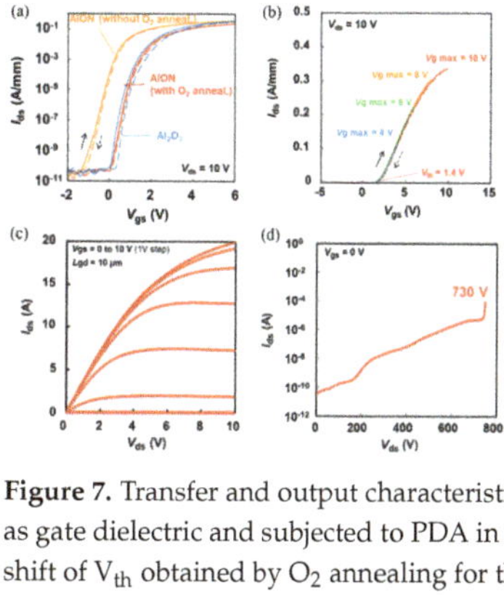

Figure 7. Transfer and output characteristics of recessed-gate AlGaN/GaN MIS-HEMTs using AlON as gate dielectric and subjected to PDA in O_2 atmosphere, reported by Ueda et al. [192]. The positive shift of V_{th} obtained by O_2 annealing for the AlON-gated transistor is shown in (**a**), while (**b**) reports the transfer curves without hysteresis obtained after applying a maximum gate voltage up to 10 V. The output characteristics of AlON-gated MIS-HEMTs in the on-state and off-state are shown in (**c**,**d**), respectively.

4. Nitride-Based Dielectrics

Despite the potentiality of Al_2O_3 and Al_2O_3-based dielectric materials, other insulators have emerged as suitable candidates for insulated-gate GaN-based transistors [1,29]. Among them, nitride-based dielectrics are of particular interest compared to oxide-based insulators because of the suppression of the Ga-O bonds that tend to induce interface traps [187].

SiN_x deposited by in situ MOCVD or LPCVD has been widely demonstrated to be very promising both as a gate dielectric and a surface passivation [29,49]. In particular, in situ SiN_x enables the dielectric deposition without exposing the (Al)GaN surface to air, which prevents the oxidation of the surface and passivates the surface states, possibly reducing the interface traps. Ogawa et al. [193] demonstrated that the in situ process of SiN_x can realize an oxide free SiN_x/AlGaN interface. Takizawa et al. [194] reported high-resolution TEM analysis revealing abrupt interfaces between SiN_x and AlGaN. Jiang et al. [78] systematically investigated MIS structures and MIS-HEMTs using in situ MOCVD-SiN_x as a gate dielectric. A D_{it} in the range of 2–3 $\times$ 10^{12} cm^{-2} eV^{-1} was obtained, which resulted in a stable V_{th} under gate bias and thermal stress. Derluyn et al. [195] reported that the reduction of surface states with in situ SiN_x passivation of HEMT structures led to higher 2DEG density and lower current collapse. Moens et al. [196] even reported on MIS-HEMTs for 650 V applications with excellent interface quality and dielectric reliability using MOCVD-grown in situ SiN_x, which demonstrated a maximum gate voltage of ~3.1 V at 10 years for a 100 ppm failure rate. LPCVD-SiN has the advantages of a large conduction band offset to GaN (~2.3 eV), a relatively high dielectric constant (~7) and a low defects density enabled by the high deposition temperature. Moreover, compared to plasma-enhanced chemical vapor deposition (PECVD)-SiN_x, LPCVD-SiN_x is free of plasma-induced damage and exhibits low oxygen contamination. In this regard, Hua et al. [197] reported on the superior properties of LPCVD-SiN_x in terms of the leakage currents, breakdown field and TDDB lifetime. Similar investigations were performed by Jauss et al. [198], who predicted a 20-year 100 ppm lifetime at 130 °C for a gate voltage of 10.1 V. However, the high deposition temperature of more than 700 °C for LPCVD-SiN_x can instead degrade the GaN surface in recessed-gate structures employed for normally off operations [48]. To overcome this issue, Hua et al. [48] successfully employed an interface protection technique consisting of a SiN_x interface layer deposited by PECVD prior to the high-temperature deposition process of LPCVD-SiN_x. With this approach, normally off hybrid MIS-FETs using high-quality LPCVD-SiN_x with a gate breakdown voltage of 21 V, a maximum gate bias of 11 V at failure rate of 63.2% for a 10-year lifetime, a stable V_{th} and a small current collapse were demonstrated. A similar approach has been also applied by Jiang et al. [78], who instead used in situ SiN_x in conjunction with PECVD SiN_x as a passivation scheme to effectively suppress the current collapse in MIS-HEMTs. Finally, it is worth mentioning that for normally off hybrid MIS-FETs, Hue et al. [42] recently developed another promising technique to protect the etched-GaN surface during the LPCVD-SiN_x high temperature deposition. This is based on an oxygen-plasma treatment followed by in situ annealing prior to the LPCVD to form a sharp and stable crystalline oxidation interlayer (COIL) protecting the surface. LPCVD-SiN_x-gated hybrid MIS-FETs with a COIL revealed a stable V_{th} and a highly reliable gate dielectric.

AlN is another promising nitride-based material which is attractive as a gate dielectric for insulated-gate GaN-based transistors due to its large bandgap, resulting in a high breakdown field, high permittivity and small mismatch to GaN, which might reduce the trap states at the AlN/(Al)GaN interface. AlN is mainly grown by MOCVD or PEALD techniques [199]. Hashizume et al. [200] were the first to report the low values of D_{it} in the range of 1 $\times$ 10^{11} cm^{-2} eV^{-1} at the MOCVD-AlN/GaN interface. Huang et al. [81] revealed an atomically sharp interface between the PEALD-AlN and AlGaN. They also demonstrated that polarization charges in the monocrystal-like AlN used as a passivation layer can effectively compensate the interface traps at the AlN/(Al)GaN interface, significantly reducing current collapse and the on-resistance degradation in ALD-AlN-passivated

AlGaN/GaN HEMTs. Polarization charges in monocrystalline thin AlN layers have also been reported to affect the V_{th} of hybrid MIS-FETs [79]. The high thermal conductivity of AlN has been also shown to be beneficial to suppress the self-heating of AlN-passivated HEMTs, thus improving the device performance [80]. AlN as passivation layer has also been demonstrated to improve the breakdown voltage of AlGaN/GaN HEMTs compared to SiN-passivated devices [201]. Very recently, Hwang et al. [202] reported a sharp interface between the GaN and PEALD-AlN. With the PEALD-AlN used as interfacial layer, they also successfully suppressed the surface oxidation of the GaN, which resulted in the improved C–V characteristics of AlN/GaN structures. AlGaN/GaN MIS-HEMTs and MIS structures using AlN deposited by a novel technique called low-temperature epitaxy (LTE) have been also recently investigated [199,203,204].

5. Summary

In this paper, we have summarized the most relevant challenges and recent progress on the development of a gate dielectric technology for insulated-gate GaN-based devices for high-frequency and high-power applications. Specifically, we first pointed out the important physical properties of the insulators which need to be considered for designing a MIS gate structure which delivers improved energy efficiency and reliable device performance. Afterwards, we highlighted that, regardless of the GaN transistor concept and the design, one of the major challenges arising from the insertion of a dielectric on (Al)GaN is represented by the trap states located at the dielectric/(Al)GaN interface or within the bulk dielectric. These trap states strongly affect the performance and the reliability of the device and need to be minimized to ensure high energy efficiency, safe operation and the long-term lifetime of the insulated-gate GaN-based transistors.

Among the various dielectrics, we focused our attention on Al_2O_3, which is one of the most promising dielectric materials due to its large bandgap and conduction band offset to (Al)GaN, its relatively high dielectric constant, its high breakdown electric field and its thermal and chemical stability against (Al)GaN. In particular, we pointed out that despite the technological progress in the ALD process, enabling the fabrication of high-quality Al_2O_3 films and of Al_2O_3-gated devices with improved and reliable performance, a large amount of defects and trap states at the Al_2O_3/(Al)GaN interface is still present and still degrades the device performance. In this regard, the main results obtained in the literature of the interface state density distribution at the Al_2O_3/(Al)GaN interface are presented and discussed in detail, and the recent progress in the performance of the Al_2O_3-gated MIS-HEMTs are reviewed.

Finally, novel Al_2O_3-based dielectric or compound materials and interface engineering approaches involving the use of Al_2O_3, which have been exploited to improve the quality and electrical performance of Al_2O_3-gate MIS structures, have been presented. Among them, AlON, or the use of nitride-based interface control layers have been demonstrated to be the most promising techniques. In addition to that, nitride-based dielectric materials have also been briefly presented as promising candidates, especially driven by their potential to function both as a gate dielectric as well as a passivation layer.

The insights of this paper help to understand the current status and the recent progress of the Al_2O_3 gate dielectric technology for insulated-gate GaN-based transistors. It also highlights that the current state of the art has made great advancements, but still requires remarkable progress in terms of gate dielectric, gate stack engineering and interface control technology. Focused efforts are still needed in order to ensure a low interface and bulk trap density, thus enabling a robust reliability under stringent and dynamic electrical stresses. Further advances in the gate dielectric technologies are necessary to overcome these obstacles and to pave the way for the massive advent of insulated-gate GaN-based technologies in the electronic market.

Funding: This research received no external funding.

Institutional Review Board Statement: Not applicable.

Informed Consent Statement: Not applicable.

Data Availability Statement: No new data were created or analyzed in this study.

Acknowledgments: Anthony Calzolaro acknowledges financial support from DFG (project no. 405782347).

Conflicts of Interest: The authors declare no conflict of interest.

References

1. Amano, H.; Baines, Y.; Beam, E.; Borga, M.; Bouchet, T.; Chalker, P.R.; Charles, M.; Chen, K.J.; Chowdhury, N.; Chu, R. The 2018 GaN Power Electronics Roadmap. *J. Phys. D Appl. Phys.* **2018**, *51*, 163001. [CrossRef]
2. Hassan, A.; Savaria, Y.; Sawan, M. GaN Integration Technology, an Ideal Candidate for High-Temperature Applications: A Review. *IEEE Access* **2018**, *6*, 78790–78802. [CrossRef]
3. Okumura, H. Present Status and Future Prospect of Widegap Semiconductor High-Power Devices. *Jpn. J. Appl. Phys.* **2006**, *45*, 7565. [CrossRef]
4. Ambacher, O.; Dimitrov, R.; Stutzmann, M.; Foutz, B.E.; Murphy, M.J.; Smart, J.A.; Shealy, J.R.; Weimann, N.G.; Chu, K.; Chumbes, M.; et al. Role of Spontaneous and Piezoelectric Polarization Induced Effects in Group-III Nitride Based Heterostructures and Devices. *Phys. Status Solidi B* **1999**, *216*, 381–389. [CrossRef]
5. Ambacher, O.; Smart, J.; Shealy, J.R.; Weimann, N.G.; Chu, K.; Murphy, M.; Schaff, W.J.; Eastman, L.F.; Dimitrov, R.; Wittmer, L. Two-Dimensional Electron Gases Induced by Spontaneous and Piezoelectric Polarization Charges in N-and Ga-Face AlGaN/GaN Heterostructures. *J. Appl. Phys.* **1999**, *85*, 3222–3233. [CrossRef]
6. Ambacher, O.; Foutz, B.; Smart, J.; Shealy, J.R.; Weimann, N.G.; Chu, K.; Murphy, M.; Sierakowski, A.J.; Schaff, W.J.; Eastman, L.F.; et al. Two Dimensional Electron Gases Induced by Spontaneous and Piezoelectric Polarization in Undoped and Doped AlGaN/GaN Heterostructures. *J. Appl. Phys.* **2000**, *87*, 334–344. [CrossRef]
7. Roccaforte, F.; Fiorenza, P.; Greco, G.; Nigro, R.L.; Giannazzo, F.; Patti, A.; Saggio, M. Challenges for Energy Efficient Wide Band Gap Semiconductor Power Devices. *Phys. Status Solidi A* **2014**, *211*, 2063–2071. [CrossRef]
8. Flack, T.J.; Pushpakaran, B.N.; Bayne, S.B. GaN Technology for Power Electronic Applications: A Review. *J. Electron. Mater.* **2016**, *45*, 2673–2682. [CrossRef]
9. Trescases, O.; Murray, S.K.; Jiang, W.L.; Zaman, M.S. GaN Power ICs: Reviewing Strengths, Gaps, and Future Directions. In Proceedings of the 2020 IEEE International Electron Devices Meeting (IEDM), San Francisco, CA, USA, 12–18 December 2020; pp. 27.4.1–27.4.4. [CrossRef]
10. Hsu, L.-H.; Lai, Y.-Y.; Tu, P.-T.; Langpoklakpam, C.; Chang, Y.-T.; Huang, Y.-W.; Lee, W.-C.; Tzou, A.-J.; Cheng, Y.-J.; Lin, C.-H.; et al. Development of GaN HEMTs Fabricated on Silicon, Silicon-on-Insulator, and Engineered Substrates and the Heterogeneous Integration. *Micromachines* **2021**, *12*, 1159. [CrossRef]
11. Chowdhury, S.; Swenson, B.L.; Mishra, U.K. Enhancement and Depletion Mode AlGaN/GaN CAVET with Mg-Ion-Implanted GaN as Current Blocking Layer. *IEEE Electron Device Lett.* **2008**, *29*, 543–545. [CrossRef]
12. Yeluri, R.; Lu, J.; Hurni, C.A.; Browne, D.A.; Chowdhury, S.; Keller, S.; Speck, J.S.; Mishra, U.K. Design, Fabrication, and Performance Analysis of GaN Vertical Electron Transistors with a Buried p/n Junction. *Appl. Phys. Lett.* **2015**, *106*, 183502. [CrossRef]
13. Ji, D.; Agarwal, A.; Li, H.; Li, W.; Keller, S.; Chowdhury, S. 880 V/2.7 m$\Omega\cdot$cm^2 MIS Gate Trench CAVET on Bulk GaN Substrates. *IEEE Electron Device Lett.* **2018**, *39*, 863–865. [CrossRef]
14. Tang, Y.; Shinohara, K.; Regan, D.; Corrion, A.; Brown, D.; Wong, J.; Schmitz, A.; Fung, H.; Kim, S.; Micovic, M. Ultrahigh-Speed GaN High-Electron-Mobility Transistors with f_T/f_{max} of 454/444 GHz. *IEEE Electron Device Lett.* **2015**, *36*, 549–551. [CrossRef]
15. Shinohara, K.; Regan, D.C.; Tang, Y.; Corrion, A.L.; Brown, D.F.; Wong, J.C.; Robinson, J.F.; Fung, H.H.; Schmitz, A.; Oh, T.C.; et al. Scaling of GaN HEMTs and Schottky Diodes for Submillimeter-Wave MMIC Applications. *IEEE Trans. Electron Devices* **2013**, *60*, 2982–2996. [CrossRef]
16. Lv, Y.; Song, X.; Guo, H.; Fang, Y.; Feng, Z. High-Frequency AlGaN/GaN HFETs with f_T/f_{max} of 149/263 GHz for D-Band PA Applications. *Electron. Lett.* **2016**, *52*, 1340–1342. [CrossRef]
17. Lee, D.S.; Laboutin, O.; Cao, Y.; Johnson, W.; Beam, E.; Ketterson, A.; Schuette, M.; Saunier, P.; Palacios, T. Impact of Al_2O_3 Passivation Thickness in Highly Scaled GaN HEMTs. *IEEE Electron Device Lett.* **2012**, *33*, 976–978. [CrossRef]
18. Chung, J.W.; Kim, T.-W.; Palacios, T. Advanced Gate Technologies for State-of-the-Art f_T in AlGaN/GaN HEMTs. In Proceedings of the 2010 International Electron Devices Meeting, San Francisco, CA, USA, 6–8 December 2010; pp. 30–32. [CrossRef]
19. Hashizume, T.; Kotani, J.; Hasegawa, H. Leakage Mechanism in GaN and AlGaN Schottky Interfaces. *Appl. Phys. Lett.* **2004**, *84*, 4884–4886. [CrossRef]
20. Greco, G.; Iucolano, F.; Roccaforte, F. Review of Technology for Normally-off HEMTs with p-GaN Gate. *Mater. Sci. Semicond. Process.* **2018**, *78*, 96–106. [CrossRef]
21. Lidow, A.; De Rooij, M.; Strydom, J.; Reusch, D.; Glaser, J. *GaN Transistors for Efficient Power Conversion*; John Wiley & Sons: Hoboken, NJ, USA, 2019.

22. Kanamura, M.; Kikkawa, T.; Iwai, T.; Imanishi, K.; Kubo, T.; Joshin, K. An over 100 W n-GaN/n-AlGaN/GaN MIS-HEMT Power Amplifier for Wireless Base Station Applications. In Proceedings of the IEEE InternationalElectron Devices Meeting, 2005. IEDM Technical Digest, Washington, DC, USA, 5 December 2005; pp. 572–575. [CrossRef]
23. Ye, P.D.; Yang, B.; Ng, K.K.; Bude, J.; Wilk, G.D.; Halder, S.; Hwang, J.C.M. GaN Metal-Oxide-Semiconductor High-Electron-Mobility-Transistor with Atomic Layer Deposited Al_2O_3 as Gate Dielectric. *Appl. Phys. Lett.* **2005**, *86*, 063501. [CrossRef]
24. Khan, M.A.; Hu, X.; Sumin, G.; Lunev, A.; Yang, J.; Gaska, R.; Shur, M.S. AlGaN/GaN Metal Oxide Semiconductor Heterostructure Field Effect Transistor. *IEEE Electron Device Lett.* **2000**, *21*, 63–65. [CrossRef]
25. Chen, K.J.; Yang, S.; Liu, S.; Liu, C.; Hua, M. Toward Reliable MIS-and MOS-Gate Structures for GaN Lateral Power Devices. *Phys. Status Solidi A* **2016**, *213*, 861–867. [CrossRef]
26. Mizue, C.; Hori, Y.; Miczek, M.; Hashizume, T. Capacitance–Voltage Characteristics of Al_2O_3/AlGaN/GaN Structures and State Density Distribution at Al_2O_3/AlGaN Interface. *Jpn. J. Appl. Phys.* **2011**, *50*, 021001. [CrossRef]
27. Hori, Y.; Yatabe, Z.; Hashizume, T. Characterization of Interface States in Al_2O_3/AlGaN/GaN Structures for Improved Performance of High-Electron-Mobility Transistors. *J. Appl. Phys.* **2013**, *114*, 244503. [CrossRef]
28. Lu, X.; Yu, K.; Jiang, H.; Zhang, A.; Lau, K.M. Study of Interface Traps in AlGaN/GaN MISHEMTs Using LPCVD SiN_x as Gate Dielectric. *IEEE Trans. Electron Devices* **2017**, *64*, 824–831. [CrossRef]
29. Hashizume, T.; Nishiguchi, K.; Kaneki, S.; Kuzmik, J.; Yatabe, Z. State of the Art on Gate Insulation and Surface Passivation for GaN-Based Power HEMTs. *Mater. Sci. Semicond. Process.* **2018**, *78*, 85–95. [CrossRef]
30. Meneghesso, G.; Meneghini, M.; Bisi, D.; Rossetto, I.; Cester, A.; Mishra, U.K.; Zanoni, E. Trapping Phenomena in AlGaN/GaN HEMTs: A Study Based on Pulsed and Transient Measurements. *Semicond. Sci. Technol.* **2013**, *28*, 074021. [CrossRef]
31. Szabó, N.; Wachowiak, A.; Winzer, A.; Ocker, J.; Gärtner, J.; Hentschel, R.; Schmid, A.; Mikolajick, T. High-k/GaN Interface Engineering toward AlGaN/GaN MIS-HEMT with Improved V_{th} Stability. *J. Vac. Sci. Technol. B* **2017**, *35*, 01A102. [CrossRef]
32. Lu, Y.; Yang, S.; Jiang, Q.; Tang, Z.; Li, B.; Chen, K.J. Characterization of V_t-Instability in Enhancement-Mode Al_2O_3-AlGaN/GaN MIS-HEMTs. *Phys. Status Solidi C* **2013**, *10*, 1397–1400. [CrossRef]
33. Zhu, J.; Hou, B.; Chen, L.; Zhu, Q.; Yang, L.; Zhou, X.; Zhang, P.; Ma, X.; Hao, Y. Threshold Voltage Shift and Interface/Border Trapping Mechanism in Al_2O_3/AlGaN/GaN MOS-HEMTs. In Proceedings of the 2018 IEEE International Reliability Physics Symposium (IRPS), Burlingame, CA, USA, 11–15 March 2018; p. P-WB. [CrossRef]
34. Winzer, A.; Schuster, M.; Hentschel, R.; Ocker, J.; Merkel, U.; Jahn, A.; Wachowiak, A.; Mikolajick, T. Analysis of Threshold Voltage Instability in AlGaN/GaN MISHEMTs by Forward Gate Voltage Stress Pulses. *Phys. Status Solidi A* **2016**, *213*, 1246–1251. [CrossRef]
35. Huang, S.; Yang, S.; Roberts, J.; Chen, K.J. Threshold Voltage Instability in Al_2O_3/GaN/AlGaN/GaN Metal-Insulator–Semiconductor High-Electron Mobility Transistors. *Jpn. J. Appl. Phys.* **2011**, *50*, 110202. [CrossRef]
36. Lagger, P.; Ostermaier, C.; Pobegen, G.; Pogany, D. Towards Understanding the Origin of Threshold Voltage Instability of AlGaN/GaN MIS-HEMTs. In Proceedings of the 2012 International Electron Devices Meeting, San Francisco, CA, USA, 10–13 December 2012; pp. 13.1.1–13.1.4. [CrossRef]
37. Lagger, P.; Steinschifter, P.; Reiner, M.; Stadtmüller, M.; Denifl, G.; Naumann, A.; Müller, J.; Wilde, L.; Sundqvist, J.; Pogany, D.; et al. Role of the Dielectric for the Charging Dynamics of the Dielectric/Barrier Interface in AlGaN/GaN Based Metal-Insulator-Semiconductor Structures under Forward Gate Bias Stress. *Appl. Phys. Lett.* **2014**, *105*, 033512. [CrossRef]
38. Nishiguchi, K.; Kaneki, S.; Ozaki, S.; Hashizume, T. Current Linearity and Operation Stability in Al_2O_3-Gate AlGaN/GaN MOS High Electron Mobility Transistors. *Jpn. J. Appl. Phys.* **2017**, *56*, 101001. [CrossRef]
39. Huang, W.; Khan, T.; Chow, T.P. Enhancement-Mode n-Channel GaN MOSFETs on p and n-GaN/Sapphire Substrates. In Proceedings of the 2006 IEEE International Symposium on Power Semiconductor Devices and IC's, Naples, Italy, 4–8 June 2006; pp. 1–4. [CrossRef]
40. Oka, T.; Ueno, Y.; Ina, T.; Hasegawa, K. Vertical GaN-Based Trench Metal Oxide Semiconductor Field-Effect Transistors on a Free-Standing GaN Substrate with Blocking Voltage of 1.6 kV. *Appl. Phys. Express* **2014**, *7*, 021002. [CrossRef]
41. Sun, M.; Zhang, Y.; Gao, X.; Palacios, T. High-Performance GaN Vertical Fin Power Transistors on Bulk GaN Substrates. *IEEE Electron Device Lett.* **2017**, *38*, 509–512. [CrossRef]
42. Hua, M.; Wei, J.; Tang, G.; Zhang, Z.; Qian, Q.; Cai, X.; Wang, N.; Chen, K.J. Normally-off LPCVD-SiN_x/GaN MIS-FET with Crystalline Oxidation Interlayer. *IEEE Electron Device Lett.* **2017**, *38*, 929–932. [CrossRef]
43. Mukherjee, K.; De Santi, C.; Borga, M.; Geens, K.; You, S.; Bakeroot, B.; Decoutere, S.; Diehle, P.; Hübner, S.; Altmann, F.; et al. Challenges and Perspectives for Vertical GaN-on-Si Trench MOS Reliability: From Leakage Current Analysis to Gate Stack Optimization. *Materials* **2021**, *14*, 2316. [CrossRef] [PubMed]
44. Hentschel, R.; Wachowiak, A.; Großer, A.; Kotzea, S.; Debald, A.; Kalisch, H.; Vescan, A.; Jahn, A.; Schmult, S.; Mikolajick, T. Extraction of the Active Acceptor Concentration in (Pseudo-) Vertical GaN MOSFETs Using the Body-Bias Effect. *Microelectron. J.* **2019**, *91*, 42–45. [CrossRef]
45. Mukherjee, K.; De Santi, C.; Borga, M.; You, S.; Geens, K.; Bakeroot, B.; Decoutere, S.; Meneghesso, G.; Zanoni, E.; Meneghini, M. Use of Bilayer Gate Insulator in GaN-on-Si Vertical Trench MOSFETs: Impact on Performance and Reliability. *Materials* **2020**, *13*, 4740. [CrossRef]
46. Pérez-Tomás, A.; Placidi, M.; Baron, N.; Chenot, S.; Cordier, Y.; Moreno, J.C.; Constant, A.; Godignon, P.; Millán, J. GaN Transistor Characteristics at Elevated Temperatures. *J. Appl. Phys.* **2009**, *106*, 074519. [CrossRef]

47. Pérez-Tomás, A.; Placidi, M.; Perpiñà, X.; Constant, A.; Godignon, P.; Jordà, X.; Brosselard, P.; Millán, J. GaN Metal-Oxide-Semiconductor Field-Effect Transistor Inversion Channel Mobility Modeling. *J. Appl. Phys.* **2009**, *105*, 114510. [CrossRef]
48. Hua, M.; Zhang, Z.; Wei, J.; Lei, J.; Tang, G.; Fu, K.; Cai, Y.; Zhang, B.; Chen, K.J. Integration of LPCVD-SiN_x Gate Dielectric with Recessed-Gate E-Mode GaN MIS-FETs: Toward High Performance, High Stability and Long TDDB Lifetime. In Proceedings of the 2016 IEEE International Electron Devices Meeting (IEDM), San Francisco, CA, USA, 3–7 December 2016; pp. 10–14. [CrossRef]
49. Yatabe, Z.; Asubar, J.T.; Hashizume, T. Insulated Gate and Surface Passivation Structures for GaN-Based Power Transistors. *J. Phys. D Appl. Phys.* **2016**, *49*, 393001. [CrossRef]
50. Roccaforte, F.; Fiorenza, P.; Greco, G.; Vivona, M.; Nigro, R.L.; Giannazzo, F.; Patti, A.; Saggio, M. Recent Advances on Dielectrics Technology for SiC and GaN Power Devices. *Appl. Surf. Sci.* **2014**, *301*, 9–18. [CrossRef]
51. Eller, B.S.; Yang, J.; Nemanich, R.J. Electronic Surface and Dielectric Interface States on GaN and AlGaN. *J. Vac. Sci. Technol. A* **2013**, *31*, 050807. [CrossRef]
52. Vetury, R.; Zhang, N.Q.; Keller, S.; Mishra, U.K. The Impact of Surface States on the DC and RF Characteristics of AlGaN/GaN HFETs. *IEEE Trans. Electron Devices* **2001**, *48*, 560–566. [CrossRef]
53. Hashizume, T.; Ootomo, S.; Hasegawa, H. Suppression of Current Collapse in Insulated Gate AlGaN/GaN Heterostructure Field-Effect Transistors Using Ultrathin Al_2O_3 Dielectric. *Appl. Phys. Lett.* **2003**, *83*, 2952–2954. [CrossRef]
54. Green, B.M.; Chu, K.K.; Chumbes, E.M.; Smart, J.A.; Shealy, J.R.; Eastman, L.F. The Effect of Surface Passivation on the Microwave Characteristics of Undoped AlGaN/GaN HEMTs. *IEEE Electron Device Lett.* **2000**, *21*, 268–270. [CrossRef]
55. Sze, S.M.; Li, Y.; Ng, K.K. *Physics of Semiconductor Devices*; John Wiley & Sons: Hoboken, NJ, USA, 2021.
56. Robertson, J.; Falabretti, B. Band Offsets of High K Gate Oxides on III-V Semiconductors. *J. Appl. Phys.* **2006**, *100*, 014111. [CrossRef]
57. Reddy, B.P.K.; Teja, K.B.R.; Kandpal, K. Investigation on High-κ Dielectric for Low Leakage AlGaN/GaN MIS-HEMT Device, Using Material Selection Methodologies. *Semiconductors* **2018**, *52*, 420–430. [CrossRef]
58. Kambayashi, H.; Satoh, Y.; Ootomo, S.; Kokawa, T.; Nomura, T.; Kato, S.; Chow, T.P. Over 100 A Operation Normally-off AlGaN/GaN Hybrid MOS-HFET on Si Substrate with High-Breakdown Voltage. *Solid-State Electron.* **2010**, *54*, 660–664. [CrossRef]
59. Lee, J.-G.; Kim, H.-S.; Seo, K.-S.; Cho, C.-H.; Cha, H.-Y. High Quality PECVD SiO_2 Process for Recessed MOS-Gate of AlGaN/GaN-on-Si Metal–Oxide–Semiconductor Heterostructure Field-Effect Transistors. *Solid-State Electron.* **2016**, *122*, 32–36. [CrossRef]
60. Liu, C.; Chor, E.F.; Tan, L.S. Investigations of HfO_2/AlGaN/GaN Metal-Oxide-Semiconductor High Electron Mobility Transistors. *Appl. Phys. Lett.* **2006**, *88*, 173504. [CrossRef]
61. Liu, C.; Chor, E.F.; Tan, L.S. Enhanced Device Performance of AlGaN/GaN HEMTs Using HfO_2 High-k Dielectric for Surface Passivation and Gate Oxide. *Semicond. Sci. Technol.* **2007**, *22*, 522. [CrossRef]
62. Shi, J.; Eastman, L.F.; Xin, X.; Pophristic, M. High Performance AlGaN/GaN Power Switch with HfO_2 Insulation. *Appl. Phys. Lett.* **2009**, *95*. [CrossRef]
63. Ye, G.; Wang, H.; Arulkumaran, S.; Ng, G.I.; Hofstetter, R.; Li, Y.; Anand, M.J.; Ang, K.S.; Maung, Y.K.T.; Foo, S.C. Atomic Layer Deposition of ZrO_2 as Gate Dielectrics for AlGaN/GaN Metal-Insulator-Semiconductor High Electron Mobility Transistors on Silicon. *Appl. Phys. Lett.* **2013**, *103*, 142109. [CrossRef]
64. Anderson, T.J.; Wheeler, V.D.; Shahin, D.I.; Tadjer, M.J.; Koehler, A.D.; Hobart, K.D.; Christou, A.; Kub, F.J.; Eddy, C.R., Jr. Enhancement Mode AlGaN/GaN MOS High-Electron-Mobility Transistors with ZrO_2 Gate Dielectric Deposited by Atomic Layer Deposition. *Appl. Phys. Express* **2016**, *9*, 071003. [CrossRef]
65. Jiang, H.; Liu, C.; Ng, K.W.; Tang, C.W.; Lau, K.M. High-Performance AlGaN/GaN/Si Power MOSHEMTs with ZrO_2 Gate Dielectric. *IEEE Trans. Electron Devices* **2018**, *65*, 5337–5342. [CrossRef]
66. Kikkawa, T.; Makiyama, K.; Ohki, T.; Kanamura, M.; Imanishi, K.; Hara, N.; Joshin, K. High Performance and High Reliability AlGaN/GaN HEMTs. *Phys. Status Solidi A* **2009**, *206*, 1135–1144. [CrossRef]
67. Deen, D.A.; Storm, D.F.; Bass, R.; Meyer, D.J.; Katzer, D.S.; Binari, S.C.; Lacis, J.W.; Gougousi, T. Atomic Layer Deposited Ta_2O_5 Gate Insulation for Enhancing Breakdown Voltage of AlN/GaN High Electron Mobility Transistors. *Appl. Phys. Lett.* **2011**, *98*, 023506. [CrossRef]
68. Fiorenza, P.; Greco, G.; Fisichella, G.; Roccaforte, F.; Malandrino, G.; Lo Nigro, R. High Permittivity Cerium Oxide Thin Films on AlGaN/GaN Heterostructures. *Appl. Phys. Lett.* **2013**, *103*, 112905. [CrossRef]
69. Chiu, Y.S.; Liao, J.T.; Lin, Y.C.; Liu, S.C.; Lin, T.M.; Iwai, H.; Kakushima, K.; Chang, E.Y. High-Permitivity Cerium Oxide Prepared by Molecular Beam Deposition as Gate Dielectric and Passivation Layer and Applied to AlGaN/GaN Power High Electron Mobility Transistor Devices. *Jpn. J. Appl. Phys.* **2016**, *55*, 051001. [CrossRef]
70. Hsu, C.-H.; Shih, W.-C.; Lin, Y.-C.; Hsu, H.-T.; Hsu, H.-H.; Huang, Y.-X.; Lin, T.-W.; Wu, C.-H.; Wu, W.-H.; Maa, J.-S.; et al. Improved Linearity and Reliability in GaN Metal–Oxide–Semiconductor High-Electron-Mobility Transistors Using Nanolaminate La_2O_3/SiO_2 Gate Dielectric. *Jpn. J. Appl. Phys.* **2016**, *55*, 04EG04. [CrossRef]
71. Hansen, P.J.; Vaithyanathan, V.; Wu, Y.; Mates, T.; Heikman, S.; Mishra, U.K.; York, R.A.; Schlom, D.G.; Speck, J.S. Rutile Films Grown by Molecular Beam Epitaxy on GaN and AlGaN/GaN. *J. Vac. Sci. Technol. B Microelectron. Nanometer Struct. Process. Meas. Phenom.* **2005**, *23*, 499–506. [CrossRef]
72. Rawat, A.; Meer, M.; kumar Surana, V.; Bhardwaj, N.; Pendem, V.; Garigapati, N.S.; Yadav, Y.; Ganguly, S.; Saha, D. Thermally Grown TiO_2 and Al_2O_3 for GaN-Based MOS-HEMTs. *IEEE Trans. Electron Devices* **2018**, *65*, 3725–3731. [CrossRef]

73. Yatabe, Z.; Hori, Y.; Ma, W.-C.; Asubar, J.T.; Akazawa, M.; Sato, T.; Hashizume, T. Characterization of Electronic States at Insulator/(Al)GaN Interfaces for Improved Insulated Gate and Surface Passivation Structures of GaN-Based Transistors. *Jpn. J. Appl. Phys.* **2014**, *53*, 100213. [CrossRef]
74. Deen, D.; Storm, D.; Meyer, D.; Katzer, D.S.; Bass, R.; Binari, S.; Gougousi, T. AlN/GaN HEMTs with High-κ ALD HfO_2 or Ta_2O_5 Gate Insulation. *Phys. Status Solidi C* **2011**, *8*, 2420–2423. [CrossRef]
75. Hashizume, T.; Ootomo, S.; Inagaki, T.; Hasegawa, H. Surface Passivation of GaN and GaN/AlGaN Heterostructures by Dielectric Films and Its Application to Insulated-Gate Heterostructure Transistors. *J. Vac. Sci. Technol. B Microelectron. Nanometer Struct. Process. Meas. Phenom.* **2003**, *21*, 1828–1838. [CrossRef]
76. Ochiai, M.; Akita, M.; Ohno, Y.; Kishimoto, S.; Maezawa, K.; Mizutani, T. AlGaN/GaN Heterostructure Metal-Insulator-Semiconductor High-Electron-Mobility Transistors with Si_3N_4 Gate Insulator. *Jpn. J. Appl. Phys.* **2003**, *42*, 2278. [CrossRef]
77. Lee, C.-T.; Chen, H.-W.; Lee, H.-Y. Metal–Oxide–Semiconductor Devices Using Ga_2O_3 Dielectrics on n-type GaN. *Appl. Phys. Lett.* **2003**, *82*, 4304–4306. [CrossRef]
78. Jiang, H.; Liu, C.; Chen, Y.; Lu, X.; Tang, C.W.; Lau, K.M. Investigation of in Situ SiN as Gate Dielectric and Surface Passivation for GaN MISHEMTs. *IEEE Trans. Electron Devices* **2017**, *64*, 832–839. [CrossRef]
79. Liu, S.; Yang, S.; Tang, Z.; Jiang, Q.; Liu, C.; Wang, M.; Chen, K.J. Al_2O_3/AlN/GaN MOS-Channel-HEMTs with an AlN Interfacial Layer. *IEEE Electron Device Lett.* **2014**, *35*, 723–725. [CrossRef]
80. Tsurumi, N.; Ueno, H.; Murata, T.; Ishida, H.; Uemoto, Y.; Ueda, T.; Inoue, K.; Tanaka, T. AlN Passivation over AlGaN/GaN HFETs for Surface Heat Spreading. *IEEE Trans. Electron Devices* **2010**, *57*, 980–985. [CrossRef]
81. Huang, S.; Jiang, Q.; Yang, S.; Zhou, C.; Chen, K.J. Effective Passivation of AlGaN/GaN HEMTs by ALD-Grown AlN Thin Film. *IEEE Electron Device Lett.* **2012**, *33*, 516–518. [CrossRef]
82. Roccaforte, F.; Greco, G.; Fiorenza, P.; Raineri, V.; Malandrino, G.; Lo Nigro, R. Epitaxial NiO Gate Dielectric on AlGaN/GaN Heterostructures. *Appl. Phys. Lett.* **2012**, *100*, 063511. [CrossRef]
83. Fiorenza, P.; Greco, G.; Giannazzo, F.; Lo Nigro, R.; Roccaforte, F. Poole-Frenkel Emission in Epitaxial Nickel Oxide on AlGaN/GaN Heterostructures. *Appl. Phys. Lett.* **2012**, *101*, 172901. [CrossRef]
84. Mehandru, R.; Luo, B.; Kim, J.; Ren, F.; Gila, B.P.; Onstine, A.H.; Abernathy, C.R.; Pearton, S.J.; Gotthold, D.; Birkhahn, R.; et al. AlGaN/GaN Metal–Oxide–Semiconductor High Electron Mobility Transistors Using Sc_2O_3 as the Gate Oxide and Surface Passivation. *Appl. Phys. Lett.* **2003**, *82*, 2530–2532. [CrossRef]
85. Luo, B.; Johnson, J.W.; Gila, B.; Onstine, A.; Abernathy, C.R.; Ren, F.; Pearton, S.J.; Baca, A.G.; Dabiran, A.M.; Wowchack, A.M.; et al. Surface Passivation of AlGaN/GaN HEMTs Using MBE-Grown MgO or Sc_2O_3. *Solid-State Electron.* **2002**, *46*, 467–476. [CrossRef]
86. Polyakov, A.Y.; Smirnov, N.B.; Gila, B.P.; Hlad, M.; Gerger, A.P.; Abernathy, C.R.; Pearton, S.J. Studies of Interface States in Sc_2O_3/GaN, MgO/GaN, and MgScO/GaN Structures. *J. Electrochem. Soc.* **2006**, *154*, H115. [CrossRef]
87. Wang, X.; Saadat, O.I.; Xi, B.; Lou, X.; Molnar, R.J.; Palacios, T.; Gordon, R.G. Atomic Layer Deposition of Sc_2O_3 for Passivating AlGaN/GaN High Electron Mobility Transistor Devices. *Appl. Phys. Lett.* **2012**, *101*, 232109. [CrossRef]
88. Balachander, K.; Arulkumaran, S.; Sano, Y.; Egawa, T.; Baskar, K. Fabrication of AlGaN/GaN Double-Insulator Metal–Oxide–Semiconductor High-Electron-Mobility Transistors Using SiO_2 and SiN as Gate Insulators. *Phys. Status Solidi A* **2005**, *202*, R32–R34. [CrossRef]
89. Anand, M.J.; Ng, G.I.; Vicknesh, S.; Arulkumaran, S.; Ranjan, K. Reduction of Current Collapse in AlGaN/GaN MISHEMT with Bilayer SiN/Al_2O_3 Dielectric Gate Stack. *Phys. Status Solidi C* **2013**, *10*, 1421–1425. [CrossRef]
90. Colon, A.; Shi, J. High-κ Insulating Materials for AlGaN/GaN Metal Insulator Semiconductor Heterojunction Field Effect Transistors. *Solid-State Electron.* **2014**, *99*, 25–30. [CrossRef]
91. Geng, K.; Chen, D.; Zhou, Q.; Wang, H. AlGaN/GaN MIS-HEMT with PECVD SiN_x, SiON, SiO_2 as Gate Dielectric and Passivation Layer. *Electronics* **2018**, *7*, 416. [CrossRef]
92. Sun, Z.; Huang, H.; Wang, R.; Liu, Y.; Sun, N.; Li, F.; Tao, P.; Ren, Y.; Song, S.; Wang, H.; et al. Effects of SiON/III-Nitride Interface Properties on Device Performances of GaN-Based Power Field-Effect Transistors. *J. Phys. D Appl. Phys.* **2020**, *54*, 025109. [CrossRef]
93. Yang, S.; Huang, S.; Schnee, M.; Zhao, Q.-T.; Schubert, J.; Chen, K.J. Fabrication and Characterization of Enhancement-Mode High-k $LaLuO_3$-AlGaN/GaN MIS-HEMTs. *IEEE Trans. Electron Devices* **2013**, *60*, 3040–3046. [CrossRef]
94. Park, K.-Y.; Cho, H.-I.; Lee, J.-H.; Bae, S.-B.; Jeon, C.-M.; Lee, J.-L.; Kim, D.-Y.; Lee, C.-S.; Lee, J.-H. Fabrication of AlGaN/GaN MIS-HFET Using an Al_2O_3 High k Dielectric. *Phys. Status Solidi C* **2003**, 2351–2354. [CrossRef]
95. Yue, Y.-Z.; Hao, Y.; Feng, Q.; Zhang, J.-C.; Ma, X.-H.; Ni, J.-Y. GaN MOS-HEMT Using Ultra-Thin Al_2O_3 Dielectric Grown by Atomic Layer Deposition. *Chin. Phys. Lett.* **2007**, *24*, 2419. [CrossRef]
96. Afanas' Ev, V.V.; Stesmans, A.; Mrstik, B.J.; Zhao, C. Impact of Annealing-Induced Compaction on Electronic Properties of Atomic-Layer-Deposited Al_2O_3. *Appl. Phys. Lett.* **2002**, *81*, 1678–1680. [CrossRef]
97. Afanas' Ev, V.V.; Houssa, M.; Stesmans, A.; Merckling, C.; Schram, T.; Kittl, J.A. Influence of Al_2O_3 Crystallization on Band Offsets at Interfaces with Si and TiN_x. *Appl. Phys. Lett.* **2011**, *99*, 072103. [CrossRef]
98. Momida, H.; Hamada, T.; Takagi, Y.; Yamamoto, T.; Uda, T.; Ohno, T. Theoretical Study on Dielectric Response of Amorphous Alumina. *Phys. Rev. B* **2006**, *73*, 054108. [CrossRef]

99. Toyoda, S.; Shinohara, T.; Kumigashira, H.; Oshima, M.; Kato, Y. Significant Increase in Conduction Band Discontinuity due to Solid Phase Epitaxy of Al_2O_3 Gate Insulator Films on GaN Semiconductor. *Appl. Phys. Lett.* **2012**, *101*, 231607. [CrossRef]
100. Yang, J.; Eller, B.S.; Zhu, C.; England, C.; Nemanich, R.J. Comparative Band Alignment of Plasma-Enhanced Atomic Layer Deposited High-k Dielectrics on Gallium Nitride. *J. Appl. Phys.* **2012**, *112*, 053710. [CrossRef]
101. Costina, I.; Franchy, R. Band Gap of Amorphous and Well-Ordered Al_2O_3 on $Ni_3Al(100)$. *Appl. Phys. Lett.* **2001**, *78*, 4139–4141. [CrossRef]
102. French, R.H. Electronic Band Structure of Al_2O_3, with Comparison to AlON and AlN. *J. Am. Ceram. Soc.* **1990**, *73*, 477–489. [CrossRef]
103. Maeda, N.; Wang, C.; Enoki, T.; Makimoto, T.; Tawara, T. High Drain Current Density and Reduced Gate Leakage Current in Channel-Doped AlGaN/GaN Heterostructure Field-Effect Transistors with Al_2O_3/Si_3N_4 Gate Insulator. *Appl. Phys. Lett.* **2005**, *87*, 073504. [CrossRef]
104. Gregušová, D.; Stoklas, R.; Čičo, K.; Heidelberger, G.; Marso, M.; Novák, J.; Kordoš, P. Characterization of AlGaN/GaN MOSHFETs with Al_2O_3 as Gate Oxide. *Phys. Status Solidi C* **2007**, *4*, 2720–2723. [CrossRef]
105. Ťapajna, M.; Jurkovič, M.; Válik, L.; Haščík, Š.; Gregušová, D.; Brunner, F.; Cho, E.-M.; Kuzmík, J. Bulk and Interface Trapping in the Gate Dielectric of GaN Based Metal-Oxide-Semiconductor High-Electron-Mobility Transistors. *Appl. Phys. Lett.* **2013**, *102*, 243509. [CrossRef]
106. Ťapajna, M.; Jurkovič, M.; Válik, L.; Haščík, Š.; Gregušová, D.; Brunner, F.; Cho, E.-M.; Hashizume, T.; Kuzmík, J. Impact of GaN Cap on Charges in Al_2O_3/(GaN/)AlGaN/GaN Metal-Oxide-Semiconductor Heterostructures Analyzed by Means of Capacitance Measurements and Simulations. *J. Appl. Phys.* **2014**, *116*, 104501. [CrossRef]
107. Esposto, M.; Krishnamoorthy, S.; Nath, D.N.; Bajaj, S.; Hung, T.-H.; Rajan, S. Electrical Properties of Atomic Layer Deposited Aluminum Oxide on Gallium Nitride. *Appl. Phys. Lett.* **2011**, *99*, 133503. [CrossRef]
108. Hori, Y.; Mizue, C.; Hashizume, T. Process Conditions for Improvement of Electrical Properties of Al_2O_3/n-GaN Structures Prepared by Atomic Layer Deposition. *Jpn. J. Appl. Phys.* **2010**, *49*, 080201. [CrossRef]
109. Suri, R. Investigation of MOS Interfaces with Atomic-Layer-Deposited High-k Gate Dielectrics on III-V Semiconductors. Ph.D. Thesis, North Carolina State University, Raleigh, NC, USA, 2010.
110. Zhang, Z.; Jackson, C.M.; Arehart, A.R.; McSkimming, B.; Speck, J.S.; Ringel, S.A. Direct Determination of Energy Band Alignments of Ni/Al_2O_3/GaN MOS Structures Using Internal Photoemission Spectroscopy. *J. Electron. Mater.* **2014**, *43*, 828–832. [CrossRef]
111. Yang, J.; Eller, B.S.; Nemanich, R.J. Surface Band Bending and Band Alignment of Plasma Enhanced Atomic Layer Deposited Dielectrics on Ga- and N-face Gallium Nitride. *J. Appl. Phys.* **2014**, *116*, 123702. [CrossRef]
112. Qin, X.; Cheng, L.; McDonnell, S.; Azcatl, A.; Zhu, H.; Kim, J.; Wallace, R.M. A Comparative Study of Atomic Layer Deposition of Al_2O_3 and HfO_2 on AlGaN/GaN. *J. Mater.Sci. Mater. Electron.* **2015**, *26*, 4638–4643. [CrossRef]
113. Winzer, A.; Szabó, N.; Ocker, J.; Hentschel, R.; Schuster, M.; Schubert, F.; Gärtner, J.; Wachowiak, A.; Mikolajick, T. Detailed Analysis of Oxide Related Charges and Metal-Oxide Barriers in Terrace Etched Al_2O_3 and HfO_2 on AlGaN/GaN Heterostructure Capacitors. *J. Appl. Phys.* **2015**, *118*, 124106. [CrossRef]
114. Hashizume, T.; Kaneki, S.; Oyobiki, T.; Ando, Y.; Sasaki, S.; Nishiguchi, K. Effects of Postmetallization Annealing on Interface Properties of Al_2O_3/GaN Structures. *Appl. Phys. Express* **2018**, *11*, 124102. [CrossRef]
115. Kaneki, S.; Ohira, J.; Toiya, S.; Yatabe, Z.; Asubar, J.T.; Hashizume, T. Highly-Stable and Low-State-Density Al_2O_3/GaN Interfaces Using Epitaxial n-GaN Layers Grown on Free-Standing GaN Substrates. *Appl. Phys. Lett.* **2016**, *109*, 162104. [CrossRef]
116. Son, J.; Chobpattana, V.; McSkimming, B.M.; Stemmer, S. Fixed Charge in High-k/GaN Metal-Oxide-Semiconductor Capacitor Structures. *Appl. Phys. Lett.* **2012**, *101*, 102905. [CrossRef]
117. Uedono, A.; Nabatame, T.; Egger, W.; Koschine, T.; Hugenschmidt, C.; Dickmann, M.; Sumiya, M.; Ishibashi, S. Vacancy-Type Defects in Al_2O_3/GaN Structure Probed by Monoenergetic Positron Beams. *J. Appl. Phys.* **2018**, *123*, 155302. [CrossRef]
118. Kubo, T.; Miyoshi, M.; Egawa, T. Post-Deposition Annealing Effects on the Insulator/Semiconductor Interfaces of Al_2O_3/AlGaN/GaN Structures on Si Substrates. *Semicond. Sci. Technol.* **2017**, *32*, 065012. [CrossRef]
119. Hung, T.-H.; Krishnamoorthy, S.; Esposto, M.; Neelim Nath, D.; Sung Park, P.; Rajan, S. Interface Charge Engineering at Atomic Layer Deposited Dielectric/III-Nitride Interfaces. *Appl. Phys. Lett.* **2013**, *102*, 072105. [CrossRef]
120. Zhou, H.; Ng, G.I.; Liu, Z.H.; Arulkumaran, S. Improved Device Performance by Post-Oxide Annealing in Atomic-Layer-Deposited Al_2O_3/AlGaN/GaN Metal–Insulator–Semiconductor High Electron Mobility Transistor on Si. *Appl. Phys. Express* **2011**, *4*, 104102. [CrossRef]
121. Nakazawa, S.; Shih, H.-A.; Tsurumi, N.; Anda, Y.; Hatsuda, T.; Ueda, T.; Kimoto, T.; Hashizume, T. Effects of Post-Deposition Annealing in O_2 on Threshold Voltage of Al_2O_3/AlGaN/GaN MOS Heterojunction Field-Effect Transistors. *Jpn. J. Appl. Phys.* **2019**, *58*, 030902. [CrossRef]
122. Choi, M.; Lyons, J.L.; Janotti, A.; Van de Walle, C.G. Impact of Native Defects in High-k Dielectric Oxides on GaN/Oxide Metal–Oxide–Semiconductor Devices. *Phys. Status Solidi B* **2013**, *250*, 787–791. [CrossRef]
123. Weber, J.R.; Janotti, A.; Van de Walle, C.G. Native Defects in Al_2O_3 and their Impact on III-V/Al_2O_3 Metal-Oxide-Semiconductor-Based Devices. *J. Appl. Phys.* **2011**, *109*, 033715. [CrossRef]
124. Liu, D.; Clark, S.J.; Robertson, J. Oxygen Vacancy Levels and Electron Transport in Al_2O_3. *Appl. Phys. Lett.* **2010**, *96*, 032905. [CrossRef]

125. Shin, B.; Weber, J.R.; Long, R.D.; Hurley, P.K.; Van de Walle, C.G.; McIntyre, P.C. Origin and Passivation of Fixed Charge in Atomic Layer Deposited Aluminum Oxide Gate Insulators on Chemically Treated InGaAs Substrates. *Appl. Phys. Lett.* **2010**, *96*, 152908. [CrossRef]
126. Kim, J.B.; Kwon, D.R.; Chakrabarti, K.; Lee, C.; Oh, K.Y.; Lee, J.H. Improvement in Al_2O_3 Dielectric Behavior by Using Ozone as an Oxidant for the Atomic Layer Deposition Technique. *J. Appl. Phys.* **2002**, *92*, 6739–6742. [CrossRef]
127. Huang, S.; Liu, X.; Wei, K.; Liu, G.; Wang, X.; Sun, B.; Yang, X.; Shen, B.; Liu, C.; Liu, S.; et al. O_3-Sourced Atomic Layer Deposition of High Quality Al_2O_3 Gate Dielectric for Normally-off GaN Metal-Insulator-Semiconductor High-Electron-Mobility Transistors. *Appl. Phys. Lett.* **2015**, *106*, 033507. [CrossRef]
128. Ťapajna, M.; Válik, L.; Gucmann, F.; Gregušová, D.; Fröhlich, K.; Haščík, Š.; Dobročka, E.; Tóth, L.; Pécz, B.; Kuzmík, J. Low-Temperature Atomic Layer Deposition-Grown Al_2O_3 Gate Dielectric for GaN/AlGaN/GaN MOS HEMTs: Impact of Deposition Conditions on Interface State Density. *J. Vac. Sci. Technol. B Nanotechnol. Microelectron. Mater. Process. Meas. Phenom.* **2017**, *35*, 01A107. [CrossRef]
129. Uenuma, M.; Takahashi, K.; Sonehara, S.; Tominaga, Y.; Fujimoto, Y.; Ishikawa, Y.; Uraoka, Y. Influence of Carbon Impurities and Oxygen Vacancies in Al_2O_3 Film on Al_2O_3/GaN MOS Capacitor Characteristics. *AIP Adv.* **2018**, *8*, 105103. [CrossRef]
130. Qin, X.; Wallace, R.M. In Situ Plasma Enhanced Atomic Layer Deposition Half Cycle Study of Al_2O_3 on AlGaN/GaN High Electron Mobility Transistors. *Appl. Phys. Lett.* **2015**, *107*, 081608. [CrossRef]
131. Yoshitsugu, K.; Horita, M.; Ishikawa, Y.; Uraoka, Y. Characterizations of Al_2O_3 Gate Dielectric Deposited on n-GaN by Plasma-Assisted Atomic Layer Deposition. *Phys. Status Solidi C* **2013**, *10*, 1426–1429. [CrossRef]
132. Ozaki, S.; Ohki, T.; Kanamura, M.; Imada, T.; Nakamura, N.; Okamoto, N.; Miyajima, T.; Kikkawa, T. Effect of Oxidant Source on Threshold Voltage Shift of AlGaN/GaN MIS-HEMTs Using ALD-Al_2O_3 Gate Insulator Films. In Proceedings of the CS MANTECH 2012 Conference, Boston, MA, USA, 23-26 April 2012.
133. Liu, Z.H.; Ng, G.I.; Arulkumaran, S.; Maung, Y.K.T.; Zhou, H. Temperature-Dependent Forward Gate Current Transport in Atomic-Layer-Deposited Al_2O_3/AlGaN/GaN Metal-Insulator-Semiconductor High Electron Mobility Transistor. *Appl. Phys. Lett.* **2011**, *98*, 163501. [CrossRef]
134. Wu, J.; Lu, X.; Ye, S.; Park, J.; Streit, D. Electrical Characterization and Reliability Analysis of Al_2O_3/AlGaN/GaN MISH Structure. In Proceedings of the 2014 IEEE International Reliability Physics Symposium, Waikoloa, HI, USA, 1–5 June 2014; p. CD-6. [CrossRef]
135. Heuken, L.; Ottaviani, A.; Fahle, D.; Zweipfennig, T.; Lükens, G.; Kalisch, H.; Vescan, A.; Heuken, M.; Burghartz, J.N. Limitations for Reliable Operation at Elevated Temperatures of Al_2O_3/AlGaN/GaN Metal–Insulator–Semiconductor High-Electron-Mobility Transistors Grown by Metal-Organic Chemical Vapor Deposition on Silicon Substrate. *Phys. Status Solidi A* **2020**, *217*, 1900697. [CrossRef]
136. Terman, L.M. An Investigation of Surface States at a Silicon/Silicon Oxide Interface Employing Metal-Oxide-Silicon Diodes. *Solid-State Electron.* **1962**, *5*, 285–299. [CrossRef]
137. Nicollian, E.H.; Goetzberger, A. The Si-Sio, Interface–Electrical Properties as Determined by the Metal-Insulator-Silicon Conductance Technique. *Bell Syst. Tech. J.* **1967**, *46*, 1033–1055. [CrossRef]
138. Shih, H.-A.; Kudo, M.; Suzuki, T. Analysis of AlN/AlGaN/GaN Metal-Insulator-Semiconductor Structure by Using Capacitance-Frequency-Temperature Mapping. *Appl. Phys. Lett.* **2012**, *101*, 043501. [CrossRef]
139. Freedsman, J.J.; Kubo, T.; Egawa, T. Trap Characterization of In-Situ Metal-Organic Chemical Vapor Deposition Grown AlN/AlGaN/GaN Metal-Insulator-Semiconductor Heterostructures by Frequency Dependent Conductance Technique. *Appl. Phys. Lett.* **2011**, *99*, 033504. [CrossRef]
140. Stoklas, R.; Gregušová, D.; Novák, J.; Vescan, A.; Kordoš, P. Investigation of Trapping Effects in AlGaN/GaN/Si Field-Effect Transistors by Frequency Dependent Capacitance and Conductance Analysis. *Appl. Phys. Lett.* **2008**, *93*, 124103. [CrossRef]
141. Yang, S.; Tang, Z.; Wong, K.-Y.; Lin, Y.-S.; Lu, Y.; Huang, S.; Chen, K.J. Mapping of Interface Traps in High-Performance Al_2O_3/AlGaN/GaN MIS-Heterostructures Using Frequency- and Temperature-Dependent C-V Techniques. In Proceedings of the 2013 IEEE International Electron Devices Meeting, Washington, DC, USA, 9–11 December 2013; pp. 6.3.1–6.3.4. [CrossRef]
142. Ramanan, N.; Lee, B.; Misra, V. Comparison of Methods for Accurate Characterization of Interface Traps in GaN MOS-HFET Devices. *IEEE Trans. Electron Devices* **2015**, *62*, 546–553. [CrossRef]
143. Winzer, A.; Szabó, N.; Wachowiak, A.; Jordan, P.M.; Heitmann, J.; Mikolajick, T. Impact of Postdeposition Annealing upon Film Properties of Atomic Layer Deposition-Grown Al_2O_3 on GaN. *J. Vac. Sci. Technol. B* **2015**, *33*, 01A106. [CrossRef]
144. Kim, S.; Hori, Y.; Ma, W.-C.; Kikuta, D.; Narita, T.; Iguchi, H.; Uesugi, T.; Kachi, T.; Hashizume, T. Interface Properties of Al_2O_3/n-GaN Structures with Inductively Coupled Plasma Etching of GaN Surfaces. *Jpn. J. Appl. Phys.* **2012**, *51*, 060201. [CrossRef]
145. Hori, Y.; Mizue, C.; Hashizume, T. Interface State Characterization of ALD-Al_2O_3/GaN and ALD-Al_2O_3/AlGaN/GaN Structures. *Phys. Status Solidi C* **2012**, *9*, 1356–1360. [CrossRef]
146. Ooyama, K.; Kato, H.; Miczek, M.; Hashizume, T. Temperature-Dependent Interface-State Response in an Al_2O_3/n-GaN Structure. *Jpn. J. Appl. Phys.* **2008**, *47*, 5426. [CrossRef]
147. Ando, Y.; Kaneki, S.; Hashizume, T. Improved Operation Stability of Al_2O_3/AlGaN/GaN MOS High-Electron-Mobility Transistors Grown on GaN Substrates. *Appl. Phys. Express* **2019**, *12*, 024002. [CrossRef]

148. Matys, M.; Adamowicz, B.; Domanowska, A.; Michalewicz, A.; Stoklas, R.; Akazawa, M.; Yatabe, Z.; Hashizume, T. On the Origin of Interface States at Oxide/III-Nitride Heterojunction Interfaces. *J. Appl. Phys.* **2016**, *120*, 225305. [CrossRef]
149. Yatabe, Z.; Hori, Y.; Kim, S.; Hashizume, T. Effects of Cl_2-Based Inductively Coupled Plasma Etching of AlGaN on Interface Properties of Al_2O_3/AlGaN/GaN Heterostructures. *Appl. Phys. Express* **2012**, *6*, 016502. [CrossRef]
150. Gregušová, D.; Stoklas, R.; Mizue, C.; Hori, Y.; Novák, J.; Hashizume, T.; Kordoš, P. Trap States in AlGaN/GaN Metal-Oxide-Semiconductor Structures with Al_2O_3 Prepared by Atomic Layer Deposition. *J. Appl. Phys.* **2010**, *107*, 106104. [CrossRef]
151. Calzolaro, A.; Szabó, N.; Großer, A.; Gärtner, J.; Mikolajick, T.; Wachowiak, A. Surface Preconditioning and Postmetallization Anneal Improving Interface Properties and V_{th} Stability under Positive Gate Bias Stress in AlGaN/GaN MIS-HEMTs. *Phys. Status Solidi A* **2020**, 2000585. [CrossRef]
152. Bao, S.-Q.-G.-W.; Ma, X.-H.; Chen, W.-W.; Yang, L.; Hou, B.; Zhu, Q.; Zhu, J.-J.; Hao, Y. Method of Evaluating Interface Traps in Al_2O_3/AlGaN/GaN High Electron Mobility Transistors. *Chin. Phys. B* **2019**, *28*, 067304. [CrossRef]
153. Wu, T.-L.; Marcon, D.; Bakeroot, B.; De Jaeger, B.; Lin, H.C.; Franco, J.; Stoffels, S.; Van Hove, M.; Roelofs, R.; Groeseneken, G.; et al. Correlation of Interface States/Border Traps and Threshold Voltage Shift on AlGaN/GaN Metal-Insulator-Semiconductor High-Electron-Mobility Transistors. *Appl. Phys. Lett.* **2015**, *107*, 093507. [CrossRef]
154. Matys, M.; Adamowicz, B.; Hori, Y.; Hashizume, T. Direct Measurement of Donor-like Interface State Density and Energy Distribution at Insulator/AlGaN Interface in Metal/Al_2O_3/AlGaN/GaN by Photocapacitance Method. *Appl. Phys. Lett.* **2013**, *103*, 021603. [CrossRef]
155. Matys, M.; Adamowicz, B.; Hashizume, T. Determination of the Deep Donor-like Interface State Density Distribution in Metal/Al_2O_3/n-GaN Structures from the Photocapacitance–Light Intensity Measurement. *Appl. Phys. Lett.* **2012**, *101*, 231608. [CrossRef]
156. Miczek, M.; Mizue, C.; Hashizume, T.; Adamowicz, B. Effects of Interface States and Temperature on the C-V Behavior of Metal/Insulator/AlGaN/GaN Heterostructure Capacitors. *J. Appl. Phys.* **2008**, *103*, 104510. [CrossRef]
157. Fang, Z.-Q.; Look, D.C.; Wang, X.-L.; Han, J.; Khan, F.A.; Adesida, I. Plasma-Etching-Enhanced Deep Centers in n-GaN Grown by Metalorganic Chemical-Vapor Deposition. *Appl. Phys. Lett.* **2003**, *82*, 1562–1564. [CrossRef]
158. Ando, Y.; Nagamatsu, K.; Deki, M.; Taoka, N.; Tanaka, A.; Nitta, S.; Honda, Y.; Nakamura, T.; Amano, H. Low Interface State Densities at Al_2O_3/GaN Interfaces Formed on Vicinal Polar and Non-Polar Surfaces. *Appl. Phys. Lett.* **2020**, *117*, 102102. [CrossRef]
159. Long, R.D.; Jackson, C.M.; Yang, J.; Hazeghi, A.; Hitzman, C.; Majety, S.; Arehart, A.R.; Nishi, Y.; Ma, T.P.; Ringel, S.A.; et al. Interface Trap Evaluation of Pd/Al_2O_3/GaN Metal Oxide Semiconductor Capacitors and the Influence of near-Interface Hydrogen. *Appl. Phys. Lett.* **2013**, *103*, 201607. [CrossRef]
160. Bisi, D.; Meneghini, M.; Van Hove, M.; Marcon, D.; Stoffels, S.; Wu, T.-L.; Decoutere, S.; Meneghesso, G.; Zanoni, E. Trapping Mechanisms in GaN-Based MIS-HEMTs Grown on Silicon Substrate. *Phys. Status Solidi A* **2015**, *212*, 1122–1129. [CrossRef]
161. Liu, Z.; Huang, S.; Bao, Q.; Wang, X.; Wei, K.; Jiang, H.; Cui, H.; Li, J.; Zhao, C.; Liu, X.; et al. Investigation of the Interface between LPCVD-SiN_x Gate Dielectric and III-Nitride for AlGaN/GaN MIS-HEMTs. *J. Vac. Sci. Technol. B* **2016**, *34*, 041202. [CrossRef]
162. Johnson, D.W.; Lee, R.T.; Hill, R.J.; Wong, M.H.; Bersuker, G.; Piner, E.L.; Kirsch, P.D.; Harris, H.R. Threshold Voltage Shift Due to Charge Trapping in Dielectric-Gated AlGaN/GaN High Electron Mobility Transistors Examined in Au-Free Technology. *IEEE Trans. Electron Devices* **2013**, *60*, 3197–3203. [CrossRef]
163. Meneghesso, G.; Meneghini, M.; Bisi, D.; Rossetto, I.; Wu, T.-L.; Van Hove, M.; Marcon, D.; Stoffels, S.; Decoutere, S.; Zanoni, E. Trapping and Reliability Issues in GaN-Based MIS HEMTs with Partially Recessed Gate. *Microelectron. Reliab.* **2016**, *58*, 151–157. [CrossRef]
164. Calzolaro, A. Fabrication and Characterization of AlGaN/GaN Metal-Insulator-Semiconductor High-Electron-Mobility-Transistors for High Power Applications. Ph.D. Thesis, Technische Universität Dresden, Dresden, Germany, 2022.
165. Nakazawa, S.; Shiozaki, N.; Negoro, N.; Tsurumi, N.; Anda, Y.; Ishida, M.; Ueda, T. Improved Hysteresis in a Normally-off AlGaN/GaN MOS Heterojunction Field-Effect Transistor with a Recessed Gate Structure Formed by Selective Regrowth. *Jpn. J. Appl. Phys.* **2017**, *56*, 091003. [CrossRef]
166. Wu, T.-L.; Franco, J.; Marcon, D.; De Jaeger, B.; Bakeroot, B.; Stoffels, S.; Van Hove, M.; Groeseneken, G.; Decoutere, S. Toward Understanding Positive Bias Temperature Instability in Fully Recessed-Gate GaN MISFETs. *IEEE Trans. Electron Devices* **2016**, *63*, 1853–1860. [CrossRef]
167. Kachi, T. Recent Progress of GaN Power Devices for Automotive Applications. *Jpn. J. Appl. Phys.* **2014**, *53*, 100210. [CrossRef]
168. Kikuta, D.; Narita, T.; Kutsuki, K.; Uesugi, T.; Kachi, T. Reliability Evaluation of Al_2O_3 Deposited by Ozone-Based Atomic Layer Deposition on Dry-Etched n-Type GaN. *Jpn. J. Appl. Phys.* **2013**, *52*, 08JN19. [CrossRef]
169. Tajima, M.; Hashizume, T. Impact of Gate and Passivation Structures on Current Collapse of AlGaN/GaN High-Electron-Mobility Transistors under Off-State-Bias Stress. *Jpn. J. Appl. Phys.* **2011**, *50*, 061001. [CrossRef]
170. Kambayashi, H.; Nomura, T.; Ueda, H.; Harada, K.; Morozumi, Y.; Hasebe, K.; Teramoto, A.; Sugawa, S.; Ohmi, T. High Quality SiO_2/Al_2O_3 Gate Stack for GaN Metal–Oxide–Semiconductor Field-Effect Transistor. *Jpn. J. Appl. Phys.* **2013**, *52*, 04CF09. [CrossRef]
171. Guo, A.; del Alamo, J.A. Positive-Bias Temperature Instability (PBTI) of GaN MOSFETs. In Proceedings of the 2015 IEEE International Reliability Physics Symposium, Monterey, CA, USA, 19–23 April 2015; pp. 6C.5.1–6C.5.7. [CrossRef]

172. Guo, A.; del Alamo, J.A. Negative-Bias Temperature Instability of GaN MOSFETs. In Proceedings of the 2016 IEEE International Reliability Physics Symposium (IRPS), Pasadena, CA, USA, 17–21 April 2016; pp. 4A-1-1–4A-1-6. [CrossRef]
173. Van Hove, M.; Kang, X.; Stoffels, S.; Wellekens, D.; Ronchi, N.; Venegas, R.; Geens, K.; Decoutere, S. Fabrication and Performance of Au-Free AlGaN/GaN-on-Silicon Power Devices With Al_2O_3 and Si_3N_4/Al_2O_3 Gate Dielectrics. *IEEE Trans. Electron Devices* **2013**, *60*, 3071–3078. [CrossRef]
174. Capriotti, M.; Alexewicz, A.; Fleury, C.; Gavagnin, M.; Bethge, O.; Visalli, D.; Derluyn, J.; Wanzenböck, H.D.; Bertagnolli, E.; Pogany, D.; et al. Fixed Interface Charges between AlGaN Barrier and Gate Stack Composed of in Situ Grown SiN and Al_2O_3 in AlGaN/GaN High Electron Mobility Transistors with Normally off Capability. *Appl. Phys. Lett.* **2014**, *104*, 113502. [CrossRef]
175. Liu, X.; Liu, Z.; Pannirselvam, S.; Pan, J.; Liu, W.; Jia, F.; Lu, Y.; Liu, C.; Yu, W.; He, J.; et al. Band Alignment of HfAlO/GaN (0001) Determined by X-Ray Photoelectron Spectroscopy: Effect of in Situ SiH_4 Passivation. *J. Alloys Compd.* **2015**, *636*, 191–195. [CrossRef]
176. Liu, X.; Chin, H.-C.; Tan, L.S.; Yeo, Y.-C. High-Permittivity Dielectric Stack on Gallium Nitride Formed by Silane Surface Passivation and Metal–Organic Chemical Vapor Deposition. *IEEE Electron Device Lett.* **2009**, *31*, 8–10. [CrossRef]
177. Hatano, M.; Taniguchi, Y.; Kodama, S.; Tokuda, H.; Kuzuhara, M. Reduced Gate Leakage and High Thermal Stability of AlGaN/GaN MIS-HEMTs Using ZrO_2/Al_2O_3 Gate Dielectric Stack. *Appl. Phys. Express* **2014**, *7*, 044101. [CrossRef]
178. Partida-Manzanera, T.; Zaidi, Z.H.; Roberts, J.W.; Dolmanan, S.B.; Lee, K.B.; Houston, P.A.; Chalker, P.R.; Tripathy, S.; Potter, R.J. Comparison of Atomic Layer Deposited Al_2O_3 and $(Ta_2O_5)_{0.12}(Al_2O_3)_{0.88}$ Gate Dielectrics on the Characteristics of GaN-Capped AlGaN/GaN Metal-Oxide-Semiconductor High Electron Mobility Transistors. *J. Appl. Phys.* **2019**, *126*, 034102. [CrossRef]
179. Kikuta, D.; Itoh, K.; Narita, T.; Mori, T. Al_2O_3/SiO_2 Nanolaminate for a Gate Oxide in a GaN-Based MOS Device. *J. Vac. Sci. Technol. A Vac. Surf. Film* **2017**, *35*, 01B122. [CrossRef]
180. Mitrovic, I.Z.; Das, P.; Jones, L.; Gibbon, J.; Dhanak, V.R.; Mahapatra, R.; Manzanera, T.P.; Roberts, J.W.; Potter, R.J.; Chalker, P.R.; et al. Band Line-up of High-k Oxides on GaN. *ECS Trans.* **2020**, *97*, 67. [CrossRef]
181. Le, S.P.; Ui, T.; Nguyen, T.Q.; Shih, H.-A.; Suzuki, T. Low-Frequency Noise in AlTiO/AlGaN/GaN Metal-Insulator-Semiconductor Heterojunction Field-Effect Transistors. *J. Appl. Phys.* **2016**, *119*, 204503. [CrossRef]
182. Le, S.P.; Nguyen, D.D.; Suzuki, T. Insulator-Semiconductor Interface Fixed Charges in AlGaN/GaN Metal-Insulator-Semiconductor Devices with Al_2O_3 or AlTiO Gate Dielectrics. *J. Appl. Phys.* **2018**, *123*, 034504. [CrossRef]
183. Roberts, J.W.; Chalker, P.R.; Lee, K.B.; Houston, P.A.; Cho, S.J.; Thayne, I.G.; Guiney, I.; Wallis, D.; Humphreys, C.J. Control of Threshold Voltage in E-Mode and D-Mode GaN-on-Si Metal-Insulator-Semiconductor Heterostructure Field Effect Transistors by in-Situ Fluorine Doping of Atomic Layer Deposition Al_2O_3 Gate Dielectrics. *Appl. Phys. Lett.* **2016**, *108*, 072901. [CrossRef]
184. Zhang, Y.; Sun, M.; Joglekar, S.J.; Fujishima, T.; Palacios, T. Threshold Voltage Control by Gate Oxide Thickness in Fluorinated GaN Metal-Oxide-Semiconductor High-Electron-Mobility Transistors. *Appl. Phys. Lett.* **2013**, *103*, 033524. [CrossRef]
185. Cai, Y.; Zhou, Y.; Chen, K.J.; Lau, K.M. High-Performance Enhancement-Mode AlGaN/GaN HEMTs Using Fluoride-Based Plasma Treatment. *IEEE Electron Device Lett.* **2005**, *26*, 435–437. [CrossRef]
186. Chu, R.; Corrion, A.; Chen, M.; Li, R.; Wong, D.; Zehnder, D.; Hughes, B.; Boutros, K. 1200-V Normally off GaN-on-Si Field-Effect Transistors with Low Dynamic on-Resistance. *IEEE Electron Device Lett.* **2011**, *32*, 632–634. [CrossRef]
187. Yang, S.; Liu, S.; Liu, C.; Hua, M.; Chen, K.J. Gate Stack Engineering for GaN Lateral Power Transistors. *Semicond. Sci. Technol.* **2015**, *31*, 024001. [CrossRef]
188. Yang, S.; Tang, Z.; Wong, K.-Y.; Lin, Y.-S.; Liu, C.; Lu, Y.; Huang, S.; Chen, K.J. High-Quality Interface in Al_2O_3/GaN/GaN/AlGaN/GaN MIS Structures With In Situ Pre-Gate Plasma Nitridation. *IEEE Electron Device Lett.* **2013**, *34*, 1497–1499. [CrossRef]
189. Chen, K.J.; Yang, S.; Tang, Z.; Huang, S.; Lu, Y.; Jiang, Q.; Liu, S.; Liu, C.; Li, B. Surface Nitridation for Improved Dielectric/III-Nitride Interfaces in GaN MIS-HEMTs. *Phys. Status Solidi A* **2015**, *212*, 1059–1065. [CrossRef]
190. Asahara, R.; Nozaki, M.; Yamada, T.; Ito, J.; Nakazawa, S.; Ishida, M.; Ueda, T.; Yoshigoe, A.; Hosoi, T.; Shimura, T.; et al. Effect of Nitrogen Incorporation into Al-Based Gate Insulators in AlON/AlGaN/GaN Metal–Oxide–Semiconductor Structures. *Appl. Phys. Express* **2016**, *9*, 101002. [CrossRef]
191. Wang, Q.; Cheng, X.; Zheng, L.; Shen, L.; Li, J.; Zhang, D.; Qian, R.; Yu, Y. Interface Engineering of an AlNO/AlGaN/GaN MIS Diode Induced by PEALD Alternate Insertion of AlN in Al_2O_3. *RSC Adv.* **2017**, *7*, 11745–11751. [CrossRef]
192. Ueda, T. GaN Power Devices: Current Status and Future Challenges. *Jpn. J. Appl. Phys.* **2019**, *58*, SC0804. [CrossRef]
193. Ogawa, E.; Hashizume, T.; Nakazawa, S.; Ueda, T.; Tanaka, T. Chemical and Potential Bending Characteristics of SiN_x/AlGaN Interfaces Prepared by in Situ Metal-Organic Chemical Vapor Deposition. *Jpn. J. Appl. Phys.* **2007**, *46*, L590. [CrossRef]
194. Takizawa, T.; Nakazawa, S.; Ueda, T. Crystalline SiN_x Ultrathin Films Grown on AlGaN/GaN Using in Situ Metalorganic Chemical Vapor Deposition. *J. Electron. Mater.* **2008**, *37*, 628–634. [CrossRef]
195. Derluyn, J.; Boeykens, S.; Cheng, K.; Vandersmissen, R.; Das, J.; Ruythooren, W.; Degroote, S.; Leys, M.R.; Germain, M.; Borghs, G. Improvement of AlGaN/GaN High Electron Mobility Transistor Structures by in Situ Deposition of a Si_3N_4 Surface Layer. *J. Appl. Phys.* **2005**, *98*, 054501. [CrossRef]
196. Moens, P.; Liu, C.; Banerjee, A.; Vanmeerbeek, P.; Coppens, P.; Ziad, H.; Constant, A.; Li, Z.; De Vleeschouwer, H.; Roig-Guitart, J.; et al. An Industrial Process for 650V Rated GaN-on-Si Power Devices Using in-Situ SiN as a Gate Dielectric. In Proceedings of the 2014 IEEE 26th International Symposium on Power Semiconductor Devices & IC's (ISPSD), Waikoloa, HI, USA, 15–19 June 2014; pp. 374–377. [CrossRef]

197. Hua, M.; Liu, C.; Yang, S.; Liu, S.; Fu, K.; Dong, Z.; Cai, Y.; Zhang, B.; Chen, K.J. Characterization of Leakage and Reliability of SiN_x Gate Dielectric by Low-Pressure Chemical Vapor Deposition for GaN-Based MIS-HEMTs. *IEEE Trans. Electron Devices* **2015**, *62*, 3215–3222. [CrossRef]
198. Jauss, S.A.; Hallaceli, K.; Mansfeld, S.; Schwaiger, S.; Daves, W.; Ambacher, O. Reliability Analysis of LPCVD SiN Gate Dielectric for AlGaN/GaN MIS-HEMTs. *IEEE Trans. Electron Devices* **2017**, *64*, 2298–2305. [CrossRef]
199. Whiteside, M.; Arulkumaran, S.; Dikme, Y.; Sandupatla, A.; Ng, G.I. Demonstration of AlGaN/GaN MISHEMT on Si with Low-Temperature Epitaxy Grown AlN Dielectric Gate. *Electronics* **2020**, *9*, 1858. [CrossRef]
200. Hashizume, T.; Alekseev, E.; Pavlidis, D.; Boutros, K.S.; Redwing, J. Capacitance-Voltage Characterization of AlN/GaN Metal–Insulator–Semiconductor Structures Grown on Sapphire Substrate by Metalorganic Chemical Vapor Deposition. *J. Appl. Phys.* **2000**, *88*, 1983–1986. [CrossRef]
201. Uemoto, Y.; Shibata, D.; Yanagihara, M.; Ishida, H.; Matsuo, H.; Nagai, S.; Batta, N.; Li, M.; Ueda, T.; Tanaka, T.; et al. 8300V Blocking Voltage AlGaN/GaN Power HFET with Thick Poly-AlN Passivation. In Proceedings of the 2007 IEEE International Electron Devices Meeting, Washington, DC, USA, 10–12 December 2007; pp. 861–864. [CrossRef]
202. Hwang, I.-H.; Kang, M.-J.; Cha, H.-Y.; Seo, K.-S. Crystalline AlN Interfacial Layer on GaN Using Plasma-Enhanced Atomic Layer Deposition. *Crystals* **2021**, *11*, 405. [CrossRef]
203. Whiteside, M.; Ng, G.I.; Arulkumaran, S.; Ranjan, K.; Dikme, Y. Low Temperature Epitaxy Grown AlN Metal-Insulator-Semiconductor Diodes on AlGaN/GaN HEMT Structure. In Proceedings of the 2019 Electron Devices Technology and Manufacturing Conference (EDTM), Singapore, 12–15 March 2019; pp. 103–105. [CrossRef]
204. Whiteside, M.; Arulkumaran, S.; Dikme, Y.; Sandupatla, A.; Ng, G.I. Improved Interface State Density by Low Temperature Epitaxy Grown AlN for AlGaN/GaN Metal-Insulator-Semiconductor Diodes. *Mater. Sci. Eng. B* **2020**, *262*, 114707. [CrossRef]

materials

MDPI

Review

Challenges and Perspectives for Vertical GaN-on-Si Trench MOS Reliability: From Leakage Current Analysis to Gate Stack Optimization

Kalparupa Mukherjee [1,*], Carlo De Santi [1], Matteo Borga [2], Karen Geens [2], Shuzhen You [2], Benoit Bakeroot [3], Stefaan Decoutere [2], Patrick Diehle [4], Susanne Hübner [4], Frank Altmann [4], Matteo Buffolo [1], Gaudenzio Meneghesso [1], Enrico Zanoni [1] and Matteo Meneghini [1]

[1] Department of Information Engineering, University of Padua, 35131 Padova, Italy; desantic@dei.unipd.it (C.D.S.); matteo.buffolo@dei.unipd.it (M.B.); gauss@dei.unipd.it (G.M.); zanoni@dei.unipd.it (E.Z.); matteo.meneghini@unipd.it (M.M.)

[2] Imec, Kapeldreef 75, 3001 Leuven, Belgium; matteo.borga@imec.be (M.B.); karen.geens@imec.be (K.G.); shuzhen.you@imec.be (S.Y.); stefaan.decoutere@imec.be (S.D.)

[3] CMST Imec/UGent, 9052 Ghent, Belgium; Benoit.Bakeroot@UGent.be

[4] Fraunhofer Institute for Microstructure of Materials and Systems IMWS, Walter-Huelse-Strasse 1, 06120 Halle, Germany; patrick.diehle@imws.fraunhofer.de (P.D.); susanne.huebner@imws.fraunhofer.de (S.H.); frank.altmann@imws.fraunhofer.de (F.A.)

* Correspondence: mukherjee@dei.unipd.it

Abstract: The vertical Gallium Nitride-on-Silicon (GaN-on-Si) trench metal-oxide-semiconductor field effect transistor (MOSFET) is a promising architecture for the development of efficient GaN-based power transistors on foreign substrates for power conversion applications. This work presents an overview of recent case studies, to discuss the most relevant challenges related to the development of reliable vertical GaN-on-Si trench MOSFETs. The focus lies on strategies to identify and tackle the most relevant reliability issues. First, we describe leakage and doping considerations, which must be considered to design vertical GaN-on-Si stacks with high breakdown voltage. Next, we describe gate design techniques to improve breakdown performance, through variation of dielectric composition coupled with optimization of the trench structure. Finally, we describe how to identify and compare trapping effects with the help of pulsed techniques, combined with light-assisted de-trapping analyses, in order to assess the dynamic performance of the devices.

Keywords: vertical GaN; quasi-vertical GaN; reliability; trapping; degradation; MOS; trench MOS; threshold voltage

check for updates

Citation: Mukherjee, K.; De Santi, C.; Borga, M.; Geens, K.; You, S., Bakeroot, B.; Decoutere, S.; Diehle, P.; Hübner, S.; Altmann, F.; et al. Challenges and Perspectives for Vertical GaN-on-Si Trench MOS Reliability: From Leakage Current Analysis to Gate Stack Optimization. *Materials* **2021**, *14*, 2316. https://doi.org/10.3390/ma14092316

Academic Editor: Fabrizio Roccaforte

Received: 23 March 2021
Accepted: 27 April 2021
Published: 29 April 2021

Publisher's Note: MDPI stays neutral with regard to jurisdictional claims in published maps and institutional affiliations.

1. Introduction

A central challenge of power electronics today is to address the continuously rising demands for safe and reliable control, conversion and distribution of energy, while maximizing the efficiency. Switched-mode power conversion strategies, with myriad applications [1–5], are now universally preferred over the simpler linear conversion methods due to the advantages of better flexibility, safety, and importantly, higher efficiency. The core requirement for efficient power conversion thus translates directly to highly efficient power transistors that can sustain repeated OFF/ON switching transitions with minimal switching and resistive losses. Higher operational frequencies are desirable, since they reduce the amount of energy transferred/cycle, which in turn reduces the size of the passive circuit components in the converters. Since higher frequencies will inevitably correspond to increased switching losses, the upper limit on the operational frequency (currently, in the MHz range) is majorly determined by the switching capabilities of the available power transistors.

Silicon-based transistors have evolved over the years to meet the market needs; however further optimization is now bounded by the theoretical limits of Si. In this regard, wide-bandgap (WBG) semiconductors have found great consensus in being promising substitutes to Si transistors, derived from their superior figures of merit (FOMs). According to Baliga's FOM (BFOM) (= $\varepsilon\mu E_G^3$, $\frac{V_{BR}^2}{R_{on}}$) [6], materials such as GaN and SiC present comprehensive improvements in the breakdown voltage (V_{BR}) vs. on-resistance (R_{on}) tradeoff. Comparing other FOMs provide easy estimations of the relevant metrics (a) conduction and switching losses from the on-resistance × output capacitance product ($R_{on} \times C_{oss}$) [7] and (b) power density from $\frac{1}{\sqrt{Q_g R_{on} A_{package} R_{th}}}$ [8] where $A_{package}$ is the package size and R_{th} is the thermal resistance. The gate charge Q_g represents the switching loss incurred by the charging and discharging cycles of the gate terminal. Here too, GaN emerges as the dominant choice over Si, as reviewed by Vecchia et al., in [1].

Thus, combining the improved transport, breakdown and thermal properties, the use of WBG materials enables cost and size-effective power transistors (converters) operating at high voltages and temperatures with higher speeds (lower switching losses), and with higher overall efficiency (lower conduction and switching losses).

Although GaN (BFOM = 3175 [9]) is superior to SiC (BFOM = 840 [9]) in most material properties, SiC has better thermal conductivity and is generally considered to be more relevant to the high voltage (>1200 V) application domain, while the commercial marketability of GaN is usually assumed to be in the low to mid voltage $\leq$650 V (power capability $\approx$ kW) domain [1,9,10]. This is primarily because of the current and voltage limitations [1,9,11] of the lateral configuration initially adopted for design of GaN power transistors. These devices were built to capitalize on the high-mobility high-density 2DEG formed at the AlGaN/GaN hetero-interface and indeed, several works on lateral GaN transistors have displayed impressive performances in the mid-voltage range [12–14], as a result of revolutionary improvements in GaN epitaxy and design over the last couple of decades.

However, to establish GaN power transistors as serious contenders in application markets such as Electric Vehicle/Hybrid Electric Vehicle (EV/HEV) [4] or power grids, voltage capabilities up to 1700–1800 V are required. To this aim, the research focus is now shifting to vertical GaN structures [2,3,15]. In addition to better heat management and normally off capabilities, vertical architectures overcome the breakdown voltage vs. device area tradeoff of lateral devices. With proper optimization, vertical transistors are also expected to present better reliability performance, since the electric field is moved within the bulk, eliminating surface issues.

Fully vertical GaN-on-GaN diode and transistor demonstrators have reported excellent performances (up to 3-4 kV capability [16–24]). However, GaN substrates are small and expensive, with wafer costs per unit area for GaN-on-GaN ranging up to \$ 100/cm^2 for 2-inch wafers [25,26]. Thus, currently these devices have limited commercial viability. Economically, the GaN-on-Si technology appears to be the most worthwhile for further development, with 8-inch wafers costing only \$1 per unit area, potentially lowering wafer costs by 100 times. [25,26]. However, owing to the mismatches in lattice constant and thermal expansion coefficient between GaN and Si, the growth of thick GaN layers on Si are subject to high dislocation/defect densities, which makes the epitaxy especially challenging. Although some innovative techniques have been successful in fabricating fully vertical GaN-on-Si diodes [27–31], and a fully vertical GaN-on-Si power transistor (V_{BR} = 520 V, R_{on} = 5 mΩ.cm^2) was recently demonstrated by Khadar et al. in [32] using substrate removal techniques, fully vertical GaN-on-Si technology is still in a very nascent stage. Recent results demonstrate the possibility of using engineered substrates (QST®), with a matched coefficient of thermal expansion, to enable low-cost vertical GaN FETs on large diameter wafers (8–12 inch) [33].

For the development of the gate module and for the optimization of the drift region of vertical GaN devices, an important step is the development of quasi-vertical GaN-on-Si devices [3,27,34–36], based on the idea of maintaining the source and drain electrodes on the same side of the wafer. This approach allows us to understand, study and overcome the challenges related to the development of vertical GaN transistors, before moving to the full vertical layout. Quasi-vertical structures can build on the recent advancements into GaN-on-Si epitaxy achieved during research into lateral GaN devices, while providing better field management due to the vertical stack. Among the several available quasi-vertical configurations such as CAVETs [24,37], OG-FETs [38,39] or Fin FETs [22,40], the trench MOSFET [2,3,32,34,41–46] is a popular choice with high cell density. It is inherently a normally off device with low R_{on}, and needs no regrowth of AlGaN/GaN channels. Figure 1 presents the schematic of a typical quasi-vertical GaN-on-Si trench MOSFET.

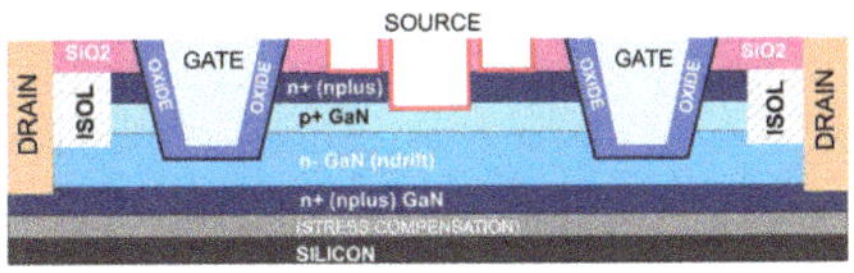

Figure 1. Schematic of a quasi-vertical n^+-p^+-n^--n^+ GaN-on-Si trench MOS device.

In the ON-state, the current in the quasi-vertical structure is sourced from the top n^+ layer, and conducted vertically through the p^+ GaN layer along the gate trench sidewalls. The current is then collected laterally through the bottom n^+ layer, before being transported back to the surface through the drain metallization. A high doping of the n^+ current-spreading layer ensures better current distribution, to minimize current crowding around the contact in the ON-state.

To design a reliable GaN-on-Si trench MOSFET, careful optimization of several interlinked physical parameters is required. As discussed earlier, the first design consideration, as for any power transistor, is to achieve a high V_{BR} and low R_{on} simultaneously. In this regard, the thickness and doping of the p-body and drift layer are the central constraints. The parameters need to be carefully engineered to ensure good reverse blocking capability in the OFF-state in addition to forward conduction in the ON-state. Regarding the M-O-S stack, gate design parameters such as dielectric composition and thickness are important in controlling the threshold voltage, leakage and gate capacitance of the device. The dielectric choice, in addition to structural optimization of the trench to minimize field crowding, controls the gate breakdown capability. Finally, the leakage and trapping needs to be minimized throughout the quasi-vertical stack.In this work, we will discuss recent case studies that address the impacts of different design choices on the performance of quasi-vertical trench MOSFETs, while demonstrating testing strategies used to identify and compare degradation mechanisms in such devices. In Section 2, p^+-n^--n^+ diode test structures are characterized; leakage modeling is used to identify the dominant mechanisms under reverse bias, and technology computer-aided design (TCAD) simulations are employed to compare the effects of high vs. low p-body doping, to present a trade-off useful for breakdown optimization. In Section 3, the optimization of the gate stack through the use of a bilayer dielectric is discussed. Specifically, the trapping and breakdown performance of bilayer (SiO_2 + Al_2O_3) vs. unilayer (Al_2O_3) dielectrics are compared, and the effects of trench optimization are visualized by scanning electron microscopy (SEM) and transmission electron microscopy (TEM) analysis. In Section 4, methodologies for the assessment of the dynamic performance of the devices are presented. In addition, light-assisted experimental techniques are discussed, which improve the detection and understanding of trapping phenomena under low and high positive gate stresses.

2. OFF-State-Leakage and Doping Constraints of Quasi-Vertical GaN-on-Si Diodes from IMEC, Leuven, Belgium

In this section, we discuss the factors influencing the leakage current and the breakdown voltage of vertical GaN-on-Si stack, specifically designed for vertical trench-MOSFETs.

The growth of thick, mostly insulating GaN drift layers on Si was made possible during the last years thanks to the intense research on lateral power GaN devices; the main goal has been to improve the OFF-state blocking capability. For the move into vertical GaN devices, the drift layer modulation needs to be more rigorous, since in addition to sustaining high reverse biases in the OFF-state, it also needs to have a low resistivity in the ON-state. The ideal drift layer is thick, to sustain a large breakdown voltage, lightly doped, to ensure high mobility, thus allowing a good ON/OFF ratio, and has a low defect density, to minimize the defect-related leakage components [3]. Unintentionally doped drift layers are weakly n-type (10^{16} carriers/cm^3 or above, [3,47,48]), due to residual impurities introduced during the growth process, such as silicon and oxygen [3,47–52].

In a vertical trench-MOSFET, the n^- drift region is in direct contact with the p-body, that may have Mg concentrations in excess of 10^{18}–10^{19} cm^{-3}. To optimize the breakdown voltage of vertical power FETs, it is therefore important to minimize both the leakage through the drift region, and to ensure that the p-body/drift region junction can sustain the high vertical field when the device is in the OFF-state [3,31,53–55].

Magnesium doping in GaN has been reported to form acceptor states located 0.16 eV above the valence band [56,57]. This relatively deep energy level results in incomplete thermal ionization of Mg acceptors at room temperature. Since the presence of hydrogen during MOCVD growth of p-type GaN can passivate the Mg-dopant through the formation of Mg-H bonds [58], a post-growth annealing treatment (while ensuring energies are lower than the threshold to create native defects) is necessary to ensure a high conductivity and hole density.

There are several possible leakage paths in the OFF-state [53,59,60]. In the quasi-vertical layout, parasitic leakage along the etch sidewalls and the bulk regions might be dominant and needs to be minimized. Other leakage paths may be present along the passivation layers, or vertically along the entire stack, reaching the substrate. To minimize the vertical leakage, the prevalent leakage mechanisms among different technology variations need to be understood, to enable directed improvements.

In performing leakage analysis of reverse biased p^+n diodes, the conduction mechanisms through dielectrics subjected to high electric fields have been found to be applicable [53,59,61]. For low to medium reverse bias, the relevant mechanisms are usually electrode-limited related to the quality of the metal-semiconductor contacts. However, these mechanisms are usually not relevant in good vertical designs. As such, bulk conduction mechanisms are more relevant, in particular, variable range hopping (VRH) [54,55,59,62–71], Poole-Frenkel emission [59,63–69,72–78], and space charge limited conduction (SCLC) [31,47,53,79].

For investigating the doping and leakage issues under OFF-state within the vertical stack, it is useful to consider the simpler quasi-vertical diode structures, which form the fundamental block of the full MOSFET. The test vehicles used for the following study were aimed at understanding the p^+-n^--n^+ stack; the schematic is presented in Figure 2. Fabricated on a 200 mm Si substrate, the diodes have Mg doping with $N_A = 6 \times 10^{19}$ cm^{-3} within the p^+ layer, and a weakly n-type drift layer with $n = 4 \times 10^{16}$ cm^{-3}. The cathode is at the buried n^+ layer below the n drift region. The reverse breakdown voltage was measured to be 170 V on these specific structures, having a drift layer thickness equal to 750 nm [45,55].

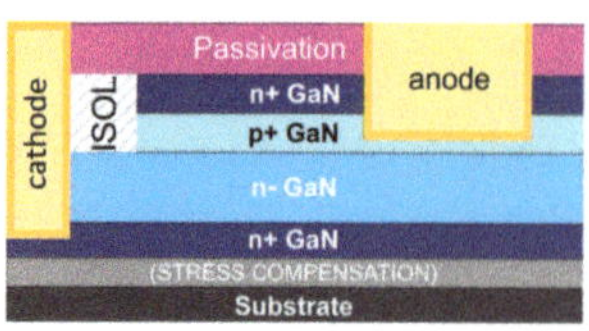

Figure 2. Schematic of the quasi-vertical p^+-n GaN-on-Si diodes.

2.1. Leakage Modeling

Since individual leakage mechanisms have distinct temperature dependencies, temperature dependent I-V behavior is obtained. Reverse biased diode characteristics over a range of temperatures (T) from 50 °C to 130 °C are displayed in Figure 3a. The maximum cathode voltage ($V_{Cathode}$) was limited to $V_{BR}/2$ to avoid degrading the samples, and obtain clean trends with T for medium voltages. The presence of two different natures of variation with T is found, hence two regions were identified to be modelled separately.

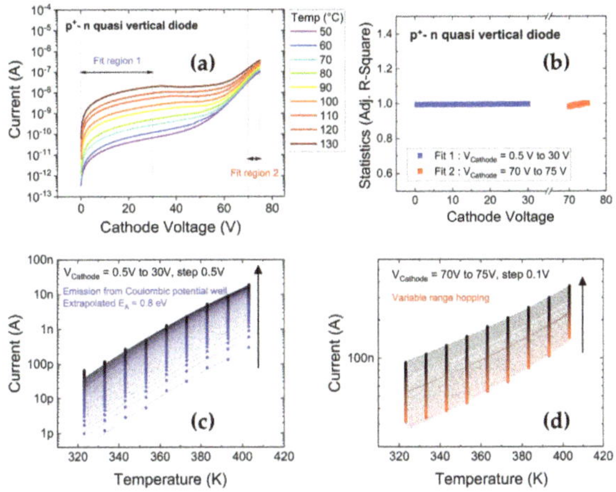

Figure 3. Modeling of the reverse-biased characteristics of the p$^+$-n diodes under test [55]. (**a**) Reverse diode characteristics from T = 50 °C to 130 °C. The two distinct regions identified in (**a**) are fitted using the Coulombic potential well model in (**c**) for $V_{Cathode}$ from 0.5 V to 30 V (in direction of arrow), and using the variable range hopping model in (**d**) for $V_{Cathode}$ from 70 V to 75 V (in direction of arrow). (**b**) Displays the good conformity of the fits with adjusted $R^2 \approx 1$ using the statistical parameter of adjusted R-square (coefficient of determination).

The first region, from $V_{Cathode}$ = 0 V to 30 V, with a strong increase in current with temperature, was found to best represent conduction from Coulombic traps through thermionic emission [54,72]. The corresponding fit data is presented in Figure 3c. This mechanism is based on the assumption that the potential around traps at low electric fields can be considered Coulombic, while at higher fields, according to the Poole-Frenkel effect, a lowering of the potential barrier is expected with a square root dependency on field, strengthening the emission process of the trap [59,72–74]. This is expressed in the following formula, and the parameters are defined in [55]:

$$I_{TE} = AT^2 \exp\left(\frac{-E_A}{kT}\right), \tag{1}$$

$$e_n \propto exp\left(-\frac{E_T - \beta F^{\frac{1}{2}}}{k_B T}\right), \tag{2}$$

$$\beta = \sqrt{\frac{q^3}{\pi \varepsilon}}, \tag{3}$$

The slope extracted from the fitting (not shown) revealed an activation energy E_A of ≈0.85 eV, usually associated with the presence of carbon acceptors [80,81], with an effective lowering in E_A (ΔE_A) = 70 meV, the corresponding Poole-Frenkel coefficient β (=1.77 × 10^{-5} eV $V^{-1/2}$ $m^{1/2}$) was found to be close to the theoretical value [55]

The second region, from $V_{Cathode}$ = 70 V to 75 V, was modelled using variable range hopping (VRH), the leakage evolution fit to the VRH model is presented in Figure 3d. The corresponding equation is written as in Equation (4), and the parameters are described in [55]:

$$I_{VRH} = I_0 exp\left[-1.76\left(\frac{T_0}{T}\right)^{\frac{1}{4}} + C_{VRH}\left(\frac{T_0}{T}\right)^{\frac{3}{4}} F^2\right] \quad (4)$$

VRH describes the conduction of electrons across multiple trap states distributed within the bandgap. With the high occurrence of substantial defect densities in GaN epitaxial layers, VRH is commonly observed in GaN diodes [59,62–70], ascribed to the hopping of charged carriers through localized defect states in depletion regions.

For both the fits in Figure 3c,d, the adjusted R-Square (Adj. R-Square) [82] is found to be close to 1, as presented in Figure 3b, attesting to the good conformity of the fits. The R-square, also referred to as the coefficient of determination, always lies between 0 to 1, corresponding to whether the fit line is able to describe 0% or 100% of the variability of the data around the mean. Adj. R-Square is a modification which takes the number of predictors (within the fitted line) into account.

2.2. Simulation of Doping Constraints in Diode Breakdown

The investigation of breakdown issues is especially suited to using TCAD simulations, which provide versatile, non-destructive and rapid optimization solutions. A representative and simplified (fully vertical) model of the test devices was built using the Sentaurus tool from Synopsys in order to investigate the nature of breakdown, relative to the chosen concentration of p-doping in GaN diodes [55]. The drift diffusion transport model is used, along with appropriate polarization, mobility and recombination models. The n^+ layers are doped with $N_D = 5 \times 10^{18}$ cm^{-3}, and the n^- drift layer doping is fixed at $N_D = 4 \times 10^{16}$ cm^{-3}. For the p-body doping, Mg is defined as the dopant species. As discussed earlier, the Mg acceptors are not expected to be completely ionized at room temperature. Hence, to correctly estimate the effects of p-doping, using the incomplete ionization model is more physical. This model takes the parameters of the individual acceptor species into account, in particular, the ionization energy. Based on this, the simulator internally computes the effective doping concentration under different conditions. For example, a defined Mg concentration of $N_A = 6 \times 10^{19}$ cm^{-3} with an ionization energy of 0.16 eV, leads to an effective base doping within the p-GaN region of $\approx 4 \times 10^{18}$ cm^{-3} (6%), except within the depletion regions around the p-n junctions, where the defined N_A is almost completely ionized.

Since the measured breakdown voltage of the test diodes is 170 V, the electric field evolution within the vertical diode is visualized at 160 V with different N_A values in Figure 4. In Figure 4a,b, the chosen N_A values are relatively low = 4×10^{17} cm^{-3} (see Figure 4a), 6×10^{17} cm^{-3} and 1×10^{18} cm^{-3}. In this scenario, the p-GaN region is observed to be severely depleted, with reach through occurring for the $N_A = 4 \times 10^{17}$ cm^{-3} case, once the depletion regions from the n^+-p and p-n^- junctions intersect. Thus, a lower bound for setting the p-doping is identified owing to this constraint. In a real growth scenario, this constraint could be considerably tighter. If the reduction in Mg concentrations due to hydrogen passivation or other impurities were considered, the breakdown could occur faster (at lower voltages) for equivalent N_A settings.

In Figure 4c,d the higher N_A values are considered, including the representative value for the structures under test with $N_A = 6 \times 10^{19}$ cm^{-3} (see Figure 4c). For these cases, the applied voltage drops almost entirely across the lightly doped n^- GaN region, leading to smaller depletion of the p^+ GaN layer. On the other hand, the peak electric field at the p^+ to n^- interface is significantly higher. In this scenario, breakdown in expected to be field-triggered, in fact, for $N_A = 6 \times 10^{19}$ cm^{-3}, we are approaching critical field for GaN ($\approx$ 3 MV/cm [83]) at the 160 V condition, which is found to agree reasonably well with the measured breakdown voltage of 170 V. Thus, the higher bound for N_A settings is identified.

Based on the results in Section 2, we infer that the density of defects within the drift region need to be optimized to control the leakage current and its temperature sensitivity. The contribution of the residual carbon concentration is found to be relevant to the low voltage regimes, and needs to be optimized to improve the leakage performance. Regarding p-doping-induced constraints on the breakdown voltage, for a lightly doped drift layer, keeping the p-doping low can reduce the peak electric field, pushing V_{BR} to higher voltages. However, the trade-off dictates that the value still needs to be high enough to avoid complete depletion of the p GaN layer unexpectedly at low voltages.

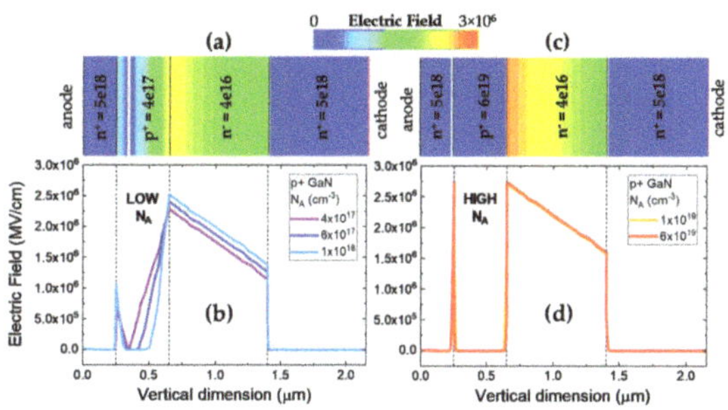

Figure 4. TCAD modeling of vertical p$^+$-n diodes under different p-doping conditions describing the expected breakdown processes (**a**) TCAD structure visualized at $N_A = 4 \times 10^{17}$ cm^{-3}; (**b**) Electric field evolution for low p doping values illustrates complete depletion (punch-through) of the p-GaN region; (**c**) TCAD structure visualized at $N_A = 6 \times 10^{19}$ cm^{-3}; (**d**) Electric field evolution for high p doping values illustrates high electric fields (approaching critical field for GaN) at the p$^+$-n$^-$ interface.

3. OFF-State and ON-State–Optimization of the M-O-S Stack in Quasi-Vertical MOSFETs from IMEC, Leuven, Belgium

This section describes recent results on the degradation and optimization of the MOS gate stack used for GaN-on-Si vertical MOSFETs.

The reliability of the gate stack is highly influenced by the choice of the oxide in trench MOSFETs, since the insulator is vulnerable to repeated stressing during the operation of the power devices over time [46,84,85]. Specifically, the properties of the insulator can greatly affect the leakage, breakdown and trapping performance of the M-O-S stack under positive gate stresses. One of the essential requirements for a gate oxide is to have high band offsets with GaN, which is critical to limit the leakage current [86–88]. In this regard, while materials such as silicon nitride or hafnium oxide (band offsets around 1 eV) are less favored, Al_2O_3 [89,90] and SiO_2 [32,34] have emerged as popular choices with conduction band offsets (ΔE_C) of 2.1 and 2.5 eV, respectively. Al_2O_3 presents good metrics [86–88]: in addition to having a high bandgap (8.9 eV), high k (dielectric constant = 9.0), and reasonably high breakdown strength (~10 MV/cm), improvements in deposition techniques now allow Al_2O_3/GaN interfaces to be formed with very low interface state densities [88,91,92]. SiO_2 also has a high bandgap (9.1 eV), and its advantage is high chemical stability, which extends to high operational stability in the devices.

Since the reliability of the MOS framework is still not completely understood, there has been limited effort in exploring alternatives to the conventional MOS structure with an unilayer dielectric. In particular, the approach of using bilayer dielectrics (with a thin interface dielectric followed by a thicker insulator), which has been found to be advantageous for Si MOSFET design, could potentially be very valuable for GaN-based MOSFETs as well. However, inherent reliability risks could be worsened with increasing complexity in the dielectric deposition process. To truly capitalize on the effects of improved dielectrics, the bulk GaN etch process, in particular the formation of the trench itself, needs to be highly optimized. The shape of the trench is usually optimized [93–97] to find the best combination of V_{BR} and R_{on}; deep trenches with rounded corners have been reported to display good metrics [93,98,99]. However, for higher trench depths (over-etch) extending beyond the p-body, the peak field under the OFF-state could be aggravated [93]. The overall etching process is aimed at creating smooth sidewalls, and preventing irregularities such as pits or voids, especially at the bottom trench corners where the peak fields are expected [93–97].

3.1. Optimising Dielectric Composition

This section demonstrates the advantages of employing a bilayer insulator composition in quasi-vertical MOSFETs through DC and pulsed measurements, and TCAD simulations [46]. The devices under test are GaN-on-Si trench MOSFETs, structurally similar to Figure 1. During Atomic Layer Etch (ALE) processing steps, an O_2 plasma is used to oxidize the GaN after which a BCl3 dry etch step is executed to remove the oxidized GaN layer. The amount of ALE cycles has been optimized to ensure a good profile of the gate trench, removing in total ~25 nm. In this section, we discuss the effects of the dielectric composition around the gate trench, as illustrated in Figure 5. The Al_2O_3 deposition is performed using atomic layer deposition (ALD) at 300 °C, while the SiO_2 in the bi-layer is deposited using plasma-enhanced chemical vapor deposition (PECVD) at a deposition temperature of 400 °C. The focal idea was to compare the robustness of devices fabricated with a bilayer dielectric composed of SiO_2 and Al_2O_3 to devices with a traditional unilayer dielectric of Al_2O_3. Effectively, the bilayer stack should combine the merits of SiO_2 as a bulk insulator with the ability of Al_2O_3 to create a high-quality interface to GaN.

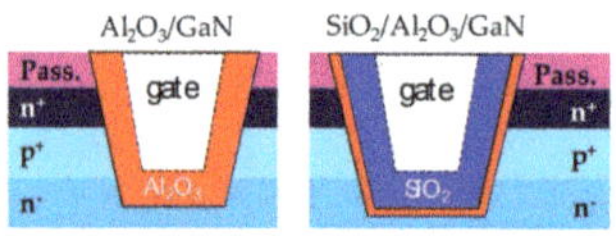

Figure 5. Dielectric composition of the devices under test. The first configuration is an unilayer of 35 nm Al_2O_3 at the GaN interface, while the second has a bilayer composition: 35 nm of SiO_2, then 2.5 nm of Al_2O_3 at the GaN interface.

As expected, the gate-source and gate-drain diode leakage of the bilayer devices was found to be lower by a couple of orders of magnitude [46]. This is attributed to the intrinsically higher breakdown field of SiO_2, as well as the additional barrier (conduction band discontinuity at the Al_2O_3/SiO_2 interface of 0.4 eV [86]) to thermionic leakage from the channel to the gate, introduced by the bilayer configuration.

To evaluate the reliability of the two stacks under the ON-state, forward gate breakdown step stress tests were performed, where the gate voltage was incremented from 0 V in steps of 3V, while V_{DS} was constant at 1 V. Very little dispersion in breakdown voltage was observed across several devices, and the gate breakdown voltage for the unilayer and bilayer configurations were found to be 9 V and 27 V [46], the bilayer devices displaying an improvement of three times.

In Figure 6, the schematic of the simulated device (Figure 6a), and the electric field distribution within the unilayer and bilayer oxides are visualized at their respective gate breakdown voltages.

In the ON-state, the channel exists continuously along the trench sidewalls. Thus, the applied gate voltage falls entirely within the oxide layer, and the internal field grows rapidly, as illustrated in Figure 6b,c. This condition can then be used to estimate the critical electric field for the two gate dielectric compositions. From theoretical considerations, the unilayer Al_2O_3 devices are expected to have an average critical electric field value of 2.6 MV/cm (9 V/35 nm), while the bilayer devices are estimated to have a critical electric field value of 7.5 MV/cm (26.2 V/35 nm) for the SiO_2 layer, and 3.2 MV/cm (0.80 V/2.5 nm) for the Al_2O_3 layer [46]. These values are well substantiated by the simulated electric fields in Figure 6 obtained at the respective breakdown voltages.

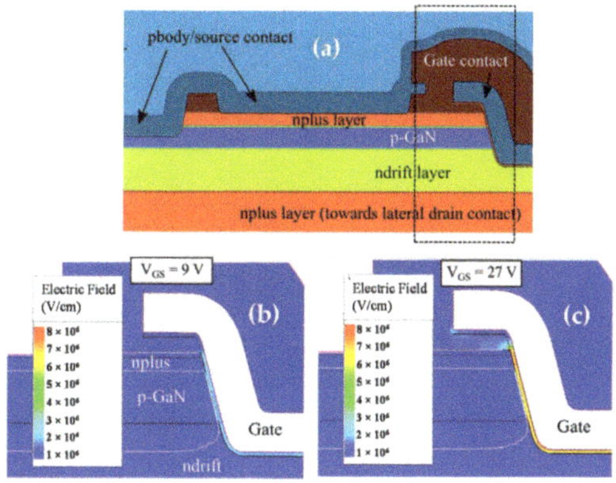

Figure 6. (**a**) Schematic of the simulated quasi-vertical trench MOSFET. Electric field distribution around the trench edges at the measured ON-state breakdown voltage visualized for (**b**) unilayer: Al_2O_3/GaN devices and (**c**) bilayer: SiO_2/Al_2O_3/GaN devices.

The second set of measurements were aimed at comparing OFF-state performance of the dielectric stacks. Figure 7 presents the results of drain step stress until breakdown, coupled with electroluminescence (EL) studies, performed at V_{GS} =0 V on 35 devices from each wafer. During each stress step, an EL image was simultaneously generated with an acquisition time of 40 s [46]. In the OFF-state, the applied stress voltage is distributed across the depleted drift layer, in addition to the dielectric stack, resulting in correspondingly higher breakdown voltages for both unilayer and bilayer devices. The V_{BR} distribution for the tested devices is compared in Figure 7a, wherein the bilayer emerges as clearly superior, with an average V_{BR} improvement of 10 V.

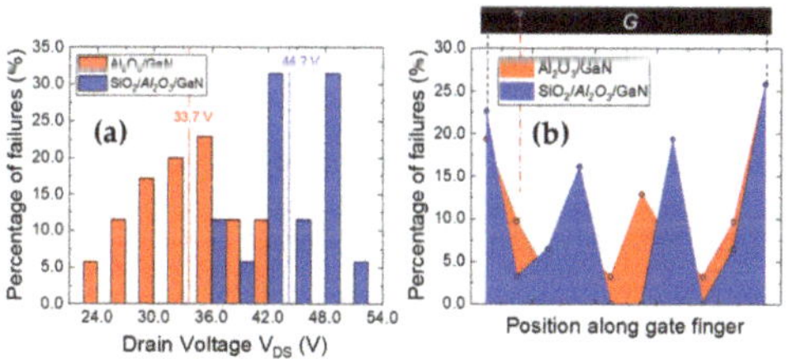

Figure 7. OFF-state drain step stress performance at V_{GS} = 0 V for a 35-device sample set (**a**) Comparison of the experimental breakdown values for both unilayer and bilayer cases (**b**) Localization of the failure spots along the gate finger, collected from observed EL spots (an example of an EL spot shown for reference at top) at corresponding V_{BR} values.

An example of an EL spot observed along the gate finger at V_{BR}, reflecting the region of breakdown in the devices, is shown in Figure 7b, along with a collated map of the breakdown spots for all tested devices, identified through EL acquisitions obtained during the step stress process, and on reaching failure. The results clearly indicate a preferential failure occurrence at the corners of the gate fingers, independent of the dielectric deposition.

The measurements displayed in Figure 7 were performed using microprobes fitted with an optimized current limiting circuit, in order to protect the failed devices from thermal runaway, and to preserve them for further post-failure analyses by TEM and Energy Dispersive X-ray Spectroscopy (EDX) [100,101] to identify the cause of breakdown [102].

Compared to the size of the original defect, an observed EL spot represents a relatively wide area in which the original defect could be present. Screening is necessary to precisely localize the defect within the observed EL spot area, which can be done by performing alternating focused ion beam (FIB) milling and SEM imaging [100]. After screening of the defect, TEM investigations were performed at various lamella thicknesses starting from 1.5 μm down to 50 nm to search for of a particular defect. Figure 8 exemplary displays the results of a defect analysis of a stressed bilayer device at the location of a particular EL spot, with a focus on the gate trench corners.

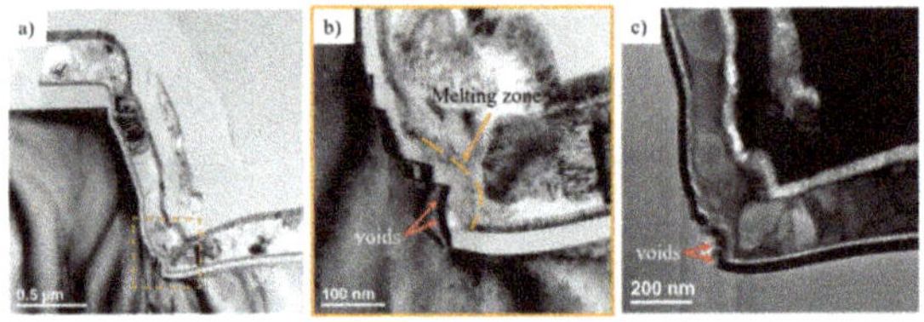

Figure 8. TEM analysis of defect at gate trench of a bilayer device at the position of an EL spot (**a**,**b**) BF-TEM and (**c**) ADF-STEM images of an approx. 50 nm thin lamella.

Device failure was identified to have been caused by an electrical breakdown of the gate isolation at the bottom edges of the trench, and was correlated with the presence of several abrupt steps of the gate trench sidewall [102]. While the defect structure was found to coincide with a melted area and several voids (see Figure 8b,c) as a consequence of gate shorts [102], EDX analysis on failed devices (not shown here, but reported in [102]) revealed that the breakdown of the gate isolation resulted in minor migrations of silicon and oxygen, and a dominant migration of nitrogen into the gate oxide.

To complete the investigation into the relative merits/demerits of the bilayer composition, trapping analyses using double pulsed [44,103] and on-the-fly transient [44,104,105] measurements were performed on several devices from both wafers, as presented in Figure 9. More details on the test methods will be provided in Section 4. The shift in the threshold voltage (ΔV_{th}) is compared for identical positive gate overdrive stresses.

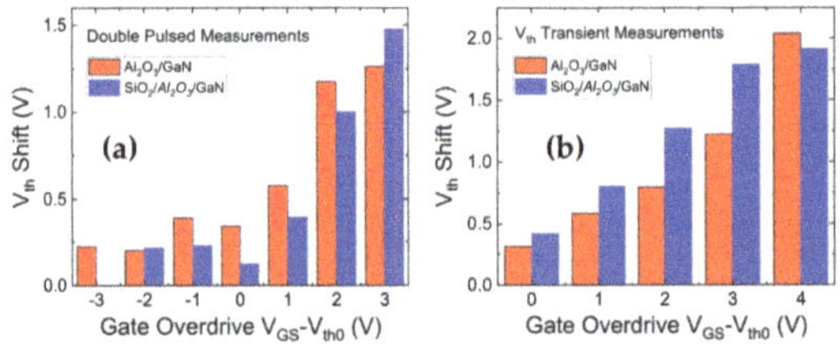

Figure 9. Comparison of bilayer vs. unilayer V_{th} shifts relative to the unstressed threshold voltage using (**a**). Double pulsed characteristics and (**b**) V_{th} transient tests.

The V_{th} shifts are comparable or slightly higher for the bilayer case, which could be due to additional trapping sites generated at the additional interface within the dielectric. However, the trapping performance for both the compositions is primarily comparable, which implies that most of the trapping can be presumed to occur at the interface and/or border traps near the shared GaN/Al_2O_3 region [44,106,107].

3.2. Optimising Trench Fabrication

In Section 3.1, the cause of breakdown was correlated to non-idealities around the trench edges. In this section, the cross-sectional analyses to identify the underlying issue, and to visualize improvements in the gate trench etch process, are summarized [102], in an effort to understand how to improve breakdown performance.

The investigated devices are GaN-on-Si trench MOSFETs with bilayer gate dielectric compositions. The fabrication process of the gate trench involved a bulk GaN etch process followed by an ALE and wet cleaning process. The first set of devices (Wafer A) are from the bilayer wafer presented in Section 3.1 (see Figure 8). The second set of devices (Wafer B) are taken from a wafer with an optimized ALE processing and wet cleaning sequence.

During the initial FIB-SEM investigation to isolate the defective/shorted gate, irregularities of the trench structure of Wafer A were observed. Hence, slice-and-view FIB-SEM analysis [100,101,108,109] was undertaken to study the trench at different locations along the gate finger, as presented in Figure 10.

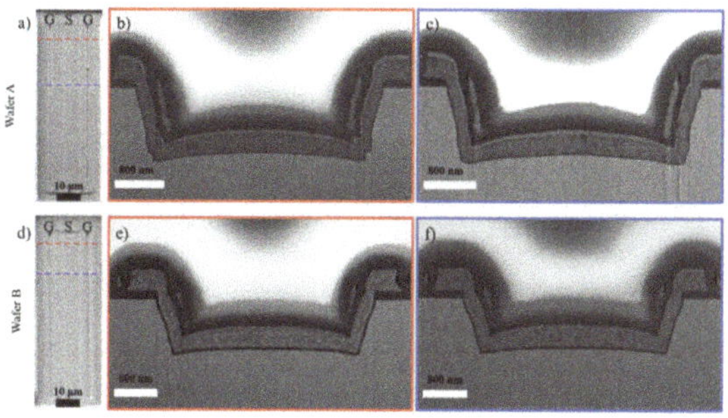

Figure 10. Slice and View analysis by FIB-SEM along the gate finger of devices from (**a–c**) Wafer A and from (**d–f**) Wafer B. (**a**) and (**d**) SEM top view images of the devices. The positions of the cross sections are marked by colored, dashed lines. (**b**,**c**) and (**e**,**f**) SEM cross sectional images. The colored frames correspond to the colored dashed lines in (**a**) and (**d**).

Several steep steps of varying shape and length were observed at each cross section along the trench sidewalls, dominantly at the lower trench corners. Since these irregularities are associated with accelerated degradations, drawing from these observations, the ALE and wet cleaning processes were improved during the fabrication of Wafer B. As displayed in Figure 10d–f, the newly fabricated trench gates have clean sidewalls, with no observed roughness or steps. Further TEM analysis [102] also corroborated these observations.

From the results in Section 3, we can improve the general understanding of the degradation mechanisms that occur within the gate stack, when subjected to prolonged gate and drain stresses. Bilayer dielectric compositions, utilizing the good interface properties of Al_2O_3 to GaN and the improved stability of the SiO_2 material, were found to be highly advantageous to breakdown performance of GaN trench MOSFETs, without significant worsening of trapping effects. However, before improving other design parameters, the fundamental GaN etch process must be robust. Microstructural defects formed during fabrication of the gate trench sidewalls can manifest in worsened reliability and faster breakdown, hence optimization techniques to minimize etch roughness are critical.

4. ON-State-Light Assisted Analysis of Trapping Mechanisms in Quasi-Vertical MOSFETs from IMEC, Leuven, Belgium

For reliable ON-state operation of GaN MOSFETs, it is fundamental to understand and minimize the trapping states for the insulator/GaN interface. Since III-V semiconductors have no native oxides, developing high quality oxide films on GaN is difficult. The progress in the application of the atomic layer deposition technique has allowed the successful deposition of low-defect Al_2O_3 films on GaN, improving the performances of MOS structures. However, identifying relevant trapping sites and the induced threshold voltage V_{th} instabilities [44,89,106,107,110,111] due to limited controllability of the GaN surface potential continues to be a primary task to the adoption of GaN vertical MOSFETs in real applications.

In Section 3.1, the trap impacts on threshold voltage were found to be comparable between bilayer and unilayer dielectric cases, indicating that states at or near the GaN/Al_2O_3 interface are presumably the major contributing factor to bias threshold instability (BTI) observations.

In this section, we focus on unilayer Al_2O_3-only trench MOSFET devices with an average V_{th} of 2 V, with device structure similar to Figure 1, to understand the trapping mechanisms through characterization of induced V_{th} shifts [44]. Within the Al_2O_3/GaN system, three fundamental trapping locations have been identified [106,107]. Trap states within the bulk dielectric and near-interface or border sites depend strongly on the properties of the deposited Al_2O_3, while the states along the Al_2O_3/GaN interface (quantified by the interface state density D_{it}) correlate to the quality of the dielectric/semiconductor boundary, and of the process. For a wide band-gap material such as GaN, it is often difficult to isolate the effects of energetically deep trap states. This is where light energy, and especially the application of UV light with energies approaching/higher than the GaN band-gap, is valuable. In the following results, we investigated V_{th} shifts under positive gate stress, by combining analytical techniques to identify trap processes and associated recovery dynamics. In each case, light energy is used to support the analyses, and provide further insight into the physical origin of the trap states.

The first set of measurements to test the dynamic performance of the devices, as summarized in Figure 11, are double pulsed measurements. The double pulse measurement system is a powerful high voltage, high speed setup to analyze the dynamic performance of devices by synchronously pulsing the gate and drain voltages. The pulsing setup switches between the quiescent (stress conditions) and measurement phases within relatively short time scales (μs). The V_G stress settings are incremented from $V_{G,Stress}$ = 0 V to 5 V, $V_{D,Stress}$ = 0 V for a quiescent time of t_Q = 100 μs, and the I_D-V_G measurement settings were V_{GS} = −1 to 7 V, V_{DS} = 8 V for a measurement time t_{meas} = 1 μs. In Figure 11a, the measurements were performed in dark conditions, displaying a positive shift in V_{th} (PBTI) of 1.2 V for Q (5,0) (V_{th} calculated as the voltage intercept at I_D = 5 mA/mm). The V_{th} shift can be attributed to the fast-pulsed stressing configuration, with no recovery intervals between the progressively stronger stress conditions.

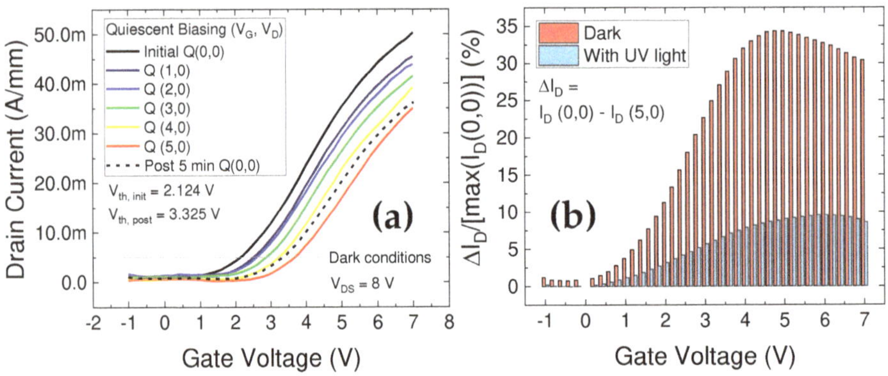

Figure 11. Double pulsed characteristics; (**a**) Measurements under dark conditions show a ΔV_{th} = 1.2 V and very little recovery in the measured I_DV_G, 5 min after the stress at Q (5,0); (**b**) Comparison of current level shifts measured under no light and UV light. Under UV illumination, shifts are lower under during stress conditions, and post-stress recovery is faster.

After a rest period of 5 min following the positive gate stress at Q (5,0), the I_D-V_G measured for Q (0,0) condition still showed substantial degradation from the initial I_D-V_G characteristic at Q (0,0), indicating semi-permanent trapping processes. This can also be visualized by plotting the $\Delta I_D/I_{D,\,max}$ ratio in Figure 11b for the high stress Q (5,0) condition.

The shift in the current levels under stress was 30% of the pre-stressed current maximum, while 5 min of recovery reduced it to 25–27%. On the other hand, repeating the same stress-recovery cycles as in Figure 11a, but under the presence of UV light displayed substantial improvement. As highlighted in Figure 11b, under UV light, for the highest stress condition of Q (5,0), the shift in the current levels was less than 10%. Furthermore, letting the device recover for 5 min thereafter, the deviation in the I_D-V_G at Q (0,0) from the unstressed initial I_D-V_G at Q (0,0) was found to be negligible ($\Delta I_D/I_{D,\,max} \approx$ 2–3%, not shown).

Based on these observations, a powerful transient setup was employed to take a closer look at the evolution of induced V_{th} shifts under longer gate stress durations, in the presence of different monochromatic light energies. This versatile setup accurately evaluates V_{th} transients in the 10 μs–100 s range where a typical measurement consists of 100 s of stress and 100 s of recovery. Twenty-two fast I_D–V_G measurements of 10 μs each are performed during the stress/recovery phases to compare the evolution of V_{th}. During initial measurements using this technique, small negative V_{th} shifts were observed at low stress voltages [89,112], and high positive V_{th} shifts were observed for gate stresses of 4 V and higher [44]. To investigate the effects of light-assisted de-trapping, the recovery was repeated under different wavelengths of light, following 100 s of trap filling at $V_{G,Stress}$ = 5 V, and $V_{D,Stress}$= 0 V.

Figure 12 presents the results of the light-assisted V_{th} transient technique. In Figure 12a, a positive V_{th} shift of 0.75 V is seen after 100 s of stress at $V_{G,Stress}$ = 5 V. The recovery transient (at $V_{G,Rec}$ = 0 V and $V_{D,Rec}$ = 0 V) in response to this stress, was measured under dark and under monochromatic light energies from 1.6 eV to 3.1 eV, as illustrated in Figure 12b,c. Under dark conditions, the recovery is slow and hence incomplete [113] at the end of the 100 s of recovery phase. For low photon energies, such as 760 nm, only 50% (0.35 V) of the stress-induced PBTI was recoverable within 100 s. For higher photon energies, de-trapping was found to be gradually accelerated. The threshold energy (associated to the lowest energetic position of deep bulk states) for improved de-trapping was identified to be 2.95 eV (420 nm), while complete recovery of the 0.75 V of positive V_{th} shift was observed within the 100 s window for the 3.1 eV (395 nm) case. As can be noticed in Figure 12b, all photon energies below 2.7 eV did not induce any significant changes, with small/negligible recovery. Small variations observed below this threshold in Figure 12c may be ascribed to small (5–10%) measurement inconsistencies and/or noise.

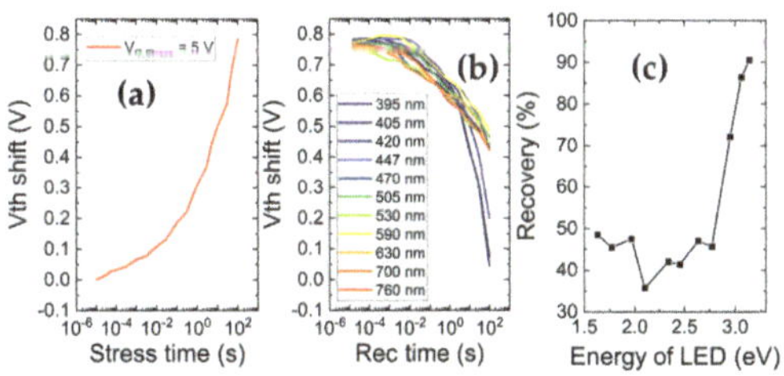

Figure 12. V_{th} transient measurements (**a**) Shift in V_{th} ($V_{th}-V_{th@10\,\mu s}$) during stress phase of 100 s at $V_{G,Stress}$ = 5 V. (**b**) V_{th} evolution during recovery phase of 100 s at $V_{G,Stress}$ = 0 V under varying light wavelengths from 760 nm to 395 nm, following equivalent stress phases as described in (**a**), (**c**) absolute V_{th} shift during recovery ($V_{th@100s}-V_{th@10\mu s}$ during recovery) versus the light energy.

A direct takeaway from this would be the presence of trap states located energetically between 2.9 eV and 3.1 eV from the conduction band of the oxide, which equates to 0.8 to 1.0 eV from the conduction band of the semiconductor, considering a conduction band offset of 2.16 eV [86] at the Al_2O_3/GaN interface.

The final light-assisted technique to identify trap distributions is the photo assisted CV method [44,114]. This measurement approach evaluates the distribution of interface states located along the gate dielectric interface to GaN. In this method, capacitance-voltage measurements, obtained under a photo-assisted de-trapped condition and a bias-induced trapped condition, are compared to quantify the interface state density. The use of UV light allows us to empty all defects at the interface (when the device is in depletion) to probe interface states deep within the bandgap. The results of the photo-assisted CV experiment are displayed in Figure 13.

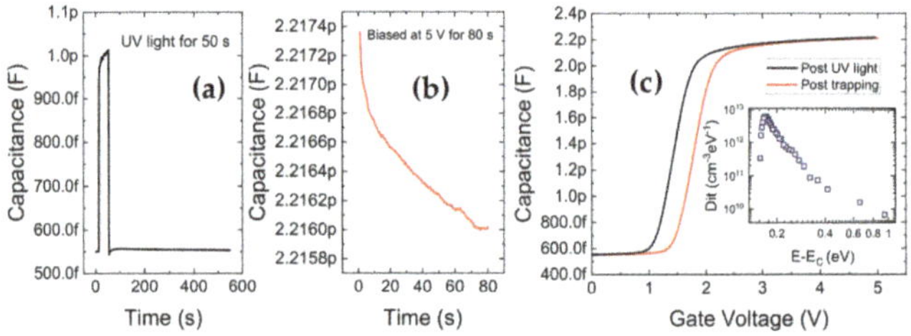

Figure 13. Photoassisted CV method for D_{it} extraction; (**a**) Capacitance-time transient during exposure to UV light at V_G = 0 V; (**b**) Capacitance-time transient during filling of traps at V_G = 5 V. (**c**) C-V comparison between detrapped (after UV light) and trapped state. (inset) Electron D_{it} vs. E_G.

The devices are biased in depletion condition for a short time and then exposed to UV light in order to empty all traps at the interface, as shown in Figure 13a. In the presence of UV light, electron-hole pairs are generated, accompanied by an increasing capacitance transient duc to the release of trapped charge inside the depleted region. The duration of UV exposure is 50 s, until the capacitance level saturates. This is followed by a longer time interval in the dark (500 s) to allow enough time for the excess photo-generated carriers to leave the system and reach thermal equilibrium. Then, the de-trapped capacitance-voltage curve from depletion to accumulation is measured from V_G = 0 V to 5 V (see Figure 13c). Bias at the end voltage (5 V) is maintained for a moderate filling time (80 s), to induce charge trapping at insulator and interface states, as shown in Figure 13b. Finally, the second C-V curve of the trapped device is measured from accumulation to depletion. The difference in C-V slope of the trapped and de-trapped curves allows the extraction of D_{it} versus energy, while the fixed shift in the curves is proportional to the amount of charge trapped in the bulk of the oxide and/or in near-interfacial or border traps. The D_{it} profile (inset of Figure 13c) reveals shallow traps located around 0.3 eV from the conduction band.

Based on the observations in Section 4, the following inferences regarding relevant trapping mechanisms under forward gate stress can be drawn, as also summarized in Figure 14.

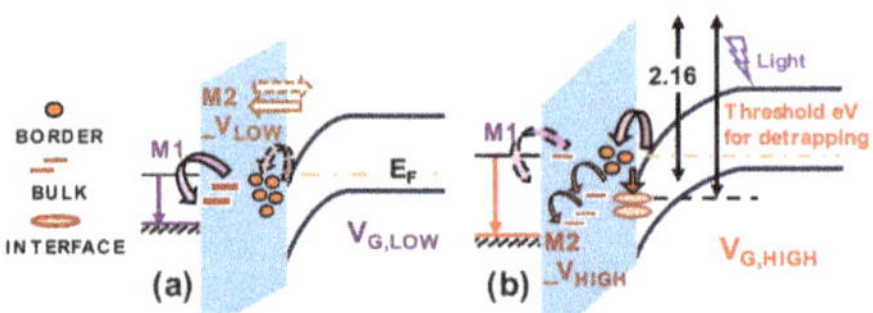

Figure 14. Energy band diagrams illustrating trapping locations in the Metal/Al_2O_3/GaN system (**a**) mechanisms activated at low V_G stress. M1: negative ΔV_{th} due to detrapped electrons from oxide towards metal. M2_V_{LOW}: moderate and recoverable positive ΔV_{th} due to injection of electrons from GaN accumulation into the border oxide traps; (**b**) mechanisms strengthened at high gate stress, M2_V_{HIGH}: strong positive ΔV_{th} due to electrons injection into energetically deeper interface traps or bulk states in the dielectric. M2_V_{HIGH} causes semi-permanent trapping which requires external light energy (inducing de-trapping) for achieving fast recovery of V_{th} [44].

The small NBTI observed during V_{th} transients at low gate stresses ($\leq$2 V) is attributed to de-trapping of electrons within the gate oxide to the metal (M1 in Figure 14). When medium gate stresses are applied ($\approx$3–4 V), small amounts of PBTI can be attributed to electron trapping from the semiconductor towards border states in the dielectric (M2_V_{LOW} in Figure 14). V_{th} shifts owing to this process are recoverable once stress is removed and the Fermi level is restored, even under dark conditions if enough recovery time is provided. For high gate stresses ($\geq$4V), strong PBTI is induced, and this contribution suffers from low recovery under dark conditions, even for long recovery times (~ days). The mechanism responsible for this semi-permanent V_{th} degradation (M2_V_{HIGH} in Figure 14) is due to the worsening of M2 under high fields, resulting in electron transport from the channel to energetically deeper trap states along the interface, or further within the bulk of the dielectric. To enable de-trapping from these deeper trap states, light energy $\geq$2.9 eV is required.

5. Conclusions

In this paper, we have summarized some of the most relevant challenges for the development of reliable GaN-on-Si vertical trench MOSFETs, for application in power electronics. Specifically, we presented the results of recent case studies, aimed at investigating (a) the origin of OFF-state leakage current, (b) the role of p-body doping in determining the breakdown voltage of the vertical stack, (c) the substantial improvement of reliability that can be obtained through the use of a bi-layer gate insulator, (d) specific failure mechanisms related to the optimization of the trench etching and cleaning procedure, and (e) a set of advanced results on the physics of interface trapping phenomena, obtained through the use of pulsed/transient measurements carried out in dark and under light. The obtained insights help understanding the current issues faced by the GaN for power community, and demonstrates strategies for identifying and analyzing the structural, leakage and trapping constraints to realize efficient and economical GaN-on-Si devices. If the pace of development and innovation within GaN-on-Si technologies is sustained, the benefits could prove to be revolutionary for the power semiconductor industry.

Funding: This project has received funding from the ECSEL Joint Undertaking (JU) under grant agreement No. 826392. The JU receives support from the European Union's Horizon 2020 research and innovation programme and Austria, Belgium, Germany, Italy, Norway, Slovakia, Spain, Sweden, Switzerland. This research activity was partly funded by project "Novel vertical GaN-devices for next generation power conversion", NoveGaN (University of Padova), through the STARS CoG Grants call. Part of this work was supported by MIUR (Italian Minister for Education) under the initiative "Departments of Excellence" (Law 232/2016).

Institutional Review Board Statement: Not applicable.

Informed Consent Statement: Not applicable.

Data Availability Statement: No new data were created or analyzed in this study.

Conflicts of Interest: The authors declare no conflict of interest.

References

1. Vecchia, M.D.; Ravyts, S.; Broeck, G.V.D.; Driesen, J. Gallium-Nitride Semiconductor Technology and Its Practical Design Challenges in Power Electronics Applications: An Overview. *Energies* **2019**, *12*, 2663. [CrossRef]
2. Flack, T.J.; Pushpakaran, B.N.; Bayne, S.B. GaN Technology for Power Electronic Applications: A Review. *J. Electron. Mater.* **2016**, *45*, 2673–2682. [CrossRef]
3. Hu, J.; Zhang, Y.; Sun, M.; Piedra, D.; Chowdhury, N.; Palacios, T. Materials and processing issues in vertical GaN power electronics. *Mater. Sci. Semicond. Process.* **2018**, *78*, 75–84. [CrossRef]
4. Keshmiri, N.; Wang, D.; Agrawal, B.; Hou, R.; Emadi, A. Current Status and Future Trends of GaN HEMTs in Electrified Transportation. *IEEE Access* **2020**, *8*, 70553–70571. [CrossRef]
5. Chowdhury, S.; Mishra, U.K. Lateral and Vertical Transistors Using the AlGaN/GaN Heterostructure. *IEEE Trans. Electron Devices* **2013**, *60*, 3060–3066. [CrossRef]
6. Baliga, B.J. Semiconductors for High-voltage, Vertical Channel Field-effect Transistors. *J. Appl. Phys.* **1982**, *53*, 7. [CrossRef]
7. Kim, I.-J.; Matsumoto, S.; Sakai, T.; Yachi, T. New power device figure of merit for high-frequency applications. In Proceedings of the International Symposium on Power Semiconductor Devices and IC's: ISPSD '95, Yokohama, Japan, 23–25 May 1995. [CrossRef]
8. Wang, H.; Wang, F.; Zhang, J. Power Semiconductor Device Figure of Merit for High-Power-Density Converter Design Applications. *IEEE Trans. Electron Devices* **2007**, *55*, 466–470. [CrossRef]
9. Roccaforte, F.; Greco, G.; Fiorenza, P.; Iucolano, F. An Overview of Normally-Off GaN-Based High Electron Mobility Transistors. *Materials* **2019**, *12*, 1599. [CrossRef]
10. Shenai, K. Future Prospects of Widebandgap (WBG) Semiconductor Power Switching Devices. *IEEE Trans. Electron Devices* **2014**, *62*, 248–257. [CrossRef]

11. Meneghesso, G.; Meneghini, M.; Bisi, D.; Rossetto, I.; Cester, A.; Mishra, U.K.; Zanoni, E. Trapping phenomena in AlGaN/GaN HEMTs: A study based on pulsed and transient measurements. *Semicond. Sci. Technol.* **2013**, *28*, 074021. [CrossRef]
12. Ueda, T. GaN power devices: Current status and future challenges. *Jpn. J. Appl. Phys.* **2019**, *58*, SC0804. [CrossRef]
13. Hu, J.; Stoffels, S.; Lenci, S.; De Jaeger, B.; Ronchi, N.; Tallarico, A.N.; Wellekens, D.; You, S.; Bakeroot, B.; Groeseneken, G.; et al. Statistical Analysis of the Impact of Anode Recess on the Electrical Characteristics of AlGaN/GaN Schottky Diodes With Gated Edge Termination. *IEEE Trans. Electron Devices* **2016**, *63*, 3451–3458. [CrossRef]
14. Uemoto, Y.; Hikita, M.; Ueno, H.; Matsuo, H.; Ishida, H.; Yanagihara, M.; Ueda, T.; Tanaka, T.; Ueda, D. Gate Injection Transistor (GIT)—A Normally-Off AlGaN/GaN Power Transistor Using Conductivity Modulation. *IEEE Trans. Electron Devices* **2007**, *54*, 3393–3399. [CrossRef]
15. Meneghini, M.; Fabris, E.; Sun, M.; Palacios, T.; Meneghesso, G.; Zanoni, E.; Ruzzarin, M.; De Santi, C.; Nomoto, K.; Hu, Z.; et al. Degradation Mechanisms of GaN-Based Vertical Devices: A Review. *Phys. Status Solidi (A)* **2020**, *217*. [CrossRef]
16. Ohta, H.; Kaneda, N.; Horikiri, F.; Narita, Y.; Yoshida, T.; Mishima, T.; Nakamura, T. Vertical GaN p-n Junction Diodes with High Breakdown Voltages Over 4 kV. *IEEE Electron Device Lett.* **2015**, *36*, 1180–3182. [CrossRef]
17. Kizilyalli, I.C.; Edwards, A.P.; Aktas, O.; Prunty, T.; Bour, D. Vertical Power p-n Diodes Based on Bulk GaN. *IEEE Trans. Electron Devices* **2015**, *62*, 414–422. [CrossRef]
18. Oka, T. Recent development of vertical GaN power devices. *Jpn. J. Appl. Phys.* **2019**, *58*, SB0805. [CrossRef]
19. Ramanathan, D.; Coles, C.; Meier, W. Not All GaN Transistors Are Built Equal: The Benefits of Vertical GaN-on-GaN. In Proceedings of the PCIM Europe Digital Days 2020; International Exhibition and Conference for Power Electronics, Intelligent Motion, Renewable Energy and Energy Management, Nuremburg, Germany, 7–8 July 2020; pp. 1–4.
20. De Santi, C.; Fabris, E.; Meneghesso, G.; Zanoni, E.; Meneghini, M. GaN Vertical p–i–n Diodes in Avalanche Regime: Time-Dependent Behavior and Degradation. *IEEE Electron Device Lett.* **2020**, *41*, 1300–1303. [CrossRef]
21. Fukushima, H.; Usami, S.; Ogura, M.; Ando, Y.; Tanaka, A.; Deki, M.; Kushimoto, M.; Nitta, S.; Honda, Y.; Amano, H. Deeply and vertically etched butte structure of vertical GaN p–n diode with avalanche capability. *Jpn. J. Appl. Phys.* **2019**, *58*, SCCD25. [CrossRef]
22. Zhang, Y.; Sun, M.; Piedra, D.; Hu, J.; Liu, Z.; Lin, Y.; Gao, X.; Shepard, K.; Palacios, T. 1200 V GaN vertical fin power field-effect transistors. In Proceedings of the 2017 IEEE International Electron Devices Meeting (IEDM), San Francisco, CA, USA, 2–6 December 2017.
23. Panasonic GaN Power Transistors. New Contender for the Power Transistor Throne: How GaN Is Threatening the MOSFET's Crown, April 2018. Available online: https://eu.industrial.panasonic.com/sites/default/pidseu/files/pan_18047_whitepaper_gan_web.pdf (accessed on 20 February 2021).
24. Ji, D.; Agarwal, A.; Li, H.; Li, W.; Keller, S.; Chowdhury, S. 880 V/2.7 M Cm2 MIS Gate Trench CAVET on Bulk GaN Substrates. *IEEE Electron Device Lett.* **2018**, *39*, 863–865. [CrossRef]
25. Hu, J.; Stoffels, S.; Lenci, S.; Bakeroot, B.; De Jaeger, B.; Van Hove, M.; Ronchi, N.; Venegas, R.; Liang, H.; Zhao, M.; et al. Performance Optimization of Au-Free Lateral AlGaN/GaN Schottky Barrier Diode with Gated Edge Termination on 200-mm Silicon Substrate. *IEEE Trans. Electron Devices* **2016**, *63*, 997–1004. [CrossRef]
26. Zhang, Y.; Sun, M.; Wong, H.-Y.; Lin, Y.; Srivastava, P.; Hatem, C.; Azize, M.; Piedra, D.; Yu, L.; Sumitomo, T.; et al. Origin and Control of OFF-State Leakage Current in GaN-on-Si Vertical Diodes. *IEEE Trans. Electron Devices* **2015**, *62*, 2155–2161. [CrossRef]
27. Zhang, Y.; Piedra, D.; Sun, M.; Hennig, J.; Dadgar, A.; Yu, L.; Palacios, T. High-Performance 500 V Quasi- and Fully-Vertical GaN-on-Si pn Diodes. *IEEE Electron Device Lett.* **2016**, *38*, 248–251. [CrossRef]
28. Zhang, X.; Zou, X.; Lu, X.; Tang, C.W.; Lau, K.M. Fully- and Quasi-Vertical GaN-on-Si p-i-n Diodes: High Performance and Comprehensive Comparison. *IEEE Trans. Electron Devices* **2017**, *64*, 809–815. [CrossRef]
29. Zhang, Y.; Yuan, M.; Chowdhury, N.; Cheng, K.; Palacios, T. 720-V/0.35-m Ω·Ω cm2Fully Vertical GaN-on-Si Power Diodes by Selective Removal of Si Substrates and Buffer Layers. *IEEE Electron Device Lett.* **2018**, *39*, 715–718. [CrossRef]
30. Zou, X.; Zhang, X.; Lu, X.; Tang, C.W.; Lau, K.M. Fully Vertical GaN p-i-n Diodes Using GaN-on-Si Epilayers. *IEEE Electron Device Lett.* **2016**, *37*, 636–639. [CrossRef]
31. Mase, S.; Hamada, T.; Freedsman, J.J.; Egawa, T. Effect of Drift Layer on the Breakdown Voltage of Fully-Vertical GaN-on-Si p-n Diodes. *IEEE Electron Device Lett.* **2017**, *38*, 1720–1723. [CrossRef]
32. Khadar, R.A.; Liu, C.; Soleimanzadeh, R.; Matioli, E. Fully Vertical GaN-on-Si power MOSFETs. *IEEE Electron Device Lett.* **2019**, *40*, 443–446. [CrossRef]
33. Zubair, A.; Perozek, J.; Niroula, J.; Aktas, O.; Odnoblyudov, V.; Palacios, T. First Demonstration of GaN Vertical Power FinFETs on Engineered Substrate. In Proceedings of the 2020 Device Research Conference (DRC), Columbus, OH, USA, 21–24 June 2020; pp. 1–2.
34. Liu, C.; Khadar, R.A.; Matioli, E. GaN-on-Si Quasi-Vertical Power MOSFETs. *IEEE Electron Device Lett.* **2017**, *39*, 71–74. [CrossRef]
35. Liu, C.; Khadar, R.A.; Matioli, E. Vertical GaN-on-Si MOSFETs With Monolithically Integrated Freewheeling Schottky Barrier Diodes. *IEEE Electron Device Lett.* **2018**, *39*, 1034–1037. [CrossRef]
36. Liu, C.; Khadar, R.A.; Matioli, E. 645 V quasi-vertical GaN power transistors on silicon substrates. In Proceedings of the 2018 IEEE 30th International Symposium on Power Semiconductor Devices and ICs (ISPSD), Chicago, IL, USA, 13–17 May 2018; pp. 240–243.

37. Chowdhury, S.; Swenson, B.L.; Mishra, U.K. Enhancement and Depletion Mode AlGaN/GaN CAVET With Mg-Ion-Implanted GaN as Current Blocking Layer. *IEEE Electron Device Lett.* **2008**, *29*, 543–545. [CrossRef]
38. Ji, D.; Li, W.; Agarwal, A.; Chan, S.H.; Haller, J.; Bisi, D.; Labrecque, M.; Gupta, C.; Cruse, B.; Lal, R.; et al. Improved Dynamic RON of GaN Vertical Trench MOSFETs (OG-FETs) Using TMAH Wet Etch. *IEEE Electron Device Lett.* **2018**, *39*, 1030–1033. [CrossRef]
39. Gupta, C.; Chan, S.H.; Enatsu, Y.; Agarwal, A.; Keller, S.; Mishra, U.K. OG-FET: An in-situ Oxide, GaN interlayer based vertical trench MOSFET. *IEEE Electron Device Lett.* **2016**, *37*, 1. [CrossRef]
40. Xiao, M.; Gao, X.; Palacios, T.; Zhang, Y. Leakage and breakdown mechanisms of GaN vertical power FinFETs. *Appl. Phys. Lett.* **2019**, *114*, 163503. [CrossRef]
41. Li, W.; Xing, H.G.; Nomoto, K.; Lee, K.; Islam, S.; Hu, Z.; Zhu, M.; Gao, X.; Pilla, M.; Jena, D. Development of GaN Vertical Trench-MOSFET With MBE Regrown Channel. *IEEE Trans. Electron Devices* **2018**, *65*, 2558–2564. [CrossRef]
42. Otake, H.; Chikamatsu, K.; Yamaguchi, A.; Fujishima, T.; Ohta, H. Vertical GaN-Based Trench Gate Metal Oxide Semiconductor Field-Effect Transistors on GaN Bulk Substrates. *Appl. Phys. Express* **2008**, *1*, 011105. [CrossRef]
43. Gupta, C.; Chan, S.H.; Lund, C.; Agarwal, A.; Koksaldi, O.S.; Liu, J.; Enatsu, Y.; Keller, S.; Mishra, U.K. Comparing electrical performance of GaN trench-gate MOSFETs with a-Plane and m-Plane Sidewall Channels. *Appl. Phys. Express* **2016**, *9*, 121001. [CrossRef]
44. Mukherjee, K.; Borga, M.; Ruzzarin, M.; De Santi, C.; Stoffels, S.; You, S.; Geens, K.; Liang, H.; Decoutere, S.; Meneghesso, G.; et al. Analysis of threshold voltage instabilities in semi-vertical GaN-on-Si FETs. *Appl. Phys. Express* **2020**, *13*, 024004. [CrossRef]
45. Borga, M.; Mukherjee, K.; De Santi, C.; Stoffels, S.; Geens, K.; You, S.; Bakeroot, B.; Decoutere, S.; Meneghesso, G.; Zanoni, E.; et al. Modeling of gate capacitance of GaN-based trench-gate vertical metal-oxide-semiconductor devices. *Appl. Phys. Express* **2020**, *13*, 024006. [CrossRef]
46. Mukherjee, K.; De Santi, C.; Borga, M.; You, S.; Geens, K.; Bakeroot, B.; Decoutere, S.; Meneghesso, G.; Zanoni, E.; Meneghini, M. Use of Bilayer Gate Insulator in GaN-on-Si Vertical Trench MOSFETs: Impact on Performance and Reliability. *Materials* **2020**, *13*, 4740. [CrossRef]
47. Zhang, Y.; Dadgar, A.; Palacios, T. Gallium nitride vertical power devices on foreign substrates: A review and outlook. *J. Phys. D: Appl. Phys.* **2018**, *51*, 273001. [CrossRef]
48. Dadgar, A. Sixteen years GaN on Si. *Phys. Status Solidi (B)* **2015**, *252*, 1063–1068. [CrossRef]
49. Zhu, T.; Oliver, R.A. Unintentional doping in GaN. *Phys. Chem. Chem. Phys.* **2012**, *14*, 9558–9573. [CrossRef] [PubMed]
50. Wickenden, A.; Koleske, D.; Henry, R.; Twigg, M.; Fatemi, M. Resistivity control in unintentionally doped GaN films grown by MOCVD. *J. Cryst. Growth* **2004**, *260*, 54–62. [CrossRef]
51. Chung, B.; Gershenzon, M. The influence of oxygen on the electrical and optical properties of GaN crystals grown by metalorganic vapor phase epitaxy. *J. Appl. Phys.* **1992**, *72*, 651–659. [CrossRef]
52. Seager, C.H.; Wright, A.F.; Yu, J.; Götz, W. Role of carbon in GaN. *J. Appl. Phys.* **2002**, *92*, 6553–6560. [CrossRef]
53. Zhang, Y.; Wong, H.-Y.; Sun, M.; Joglekar, S.; Yu, L.; Braga, N.A.; Mickevicius, R.V.; Palacios, T. Design space and origin of off-state leakage in GaN vertical power diodes. In Proceedings of the 2015 IEEE International Electron Devices Meeting (IEDM), Washington, DC, USA, 7–9 December 2015; pp. 35.1.1–35.1.4.
54. Rackauskas, B.; Dalcanale, S.; Uren, M.J.; Kachi, T.; Kuball, M. Leakage mechanisms in GaN-on-GaN vertical pn diodes. *Appl. Phys. Lett.* **2018**, *112*, 233501. [CrossRef]
55. Mukherjee, K.; De Santi, C.; Buffolo, M.; Borga, M.; You, S.; Geens, K.; Bakeroot, B.; Decoutere, S.; Gerosa, A.; Meneghesso, G.; et al. Understanding the Leakage Mechanisms and Breakdown Limits of Vertical GaN-on-Si p^+n^-n Diodes: The Road to Reliable Vertical MOSFETs. *Micromachines* **2021**, *12*, 445. [CrossRef]
56. Podor, B. Thermal ionization energy of Mg acceptors in GaN: Effects of doping level and compensation. In Proceedings of the SPIE 4412, International Conference on Solid State Crystals 2000: Growth, Characterization, and Applications of Single Crystals, Zakopane, Poland, 10 August 2001; pp. 299–303.
57. Sabui, G.; Parbrook, P.J.; Arredondo-Arechavala, M.; Shen, Z.J. Modeling and simulation of bulk gallium nitride power semiconductor devices. *AIP Adv.* **2016**, *6*, 055006. [CrossRef]
58. Yang, J.; Zhao, D.; Jiang, D.; Chen, P.; Zhu, J.; Liu, Z.; Le, L.; He, X.; Li, X.; Zhang, Y.T.; et al. Influence of hydrogen impurities on p-type resistivity in Mg-doped GaN films. *J. Vac. Sci. Technol. A* **2015**, *33*, 021505. [CrossRef]
59. Musolino, M.; Van Treeck, D.; Tahraoui, A.; Scarparo, L.; De Santi, C.; Meneghini, M.; Zanoni, E.; Geelhaar, L.; Riechert, H. A physical model for the reverse leakage current in (In,Ga)N/GaN light-emitting diodes based on nanowires. *J. Appl. Phys.* **2016**, *119*, 44502. [CrossRef]
60. Han, D.-P.; Oh, C.-H.; Kim, H.; Shim, J.-I.; Kim, K.-S.; Shin, D.-S. Conduction Mechanisms of Leakage Currents in InGaN/GaN-Based Light-Emitting Diodes. *IEEE Trans. Electron Devices* **2014**, *62*, 587–592. [CrossRef]
61. Chiu, F.-C. A Review on Conduction Mechanisms in Dielectric Films. *Adv. Mater. Sci. Eng.* **2014**, *2014*, 1–18. [CrossRef]
62. Hill, R.M. Hopping conduction in amorphous solids. *Philos. Mag.* **1971**, *24*, 1307–1325. [CrossRef]
63. Shan, Q.; Meyaard, D.S.; Dai, Q.; Cho, J.; Schubert, E.F.; Son, J.K.; Sone, C. Transport-mechanism analysis of the reverse leakage current in GaInN light-emitting diodes. *Appl. Phys. Lett.* **2011**, *99*, 253506. [CrossRef]
64. Zhou, S.; Lv, J.; Wu, Y.; Zhang, Y.; Zheng, C.; Liu, S. Reverse leakage current characteristics of InGaN/GaN multiple quantum well ultraviolet/blue/green light-emitting diodes. *Jpn. J. Appl. Phys.* **2018**, *57*, 051003. [CrossRef]

65. Kuksenkov, D.V.; Temkin, H.; Osinsky, A.; Gaska, R.; Khan, M.A. Origin of conductivity and low-frequency noise in reverse-biased GaN p-n junction. *Appl. Phys. Lett.* **1998**, *72*, 1365–1367. [CrossRef]
66. Kim, J.; Kim, J.-Y.; Tak, Y.; Kim, J.; Hong, H.-G.; Yang, M.; Chae, S.; Park, J.; Park, Y.; Chung, U.-I. Investigation of Reverse Leakage Characteristics of InGaN/GaN Light-Emitting Diodes on Silicon. *IEEE Electron Device Lett.* **2012**, *33*, 1741–1743. [CrossRef]
67. Jung, E.; Lee, J.K.; Kim, M.S.; Kim, H. Leakage Current Analysis of GaN-Based Light-Emitting Diodes Using a Parasitic Diode Model. *IEEE Trans. Electron Devices* **2015**, *62*, 3322–3325. [CrossRef]
68. Look, D.C.; Reynolds, D.C.; Kim, W.; Aktas, O.; Botchkarev, A.; Salvador, A.; Morkoç, H. Deep-center hopping conduction in GaN. *J. Appl. Phys.* **1996**, *80*, 2960–2963. [CrossRef]
69. Ferdous, M.S.; Wang, X.; Fairchild, M.N.; Hersee, S.D. Effect of threading defects on InGaN/GaN multiple quantum well light emitting diodes. *Appl. Phys. Lett.* **2007**, *91*, 231107. [CrossRef]
70. Zhao, L.; Chen, L.; Yu, G.; Yan, D.; Yang, G.; Gu, X.; Liu, B.; Lu, H. Tunneling-Hopping Transport Model for Reverse Leakage Current in InGaN/GaN Blue Light-Emitting Diodes. *IEEE Photon. Technol. Lett.* **2017**, *29*, 1447–1450. [CrossRef]
71. Tsou, C.-W.; Ji, M.-H.; Bakhtiary-Noodeh, M.; Detchprohm, T.; Dupuis, R.D.; Shen, S.-C. Temperature-Dependent Leakage Current Characteristics of Homojunction GaN p-i-n Rectifiers Using Ion-Implantation Isolation. *IEEE Trans. Electron Devices* **2019**, *66*, 4273–4278. [CrossRef]
72. Mazzola, M.S.; Saddow, S.E.; Neudeck, P.G.; Lakdawala, V.K.; We, S. Observation of the D-center in 6H-SiC p-n diodes grown by chemical vapor deposition. *Appl. Phys. Lett.* **1994**, *64*, 2730–2732. [CrossRef]
73. Hill, R.M. Poole-Frenkel conduction in amorphous solids. *Philos. Mag.* **1971**, *23*, 59–86. [CrossRef]
74. Frenkel, J. On Pre-Breakdown Phenomena in Insulators and Electronic Semi-Conductors. *Phys. Rev.* **1938**, *54*, 647–648. [CrossRef]
75. Simmons, J.G. Conduction in thin dielectric films. *J. Phys. D Appl. Phys.* **1971**, *4*, 613–657. [CrossRef]
76. Lee, M.; Lee, H.U.; Song, K.M.; Kim, J. Significant improvement of reverse leakage current characteristics of Si-based homoepitaxial InGaN/GaN blue light emitting diodes. *Sci. Rep.* **2019**, *9*, 1–6. [CrossRef]
77. Hirsch, L.; Barrieere, A.S.; A. Electrical characterization of InGaN/GaN light emitting diodes grown by molecular beam epitaxy. *J. Appl. Phys.* **2003**, *94*, 5014. [CrossRef]
78. Kim, J.; Kim, J.; Tak, Y.; Chae, S.; Kim, J.-Y.; Park, Y. Effect of V-Shaped Pit Size on the Reverse Leakage Current of InGaN/GaN Light-Emitting Diodes. *IEEE Electron Device Lett.* **2013**, *34*, 1409–1411. [CrossRef]
79. Zhang, Y.; Sun, M.; Piedra, D.; Azize, M.; Zhang, X.; Fujishima, T.; Palacios, T. GaN-on-Si Vertical Schottky and p-n Diodes. *IEEE Electron Device Lett.* **2014**, *35*, 618–620. [CrossRef]
80. Demchenko, D.O.; Diallo, I.C.; Reshchikov, M.A. Yellow Luminescence of Gallium Nitride Generated by Carbon Defect Complexes. *Phys. Rev. Lett.* **2013**, *110*, 087404. [CrossRef]
81. Huber, M.; Silvestri, M.; Knuuttila, L.; Pozzovivo, G.; Andreev, A.; Kadashchuk, A.; Bonanni, A.; Lundskog, A. Impact of residual carbon impurities and gallium vacancies on trapping effects in AlGaN/GaN metal insulator semiconductor high electron mobility transistors. *Appl. Phys. Lett.* **2015**, *107*, 032106. [CrossRef]
82. Karch, J. Improving on Adjusted R-Squared. *Collabra Psychol.* **2020**, *6*. [CrossRef]
83. Cheng, K.; Liang, H.; Van Hove, M.; Geens, K.; De Jaeger, B.; Srivastava, P.; Kang, X.; Favia, P.; Bender, H.; Decoutere, S.; et al. AlGaN/GaN/AlGaN Double Heterostructures Grown on 200 mm Silicon (111) Substrates with High Electron Mobility. *Appl. Phys. Express* **2011**, *5*, 011002. [CrossRef]
84. Gao, J.; Hao, M.; Li, W.; Xu, Z.; Mandal, S.; Nemanich, R.; Chowdhury, S. Al2 O3 Insertion Layer for Improved PEALD SiO2 /(Al)GaN Interfaces. *Phys. Status Solidi (A)* **2018**, *215*. [CrossRef]
85. Posthuma, N.E.; You, S.; Stoffels, S.; Liang, H.; Zhao, M.; Decoutere, S. Gate Architecture Design for Enhancement Mode P-GaN Gate HEMTs for 200 and 650V Applications. In Proceedings of the 2018 IEEE 30th International Symposium on Power Semiconductor Devices and ICs (ISPSD), Chicago, IL, USA, 13–17 May 2018; pp. 188–191.
86. Robertson, J.; Falabretti, B. Band offsets of high K gate oxides on III-V semiconductors. *J. Appl. Phys.* **2006**, *100*, 014111. [CrossRef]
87. Robertson, J. High dielectric constant oxides. *Eur. Phys. J. Appl. Phys.* **2004**, *28*, 265–291. [CrossRef]
88. Yatabe, Z.; Hori, Y.; Ma, W.-C.; Asubar, J.T.; Akazawa, M.; Sato, T.; Hashizume, T. Characterization of electronic states at insulator/(Al)GaN interfaces for improved insulated gate and surface passivation structures of GaN-based transistors. *Jpn. J. Appl. Phys.* **2014**, *53*. [CrossRef]
89. Ruzzarin, M.; Meneghini, M.; Bisi, D.; Sun, M.; Palacios, T.; Meneghesso, G.; Zanoni, E. Instability of Dynamic- and Threshold Voltage in GaN-on-GaN Vertical Field-Effect Transistors. *IEEE Trans. Electron Devices* **2017**, *64*, 3126–3131. [CrossRef]
90. Ruzzarin, M.; Meneghini, M.; De Santi, C.; Meneghesso, G.; Zanoni, E.; Sun, M.; Palacios, T. Degradation of vertical GaN FETs under gate and drain stress. In Proceedings of the 2018 IEEE International Reliability Physics Symposium (IRPS), Burlingame, CA, USA, 11–15 March 2018.
91. Hori, Y.; Mizue, C.; Hashizume, T. Process Conditions for Improvement of Electrical Properties of Al_2O_3/n-GaN Structures Prepared by Atomic Layer Deposition. *Jpn. J. Appl. Phys.* **2010**, *49*, 080201. [CrossRef]
92. Kim, S.; Hori, Y.; Ma, W.-C.; Kikuta, D.; Narita, T.; Iguchi, H.; Uesugi, T.; Kachi, T.; Hashizume, T. Interface Properties of Al_2O_3/n-GaN Structures with Inductively Coupled Plasma Etching of GaN Surfaces. *Jpn. J. Appl. Phys.* **2012**, *51*, 60201. [CrossRef]
93. Saxena, R.S.; Kumar, M.J. Trench Gate Power MOSFET: Recent Advances and Innovations. *arXiv* **2012**, arXiv:1208.5553.

94. Zhang, Y.; Sun, M.; Liu, Z.; Piedra, D.; Hu, J.; Gao, X.; Palacios, T. Trench formation and corner rounding in vertical GaN power devices. *Appl. Phys. Lett.* **2017**, *110*, 193506. [CrossRef]
95. Osipov, K.; John, W.; Kemf, N.; Chevtchenko, S.; Kurpas, P.; Matalla, M.; Krüger, O.; Wuerfl, J. Fabrication Technology of GaN/AlGaN HEMT Slanted Sidewall Gates Using Thermally Reflowed ZEP Resist and CHF3/SF6 Plasma Etching. In Proceedings of the International Conference on Compound Semiconductor Manufacturing Technology, CS MANTECH, New Orleans, LA, USA, 13–16 May 2013; pp. 139–142.
96. Thies, A.; Kemf, N.; Chevtchenko, S.; Krüger, O. Formation of Slanted Gates for Gan-Based Hemts by Combined Plasma and Wet Chemical Etching of Silicon Nitride. In Proceedings of the International Conference on Compound Semiconductor Manufacturing Technology, CS MANTECH, New Orleans, LA, USA, 13–16 May 2013; pp. 395–398.
97. Gupta, C.; Ji, D.; Chan, S.H.; Agarwal, A.; Leach, W.; Keller, S.; Chowdhury, S.; Mishra, U.K. Impact of Trench Dimensions on the Device Performance of GaN Vertical Trench MOSFETs. *IEEE Electron Device Lett.* **2017**, *38*, 1559–1562. [CrossRef]
98. Zeng, J.; Mawby, P.; Towers, M.; Board, K. Modelling of the quasisaturation behaviour in the high-voltage MOSFET with vertical trench gate. *IEE Proc. Circuits Devices Syst.* **1996**, *143*, 28. [CrossRef]
99. Shimizu, R.; Kuribayashi, H.; Hiruta, R.; Sudoh, K.; Iwasaki, H. Mechanism and Control Technology of Trench Corner Rounding by Hydrogen Annealing for Highly Reliable Trench MOSFET. In Proceedings of the 18th International Symposium on Power Semiconductor Devices & IC's, Naples, Italy, 4–8 June 2006.
100. Simon-Najasek, M.; Huebner, S.; Altmann, F.; Graff, A. Advanced FIB sample preparation techniques for high resolution TEM investigations of HEMT structures. *Microelectron. Reliab.* **2014**, *54*, 1785–1789. [CrossRef]
101. Graff, A.; Simon-Najasek, M.; Poppitz, D.; Altmann, F. Physical failure analysis methods for wide band gap semiconductor devices. In Proceedings of the 2018 IEEE International Reliability Physics Symposium (IRPS), Burlingame, CA, USA, 11–15 March 2018.
102. Diehle, P.; Hubner, S.; De Santi, C.; Mukherjee, K.; Zanoni, E.; Meneghini, M.; Geens, K.; You, S.; Decoutere, S.; Altmann, F. Root cause analysis of gate shorts in semi-vertical GaN MOSFET devices. In Proceedings of the 2020 13th International Conference on Advanced Semiconductor Devices and Microsystems (ASDAM), Smolenice, Slovakia, 11–14 October 2020; pp. 10–13.
103. Bisi, D.; Stocco, A.; Meneghini, M.; Rampazzo, F.; Cester, A.; Meneghesso, G.; Zanoni, E. High-Voltage Double-Pulsed Measurement System for GaN-Based Power HEMTs. In Proceedings of the 2014 IEEE International Reliability Physics Symposium, Waikoloa, HI, USA, 1–5 June 2014; pp. CD.11.1–CD.11.4.
104. Lagger, P.; Reiner, M.; Pogany, D.; Ostermaier, C. Comprehensive Study of the Complex Dynamics of Forward Bias-Induced Threshold Voltage Drifts in GaN Based MIS-HEMTs by Stress/Recovery Experiments. *IEEE Trans. Electron Devices* **2014**, *61*, 1022–1030. [CrossRef]
105. Stockman, A.; Canato, E.; Tajalli, A.; Meneghini, M.; Meneghesso, G.; Zanoni, E.; Moens, P.; Bakeroot, B. On the origin of the leakage current in p-gate AlGaN/GaN HEMTs. In Proceedings of the 2018 IEEE International Reliability Physics Symposium (IRPS), Burlingame, CA, USA, 11–15 March 2018. [CrossRef]
106. Ren, B.; Sumiya, M.; Liao, M.; Koide, Y.; Liu, X.; Shen, Y.; Sang, L. Interface trap characterization of Al_2O_3/GaN vertical-type MOS capacitors on GaN substrate with surface treatments. *J. Alloys Compd.* **2018**, *767*, 600–605. [CrossRef]
107. Kaneki, S.; Ohira, J.; Toiya, S.; Yatabe, Z.; Asubar, J.T.; Hashizume, T. Highly-stable and low-state-density Al_2O_3/GaN interfaces using epitaxial n-GaN layers grown on free-standing GaN substrates. *Appl. Phys. Lett.* **2016**, *109*, 162104. [CrossRef]
108. Rosenkranz, R. Failure localization with active and passive voltage contrast in FIB and SEM. *J. Mater. Sci. Mater. Electron.* **2011**, *22*, 1523–1535. [CrossRef]
109. Simon-Najasek, M.; Jatzkowski, J.; Große, C.; Altmann, F. A New Technique for Non-Invasive Short-Localisation in Thin Dielectric Layers by Electron Beam Absorbed Current (EBAC) Imaging. In Proceedings of the 38th International Symposium for Testing and Failure Analysis, Pheonix, AZ, USA, 11–15 November 2012; pp. 61–66. [CrossRef]
110. Bisi, D.; Chan, S.H.; Liu, X.; Yeluri, R.; Keller, S.; Meneghini, M.; Meneghesso, G.; Zanoni, E.; Mishra, U.K. On trapping mechanisms at oxide-traps in Al_2O_3/GaN metal-oxide-semiconductor capacitors. *Appl. Phys. Lett.* **2016**, *108*, 112104. [CrossRef]
111. Lagger, P.; Steinschifter, P.; Reiner, M.; Stadtmüller, M.; Denifl, G.; Naumann, A.; Muller, J.; Wilde, L.; Sundqvist, J.; Pogany, D.; et al. Role of the dielectric for the charging dynamics of the dielectric/barrier interface in AlGaN/GaN based metal-insulator-semiconductor structures under forward gate bias stress. *Appl. Phys. Lett.* **2014**, *105*, 033512. [CrossRef]
112. Sato, M.; Kamiyama, S.; Matsuki, T.; Ishikawa, D.; Ono, T.; Morooka, T.; Yugami, J.; Ikeda, K.; Ohji, Y. Study of a Negative Threshold Voltage Shift in Positive Bias Temperature Instability and a Positive Threshold Voltage Shift the Negative Bias Temperature Instability of Yttrium-Doped HfO2Gate Dielectrics. *Jpn. J. Appl. Phys.* **2010**, *49*, 04DC24. [CrossRef]
113. Choi, M.; Janotti, A.; Van De Walle, C.G. Native point defects and dangling bonds in α-Al_2O_3. *J. Appl. Phys.* **2013**, *113*, 044501. [CrossRef]
114. Swenson, B.L.; Mishra, U.K. Photoassisted high-frequency capacitance-voltage characterization of the Si3N4/GaN interface. *J. Appl. Phys.* **2009**, *106*, 064902. [CrossRef]

materials

Review

Diamond/GaN HEMTs: Where from and Where to?

Joana C. Mendes [1,*], Michael Liehr [2] and Changhui Li [2]

[1] Instituto de Telecomunicações e Departamento de Eletrónica, Telecomunicações e Informática, Universidade de Aveiro, 3810-193 Aveiro, Portugal
[2] W&L Coating Systems GmbH, Bingenheimer Str. 32, D-61203 Reichelsheim, Germany; m.liehr@wl-coating.com (M.L.); c.li@wl-coating.com (C.L.)
* Correspondence: joanacatarina.mendes@ua.pt

Abstract: Gallium nitride is a wide bandgap semiconductor material with high electric field strength and electron mobility that translate in a tremendous potential for radio-frequency communications and renewable energy generation, amongst other areas. However, due to the particular architecture of GaN high electron mobility transistors, the relatively low thermal conductivity of the material induces the appearance of localized hotspots that degrade the devices performance and compromise their long term reliability. On the search of effective thermal management solutions, the integration of GaN and synthetic diamond with high thermal conductivity and electric breakdown strength shows a tremendous potential. A significant effort has been made in the past few years by both academic and industrial players in the search of a technological process that allows the integration of both materials and the fabrication of high performance and high reliability hybrid devices. Different approaches have been proposed, such as the development of diamond/GaN wafers for further device fabrication or the capping of passivated GaN devices with diamond films. This paper describes in detail the potential and technical challenges of each approach and presents and discusses their advantages and disadvantages.

Keywords: diamond; GaN; HEMT; thermal management; GaN-on-diamond; CVD

check for updates

Citation: Mendes, J.C.; Liehr, M.; Li, C. Diamond/GaN HEMTs: Where from and Where to? *Materials* **2022**, *15*, 415. https://doi.org/10.3390/ma15020415

Academic Editor: Fabrizio Roccaforte

Received: 8 October 2021
Accepted: 30 December 2021
Published: 6 January 2022

1. Introduction

Gallium nitride (GaN) is a wide bandgap compound III-V semiconductor with high breakdown electric field, high electron mobility, and high electron saturation velocity that translate in a tremendous potential for high power and high frequency applications (Table 1). The GaN high electron mobility transistor (HEMT) is a device that takes advantage of the two-dimensional electron gas (2DEG) that spontaneously forms at an aluminum gallium nitride (AlGaN)/GaN heterojunction thanks to the strong internal piezoelectric and spontaneous polarization. This 2DEG typically exhibits high values of sheet carrier density ($\approx 10^{13}$ cm^{-2}) and carrier mobility (1000–2000 cm^2/(V·s)) [1].

GaN HEMTs can be used in power switching for information technologies, automotive, healthcare, and industrial manufacturing applications [1–3]. Thanks to the large bandgap, leakage currents in GaN power devices are orders of magnitude smaller than in silicon (Si), allowing for operation at higher temperature without thermal runaway and reducing the cooling requirements. The high breakdown electric field allows shorter drift distances for a given blocking voltage, when compared to Si devices, yielding to a drastic reduction in the specific on-resistance that in turn translates into smaller device area and correspondingly lower capacitance. This reduces switching losses and enables higher switching frequencies.

GaN HEMTS have also paved their way into mobile and satellite communications and radar systems [4]. In addition to the properties listed above, the high breakdown electric field of GaN allows higher matching impedances and circuits with broader bandwidth and high power-added efficiency (*PAE*) [5]. The ability of GaN to withstand higher temperatures further increases the power density of a given HEMT device and power amplifiers with absolute power levels of tens to hundreds of Watts have been reported [6].

Table 1. Selected properties of relevant semiconductors [1,6–13].

Material Property	Si	SiC	GaN	Diamond
Bandgap (eV)	1.12	2.9 (6H)/ 3.2 (4H)	3.4	5.47
Breakdown field ($\times 10^6$ V/cm)	0.25	2.5 (6H)/ 3 (4H)	3–3.75	20
Electron mobility ($cm^2/(V{\cdot}s)$)	1350	415 (6H)/ 950 (4H)	1100–1300	2400
Thermal conductivity ($W/(m{\cdot}K)$)	150	380–450	130–210	2200 (single crystal) >1800 (polycrystalline)
Coefficient of thermal expansion ($\times 10^{-6}$ K^{-1})	2.6	3.08	5.6 (a_0)/ 3.2 (c_0)	0.8

Despite the fact that commercial HEMT devices based on a different combination of processes and design technologies are currently available from a broad range of manufacturers [4], GaN-based technologies still face some challenges that affect their overall performance and limit their potential benefits.

1. Due to the intrinsic nature of the 2DEG, GaN HEMTs are normally-on (depletion mode) devices. For power switching applications, normally-off devices are preferred due to static power consumption, simplification of circuit design, and safety concerns. Normally-off transistors can be obtained with different techniques [14], however their performance is typically worse than that of their normally-on counterparts [15,16].
2. The existence of electrically active surface traps located at the passivation/top-layer interface [17] and of bulk traps present in the GaN and buffer layers [18] induces effects such as current collapse [19–21], dynamic on-resistance (or knee walkout) [22], degradation of cut-off frequency [23], and DC-RF dispersion [24], which compromise the reliability of the devices and prevent harnessing the full potential of GaN HEMT power devices [25].
3. Due to the intrinsic nature of GaN HEMTs, harsh and localized self-heating in the conducting channel may occur [26]; this effect increases with the device power density and further compromises reliability [22,27]. On one side, the electrical behavior of the traps mentioned in the above paragraph is temperature-dependent [25]. On the other side, additional phonon scattering in the channel degrades the 2DEG effective carrier mobility, leading to degraded DC and RF performance [28]. Finally, since the relation between the mean time-to-failure (MTTF) of an electronic component and its operating temperature is semi-exponential [29], even a small temperature reduction can have a great impact on the lifetime of HEMTs with thermally-activated degradation mechanisms [30].

From what was described above, it can be concluded that the capability of efficiently transferring the heat away from the localized hotspots and the consequent control of the device temperature is fundamental to achieve high levels of stability and reliability in HEMT applications [31]. Diamond has the highest thermal conductivity (κ) of any bulk material, and the integration of diamond films and GaN HEMTs as substrates or packaging has already proven to enhance the extraction of the heat generated during the devices operation, leading to a substantial decrease in the junction temperature as well as to an increase in the maximum power density the HEMTs can safely handle. This anticipates a superior high-frequency handling capacity, higher energy efficiency and flexibility, and a better utilization of the electromagnetic spectrum.

Diamond has been successfully integrated with HEMT devices following different approaches. This manuscript aims at describing each of them in detail, pointing out the technical challenges and benefits, and providing the reader with a critical discussion of the feasibility of each approach. The manuscript is organized as follows: Section 2 discusses the critical aspects that impact the thermal management of GaN HEMTs; Section 3 describes the different strategies that have been followed by different groups to integrate diamond and GaN into high performance devices; Section 4 describes the challenges faced by each integration technology and discusses their feasibility; Section 5 draws the main conclusions and Appendix A summarizes the performance of the different GaN/diamond HEMTs reported so far.

2. Thermal Management of GaN HEMTs

2.1. GaN HEMT

The concept of mobility enhancement through modulation doping of an aluminum gallium arsenide (AlGaAs)/gallium arsenide (GaAs) multilayer heterojunction was introduced by Dingle et al. in 1978 [32]. Since the energy of the GaAs conduction band is lower than the energy of the AlGaAs donor states, electrons from the later move into the GaAs regions, forming a 2DEG. By introducing a rectifying contact on top of the heterojunction, Mimura et al. [33] were able to control, by field effect, the concentration of the 2DEG. Soon after the deposition of high quality GaN films on sapphire substrates by Metalorganic Chemical Vapor Deposition (MOCVD) [34] and using the same principle, the AlGaN/GaN HEMT was reported by Khan and co-workers [35]. In a conventional AlGaN/GaN HEMT, the current flowing in the 2DEG channel between the source and drain Ohmic electrodes is modulated by a negative bias applied to the gate Schottky contact.

The cross-section view of a general GaN HEMT structure is shown in Figure 1 and the κ of the materials typically used are listed in Table 2. The actual structure, composition, and thickness of each layer in a particular HEMT depend on its specific purpose and/or the vendor's manufacturing practices. Since the review and discussion of the literature presenting different devices modifications is out of the scope of this work, only the layers considered relevant to the thermal transport are represented in Figure 1. Other possible layers (not shown) would include the spacer and cap layers.

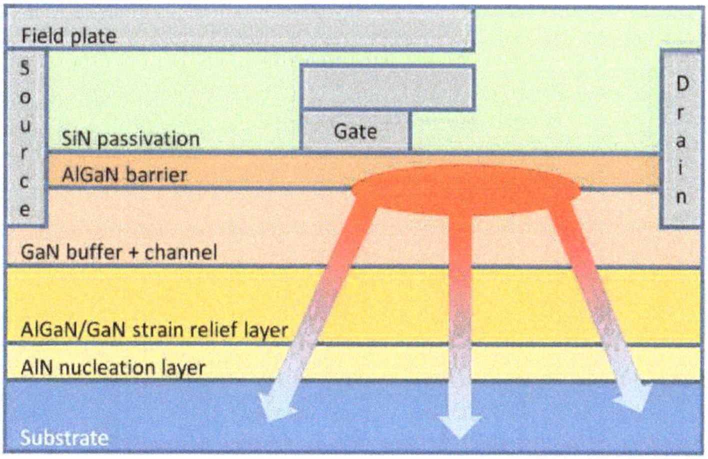

Figure 1. Simplified representation of an AlGaN/ GaN HEMT structure; drawing not to scale.

1. Substrate. Homoepitaxial growth of GaN-based films is hampered by the limited availability of GaN substrates in standard wafer sizes. As a consequence, the different layers are typically deposited by either Molecular Beam Epitaxy (MBE) or MOCVD onto sapphire, Si, or silicon carbide (SiC) substrates. Epitaxial films with dislocation densities of 10^8 cm^{-2} are typically obtained [36]; dislocation densities lower than 10^7 cm^{-2} involve hydride vapor phase epitaxy (HVPE) growth.

2. Nucleation layer. The deposition of high quality epitaxial GaN films with smooth surfaces and low dislocation density is not a straightforward task due to lattice mismatch and to the difference in the coefficients of thermal expansion (*CTE*s) of GaN and substrate [37]. A nucleation layer, typically 40–200 nm of aluminum nitride (AlN) [38], is thus initially deposited on the substrate surface for strain accommodation and increased interface resistivity [39,40].
3. Strain relief layer. AlGaN/GaN transition layers, often up to 1 μm thickness, further accommodate the lattice mismatch during the growth of GaN on the foreign substrate [41].
4. GaN layer. A 0.6–1.5 μm-thick GaN buffer layer that provides electrical isolation to reduce substrate leakage and prevents the propagation of threading dislocations and contaminants that might migrate from the substrate into the top high quality channel region follows.
5. Barrier layer. The typically 5–25 nm-thick barrier layer can be made of AlN [42], indium gallium nitride (InGaN) [43], or other high bandgap alloys, but $Al_xGa_{1-x}N$ with an aluminum (Al) fractional content in the range of 22–32% [6] is the most widely reported material.
6. Passivation layer. A thin dielectric layer, typically silicon nitride (Si_3N_4), compensates the surface/interface states responsible for the current collapse issue by introducing shallow donors [44].
7. Field plate. The source and gate-connected field plates are usually employed to reduce the strength of the electric field near the gate terminal, reducing the gate tunneling injection current responsible for charging the surface traps [45].

Table 2. Typical thermal conductivity of the materials present in an AlGaN/GaN HEMT [1,7,9–11,46–48].

Material	Thermal Conductivity (W/(m·K))
Sapphire	46
Si	150
SiC	380–450
Single-crystal diamond	2200
Polycrystalline diamond	>1800
GaN	130–210
AlN	130
$Al_xGa_{1-x}N$	10–11
$In_xGa_{1-x}N$	5–11
SiN	1.6

2.2. Getting the Heat Out

Thanks to the high breakdown voltage and saturation velocity of carriers in the 2DEG, AlGaN/GaN HEMTs are able to handle substantial power densities, which result in self-heating and highly-localized power densities (in some cases as high as 10^5 W/cm^2) [49]. The resulting high temperatures decrease the MTTF, the performance, and the reliability of HEMT devices—and this makes the implementation of efficient thermal management techniques mandatory. This is particularly important for devices that have inherently small features yet process large power densities, such as high-power RF/millimeter-wave transistors and single-mode visible semiconductor lasers.

Most of the heat generated during device operation will diffuse from the hotspot through the different layers represented in Figure 1 until it reaches the heat sink attached to the back of the substrate. A few intrinsic obstacles hinder the transfer of the heat towards the heat sink. The AlGaN/GaN strain relief transition layers improve the electrical performance at the top of the GaN layer, however, if the concentration of Al is higher than 5%, the κ of the AlGaN decreases to around 1/10th that of bulk GaN, hampering the transfer of heat from the GaN buffer layer to the underlying substrate [41].

The thermal transport across the strain-relief layer/nucleation layer/substrate interfaces also plays an important role in determining the overall κ of the GaN HEMT material system. The existence of an interface between two solids results in the scattering of thermal energy carriers (electrons and phonons), which translates in the appearance of a thermal boundary resistance (*TBR*) between the different materials and in a temperature discontinuity across the interface [50,51].

Despite its low thickness in comparison to the GaN layer, the nucleation layer profoundly impacts the transport of heat from the strain relief layer towards the underlying substrate. Instead of being composed of high quality AlN, this layer contains dislocations, grain boundaries, and point defects (impurity atoms and vacancies) [52,53], within the itself or near the interfaces, that hinder heat transport by increasing phonon scattering rates and reducing the phonon mean free path [54].

The interfacial and nucleation layer thermal resistances contribute with an effective *TBR* between the strain relief layers and the substrate between 10 and 70 $m^2 \cdot K/GW$ [38] that may cause an additional 30–50% channel temperature rise in AlGaN/GaN HEMTs [38,55].

2.3. Why GaN-Diamond HEMTs?

In the search of good thermal conductors, carbon-based materials, such as highly oriented pyrolytic graphite (HOPG), graphene, and diamond, are obvious candidates.

HOPG is an anisotropic material with high in-plane κ ($\kappa_{in\text{-}plane} \approx 2000$ W/(m·K)) but a much lower out-of-plane κ ($\kappa_{out\text{-}of\text{-}plane} \approx 6$–9 W/(m·K)) [56]. Despite being also an electrical conductor, graphite already found its place for thermal management of electronic components at system level, and high κ graphite films for assembling integrated circuits or CPUs, for example, are available from a few vendors. Other examples include graphite heat sinks [57] and composite graphite/metal laminates [58].

Graphene is another anisotropic material with even higher in-plane κ ($\kappa_{in\text{-}plane} > 3000$ W/(m·K)) [59]. However, the particular value of κ depends significantly on the preparation method and can be reduced greatly up to one order of magnitude compared to that of pristine graphene because of poor alignment and structural defects in the material [60]. Due to its intrinsic 2D nature, electrically conductive graphene films are more suited for integration at device level. In 2012 Yan et al. [61] reported the use of graphene quilts for the thermal management of GaN HEMTs, obtaining a ≈20 °C decrease in the hot spot temperature in transistors operating at ≈13 W/mm.

Diamond is an isotropic material with high κ (2200 W/(m·K), increasing to 3300 W/(m·K) in the case of isotopically pure material) while being electrically insulating, with a breakdown field 60 times greater that of Si (2×10^7 V/cm [13])—Table 1. Single-crystal diamond (SCD), grown by high pressure high temperature (HPHT) method, has the best thermal and electric properties; however its area is limited to a few mm^2. Alternatively, polycrystalline diamond (PCD) films can be grown by chemical vapor deposition (CVD) on large-area substrates such as Si, overcoming the area limitation while still guaranteeing κ values in the range 1000–1800 W/(m·K) [62].

3. Integration of Diamond and GaN

The integration of diamond and GaN for the fabrication of HEMTs with superior thermal handling capability has been an active research topic involving academic and industrial institutions for more than 20 years. Generally speaking, the integration of both materials can be done in two ways: replacing the GaN substrate with diamond or placing the diamond on top of the device, close to the gate hotspot.

The fabrication of GaN-on-diamond wafers can be made using three fundamentally different approaches: (i) depositing diamond films by CVD directly on the back of GaN wafers, following the substrate removal (hereafter referred to as GaN-on-diamond); (ii) bonding

GaN HEMT wafers and diamond substrates (bonded wafers); and (iii) growing the epitaxial GaN layers directly on diamond substrates (GaN epitaxy). Placing the diamond on top of the HEMT device can be done simply by growing the diamond films directly on the passivated surface of the device (capping diamond).

The historical evolution of each approach is presented in detail in the following paragraphs.

3.1. GaN-on-Diamond

The GaN-on-diamond concept was initially introduced in 2003 [49]. The original idea relied on the deposition of the PCD film directly on the back of the GaN layer. The first GaN-on-diamond wafer pieces were produced in 2004 by etching the Si substrate of a GaN wafer initially deposited by MOCVD, followed by depositing a 50 nm-thick dielectric layer, and then by growing a 25 µm-thick PCD film by hot filament CVD (HFCVD). The PCD was deposited on the Ga-face, leaving behind an N-face GaN-on-diamond wafer after the etching of the temporary Si carrier. The fabrication process remains fundamentally the same till today and is represented in Figure 2. The required steps involved (i) performing the GaN epitaxy on a Si substrate, (ii) bonding the GaN HEMT epilayers onto a temporary Si carrier, (iii) etching away the original host Si substrate, (iv) depositing a 50 nm-thick layer of SiN onto the exposed rear face of the GaN, and (v) depositing a 25 µm-thick PCD film by HFCVD onto the dielectric. Finally, by (vi) removing the temporary Si carrier, a free-standing GaN-on-diamond wafer was obtained.

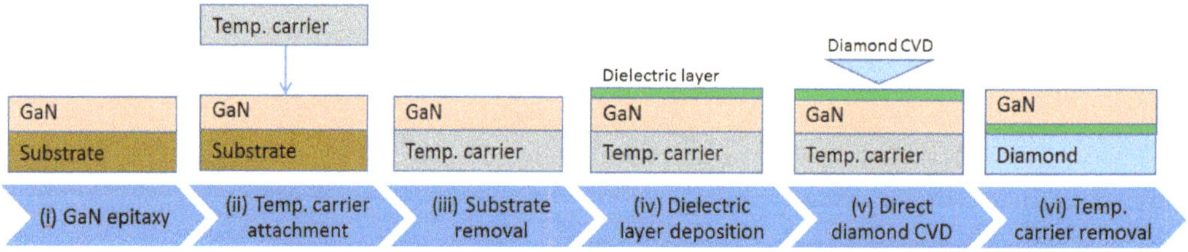

Figure 2. GaN-on-diamond fabrication steps.

By 2006, the process had been optimized to fabricate a Ga-face (i.e., right side up) GaN-on-Diamond HEMT epitaxial wafer and the operation of an AlGaN/GaN HEMT with a 25 µm-thick PCD film located 1.2 µm below the electron channel was reported [63]. The scanning electron microscope (SEM) cross-section of the GaN-on-diamond wafer is shown in Figure 3a. These first unpassivated HEMTs had a high contact resistance that translated in low current capability (maximum drain current $I_{D\,max}$ = 306 mA/mm) and low peak transconductance ($g_{m\,peak}$ = 70 mS/mm). Further iterations improved the performance of the HEMTs [64,65], nevertheless, and in spite of having half the thermal resistance R_{th} of GaN-on-SiC HEMTs (≈6 against ≈12 K·mm/W, respectively), in 2007 GaN-on-diamond HEMTs were still outperformed by GaN-on-SiC technology—Figure 3b. By 2009 GaN HEMTs with 75 µm of PCD showed cut-off (f_T) and maximum oscillation (f_{max}) frequencies of 85 and 95 GHz, respectively [66], and soon after that the first RF power amplifier module [67] was reported.

In 2011 DARPA introduced the Near Junction Thermal Transport (NJTT) concept that aimed at extracting the heat from within 1 µm of a transistor's active region. The low κ AlN/AlGaN nucleation/transition layers under the GaN channel were eliminated [41,68,69], allowing the deposition of the films within a few hundreds of nanometers away from the hot junction, and the thermal conductance between the GaN and the diamond was more than doubled. Continuous wave (CW) load-pull tests performed at 10 GHz showed that the GaN-on-diamond HEMTs reached over 7 W/mm output power density (P_D) and 46% *PAE* at 40 V drain bias (V_D) [70]; however, and despite showing a 25% lower temperature rise for the same thermal power than their GaN-on-SiC counterparts, these devices had a high gate leakage current associated with residual surface defects in the gate region which still prevented the harnessing of the full potential of the technology.

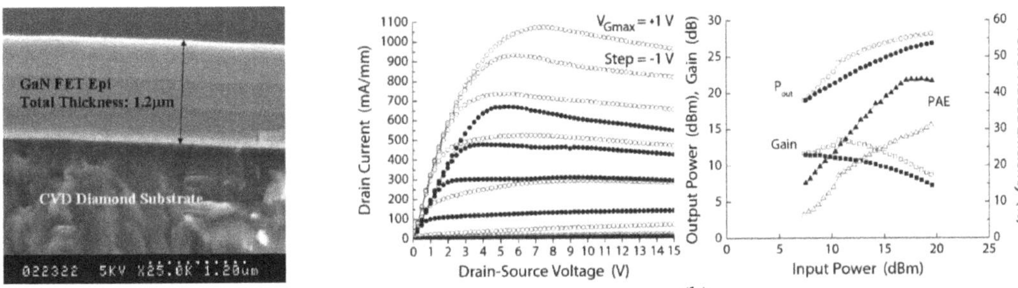

(**a**) (**b**)

Figure 3. (**a**) SEM cross-section of a GaN-on-diamond wafer (© 2006 IEEE. Reprinted, with permission, from G. H. Jessen et al., "AlGaN/GaN HEMT on diamond technology demonstration," Tech. Dig.—IEEE Compd. Semicond. Integr. Circuit Symp. CSIC, pp. 271–274, 2006 [63]). (**b**) Electrical characteristics of similar GaN-on-diamond (solid symbols) and GaN-on-SiC (open symbols) HEMTs; output power and *PAE* measured at 10 GHz CW class B operation and 20 V V_D (© 2007 IEEE. Reprinted, with permission, from J. G. Felbinger et al., "Comparison of GaN HEMTs on diamond and SiC substrates," IEEE Electron Device Lett. [65]).

The κ of the diamond films was further improved by replacing the HFCVD process with microwave plasma CVD (MPCVD) and by the end of the NJTT program, in 2014, the GaN-on-diamond technology had allowed the reduction of the junction temperature by 40–45% and the tripling of the areal RF power density in comparison with GaN-on-SiC [71].

At the same time the bottleneck of the heat extraction was recognized to be the *TBR* between GaN and diamond ($TBR_{GaN/diamond}$) [72] and most of following research focused on decreasing it, whether by decreasing the dielectric thickness, by using a different dielectric, or by optimizing the diamond nucleation layer [10,72–84]. The impact of the thickness of the GaN buffer layer on the R_{th} of the HEMT devices [85–90] and the effects of the stress caused by the difference in the *CTE*s of GaN and diamond [91–96] were also evaluated by different research groups. A more detailed description and discussion of the main findings is included in Section 4.1. The mechanical and thermo-mechanical integrity of the diamond/GaN interface, which impacts profoundly the reliability of the devices, was also addressed [97–99]. As a general finding, Liu et al. concluded that the GaN/diamond interface has a high mechanical stability, showing the potential of this material system for the fabrication of reliable devices [97].

The performance of more recent HEMT devices was thoroughly analyzed by Ranjan and co-workers [28,100], who evaluated the effect of the bias conditions on the self-heating and transport properties of GaN-on-Si and GaN-on-diamond (with 30 nm of SiN) HEMTs. A ≈4 times improvement in the DC and RF performances of the later was observed. The DC P_D of GaN-on-diamond HEMTs was 27.56 W/mm for 55 V applied V_D, whereas GaN-on-Si devices were burnt at ≈9 W/mm for 20 V V_D. The reduction of $I_{D\,max}$ due to channel self-heating for GaN-on-diamond and GaN-on-Si HEMTs was 10% and 33%, and for V_D = 10 V f_T/f_{max} were 10.2/31.4 GHz and 7/18.2 GHz, respectively. The GaN-on-diamond HEMT had an almost constant small signal gain for V_D between 10 and 40 V. Finally, the increase rate of the gate current with V_D was 3.3 times smaller for GaN-on-diamond devices. Figure 4a–c show some of the devices' electrical characteristics.

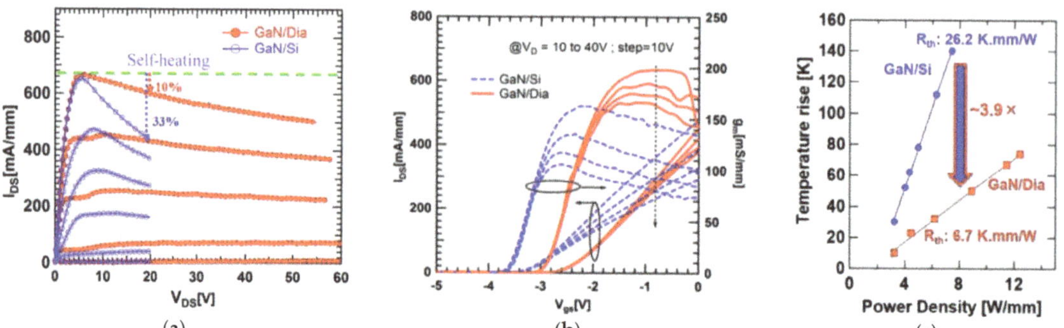

(a) (b) (c)

Figure 4. (**a**) I_D versus V_D characteristics, (**b**) Transfer characteristics for different V_D, and (**c**) Temperature rise versus P_D for similar GaN-on-diamond and GaN-on-Si HEMTs (reprinted from [28]; permission conveyed through CCBY 4.0: https://creativecommons.org/licenses/by/4.0/~ (accessed on 7 October 2021)).

In 2019, HEMTs fabricated on latest generation Element Six GaN-on-diamond wafers (with a 30 nm-thick dielectric layer) showed 2.95 K·mm/W R_{th}, 56 W/mm DC power capability, and average/maximum channel temperature of 176/205 °C [101]. Nevertheless, these devices had high leakage currents which ultimately limited their breakdown voltage, showing there is still some room for optimizing the fabrication process and maximizing the performance of GaN-on-diamond HEMTs. In the same year, researchers from RFHIC [102] reported the fabrication of 4" GaN-on-diamond wafers with a $TBR_{GaN/diamond}$ of 31.0 m^2·K/GW and an uniformity of ±10%. The development of an inner slot via hole process allowed the opening of 10 μm-diameter holes in the diamond using a laser drilling process—Figure 5. On-wafer pulsed load-pull tests performed at 2 GHz revealed 18.1 W/mm P_D for an encapsulated 10 × 200 μm gate HEMT.

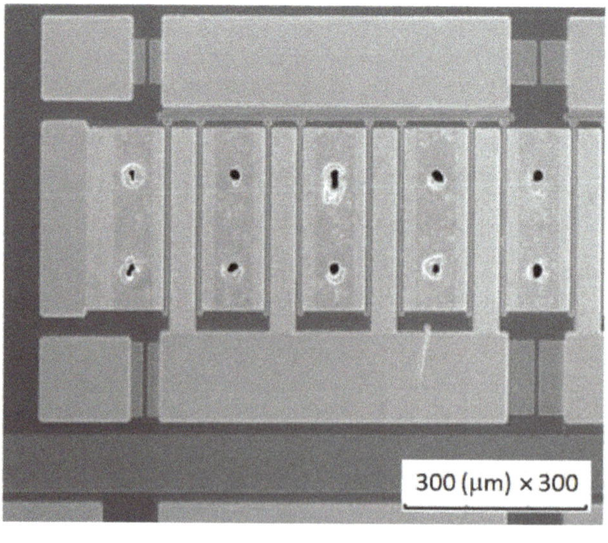

Figure 5. Inner slot via hole shape on source pad (© 2019 IEEE. Reprinted, with permission, from W. S. Lee, et al., "A GaN/Diamond HEMTs with 23 W/mm for Next Generation High Power RF Application," in IEEE MTT-S International Microwave Symposium Digest, 2019, vol. 2019-June, pp. 1395–1398 [102]).

The evolution of the relevant technological parameters (dielectric material and thickness, diamond film κ, and value of $TBR_{GaN/diamond}$) since 2006 is listed in Table 3. It can be seen that most of the GaN-on-diamond HEMTs fabricated by this method feature a ≈30 nm-thick SiN layer, and a $TBR_{GaN/diamond}$ in the range of 20–30 m^2·K/GW. Section 4.1 describes in more detail the impact of the dielectric layer on the experimental $TBR_{GaN/diamond}$ values and the electrical parameters of GaN-on-diamond HEMTs fabricated so far using this method are summarized in Table A1 in the Appendix A.

Table 3. Evolution of GaN-on-diamond technology.

Ref.	Year	Dielectric Layer		Diamond Film		$TBR_{GaN/diamond}$ (m^2·K/GW)
		Thickness (nm)	Material	CVD Type/Thickness (μm)	κ (W/(m·K))	
[63]	2006	–	–	HFCVD/25	–	–
[66]	2009	–	–	75	–	–
[67]	2010	–	–	MPCVD/100	>1500	–
[69]	2013	40	–	30	–	36 ± 12 [b]
[70]	2013	50	–	HFCVD/100	–	18 [a]
[103]	2014	30	–	100	–	29 ± 2 [b]
[72]	2014	25	–	HFCVD/95	710	27 ± 3 [a]
		50		MPCVD/120	1200	36 [a]
[77]	2014	50	–	MPCVD/110	1200	17 [b]
		90				41 [b]
[104]	2014	–	–	MPCVD	1600	19 ± 3 [b]
[105]	2014	–	–	100	–	47.6 [b]
						19 [b]
[78]	2015	34	SiN	HFCVD/100	650	25 ± 3 [b]
		100		MPCVD/100	1500	50 ± 5 [b]
		28		MPCVD/100	1500	12 ± 2 [b]
[79]	2016	40	SiN	MPCVD/100	1370	25.5 ± 0.5 [b]
						31.0 ± 0.7 [b]
[80]	2017	31	SiN	HFCVD/100	–	31.8 ± 5.3 [b]
		22	SiN	MPCVD/100		19.8 ± 4.1 [b]
						17.4 ± 3.0 [b]
[81]	2017	30	SiN	MPCVD/100	–	23 ± 3 [b]
[82]	2017	5	SiN	MPCVD/1	100–700	6.5 [b]
			AlN			15.9 [b]
		No interlayer				61.1 [b]
[10]	2018	5	SiN	MPCVD/1	–	9.5 + 3.8/−1.7 [b]
			AlN			18.2 + 1.5/−3.6 [b]
		No interlayer				41.4 + 14.0/−12.3 [b]
[83]	2019	100	SiN	MPCVD/2	–	38.5 ± 2.4 [b]
		100	AlN			56.4 ± 5.5 [b]
[102]	2019	35	SiN	MPCVD/120	–	31.0 ± 3.1 [b]

Table 3. *Cont.*

Ref.	Year	Dielectric Layer: Thickness (nm)	Dielectric Layer: Material	Diamond Film: CVD Type/Thickness (μm)	Diamond Film: κ (W/(m·K))	$TBR_{GaN/diamond}$ (m^2·K/GW)
[99]	2019	50	SiN	MPCVD/100	-	33
		36	SiN			22
		41	SiN			15
[88]	2019	36	SiN	MPCVD/75	-	20 [c]
		17				13 [c]
[101]	2019	30	SiN	MPCVD/100	-	18 [b]
[106]	2020	20	$Al_{0.32}Ga_{0.68}N$	MPCVD/35	-	30 ± 5 [b]

Values obtained with [a] Raman thermography, [b] transient thermoreflectance, and [c] from luminescence spectra.

3.2. Bonded Wafers

In 2013 BAE Systems proposed a "device-first" technology that allowed the placement of the diamond heat spreader within 1 μm of the device hotspot [107,108]. After the complete fabrication of the devices, the wafer was bonded to a temporary carrier and the substrate and the GaN buffer layer were removed. The back side of the HEMTs was then bonded at room temperature (RT) to a 1" square PCD diamond substrate, fabricated in a different step, using an adhesive and pressing the two materials together. Thanks to the low roughness of the GaN back surface after the removal of the buffer layers (<1 nm), and depending on the κ of the adhesive used, $TBR_{GaN/diamond}$ was estimated to be in the range 15–60 m^2·K/GW. The generic process flow is schematically represented in Figure 6.

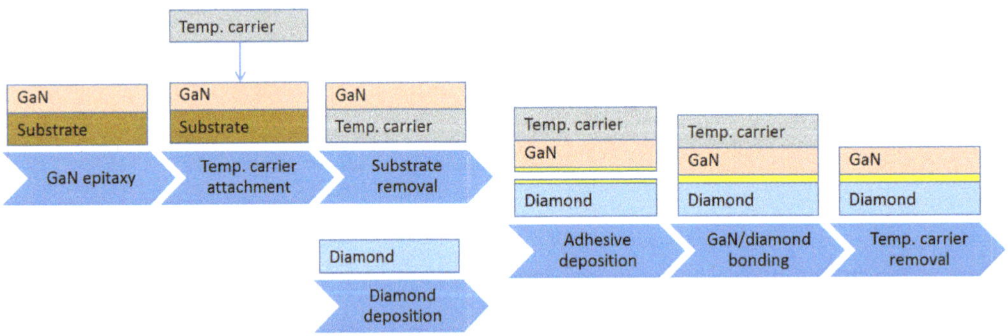

Figure 6. Bonding of GaN and diamond substrates.

In 2014, functional GaN HEMTs originally fabricated in a SiC substrate were bonded to a 1" PCD wafer at a temperature lower than 150 °C [109] by means of a 35 nm-thick layer of Si-containing bonding material [110]. The experimental value of $TBR_{GaN/diamond}$ was 34 m^2·K/GW and the yield of the bonding process was ≈70% (Figure 7a). Even when dissipating 3 times more power, the temperature of the GaN-on-diamond HEMTs was lower than that of their GaN-on-SiC counterparts. However, original GaN-on-SiC devices outperformed GaN-on-diamond ones: $I_{D\,max}$ and $g_{m\,peak}$ were reduced by 16% and 11%, respectively, after the GaN-on-SiC HEMTs were transferred to diamond. The degradation of the DC characteristics was attributed to changes in residual mechanical stress in the device epitaxial layers during the substrate transfer process, as well as to the mechanical and chemical treatments applied. The RF characteristics of the GaN-on-diamond devices also degraded in comparison with GaN-on-SiC ones: at 10 GHz and for V_D = 20 V PAE/P_D were 38%/3.4 W/mm and 48%/4.6 W/mm for both devices, respectively, when tuned for maximum power, and 42%/3.0 W/mm and 57%/4.1 W/mm (when tuned for efficiency). According to the authors, this was primarily due to the omission of air-bridge structures in the GaN-on-diamond devices with the unconnected device channels acting as RF parasitics during power measurements.

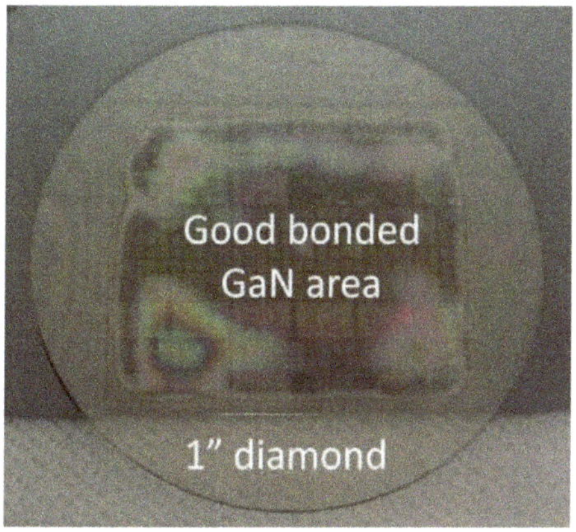

(**a**)

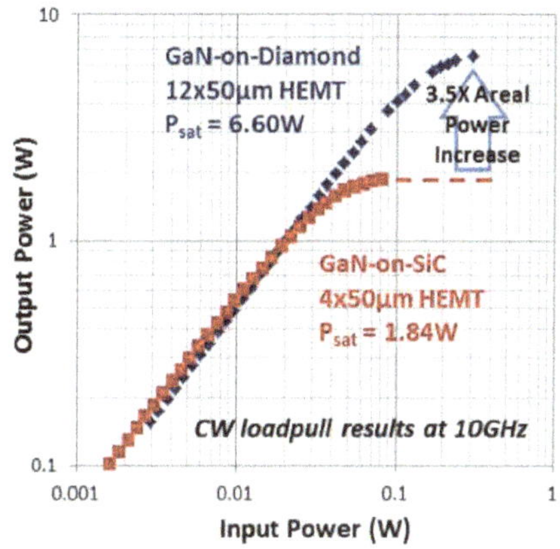

(**b**)

Figure 7. (**a**) GaN HEMTs bonded to a 1″ PCD substrate (© 2014 IEEE. Reprinted, with permission, from K. K. Chu et al., "S2-T4. Low temperature substrate bonding technology for high power GaN-on-diamond", Lester Eastman Conf. 2014—High Perform. Devices, LEC 2014, pp. 1–4, 2014 [109]). (**b**) Comparison of input-output power curves obtained at 10 GHz for GaN/diamond HEMT with 3 times larger gate periphery than GaN-on-SiC HEMT (© 2015 IEEE. Reprinted, with permission, from K. K. Chu et al., "High-Performance GaN-on-Diamond HEMTs Fabricated by Low-Temperature Device Transfer Process," 2015 IEEE Compd. Semicond. Integr. Circuit Symp. CSICS 2015, pp. 7–10, 2015 [111]).

Upon solving these issues, the *PAE* of 12 × 50 µm GaN/diamond HEMTs increased to 51% and P_D to 11.0 W/mm at 10 GHz. For the same V_D, the P_D of 4 × 50 µm GaN-on-SiC HEMTs was only 9.2 W/mm, showing a 3.5 times areal power increase with GaN-on-diamond HEMTs (Figure 7b) [111]. Even under these conditions, the temperature at the center gates was slightly lower for the GaN-on-diamond HEMT than for the GaN-on-SiC HEMT (195 against 202 °C, respectively). The main challenge of this process was identified as the ability to achieve large area bonding with very low $TBR_{\text{GaN/diamond}}$ [110].

In 2017, Liu and co-workers [112] bonded HEMT devices previously fabricated on SiC substrates to a 3″ commercial PCD wafer at a temperature of 180 °C and obtained a functional device yield over 80%. The experimental $TBR_{\text{GaN/diamond}}$ (51 $m^2{\cdot}K/GW$) was still relatively high and the DC current showed a 12−19% reduction due to self-heating; nevertheless, the peak junction temperature of a 10 × 125 µm HEMT with compressed gate pitch of 20 µm decreased from 241 to 191 °C after being transferred from the SiC to the diamond substrate, suggesting a 20% reduction in R_{th}. CW load-pull measurements were performed at 10 GHz and class AB operation on 4 × 125 µm/40 µm gate pitch HEMTs; after being transferred to the PCD substrate, the same HEMT delivered 5.5 W/mm P_D with a *PAE* of 50.5% (against 4.8 W/mm and 50.9% when on the original SiC substrate).

The RT bonding of GaN and PCD films [113] or SCD substrates [114,115] has also been achieved using surface-activated-bonding (SAB). SAB is a direct solid state covalent bonding method that takes place in ultra-high vacuum conditions without obvious interfacial chemical reactions. The previous bombardment of the to-be-bonded surfaces with an argon (Ar) ion beam induces the surface activation, generating surface dangling bonds and making it possible to bond the surfaces at RT [116]. A few nm-thick Si interlayer is typically sputtered on the diamond and GaN surfaces to improve the adhesion. The two activated surfaces are then pressed together at RT and the dangling bonds form covalent bonds at the interface. Transmission electron microscope (TEM) images of the resulting uniform GaN/diamond interface can be seen in Figure 8a (from [113]). After the bonding process the Si layer was ≈24 nm-thick. The bombardment of the diamond surface with Ar ions induced its amorphization, creating an additional ≈3 nm-thick amorphous diamond layer. The thickness of the Si layer was further reduced to ≈10 nm [115] (Figure 8b), creating an interface with a $TBR_{GaN/diamond}$ of ≈19 $m^2 \cdot K/GW$. In a slightly different experimental setup, the surfaces activation was performed with a mixed beam of Si and Ar ions; the thickness of both Si interlayer and amorphous diamond layer was reduced to ≈2 nm (Figure 8c) and the $TBR_{GaN/diamond}$ was as low as ≈11 $m^2 \cdot K/GW$. Despite these interesting preliminary results, functional HEMTs on this GaN/diamond material stack are yet to be reported.

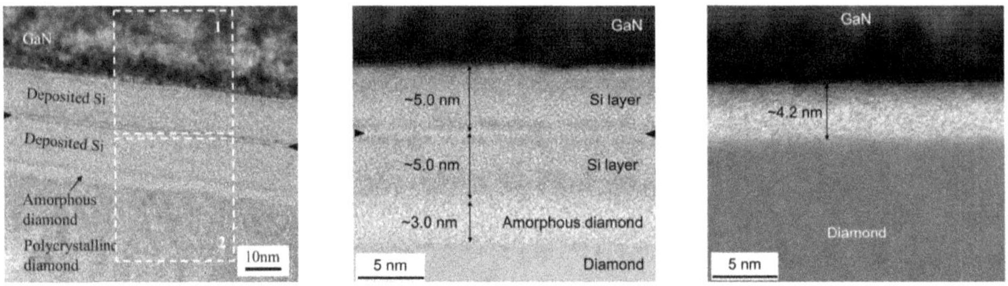

(**a**) (**b**) (**c**)

Figure 8. Cross-section TEM image of GaN/diamond interface obtained by SAB at room temperature after (**a**) sputtering a ≈12 nm (Reprinted from Scr. Mater., vol. 150, F. Mu et al., "Room temperature GaN-diamond bonding for high-power GaN-on-diamond devices", pp. 148–151 [113], Copyright 2018, with permission from Elsevier) and (**b**) ≈5 nm (Reprinted with permission from Z. Cheng et al., "Interfacial Thermal Conductance across Room-Temperature Bonded GaN-Diamond Interfaces for GaN-on-Diamond Devices," ACS Appl. Mater. Interfaces, vol. 12, pp. 8376–8384, 2020 [115]. Copyright 2020 American Chemical Society) Si nanolayer on both surfaces; (**c**) activating both surfaces with a mixed beam of Si and Ar ions (Reprinted with permission from Z. Cheng et al., "Interfacial Thermal Conductance across Room-Temperature Bonded GaN-Diamond Interfaces for GaN-on-Diamond Devices," ACS Appl. Mater. Interfaces, vol. 12, pp. 8376–8384, 2020 [115]. Copyright 2020 American Chemical Society).

Using a similar SAB method with an intermediate Si nanolayer, in 2019 Mitsubishi Electric Corp. announced the successful transfer of a GaN-on-Si multi-cell HEMT to a SCD substrate [114]. No voids were identified in the ≈6 nm-thick bonding interface and the improved DC characteristics of the GaN-on-diamond devices showed that the GaN HEMTs layers were successfully transferred to the diamond substrate (Figure 9). This achievement is expected to improve the *PAE* of high-power amplifiers in mobile communication base stations and satellite communications systems, thereby helping to reduce power consumption. Mitsubishi Electric targeted the commercial launch of GaN-on-diamond HEMTS for 2025.

Wang et al. [117] reported bonding GaN and SCD/PCD diamond substrates at RT and in atmospheric air. Following the surface activation with Ar ions, a double molybdenum (Mo)/gold (Au) layer (5 nm/11 nm) was sputtered on the surfaces of both materials and they were pressed together with an applied load of 2000 N. The bonded surfaces showed a voidage as low as 1.5% and the bonding strength was evaluated as 6.8 MPa. Even though no thermal measurements were made, after 1000 cycles of thermal cycling between −45 and 125 °C the bonding area remained at 73%, suggesting that the Mo/Au nanolayer can effectively balance the difference in the *CTE*s of GaN and PCD wafers.

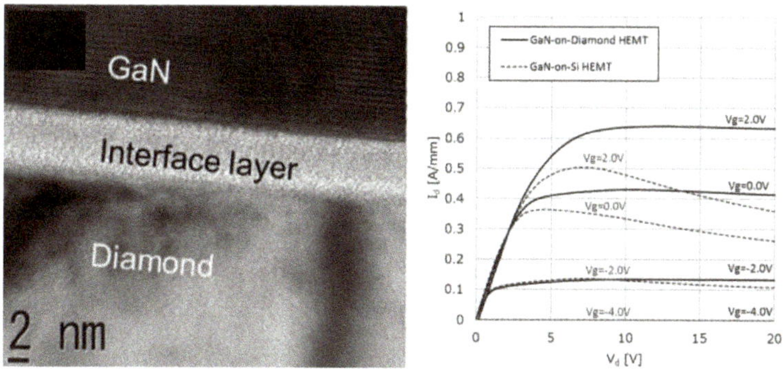

(a) (b)

Figure 9. (**a**) Cross-section TEM image of GaN/diamond interface obtained by SAB. (**b**) I_D-V_D characteristics of GaN-on-Si (dashed lines) and GaN/diamond (solid lines) HEMTs (reprinted from [114], Copyright 2019 The Japan Society of Applied Physics).

Minoura and his colleagues from Fujitsu Limited [118] bonded AlGaN/GaN and indium aluminum gallium nitride (InAlGaN)/GaN-on-SiC HEMTs to SCD substrates by a modified SAB method. A thin (<10 nm) titanium (Ti) layer was previously deposited on the surface of the SCD substrate to prevent the amorphization of the diamond surface during the bombardment with the Ar ions. The TEM image of the SiC/diamond interface is shown in Figure 10a. Using this method, the *TRB* at the diamond/SiC interface was 66 $m^2 \cdot K/GW$ and the R_{th} of the AlGaN/GaN HEMT bonded to the SCD substrate was about 1/3 compared to that without diamond. The P_D of InAlGaN/GaN HEMTs measured with pulsed load-pull measurements (for V_D = 50 V, a pulse width of 10 µs, and 10% duty cycle) increased from 14.8 to 19.8 W/mm with the bonding of the SCD heat spreader structure. With 1% duty cycle, the SCD-bonded HEMT showed a P_D of 22.3 W/mm. Figure 10b,c show the variation of the normalized P_D with the duty-cycle and the P_D as a function of V_D, respectively.

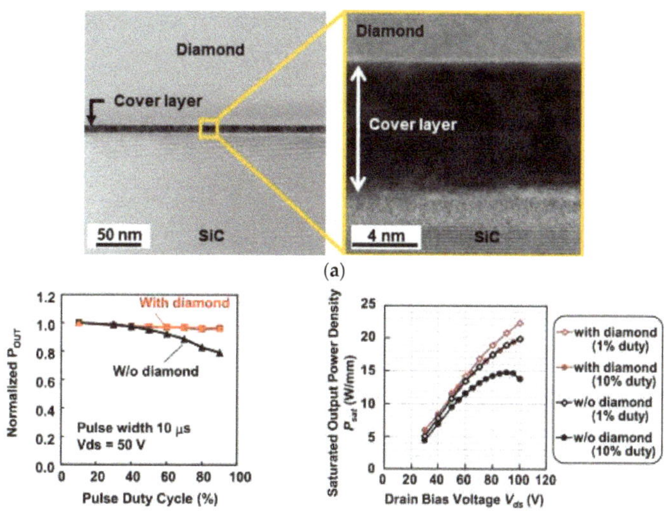

(b) (c)

Figure 10. (**a**) Cross-section TEM image of SiC/SCD interface obtained by SAB after protecting the diamond surface with a Ti film. (**b**) Normalized P_D vs. duty-cycle. (**c**) P_D as a function of V_D (reprinted from [118], Copyright 2020 The Japan Society of Applied Physics).

SAB of GaN HEMTs and diamond was also achieved by sputtering a 450 nm-thick AlN layer on both GaN AlN nucleation layer (after removal of the Si substrate) and diamond surfaces, followed by surface activation with an Ar^+-based plasma and by thermocompression at 160 °C [119]. The strain relief layers and the AlN nucleation layer prevented the flow of heat from the top GaN layer into the diamond substrate; as a consequence, the temperature of the HEMT on the diamond substrate was higher than that on the Si substrate [120].

Van der Waals (VdW) bonding, a process first employed for GaAs thin films [121], has also been used to bond GaN devices and SCD/PCD substrates at temperatures below 300 °C [122,123]. This technique guarantees a good thermal interface ($TBR_{GaN/SCD}$ was estimated to be as low as 10 $m^2 \cdot K/GW$ [122]) without the observation of stress or degradation. RF-devices operating at 3 GHz with improved efficiency (*PAE* of 54.2% against 50.6% for GaN-on-Si HEMTs)) were demonstrated with SCD; on the other hand, the bonding obtained with PCD was not reproducible [123].

The evolution of the relevant technological parameters of GaN/diamond bonded wafers is summarized in Table 4. The *TBR* of the GaN/diamond interface obtained after the SAB and VdW bonding of GaN and SCD can be as low as 11 and 10 $m^2 \cdot K/GW$, respectively. Functional HEMTs have been fabricated on GaN bonded to SCD and PCD substrates. The electrical parameters of HEMTs fabricated using this method are summarized in Table A2 in the Appendix A.

Table 4. Evolution of GaN/diamond bonded wafers technology.

Ref.	Year	Adhesive Layer		Diamond Substrate	Bonding Process	*TBR* ($m^2 \cdot K/GW$)
		Thickness (nm)	Material			
[107]	2013	–	Si-based	1″ × 1″ PCD	Pressing at RT	–
[109]	2014	35	Si-based	1″ PCD wafer	Pressing <150 °C	34 ± 5 [a]
[112]	2017	30–40	–	3″ PCD wafer	Pressing 180 °C	51 [b]
[113]	2018	24	Si	900 μm-thick PCD on Si	SAB RT	–
[114]	2019	6	Si	1 cm × 1 cm SCD	SAB RT	–
[115]	2020	10	Si	SCD	SAB RT	19 [a]
		2				11 [a]
[117]	2020	2 × 5/11	Mo/Au	PCD/SCD	SAB RT	–
[118]	2020	10	Ti/Si	5 mm × 5 mm SCD	SAB RT	66 [c]
[119]	2020	2 × 450	AlN	–	SAB 160 °C	–
[122]	2018	No interlayer		SCD	VdW bonding	10 [d]

Values obtained with [a] transient thermoreflectance, [b] on-wafer IR imaging system, [c] periodic heating method, and [d] estimated from simulations.

3.3. GaN Epitaxy

The last technique to fabricate GaN-on-diamond wafers involves the epitaxial deposition of GaN on the diamond substrate. However, the deposition of epitaxial GaN films on SCD on PCD substrates is inherently difficult, however, due to a series of factors [124]. First, the large lattice mismatch between the two materials (around 13%) can generate defects and lead to poor crystal quality. It should be mentioned, however, that the lattice mismatch between GaN and sapphire is even higher (16% [36]), so this fact alone is not determinant. In addition to the lattice mismatch, the large difference in the *CTE*s of diamond and GaN (0.8×10^{-6} K^{-1} against 5.6×10^{-6} K^{-1}, respectively, whereas the *CTE* of sapphire is 7.5×10^{-6} K^{-1}) can lead to a highly stressed interface that will further impact the dislocating density and lead to the possible cracking of the GaN layer. Finally, in the case of PCD substrates, the absence of a fixed epitaxial relationship between the GaN and the diamond increases the difficulty of nucleating a continuous epitaxial GaN layer.

Nevertheless, different groups have tried to grow GaN on diamond substrates. First reports date from 2003, when Hageman and co-workers [125] reported the growth of 2.5 μm-thick polycrystalline and hexagonal GaN films on (100) natural SCD by a dual step. A thin GaN layer was initially grown by MOCVD on a 10 nm-thick AlN nucleation layer previously deposited on the SCD substrate. Subsequently, the MOCVD pre-grown samples were used as templates for the growth of thick GaN layers using HVPE. The group later reported the growth of 0.07−1.55 μm-thick GaN films on (001) SCD HPHT substrates by MOCVD [126].

Using a different method, Dussaigne et al. [127] reported the growth of 1 μm-thick GaN epilayers on 100 nm-thick AlN layers previously deposited on (111) HPHT SCD substrates by ammonia (NH_3)-source MBE. The films displayed a low RMS roughness of 1.3 nm and some cracks due to the difference in the *CTE*s of diamond and GaN; cracks were not formed for GaN epilayers with thicknesses lower than 250 nm. In a subsequent work [128], 200 nm-thick AlN and GaN strain engineered interlayers were deposited on top of the AlN buffer layer. An 800 nm-thick GaN layer with 8×10^9 cm^{-2} dislocation density was then grown, followed by 24 nm of AlGaN (with an Al content of 28%). A 2DEG formed at the interface, allowing the fabrication of HEMT structures with $I_{D\,max}$ = 730 mA/mm, $g_{m\,peak}$ = 137.5 mS/mm, and f_T/f_{max} = 21/42.5 GHz [129].

Following a different approach, researchers from NTT Corporation reported the formation of a 2DEG in an AlGaN/GaN heterostructure formed on a 600 nm-thick GaN layer epitaxially grown on a type Ib (111) SCD substrate by metalorganic vapor-phase epitaxy (MOVPE) [130] with similar dislocation density (8.4×10^9 cm^{-2} [131]). The AlN/GaN stress-relief layers were deposited on a 180 nm-thick single crystal AlN buffer layer deposited on the diamond substrate and the fabricated AlGaN/GaN HEMTs showed $I_{D\,max}$ = 220 mA/mm and f_T/f_{max} = 3/7 GHz. The R_{th} of GaN-on-diamond HEMTs was 4.1 K·mm/W, against 7.9 K·mm/W for similar GaN-on-SiC structures. By depositing the GaN layer on type IIa (111) SCD diamond substrates, R_{th} was further decreased to 1.5 K·mm/W and RF operation with 2.13 W/mm output power density and 46% *PAE* was achieved [131]. The impact of the misorientation angle of the (111) SCD surface on the surface morphology of the HEMT structures was also studied by the authors [132,133].

Other groups reported growth of GaN on PCD and NCD substrates. The benefits of this approach would be twofold: large-area PCD substrates are available from a few vendors, and the price/area is significantly lower than the one of HPHT SCD substrates. However, only polycrystalline GaN films could be initially deposited. Van Dreumel et al. [134] deposited polycrystalline GaN layers by MOCVD on the surface of nanocrystalline diamond (NCD) films previously deposited by HFCVD on Si substrates, while Zhang et al. [135] deposited fine grained polycrystalline GaN films by MOCVD on the nucleation surfaces of freestanding PCD films prepared by glow discharge. The final goal in this last work was the fabrication of diamond/GaN surface acoustic wave devices, though, and not the fabrication of AlGaN/GaN HEMTs.

More recently, Webster et al. [124] reported the ability to grow epitaxially-oriented GaN films on thick PCD substrates by MOVPE. An AlN layer was previously deposited at 650 °C as nucleation layer for the subsequent growth of a 1.5 μm-thick GaN layer at higher temperature, on top of which the AlGaN/GaN HEMTs structures were fabricated. SiN stripes were then deposited on the AlGaN/GaN stack and the unmasked regions were etched down to the PCD substrate. After an epitaxial layer overgrowth (ELO) re-growth cycle the dislocation density of the GaN layer was reduced by two orders of magnitude (from $\approx 7 \times 10^{-9}$ to $<10^8$ cm^{-2}). With this technique the team was able to grow GaN with a significant degree of epitaxial orientation on an area up to 15 μm^2 on a PCD substrate—Figure 11. The appearance of some cracks on the GaN surface was attributed to the large difference between *CTE*s of diamond and GaN.

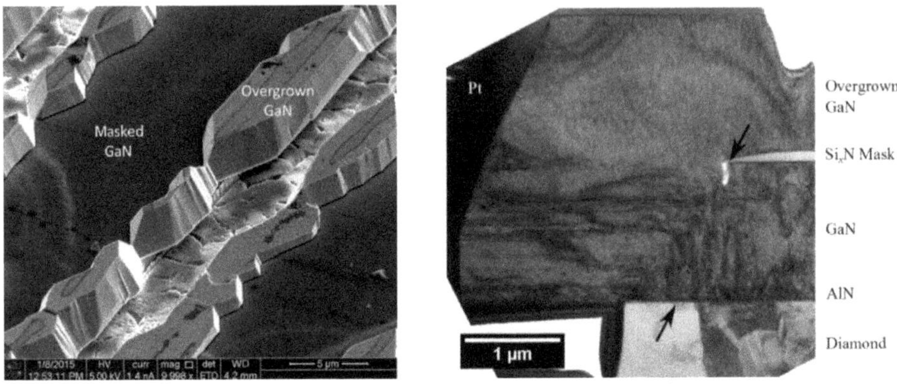

(a) **(b)**

Figure 11. (**a**) SEM image of masked and ELO growth of GaN. (**b**) TEM image showing original and ELO growth of GaN; arrows indicate boundary between original (right of boundary) and ELO (left of boundary) growth (reprinted from [124]; permission conveyed through CCBY 3.0: https://creativecommons.org/licenses/by/3.0/ (accessed on 7 October 2021)).

Later the team proposed a different approach [136]; the (111) Si substrate from commercial diamond-on-Si substrates was etched, exposing the back surface of the PCD substrate that was shown to feature a thin Si_xC layer formed during the deposition of the PCD on the Si substrate. According to the authors, this Si_xC layer provides sufficient crystallographic information for the epitaxial growth process to occur. Following the etching of the Si substrate, the PCD films exhibited some curvature, due to the thermal stress caused by the high diamond deposition temperatures and the difference in the *CTE*s of both materials. AlN or $Al_{0.75}Ga_{0.25}N$ nucleation layers were grown on the SiC layer and GaN was deposited by MOVPE. Interestingly, the growth on the nucleation surface of the curved diamond substrate was shown to effectively remove the surface defects induced by the non-single crystalline nature of the Si_xC layer and to reduce the tensile stress induced by the *CTE* mismatch, allowing the growth of crack-free GaN epitaxial layers up to 1.1 µm thick. As the thickness of the GaN layers increased, the dislocation density was significantly reduced (Figure 12). Upon optimization of the whole process, the tensile strain inherent in GaN epitaxial layers grown on PCD can be further reduced, and thicker crack-free GaN layers can be deposited. Since the increase of the thickness of the GaN epitaxial layer is usually accompanied by reductions in the dislocation density, the use of thicker GaN layers is expected to have a positive impact on the performance of HEMT devices fabricated using PCD substrates.

In 2020 Ahmed et al. [137] also proposed the use of ELO to integrate GaN and PCD diamond; the processing steps are represented in Figure 13. Unlike the procedure described in [124], where the GaN was deposited directly on the PCD by MOVPE, the AlGaN/GaN stack capped with a thin SiN layer was initially deposited on a Si substrate by MOCVD. 500 nm-thick stripes of PCD were then selectively deposited on the SiN by HFCVD following the procedure described in [138]; the exposed SiN was then removed by reactive ion etching (RIE) and the GaN surface was made accessible (step 4 in Figure 13a). The wafer was then returned to the MOCVD reactor for the ELO of GaN (step 5 in Figure 13a). Figure 13b,c show the SEM images of the final structure before and after the regrowth step, respectively. A continuous GaN film formed across 5 µm-wide diamond stripes separated by 5 µm-wide GaN windows oriented along the $\langle 1\bar{1}00 \rangle$ direction of the initial GaN. The dislocation density on ELO GaN ($\approx 10^7$ cm^{-2}) was $\approx$2 orders of magnitude lower than on initial GaN ($\approx 10^9$ cm^{-2}). Although coalescence was achieved, several voids, pinholes, and cracks were visible in some locations along the coalescence regions; the cracks were generated due to the release of thermal stress energy upon cooling of the wafer following

GaN growth. In order to obtain the final GaN/PCD final devices, some steps are further required: MOCVD growth of the HEMT structure, removal of Si substrate and exposure of diamond strip.es surface, thick diamond CVD, and transistor fabrication.

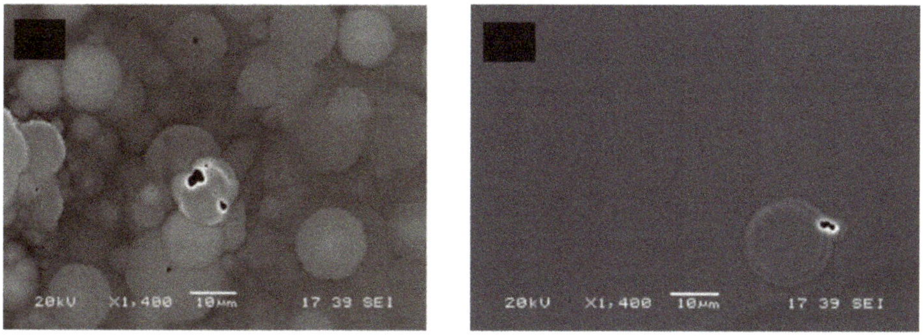

(**a**) (**b**)

Figure 12. SEM images showing improvement in surface morphology of thicker GaN films. (**a**) 300 nm- and (**b**) 1.1 µm-thick GaN films (republished with permission of IOP Publishing, from "Growth of GaN epitaxial films on polycrystalline diamond by metal-organic vapor phase epitaxy," Q. Jiang, D. W. E. Allsopp, and C. R. Bowen, vol. 50, no. 16, 2017 [136]; permission conveyed through Copyright Clearance Center, Inc.).

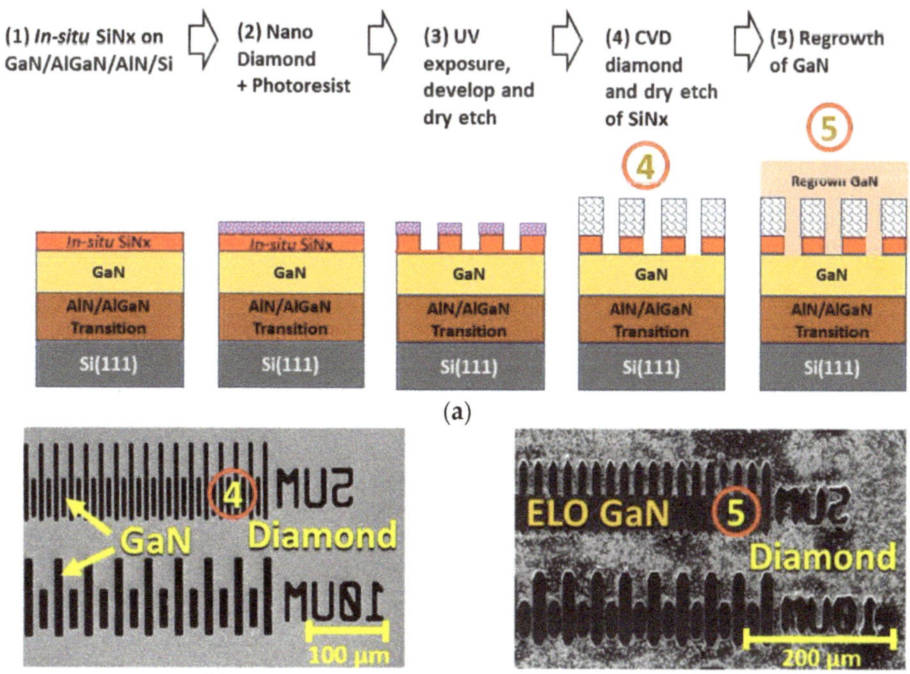

(**b**) (**c**)

Figure 13. (**a**) Processing steps of epitaxial growth of GaN on PCD. SEM images of (**b**) selectively deposited PCD stripes (after step 4) and (**c**) same features after ELO of GaN (after step 5) (reprinted with permission from R. Ahmed et al., "Integration of GaN and Diamond Using Epitaxial Lateral Overgrowth," ACS Appl. Mater. Interfaces, vol. 12, no. 35, pp. 39397–39404, 2020 [137]. Copyright 2020 American Chemical Society).

The experimental results obtained so far are compiled in Table 5. As a summary one can say that research efforts focused on both SCD and PCD substrates, however functional HEMTs have only been fabricated on SCD substrates. Despite promising results, the interest on these substrates faded away and more recent research focused on the ELO of GaN films on PCD substrates with reduced dislocation density. The electrical parameters of HEMTs fabricated on GaN films deposited on SCD substrates are summarized in Table A3 in the Appendix A.

Table 5. GaN epitaxy on diamond substrates.

Ref.	Year	Diamond	GaN			
			Deposition Method	Type	Thickness (μm)	Disl. Dens. (cm^{-2})
[125]	2003	Natural SCD	MOCVD on 10 nm AlN layer + HVPE	Polycryst.	2.5	–
[126]	2011	(011) SCD	MOCVD on AlN layer	Polycryst.	0.07–1.55	–
[127]	2009	(111) SCD	NH_3-MBE on 100 nm AlN layer	Epilayer	1	–
[128,129]	2010	(111) SCD	NH_3-MBE on 200 nm AlN + GaN strain engineered interlayers	Epilayer	0.8	8.4×10^9
[130]	2011	Ib (111) SCD	MOVPE on 180 nm single crystal AlN + 400 nm AlN/GaN	Epilayer	0.6	–
[131]	2012	IIa (111) SCD	MOVPE on 180 nm single crystal AlN + 500 nm AlN/GaN	Epilayer	0.6	8.4×10^9
[132,133]	2012	Ib (111) SCD	MOVPE on 180 nm single crystal AlN + 500 nm AlN/GaN	Epilayer	0.6	–
[134]	2009	NCD	MOCVD on 50 nm GaN	Polycryst.	3	–
[135]	2010	PCD	MOCVD	Polycryst.	0.8	–
[124]	2015	PCD	MOVPE on 70 nm AlN layer + deposition of SiN stripes/etching + ELO	15 μm wide epilayer	1.5	$\approx 7 \times 10^{-9} \rightarrow <10^8$
[136]	2017	PCD thin films	Etching of Si substrate + MOVPE on 10–40 nm AlN/$Al_{0.75}Ga_{0.25}N$ layers	Epilayer	0.2–1.1	–
[137]	2020	Post-deposited PCD	MOCVD of SiN-capped AlGaN/GaN stack on Si + selective deposition of PCD stripes + ELO	5 μm wide epilayer	≈1–5	$\approx 10^9 \rightarrow \approx 10^7$

3.4. Capping Diamond

The capping diamond concept was initially proposed in 1991 [139]. This approach relies on the deposition of the PCD film directly on the device die, immediately on top of the passivation layers, bringing the diamond film in close contact with the hot spots—Figure 14. Initial calculations predicted a 50 °C reduction in the channel temperature of a 2 W/mm GaAs device caused by "thermally shorting" the source, gate and drain contacts with a highly thermally conductive diamond film [140].

Despite no experimental works reported the direct growth of PCD films on GaAs FETs, the concept was picked up by the GaN community and the first successfully diamond-coated working GaN device was reported 10 years later, in 2001, by researchers from the Fraunhofer Institute and from DaimlerChrysler high frequency electronics labs [46]. Diamond films with thicknesses between 0.7 and 2 μm were deposited directly on the SiN passivation layer of GaN FETs using an MPCVD ellipsoid reactor operating at 2.45 GHz at a temperature lower than 500 °C—Figure 15a. Since the regular ultrasonic seeding with diamond particles was found to damage the SiN protective layer, an alternate seeding method based on the sedimentation of fine diamond particles from an agitated emulsion was used. The output and transfer characteristics of the devices were measured before and after the deposition of the 0.7 μm-thick PCD film, showing that the coated FETs remained fully operational—Figure 15b.

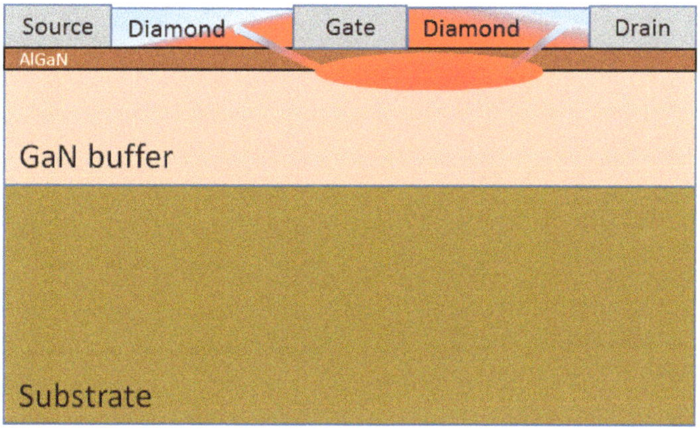

Figure 14. Capping diamond concept.

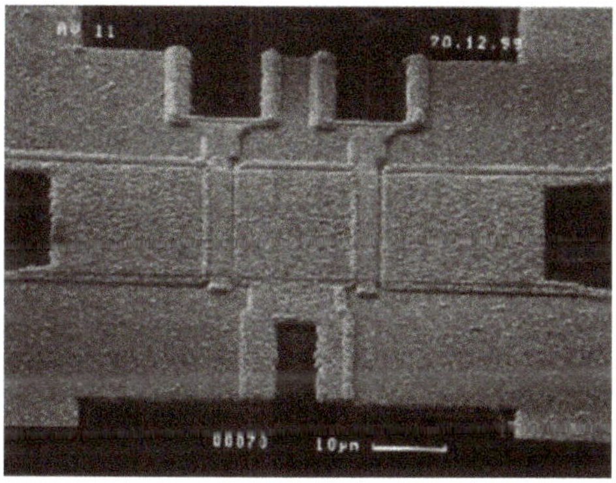

(a)

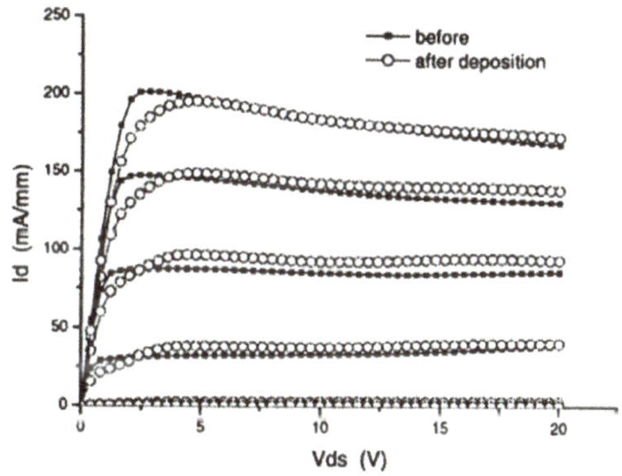

(b)

Figure 15. (**a**) Reflection electron microscope image (REM) and (**b**) Transfer characteristics of a two-finger GaN FET coated with a 0.7 µm-thick PCD film (reprinted from Diam. Relat. Mat., vol. 10, no. 3–7, M. Seelmann-Eggebert, et al., "Heat-spreading diamond films for GaN-based high-power transistor devices," pp. 744–749 [46], Copyright 2001, with permission from Elsevier).

Despite these encouraging first results, the next successful reports describing the deposition of diamond films directly on HEMT devices date from 2010 [141]. In a joint work by the University of Maryland and the Naval Research Laboratory (USA), the authors reported the deposition of a 0.5 µm-thick NCD film on a GaN-on-Si HEMT by MPCVD. A 50 nm-thick silicon dioxide (SiO_2) passivation layer was previously deposited on the surface of the device and the deposition temperature was increased to 750 °C. The NCD-coated device exhibited 20% lower temperature in comparison with the SiO_2-passivated one, however, after NCD coating, $I_{D\,max}$ and $g_{m\,peak}$ decreased from 176 mA/mm and 145.4 mS/mm to 157 mA/mm and 113.9 mS/mm, respectively. This effect occurred due to a reduction in pinch-off voltage following the increased drain-gate coupling through the SiO_2 and NCD film. The NCD film was unintentionally contaminated with boron impurities (1×10^{10} cm^{-2}) and an increase of the leakage current was observed. In order to protect the thermally sensitive Schottky gate contact, a "gate after diamond" process was further developed by the team [142]. This approach involved the deposition of a 0.5 µm-thick NCD layer with κ in excess of 400 W/(m·K) on a 50 nm-thick SiO_2 layer after completion of the mesa and ohmic processes, but before the gate metallization step. An oxygen (O_2)-based plasma etch was then used to recess etch the diamond in the gate region before metal deposition. Similarly to what happened in the previous approach, $I_{D\,max}$ decreased from ≈360 to ≈270 mA/mm after the deposition of the NCD film. The process was further optimized to allow the deposition of the NCD layer directly on the GaN surface [143,144], which improved the performance of the 2DEG in comparison with a reference HEMT passivated with SiN (2DEG density/mobility of 1.02×10^{13} cm^{-2}/1280 cm^2/(V·s) against 8.92×10^{12} cm^{-2}/1220 cm^2/(V·s), respectively), increased $I_{D\,max}$ and $g_{m\,peak}$ from 380 to 445 mA/mm and from 114 to 127 mS/mm, respectively, and decreased the device on-resistance from 14.6 to 11.9 Ω·mm. At 5 W/mm DC power, the temperature of the NCD-capped HEMT was ≈20% lower than that of the SiN-capped one and the R_{th} was 0.98 K·mm/W, ≈3.75 times lower. The use of a p-type diamond gate electrode was also proposed [145]. Again $I_{D\,max}$ increased from ≈290 mA/mm (with a nickel (Ni)/Au gate) to ≈430 mA/mm (with the NCD gate), the on-resistance decreased from 29.4 to 12.1 Ω·mm and the leakage gate current decreased by nearly one order of magnitude at a gate voltage of −10 V. In a later process development, the SiO_2 layer was replaced with a 10 nm-thick SiN layer [146]. The capping of a SiN-passivated GaN-on-SiC HEMT with 500 nm NCD lead to an ≈8% reduction in the peak temperature, less than the 20% obtained with the gate-after-diamond approach due to the SiN passivation layer and the resulting increased $TBR_{GaN/diamond}$ [147].

Following a different approach in 2011 Alomari et al. [148] reported the deposition of a 0.5 µm-thick NCD film on $In_{0.17}AlN_{0.83}$/GaN HEMTs by bias-enhanced nucleation (BEN) and HFCVD at temperatures of 750–800 °C—Figure 16a. The HEMT structures included the dielectric passivation and stress control layers based on SiO_2 and Si_3N_4 and a thin sputtered Si layer for BEN step (Figure 16b). The DC characteristics of the devices remained fundamentally similar after the NCD deposition. f_T and f_{max} were 4.2 and 5 GHz, respectively, and the RF-tested device showed high gate leakage. In 2014 the NCD layer thickness had been increased to 2.8 µm [149].

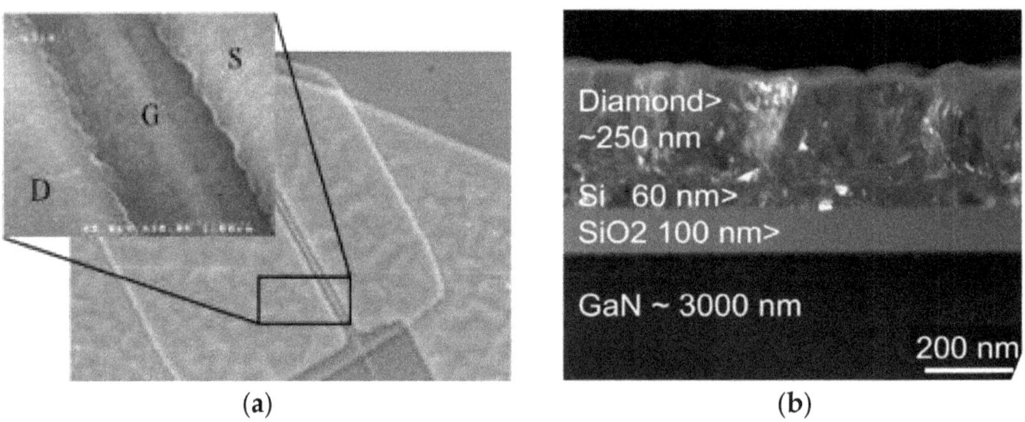

Figure 16. (**a**) SEM top view of a GaN HEMT coated with a 0.5 µm-thick NCD layer. (**b**) TEM cross section of the NCD film showing no voids at the passivation/growth interface (reprinted from Diam. Relat. Mater., vol. 20, no. 4, M. Alomari et al., "Diamond overgrown InAlN/GaN HEMT," pp. 604–608 [148], Copyright 2011, with permission from Elsevier).

Zhou et al. [150] deposited 155–1000 nm-thick PCD films onto Si_3N_4-passivated AlGaN/GaN-on-Si HEMT structures and used transient thermoreflectance experimental results, together with device thermal simulations, to evaluate the impact of the diamond capping layer thickness on the maximum device temperature. A 12% maximum reduction in peak channel temperature could be achieved with a 16 × 125 µm/50 µm gate-pitch AlGaN/GaN-on-Si HEMT with a 1 µm-thick PCD film deposited on the source-drain opening. If the $TBR_{GaN/diamond}$ was not included, a further 10% temperature reduction could be expected. Little further thermal benefit was predicted when using PCD films thicker than 2 μm, with only a maximum 15% temperature reduction. If PCD could be grown on both source-drain opening and metal contacts, a 1.5% better thermal benefit would be achieved for thicker films by increasing the area of the heat spreader. The quality of the initial few μm of the capping diamond layer was also shown to play an important role in reducing the channel temperature [151].

In 2019, researchers from the Power and Wide-Band-Gap Electronics Research Laboratory and Lake Diamond SA deposited 3 μm-thick PCD heat spreaders on the top of vertical GaN PiN diodes using a 30 nm-thick SiN adhesion layer [152]. With 0.9 W of dissipated power, the temperature of the coated device was 64% lower than the temperature of a reference uncoated device.

The appearance of biaxial strain in the Al(Ga)N layers due to the high diamond deposition temperatures, the lattice mismatch between the layers, and the different *CTE*s of the stack materials was studied in detail by Siddique et al. [91]. In their study, a 46 nm-thick SiN layer was used to protect the III-nitride layers beneath the AlGaN barrier layer during the diamond CVD. Figure 17 shows scanning transmission electron microscope (STEM) images of the SiN layer (a) before and (b) after the diamond deposition; even though the SiN layer was partially degraded during the deposition of the diamond and was thinned down to 20 nm in some regions, it remained continuous across the entirety of the AlGaN barrier layer. The deposition of the diamond film was seen to increase the biaxial tensile stress in the AlN and GaN layers by 3%, which caused a 4.5% reduction of the total 2DEG sheet charge density (from 1.04×10^{13} to 0.99×10^{13} cm^{-2}). A 52.8 $m^2{\cdot}K/GW$ $TBR_{GaN/diamond}$ was measured (including the SiN, GaN cap, AlGaN, and AlN layers).

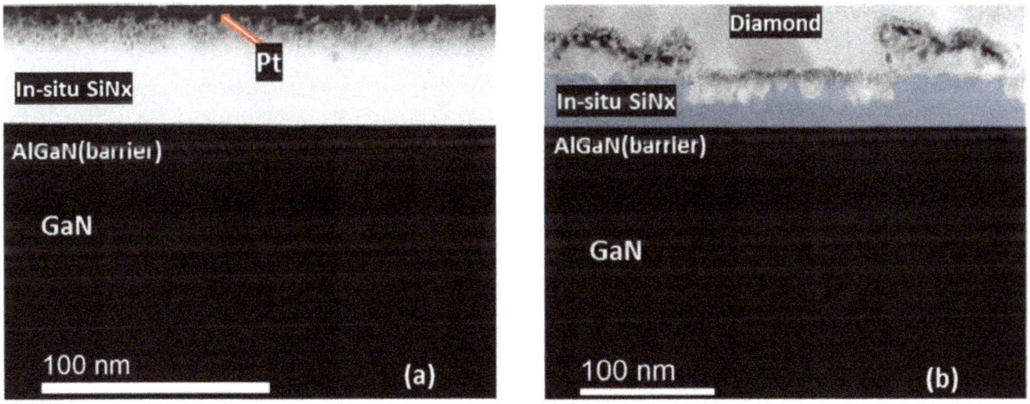

Figure 17. Bright-field STEM images showing the Al(Ga)N/SiN layer interfacial region (**a**) before and (**b**) after diamond growth (reprinted with permission from A. Siddique et al., "Structure and Interface Analysis of Diamond on an AlGaN/GaN HEMT Utilizing an in Situ SiNx Interlayer Grown by MOCVD," ACS Appl. Electron. Mater., vol. 1, pp. 1387–1399, 2019 [91]. Copyright 2019 American Chemical Society).

In order to prevent the modulation of the 2DEG due to the stress induced in the channel by the capping diamond layer, Arivazhagan et al. [153] proposed depositing the diamond layer on the drain electrode alone (instead of directly on the device channel). The impact of this approach was evaluated using thermo-electrical simulations. The results suggest that the deposition of the PCD film on the drain electrode will lower self-heating and allow higher $I_{D\,max}$ than in conventional GaN-on-Si HEMTs—without inducing any change in the 2DEG sheet charge density or in the device threshold voltage.

Despite the technological difficulties inherent to this technological approach, a breakthrough has been recently achieved by Fujitsu Limited. In 2021 Yaita et al. reported a diamond-coated GaN-on-SiC HEMT with improved DC characteristics [154]. A 2.5 μm-thick PCD film was deposited by HFCVD on a 100 nm-thick SiN capping layer at 700 °C. In order to prevent the degradation by the elevated temperatures required to deposit the diamond film, the Schottky gate contact was replaced by a 40 nm-thick SiN insulating layer, followed by a Ni/Au gate contact. In addition, metal heat spreaders were attached to the deposited PCD. The I_D and g_m of the diamond-coated HEMT increased from 0.9 to 1.1 A/mm and from 102 to 148 mS/mm, respectively (Figure 18a). At the same time, for 25 W/mm P_D the hotspot temperature lowered by more than 100 °C, which corresponds to a decrease in R_{th} from 12.7 to 7.4 K·mm/W, and to a ≈40% reduction in the amount of heat generated by the diamond-coated GaN HEMT (Figure 18b). Fujitsu aims to commercialize improved-heat-dissipation GaN HEMT amplifiers in year 2022 for use in weather radar systems and next-generation wireless communication systems [155].

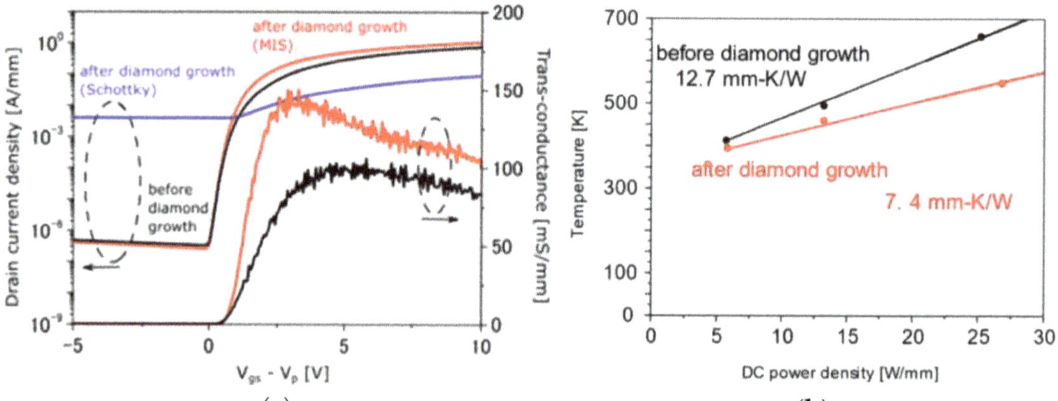

Figure 18. (**a**) Transfer curve and transconductance vs. gate voltage. (**b**) Hotspot temperature calculated from the Raman peak shift (reprinted from [154], Copyright 2021 The Japan Society of Applied Physics.

Experimental results presented so far have been limited to the scope of single or dual-gate HEMTs. Zhang and co-workers implemented 3D thermal simulations to evaluate the impact of capping multi-finger HEMTs with PCD layers [156]. They observed that the capping diamond layer reduced the junction temperature and the temperature non-uniformity in the near-junction region across the channel; the efficiency of the capping diamond layer was also observed to increase with increasing thickness but with a decreasing trend. The largest thermal benefit could be expected under challenging conditions, such as high P_D, narrow gate pitch, high *TBR*, as well as for traditional GaN-on-Si HEMTs. In the particular case of a 12-finger GaN-on-diamond HEMT operating at of 6 W/mm P_D per gate, 20 μm gate pitch, and assuming a similar *TBR* between both GaN/diamond interfaces of 15 m^2·K/GW, the inclusion of a 20 μm-thick PCD capping layer would reduce the junction temperature from 195.8 to 172.2 °C, which corresponds to a net decrease of 23.6 °C. By reducing the thickness of the PCD layer to 1 μm, the reduction of junction temperature would still be as high as 14.9 °C.

The impact of capping double-channel HEMTs was also evaluated with thermo-electrical simulations [157]. The authors observed that the PCD layer provides a lateral heat conduction path close to the hot spot located at near the drain side of the gate edge, modulating the channel lattice temperature distribution and making it become more uniform. For a P_D of 46 W/mm the peak lattice temperature was reduced by 64 °C, indicating that the PCD layer plays an important role in device heat dissipation. Similar to what was observed in [156], the effect of lattice temperature reduction increases with increasing PCD layer thickness. Taking the effect of temperature reduction and cost into consideration, the authors proposed the optimum thickness for the PCD layer to be 1 μm.

The impact of capping a pulsed-mode AlGaN/GaN HEMT with a PCD layer has been also evaluated with thermal 3D simulations [158]. The PCD capped layer not only reduced significantly the peak junction temperature but also suppressed its oscillation in the pulse mode operation, smoothing the temporal variation of the junction temperature. Again the cooling performance of the PCD capped layer was observed to increase with rising thickness but with a decreasing trend. The overall efficiency of this approach was shown to be more effective under harsh thermal conditions, including smaller duty cycle, higher TRB, and lower κ substrate.

The evolution of the relevant experimental results has been compiled in Table 6. While experimental results and functional HEMTs were reported between 2001 and 2014, most recent work focused on evaluating the biaxial strain induced by the diamond deposition temperatures (typically in excess of 500 °C) and on anticipating the thermal benefit of the diamond layer using thermal simulations. The electrical parameters of diamond-capped HEMTS are summarized in Table A4 in the Appendix A.

Table 6. Evolution of capping diamond technology.

Ref.	Year	Passivation Layer		Diamond Film			*TBR* (m^2·K/GW)
		Thickness (nm)	Material	Thickness/Type (μm)	Dep. Temp. (°C)	CVD Type	
[46]	2001	–	SiN	0.7–2/PCD	<500	MPCVD	–
[141,142]	2010, 2012	50	SiO_2	0.5/NCD	750	MPCVD	–
[143]	2013	No interlayer	–	–/NCD	750	MPCVD	–
[146]	2017	10	SiN	0.5/NCD	750	MPCVD	–
[148]	2011	–	SiO_2/Si_3N_4	0.5/NCD	750–800	HFCVD	–
[149]	2014	–	Si_3N_4	2.8/NCD	750–800	HFCVD	–
[150]	2017	50	Si_3N_4	0.155–1/PCD	650	MPCVD	45 + 13/−11—91 + 13/−9 [a]
[91]	2019	46	SiN	1.46/PCD	720–750	HFCVD	52.8 + 5.1/−3.2 [a]
[152]	2019	30	SiN	3/PCD	820	MPCVD	–
[154]	2021	100	SiN	2.5/PCD	700	HFCVD	–

[a] Values obtained with transient thermoreflectance.

4. GaN/Diamond HEMTs: Where to Go?

Each of the previously described approaches has some advantages over the others and, simultaneously, shows room for improvement if particular technological developments can be achieved. This is will be the topic of the current section.

4.1. *Challenges of Fabricating GaN-on-Diamond Wafers*

The deposition of diamond films on the back of GaN wafers has been routinely performed by Element Six for more than 10 years. The fabrication of GaN-on-diamond wafers involves the deposition of 100 μm-thick diamond films at temperatures higher than 700 °C. The large deposition temperatures induce a large residual stress at the GaN/diamond interface that may cause the bowing of the wafer and the cracking of the GaN layers. In addition, the $TBR_{GaN/diamond}$ and the low quality of the initial layers of the deposited diamond films are currently the bottleneck of the heat extraction. Each of these issues will be discussed in detail in the following paragraphs.

4.1.1. Decreasing Thermal Stress

One of the biggest issues of this approach is related with the high diamond deposition temperature (>700 °C) which induces a large residual stress at the GaN/diamond interface because of the mismatch between the *CTE*s of both materials [159,160]. The residual stress depends on the GaN thickness, on the diamond growth temperature, and on the sacrificial carrier wafer [91,92]. Residual stresses larger than 1 GPa at the free surface of the GaN have been reported [93]; these elevated stress conditions induce layer cracking and wafer bow, and impact the electrical performance of the devices [161–164]. A large thermal stress at the GaN-diamond interface causes a significant reliability concern when considering the function and lifetime of GaN-on-diamond devices.

To avoid the issues related with the residual stress, a slightly modified approach has been recently proposed [94]. The GaN was initially grown on an AlN nucleation layer deposited on a Si substrate. The Si substrate was selectively etched, leaving behind 0.5 mm-diameter GaN membranes with the AlN nucleation layer exposed onto which 50 μm-thick diamond films were further grown. The high quality diamond/AlN interface showed no visible voids or cracks, confirming the strong bond between both materials. This approach presents no technological barrier to the incorporation of an AlN "initiation" layer into the GaN buffer close to the device channel; however, due to the low thickness of the GaN membranes and the simultaneously high temperatures required for diamond growth, the bowing of the membranes during the diamond deposition step cannot be neglected. This phenomenon was recently analyzed in detail in [95] using commercial GaN-on-Si wafers as the starting material. The processing steps are described in Figure 19a. The GaN membrane diameter was shown to impact the maximum displacement from the original plane (bowing) before and after diamond deposition, as well as the membrane stress. The 5 mm-diameter membranes allowed a larger displacement with a subsequent lower stress value, however the larger bowing "pushed" the GaN surface further into the plasma during diamond deposition, exposing it to high temperatures and resulting in thermal runaway and damage to the GaN/III-N film—Figure 19b. In addition, the large membrane bows present a big challenge for device manufacturing using contact lithography. A possible way to prevent the bowing would be the use of pre-stressed GaN-on-Si wafers as the starting material.

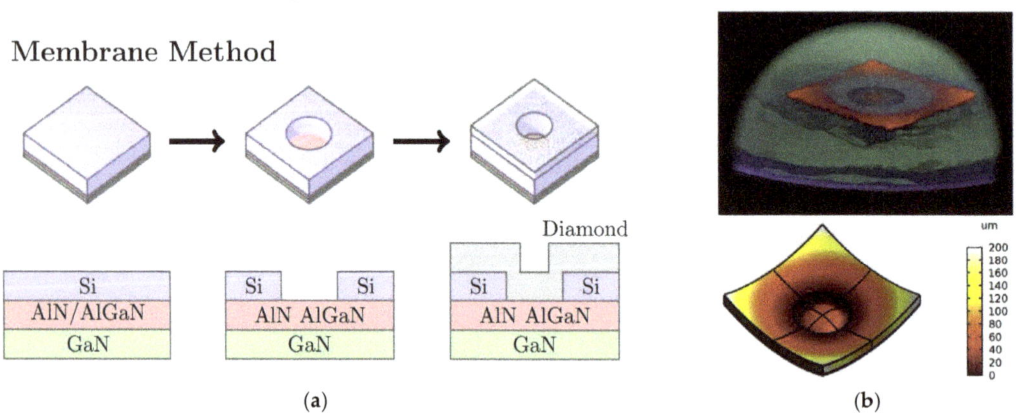

Figure 19. (**a**) Fabrication steps of a GaN-on-diamond membrane. (**b**) Image of the 5 mm-diameter membrane inside the MPCVD system and corresponding thermally-induced mechanical displacement (reprinted from [95]; permission conveyed through CCBY 4.0: https://creativecommons.org/licenses/by/4.0/ (accessed on 7 October 2021)).

Following a radically different approach, represented in Figure 20, Jia et al. [96] reported a low stress GaN-on-diamond wafer fabricated by dual sided deposition of diamond. In this process, the temporary carrier represented in Figure 2 was a 2 μm-thick Si layer followed by 100 μm of low-quality PCD deposited by CVD. A thin SiN layer was deposited on the surface after the substrate removal, followed by high-quality PCD. The final free-standing GaN-on-diamond wafer was obtained by removing the low-quality PCD and Si layers. The surface and quality of the GaN layer after the etching of the low quality

PCD and Si layers remained the same while the stress was reduced to 0.5 GPa. However, it should be mentioned that the AlN buffer layer was not removed and this may have helped to reduce the internal stress caused by diamond growth.

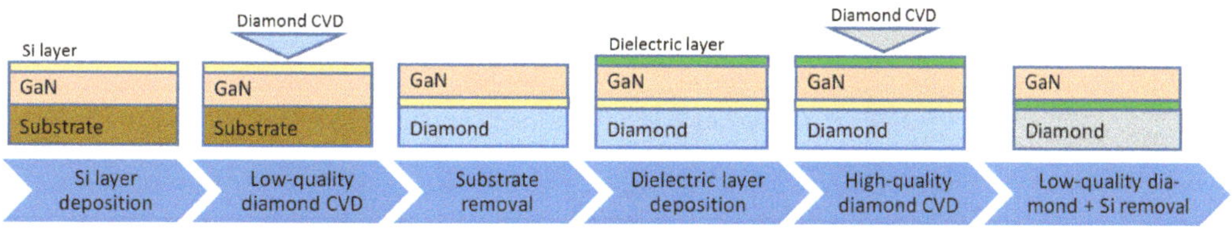

Figure 20. Fabrication of GaN-on-diamond wafers by dual-sided diamond deposition.

4.1.2. Optimizing the Thermal Barrier Resistance at the Diamond/GaN Interface

The minimization of the $TBR_{GaN/diamond}$ will have a positive impact on the thermal performance and reliability of GaN-on-diamond HEMTs. In theory this can be achieved by reducing the thickness of dielectric layer, thereby reducing its bulk R_{th}, or by improving the interface between GaN/dielectric and dielectric/diamond to reduce the interface effects. As an example, for a particular GaN-on-diamond HEMT, the reduction in the $TBR_{GaN/diamond}$ related with the SiN layer from 13 to 3 m^2·K/GW would reduce the total R_{th} of the device from 2.8 to 1.9 K·mm/W, which corresponds to a 35% decrease in the peak operating temperature rise at a given power level [73]. However, one should also keep in mind that, despite smoother GaN/SiN/diamond interfaces lead to lower $TBR_{GaN/diamond}$, they also show reduced interfacial fracture toughness, in comparison with rougher interfaces [99], which could negatively impact the devices reliability.

The standard fabrication of GaN-on-diamond wafers requires a 30 nm-thick SiN layer to protect the surface of the GaN from the diamond deposition conditions. Despite its low thickness, this layer may contribute with a *TBR* of ≈30 m^2·K/GW, adding more than 20% to the total device R_{th} [74]. Some groups decreased the thickness of the SiN layer to 5 nm and obtained *TBR* values of 9.5 and 6.5 m^2·K/GW [10,82] (close to the theoretical minimum of 5.5 m^2·K/GW calculated by the diffuse mismatch model [82]).

An apparently simple way of decreasing $TBR_{GaN/diamond}$ would be to remove the SiN layer. However, depositing diamond directly onto GaN is not a straightforward task. At typical diamond CVD temperatures (700 °C and above), the atomic hydrogen (H) can etch the surface of the GaN substrate (forming NH_3 and liquid gallium (Ga)). This etching can be prevented if the density of diamond seeds on the GaN surface is so high that the lateral diamond growth rate exceeds the GaN etching rate. In this case, a protective diamond layer grows to cover the GaN surface before significant etching can occur. However, even under these conditions the interface between GaN and diamond is rather weak because Ga does not form a carbide. This means the diamond adheres to the GaN surface mainly via weak VdW interactions, rather than by strong covalent bonds. This becomes a serious problem when the coated samples are cooled down to RT due to the difference in the *CTE*s of both materials, which causes compressive stress to accumulate in the diamond layer and can lead to delamination of the entire diamond layer [75].

Despite these difficulties, different groups have reported direct growth of diamond films on GaN [10,82], however the experimental values of $TBR_{GaN/diamond}$ are higher than the ones obtained with SiN (41/30 against 9.5/5.5 m^2·K/GW, respectively [10,82]) and significantly larger than the minimum theoretical value for the GaN/diamond interface (3 m^2·K/GW [82]). During the diamond deposition extensive deterioration of the GaN surface occurs, which results in the appearance of voids measuring up to 50 nm. This results in an extremely rough surface that increases the scattering of the phonons. Smith et al. [76] and Waller et al. [74] reported $TBR_{GaN/diamond}$ as high as 220 m^2·K/GW. In both works the GaN surface was seeded using a two-step electrostatic spray technique (see more details in Section 4.1.3.1).

A different possibility is to replace the SiN layer with a layer of material with higher κ, such as AlN [10,82] (30 [46] against 285 W/(m·K) [165], respectively, in crystalline forms). However, AlN thin films decompose in hydrogen (H_2) and H_2/CH_4 plasmas at low pressures (25–5 Torr) and high temperatures (650–1070 °C) [166], and direct growth of diamond has proven to be very difficult. A possible solution would be the pre-treatment of the AlN surface, prior to exposure to CVD diamond growth conditions. Previous exposure of the AlN substrates to carbon tetrafluoride (CF_4) plasma allowed increasing the seeding density by nearly 3 orders of magnitude in comparison with untreated substrates [167]. Mandal et al. [84] reported that exposure of the AlN surface to H_2/nitrogen (N_2) plasma was necessary for the deposition of thick (>100 μm) and adherent diamond layers. However, the real usefulness of replacing SiN with AlN is doubtful, since AlN and SiN thin films are amorphous and feature similar κ values (≈1–5 nm) [82]. In fact, two different groups reported higher values of $TBR_{GaN/diamond}$ obtained with 5 nm-thick AlN layers than with similar SiN layers (18.2/9.5 against 15.9/6.5 $m^2 \cdot K/GW$, respectively [10,82]). In both works this difference was attributed to discontinuities in the AlN layer itself which resulted in the etching of the GaN surface (Figure 21); however it is not clear if the as-deposited AlN layer was discontinuous or if it was etched during the deposition of diamond. Jia et. al [83] observed the same tendency with thicker dielectric layers: the *TBR* obtained with 100 nm-thick AlN layers and SiN layers was 56.4 and 38.5 $m^2 \cdot K/GW$, respectively.

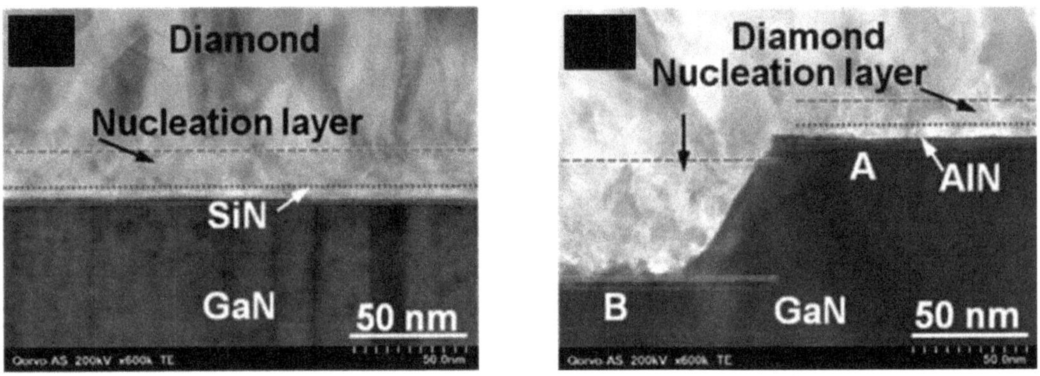

(a) (b)

Figure 21. TEM cross-sections of GaN/diamond interfaces with a 5 nm-thick barrier layer of (**a**) SiN and (**b**) AlN. Uniform and smooth GaN/dielectric and dielectric/diamond interfaces are obtained with the SiN layer. With AlN, some regions (A) show smooth interfaces, however in other regions (B) ≈60 nm of GaN has been etched away (reprinted with permission from Y. Zhou et al., "Barrier layer optimization for enhanced GaN-on-diamond device cooling," ACS Appl. Mater. Interfaces, vol. 9, no. 39, pp. 34416–34422, 2017 [82]. Copyright 2017 American Chemical Society).

Seeding the pre-treated AlN surfaces with H-terminated detonation nanodiamond (DND) seeds resulted in an average *TBR* of 16 $m^2 \cdot K/GW$. A breakthrough was recently reported by Smith and co-workers [76], who used a two-step electrostatic spray technique to seed 130 nm-thick AlN films deposited on Si substrates. Using this method the experimental *TBR* was as low as 1.47 $m^2 \cdot K/GW$, close to 0.8 $m^2 \cdot K/GW$, the theoretical minimum *TBR* achievable at the AlN−diamond interface from a diffuse mismatch model, relying only upon the density of states in these two materials [84] (see more details in Section 4.1.3.1).

It should be mentioned that in the experiments by Mandal et al. [84] and Smith et al. [76] the *TBR* values refer to the interface between the diamond and the AlN substrate: as a consequence, no direct comparison can be made with the $TBR_{GaN/diamond}$ previously reported by other groups (which includes the interfaces GaN/dielectric layer and dielectric layer/diamond, as well as the R_{th} of the dielectric layer itself). Nevertheless, the replacement of the low κ amorphous SiN layer with crystalline AlN is a promising approach. Ideally, the AlN layer should be integrated just below the GaN channel. This layer would act as an etch stop during the device epitaxy, as well as a seed layer for the diamond growth [168]. This is the approach proposed by Field and co-workers [106]; since integrating a thin AlN or high Al content AlGaN layers at this point in the epitaxy is challenging because of alloying with surrounding layers, they used a relatively low Al content crystalline $Al_{0.32}Ga_{0.68}N$ layer as the etch stop and interlayer and grew diamond following the same procedure as in [84]. Due to the sample layout, a 10 nm-thick SiC layer was formed between the $Al_{0.32}Ga_{0.68}N$/diamond interface, which improved the heat transport across the two materials. A *TBR* of 30 $m^2 \cdot K/GW$ was measured, a value still much higher than the theoretical minimum of 4 $m^2 \cdot K/GW$ obtained for this interface [106].

4.1.3. Optimizing Diamond CVD for Thermal Management Applications

The initial layers of a diamond film typically feature small grains and feature a correspondingly low κ. This effect, combined with the *TBR* at the diamond/dielectric interface, may contribute to an additional *TBR* of 10 $m^2 \cdot K/GW$ [79]. The importance of the quality of the diamond nucleation layer was highlighted in a recent work by Song et al. [169], who showed that the R_{th} of a 12 finger/30 μm gate pitch GaN-on-diamond HEMT dissipating 5 W/mm would lower from 13.0 to ≈11.0 K·mm/W in the absence of phonon scattering by external defects in the GaN layer and interface (a value ≈49% lower than that of a state-of-the-art similar GaN-on-SiC structure). If the κ of the diamond nucleation layer were the same as its bulk conductivity, the R_{th} of this devices would further decrease to ≈10.0 K·mm/W (≈54% lower than the R_{th} of a similar GaN-on-SiC HEMT). Following the same trend, in the case of a HEMT with one finger gate, if the *TBR* at the GaN/diamond interface decreased from 13 to 3 $m^2 \cdot K/GW$, the device R_{th} would go from 2.8 to 1.9 K·mm/W, which corresponds to a ≈30% reduction in the peak operating temperature at a given power level [73].

The morphology and properties of the diamond nucleation layer are intrinsically related with the process of depositing diamond films on non-diamond substrates. The deposition of diamond on a foreign substrate requires a seeding step, during which the substrate surface is enriched with diamond nanoparticles (DNP). Different techniques can be used for this purpose, such as ultrasonic agitation in a suspension containing DNP or spin-coating of a solution saturated with the same. Once exposed to diamond growth conditions, the diamond seeds grow three-dimensionally and eventually coalesce, forming a closed diamond film. At this stage the individual crystallites start growing perpendicularly to the surface, following the Van der Drift model [170], until growth terminates. The incubation time for the onset of the formation of diamond crystallites can be 15–45 min, depending on the growth parameters.

The growth of the diamond crystals from individual diamond seeds translates in the existence of a so-called diamond nucleation region which contains a high concentration of defects and grain boundaries that increase the phonon scattering and consequently decrease the thermal conductance. The thickness of this nucleation layer ranges typically between 10 and 50 nm, depending on the seeding method and deposition conditions, and its κ can be as low as 3 W/(m·K) [78]. The appearance of voids at the diamond/substrate interface at the locations where the enlarged diamond seeds touch one another is also common. This effect is represented schematically in Figure 22a [84]. Figure 22b shows a high-angle annular dark-field STEM (HAAD-STEM) image of the interface where such voids can be easily identified.

The size of the diamond grains is typically a few nm close to the substrate and increases with the thickness of the film. The evolution of the grain size has been studied computationally [171,172] and experimentally [173]; it has been shown that, depending on the growth conditions, the lateral size of the grains and their aspect ratio are strongly changing with the film thickness. As a consequence, the grain boundary density varies with the depth of the PCD layer, translating into an inhomogeneous $\kappa_{in\text{-}plane}$. On the other hand, in columnar PCD films $\kappa_{out\text{-}of\text{-}plane}$ is typically higher than $\kappa_{in\text{-}plane}$ [174]. However, this condition does not necessarily hold true in the nucleation region, where $\kappa_{in\text{-}plane}$ can be higher than $\kappa_{out\text{-}of\text{-}plane}$ and vice-versa for a given PCD film thickness [175]. The grain size dependence of κ, which is especially pronounced near the nucleation region, is therefore a critical parameter for maximizing the heat-spreading capabilities of PCD films on hybrid diamond/GaN devices. The dependence of the in-plane and out-of-plane κ with the PCD film thickness is shown in Figure 23.

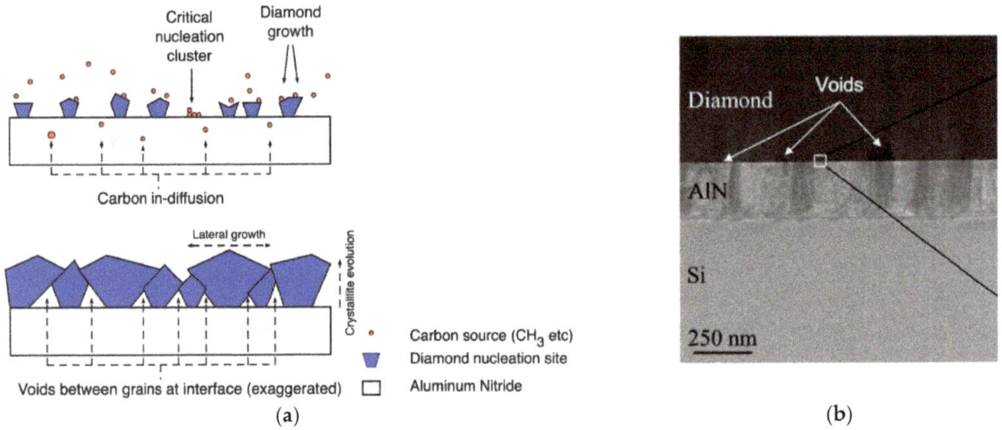

Figure 22. (**a**) Schematic of diamond film growth with low nucleation density; (**b**) HAAD-STEM image of the interface where the voids are clearly seen (reprinted from [84]; permission conveyed through CCBY 4.0: https://creativecommons.org/licenses/by/4.0/ (accessed on 7 October 2021)).

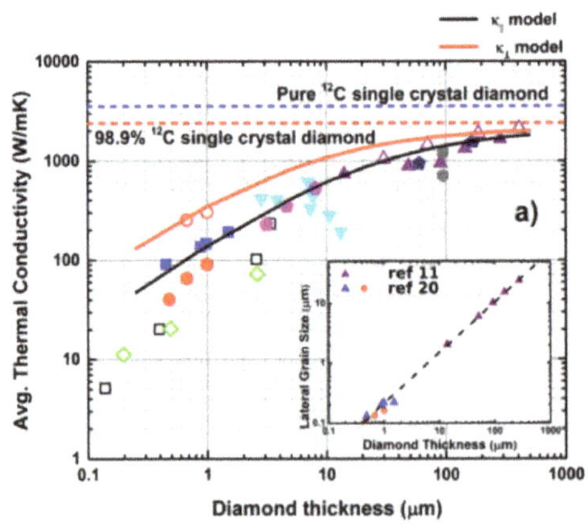

Figure 23. Evolution of in-plane and out-of-plane κ with diamond film thickness (© 2016 IEEE. Reprinted, with permission, from J. Anaya, et al., "Thermal management of GaN-on-diamond high electron mobility transistors: Effect of the nanostructure in the diamond near nucleation region," in 15th IEEE Intersociety Conference on Thermal and Thermomechanical Phenomena in Electronic Systems (ITherm), 2016, pp. 1558 1–8 [175]).

4.1.3.1. Impact of the Seeding Procedure

It is obvious that the deposition conditions (CH_4/H_2 ratio, substrate temperature, and pressure) impact directly the growth rate as well as the macro-level characteristics/quality of the diamond deposits. Similarly, the seeding step has a considerable impact on the quality of the diamond film in the nucleation region, close to the interface with the substrate.

Increasing the seeding density has been the motto of researchers during the past years. Given their small size, which allows for homogeneous and high density seeding, DND seeds are the most frequently used diamond particles. The seeding methods that lead to the highest seeding densities include the ultrasonic agitation in a suspension containing DND seeds and the enhancement of the electrostatic attraction between the DND seeds and the substrate.

1. Ultrasonic agitation in a suspension containing DND seeds. Ultrasonic seeding has been widely used since the early 1990s [176]. Seeding can be performed with different sized diamond grit, as well as with a mixture of diamond grit and metal particles. As an example, the adhesion of 0.25 μm tungsten (W) and Ti particles to a nanodiamond suspension was reported to increase the seeding density and the adhesion of diamond films deposited on Si and Si_3N_4 substrates [177]. A significant improvement was achieved with DND particles colloidal solutions, which enabled seeding densities in excess of 10^{12} cm^{-2} [178].
2. Enhancement of electrostatic attraction between DND seeds and substrate. By properly tuning the ζ-potentials of seeds and/or substrates, one can benefit from enhanced electrostatic attraction between the DND seeds and the substrate. This can be achieved by terminating the DND seeds with oxygen (O) or H atoms [179] or by preparing polymer/nanodiamond colloids [180,181]. This effect can be further enhanced by exposing the substrate to a plasma treatment in order to guarantee that the ζ-potentials of diamond seeds and substrate have opposite signs [84].

Despite leading to the highest reported seeding densities, DND particles may not be the best choice for thermal management applications, which rely on the minimization of the *TBR* between diamond and substrate. DND particles possess an amorphous shell [182], which may contribute to the increase of sp^2 bonds close to the interface, which in turn will enhance phonon scattering, thus compromising the thermal transport across the interface. In addition, while it is true that a high seeding density guarantees a lower coalescence time, it is not absolutely clear that this is an advantage for thermal applications. A very high density of seeds means that they are not allowed to grow significantly before they coalesce with each other; as a consequence the amount of defects and grain boundaries further increases—and so does the scattering of the phonons.

If seeding is performed with larger sized particles, the grain/grain boundaries ratio might be maximized. Following this reasoning, in 2017 Liu and co-workers evaluated the effect of seeding GaN substrates with a 30 nm-thick SiN protective layer with 30 and 100 nm diamond seeds [81]. Unlike what happened with the smaller 30 nm particles, seeding with 100 nm particles damaged the SiN layer, which resulted in the etching of the GaN surface and appearance of pin holes during diamond growth. More recently, Bai et al. [183] evaluated the impact of seeding with 4 and 20 nm seeds on the κ of diamond films deposited in Si substrates. Seeding with the larger 20 nm seeds resulted in a smaller seeding density (7×10^9 cm^{-2} in contrast with 3×10^{11} cm^{-2} with the 4 nm seeds) but resulted in larger-sized grains near the interface region, and in a correspondingly higher in-plane κ as measured by Raman thermography.

In order to overcome the limitations of the standard seeding procedures, Smith et al. [76] proposed a two-step electrostatic spray technique to seed 130 nm-thick AlN films deposited on Si substrates. Using this method, the surface of the AlN was initially seeded with 2 μm diamond particles with smooth facets, which guaranteed a large contact area with the AlN surface and favored the thermal transport across the AlN/diamond interface. Following this step, the substrate was electrostatically sprayed again with 3.3 nm DNP, which filled in the gaps between the larger seeds and prevent the formation of voids. The schematic diagram explaining the rationale for the two-step seeding is shown in Figure 24. The advantages of this method are twofold: the microparticles of diamond guarantee a lower grain boundary ratio (when compared to conventional seeding with DND particles) and the DNP fill the voids between the larger particles, protecting the AlN surface from the plasma. This layer of electrosprayed seeds replaces the highly defective diamond nucleation layer characteristic of heteroepitaxial diamond films. Using this method, an extremely low *TBR* of 1.47 $m^2 \cdot K/GW$ (close to the theoretical minimum of 0.8 $m^2 \cdot K/GW$ [84]) was obtained at a diamond/AlN interface.

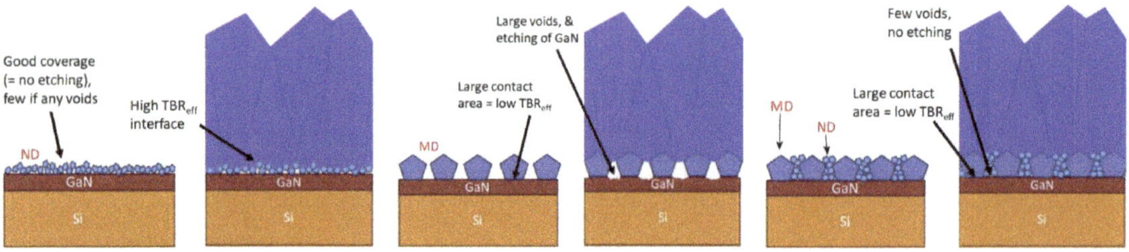

Figure 24. Rationale behind the two-step seeding: seeding with (**a**) ND particles alone, (**b**) diamond microparticles alone, and (**c**) diamond microparticles followed by ND (reprinted from Carbon, vol 167, E. J. W. Smith et al., "Mixed-size diamond seeding for low-thermal-barrier growth of CVD diamond onto GaN and AlN", pp. 620–626 [76], Copyright 2020, with permission from Elsevier).

4.1.4. Optimizing the Thickness of the GaN Epilayers

The determination of the GaN buffer layer thickness typically takes into account the electrical performance of the device and material quality requirements, instead of thermal requirements [184]. In latest generation GaN-on-diamond devices the low thermally conductive nucleation and strain relief layers are removed before the deposition of the diamond film, and as a consequence the thickness of the GaN buffer layer has a non-negligible impact on the total R_{th} of the device. In fact, in a joint work between the University of Bristol and Element Six [88], the thickness of the GaN buffer layer was reduced from 700 to 354 nm and the thickness of the dielectric layer to 17 nm, leading to a device R_{th} of 9 K·mm/W, a value significantly lower than the R_{th} of GaN-on-SiC devices (16 K·mm/W). Besides lowering the maximum temperature, the thinning of the GaN layer decreased self-heating, resulting in a smaller change in output conductance and providing a means to reduce the thermally-generated device non-linearities.

Taking into account these results, it might initially be assumed that the GaN layer should be as thin as possible. If it is too thick, the R_{th} associated with the GaN layer and consequently the R_{th} of the device increase. However, if the GaN layer is too thin (especially when the heat source length is comparable to the device length and for small $TBR_{GaN/diamond}$ values), the concentrated heat flux coming out from the heat source reaches the GaN/diamond interface without spreading, causing the region right under the heat source to heat up significantly and leading to an increased R_{th} [85]. As an example, the peak channel temperature of a 4 × 125 μm/40 μm gate pitch GaN-on-diamond HEMT dissipating 10 W/mm is shown in Figure 25 for different values of $TBR_{GaN/diamond}$ and GaN layer thicknesses (from [185]). It can be seen that thinner GaN layers may lead to lower or higher peak channel temperatures, depending on the particular value of the $TBR_{GaN/diamond}$.

Simulations have systematically shown that the device R_{th} monotonically decreases with increasing GaN layer thickness until it reaches a minimum. The higher the κ of the diamond, the lower the $TBR_{GaN/diamond}$ [86,87,89], and the smaller the hotspot area [85], the more important this dependence becomes.

In 2020, Song and co-workers [90] have shown that, while the device R_{th} is fairly low at a typical GaN thickness of 1 μm (≈12.9 and ≈16.4 K·mm/W for $TBR_{GaN/diamond}$ of 6.5 and 30 $m^2 \cdot K/GW$, respectively), a reduction in the GaN thickness below 1 μm may result in a substantial increase in the device R_{th}, in particular when $TBR_{GaN/diamond}$ is high (≈31% and ≈118% increase for $TBR_{GaN/diamond}$ of 6.5 and 30 $m^2 \cdot K/GW$, respectively, and 0.1 μm GaN thickness). For the same $TBR_{GaN/diamond}$ values, the GaN thickness that minimizes the R_{th} of the device is ≈3.6 and ≈5.8 μm, respectively, and the minimum R_{th} is ≈5% and ≈19% the R_{th} with 1 μm of GaN.

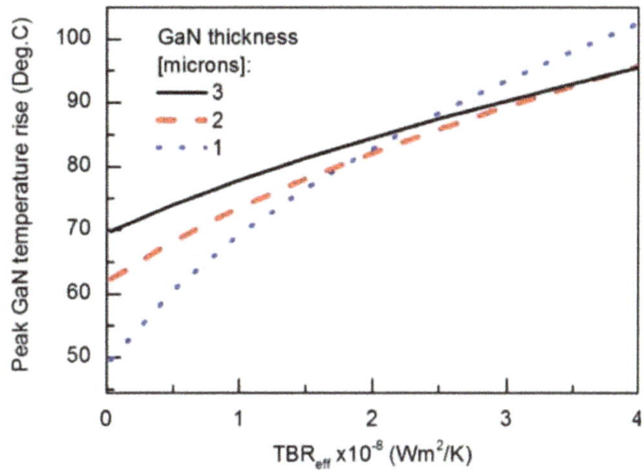

Figure 25. Peak channel temperature of a 4 × 125 μm/40 μm gate pitch GaN-on-diamond HEMT dissipating 10 W/mm for different $TBR_{GaN/diamond}$ values and GaN layer thicknesses (Copyright © 2014 IEICE. Reprinted, with permission, from M. Kuball, J. A. Calvo, R. B. Simon, and J. W. Pomeroy, "Novel thermal management and its analysis in GaN electronics," Asia-Pacific Microw. Conf., pp. 920–922, 2014 [185]).

From what was presented above, it can easily be concluded that in order to take the most benefit of the diamond substrate, the impact of the GaN buffer layer thickness on the overall R_{th} of the device should also be taken into account in the design phase, in addition to the traditionally considered electrical performance and material quality requirements.

4.2. Challenges of Bonding GaN and Diamond Wafers

SAB and VdW bonding are the most promising methods for bonding diamond GaN and diamond substrates. Both methods have a few advantages in comparison to the fabrication of GaN-on-diamond wafers by direct diamond CVD. To begin with, the GaN wafers without the nucleation and strain relief layers can be bonded to SCD plates that feature an extremely high κ, whereas the κ of the films close to the GaN on the GaN-on-diamond wafers is quite low. In addition, only an extremely thin Si layer (or no layer, in the case of VdW bonding) is required to bond the two materials, which allows the $TBR_{GaN/diamond}$ to be significantly reduced. Finally, the bonding is performed at a temperature significantly lower than the 700–800 °C required to deposit good quality PCD (RT in the case of SAB, 300 °C in the case of VdW bonding), which means there is no residual stress (or that it is minimal) at the diamond/GaN interface.

But the bonding of the wafers also presents some drawbacks. The area of available SCD substrates is quite small (only a few mm^2), which seriously compromises the scalability of the process. Bonding GaN wafers with large area PCD substrates could in principle overcome this limitation; however, bonding with PCD substrates is not reproducible in the case of VdW bonding. In a recent work, the SAB of GaN and PCD with a Mo/Au interlayer has been reported and the mechanical strength of the interface was evaluated, but the thermal characterization is still lacking.

Despite these limitations, the potential of the technique has been recognized by Fujitsu and Mitsubishi, and it is expected that commercial devices will be available in a near future. The SAB of GaN HEMTs and large area PCD substrates is also regarded as an area worth of investigation, since the successful bonding of GaN and PCD would allow the fabrication of large area GaN/diamond wafers with improved heat extraction.

4.3. Challenges of Epitaxially Growing GaN on Diamond

The GaN epitaxy allows the deposition of the GaN layers directly onto high thermally conductive diamond substrates without the need of any dielectric layer. However, the required AlN nucleation and stress-relief AlGaN/GaN stacked layers with lower κ will themselves hamper the flow of the heat from the top HEMT structures to the back of the diamond substrates. In addition, if the epitaxy is performed on SCD substrates, the area will be limited to a few mm^2. This means that the true benefit of GaN epitaxy is doubtful when compared with the direct bonding of the HEMT structures and the diamond substrates.

A breakthrough has been achieved with the ELO of the GaN layers on PCD substrates, thus overcoming the small area availability of the SCD substrates. Nevertheless, if the GaN is deposited directly on the PCD substrate, as proposed by Webster et al. [124], the low thermally conductive stress-relief layers will once again compromise the flow of heat. In the approach followed by Ahmed et al. [137], on the other side, the GaN is initially deposited on a Si substrate and selectively deposited PCD stripes replace the SiN stripes used in [124] for the ELO step. The final integration of the ELO grown GaN epilayers with the thick diamond heat spreaders will require a few more steps: (i) the MOCVD of the HEMT structures, (ii) the removal of the Si substrate and AlGaN/GaN stress-relief layers, and (iii) the direct CVD of the diamond film. While (ii) allows placing the diamond directly in contact with the GaN epilayers, (iii) will require the deposition of a dielectric layer on the exposed GaN, as in the case of direct diamond CVD. The structure of the final GaN-on-diamond wafers fabricated using this method will be similar to the structure of GaN-on-diamond wafers described in Section 3.1, but with a difference, since the diamond stripes embedded in the ELO GaN can be overgrown by the thick diamond without the need of a protective dielectric layer. Despite this improvement, the evaluation of the potential of this approach needs to take into consideration that the fabrication procedure is significantly more complex than that of standard GaN-on-diamond wafers.

4.4. Challenges of Capping GaN HEMTs with Diamond

The capping of the HEMT devices with a diamond film is the technically simplest approach. In theory, the GaN HEMTs can be capped with the diamond film after the passivation steps without any changes in the fabrication procedure. However, due to the nature of the diamond CVD process, the diamond-capped HEMTs face some of the same issues as the GaN-on-diamond wafers.

On one side, the passivation layer, though preventing the degradation of III-nitride layers in the harsh CVD environment, contributes with a non-negligible *TBR* which hampers the flow of heat towards the diamond heat spreader. On the other side, the quality—and hence the κ—of the diamond film close to the interface also play a critical role. In this sense, the discussion presented in Sections 4.1.2 and 4.1.3 holds valid for the capping diamond approach. In addition, the top AlGaN barrier layer features a low κ (Table 2) and introduces a non-negligible R_{th} between the hot spot and the capping diamond layer.

Finally, thermal stresses will inevitably accumulate at the GaN/diamond interface due to the *CTE*s mismatch. Any change in the stress–strain state in AlGaN/GaN heterostructures, especially in the fully strained pseudomorphically grown AlGaN barrier layer, would have a significant impact on the 2DEG characteristics. Therefore, the proper understanding of the impact the stress-strain state induced by the diamond CVD on the 2DEG characteristics will require a thorough evaluation of the barrier layer stress.

Nevertheless, these limitations did not prevent Fujitsu from developing diamond capped GaN HEMTs with improved heat dissipating capabilities, meaning that capping HEMTs with diamond may provide a valuable way of improving the thermal management of these devices.

4.5. What Is the Best Approach?

From what was described in the previous sections, it can be concluded that there is no universally best approach, and each one has its own pros and cons.

GaN-on-diamond is undoubtedly the more mature technology. Large-area deposition capability is a significant advantage and, despite being relatively complex, the fabrication process has been optimized and the main issues have, at least partially, been solved. Other companies, in addition to Element Six, have been involved in the process. In 2019 RFHIC reported a manufacturing procedure that allows the fabrication of 4″ GaN-on-diamond HEMTs using a laser drilling process [102]. GaN-on-diamond HEMTs and RF power amplifiers can be currently purchased from Qorvo and Akash Systems, Inc. However, and despite the maturity and success of this technology, some room for improvement still exists (namely the decrease of the gate leakage current), and exciting improvements are expected in the coming years.

The bonding of GaN and diamond SCD substrates is also reaching a high level of maturity, and companies such as Mitsubishi Electric Corp. and Fujitsu Limited have reported the successful transfer of GaN and GaN-on-SiC HEMTS to diamond substrates. This anticipates a bright future for bonded GaN/diamond devices. On one side, these devices show a potential for decreasing the $TBR_{\text{GaN/diamond}}$ below the minimum achievable with GaN-on-diamond technology, since the 30 nm-thick SiN dielectric layer can be replaced by a 2–10 nm-thick Si-based layer. However, since the reported GaN HEMTs have been bonded to SCD substrates, this technique will have a significantly lower yield than the GaN-on-diamond approach and the full scaling up of the technology will be more challenging. If, however, future research deems the bonding of GaN HEMTs and large area PCD substrates feasible and reproducible this technique may compete with the GaN-on-diamond technology.

The epitaxy of GaN on diamond substrates, though feasible, may not bring any realistic advantage. On one side, the best quality GaN films have been grown on low-area SCD substrates. The ELO of GaN has made growth of GaN films with low dislocation density possible on PCD substrates, however functional HEMTs are yet to be demonstrated. On the other side, in the majority of approaches reported so far the nucleation and strain relief layers are part of the final HEMT material stack, and they will hinder the transport of heat to the diamond substrate.

Despite the low κ of the AlGaN barrier layer the capping of passivated HEMTs with a thin diamond film is expected to decrease the peak temperature of the devices between 8% and 20% in comparison with GaN-on-SiC and GaN-on-Si HEMTs, respectively. Despite being a relatively modest number—if compared with the improvement obtained with latest generation GaN-on-diamond devices—this approach is technologically simple, since the diamond can be deposited directly on the passivation layer, and allows for large area growth. A bright future is anticipated also for this approach, as proven by the work recently reported by Fujitsu researchers.

The comparative advantages and disadvantages of each approach are further summarized in Table 7.

Table 7. Current advantages and disadvantages of each approach.

	GaN-on-Diamond	Bonded Wafers		GaN Epitaxy		Capping Diamond
		SCD	PCD	SCD	PCD	
Large area	Yes	No	Yes	No	Yes	Yes
$\kappa_{diamond}$ at interface	Low	High	High	High	ELO GaN-after-PCD: high PCD-after-ELO GaN: evaluation required	Low
$TBR_{GaN/diamond}$	Large	Small	VdW: not reproducible SAB: evaluation required	Evaluation required	ELO GaN-after-PCD: evaluation required PCD-after-ELO GaN: evaluation required	Large
Removal of AlGaN/GaN stress-relief layers	Possible	Possible	Possible	Not possible	ELO GaN-after-PCD: not possible PCD-after-ELO GaN: possible	Not relevant
AlGaN top barrier layer	Not relevant	Not relevant	Not relevant	Not relevant	Not relevant	Present
Induced thermal stress	Relevant	Not relevant	Not relevant	Not relevant	ELO GaN-after-PCD: not relevant PCD-after-ELO GaN: evaluation required	Relevant
Manufacturing complexity	Fair	Fair	Fair	Simple	ELO GaN-after-PCD: fair PCD-after-ELO GaN: complex	Simple

Advantage
Limitation
Severe limitation

5. Conclusions

The integration of diamond and GaN devices has been an active research topic for 20 years. The involvement of companies like Fujitsu and Mitsubishi, for instance, is representative of the impact that hybrid GaN/diamond electronic devices can have on some applications.

The integration of diamond and GaN has been achieved by different methods: the direct CVD of the diamond films on the back of GaN wafers, the bonding of HEMTs and diamond substrates, the direct epitaxy of the GaN layers on diamond substrates, and the diamond capping of passivated HEMTs. The technological advances, the room for improvement, and the advantages/disadvantages of each method have been presented and discussed.

Generally speaking, the fabrication of diamond-on-GaN wafers by direct diamond CVD on the back of the GaN wafers has been quite successful and commercial RF power amplifiers fabricated on GaN-on-diamond wafers are currently available for satellite communications. The bonding of GaN HEMTs and SCD substrates and the capping of GaN HEMTs have also been raising interest from companies such as Mitsubishi Electric Corp. and Fujitsu Limited. Recent advances in the epitaxial growth of GaN layers on PCD substrates anticipate interesting technological developments in a near future. Far from having reached the limits of the technology, it can be thus said that the integration of diamond and GaN will remain an active research topic in the years to come, involving academic and industrial players, with the ultimate goal of increasing the power density and reliability of GaN HEMTs.

Author Contributions: Conceptualization, J.C.M.; investigation, J.C.M.; writing—original draft preparation, J.C.M.; writing—review and editing, J.C.M., M.L. and C.L. All authors have read and agreed to the published version of the manuscript.

Funding: This research was co-funded by EU funds under the project UIDB/50008/2020-UIDP/50008/2020. Joana C. Mendes was hired by Instituto de Telecomunicações under the decree law Nr. 57/2016.

Institutional Review Board Statement: Not applicable.

Informed Consent Statement: Not applicable.

Data Availability Statement: Not applicable.

Conflicts of Interest: The authors declare no conflict of interest.

Appendix A. Performance of Diamond/GaN HEMT Transistors

Table A1. Relevant electrical parameters of HEMTs fabricated on GaN-on-diamond wafers.

Ref.	Year	HEMT Dimensions [a]		DC Measurements			Small-Signal			Large Signal			R_{th} (K·mm/W)
				$I_{D\,max}$ (mA/mm)	$g_{m\,peak}$ (mS/mm)	$P_{D,\,DC}$ (W/mm)	V_D (V)	f_T (GHz)	f_{max} (GHz)	Conditions	*PAE* (%)	P_D (W/mm)	
[63]	2006	$NF/W_G/L_{SD}/L_G$	2/150/4.5/1.5	306	70	–	–	8	11.46	–	–	–	–
[64]	2007	$NF/W_G/L_{SD}/L_G$	1/50/4.5/1.2	800	180	–	–	–	–	–	–	–	–
		$NF/W_G/L_{SD}/L_G$	2/37.5/4.5/1.2	–	–	–	10	12.3	21.8	–	–	–	–
[65]	2007	$NF/W_G/L_{SD}/L_G$	2/125/5.3/0.25	670	187	–	–	27.4	–	Class B; V_D = 25 V	47	2.79	6
										Class B; V_D = 20 V	44	1.92	–
[66]	2009	$NF/W_G/L_{GD}/L_{SG}/L_G$	2/50/2.5/0.5/0.04	580	220	–	–	85	91	–	–	–	–
[70,186]	2013	$NF/W_G/L_{SD}/L_G$	1/50/4/0.25	1100	300	–	–	–	–	–	–	–	–
		$NF/W_G/L_{SD}/L_G$	2/100/4/0.25	–	–	–	30	30	–	10 GHz; V_D = 40 V	>46	>7	–
[28]	2019	$NF/W_G/L_{GD}/L_{SG}/L_G$	2/100/3/2/2	662	199	27.56	10	10.2	31.4	–	–	–	6.7
[101]	2019	–	–	–	–	56	–	–	–	–	–	–	2.95
[102]	2019	$NF/W_G/L_G$	2/300/0.5	–	–	–	–	–	–	Pulsed; PW = 50 µs; DC = 10%; V_D = 100 V	–	22.5	–
		$NF/W_G/L_G$	10/50/0.5	–	–	–	–	–	–	Pulsed; PW = 50 µs; DC = 10%; V_D = 100 V	–	23.2	–
		$NF/W_G/L_G$	10/200/0.5 [b]		–	–	–	–	–	Pulsed; PW = 50 µs; DC = 10%; V_D = 80 V	–	18.1	–

[a] *NF*—number of fingers; L_G—gate length (µm); W_G—gate width (µm); L_{SD}—source-drain spacing (µm); L_{GD}—gate-drain spacing (µm); L_{SG}—source-gate spacing (µm); PW—pulse width; DC—duty cycle). [b] Fully packaged device.

Table A2. Relevant electrical parameters of HEMTs fabricated on GaN/diamond bonded wafers.

Ref.	Year	HEMT Dimensions [a]		DC Measurements		Small-Signal			Large Signal		
				$I_{D\,max}$ (mA/mm)	$g_{m\,peak}$ (mS/mm)	V_D (V)	f_T (GHz)	f_{max} (GHz)	Conditions	*PAE* (%)	P_D (W/mm)
[109]	2014	NF/W_G	2/100	1000	330	–	–	–	10 GHz; V_D = 20 V; tuned for P_D	38	3.4
									10 GHz; V_D = 20 V; tuned for *PAE*	42	3.0

Table A2. *Cont.*

Ref.	Year	HEMT Dimensions [a]		DC Measurements		Small-Signal			Large Signal		
				$I_{D\,max}$ (mA/mm)	$g_{m\,peak}$ (mS/mm)	V_D (V)	f_T (GHz)	f_{max} (GHz)	Conditions	*PAE* (%)	P_D (W/mm)
[109]	2014	NF/W_G	2/100	1000	330	–	–	–	10 GHz; $V_D = 40$ V; tuned for P_D	30	6.0
									10 GHz; $V_D = 40$ V; tuned for *PAE*	33	5.4
[111]	2016	NF/W_G	12/50	1200	390	–	–	–	10 GHz; $V_D = 40$ V	51	11.0
[112]	2017	NF/W_G	10/125	1000	–	–	–	–	–	–	–
		NF/W_G	4/125						Class AB; 10 GHz; $V_D = 28$ V	50.5	5.5
[114]	2019	NF/W_G	4 × 8/180	640	–	–	–	–	–	–	–
[118]	2020	W_G/P_g	1000/30	–	–	–	–	–	Pulsed; PW = 10 μs; DC = 10%; $V_D = 50$ V	–	19.8
									Pulsed; PW = 10 μs; DC = 1%; $V_D = 50$ V		22.3
[119,120]	2020	$NF/W_G/L_G$	2/75/0.08	690	325	4	85	106	–	–	–
[122]	2018	$NF/W_G/L_G$	2/300/0.5	–	–	–	–	–	3 GHz; $V_D = 50$ V; tuned for P_D	46.5	6.63
									3 GHz; $V_D = 50$ V; tuned for *PAE*	54.2	5.39
									Pulsed; $V_D = 50$ V; tuned for P_D	53.5	7.44
									Pulsed; $V_{DS} = 50$ V; tuned for *PAE*	59.1	5.91

[a] *NF*—number of fingers; L_G—gate length (μm); W_G—gate width (μm); P_G—gate pitch (μm); PW—pulse width.

Table A3. Relevant electrical parameters of HEMTs fabricated on GaN layers deposited on diamond substrates.

Ref.	Year	HEMT Dimensions [a]		DC Measurements		Small-Signal			Large Signal			R_{th} (K·mm/W)
				$I_{D\,max}$ (mA/mm)	$g_{m\,peak}$ (mS/mm)	V_D (V)	f_T (GHz)	f_{max} (GHz)	Conditions	*PAE* (%)	P_D (W/mm)	
[129]	2010	$NF/W_G/L_G/L_{SD}$	1/50/0.2/4.5	730	137.5	10	21	42.5	–	–	–	–
[130]	2011	$NF/W_G/L_G/L_{SD}$	1/100/3/20	220	–	10	3	7	–	–	–	–
		$NF/W_G/L_G$	1/630/6	–	–	–	–	–	–	–	–	4.1
[131]	2012	$NF/W_G/L_{GD}/L_{SG}/L_G/L_{SD}$	2/100/2.3/2.3/0.4/5	770	160	15	25	18	–	–	–	–
		$NF/W_G/L_{GD}/L_{SG}/L_G/L_{SD}$	2/400/2.3/2.3/0.4/5	–	–	–	–	–	1 GHz; $V_D = 50$ V	46	2.13	1.5

Table A3. *Cont.*

Ref.	Year	HEMT Dimensions [a]		DC Measurements		Small-Signal			Large Signal			R_{th} (K·mm/W)
				$I_{D\,max}$ (mA/mm)	$g_{m\,peak}$ (mS/mm)	V_D (V)	f_T (GHz)	f_{max} (GHz)	Conditions	*PAE* (%)	P_D (W/mm)	
[132]	2012	$NF/W_G/L_{GD}/L_{SG}/L_G/L_{SD}$	1/100/7.5/7.5/5/20	275	60	–	–	–	–	–	–	–
[133]	2012	$NF/L_G/L_{GD}/L_{SG}/L_{SD}$	1/0.4/2.3/2.3/5	800	160	–	–	–	–	–	–	–

[a] *NF*—number of fingers; L_G—gate length (μm); W_G—gate width (μm); L_{SD}—source-drain spacing (μm); L_{GD}—gate-drain spacing (μm); L_{SG}—source-gate spacing (μm).

Table A4. Relevant electrical parameters of HEMTs featuring capping diamond.

Ref.	Year	HEMT Dimensions [a]		DC Measurements		Small-Signal		R_{th} (K·mm/W)
				$I_{D\,max}$ (mA/mm)	$g_{m\,peak}$ (mS/mm)	f_T (GHz)	f_{max} (GHz)	
[46]	2001	*NF*	2	190	100	–	–	–
[141]	2010	–	–	150	113.9	–	–	–
[142]	2012	–	–	270	–	–	–	–
[143,144]	2013, 2014	–	–	445	127	–	–	0.96
[145]	2013	$NF/W_G/L_G/L_{SD}$	1/100/3/20	430	–	–	–	–
[148]	2011	$NF/W_G/L_G$	1/50/0.25	400	170	4.2	5	–
[154]	2021	–	–	1100	148	–	–	7.4

[a] *NF*—number of fingers; L_G—gate length (μm); W_G—gate width (μm); L_{SD}—source-drain spacing (μm).

References

1. Roccaforte, F.; Fiorenza, P.; Nigro, R.L.; Giannazzo, F.; Greco, G. Physics and technology of gallium nitride materials for power electronics. *Riv. Del Nuovo Cim.* **2018**, *41*, 625–681.
2. Longobardi, G. GaN for power devices: Benefits, applications, and normally-off technologies. In Proceedings of the 2017 International Semiconductor Conference, Sinaia, Romania, 9 November 2017, pp. 11–18.
3. Kuball, M. GaN-on-Diamond Technology Recent UK Successes. In Proceedings of the Diamond D-Day—Workshop hosted by the Centre for Device Thermography and Reliability of the University of Bristol and the GaNDaME Programme, Bristol, UK, 29 January 2019.
4. Amano, H.; Baines, Y.; Beam, E.; Borga, M.; Bouchet, T.; Chalker, P.R.; Charles, M.; Chen, K.J.; Chowdhury, N.; Chu, R.; et al. The 2018 GaN power electronics roadmap. *J. Phys. D Appl. Phys.* **2018**, *51*, 163001. [CrossRef]
5. Lin, Y.Q.; Patterson, A. Design Solutions for 5G Power Amplifiers using 0.15 μm and 0.25 μm GaN HEMTs. In Proceedings of the 2020 International Symposium on VLSI Design, Automation and Test (VLSI-DAT), Hsinchu, Taiwan, 10–13 August 2020; pp. 31–33.
6. Iucolano, F.; Boles, T. GaN-on-Si HEMTs for wireless base stations. *Mater. Sci. Semicond. Process* **2019**, *98*, 100–105. [CrossRef]
7. Chow, T.P.; Omura, I.; Higashiwaki, M.; Kawarada, H.; Pala, V. Smart power devices and ICs using GaAs and wide and extreme bandgap semiconductors. *IEEE Trans. Electron Devices* **2017**, *64*, 856–873. [CrossRef]
8. Millán, J. Wide band-gap power semiconductor devices. *IET Circ. Devices Syst.* **2007**, *1*, 372–379. [CrossRef]
9. Gracio, J.J.; Fan, Q.H.; Madaleno, J.C. Diamond growth by chemical vapour deposition. *J. Phys. D Appl. Phys.* **2010**, *43*, 374017. [CrossRef]
10. Yates, L.; Anderson, J.; Gu, X.; Lee, C.; Bai, T.; Mecklenburg, M.; Aoki, T.; Goorsky, M.S.; Kuball, M.; Piner, E.L.; et al. Low Thermal Boundary Resistance Interfaces for GaN-on-Diamond Devices. *ACS Appl. Mater. Interfaces* **2018**, *10*, 24302–24309. [CrossRef]
11. Diamond Materials, The CVD Diamond Booklet. Available online: https://www.diamond-materials.com/site/assets/files/1095/cvd_diamond_booklet.pdf (accessed on 7 October 2021).
12. Watanabe, H.; Yamada, N.; Okaji, M. Linear thermal expansion coefficient of silicon from 293 to 1000 K. *Int. J. Thermophys.* **2004**, *25*, 221–236. [CrossRef]
13. Stockmeier, M.; Müller, R.; Sakwe, S.A.; Wellmann, P.J.; Magerl, A. On the lattice parameters of silicon carbide. *J. Appl. Phys.* **2009**, *105*, 033511. [CrossRef]
14. He, J.; Cheng, W.C.; Wang, Q.; Cheng, K.; Yu, H.; Chai, Y. Recent Advances in GaN-Based Power HEMT Devices. *Adv. Electron. Mater.* **2021**, *7*, 2001045. [CrossRef]
15. Palacios, T.; Suh, C.-S.; Chakraborty, A.; Keller, S.; DenBaars, S.P.; Mishra, U.K. High-Performance E-Mode AlGaN/GaN HEMTs. *IEEE Electron Device Lett.* **2006**, *27*, 428–430. [CrossRef]
16. Lu, B.; Saadat, O.I.; Palacios, T. High-Performance Integrated Dual-Gate AlGaN/GaN Enhancement-Mode Transistor. *IEEE Electron Device Lett.* **2010**, *31*, 990–992. [CrossRef]
17. Longobardi, G.; Udrea, F.; Sque, S.; Croon, J.; Hurkx, F.; Šonský, J. Investigation of surface charges and traps in gallium nitride/aluminium gallium nitride/gallium nitride high-voltage transistors via measurements and technology computer-aided design simulations of transfer characteristics of metal-insulator-semiconductor. *IET Power Electron.* **2015**, *8*, 2322–2328. [CrossRef]
18. Hu, J.; Stoffels, S.; Lenci, S.; Groeseneken, G.; Decoutere, S. On the identification of buffer trapping for bias-dependent dynamic RON of AlGaN/GaN schottky barrier diode with AlGaN:C back barrier. *IEEE Electron Device Lett.* **2016**, *37*, 310–313. [CrossRef]
19. Mizutani, T.; Ohno, Y.; Akita, M.; Kishimoto, S.; Maezawa, K. A study on current collapse in AlGaN/GaN HEMTs induced by bias stress. *IEEE Trans. Electron Devices* **2003**, *50*, 2015–2020. [CrossRef]
20. Meneghesso, G.; Meneghini, M.; Zanoni, E. Reliability and instabilities in GaN-based HEMTs. In Proceedings of the 2014 IEEE International Conference on Electron Devices and Solid-State Circuits, EDSSC 2014, Chengdu, China, 18–20 June 2014; pp. 5–6.
21. Wang, M.; Yan, D.; Zhang, C.; Xie, B.; Wen, C.P.; Wang, J.; Hao, Y.; Wu, W.; Shen, B. Investigation of surface- and buffer-induced current collapse in GaN high-electron mobility transistors using a soft switched pulsed I-V measurement. *IEEE Electron Device Lett.* **2014**, *35*, 1094–1096. [CrossRef]
22. Jin, D.; del Alamo, J.A. Mechanisms responsible for dynamic ON-resistance in GaN high-voltage HEMTs. In Proceedings of the 2012 24th International Symposium on Power Semiconductor Devices and ICs, Bruges, Belgium, 3–7 June 2012; pp. 333–336.
23. Ubochi, B.; Faramehr, S.; Ahmeda, K.; Igić, P.; Kalna, K. Operational frequency degradation induced trapping in scaled GaN HEMTs. *Microelectron. Reliab.* **2017**, *71*, 35–40. [CrossRef]
24. McGovern, P.; Tasker, P.J.; Powell, J.R.; Hilton, K.P.; Glasper, J.L.; Balmer, R.S.; Martin, T.; Uren, M.J. Analysis of DC-RF Dispersion in AlGaN/GaN HFETs Using Pulsed I-V and Time-Domain Waveform Measurements. In Proceedings of the IEEE MTT-S International Microwave Symposium Digest, Long Beach, CA, USA, 17 June 2005; pp. 513–516.
25. Gomes, J.L.; Nunes, L.C.; Goncalves, C.F.; Pedro, J.C. An Accurate Characterization of Capture Time Constants in GaN HEMTs. *IEEE Trans. Microw. Theory Technol.* **2019**, *67*, 2465–2474. [CrossRef]
26. Gaska, R.; Osinsky, A.; Yang, J.W.; Shur, M.S. Self-heating in high-power AlGaN-GaN HFETs. *IEEE Electron Device Lett.* **1998**, *19*, 89–91. [CrossRef]
27. Azarifar, M.; Donmezer, N. A multiscale analytical correction technique for two-dimensional thermal models of AlGaN/GaN HEMTs. *Microelectron. Reliab.* **2017**, *74*, 82–87. [CrossRef]

28. Ranjan, K.; Arulkumaran, S.; Ng, G.I.; Sandupatla, A. Investigation of Self-Heating Effect on DC and RF Performances in AlGaN/GaN HEMTs on CVD-Diamond. *IEEE J. Electron Devices Soc.* **2019**, *7*, 1264–1269. [CrossRef]
29. Kraus, A.D.; Bar-cohen, A. *Thermal Analysis and Control of Electronic Equipment*; Hemisphere Pub. Corp.: Washington, DC, USA, 1983.
30. Joshin, K.; Kikkawa, T. High-power and high-efficiency GaN HEMT amplifiers. In Proceedings of the 2008 IEEE Radio and Wireless Symposium, Orlando, FL, USA, 22–24 January 2008; pp. 65–68.
31. Darwish, A.M.; Bayba, A.J.; Hung, H.A. Thermal resistance calculation of AlGaN-GaN devices. *IEEE Trans. Microw. Theory Technol.* **2004**, *52*, 2611–2620. [CrossRef]
32. Dingle, R.; Störmer, H.L.; Gossard, A.C.; Wiegmann, W. Electron mobilities in modulation-doped semiconductor heterojunction superlattices. *Appl. Phys. Lett.* **1978**, *33*, 665–667. [CrossRef]
33. Mimura, T.; Hiyamizu, S.; Fujii, T.; Nanbu, K. A New Field-Effect Transistor with Selectively Doped {GaAs}/n-{AlxGa}1-{xAs} Heterojunctions. *Jpn. J. Appl. Phys.* **1980**, *19*, L225–L227. [CrossRef]
34. Nakamura, S.; Harada, Y.; Seno, M. Novel metalorganic chemical vapor deposition system for GaN growth. *Appl. Phys. Lett.* **1991**, *58*, 2021–2023. [CrossRef]
35. Khan, M.A.; Bhattarai, A.; Kuznia, J.N.; Olson, D.T. High electron mobility transistor based on a GaN-AlxGa 1-xN heterojunction. *Appl. Phys. Lett.* **1993**, *63*, 1214–1215. [CrossRef]
36. Keller, S. Substrates and Materials. In *Power GaN Devices: Materials, Applications and Reliability*; Meneghini, M., Meneghesso, G., Zanoni, E., Eds.; Springer International Publishing: Cham, Switzerland, 2017; pp. 27–52.
37. Amano, H.; Sawaki, N.; Akasaki, I.; Toyoda, Y. Metalorganic vapor phase epitaxial growth of a high quality GaN film using an AlN buffer layer. *Appl. Phys. Lett.* **1986**, *48*, 353–355. [CrossRef]
38. Manoi, A.; Pomeroy, J.W.; Killat, N.; Kuball, M. Benchmarking of thermal boundary resistance in AlGaN/GaN HEMTs on SiC substrates: Implications of the nucleation layer microstructure. *IEEE Electron Device Lett.* **2010**, *31*, 1395–1397. [CrossRef]
39. Chang, S.; Zhao, M.; Spampinato, V.; Franquet, A.; Chang, L. The influence of AlN nucleation layer on RF transmission loss of GaN buffer on high resistivity Si (111) substrate. *Semicond. Sci. Technol.* **2020**, *35*, 035029. [CrossRef]
40. Chen, J.-T.; Pomeroy, J.W.; Rorsman, N.; Xia, C.; Virojanadara, C.; Forsberg, U.; Kuball, M.; Janzén, E. Low thermal resistance of a GaN-on-SiC transistor structure with improved structural properties at the interface. *J. Cryst. Growth* **2015**, *428*, 54–58. [CrossRef]
41. Pomeroy, J.; Bernardoni, M.; Sarua, A.; Manoi, A.; Dumka, D.C.; Fanning, D.M.; Kuball, M. Achieving the best thermal performance for GaN-on-diamond. In Proceeding of the IEEE Compound Semiconductor Integrated Circuit Symposium, CSICS, Monterey, CA, USA, 13–16 October 2013; pp. 1–4.
42. Smorchkova, I.P.; Chen, L.; Mates, T.; Shen, L.; Heikman, S.; Moran, B.; Keller, S.; DenBaars, S.P.; Speck, J.S.; Mishra, U.K. AlN/GaN and (Al,Ga)N/AlN/GaN two-dimensional electron gas structures grown by plasma-assisted molecular-beam epitaxy. *J. Appl. Phys.* **2001**, *90*, 5196–5201. [CrossRef]
43. Lee, D.S.; Gao, X.; Guo, S.; Kopp, D.; Fay, P.; Palacios, T. 300-GHz InAlN/GaN HEMTs With InGaN Back Barrier. *IEEE Electron Device Lett.* **2011**, *32*, 1525–1527. [CrossRef]
44. Yang, S.; Tang, Z.; Hua, M.; Zhang, Z.; Wei, J.; Lu, Y.; Chen, K.J. Investigation of SiNx and AlN Passivation for AlGaN/GaN High-Electron-Mobility Transistors: Role of Interface Traps and Polarization Charges. *IEEE J. Electron Devices Soc.* **2020**, *8*, 358–364. [CrossRef]
45. Brannick, A.; Zakhleniuk, N.A.; Ridley, B.K.; Shealy, J.R.; Schaff, W.J.; Eastman, L.F. Influence of Field Plate on the Transient Operation of the AlGaN/GaN HEMT. *IEEE Electron Device Lett.* **2009**, *30*, 436–438. [CrossRef]
46. Seelmann-Eggebert, M.; Meisen, P.; Schaudel, F.; Koidl, P.; Vescan, A.; Leier, H. Heat-spreading diamond films for GaN-based high-power transistor devices. *Diam. Relat. Mater.* **2001**, *10*, 744–749. [CrossRef]
47. Mitterhuber, L.; Hammer, R.; Dengg, T.; Spitaler, J. Thermal Characterization and Modelling of AlGaN-GaN Multilayer Structures for HEMT Applications. *Energies* **2020**, *13*, 2363. [CrossRef]
48. Kucukgok, B.; Wu, X.; Wang, X.; Liu, Z.; Ferguson, I.T.; Lu, N. The structural properties of InGaN alloys and the interdependence on the thermoelectric behavior. *AIP Adv.* **2016**, *6*, 25305. [CrossRef]
49. Ejeckam, F.; Francis, D.; Faili, F.; Twitchen, D.; Bolliger, B.; Babic, D.; Felbinger, J. S2-T1: GaN-on-diamond: A brief history. In Proceedings of the Lester Eastman Conference 2014—High Perform. Devices, LEC, Ithaca, NY, USA, 5–7 August 2014; pp. 1–5.
50. Swartz, E.T.; Pohl, R.O. Thermal boundary resistance. *Rev. Mod. Phys.* **1989**, *61*, 605–668. [CrossRef]
51. Bao, W.; Wang, Z.; Tang, D. Phonon transport across GaN/AlN interface: Interfacial phonon modes and phonon local non-equilibrium analysis. *Int. J. Heat Mass Transf.* **2022**, *183*, 122090. [CrossRef]
52. Zhao, Y.; Zhu, C.; Wang, S.; Tian, J.Z.; Yang, D.J.; Chen, C.K.; Cheng, H.; Hing, P. Pulsed photothermal reflectance measurement of the thermal conductivity of sputtered aluminum nitride thin films. *J. Appl. Phys.* **2004**, *96*, 4563–4568. [CrossRef]
53. Zou, J.; Kotchetkov, D.; Balandin, A.A.; Florescu, D.I.; Pollak, F.H. Thermal conductivity of GaN films: Effects of impurities and dislocations. *J. Appl. Phys.* **2002**, *92*, 2534–2539. [CrossRef]
54. Riedel, G.J.; Pomeroy, J.W.; Hilton, K.P.; Maclean, J.O.; Wallis, D.J.; Uren, M.J.; Martin, T.; Forsberg, U.; Lundskog, A.; Kakanakova-Georgieva, A.; et al. Reducing thermal resistance of AlGaN/GaN electronic devices using novel nucleation layers. *IEEE Electron Device Lett.* **2009**, *30*, 103–106. [CrossRef]
55. Sarua, A.; Ji, H.; Hilton, K.P.; Wallis, D.J.; Uren, M.J.; Martin, T.; Kuball, M. Thermal boundary resistance between GaN and substrate in AlGaN/GaN electronic devices. *IEEE Trans. Electron Devices* **2007**, *54*, 3152–3158. [CrossRef]
56. Inagaki, M.; Kaburagi, Y.; Hishiyama, Y. Thermal management material: Graphite. *Adv. Eng. Mater.* **2014**, *16*, 494–506. [CrossRef]

57. Chu, R.C.; Ellsworth, J.M.J.; Marotta, E.; Singh, P. Graphite-Based Heat Sinks and Apparatus for the Manufacture Thereof. U.S. Patent 6,907,917 B2, 21 June 2005.
58. Huang, Y.; Ouyang, Q.; Guo, Q.; Guo, X.; Zhang, G.; Zhang, D. Graphite film/aluminum laminate composites with ultrahigh thermal conductivity for thermal management applications. *Mater. Des.* **2016**, *90*, 508–515. [CrossRef]
59. Balandin, A.A. Thermal properties of graphene and nanostructured carbon materials. *Nat. Mater.* **2011**, *10*, 569–581. [CrossRef] [PubMed]
60. Huang, P.; Li, Y.; Yang, G.; Li, Z.-X.; Li, Y.-Q.; Hu, N.; Fu, S.-Y.; Novoselov, K.S. Graphene film for thermal management: A review. *Nano Mater. Sci.* **2021**, *3*, 1–16. [CrossRef]
61. Yan, Z.; Liu, G.; Khan, J.M.; Balandin, A.A. Graphene quilts for thermal management of high-power GaN transistors. *Nat. Commun.* **2012**, *3*, 827–828. [CrossRef] [PubMed]
62. Mukherjee, D.; Mendes, J.C.; Alves, L.N. Diamond for Future (Power) Electronics. *Power Electron. Eur.* **2020**, *4*, 6–8.
63. Jessen, G.H.; Gillespie, J.K.; Via, G.D.; Crespo, A.; Langley, D.; Wasserbauer, J.; Faili, F.; Francis, D.; Babic, D.; Ejeckam, F.; et al. AlGaN/GaN HEMT on diamond technology demonstration. In Proceedings of the Technical Digest—IEEE Compound Semiconductor Integrated Circuit Symposium, CSIC, San Antonio, TX, USA, 12–15 November 2006; pp. 271–274.
64. Dumka, D.C.; Saunier, P. AlGaN/GaN HEMTs on diamond substrate. In Proceedings of the 65th DRC Device Research Conference, South Bend, Indiana, 18–20 June 2007; pp. 31–32.
65. Felbinger, J.G.; Chandra, M.V.S.; Sun, Y.; Eastman, L.F.; Wasserbauer, J.; Faili, F.; Babic, D.; Francis, D.; Ejeckam, F. Comparison of GaN HEMTs on diamond and SiC substrates. *IEEE Electron Device Lett.* **2007**, *28*, 948–950. [CrossRef]
66. Diduck, Q.; Felbinger, J.; Eastman, L.F.; Francis, D.; Wasserbauer, J.; Faili, F.; Babić, D.I.; Ejeckam, F. Frequency performance enhancement of AlGaN/GaN HEMTs on diamond. *Electron. Lett.* **2009**, *45*, 758–759. [CrossRef]
67. Babić, D.I.; Diduck, Q.; Yenigalla, P.; Schreiber, A.; Francis, D.; Faili, F.; Ejeckam, F.; Felbinger, J.G.; Eastman, L.F. GaN-on-diamond field-effect transistors: From wafers to amplifier modules. In Proceedings of the International Convention on Information and Communications Technology, Electronics and Microelectronics (MIPRO), Opatija, Hrvatska, 24–28 May 2010; pp. 60–66.
68. Dumka, D.C.; Chowdhury, U.; Jimenez, J.L.; Chou, T.-M.; Ketterson, A.; Fanning, D.M.; Murdock, B.; Faili, F.; Francis, D.; Ejeckam, F. Near-Junction Thermal Management in High Power GaN HEMTs. In Proceedings of the Government Microcircuit Applications and Critical Technology Conf. (GOMACTech 2012), Las Vegas, NV, USA, 19–22 March 2012; Volume 556, pp. 6–9.
69. Cho, J.; Li, Z.; Bozorg-Grayeli, E.; Kodama, T.; Francis, D.; Ejeckam, F.; Faili, F.; Asheghi, M.; Goodson, K.E. Improved thermal interfaces of GaN-Diamond composite substrates for HEMT applications. *IEEE Trans. Compon. Packag. Manuf. Technol.* **2013**, *3*, 79–85. [CrossRef]
70. Dumka, D.C.; Chou, T.M.; Jimenez, J.L.; Fanning, D.M.; Francis, D.; Faili, F.; Ejeckam, F.; Bernardoni, M.; Pomeroy, J.W.; Kuball, M. Electrical and thermal performance of AlGaN/GaN HEMTs on diamond substrate for RF applications. In Proceedings of the 2013 IEEE Compound Semiconductor Integrated Circuit Symposium (CSICS), Monterey, CA, USA, 13–16 October 2013; pp. 1–4.
71. Altman, D.; Tyhach, M.; McClymonds, J.; Kim, S.; Graham, S.; Cho, J.; Goodson, K.; Francis, D.; Faili, F.; Ejeckam, F.; et al. Analysis and characterization of thermal transport in GaN HEMTs on Diamond substrates. In Proceedings of the Intersociety Conference on Thermal and Thermomechanical Phenomena in Electronic Systems (ITherm), Orlando, FL, USA, 27–30 May 2014; pp. 1199–1205.
72. Pomeroy, J.W.; Bernardoni, M.; Dumka, D.C.; Fanning, D.M.; Kuball, M. Low thermal resistance GaN-on-diamond transistors characterized by three-dimensional Raman thermography mapping. *Appl. Phys. Lett.* **2014**, *104*, 083513. [CrossRef]
73. el Helou, A.; Tadjer, M.; Hobart, K.; Raad, P.E. Effects of Thermal Boundary Resistance on the Thermal Performance of GaN HEMT on Diamond. In Proceedings of the 2020 IEEE International Reliability Physics Symposium (IRPS), Dallas, TX, USA, 28 April–30 May 2020; pp. 2–5.
74. Waller, W.M.; Pomeroy, J.W.; Field, D.; Smith, E.J.W.; May, P.W.; Kuball, M. Thermal boundary resistance of direct van der Waals bonded GaN-on-diamond. *Semicond. Sci. Technol.* **2020**, *35*, 095021. [CrossRef]
75. Wu, Q.; Wu, Q.; Xu, Y.; Zhou, J.; Kong, Y.; Chen, T.; Wang, Y.; Lin, F.; Fu, Y.; Jia, Y.; et al. Performance Comparison of GaN HEMTs on Diamond and SiC Substrates Based on Surface Potential Model. *ECS J. Solid State Sci. Technol.* **2017**, *6*, Q171–Q178. [CrossRef]
76. Smith, E.J.W.; Piracha, A.H.; Field, D.; Pomeroy, J.W.; Mackenzie, G.R.; Abdallah, Z.; Massabuau, F.C.P.; Hinz, A.M.; Wallis, D.J.; Oliver, R.A.; et al. Mixed-size diamond seeding for low-thermal-barrier growth of CVD diamond onto GaN and AlN. *Carbon N. Y.* **2020**, *167*, 620–626. [CrossRef]
77. Pomeroy, J.W.; Simon, R.B.; Sun, H.; Francis, D.; Faili, F.; Twitchen, D.J.; Kuball, M. Contactless thermal boundary resistance measurement of GaN-on-diamond wafers. *IEEE Electron Device Lett.* **2014**, *35*, 1007–1009. [CrossRef]
78. Sun, H.; Simon, R.B.; Pomeroy, J.W.; Francis, D.; Faili, F.; Twitchen, D.J.; Kuball, M. Reducing GaN-on-diamond interfacial thermal resistance for high power transistor applications. *Appl. Phys. Lett.* **2015**, *106*, 111906. [CrossRef]
79. Sun, H.; Pomeroy, J.W.; Simon, R.B.; Francis, D.; Faili, F.; Twitchen, D.J.; Kuball, M. Temperature-Dependent Thermal Resistance of GaN-on-Diamond HEMT Wafers. *IEEE Electron Device Lett.* **2016**, *37*, 621–624. [CrossRef]
80. Cho, J.; Francis, D.; Altman, D.H.; Asheghi, M.; Goodson, K.E. Phonon conduction in GaN-diamond composite substrates. *J. Appl. Phys.* **2017**, *121*, 055105. [CrossRef]
81. Liu, D.; Francis, D.; Faili, F.; Middleton, C.; Anaya, J.; Pomeroy, J.W.; Twitchen, D.J.; Kuball, M. Impact of diamond seeding on the microstructural properties and thermal stability of GaN-on-diamond wafers for high-power electronic devices. *Scr. Mater.* **2017**, *128*, 57–60. [CrossRef]

82. Zhou, Y.; Anaya, J.; Pomeroy, J.; Sun, H.; Gu, X.; Xie, A.; Beam, E.; Becker, M.; Grotjohn, T.A.; Lee, C.; et al. Barrier layer optimization for enhanced GaN-on-diamond device cooling. *ACS Appl. Mater. Interfaces* **2017**, *9*, 34416–34422. [CrossRef]
83. Jia, X.; Wei, J.-J.; Kong, Y.; Li, C.-M.; Liu, J.; Chen, L.; Sun, F.; Wang, X. The influence of dielectric layer on the thermal boundary resistance of GaN-on-diamond substrate. *Surf. Interface Anal.* **2019**, *51*, 783–790. [CrossRef]
84. Mandal, S.; Yuan, C.; Massabuau, F.; Pomeroy, J.W.; Cuenca, J.; Bland, H.; Thomas, E.; Wallis, D.; Batten, T.; Morgan, D.; et al. Thick, Adherent Diamond Films on AlN with Low Thermal Barrier Resistance. *ACS Appl. Mater. Interfaces* **2019**, *11*, 40826–40834. [CrossRef]
85. Park, K.; Bayram, C. Thermal resistance optimization of GaN/substrate stacks considering thermal boundary resistance and temperature-dependent thermal conductivity. *Appl. Phys. Lett.* **2016**, *109*, 151904. [CrossRef]
86. Guo, H.; Kong, Y.; Chen, T. Thermal simulation of high power GaN-on-diamond substrates for HEMT applications. *Diam. Relat. Mater.* **2017**, *73*, 260–266. [CrossRef]
87. Guo, H.; Kong, Y.; Chen, T. Impact of thermal boundary resistance on the thermal design of GaN-on-diamond HEMTs. In Proceedings of the 2019 IEEE 69th Electronic Components and Technology Conference (ECTC), Las Vegas, NV, USA, 28–31 May 2019; Volume 2019-May, pp. 1842–1847.
88. Middleton, C.; Chandrasekar, H.; Singh, M.; Pomeroy, J.W.; Uren, M.J.; Francis, D.; Kuball, M. Impact of thinning the GaN buffer and interface layer on thermal and electrical performance in GaN-on-diamond electronic devices. *Appl. Phys. Express* **2019**, *12*, 024003. [CrossRef]
89. Zou, B.; Sun, H.; Guo, H.; Dai, B.; Zhu, J. Thermal characteristics of GaN-on-diamond HEMTs: Impact of anisotropic and inhomogeneous thermal conductivity of polycrystalline diamond. *Diam. Relat. Mater.* **2019**, *95*, 28–35. [CrossRef]
90. Song, C.; Kim, J.; Cho, J. The effect of GaN epilayer thickness on the near-junction thermal resistance of GaN-on-diamond devices. *Int. J. Heat Mass Transf.* **2020**, *158*, 119992. [CrossRef]
91. Siddique, A.; Ahmed, R.; Anderson, J.; Nazari, M.; Yates, L.; Graham, S.; Holtz, M.; Piner, E.L. Structure and Interface Analysis of Diamond on an AlGaN/GaN HEMT Utilizing an in Situ SiNx Interlayer Grown by MOCVD. *ACS Appl. Electron. Mater.* **2019**, *1*, 1387–1399. [CrossRef]
92. Hancock, B.L.; Nazari, M.; Anderson, J.; Piner, E.; Faili, F.; Oh, S.; Twitchen, D.; Graham, S.; Holtz, M. Ultraviolet micro-Raman spectroscopy stress mapping of a 75-mm GaN-on-diamond wafer. *Appl. Phys. Lett.* **2016**, *108*, 211901. [CrossRef]
93. Hines, N.J. Advanced Solutions for Thermal Management and Reliabiliy of Gallium Nitride-Based High Electron Mobility Transistors on Diamond Substrates. Doctoral Thesis, Georgia Institute of Technology, Atlanta, GA, USA, 2019.
94. Smith, M.D.; Cuenca, J.A.; Field, D.E.; Fu, Y.C.; Yuan, C.; Massabuau, F.; Mandal, S.; Pomeroy, J.W.; Oliver, R.A.; Uren, M.J.; et al. GaN-on-diamond technology platform: Bonding-free membrane manufacturing process. *AIP Adv.* **2020**, *10*, 035306. [CrossRef]
95. Cuenca, J.A.; Smith, M.D.; Field, D.E.; Massabuau, F.C.-P.; Mandal, S.; Pomeroy, J.; Wallis, D.J.; Oliver, R.A.; Thayne, I.; Kuball, M.; et al. Thermal stress modelling of diamond on GaN/III-Nitride membranes. *Carbon* **2021**, *174*, 647–661. [CrossRef]
96. Jia, X.; Wei, J.; Huang, Y.; Shao, S.; An, K.; Kong, Y.; Liu, J.; Chen, L.; Li, C. Fabrication of low stress GaN-on-diamond structure via dual-sided diamond film deposition. *J. Mater. Sci.* **2021**, *56*, 6903–6911. [CrossRef]
97. Liu, D.; Sun, H.; Pomeroy, J.W.; Francis, D.; Faili, F.; Twitchen, D.J.; Kuball, M. GaN-on-diamond electronic device reliability: Mechanical and thermo-mechanical integrity. *Appl. Phys. Lett.* **2015**, *107*, 251902. [CrossRef]
98. Zhai, W.; Zhang, J.; Chen, X.; Bu, R.; Wang, H.; Hou, X. FEM thermal and stress analysis of bonded GaN-on-diamond substrate. *AIP Adv.* **2017**, *7*, 095105. [CrossRef]
99. Liu, D.; Fabes, S.; Li, B.S.; Francis, D.; Ritchie, R.O.; Kuball, M. Characterization of the Interfacial Toughness in a Novel 'GaN-on-Diamond' Material for High-Power RF Devices. *ACS Appl. Electron. Mater.* **2019**, *1*, 354–369. [CrossRef]
100. Ranjan, K.; Ing, N.G.; Arulkumaran, S.; Vicknesh, S.; Foo, S.C. Enhanced DC and RF performance of AlGaN/GaN HEMTs on CVD-Diamond in high power CW operation. In Proceedings of the 2019 Electron Devices Technology and Manufacturing Conference (EDTM 2019), Singapore, 12–15 March 2019; pp. 97–99.
101. Tadjer, M.J.; Anderson, T.J.; Ancona, M.G.; Raad, P.E.; Komarov, P.; Bai, T.; Gallagher, J.C.; Koehler, A.D.; Goorsky, M.S.; Francis, D.A.; et al. GaN-On-Diamond HEMT Technology with TAVG = 176 C at PDC,max = 56 W/mm Measured by Transient Thermoreflectance Imaging. *IEEE Electron Device Lett.* **2019**, *40*, 881–884. [CrossRef]
102. Lee, W.S.; Lee, K.W.; Lee, S.H.; Cho, K.; Cho, S. A GaN/Diamond HEMTs with 23 W/mm for Next Generation High Power RF Application. In Proceedings of the IEEE MTT-S International Microwave Symposium Digest, Boston, MA, USA, 2–7 June 2019; pp. 1395–1398.
103. Cho, J.; Won, Y.; Francis, D.; Asheghi, M.; Goodson, K.E. Thermal interface resistance measurements for GaN-on-diamond composite substrates. In Proceedings of the 2014 IEEE Compound Semiconductor Integrated Circuit Symposium (CSICS), La Jolla, CA, USA, 19–22 October 2014; pp. 1–4.
104. Bolliger, B.; Ejeckam, F.; Dodson, J.; Faili, F.; Francis, D.; Lowe, F.; Twitchen, D. CVD diamond for enhanced GaN devices. In Proceedings of the 2014 Asia-Pacific Microwave Conference, Sendai, Japan, 4–7 November 2014.
105. Tyhach, M.; Altman, D.; Bernstein, S.; Korenstein, R.; Cho, J.; Goodson, K.E.; Francis, D.; Faili, F.; Ejeckam, F.; Kim, S.; et al. S2-T3: Next generation gallium nitride HEMTs enabled by diamond substrates. In Proceedings of the 2014 Lester Eastman Conference on High Performance Devices (LEC), Ithaca, NY, USA, 5–7 August 2014.

106. Field, D.E.; Cuenca, J.A.; Smith, M.; Fairclough, S.M.; Massabuau, F.C.-P.; Pomeroy, J.W.; Williams, O.; Oliver, R.A.; Thayne, I.; Kuball, M. Crystalline Interlayers for Reducing the Effective Thermal Boundary Resistance in GaN-on-Diamond. *ACS Appl. Mater. Interfaces* **2020**, *12*, 54138–54145. [CrossRef]
107. Chao, P.C.; Chu, K.; Creamer, C. A new high power GaN-on-diamond HEMT with low-temperature bonded substrate technology. In Proceedings of the 2013 International Conference on Compound Semiconductor Manufacturing Technology, CS MANTECH 2013, New Orleans, LA, USA, 13–16 May 2013; pp. 179–182.
108. Chu, K.K.; Yurovchak, T.; Chao, P.C.; Creamer, C.T. Thermal modeling of high power GaN-on-diamond HEMTs fabricated by low-temperature device transfer process. In Proceedings of the 2013 IEEE Compound Semiconductor Integrated Circuit Symposium (CSICS), Monterey, CA, USA, 13–16 October 2013; pp. 6–9.
109. Chu, K.K.; Chao, P.C.; Diaz, J.A.; Yurovchak, T.; Creamer, C.T.; Sweetland, S.; Kallaher, R.L.; McGray, C. S2-T4: Low-temperature substrate bonding technology for high power GaN-on-diamond HEMTs. In Proceedings of the 2014 Lester Eastman Conference on High Performance Devices (LEC), Ithaca, NY, USA, 5–7 August 2014; pp. 1–4.
110. Chao, P.C.; Chu, K.; Diaz, J.; Creamer, C.; Sweetland, S.; Kallaher, R.; McGray, C.; Via, G.D.; Blevins, J. GaN-on-Diamond HEMTs with 11 W/mm Output Power at 10 GHz. *MRS Adv.* **2016**, *1*, 147–155. [CrossRef]
111. Chu, K.K.; Chao, P.C.; Diaz, J.A.; Yurovchak, T.; Schmanski, B.J.; Creamer, C.T.; Sweetland, S.; Kallaher, R.L.; McGray, C.; Via, G.D.; et al. High-Performance GaN-on-Diamond HEMTs Fabricated by Low-Temperature Device Transfer Process. In Proceedings of the 2015 IEEE Compound Semiconductor Integrated Circuit Symposium (CSICS), New Orleans, LA, USA, 11–14 October 2015; pp. 7–10.
112. Liu, T.; Kong, Y.; Wu, L.; Guo, H.; Zhou, J.; Kong, C.; Chen, T. 3-inch GaN-on-Diamond HEMTs with Device-First Transfer Technology. *IEEE Electron Device Lett.* **2017**, *38*, 1417–1420. [CrossRef]
113. Mu, F.; He, R.; Suga, T. Room temperature GaN-diamond bonding for high-power GaN-on-diamond devices. *Scr. Mater.* **2018**, *150*, 148–151. [CrossRef]
114. Hiza, S.; Fujikawa, M.; Takiguchi, Y.; Nishimura, K.; Yagyu, E.; Matsumae, T.; Kurashima, Y.; Takagi, H.; Yamamuka, M. High-Power GaN-on-Diamond HEMTs Fabricated by Surface-Activated Room-Temperature Bonding. In Proceedings of the 2019 International Conference on Solid State Devices and Materials, Aichi, Japan, 2–5 September 2019; Volume 4, pp. 467–468.
115. Cheng, Z.; Mu, F.; Yates, L.; Suga, T.; Graham, S. Interfacial Thermal Conductance across Room-Temperature Bonded GaN-Diamond Interfaces for GaN-on-Diamond Devices. *ACS Appl. Mater. Interfaces* **2020**, *12*, 8376–8384. [CrossRef] [PubMed]
116. Suga, T.; Mu, F. Direct Bonding of GaN to Diamond Substrate at Room Temperature. In Proceedings of the 2020 IEEE 70th Electronic Components and Technology Conference (ECTC), Orlando, FL, USA, 3–30 June 2020; pp. 1328–1331.
117. Wang, K.; Ruan, K.; Hu, W.; Wu, S.; Wang, H. Room temperature bonding of GaN on diamond wafers by using Mo/Au nano-layer for high-power semiconductor devices. *Scr. Mater.* **2020**, *174*, 87–90. [CrossRef]
118. Minoura, Y.; Ohki, T.; Okamoto, N.; Yamada, A.; Makiyama, K.; Kotani, J.; Ozaki, S.; Sato, M.; Nakamura, N. Surface activated bonding of SiC/diamond for thermal management of high-output power GaN HEMTs. *Jpn. J. Appl. Phys.* **2020**, *59*, SGGD03. [CrossRef]
119. Abou Daher, M.; Lesecq, M.; Tilmant, P.; Defrance, N.; Rousseau, M.; Cordier, Y.; De Jaeger, J.C.; Tartarin, J.G. AlGaN/GaN high electron mobility transistors on diamond substrate obtained through aluminum nitride bonding technology. *J. Vac. Sci. Technol. B* **2020**, *38*, 033201. [CrossRef]
120. Abou Daher, M.; Lesecq, M.; Defrance, N.; Okada, E.; Boudart, B.; Guhel, Y.; Tartarin, J.G.; de Jaeger, J.C. Electrical and thermal analysis of AlGaN/GaN HEMTs transferred onto diamond substrate through an aluminum nitride layer. *Microw. Opt. Technol. Lett.* **2021**, *63*, 1–5. [CrossRef]
121. Yablonovitch, E.; Hwang, D.M.; Gmitter, T.J.; Florez, L.T.; Harbison, J.P. Van der Waals bonding of GaAs epitaxial liftoff films onto arbitrary substrates. *Appl. Phys. Lett.* **1990**, *56*, 2419–2421. [CrossRef]
122. Gerrer, T.; Cimalla, V.; Waltereit, P.; Müller, S.; Benkhelifa, F.; Maier, T.; Czap, H.; Ambacher, O.; Quay, R. Transfer of AlGaN/GaN RF-devices onto diamond substrates via van der Waals bonding. *Int. J. Microw. Wirel. Technol.* **2018**, *10*, 666–673. [CrossRef]
123. Cimalla, V. Low Temperature Bonding of GaN electronics on diamond heat spreaders. In Proceedings of the Diamond D-Day—Workshop hosted by the Centre for Device Thermography and Reliability of the University of Bristol and the GaNDaME Programme, Bristol, UK, 29 January 2019.
124. Webster, R.F.; Cherns, D.; Kuball, M.; Jiang, Q.; Allsopp, D. Electron microscopy of gallium nitride growth on polycrystalline diamond. *Semicond. Sci. Technol.* **2015**, *30*, 114007. [CrossRef]
125. Hageman, P.R.; Schermer, J.J.; Larsen, P.K. GaN growth on single-crystal diamond substrates by metalorganic chemical vapor deposition and hydride vapour deposition. *Thin Solid Films* **2003**, *443*, 9–13. [CrossRef]
126. van Dreumel, G.W.G.; Tinnemans, P.T.; van Den Heuvel, A.A.J.; Bohnen, T.; Buijnsters, J.G.; ter Meulen, J.J.; van Enckevort, W.J.P.; Hageman, P.R.; Vlieg, E. Realising epitaxial growth of GaN on (001) diamond. *J. Appl. Phys.* **2011**, *110*, 013503. [CrossRef]
127. Dussaigne, A.; Malinverni, M.; Martin, D.; Castiglia, A.; Grandjean, N. GaN grown on (111) single crystal diamond substrate by molecular beam epitaxy. *J. Cryst. Growth* **2009**, *311*, 4539–4542. [CrossRef]
128. Dussaigne, A.; Gonschorek, M.; Malinverni, M.; Py, M.A.; Martin, D.; Mouti, A.; Stadelmann, P.; Grandjean, N. High-mobility AlGaN/GaN two-dimensional electron gas heterostructure grown on (111) single crystal diamond substrate. *Jpn. J. Appl. Phys.* **2010**, *49*, 061001. [CrossRef]

129. Alomari, M.; Dussaigne, A.; Martin, D.; Grandjean, N.; Gaquière, C.; Kohn, E. AlGaN/GaN HEMT on (111) single crystalline diamond. *Electron. Lett.* **2010**, *46*, 299–301. [CrossRef]
130. Hirama, K.; Taniyasu, Y.; Kasu, M. AlGaN/GaN high-electron mobility transistors with low thermal resistance grown on single-crystal diamond (111) substrates by metalorganic vapor-phase epitaxy. *Appl. Phys. Lett.* **2011**, *98*, 162112. [CrossRef]
131. Hirama, K.; Kasu, M.; Taniyasu, Y. RF high-power operation of AlGaN/GaN HEMTs epitaxially grown on diamond. *IEEE Electron Device Lett.* **2012**, *33*, 513–515. [CrossRef]
132. Hirama, K.; Kasu, M.; Taniyasu, Y. Growth and device properties of AlGaN/GaN high-electron mobility transistors on a diamond substrate. *Jpn. J. Appl. Phys.* **2012**, *51*, 01AG09. [CrossRef]
133. Hirama, K.; Taniyasu, Y.; Kasu, M. Epitaxial growth of AlGaN/GaN high-electron mobility transistor structure on diamond (111) surface. *Jpn. J. Appl. Phys.* **2012**, *51*, 090114. [CrossRef]
134. van Dreumel, G.W.G.; Buijnsters, J.G.; Bohnen, T.; ter Meulen, J.J.; Hageman, P.R.; van Enckevort, W.J.P.; Vlieg, E. Growth of GaN on nano-crystalline diamond substrates. *Diam. Relat. Mater.* **2009**, *18*, 1043–1047. [CrossRef]
135. Zhang, D.; Bai, Y.Z.; Qin, F.W.; Bian, J.M.; Jia, F.C.; Wu, Z.L.; Zhao, J.J.; Jiang, X. Preparation and characteristics of GaN films on freestanding CVD thick diamond films. *Chin. Phys. Lett.* **2010**, *27*, 018102.
136. Jiang, Q.; Allsopp, D.W.E.; Bowen, C.R. Growth of GaN epitaxial films on polycrystalline diamond by metal-organic vapor phase epitaxy. *J. Phys. D Appl. Phys.* **2017**, *50*, 165103. [CrossRef]
137. Ahmed, R.; Siddique, A.; Anderson, J.; Gautam, C.; Holtz, M.; Piner, E. Integration of GaN and Diamond Using Epitaxial Lateral Overgrowth. *ACS Appl. Mater. Interfaces* **2020**, *12*, 39397–39404. [CrossRef]
138. Ahmed, R.; Siddique, A.; Anderson, J.; Engdahl, C.; Holtz, M.; Piner, E. Selective Area Deposition of Hot Filament CVD Diamond on 100 mm MOCVD Grown AlGaN/GaN Wafers. *Cryst. Growth Des.* **2019**, *19*, 672–677. [CrossRef]
139. Rotter, S. Diamond and Diamond-Like-Carbon Films for Electronics. In *Workshop on Compound Semiconductor Devices and Integrated Circuits 91 (WOCSDICE 91)*; Sandia National Labs.: Albuquerque, NM, USA, 1991.
140. Rotter, S. A proposed novel application of thin overlay diamond films for improved thermal management budget of short gate high power devices. In Proceedings of the Fourth International Conference on New Diamond Science and Technology, Kobe, Japan, 18–22 July 1994; pp. 665–668.
141. Tadjer, M.J.; Anderson, T.J.; Hobart, K.D.; Feygelson, T.I.; Mastro, M.A.; Caldwell, J.D.; Hite, J.K.; Eddy, C.R., Jr.; Kub, F.J.; Butler, J.E.; et al. Reduced Self-heating in AlGaN/GaN HEMTs Using Nanocrystalline Diamond Heat Spreading Films. In Proceedings of the 68th Device Research Conference, Notre Dame, IN, USA, 21–23 June 2010; pp. 125–126.
142. Tadjer, M.J.; Anderson, T.J.; Hobart, K.D.; Feygelson, T.I.; Caldwell, J.D.; Eddy, C.R.; Kub, F.J.; Butler, J.E.; Pate, B.; Melngailis, J. Reduced self-heating in AlGaN/GaN HEMTs using nanocrystalline diamond heat-spreading films. *IEEE Electron Device Lett.* **2012**, *33*, 23–25. [CrossRef]
143. Anderson, T.J.; Hobart, K.D.; Tadjer, M.J.; Koehler, A.D.; Feygelson, T.I.; Hite, J.K.; Pate, B.B.; Kub, F.J.; Eddy, C.R., Jr. Nanocrystalline diamond for near junction heat spreading in GaN power HEMTs. In Proceedings of the Technical Digest—IEEE Compound Semiconductor Integrated Circuit Symposium, CSIC, Monterey, CA, USA, 13–16 October 2013; pp. 8–11.
144. Anderson, T.J.; Hobart, K.D.; Tadjer, M.J.; Koehler, A.D.; Feygelson, T.I.; Pate, B.B.; Hite, J.K.; Kub, F.J.; Eddy, C.R., Jr. Nanocrystalline Diamond for Near Junction Heat Spreading in GaN Power HEMTs. *ECS Trans.* **2014**, *61*, 45–49. [CrossRef]
145. Anderson, T.J.; Koehler, A.D.; Hobart, K.; Tadjer, M.J.; Feygelson, T.I.; Hite, J.K.; Pate, B.B.; Kub, F.J.; Eddy, C.R., Jr. Nanocrystalline Diamond-Gated AlGaN/GaN HEMT. *IEEE Electron Device Lett.* **2013**, *34*, 1382–1384. [CrossRef]
146. Anderson, T.J.; Hobart, K.D.; Tadjer, M.J.; Koehler, A.D.; Imhoff, E.A.; Hite, J.K.; Feygelson, T.I.; Pate, B.B.; Eddy, C.R., Jr.; Kub, F.J. Nanocrystalline Diamond Integration with III-Nitride HEMTs. *ECS J. Solid State Sci. Technol.* **2017**, *6*, Q3036–Q3039. [CrossRef]
147. El Helou, A.; Komarov, P.; Tadjer, M.J.; Anderson, T.J.; Francis, D.A.; Feygelson, T.; Pate, B.B.; Hobart, K.D.; Raad, P.E. High-Resolution Thermoreflectance Imaging Investigation of Self-Heating in AlGaN/GaN HEMTs on Si, SiC, and Diamond Substrates. *IEEE Trans. Electron Devices* **2020**, *67*, 5415–5420. [CrossRef]
148. Alomari, M.; Dipalo, M.; Rossi, S.; Diforte-Poisson, M.A.; Delage, S.; Carlin, J.F.; Grandjean, N.; Gaquiere, C.; Toth, L.; Pecz, B.; et al. Diamond overgrown InAlN/GaN HEMT. *Diam. Relat. Mater.* **2011**, *20*, 604–608. [CrossRef]
149. Kohn, E.; Alomari, M.; Gao, Z.; Rossi, S.; Dussaigne, A.; Carlin, J.-F.; Grandjean, N.; Aretouli, K.; Adikimenakis, A.; Konstantinidis, G.; et al. Direct Integration of Diamond Heat Spreader with GaN-Based HEMT Device Structures. In Proceedings of the High Performance Devices. Lester Eastman Conference (LEC), Ithaca, NY, USA, 5–7 August 2014; pp. 1–4.
150. Zhou, Y.; Ramaneti, R.; Anaya, J.; Korneychuk, S.; Derluyn, J.; Sun, H.; Pomeroy, J.; Verbeeck, J.; Haenen, K.; Kuball, M. Thermal characterization of polycrystalline diamond thin film heat spreaders grown on GaN HEMTs. *Appl. Phys. Lett.* **2017**, *111*, 041901. [CrossRef]
151. Wang, A.; Tadjer, M.J.; Anderson, T.J.; Baranyai, R.; Pomeroy, J.W.; Feygelson, T.I.; Hobart, K.D.; Pate, B.B.; Calle, F.; Kuball, M. Impact of intrinsic stress in diamond capping layers on the electrical behavior of AlGaN/GaN HEMTs. *IEEE Trans. Electron Devices* **2013**, *60*, 3149–3156. [CrossRef]
152. Soleimanzadeh, R.; Khadar, R.A.; Naamoun, M.; Van Erp, R.; Matioli, E. Near-junction heat spreaders for hot spot thermal management of high power density electronic devices. *J. Appl. Phys.* **2019**, *126*, 165113. [CrossRef]
153. Arivazhagan, L.; Jarndal, A.; Nirmal, D. GaN HEMT on Si substrate with diamond heat spreader for high power applications. *J. Comput. Electron.* **2021**, *20*, 873–882. [CrossRef]

154. Yaita, J.; Yamada, A.; Kotani, J. Growth of microcrystalline diamond films after fabrication of GaN high-electron-mobility transistors for effective heat dissipation. *Jpn. J. Appl. Phys.* **2021**, *60*, 076502. [CrossRef]
155. Fujitsu. Fujitsu Successfully Grows Diamond Film to Boost Heat Dissipation Efficiency of GaN HEMT. Press Release 2019. Available online: https://www.fujitsu.com/global/about/resources/news/press-releases/2019/1205-01.html%202019 (accessed on 7 October 2021).
156. Zhang, H.; Guo, Z.; Lu, Y. Enhancement of Hot Spot Cooling by Capped Diamond Layer Deposition for Multifinger AlGaN/GaN HEMTs. *IEEE Trans. Electron Devices* **2020**, *67*, 47–52. [CrossRef]
157. Zhu, T.; Zheng, X.-F.; Cao, Y.-R.; Wang, C.; Mao, W.; Wang, Y.-Z.; Mi, M.-H.; Wu, M.; Mo, J.-H.; Ma, X.-H.; et al. Study on the Effect of Diamond Layer on the Performance of Double-Channel AlGaN/GaN HEMTs. *Semicond. Sci. Technol.* **2020**, *35*, 055006. [CrossRef]
158. Zhang, H.; Guo, Z. Thickness Dependence and Anisotropy of Capped Diamond Thermal Conductivity on Cooling of Pulse-Operated GaN HEMTs. *IEEE Trans. Compon. Packag. Manuf. Technol.* **2021**, *11*, 233–240. [CrossRef]
159. Moelle, C.; Klose, S.; Szücs, F.; Fecht, H.J.; Johnston, C.; Chalker, P.R.; Werner, M. Measurement and calculation of the thermal expansion coefficient of diamond. *Diam. Relat. Mater.* **1997**, *6*, 839–842. [CrossRef]
160. Reeber, R.R.; Wang, K. Lattice parameters and thermal expansion of GaN. *J. Mater. Res.* **2000**, *15*, 40–44. [CrossRef]
161. Tripathy, S.; Chua, S.J.; Chen, P.; Miao, Z.L. Micro-Raman investigation of strain in GaN and AlxGa 1-xN/GaN heterostructures grown on Si(111). *J. Appl. Phys.* **2002**, *92*, 3503. [CrossRef]
162. Ahmad, I.; Holtz, M.; Faleev, N.N.; Temkin, H. Dependence of the stress-temperature coefficient on dislocation density in epitaxial GaN grown on α-Al2O3 and 6H-SiC substrates. *J. Appl. Phys.* **2004**, *95*, 1692–1697. [CrossRef]
163. Choi, S.; Heller, E.; Dorsey, D.; Vetury, R.; Graham, S. The impact of mechanical stress on the degradation of AlGaN/GaN high electron mobility transistors. *J. Appl. Phys.* **2013**, *114*, 164501. [CrossRef]
164. Kang, B.S.; Kim, S.; Kim, J.; Ren, F. Effect of external strain on the conductivity of AlGaN/GaN high-electron-mobility transistors. *Appl. Phys. Lett.* **2003**, *83*, 4845. [CrossRef]
165. Slack, G.A.; Tanzilli, R.A.; Pohl, R.O.; Vandersande, J.W. The intrinsic thermal conductivity of AIN. *J. Phys. Chem. Solids* **1987**, *48*, 641–647. [CrossRef]
166. Pobedinskas, P.; Degutis, G.; Dexters, W.; Hardy, A.; van Bael, M.K.; Haenen, K. Influence of hydrogen and hydrogen/methane plasmas on AlN thin films. *Appl. Phys. Lett.* **2014**, *104*, 081917. [CrossRef]
167. Pobedinskas, P.; Degutis, G.; Dexters, W.; Janssen, W.; Janssens, S.D.; Conings, B.; Ruttens, B.; D'Haen, J.; Boyen, H.-G.; Hardy, A.; et al. Surface plasma pretreatment for enhanced diamond nucleation on AlN. *Appl. Phys. Lett.* **2013**, *102*, 201609. [CrossRef]
168. Smith, M.D.; Li, X.; Uren, M.J.; Thayne, I.G.; Kuball, M. Polarity dependence in Cl2-based plasma etching of GaN, AlGaN and AlN. *Appl. Surf. Sci.* **2020**, *521*, 146297. [CrossRef]
169. Song, C.; Kim, J.; Lee, H.; Cho, J. Fundamental limits for near-junction conduction cooling of high power GaN-on-diamond devices. *Solid State Commun.* **2019**, *295*, 12–15. [CrossRef]
170. van der Drift, A. Evolutionary selection, a principle governing growth orientation in vapour-deposited layers. *Philips Res. Rep.* **1967**, *22*, 267–288.
171. Ophus, C.; Ewalds, T.; Luber, E.J.; Mitlin, D. The role of self-shadowing on growth and scaling laws of faceted polycrystalline thin films. *Acta Mater.* **2010**, *58*, 5150–5159. [CrossRef]
172. Smereka, P.; Li, X.; Russo, G.; Srolovitz, D.J. Simulation of faceted film growth in three dimensions: Microstructure, morphology and texture. *Acta Mater.* **2005**, *53*, 1191–1204. [CrossRef]
173. Liu, T.; Raabe, D.; Mao, W.-M. A review of crystallographic textures in chemical vapor-deposited diamond films. *Front. Mater. Sci. China* **2010**, *4*, 1–16. [CrossRef]
174. Graebner, J.E.; Jin, S.; Kammlott, G.W.; Bacon, B.; Seibles, L.; Banholzer, W. Anisotropic thermal conductivity in chemical vapor deposition diamond. *J. Appl. Phys.* **1992**, *71*, 5353–5356. [CrossRef]
175. Anaya, J.; Sun, H.; Pomeroy, J.; Kuball, M. Thermal management of GaN-on-diamond high electron mobility transistors: Effect of the nanostructure in the diamond near nucleation region. In Proceedings of the 15th IEEE Intersociety Conference on Thermal and Thermomechanical Phenomena in Electronic Systems (ITherm), Las Vegas, NV, USA, 31 May–3 June 2016; pp. 1558–1565.
176. Iijima, S.; Aikawa, Y.; Baba, K. Early formation of chemical vapor deposition diamond films. *Appl. Phys. Lett.* **1990**, *57*, 2646–2648. [CrossRef]
177. Liu, J.; Hei, L.F.; Chen, G.C.; Li, C.M.; Song, J.H.; Tang, W.Z.; Lu, F.X. Influence of seeding pre-treatments on mechanical properties of ultrananocrystalline diamond films on silicon and Si3N4 substrates. *Thin Solid Films* **2014**, *556*, 385–389. [CrossRef]
178. Williams, O.A.; Douhéret, O.; Daenen, M.; Haenen, K.; Osawa, E.; Takahashi, M. Enhanced diamond nucleation on monodispersed nanocrystalline diamond. *Chem. Phys. Lett.* **2007**, *445*, 255–258. [CrossRef]
179. Mandal, S.; Thomas, E.L.H.; Middleton, C.; Gines, L.; Griffiths, J.T.; Kappers, M.J.; Oliver, R.A.; Wallis, D.J.; Goff, L.E.; Lynch, S.A.; et al. Surface Zeta Potential and Diamond Seeding on Gallium Nitride Films. *ACS Omega* **2017**, *2*, 7275–7280. [CrossRef]
180. Kim, J.H.; Lee, S.K.; Kwon, O.M.; Hong, S.I.; Lim, D.S. Thickness controlled and smooth polycrystalline CVD diamond film deposition on SiO2 with electrostatic self assembly seeding process. *Diam. Relat. Mater.* **2009**, *18*, 1218–1222. [CrossRef]
181. Scorsone, E.; Saada, S.; Arnault, J.C.; Bergonzo, P. Enhanced control of diamond nanoparticle seeding using a polymer matrix. *J. Appl. Phys.* **2009**, *106*, 014908. [CrossRef]

182. Ōsawa, E. Monodisperse single nanodiamond particulates. *Pure Appl. Chem.* **2008**, *80*, 1365–1379. [CrossRef]
183. Bai, T.; Wang, Y.; Feygelson, T.I.; Tadjer, M.J.; Hobart, K.D.; Hines, N.J.; Yates, L.; Graham, S.; Anaya, J.; Kuball, M.; et al. Diamond Seed Size and the Impact on Chemical Vapor Deposition Diamond Thin Film Properties. *ECS J. Solid State Sci. Technol.* **2020**, *9*, 053002. [CrossRef]
184. Nochetto, H.C.; Jankowski, N.R.; Bar-Cohen, A. The impact of GaN/substrate thermal boundary resistance on a HEMT device. In Proceedings of the ASME 2011 International Mechanical Engineering Congress and Exposition. Volume 10: Heat and Mass Transport Processes, Parts A and B., Denver, CO, USA, 11–17 November 2011; Volume 10, pp. 241–249.
185. Kuball, M.; Calvo, J.A.; Simon, R.B.; Pomeroy, J.W. Novel thermal management and its analysis in GaN electronics. In Proceedings of the 2014 Asia-Pacific Microwave Conference, Sendai, Japan, 4–7 November 2014; pp. 920–922.
186. Dumka, D.C.; Chou, T.M.; Faili, F.; Francis, D.; Ejeckam, F. AlGaN/GaN HEMTs on diamond substrate with over 7 W/mm output power density at 10 GHz. *Electron. Lett.* **2013**, *49*, 1298–1299. [CrossRef]

materials

Review

Ga_2O_3 and Related Ultra-Wide Bandgap Power Semiconductor Oxides: New Energy Electronics Solutions for CO_2 Emission Mitigation

Zeyu Chi [1,†], Jacob J. Asher [2,†], Michael R. Jennings [2], Ekaterine Chikoidze [1] and Amador Pérez-Tomás [3,*]

1 Groupe d'Etude de la Matière Condensée (GEMaC), UVSQ-CNRS, Université Paris-Saclay, 45 Av. des Etats-Unis, CEDEX, 78035 Versailles, France; zeyu.chi@uvsq.fr (Z.C.); ekaterine.chikoidze@uvsq.fr (E.C.)
2 Bay Campus, College of Engineering, Swansea University, Fabian Way, Crymlyn Burrows, Swansea SA1 8EN, UK; j.j.asher.796600@swansea.ac.uk (J.J.A.); m.r.jennings@swansea.ac.uk (M.R.J.)
3 Catalan Institute of Nanoscience and Nanotechnology (ICN2), CSIC and BIST, Campus UAB, Bellaterra, 08193 Barcelona, Spain
* Correspondence: amador.perez@icn2.cat
† These authors contributed equally to this work.

Abstract: Currently, a significant portion (~50%) of global warming emissions, such as CO_2, are related to energy production and transportation. As most energy usage will be electrical (as well as transportation), the efficient management of electrical power is thus central to achieve the XXI century climatic goals. Ultra-wide bandgap (UWBG) semiconductors are at the very frontier of electronics for energy management or energy electronics. A new generation of UWBG semiconductors will open new territories for higher power rated power electronics and solar-blind deeper ultraviolet optoelectronics. Gallium oxide—Ga_2O_3 (4.5–4.9 eV), has recently emerged pushing the limits set by more conventional WBG (~3 eV) materials, such as SiC and GaN, as well as for transparent conducting oxides (TCO), such asIn_2O_3, ZnO and SnO_2, to name a few. Indeed, Ga_2O_3 as the first oxide used as a semiconductor for power electronics, has sparked an interest in oxide semiconductors to be investigated (oxides represent the largest family of UWBG). Among these new power electronic materials, $Al_xGa_{1-x}O_3$ may provide high-power heterostructure electronic and photonic devices at bandgaps far beyond all materials available today (~8 eV) or $ZnGa_2O_4$ (~5 eV), enabling spinel bipolar energy electronics for the first time ever. Here, we review the state-of-the-art and prospects of some ultra-wide bandgap oxide semiconductor arising technologies as promising innovative material solutions towards a sustainable zero emission society.

Keywords: energy electronics; ultra-wide bandgap; power electronics; diodes; transistors; gallium oxide; Ga_2O_3; spinel; $ZnGa_2O_4$

Citation: Chi, Z.; Asher, J.J.; Jennings, M.R.; Chikoidze, E.; Pérez-Tomás, A. Ga_2O_3 and Related Ultra-Wide Bandgap Power Semiconductor Oxides: New Energy Electronics Solutions for CO_2 Emission Mitigation. *Materials* **2022**, *15*, 1164. https://doi.org/10.3390/ma15031164

Academic Editors: John Buckeridge and Ichimura Masaya

Received: 17 December 2021
Accepted: 25 January 2022
Published: 2 February 2022

1. Introduction

According to the latest Intergovernmental Panel on Climate Change (IPCC) report released in August 2021 [1], climate change is widespread, rapid, and intensifying and some trends are now regarded as irreversible. Human-induced climate change is already affecting many weather and climate extremes in every region across the globe. Scientists are also observing changes across the whole Earth's climate system; in the atmosphere, in the oceans, ice floes, and on land. Many of these changes are unprecedented and some of the shifts are now in motion, while some—such as rising sea levels—are already irreversible for the coming centuries to millennia. Stabilizing the climate will require strong, rapid, and sustained reductions in greenhouse gas emissions, and reaching net zero CO_2 emissions. Limiting other greenhouse gases and air pollutants, especially methane, could be beneficial for the health of the climate as well as the population [1]. The breakdown for the different greenhouse gas emissions can be seen in Figure 1 [2], where transport and electrical production account for up to 40%. Therefore, many energy-related megatrends of our modern society must focus on themes such as energy efficiency, e-mobility, smart grid and digitalization requiring green energy management electronics or power electronic solutions [3].

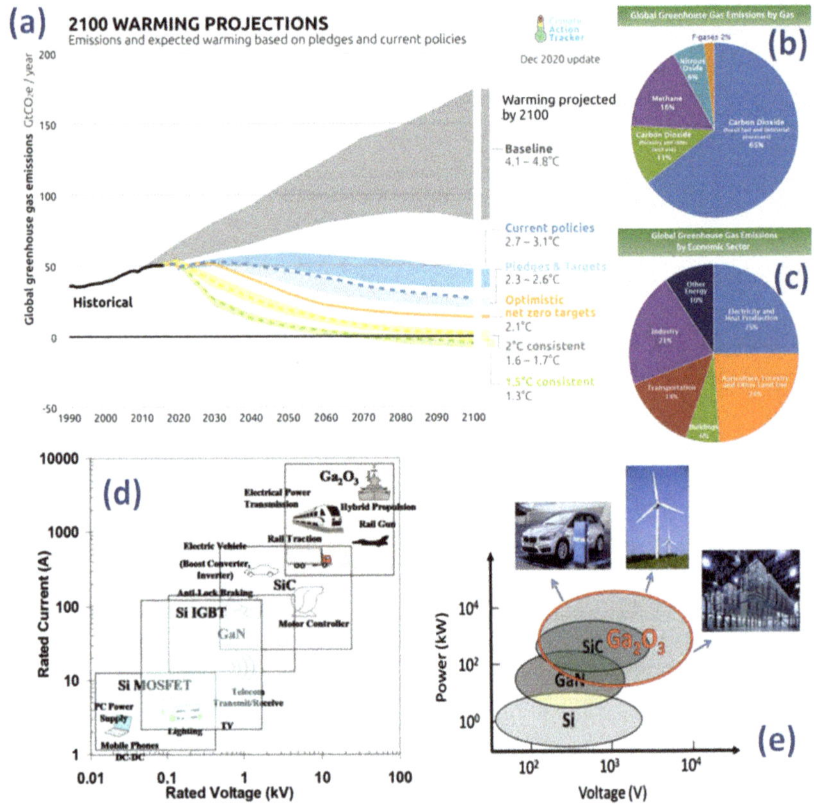

Figure 1. (**a**) Projected global warming figures for 2100. (**b**) Global warming emissions by gas. (**c**) Global greenhouse gas emissions by economic sector. (**d**) Selected applications for power semiconductors Si, SiC, GaN, and Ga_2O_3 for power electronics in terms of current and voltage requirements. (**e**) Owing to its ultra-wide bandgap, Ga_2O_3 can create additional possible applications for ultra-high power electronics including fast chargers for electric vehicles, high voltage direct current (HVDC) for data centers, and alternative energy sources. Figure sources: https://www.epa.gov/ghgemissions/global-greenhouse-gas-emissions-data (accessed on 16 December 2021). Source: (**a**) Source: IPCC (2014); based on global emissions from 2010. Details about the sources included in these estimates can be found in the Contribution of Working Group III to the Fifth Assessment Report of the Intergovernmental Panel on Climate Change. (**b**) IPCC (2014) based on global emissions from 2010. Details about the sources included in these estimates can be found in the Contribution of Working Group III to the Fifth Assessment Report of the Intergovernmental Panel on Climate Change. (**c**) Boden, T.A., Marland, G., and Andres, R.J. (2017). Global, Regional, and National Fossil-Fuel CO_2 Emissions. Carbon Dioxide Information Analysis Center, Oak Ridge National Laboratory, U.S. Department of Energy, Oak Ridge, Tenn., U.S.A. doi 10.3334/CDIAC/00001_V2017. Panels (**d**) and (**e**) adapted with permission from [4](© 2018 COPYRIGHT AIP Publishing).

Around half of the power used in the world is electrical and this is expected to increase steadily in the near future [5]. The vast majority (if not all) of this electricity will flow through, at least, one power electronic device during its generation, transmission, and final use. This is a critical aspect of power management which is sometimes overlooked, as power electronics make renewable (and non-renewable) energy impactful by increasing their efficiency [6]. As Si-based devices are replaced with other materials which are more energy efficient, this will affect the overall power consumption which will have a knock-on effect on CO_2 emissions by a significant amount [7]. Furthermore, devices made with a semiconductor having a bandgap larger than silicon can be made with less material and have lower cooling requirements, hence saving a lot of space and weight in applications such as electrical transport. This integration obviously impacts the amount of power required and, therefore, saves energy and its associated emissions. Since the 1980s, there has been a lot of work towards replacing silicon-based (E-gap of 1.12 eV) power electronics devices with wide bandgap (3–3.4 eV) semiconductor (WBG) based devices (in particular, silicon carbide (SiC) and gallium nitride (GaN)) and power devices with superior specs (higher temperature of operation, higher power handling capability, etc.) are now commercially available (typically in the range of 650 V–3.5 kV) [8,9] (Figure 1). While SiC devices and GaN transistors are already qualified in many emerging applications, silicon-based devices are still dominating in most applications. There are several reasons for this dominance, to start with, Si-based devices still have substantial potential. Their electrical and thermal performance is outstanding, their reliability is proven as can be seen from their years in application, as well as their low cost. In contrast WBG devices are starting their development, where we are still learning about materials development and device design. The benefits on the system level needs to be qualified and long-term reliability issues need to be determined; as these materials are developed, the costs for high-quality large volume production should decrease.

More recently, the frontier in the field is now given by ultra-wide bandgap semiconductors (UWBG), which have the promise of further upshifting the power rating and operation temperature. The same UWBG oxides also offer the potential for deeper ultraviolet optoelectronics [10]. Although another UWBG semiconductor, diamond, has been investigated over the last forty years, there has been limited progress and only recently have other materials, such as gallium oxide (Ga_2O_3) or aluminum nitride (AlN), yielded device demonstrations with appropriate performances. In particular, Ga_2O_3 is a newer UWBG material (4.5–5 eV) and is receiving a lot of attention as a novel semiconductor, owing to its unusual material properties. The doping (*n*-type) is very tunable with an extremely high breakdown field and unique optoelectronic properties, these alongside the possibility of growing large native substrates (over 6″) with a low cost [11]. Besides, representing the first viable oxide semiconductor for power electronics, Ga_2O_3 has opened the door to many more oxide compounds to be scrutinized (e.g., spinel $ZnGa_2O_4$) as they represent the largest family of ultra-wide bandgap semiconductors. UWBG oxide semiconductors are now at the very frontier of energy electronics, and much cutting-edge research, challenges, and opportunities are taking place [12]. These will be succinctly overviewed in this paper.

2. Oxide Semiconductors for Power Electronics

As an alternative to silicon, there is a new generation of wide bandgap semiconductors which have the capability to operate at higher voltages, temperatures, and switching frequencies with greater efficiencies compared to existing Si devices. This characteristic results in lower losses and enables significantly reduced volume due to decreased cooling requirements and smaller passive components contributing to overall lower system cost. Wide bandgap semiconductors (in the context of power electronic devices) usually represent materials whose band gap is larger than that of silicon. A (non-exhaustive) list of different wide bandgap semiconductors is presented in Figure 2. There are several families of wide bandgap semiconductors depending on their chemical composition. The III–V wide bandgap semiconductors are primarily nitrides, phosphides, and arsenides. Chalcogen semiconductors are those containing a transition metal and a chalcogen anion (S, Se, or Te), therefore forming sulfides, selenides, and tellurides. There are few halogen wide bandgap semiconductors in the form of chloride, iodides, and bromides. Silicon carbide (which exhibits a very large number of polytypes) and diamond are both carbon-based materials. SiC is a relevant wide bandgap semiconductor since it is the only compound semiconductor that can be thermally oxidized to form SiO_2 in the same fashion as silicon [13].

Group	Material	Formula	Band gap (eV)
III-V			
III-V	Boron nitride, cubic	C-BN	6.36
III-V	Boron nitride, hexagonal	H-BN	5.96
III-V	Aluminium nitride	AlN	6.28
III-V	Aluminium gallium nitride	$Al_xGa_{1-x}N$	3.44-6.28
III-V	Boron arsenide	$B_{12}As_2$	3.47
III-V	Gallium nitride	GaN	3.44
III-V	Indium gallium nitride	$In_xGa_{1-x}N$	2.0-3.44
III-V	Aluminium phosphide	AlP	2.45
Carbon Based			
IV	Diamond	C	5.47
IV	Silicon carbide, 4H-SiC	4H-SiC	3.3
IV	Silicon carbide, 6H-SiC	6H-SiC	3.0
IV	Silicon carbide, 3C-SiC	3C-SiC	2.3
Chalcogens			
II-VI	Calcium sulfide	CaS	5.38
II-VI	Magnesium Sulfide	MgS	4.45
II-VI	Zinc sulfide	H-ZnS	3.91
II-VI	Zinc sulfide	C-ZnS	3.54
II-VI	Magnesium Selenide	MgSe	3.6
II-VI	Magnesium Teluride	MgTe	3.49
II-VI	Maganese teluride	MnTe	3.2
II-VI	Maganese sulfide	MnS	3.1
II-VI	Zinc selenide	ZnSe	2.7
II-VI	Maganese selenide	MnSe	2.65
II-VI	Cadmium sulfide	CdS	2.42
II-VI	Zinc telluride	ZnTe	2.25
I-III-VI_2	Cupper Aluminum Sulfide	$CuAlS_2$	3.50
Halogens			
I-VII	Cuprous chloride	CuCl	3.4
I-VII	Cuprous bromide	CuBr	2.91
I-VII	Cuprous iodide	CuI	2.95

Group	Material	Formula	Band gap (eV)
Oxides			
Insulator	Aluminum oxide	Al_2O_3	9.00
Insulator	Magnesium oxide	MgO	7.80
Insulator	Lanthanum aluminate	$LaAlO_3$	5.6
Insulator	lithium metagallate	α-$LiGaO_2$	5.6
Insulator	lithium metagallate	β-$LiGaO_2$	5.6
Semi	Gallium Oxide	α-Ga_2O_3	5.2
Semi	Gallium Oxide	β-Ga_2O_3	4.9
Semi	Magnesium gallate	$MgGa_2O_4$	4.9
Semi	Zinc Germanate	Zn_2GeO_4	4.68
Semi	Zinc Gallate	$ZnGa_2O_4$	4.60
Semi	Indium Germanate	$In_2Ge_2O_7$	4.43
Semi	Silver metagallate	$AgGaO_2$	4.12
Insulator	Lithium niobate	$LiNbO_3$	4
Semi	Zinc Alluminate	$ZnAl_2O_4$	3.8
Semi	Tin dioxide	SnO_2	3.7
Semi	Nickel oxide	NiO	3.6
Semi	Zinc oxide	ZnO	3.37
Semi	Strontium titanate	$SrTiO_3$	3.3
Semi	Titanium dioxide	a-TiO_2	3.2
Semi	Titanium dioxide	r-TiO_2	3.02
Insulator	Barium titanate	$BaTiO_3$	3
Semi	Indium Oxide	In_2O_3	2.9
Semi	p-type delafossite	$CuAlO_2$	3.5
Semi	p-type delafossite	$CuGaO_2$	3.6
Semi	p-type delafossite	$SrCu_2O_2$	3.2

Figure 2. Wide bandgap semiconductors (in the context of power electronic devices) usually representmaterialswhosebandgap is larger than that of silicon. In practice, wide bandgap materials of choice have a bandgap of around ~3 eV, with silicon carbide and gallium nitride in a prominent position. Recently, a new family of semiconductor materials with even larger bandgaps (known as ultra-wide bandgap semiconductors) is being investigated for the new generation of optoelectronic and power electronic applications. As a rule of thumb, an ultra-wide bandgap semiconductor is one whose bandgap is larger than that of GaN (i.e., 3.4 eV). Perhaps the most investigated ultra-wide bandgap semiconductors are diamond, some nitrides (AlGaN, AlN, and BN), and a few oxides. Among these oxides, gallium oxide is the only oxide semiconductor with ultra-large bandgap where it is possible to modulate the conductivity (i.e., doping) to define power electronic devices.

A special case of chalcogenides would be oxides; although group 16 is defined as chalcogens, the term chalcogenide is more commonly reserved for sulfides, selenides, and tellurides only. Oxides are ubiquitous in nature due to the large abundance of oxygen in the earth and the large oxygen electronegativity (i.e., the atom tendency to attract electrons and thus form bonds) that easily creates largely covalent stable chemical bonds with almost all elements to give the corresponding oxides. Indeed, almost the entire Earth's crust parts are oxides as the individual crust elements are inclemently oxidized by the oxygen present in the atmosphere or in the water [14]. Besides, the Earth's mantle (which represents 60–70% and ~80% of the Earth's mass and volume, respectively) is predominantly a layer of silicate (i.e., compounds containing silicon and oxygen including silica, orthosilicates, metasilicates, pyrosilicates, etc.) and magnesium oxide (MgO)-rich rock between the crust and the outer core [14]. The upper mantle is dominantly peridotite, composed primarily of variable proportions of the minerals olivine ($(Mg,Fe)_2SiO_4$), pyroxenes ($XY(Si,Al)_2O_6$), and aluminous phases, such as feldspar ($NaAlSi_3O_8$–$CaAl_2Si_2O_8$) and spinel ($MgAl_2O_4$). The lower mantle is composed primarily of bridgmanite ($(Mg, Fe)SiO_3$) and ferropericlase ($(Mg, Fe)O$), with significant amounts of calcium perovskite ($CaSiO_3$) and calcium-ferrite oxides [15].

Thus, in general, oxides can be regarded as naturally abundant and stable compounds. Since the early days of solid-state physics, (undoped) oxides have been considered to be insulators (or more precisely, highly resistive wide bandgap semiconductors). The bandgap of many common oxides, such as Al_2O_3, SnO_2, TiO_2, In_2O_3, Cu_2O, WO_3, ZnO, or NiO, is much wider than that of silicon (1.12 eV). Therefore, they are intrinsically poor conductors at room temperature if they are not properly doped into a degenerated state. Recently, much effort has been put into increasing the conductivity of some of these oxides (in particular those where *s* and *p* electrons propagate with a large mobility) while maintaining the optical transparency. Good examples are the doping of Al in ZnO, Sn in In_2O_3, and F in SnO_2, which are known as transparent conducting oxides (TCOs).

In practice, wide bandgap materials of choice have a bandgap of around ~3 eV, with silicon carbide and gallium nitride in a prominent position. Recently, a new family of semiconductor materials with even larger bandgaps (known as ultra-wide bandgap semiconductors) is being investigated for the new generation of optoelectronic and power electronic applications. As a rule of thumb, an ultra-wide bandgap semiconductor is one with a band gap larger than that of GaN (i.e., 3.4 eV). Perhaps the most investigated ultra-wide bandgap semiconductors are diamond, some nitrides (AlGaN, AlN, and BN), and few oxides. Among oxides, gallium oxide (Ga_2O_3) is the only oxide semiconductor with ultra-large bandgap where it is possible to modulate the conductivity (i.e., doping) to define power electronic devices. SiC and GaN power devices have already attracted much attention in higher efficiency electrical power conversion [4]. The major advantage of β-Ga_2O_3 is that the single crystal structure can be synthesized via several standard melt growth methods, e.g., the Czochralski (CZ) technique. This is a huge advantage of Ga_2O_3 over SiC, GaN, and diamond for scaling up production, hence we would expect the cost of β-Ga_2O_3 power electronics to decrease and be more in line with silicon with respect to their SiC and GaN counterparts [16,17].

3. Gallium Oxide (Ga_2O_3)

Ga_2O_3 has, at least, six polymorphs of which only one is thermodynamically stable at high temperatures (β phase, monoclinic), while the others are metastable and tend to convert to β upon high-temperature treatments including the phases α, corundum, δ, cubic, and ε, hexagonal, γ, defective-spinel, and orthorhombic κ polymorph [18]. The basic principles of polymorphism in crystals are clear: the lattices adapt to the minimum energy with respect to the temperature and pressure. Nearly all Ga_2O_3-containing devices utilize the monoclinic β phase, the most stable and best-characterized polymorph. As a well-known representative of a binary metal-oxide, gallium oxide cannot therefore be regarded as a new material, but as a revisited and rejuvenated one. For example, early crystallographic studies for single crystals [19] together with diverse luminescence studies of doped β-Ga_2O_3 were reported as early as the1960s [20]. Lorenz et al. [21] already published in 1966 that *n*-type Ga_2O_3 exhibits mobilities in the range of 100 $cm^2V^{-1}s^{-1}$ and an adequate device doping of 10^{18} cm^{-3} can be achieved just by controlling the native oxygen vacancies' density. Its deep-ultraviolet intrinsic bandgap of around 4.5–4.9 eV and excellent photoconductivity are also well-known from early contemporary studies [22]. It was not until this decade that the potential of Ga_2O_3 for a certain class of extreme or power electronics was realized due to further availability of large-area single crystals with high quality and the control of doping. In the past, Ga_2O_3 was somehow ignored as an ultra-wide bandgap material, as it was eclipsed by the potential of diamond which has never been fully realized [23].

Previously, SiC and GaN were the wide bandgap materials of choice [6]. However, from an ultra-high energy electronics perspective, Ga_2O_3 transistors and diodes exhibit the potential of delivering outstanding performances in the form of high breakdown voltage, high power and low losses because of superior material properties, thus extending the power handling limits given by the SiC and GaN integration into the mainstream [4]. Indeed, an ultra-large breakdown electric field, (which is usually assumed to be of the order of E_c~8 MVcm^{-1}), is a prime material advantage of Ga_2O_3. However, this value may be well underestimated; it was very recently suggested that the critical electric field of Ga_2O_3 could be as large as 13.2 MVcm^{-1}, if the residual donors are efficiently removed [24].

A high critical field crucially promotes the suitability of a semiconductor material for power devices that would be able to manage a large amount of electrical energy per unit area. Baliga's figure of merit [25] for power electronics is proportional to $E_c{}^3$, whilst only being linearly proportional to the bulk electron mobility (μ). Although Ga_2O_3 presents a similar conduction band dispersion (i.e., effective mass) than GaN, a relatively small bound limit of μ ~300 cm^{2}V^{-1}s^{-1} is frequently given [26]. This is due to a massive Fröhlich interaction which is common to many conducting oxides. Balancing critical field and mobility, the on-state losses can be still an order of magnitude lower than those for SiC and GaN for a given breakdown voltage (Figure 3). Comparing these values to other power semiconductors (see Figure 3), β-Ga_2O_3 appears favorable, surpassing SiC and GaN. A major additional technological advantage of the β-Ga_2O_3 is that the single crystal structure can be synthesized via several standard melt growth methods including the Czochralski (CZ) technique [27]. This, in practice, would imply SiC performances (or better ones) at a fraction of cost.

WBG	Bandgap E_g [eV]	Permitt. ε_r [ε_0]	Mobility μ_n [cm^2/Vs]	Crit. Field E_c [MV/cm]	BFOM x10^6 [V^2/Wcm^2]	Ther. Cond. k [W/mK]
Si	1.12	11.9	1240	0.3	8.8	145
4H-SiC	3.20	9.7	980	3.1	6270	350
GaN	3.40	10.4	1000	4.9	27900	140
β-Ga_2O_3	4.90	10.0	150	10.3	36300	27
Diamond	5.50	5.7	2000	13.0	554000	3450
AlN	6.00	9.8	426	15.4	336000	319
c-BN	6.40	7.1	825	17.5	695000	2145

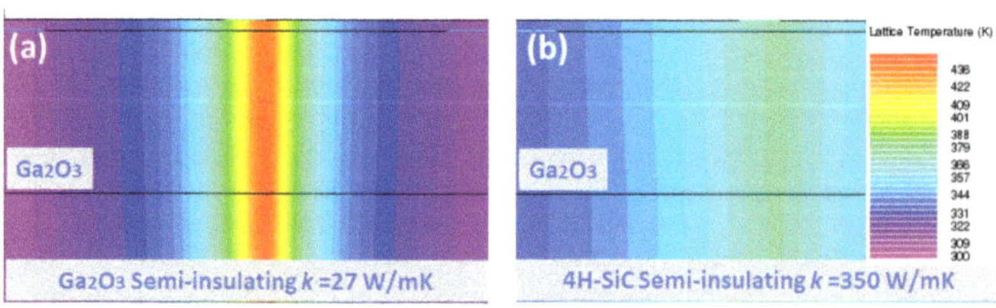

Figure 3. A summary of the main power device figure of merit (or Baliga's figure of merit. BFOM) parameters of the most popular wide bandgap semiconductors. Gallium oxide has a particularly poor thermal conductivity. However, when integrated into devices, heterojunctions with other better suited heat sinks (such as silicon carbide) area way to circumvent that limitation. As shown in the bottom panels, the simulate lattice temperature is lower on SiC (**b**) when compared with Ga_2O_3 substrates (**a**). Furthermore, thinning the Ga_2O_3 active film helps thermal performances. Adapted with permission from [11] © 2018 COPYRIGHT Society of Photo-Optical Instrumentation Engineers (SPIE).

There are certain applications, such as maritime and air transport, that are difficult to electrify as the power ratings are generally larger than, say, urban electric cars (Figure 1d,e).

For electric cars, devices delivering at or below the 1.2 kV perform well as rapid chargers or drive converters. These power ratings are well covered with "conventional" WBG, such as SiC and GaN. As the critical electric field of Ga_2O_3 has been reported to be at least two times, (or even four times larger), than that of these WBGs, the blocking voltage range of single electronics devices may be significantly extended in the future beyond what is theoretically possible today. These promises will impact directly on the size and weight of planes and ships resulting in less energy and emissions. As energy and transportation represents a major portion of the current CO_2 emissions contributing to global warming, it is expected that UWBG such as Ga_2O_3 may open new opportunities in sectors that are now difficult to decarbonize. Other prominent examples where the advantage of ultra-wide bandgap semiconductors can be exploited are as more solar-blind (UV transparent) transparent conducting electrodes [11] and electron (or hole) transport layers within solar cells or photodiodes [28].

3.1. Gallium Oxide Bulk Crystal Growth

Commonly used growth techniques of bulk β-Ga_2O_3 crystal are (Table 1): Verneuil method [21,29], Czochralski (CZ) method [30–33], floating-zone (FZ) method [34], edge-defined film fed (EFG) method [16,17], and Bridgman (horizontal or vertical, HB and VB) method [35,36],summarizing the basic features of melt growth methods reported so far.

Table 1. Overview of β-Ga_2O_3 bulk crystal growth methods.

Method	Verneuil	FZ	CZ	EFG	VB
Schematic illustration	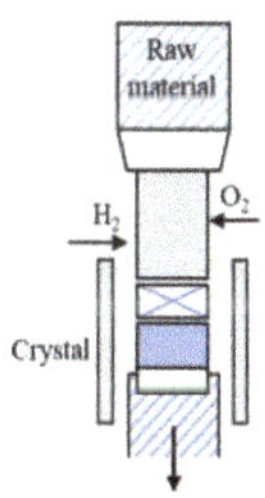	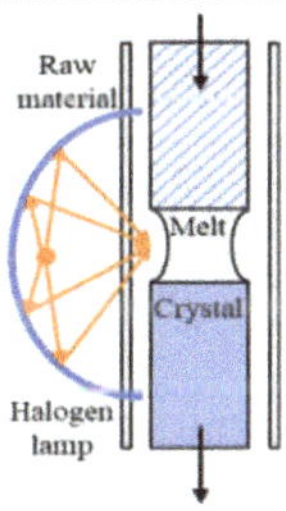	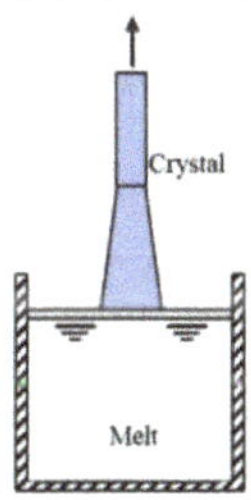	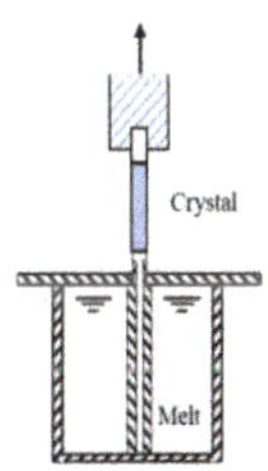	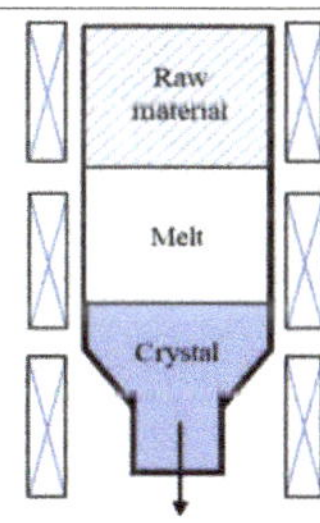
Bulk size	3/8-inch diameter 1-inch length	1-inch diameter	2-inch diameter	6-inch width 4-inch diameter	2-inch diameter
Growth rate (mm/h)	10	20–40	2	15	5
FWMH	-	22 arcsec	22–50 arcsec	17 arcsec	10–50 arcsec
Dislocation density	-	-	~10^3 cm^{-2}	10^3 cm^{-2}	10^2–2×10^3 cm^{-2}
Residual impurity	$2\times10^{18}$$cm^{-3}$	~10^{17} cm^{-3} (Si, Sn)	~10^{16} cm^{-3} (Si)	~10^{17} cm^{-3} (Si)	~several tens wt.ppm (Rh)
Intentional doping	~10^{19} cm^{-3} (Mg), ~10^{21} cm^{-3} (Zr) at 900 °C	~10^{19} cm^{-3} (Nb, Ta)	~10^{19} cm^{-3} (Sn, Si, Hf)	6-7 × 10^{18} cm^{-3} (Si, Sn)	3.6 × 10^{18} cm^{-3} (Sn)
Refs.	[21,28,37,38]	[34,38–43]	[16,17,30,33,43–45]	[16,17,46,47]	[35,36,48]

The Verneuil method, being a crucible-free technique, enables both oxidizing and reducing of growth conditions [21]. The synthesis under a reducing condition benefited electron conductivity [49]. *N*-type doping was realized by Harwig et al. [37], the free carrier concentration was determined to be ~10^{19} cm^{-3} by Mg doping, and ~10^{21} cm^{-3} by Zr doping at 900 °C. The β-Ga_2O_3 bulk crystal grown by this method has poor quality, and it was used mainly last century, as other more efficient techniques were well developed. The FZ method is also a crucible-free technique, it was recently used to grow bulk β-Ga_2O_3 crystal to investigate the scintillation features [50,51] as it can be employed in an air atmosphere, which may allow for creation of fewer oxygen defect centers being the emission origin of Ga_2O_3 [52]. Tomioka et al. [41] analyzed the residual impurities of β-Ga_2O_3 grown by the FZ method by inductively-coupled plasma mass spectroscopy; besides Si or Sn, Al, Mg, and Fe have also been detected with a concentration of ~10^{16} cm^{-3}. Al was presumed to be a neutral impurity, while Mg and Fe were considered as deep ionized acceptors and could compensate Si donors. To our knowledge, the lowest FWHM reported is ~22 arcsec for the peak β-Ga_2O_3 (400) by Hossain et al. [39], in this work, the Laue diffraction patterns also confirmed that the grown β-Ga_2O_3 crystal has a good crystallinity. However, FWHM of β-Ga_2O_3 rocking curves larger than 100 arcsec has also been measured [38,53]. However, both these techniques mentioned above suffer from small crystal size (wafer is no more than 1 inch so far, as summarized in Table 1.

Using an Ircrucible, the CZ method has been predicted to be a potential candidate for large boule, but thermal instability is an issue at high temperature that leads to decomposition of Ga_2O_3. Thus, this technique requires atmosphere control. Being a crack-free technique, the β-Ga_2O_3 crystal grown by the CZ method has small or even no boundaries. Several works reported by Galazka et al. [32,43,44] evidenced that the FWHM of the X-ray rocking curve could be as low as 22–50 arcsec on average, and the dislocation density was ~103 cm^{-2}. Moreover, Galazka et al. [31] recently reported that bulk Ga_2O_3 grown by the CZ method has an electron mobility of 80–152 $cm^2V^{-1}s^{-1}$ with a low residual Si impurity concentration of ~10^{16} cm^{-3}. Similar to the CZ method, the EFG method has the same technique issue. However, this technique is available for a 4-inch wafer and recently became commercially available. Commonly observed twin-boundaries in the EFG grown β-Ga_2O_3 were efficiently avoided by optimizing the growth process (the so-called shouldering process). Different from the traditional growth direction (010), Oshima et al. [54] demonstrated that the (001) oriented β-Ga_2O_3 grown by the EFG is more suitable than (010) for a Schottky barrier diode (SBD). A weak correlation between pits and electrical properties has been revealed [27,54]. The use of the VB method allows withstanding of high oxygen concentrations as a Pt-Rh (70–30%) alloy crucible. Additionally, this crucible also facilitates the pulling-up process as the grown crystal does not adhere to the wall. The major residual impurities are generally Rh (~several tens wt.ppm) from the crucible, Sn and Si (~several wt.ppm) from raw materials, and Fe and Zr (~several wt.ppm) from the furnace [36,48].This technique recently became *n*-type doping available by using a resistance heating VB furnace, and electron concentration and electron mobility were determined to be 3.6×10^{18} cm^{-3} and 60 $cm^2V^{-1}s^{-1}$, respectively, by 0.1 mol% Sn-doped [35,48]. As the CZ, EFG, and VB method use the crucible, they all have a high level of scalability.

3.2. Gallium Oxide Thin-Film Growth

Bulk devices and subsequent epitaxy of β-Ga_2O_3 layers could be provided by bulk growth, while high-quality epitaxial growth technologies are still required in order to study and fabricate more complex devices. Halide vapor phase epitaxy (HVPE), metal-organic vapor phase epitaxy (MOVPE), pulsed laser deposition (PLD), atomic layer deposition (ALD), molecular beam epitaxy (MBE), mist-chemical vapor deposition (CVD), and metal-organic chemical vapor deposition (MOCVD) are all involved in thin-film growth of Ga_2O_3.

Vapor phase epitaxy is a commercially promising technique for mass production of β-Ga_2O_3. Based on VPE, the halide vapor phase epitaxy (HVPE) method enables a growth rate as high as 250 μm/h [55] and the wafer size from 2 to 6 inches [56], it is thus a suitable technique for thick films with high purity for high voltage vertical switching devices. Furthermore, with the presence of chlorine catalyst in the growth chamber, this technique exhibits the growth of metastable phases of Ga_2O_3, such as α and ε [57]. The HVPE method suffers from a high level of roughness on the surface even at a relatively low growth rate [56,58]; an electrical mechanical [59] or a chemical mechanical [60] polishing can be employed to remove further deep surface pits formed during the growth. Leach et al. [61] reported a vast difference in surface morphology and XRD full-width half-maximum (FWMH), between sufficiently and insufficiently CMP polished (discriminated by the polishing times of the various polishing steps) β-Ga_2O_3 wafers grown by HVPE. Despite the poor morphology, the FWHM of the films grown on on-axis substrate were as narrow as 28 arcsec. Moreover, Murakami et al. [62] revealed that effective donor concentration without intentional doping could reach as low as 10^{13} cm^{-3}.

Metal-organic vapor phase epitaxy (MOVPE)can provide a highly scalable growth as its deposition areas are large. Triethylgallium (TEGa), trimethylgallium (TMGa), and $O_2$are most commonly the precursors for gallium and oxygen, respectively. The homoepitaxial growth of β-Ga_2O_3 by MOVPE can be strongly affected by substrate orientation. The growth rate is approximately 1.6–2.0 nm/min on the (100) plane, 0.65–1 μm/h on the (010) plane, and 1.6–4.3 nm/min on the (00-1) plane with miscut angles [63]. Recently, the growth rate can be elevated to 3.6 nm/min on the (100) plane [64] by tuning the growth pressure. A high-quality homoepitaxial growth on β-Ga_2O_3 the (100) with an FWMH of 43 arcsec has been reported by Gogova et al. [65]. The study of residual donor source is still in progress [66] while an electron concentration of 8×10^{19} cm^{-3} by Si-doping was realized by Baldini et al. [67], which is the highest doping level by this technique so far.

Pulsed laser deposition (PLD) has often been used for doped layers of Ga_2O_3 as it can transport materials from the target to the substrate stoichiometrically, thus the thickness of layers can be incisively controlled. It also has a relatively low operating temperature compared to other techniques. However, the quality of the materials deposited and the deposition rate are relatively low compared with other CVD and MBE methods. The roughness measured on the surface of Ga_2O_3 films had a root mean square between 1 and 7 nm [68–70].

A growth rate of 10.8 nm/min could be reached without oxygen, while it decreased to 6.5 nm/min by increasing oxygen pressure to 50 mbar [71]. Indeed, oxygen partial pressure and temperature are considered as the dominant parameters for properties of materials grown by the PLD [72]. The crystallinity was enhanced by increasing oxygen pressure at either low deposition temperature (250 °C [71]) or high deposition temperature (780 °C [68]). A higher oxygen partial pressure also leads to self-trapped holes at O1*s* and between two O2*s* sites [68], which could further act on the transport properties. Unlike the influence of oxygen pressure, a higher temperature does not always lead to a better film quality [73,74]. While, as expected, a higher annealing temperature could improve the crystallinity, as it helps the re-arrangement of Ga and O atoms to their optimal sites [75,76]. The highest *n*-type doping level achieved by the PLD is 1.7×10^{20} cm^{-3} by Si doping [69].

Atomic layer deposition (ALD), initially called atomic layer epitaxy (ALE), is a sub-set of the chemical vapor deposition (CVD) technique based on self-saturation, sequential surface reactions. ALD is a more general deposition containing ALE and molecular layering (ML) techniques [77]. The highly controlled thickness of films and conformal coverage are the main advantages of ALD over other techniques, it also allows a relatively lower deposition temperature compared to MBE and CVD techniques and a lower growth rate (generally less than 0.1 nm/cycle). Sn-doped Ga_2O_3 grown by ALD was investigated by Siah et al. [78], however the concentration of Sn was estimated as 2×10^{20} cm^{-3}, with the free electrons determined to be 4×10^{18} cm^{-3}. This was due to the low growth temperature.

Thus, post-annealing is generally also required to improve the crystalline quality. Additionally, the temperature during growth depends mainly on the gallium precursor chosen [79,80]. Besides the conventional ALD, the plasma-enhanced atomic layer deposition (PEALD) further permits a lower deposition temperature and better Ga_2O_3film properties with very smooth surface roughness (<1 nm) [81–83].

Molecular beam epitaxy (MBE)suits research purposes better than commercial use, as it enables the growth of high structural quality β-Ga_2O_3 with a relatively low growth rate (<1 μm/h) and high production cost, while high voltage vertical devices often require thick drift regions (dozens of microns). The orientation of growth has been found to be one factor that influences the growth rate [84]. Mazzolini et al. [85] further demonstrated the growth rate of different orientations Γ(010) (2.3 nm/min) > Γ(001) > Γ(−201) > Γ(100) of In-catalyzed β-Ga_2O_3layers;this phenomenon was believed to be associated with the surface free energy related to the binding energy of the In ad-atom. Nepal et al. [86] reported a heteroepitaxial growth on SiC with (−402) having a relatively high FWMH (694 arcsec), which can be reduced to 30–60 arcsec by homoepitaxial growth [87]. The thin films grown by MBE also benefit a smooth surface with a roughness of less than 1 nm [88,89]. The densities of the threading dislocation etch pits was determined to be ~10^5 cm^{-2} for the film grown at 850 °C [89]. An electron concentration of 10^{20} cm^{-3} has been achieved by Sn doping [90].

Techniques based on chemical vapor deposition (CVD) have also been employed for the growth of Ga_2O_3. Scalability and mass production are the most advantageous characteristics of the mist-CVD technique, as it is a vacuum free, low-cost, and solution-processed approach. This technique is also often used for epitaxial growth of α-Ga_2O_3 on sapphire [91–94]. Morimoto et al. [94] also pointed out the facilities of mist-CVD for $Ga_2O_3$by F doping. Both homoepitaxial [95,96] and heteroepitaxial [97] growth of β-Ga_2O_3have been successfully performed. It is also worth noting that the FWMH of rocking curves was 39–91 arcsec for homoepitaxial growth with growth rate of 0.5–3.2 μm/h [96,98]. An electron concentration was measured as 5×10^{20} cm^{-3} by Sn doping [98].

The metal-organic chemical vapor deposition (MOCVD) technique uses Ga-based organic material as metal precursors, such as trimethylgallium (TMGa) and triethylgallium (TEGa), which usually leads to C-contamination of the as-grown film (relatively less carbon by using TEGa than TMGa). It is well-known that such contamination can be efficiently reduced by high growth temperature, and eliminated by post-annealing. Li et al. [99] reported a high-quality homoepitaxially grown film with FWMH and surface roughness of 21.6 arcsec and 0.68 nm, respectively. The growth rate is generally from several hundred nm/h [100,101] to10 μm/h [102–104]. This technique is also available for both *n*- and *p*-type dupability [24,105] (Figure 4).

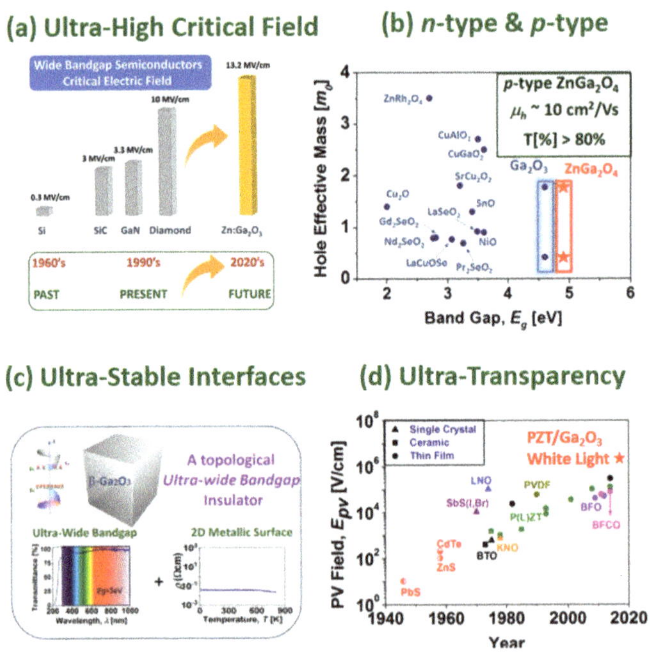

Figure 4. Ga_2O_3 and related oxides have been demonstrated to exhibit some remarkable features, such as (**a**) ultra-high critical electric field, (**b**) potential bipolar operation due to its demonstrated *n*-type and *p*-type conductivity, (**c**) ultra-stable interfaces that may host a 2D electron gas, (**d**) extended transparency into the UV-A region for transparent conducting oxide (TCO) applications (tail state density is located deeper in the ultraviolet than conventional TCOs). Panel (**a**) adapted with permission from Chikoidze et al. [24] © 2022 Elsevier Ltd. All rights reserved. Panel (**b**) adapted with permission from Chikoidze et al. [106] Copyright © 2022, American Chemical Society. Panel (**c**) adapted with permission from Chikoidze et al. [107]. © 2022 Elsevier Ltd. All rights reserved. Panel (**d**) adapted with permission from Perez-Tomas et al. [108,109] © 2022 WILEY-VCH Verlag GmbH & Co. KGaA. Adapted with permission from [12] © 2021 COPYRIGHT Society of Photo-Optical Instrumentation Engineers (SPIE).

3.3. Gallium Oxide Doping Issues and Recent Progress

β-Ga_2O_3 is very easily doped *n*-type to the degenerate state, *n*-type doped β-Ga_2O_3 with carrier concentration from 10^{16} to 10^{20} cm^{-3} [110,111] has been achieved by Sn and Ge doping by MBE, Si and Sn doping by MOVPE, and Sn doping by MOCVD [69]. A high mobility at room temperature of 145–184 $cm^2V^{-1}s^{-1}$ [100,101,112] has been reached by Si doping, and even till 10^4 $cm^2V^{-1}s^{-1}$ at 46 K [109]. Having a high critical field (5.2 MV.cm^{-1} without intentional doping [113]), the β-Ga_2O_3devices demonstrate high performance. Nevertheless, all the Ga_2O_3devices demonstrated thus far have been unipolar in nature (i.e., only *n*-type). In order to realize the full potential for WBG opto-electronics β-$Ga_2O_3$and to sustain high breakdown voltage (>6.5 kV), we need vertical geometry bipolar-junction-based devices. Therefore, the realization of *p*-type β-Ga_2O_3 is a primary challenge today for the gallium oxide scientific community (Figure 4).

There is a tendency in oxide compounds to have *n*-type conductivity, caused by vacancies in the oxygen atoms. This, as well as the fact that it is a UWBG material, intrinsic conduction is rare and even causes *p*- and *n*-type doping tends not to be symmetrical. This asymmetry is seen in gallium oxide, the hole conductivity is poor and is likely the main limitation for development of gallium oxide technology. Fundamental restrictions such as this area recurring issue in oxides, such as: (i) acceptor point defects with high formation energy; (ii) native donor defects with low energy—resting holes; and (iii) *p*-type oxides suffer from a high effective mass of the holes (this results in a low mobility), due to the top of the VB predominantly from localized O 2-p derived orbits.

Native *p*-type conductivity: Using thermodynamical calculations for the point defects on gallium oxide it can be seen that gallium oxide is "lucky", as when β-$Ga_2O_3$is at 500 °C, $P_{hole} \approx 1.33 \times 10^{-2}$ atm with a hole concentration around $p \approx 10^{15}$ cm^{-3} [114]. Comparing this to calculations for ZnO gives $P_{hole} \approx 10^3$ atm, for the same temperature. This divergence is believed to be from higher formation energy of the donor vacancies in β-Ga_2O_3 (approximately 1 eV higher per vacancy), making compensation mechanism by point defects less favorable in gallium oxide than in ZnO. As a consequence, it can be expected that *p*-type samples of β-Ga_2O_3 with higher carrier concentrations (then intrinsic) can be obtained when doping with shallow acceptor impurities.

The native hole concentration was investigated by Nanovation (SME, France) [114] where undoped β-Ga_2O_3thin film grown on c-sapphire substrates by pulsed laser deposition (PLD) showing resistivity of $\rho = 1.8 \times 10^2$ Ω.cm, hole concentration of $p = 2 \times 10^{13}$ cm^{-3} and a hole mobility of 4.2 $cm^2V^{-1}s^{-1}$ [114]. The determination of conductivity mechanism showed that Ga vacancies act as deep level acceptors with the activation energy of 0.56 eV in the low compensated sample, having $Ea = 1.2$ eV ionization energy. Later, the improvement was shown that native *p*-type conductivity by post-annealing in an oxygen atmosphere for β-Ga_2O_3 thin film was grown on c-sapphire substrates by MOCVD [115]. After oxygen annealing, the hole concentration was increased from 5.6×10^{14} cm^{-3} to 5.6×10^{17} cm^{-3} at 850 K. The author claimed that the annealing effect is related to the formation of V_{Ga}—V_O^{++} complexes as a shallow acceptor center with $E_a = 0.17$ eV activation energy.

Device applications require higher hole concentrations (at operating temperature), which could be achieved via external acceptor impurity incorporation.

There are already extensive theoretical studies (standard density functional theory (DFT and DFT with GGA+U) of acceptor impurity doping of β-Ga_2O_3 in order to identify efficient *p*-type dopant. Kyrtsos et al. [116] demonstrated by DFT calculations that dopants, such as Zn, Li, and Mg, will introduce deep acceptor level with ionization energies of more than 1 eV, thus, they cannot contribute to the *p*-type conductivity. However, this result could be influenced by the underestimation of the bandgap due to the semi-local approach. Varley et al. [117] predicted that self-trapped holes are more favorable than delocalized holes due to their energies and by theoretical calculation (self-trapping energy is 0.53 eV and barrier to trapping is 0.10 eV). This indicates that free holes are unstable and will spontaneously localize towards small polarons.

Lyons [118] examined the elements of group 5 and group 12 (Be, Mg, Ca, Sr, Zn, Cd) as acceptor impurities in β-Ga_2O_3 by hybrid DFT, all of them will exhibit the acceptor ionization levels of more than 1.3 eV. Mg was determined to be the most stable acceptor species, followed by Be. Sun et al. [119] used ab initio calculations to simulate the doping by Ge, Sn, Si, N, and Cl. Among them, N has been predicted to be a deep acceptor with an impurity level of 1.45 eV, as it has a similar atomic size as oxygen but has one less valence electron, and a higher *2p* orbital than oxygen. While all others act as donors, another ab initio calculation also demonstrated that nitrogen doping could introduce an acceptor level at 1.33 eV above the VBM.

Very recently, Goyal et al. [120] simulated a growth-annealing-quench sequence for hydrogen-assisted Mg doping in Ga_2O_3 by using the first principles defect theory and defect equilibrium calculations. The H_2O partial pressure and H exposure can strongly influence the Mg dopants concentration during the growth, by increasing the solubility limit of the acceptor, or by reducing the compensation. A conversion from *n*-type to *p*-type was achieved by annealing at O-rich/H-poor conditions. A Fermi level at +1.5 eV above the VB has been found after quenching.

Doping with two elements (co-doping) has been predicted by DFT which showed a promising method to obtain *p*-type β-Ga_2O_3, as it can break the solubility limit of mono-doping and improves the photoelectric properties of semiconductor materials which results in increasing the conductivity.

The principle is to increase carrier concentration and decrease the compensating defect formation energy. This is inherently caused by the localized nature of the O2 *p*-derived VB that leads to difficulty in introducing shallow acceptors and large hole effective mass [121].

Co-doping has been successfully used for II-VI compounds, co-doping containing N (Zn-N, N-P, Al-N, and In-N) has been demonstrated to be an effective way to improve the *p*-type conductivity [122–124], in particular, Zhang et al. [124] predicted two shallow impurity levels above the VB of about 0.149 eV and 0.483 eV in N–Zn co-doped β-Ga_2O_3. Co-doping by N-P made an acceptor level decrease ~0.8 eV, and an impurity level appears at 0.55 eV above the VB of β-Ga_2O_3. A significant loss of holes' effective mass was also evidenced [124]. There are a few experimental works reported regarding *p*-type doping of gallium oxide. Mg-doped β-$Ga_2O_3$was studied by Qian et al. [125] for the photo-blind detector, and the β-Ga_2O_3 containing 4.92 at% Mg has shown an acceptor level by XPS. A variation of bandgap has also been reported [83,126] however, the Hall effect measurement validity failed at room temperature due to the very high resistivity of the samples [127].

Suet al. [128] deposited Mg-Zn co-doped β-Ga_2O_3 on sapphire (0001), however, anti-sites' impurity defects (i.e., ZnGa and GaZn) were determined as deep acceptors (0.79 eV for ZnGa and 1.00 eV for GaZn) by absorption spectra. Feng et al. [129] demonstrated Zn doping (1.3–3.6 at%) in β-Ga_2O_3nanowires can reduce the bandgap slightly, they also proved the *p*-type conductivity by making *p-n* junction. Chikoidze et al. [24] suggested that Zn in β-Ga_2O_3 has an amphoteric nature: it can be an acceptor as Zn_{Ga} defect and at the same time, a donor being in Zn_i interstitial sites. It was shown that in (0.5%) Zn:Ga_2O_3 the auto-compensation of donor (Zn_i) -acceptor (Zn_{Ga}) defects takes place.

Islam et al. [130] reported that hydrogen annealing could vastly reduce the resistivity and reach a remarkable hole density of ~ 10^{15} cm^{-3} at room temperature. Besides, the ionization energy of acceptor is as low as 42 meV by incorporation of hydrogen in the lattice. This improvement is related to hydrogen decorated gallium vacancies $V_{Ga\text{-}H}$: during the diffusion of hydrogen into the Ga_2O_3crystal, H^+ absorbed at the surface will be attracted toward the ${V_{Ga}}^{3-}$, it stabilizes the negative charge and thus lowers the acceptor level. This mechanism leads to H^+ decorated Ga-vacancy ${V_{Ga\text{-}2H}}^{1-}$ and, therefore, the *p*-type conductivity.

Nitrogen-doped *p*-Ga_2O_3 has been experimentally achieved by non-conventional growth technique. Wu et al. [131] demonstrated a multi-step structural phase transition growth from hexagonal $P6_3mc$ GaN to rhombohedral R3C α-$GaN_xO_{3(1-x)/2}$ and realized the monolithic C2/m N-doped β-Ga_2O_3 thin layer finally with an acceptor ionization energy of 0.165 eV. The resistivity, hole concentration, and hole mobility are 17.0 Ω.cm, 1.56×10^{16}cm^{-3}, and 23.6 cm^{2}V^{-1}s^{-1}, respectively, by employing the Hall effect measurement. A performant field-effect transistor was also fabricated based on this *p*-type β-Ga_2O_3. Clearly, further experimental studies of optimal acceptor defects with room temperature activation are required.

3.4. Gallium Oxide Power Rectifiers

Once the device-grade epitaxial layers have been grown either homo- (bulk Ga_2O_3) hetero- (e.g., sapphire, silicon), or both, the simplest electronic devices one can define are rectifiers. In a Schottky rectifier, the counter-electrode (cathode) is processed to allow low resistance Ohmic contact while the anode contact is intended as a Schottky junction over a lightly doped epitaxy; it conducts electrons in the forward mode while sustaining large electric fields (by the creation of a depletion space charge region) in the reverse mode. As mentioned previously, devices using Ga_2O_3 are primarily limited to unipolar devices and Schottky diodes are made, in general, on *n*-type semiconductor layers as electrons are lighter than holes. However, it is also important to consider the appropriate metal contacts to Ga_2O_3 as they are responsible for connecting the semiconductor to the surrounding electrical circuit/system and parameters such as the Schottky barrier height are crucial. For different contacts to Ga_2O_3, such as in GaN and AlGaN, which utilize stacks of different metals [132], this decision can make an important difference to the nature of the contact. Regarding Schottky contacts to Ga_2O_3,Ni/Au is a common choice (see Table 2). Other Schottky contacts investigated include Pt, Ni, Cu, W, Ir, TiN/Au, Pt/Ti/Au, Ni/Au, ndPt/Au [133–136]. Very recently, an ultra-large Schottky barrier of ~1.8 eV was extracted for all-oxide $PdCoO_2$/β-Ga_2O_3 Schottky diodes [137]. The polar layered structure of $PdCoO_2$ generates electric dipoles, realizing a large Schottky barrier height of ~1.8 eV (well beyond the 0.7 eV expected from the basal Schottky–Mott relation) along with a large on/off ratio approaching 10^8, even at a high temperature of 350 °C (Figure 5c). As there are a number of polar oxides, this is a promising approach to increase the reverse blocking voltage of Ga_2O_3 diodes [138].

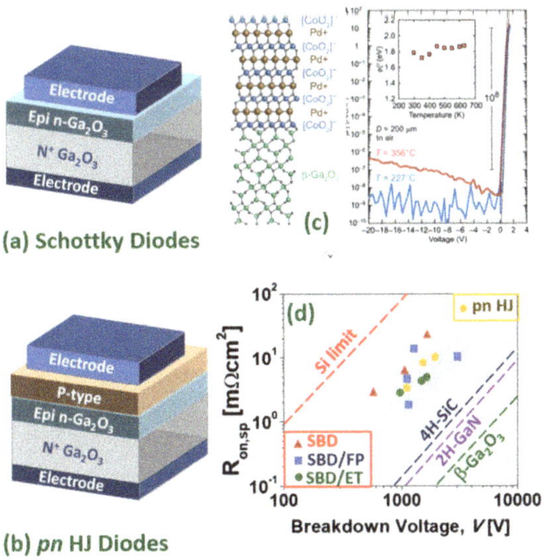

Figure 5. Schematics of (**a**) vertical Ga_2O_3 Schottky diodes and (**b**) *p-n* heterojunction diodes. (**c**) A $PdCoC_2$/Ga_2O_3 exhibiting the ultra-large Schottky barrier of 1.8 eV. (**d**) Baliga's FOM for selected Schottky and p-n HJ diodes from the literature. Panel (**c**) adapted with permission from Harada et al. [137] © 2022 AAAS 4.0 (CC BY-NC). Adapted with permission from [12] © 2021 copyright Society of Photo-Optical Instrumentation Engineers (SPIE).

Table 2. Table displaying varying SBD designs: L—lateral, V—vertical, TCO—thin conductive oxide film, FP—field plate, BET—bevel edge termination, FPET—field plate edge termination, MDS—metal-dielectric-semiconductor Schottky diode. Included here are different structures which exhibited SBD (some exhibiting Schottky contacts as opposed to useable device) using a range of different designs and metal stacks.

Device Configuration	Schottky Metal Stack	Ohmic Metal Stack	V_{br}	Ideality Factor	Ref.
V-SBD-BET	Ni/Au	Ti/Al/Ni/Au	427 V	1.07	[133]
V-SBD-FP	Ni/Au	Ti/Au	730 V V	1.02	[139]
V-SBD-FPET	Ni/Au	Ti/Au	1722 V	1.03	[140]
L-SBD-FP	Ni/Au	Ti/Au	<3 kV	~1.25	[141]
L-SBD	Ni/Au	Ti/Au	1.7 kV	-	[142]
L-SBD	Pt	Ti/Au	-	1.40	[134]
L-SBD	Ir	Ti/Au	-	1.45	[134]
V-SBD	Ni	Ti/Au	-	1.57	[134]
L-SBD	Ni	Ti/Au	-	1.33	[134]
V-SBD	Cu	Ti/Au	-	1.53	[134]
L-SBD	W	Ti/Au	-	1.4	[134]
V-SBD	Ni/Au	Sn	~210 V	3.38	[143]
L-SBD	Ptx	Ti/Al/Au	-	-	[144]
V-SBD	Pt/Au	Ti/Au	-	-	[135]
V-SBD	TiN	Ti/Au	-	1.03	[145]
V-SBD	Pt/Ti/Au	Ti/Au	-	1.03	[136]
V-SBD-TCO	SnO/Ti	Ti/Au	-	1.09	[146]
V-MDS(TiO_2)	Ni/Au	Ti/Au	1010 V	-	[147]

In the counter-electrode, highly doped regions beneath the metallization are deployed to assist ohmicity of the contacts [139]. The dopants for this have previously been discussed. Another approach to this is using thin films of highly-conducting oxides [140].

Ohmic contacts to β-Ga_2O_3 are commonly based on Ti/Au, however other metal contacts have been utilized, such as In, Ti, Ti/Al/Au, In/Au, and Ti/Al/Ni/Au. Besides, there are other metals which have exhibited pseudo Ohmic behavior including Zr, Ag, and Sn [132]. This pseudo nature meant that, initially, ohmicity was observed but, after annealing, rectifying behavior became dominant. Therefore, the Schottky/Ohmic nature is also dependent upon the Ga_2O_3's surface/interface states together with the exact choice of metal stack, explaining, in turn, the varying contact resistivity of certain metals. While delivering low contact resistance, it is worth mentioning that Au is not considered a CMOS-compatible metal. This is an issue shared with GaN-based technology [148].

For the continued development of high voltage β-Ga_2O_3devices, edge termination is an important aspect as it is with its Si, GaN, and 4H-SiCcounterparts. Edge termination in β-$Ga_2O_3$is being explored and focused specifically on field plates (FP), imparted edge termination (ET), guard ring field plates, thermally oxidized termination, beveled mesas, and trench. These techniques are all deployed to further manage the electrical field to reduce the electric field crowding at the diode edges to increase its blocking capabilities. SBD devices can be made with either a vertical architecture, using homoepitaxial Ga_2O_3 or with a lateral architecture using either homo- or heteroepitaxial (e.g., on sapphire) Ga_2O_3. In general, the vertical structure is preferred as the device pitch is reduced and the encapsulation is simpler. Hu et al. [141] demonstrated a field-plated lateral β-Ga_2O_3 SBD on a sapphire substrate with a reverse blocking voltage of more than 3 kV, an R_{on}

of 24.3 mΩcm^2 (anode–cathode spacing 24 μm), and an FOM >0.37 GWcm^{-2} (while an FOM of ~500 GWcm^{-2} was achieved as the anode-cathode spacing (and V_{br}) was reduced). Zhou et al. [149] implemented a Mg implanted ET device on a vertical β-Ga_2O_3 SBD with a reverse blocking voltage of 1.55 kV and a low specific on-resistance of 5.1 mΩcm^2 (epi thickness 10 μm) and an FOM of 0.47 GWcm^{-2}. Analogously, Lin et al. [150] implemented a guard ring with or without an FP on vertical SBDs. The terminated devices exhibited a specific on-resistance of 4.7 mΩcm^2 and a V_{br} of 1.43 kV. Wang et al. [151] implemented a thermally oxidized termination on a vertical SBD with a V_{br} of 940 V, a specific on-resistance of 3.0 mΩcm^2, and an FOM of 0.295 GWcm^{-2}. Allen et al. [152] implemented a small-angle beveled field plate (SABFP), on thinned Ga_2O_3 substrates and a non-punch-through vertical SBD design rendering a V_{br} of 1100 V, a peak electric field of 3.5 MVcm^{-1}, and an FOM of 0.6 GWcm^{-2}.

Somehow the state of the art is given by Li et al. [153]. They demonstrated an FP vertical Ga_2O_3 trench SBDs with a V_{br} of 2.89 kV (which is ~500 V higher than those without FPs). The trench SBDs exhibited a differential specific on-resistance of 10.5 (8.8) mΩcm^2 from DC (pulsed) measurements leading to an FOM of 0.80 (0.95) GWcm^{-2}. This Baliga's power FOM is approaching that for the best vertical SBD GaN devices (e.g., 1.7 GWcm^{-2} [154]) but is still several times smaller than lateral AlGaN/GaN SBD (e.g., 3.6 GWcm^{-2} [155]) and bipolar *p-n* vertical GaN diodes (e.g., ~4.6 GWcm^{-2} [156]). Both, the 2D gas formed at the AlGaN/GaN interface and the bipolar injection are effective ways of further reducing the on-resistance in these devices while keeping the breakdown voltage high. The lack of low resistivity *p*-type layer for the anode has to date, prevented a competitive homojunction p-n Ga_2O_3 diode, but *p-n* heterojunction diodes have been realized by integrating *n*-type Ga_2O_3 with *p*-type semiconductors, such as CuO (1.49 kV) [157] and NiO (1.06 kV/1.86kV) [158,159]. Nickel oxide as the *p*-type blocking layer in heterojunction power diodes resulted in a particularly promising approach with this NiO/Ga_2O_3device [160] yielding a Baliga's FOM of 0.33 GWcm^{-2} (Figure 5c,d).

Recently, extremely high-*k* dielectrics have been explored for electric field management in WBG semiconductor-based lateral and vertical device structures [160–164]. According to the TCAD simulations of Roy et al. [165], a super-dielectric Ga_2O_3 SBD with practically achievable device dimensions with extremely high FOM should be possible; e.g., 20kVcanbeachievedforan R_{on} of 10 mΩ-cm^2 with a dielectric constant of 300, a Ga_2O_3 width/dielectric width ratio of 0.2, and an aspect ratio (drift layer length (anode to cathode spacing)/drift layer width ratio) of 10 resulting in a PFOM of 40 GWcm^{-2} (surpassing the theoretical unipolar FOM of β-Ga_2O_3SBD by four times).

3.5. Gallium Oxide Power Transistors

A power MOSFET fabrication process generally includes a number of technological steps including either gate dielectrics, surface passivation, drain/source ohmic contacts, implant doping, isolation, mesa etch, or in combination. Due to the large bandgap of Ga_2O_3, the most suitable gate insulators are those with enough (conduction and valence) band-offsets to avoid current injection through the gate (e.g., SiO_2 and Al_2O_3 and perhaps other oxides such as Y_2O_3, MgO, and Mg_2AlO_4). While balancing the dielectric constant to achieve more gate capacitance and more carriers in the conductive channel [166]. Defining a contact region by implantation, such as in Si, SiC, and GaN power MOSFET technologies, is a usual choice [167], in Ga_2O_3 this is typically n^+ Si-ion implantation. While other techniques have been suggested to further decrease the contact resistivity, such as formation of surface states [168] or the adoption of a TCO as a metallic interface [169].

As in, the more mature, AlGaN/GaN HEMT technology, Ohmic contacts are typically made with a multilayer metal stack consisting of an adhesion layer (e.g., Ti, Ta), an overlayer (Al), a barrier layer (e.g., Ni, Ti, Mo), and a capping of Au [170,171]. Nevertheless, it has been argued that simpler metal structures, such as Ti/Ga_2O_3, are also efficient if there is an oxygen deficient Ga_2O_3 surface [172] (a double charged oxygen vacancy is a well-known intrinsic donor in oxides [107]). Indeed, Yao et al. [132] suggested that the surface states appear to have a more dominant role in the transformation from a Schottky to an Ohmic interface than the choice of metal.

As with power SBDs, power MOSFETs can be defined in a vertical Ga_2O_3 homoepitaxial structure (typical of SiC power MOSFETs) and lateral structure (typical of AlGaN/GaN power HEMTs) which can be either homoepitaxial or heteroepitaxial (Figure 6). Ga_2O_3 power MOSFETs are mostly unipolar *n*-type and operate in depletion mode (D-mode or normally-on) but a number of techniques have been reported to make enhancement mode (E-mode or normally-off) Ga_2O_3 devices. For example, Chabak et al. [173] reported an enhancement-mode β-Ga_2O_3 MOSFETs on a Si-doped homoepitaxial channel grown by molecular beam epitaxy and, using a gate recess process to partially remove the epitaxial channel under the 1-μm gated region to fully deplete at zero gate bias. With a breakdown voltage of 505 V (8 mm source-drain spacing), a maximum current density of 40 mA mm^{-1}, and an on/off ratio of 10^9. Hu et al. [174] achieved (in 2018) a larger blocking voltage (1.075 kV), a larger threshold voltage (1.2–2.2 V), and a larger output current (~500 A cm^{-2}) in a first demonstration of vertical E-mode MOSFET with significatively larger FOM (~80 MW cm^{-2}).

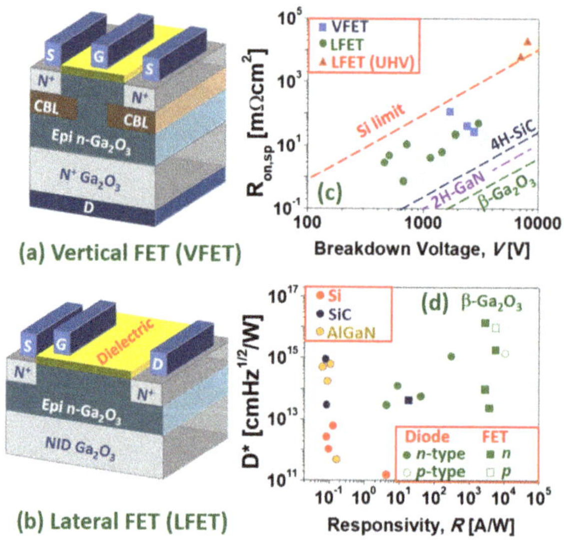

Figure 6. Schematics of (**a**) a vertical Ga_2O_3 power transistor (VFET) and (**b**) a lateral transistor (LFET). (**c**) Baliga's FOM for selected LFETs and VFETs from the literature. (**d**) Prospects of Ga_2O_3 devices as UV PDs, D* refers to specific detectivity; dots symbols referrer to diodes (either SBD or MSM), while square symbols denote transistors (data adapted from Wu et al. [131]). Adapted with permission from [12] © 2021 copyright Society of Photo-Optical Instrumentation Engineers (SPIE).

The E-mode was accomplished by doping profiling in a FinFET design (a type of 3D, non-planar transistor which has become the usual layout for the smallest CMOS 14 nm, 10 nm, and 7 nm nodes). This kind of E-mode vertical power device was later optimized to sustain up to a blocking voltage of 1.6kV [175], a threshold voltage of 2.66 kV, a maximum current density of 25.2 $mWcm^2$, and a record FOM of 280 MW cm^{-2} [176]. Among D-mode devices, the ones reported by Lv et al. [177] stand out for exhibiting a particularly large FOM. They reported (in 2019) [177] source-FP β-Ga_2O_3 MOSFETs on a Si-doped/Fe-doped semi-insulating β-Ga_2O_3 substrate exhibiting 222 mA mm^{-1} (18 mm source-drain spacing) with on-resistance of 11.7 $m\Omega cm^2$, a V_{br} of 680 V and an FOM of 50.4 $MWcm^{-2}$. Later (in 2020) [178], they adopted a T-shaped gate and source connected FP structure to increase the V_{br} up to 1.4 kV/2.9 kV (for 4.8 μm/17.8 μm source-drain spacing), with a specific on-resistances of 7.08 $m\Omega cm^2$/46.2 $m\Omega cm^2$. These yielded a record high FOM of

277 MW cm^{-2}, together with negligible gate or drain pulsed current collapse and a drain current on/off ratio of 10^9.

Other lateral D-mode devices with high FOM were reported by Tetzner et al. [179]. By using sub-μm gate lengths (combined with gate recess) and optimization of compensation-doped high-quality crystals, implantation based inter-device isolation, and SiNx-passivation, breakdown voltages of 1.8 kV and an FOM of 155 MW cm^{-2} were achieved. In 2020, Sharma et al. [180] reported Ga_2O_3 lateral D-mode field-plated MOSFETs exhibiting an ultra-high V_{br} of 8.03 kV (70 mm) by using polymer SU8 passivation. The current was rather low, however, due to plasma-induced damage of channel and access regions resulting in an impractical FOM of 7.73 kW cm^{-2} (i.e., not above the silicon limit). As reported by Kalarickal et al. [164], ultra-high-*k* ferroelectric dielectrics, such as $BaTiO_3$, can, in principle, provide an efficient field management strategy by improving the uniformity of electric field profile in the gate-drain region of lateral FETs. High average breakdown fields of 1.5 MV/cm (918 V) and 4 $MVcm^{-1}$ (201 V) were demonstrated for gate-drain spacings of 6μm and 0.6 μm, respectively, in β-Ga_2O_3, at a high channel sheet charge density of 1.8×10^{13} cm^{-2}. An elevated sheet charge density together with a high breakdown field enabled a record power FOM of 376 $MWcm^{-2}$ at a gate-drain spacing of 3 μm (Figure 6c). As in the case of SBDs, these performances for the Ga_2O_3 devices are already impressive and well beyond the silicon limit but still lag behind the best (much more mature) GaN devices in their respective power ratings [181,182].

All the above power MOSFET devices are unipolar *n*-type. These devices are sometimes referred as MISFETs so as to distinguish them from the conventional p-n junction based MOSFETs, since there are no *p*-regions in these MISFETs [175]. As mentioned in the previous sections, there are, however, several reports of *p*-type Ga_2O_3in nominally undoped, H-doped and N-doped β-Ga_2O_3. In particular, Wuetal. [131] proposed a growth mechanism of multistep structural phase transitions from hexagonal P63mc GaN to rhombohedral R3c α-$GaN_xO_{3(1-x)/2}$,and finally to monolithic C2/m N-doped β-Ga_2O_3. This improves the crystalline quality, facilitates acceptor doping, increases the acceptor activation efficiency, and thus enhances the *p*-type conductivity (acceptor ionization energy of 0.165 eV, Hall resistivity of 17.0 Ωcm, Hall hole mobility of 23.6 $cm^2V^{-1}s^{-1}$, hole concentration of 1.56×10^{16} cm^{-3}). P-type β-Ga_2O_3 films-based lateral MOSFET deep-ultraviolet (DUV) PDs were fabricated with extremely high responsivity (5.1×10^3 A/W) and detectivity (1.0×10^{16}Jones) under 250 nm light illumination (40 μW/cm^2) conditions. Figure 6d shows the responsivity and detectivity (D*) for state-of-the-art DUV PDs based on various WBG materials (adapted from [131]), in which it can be seen how β-Ga_2O_3 surpasses conventional Si-, SiC-, and AlGaN-based devices in terms of responsivity and detectivity.

4. Other Emerging Oxide Semiconductors for Power Electronics

Ga_2O_3phase engineering: Owing to the nonpolar nature of β-Ga_2O_3 crystals, modulation-doped heterostructure is one of the possible approaches to realize Ga_2O_3-based FETs [183]. Analogously, *p*-type semiconductors (e.g., *p*-type nitrides such as GaN) may be introduced to yield normally-off β-Ga_2O_3 field-effect transistors with tunable positive threshold voltages [184]. Other phases of Ga_2O_3 have also received attention due to potentially favorable growth characteristics, and to the possibility of polarization engineering made possible by the polar nature of their crystal structures. In principle, this polarization could be utilized to produce Ga_2O_3 two-dimensional electron gases (2DEGs) in analogy with GaN/AlN-based transistors [185].

Ga_2O_3alloy engineering: The aluminum gallium oxide, $Al_xGa_{1-x}O_3$, is a ternary alloy of Al_2O_3 and Ga_2O_3. It was already noted by Roy [186] in 1952 that the gallium ion closely resembles the aluminum ion and substitutes for it in several structures. Because β-$(AlGa)_2O_3$ is not the energetically favored crystalline phase for large Al compositions, the crystal converts to competing structural phases when grown on β-Ga_2O_3 substrates [187]. Thus, it has been difficult to obtain gallium oxide UWBG materials exceeding the bandgap of ~6 eV which is available to the materials in the nitride family in AlN. Very recently however, it was found that single-crystalline layers of α-$(AlGa)_2O_3$ alloys spanning bandgaps of 5.4–8.6 eV can be grown by molecular beam epitaxy [188]. By varying the alloy composition, bandgap energies from ~5.4 up to 8.6 eV with a bowing parameter of 1.1 eV are achieved, making α-$(Al_xGa_{1-x})_2O_3$ the largest bandgap epitaxial material family to date. If these layers can be controllably doped, it would pave the way for α-$(Al_xGa_{1-x})_2O_3$–based high-power heterostructure electronic and photonic devices at bandgaps far beyond all materials available today [189].

Spinel electronics: The spinel zinc gallate, $ZnGa_2O_4$, is a nearly stoichiometric mixed oxide made of Ga_2O_3 and ZnO.A potential advantage of spinel $ZnGa_2O_4$ is its great dopability prospects owing to the spinel's inherent diversity in cation coordination possibilities [106]. Normal spinels have all A cations in the tetrahedral site and all B cations in the octahedral site, e.g., Zn-tetrahedral site $Zn^{2+}(T_d)$ and Ga-octahedral site $Ga^{3+}(O_h)$, so that normal $ZnGa_2O_4$ is $Zn(^{2+}[T_d])Ga_2(^{3+}[O_h])O_4(^{2-})$. The spinel's off-stoichiometry, from the ideal 1:2:4 proportions, or the creation of cation antisite defects are known routes for doping these compounds. Dominant defects in spinels are antisite donors (e.g., Zn_{Ga}) or donor-like $Ga^{3+}(O_h)$-on-T_d and antisite acceptors (e.g., GaZn) with acceptor-like $Zn^{2+}(T_d)$-on-O_h antisite defects resulting in an intrinsic bipolar power semiconductor [190]. $ZnGa_2O_4$ is therefore a potential outstanding UWBG (~5 eV) oxide semiconductor but is only one among the many possible spinel oxides. There are over 1000 compounds that are known to crystalize in the spinel structure. The sub-family of spinel oxides is a large and important class of multi-functional oxide semiconductors with many optoelectronics applications in areas such as batteries, fuel cells, catalysis, photonics (phosphors, bio-imaging, photodetectors), spintronics (magnets, bio-magnets), or thermoelectricity [191]. Other magnesium-based Ga-spinels, such as $MgGa_2O_4$and $Zn_{1-x}Mg_xGa_2O_4$, are related oxides that are currently being investigated [192,193].

5. Conclusions

The rational use of electrical energy and information are central themes in the greatest climatic challenge of the 21st century. UWBG oxides, such as Ga_2O_3 and related materials, are promising power electronic candidates since their critical electric field is large compared to beyond silicon WBG (i.e., SiC and GaN), while still yielding a moderate mobility, high quality epi-layers, and large bulk single crystals (more than 6-inch) using low cost and scalable fabrication approaches. During the last decade, the Ga_2O_3 power diode and transistor progress has been impressive, with devices now approaching the frontier of the field. The material system also opens new optoelectronics avenues (owing its UVC spanning bandgap), and new electronics perspectives based on stabile interfaces and a natural integration with extremely high-*k* functional oxides. The advances offered by Ga_2O_3 are also opening the door to many more UWBG oxides (the largest family of wide bandgap semiconductors), such as the spinel, $ZnGa_2O_4$, along with many more that are anticipated. Therefore, the ever-increasing family of UWBG oxides is at the very frontier of a more efficient energy electronics which is adapted to tackle the 21st century climatic targets, although there still is a lot of room for performance improvements, technical innovation, and new discoveries.

Author Contributions: Conceptualization A.P.-T., E.C., M.R.J.; writing—original draft preparation, A.P.-T., Z.C., J.J.A.; writing—review and editing, A.P.-T., Z.C., J.J.A.; visualization, A.P.-T.; supervision, A.P.-T.; project administration, A.P.-T.; funding acquisition, E.C., A.P.-T., M.R.J. All authors have read and agreed to the published version of the manuscript.

Funding: This research was funded by the French National Research Agency ANR, "Accelerating the demonstration of Gallium Oxide's outstanding potential for Energy application" (GOPOWER), grant number Project-ANR-21-CE50-0015.

Institutional Review Board Statement: Not applicable.

Informed Consent Statement: Not applicable.

Data Availability Statement: Not applicable.

Acknowledgments: Authors acknowledge IRP -"GALLIA", CNRS, France. The ICN2 is funded by the CERCA programme/Generalitat de Catalunya. The ICN2 is supported by the Severo Ochoa Centres of Excellence programme, funded by the Spanish Research Agency (AEI, grant no. SEV-2017-0706).

Conflicts of Interest: The authors declare no conflict of interest.

References

1. IPCC Working Group I Report, Climate Change 2021: The Physical Science Basis. Available online: www.ipcc.ch (accessed on 9 August 2021).
2. US Environmental Protection Agency (EPA). Greenhouse Gas Emissions: Global Greenhouse Gas Emissions Data. 2021. Available online: https://www.epa.gov/ghgemissions/global-greenhouse-gas-emissions-data (accessed on 6 November 2017).
3. Leo Lorenz, Power Device Development Trends—From Silicon to Wide Bandgap? Available online: www.power-and-beyond.com (accessed on 16 December 2021).
4. Pearton, S.J.; Ren, F.; Tadjer, M.; Kim, J. Perspective: Ga_2O_3 for ultra-high power rectifiers and MOSFETS. *J. Appl. Phys.* **2018**, *124*, 220901. [CrossRef]
5. Reese, S.B.; Remo, T.; Green, J.; Zakutayev, A. How Much Will Gallium Oxide Power Electronics Cost? *Joule* **2019**, *3*, 903–907. [CrossRef]
6. Chu, S.; Cui, Y.; Liu, N. The path towards sustainable energy. *Nat. Mater.* **2017**, *16*, 16–22. [CrossRef] [PubMed]
7. Millan, J.; Godignon, P.; Perpiñà, X.; Perez-Tomas, A.; Rebollo, J. A Survey of Wide Bandgap Power Semiconductor Devices. *IEEE Trans. Power Electron.* **2013**, *29*, 2155–2163. [CrossRef]
8. Spaziani, L.; Lu, L. Silicon, GaN and SiC: There's room for all: An application space overview of device considerations. In Proceedings of the 2018 IEEE 30th International Symposium on Power Semiconductor Devices and ICs (ISPSD), Chicago, IL, USA, 13–17 May 2018; pp. 8–11.
9. Jones, E.A.; Wang, F.F.; Costinett, D. Review of Commercial GaN Power Devices and GaN-Based Converter Design Challenges. *IEEE J. Emerg. Sel. Top. Power Electron.* **2016**, *4*, 707–719. [CrossRef]
10. Tsao, J.Y.; Chowdhury, S.; Hollis, M.A.; Jena, D.; Johnson, N.M.; Jones, K.A.; Kaplar, R.J.; Rajan, S.; Van de Walle, C.G.; Bellotti, E.; et al. Ultrawide-Bandgap Semiconductors: Research Opportunities and Challenges. *Adv. Electron. Mater.* **2017**, *4*, 1600501. [CrossRef]
11. Perez-Tomas, A.; Teherani, F.H.; Bove, P.; Sandana, E.V.; Chikoidze, E.; Jennings, M.R.; Rogers, D.J.; Russell, S.A.O. Wide and ultra-wide bandgap oxides: Where paradigm-shift photovoltaics meets transparent power electronics. In Proceedings of the SPIE 10533, Oxide-Based Materials and Devices IX, San Francisco, CA, USA, 27 January–1 February 2018. [CrossRef]
12. Perez-Tomas, A.; Chikoidze, E.; Rogers, D.J. A walk on the frontier of energy electronics with power ultra-wide bandgap oxides and ultra-thin neuromorphic 2D materials. In Proceedings of the SPIE 11687, Oxide-Based Materials and Devices XII, San Francisco, CA, USA, 6–12 March 2021. [CrossRef]
13. Perez-Tomas, A.; Lodzinski, M.; Guy, O.J.; Jennings, M.R.; Placidi, M.; Llobet, J.; Gammon, P.M.; Davis, M.C.; Covington, J.A.; Burrows, S.E.; et al. Si/SiC bonded wafer: A route to carbon free SiO_2 on SiC. *Appl. Phys. Lett.* **2009**, *94*, 103510. [CrossRef]
14. Nawaz, M. Introductory Chapter: Earth Crust-Origin, Structure, Composition and Evolution. In *Earth Crust*; IntechOpen: London, UK, 2019. [CrossRef]
15. Jackson, I. *The Earth's Mantle—Composition, Structure and Evolution*; Cambridge University Press: Cambridge, UK, 1998.
16. Higashiwaki, M.; Sasaki, K.; Murakami, H.; Kumagai, Y.; Koukitu, A.; Kuramata, A.; Masui, T.; Yamakoshi, S. Recent progress in Ga_2O_3 power devices. *Semicond. Sci. Technol.* **2016**, *31*, 034001. [CrossRef]
17. Kuramata, A.; Koshi, K.; Watanabe, S.; Yamaoka, Y.; Masui, T.; Yamakoshi, S. Bulk crystal growth of Ga_2O_3. In Proceedings of the SPIE 10533, Oxide-Based Materials and Devices IX, San Francisco, CA, USA, 27 January–1 February 2018. [CrossRef]
18. Playford, H.Y.; Hannon, A.C.; Barney, E.R.; Walton, R.I. Structures of Uncharacterised Polymorphs of Gallium Oxide from Total Neutron Diffraction. *Chem. A Eur. J.* **2013**, *19*, 2803–2813. [CrossRef]
19. Geller, S. Crystal Structure of β—Ga_2O_3. *J. Chem. Phys.* **1960**, *33*, 676–684. [CrossRef]
20. Blasse, G.; Bril, A. Some observations on the luminescence of β-Ga_2O_3. *J. Phys. Chem. Solids* **1970**, *31*, 707–711. [CrossRef]
21. Lorenz, M.; Woods, J.; Gambino, R. Some electrical properties of the semiconductor β Ga_2O_3. *J. Phys. Chem. Solids* **1967**, *28*, 403–404. [CrossRef]
22. Tippins, H.H. Optical Absorption and Photoconductivity in the Band Edge of $\beta-Ga_2O_3$. *Phys. Rev.* **1965**, *140*, A316–A319. [CrossRef]
23. Nouketcha, F.L.L.; Cui, Y.; Lelis, A.; Green, R.; Darmody, C.; Schuster, J.; Goldsman, N. Investigation of Wide- and Ultrawide-Bandgap Semiconductors From Impact-Ionization Coefficients. *IEEE Trans. Electron. Devices* **2020**, *67*, 3999–4005. [CrossRef]
24. Chikoidze, E.; Tchelidze, T.; Sartel, C.; Chi, Z.; Kabouche, R.; Madaci, I.; Rubio, C.; Mohamed, H.; Sallet, V.; Medjdoub, F.; et al. Ultra-high critical electric field of 13.2 MV/cm for Zn-doped p-type β-Ga_2O_3. *Mater. Today Phys.* **2020**, *15*, 100263. [CrossRef]
25. Baliga, B. Power semiconductor device figure of merit for high-frequency applications. *IEEE Electron Device Lett.* **1989**, *10*, 455–457. [CrossRef]

26. Ma, N.; Tanen, N.; Verma, A.; Guo, Z.; Luo, T.; Xing, H.; Jena, D. Intrinsicelectronmobilitylimits in β-Ga_2O_3. *Appl. Phys. Lett.* **2016**, *109*, 212101. [CrossRef]
27. Fu, B.; Jia, Z.; Mu, W.; Yin, Y.; Zhang, J.; Tao, X. A review of β-Ga_2O_3 single crystal defects, their effects on device performance and their formation mechanism. *J. Semicond.* **2019**, *40*, 011804. [CrossRef]
28. Stuchlikova, T.H.; Stuchlik, J.; Remes, Z.; Taylor, A.; Mortet, V.; Ashcheulov, P.; Gregora, I.; Krivyakin, G.; Volodin, V. High-Temperature PIN Diodes Based on Amorphous Hydrogenated Silicon-Carbon Alloys and Boron-Doped Diamond Thin Films. *Phys. Status Solidi* **2020**, *257*, 1900247. [CrossRef]
29. Chase, A.O. Growth of beta-Ga_2, O_3 by the Verneuil Technique. *J. Am. Ceram. Soc.* **1964**, *47*, 470. [CrossRef]
30. Galazka, Z.; Ganschow, S.; Irmscher, K.; Klimm, D.; Albrecht, M.; Schewski, R.; Pietsch, M.; Schulz, T.; Dittmar, A.; Kwasniewski, A.; et al. Bulk single crystals of β-Ga_2O_3 and Ga-based spinels as ultra-wide bandgap transparent semiconducting oxides. *Prog. Cryst. Growth Charact. Mater.* **2020**, *67*, 100511. [CrossRef]
31. Galazka, Z.; Irmscher, K.; Schewski, R.; Hanke, I.M.; Pietsch, M.; Ganschow, S.; Klimm, D.; Dittmar, A.; Fiedler, A.; Schroeder, T.; et al. Czochralski-grown bulk β-Ga_2O_3 single crystals doped with mono-, di-, tri-, and tetravalent ions. *J. Cryst. Growth* **2019**, *529*, 125297. [CrossRef]
32. Galazka, Z.; Schewski, R.; Irmscher, K.; Drozdowski, W.; Witkowski, M.E.; Makowski, M.; Wojtowicz, A.J.; Hanke, I.M.; Pietsch, M.; Schulz, T.; et al. Bulk β-Ga_2O_3 single crystals doped with Ce, Ce+Si, Ce+Al, and Ce+Al+Si for detection of nuclear radiation. *J. Alloys Compd.* **2019**, *818*, 152842. [CrossRef]
33. Irmscher, K.; Galazka, Z.; Pietsch, M.; Uecker, R.; Fornari, R. Electrical properties of β-Ga_2O_3 single crystals grown by the Czochralski method. *J. Appl. Phys.* **2011**, *110*, 063720. [CrossRef]
34. Víllora, E.G.; Shimamura, K.; Yoshikawa, Y.; Aoki, K.; Ichinose, N. Large-size β-Ga_2O_3 single crystals and wafers. *J. Cryst. Growth* **2004**, *270*, 420–426. [CrossRef]
35. Hoshikawa, K.; Kobayashi, T.; Matsuki, Y.; Ohba, E. 2-inch diameter (1 0 0) β-Ga_2O_3 crystal growth by the vertical Bridgman technique in a resistance heating furnace in ambient air. *J. Cryst. Growth* **2020**, *545*, 125724. [CrossRef]
36. Hoshikawa, K.; Ohba, E.; Kobayashi, T.; Yanagisawa, J.; Miyagawa, C.; Nakamura, Y. Growth of β-Ga_2O_3 single crystals using vertical Bridgman method in ambient air. *J. Cryst. Growth* **2016**, *447*, 36–41. [CrossRef]
37. Harwig, T.; Schoonman, J. Electrical properties of β-Ga_2O_3 single crystals. II. *J. Solid State Chem.* **1978**, *23*, 205–211. [CrossRef]
38. Cui, H.; Mohamed, H.; Xia, C.; Sai, Q.; Zhou, W.; Qi, H.; Zhao, J.; Si, J.; Ji, X. Tuning electrical conductivity of β-Ga_2O_3 single crystals by Ta doping. *J. Alloys Compd.* **2019**, *788*, 925–928. [CrossRef]
39. Hossain, E.; Kulkarni, R.; Mondal, R.; Guddolian, S.; Rahman, A.A.; Thamizhavel, A.; Bhattacharya, A. Optimization of Gas Ambient for High Quality β-Ga_2O_3 Single Crystals Grown by the Optical Floating Zone Technique. *ECS J. Solid State Sci. Technol.* **2019**, *8*, Q3144–Q3148. [CrossRef]
40. Suzuki, N.; Ohira, S.; Tanaka, M.; Sugawara, T.; Nakajima, K.; Shishido, T. Fabrication and characterization of transparent conductive Sn-doped β-Ga_2O_3 single crystal. *Phys. Status Solidi* **2007**, *4*, 2310–2313. [CrossRef]
41. Tomioka, Y.; Ozaki, Y.; Inaba, H.; Ito, T. Compensation effects between impurity cations in single crystals of a wide gap semiconductor β-Ga_2O_3 prepared by the floating zone method. *Jpn. J. Appl. Phys.* **2019**, *58*, 091009. [CrossRef]
42. Zhou, W.; Xia, C.; Sai, Q.; Zhang, H. Controlling n-type conductivity of β-Ga_2O_3 by Nb doping. *Appl. Phys. Lett.* **2017**, *111*, 242103. [CrossRef]
43. Galazka, Z. β-Ga_2O_3 for wide-bandgap electronics and optoelectronics. *Semicond. Sci. Technol.* **2018**, *33*, 113001. [CrossRef]
44. Galazka, Z.; Uecker, R.; Klimm, D.; Irmscher, K.; Naumann, M.; Pietsch, M.; Kwasniewski, A.; Bertram, R.; Ganschow, S.; Bickermann, M. Scaling-Up of Bulk β-Ga_2O_3Single Crystals by the Czochralski Method. *ECS J. Solid State Sci. Technol.* **2016**, *6*, Q3007–Q3011. [CrossRef]
45. Saleh, M.; Varley, J.B.; Jesenovec, J.; Bhattacharyya, A.; Krishnamoorthy, S.; Swain, S.; Lynn, K.G. Degenerate doping in β-Ga_2O_3 single crystals through Hf-doping. *Semicond. Sci. Technol.* **2020**, *35*, 04LT01. [CrossRef]
46. Kuramata, A.; Koshi, K.; Watanabe, S.; Yamaoka, Y.; Masui, T.; Yamakoshi, S. High-quality β-Ga_2O_3single crystals grown by edge-defined film-fed growth. *Jpn. J. Appl. Phys.* **2016**, *55*, 1202A2. [CrossRef]
47. Yao, Y.; Sugawara, Y.; Ishikawa, Y. Identification of Burgers vectors of dislocations in monoclinic β-Ga_2O_3 via synchrotron X-ray topography. *J. Appl. Phys.* **2020**, *127*, 205110. [CrossRef]
48. Hoshikawa, K.; Kobayashi, T.; Ohba, E. 50 mm diameter Sn-doped (0 0 1) β-Ga_2O_3 crystal growth using the vertical Bridgeman technique in ambient air. *J. Cryst. Growth* **2020**, *546*, 125778. [CrossRef]
49. Aubay, E.; Gourier, D. Magnetic bistability and Overhauser shift of conduction electrons in gallium oxide. *Phys. Rev. B* **1993**, *47*, 15023–15036. [CrossRef]
50. He, N.; Tang, H.; Liu, B.; Zhu, Z.; Li, Q.; Guo, C.; Gu, M.; Xu, J.; Liu, J.; Xu, M.; et al. Ultra-fast scintillation properties of β-Ga_2O_3 single crystals grown by Floating Zone method. *Nucl. Instrum. Methods Phys. Res. Sect. A Accel. Spectrom. Detect. Assoc. Equip.* **2018**, *888*, 9–12. [CrossRef]
51. Usui, Y.; Oya, T.; Okada, G.; Kawaguchi, N.; Yanagida, T. Comparative study of scintillation and optical properties of Ga_2O_3 doped with ns2 ions. *Mater. Res. Bull.* **2017**, *90*, 266–272. [CrossRef]
52. Yanagida, T.; Kawaguchi, N. Optical and scintillation properties of alkaline earth doped Ga_2O_3 single crystals prepared by the floating zone method. *Jpn. J. Appl. Phys.* **2019**, *59*, SCCB20. [CrossRef]

53. Zhang, S.; Lian, X.; Ma, Y.; Liu, W.; Zhang, Y.; Xu, Y.; Cheng, H. Growth and characterization of 2-inch high quality β-Ga_2O_3 single crystals grown by EFG method. *J. Semicond.* **2018**, *39*, 083003. [CrossRef]
54. Oshima, T.; Hashiguchi, A.; Moribayashi, T.; Koshi, K.; Sasaki, K.; Kuramata, A.; Ueda, O.; Oishi, T.; Kasu, M. Electrical properties of Schottky barrier diodes fabricated on (001) β-Ga_2O_3substrates with crystal defects. *Jpn. J. Appl. Phys.* **2017**, *56*, 86501. [CrossRef]
55. Oshima, Y.; Víllora, E.G.; Shimamura, K. Quasi-heteroepitaxial growth of β-Ga_2O_3 on off-angled sapphire (0 0 0 1) substrates by halide vapor phase epitaxy. *J. Cryst. Growth* **2015**, *410*, 53–58. [CrossRef]
56. Xiong, Z.-N.; Xiu, X.-Q.; Li, Y.-W.; Hua, X.-M.; Xie, Z.-L.; Chen, P.; Liu, B.; Han, P.; Zhang, R.; Zheng, Y.-D. Growth of β -Ga_2O_3 Films on Sapphire by Hydride Vapor Phase Epitaxy. *Chin. Phys. Lett.* **2018**, *35*, 058101. [CrossRef]
57. Yao, Y.; Lyle, L.A.M.; Rokholt, J.A.; Okur, S.; Tompa, G.S.; Salagaj, T.; Sbrockey, N.; Davis, R.F.; Porter, L.M. (Invited) Growth and Characterization of α-,β-, andε-Ga_2O_3 Epitaxial Layers on Sapphire. *ECS Trans.* **2017**, *80*, 191–196. [CrossRef]
58. Xiu, X.; Zhang, L.; Li, Y.; Xiong, Z.; Zhang, R.; Zheng, Y. Application of halide vapor phase epitaxy for the growth of ultra-wide band gap Ga_2O_3. *J. Semicond.* **2019**, *40*, 011805. [CrossRef]
59. Modak, S.; Chernyak, L.; Khodorov, S.; Lubomirsky, I.; Yang, J.; Ren, F.; Pearton, S.J. Impact of Electron Injection and Temperature on Minority Carrier Transport in Alpha-Irradiated ß-Ga_2O_3 Schottky Rectifiers. *ECS J. Solid State Sci. Technol.* **2019**, *8*, Q3050–Q3053. [CrossRef]
60. De Santi, C.; Nardo, A.; Wong, M.; Goto, K.; Kuramata, A.; Yamakoshi, S.; Murakami, H.; Kumagai, Y.; Higashiwaki, M.; Meneghesso, G.; et al. Stability and degradation of isolation and surface in Ga_2O_3 devices. *Microelectron. Reliab.* **2019**, *100–101*, 113453. [CrossRef]
61. Leach, J.H.; Udwary, K.; Rumsey, J.; Dodson, G.; Splawn, H.; Evans, K.R. Halide vapor phase epitaxial growth of β-Ga_2O_3 and α-Ga_2O_3 films. *APL Mater.* **2019**, *7*, 022504. [CrossRef]
62. Murakami, H.; Nomura, K.; Goto, K.; Sasaki, K.; Kawara, K.; Thieu, Q.T.; Togashi, R.; Kumagai, Y.; Higashiwaki, M.; Kuramata, A.; et al. Homoepitaxial growth of β-Ga_2O_3layers by halide vapor phase epitaxy. *Appl. Phys. Express* **2014**, *8*, 015503. [CrossRef]
63. Bin Anooz, S.; Grüneberg, R.; Wouters, C.; Schewski, R.; Albrecht, M.; Fiedler, A.; Irmscher, K.; Galazka, Z.; Miller, W.; Wagner, G.; et al. Step flow growth of β-Ga_2O_3 thin films on vicinal (100) β-Ga_2O_3 substrates grown by MOVPE. *Appl. Phys. Lett.* **2020**, *116*, 182106. [CrossRef]
64. Bin Anooz, S.; Grüneberg, R.; Chou, T.-S.; Fiedler, A.; Irmscher, K.; Wouters, C.; Schewski, R.; Albrecht, M.; Galazka, Z.; Miller, W.; et al. Impact of chamber pressure and Si-doping on the surface morphology and electrical properties of homoepitaxial (100) β-Ga_2O_3 thin films grown by MOVPE. *J. Phys. D Appl. Phys.* **2020**, *54*, 034003. [CrossRef]
65. Gogova, D.; Schmidbauer, M.; Kwasniewski, A. Homo- and heteroepitaxial growth of Sn-doped β-Ga_2O_3 layers by MOVPE. *CrystEngComm* **2015**, *17*, 6744–6752. [CrossRef]
66. Albrecht, M.; Schewski, R.; Wouters, C.; Fielder, A.; Irmscher, K.; Galazka, Z.; Popp, A.; Anooz, S.B.; Baldini, M.; Wagner, G. Ga_2O_3 from Materials to Devices. *Appl. Phys. Res.* **2021**, *5*, 011301.
67. Baldini, M.; Albrecht, M.; Fiedler, A.; Irmscher, K.; Schewski, R.; Wagner, G. Editors' Choice—Si and Sn Doped Homoepitaxial β-Ga_2O_3 Layers Grown by MOVPE on (010)-Oriented Substrates. *ECS J. Solid State Sci. Technol.* **2016**, *6*, Q3040–Q3044. [CrossRef]
68. Cui, R.-R.; Zhang, J.; Luo, Z.-J.; Guo, X.; Ding, Z.; Deng, C.-Y. Microstructure, optical, and photoluminescence properties of β-Ga_2O_3 films prepared by pulsed laser deposition under different oxygen partial pressures. *Chin. Phys. B* **2021**, *30*, 028505. [CrossRef]
69. Leedy, K.; Chabak, K.D.; Vasilyev, V.; Look, D.C.; Boeckl, J.J.; Brown, J.L.; Tetlak, S.E.; Green, A.J.; Moser, N.A.; Crespo, A.; et al. Highly conductive homoepitaxial Si-doped Ga_2O_3 films on (010) β-Ga_2O_3 by pulsed laser deposition. *Appl. Phys. Lett.* **2017**, *111*, 012103. [CrossRef]
70. Shen, H.; Baskaran, K.; Yin, Y.; Tian, K.; Duan, L.; Zhao, X.; Tiwari, A. Effect of thickness on the performance of solar blind photodetectors fabricated using PLD grown β-Ga_2O_3 thin films. *J. Alloys Compd.* **2019**, *822*, 153419. [CrossRef]
71. Vu, T.K.O.; Lee, D.U.; Kim, E.K. The effect of oxygen partial pressure on band gap modulation of Ga_2O_3 grown by pulsed laser deposition. *J. Alloys Compd.* **2019**, *806*, 874–880. [CrossRef]
72. Pearton, S.J.; Yang, J.; Cary, P.H., IV; Ren, F.; Kim, J.; Tadjer, M.J.; Mastro, M.A. A review of Ga_2O_3 materials, processing, and devices. *Appl. Phys. Rev.* **2018**, *5*, 011301. [CrossRef]
73. Chen, Z.; Wang, X.; Zhang, F.; Noda, S.; Saito, K.; Tanaka, T.; Nishio, M.; Guo, Q. Temperature dependence of luminescence spectra in europium doped Ga_2O_3 film. *J. Lumin.* **2016**, *177*, 48–53. [CrossRef]
74. Wang, Q.; Chen, J.; Huang, P.; Li, M.; Lu, Y.; Homewood, K.P.; Chang, G.; Chen, H.; He, Y. Influence of growth temperature on the characteristics of β-Ga_2O_3 epitaxial films and related solar-blind photodetectors. *Appl. Surf. Sci.* **2019**, *489*, 101–109. [CrossRef]
75. Yadav, M.K.; Mondal, A.; Das, S.; Sharma, S.; Bag, A. Impact of annealing temperature on band-alignment of PLD grown Ga_2O_3/Si (100) heterointerface. *J. Alloys Compd.* **2019**, *819*, 153052. [CrossRef]
76. Yu, J.; Nie, Z.; Dong, L.; Yuan, L.; Li, D.; Huang, Y.; Zhang, L.; Zhang, Y.; Jia, R. Influence of annealing temperature on structure and photoelectrical performance of β-Ga_2O_3/4H-SiC heterojunction photodetectors. *J. Alloys Compd.* **2019**, *798*, 458–466. [CrossRef]
77. Puurunen, R.L. A Short History of Atomic Layer Deposition: Tuomo Suntola's Atomic Layer Epitaxy. *Chem. Vap. Depos.* **2014**, *20*, 332–344. [CrossRef]
78. Siah, S.C.; Brandt, R.E.; Lim, K.; Schelhas, L.T.; Jaramillo, R.; Heinemann, M.D.; Chua, D.; Wright, J.; Perkins, J.D.; Segre, C.U.; et al. Dopant activation in Sn-doped Ga_2O_3 investigated by X-ray absorption spectroscopy. *Appl. Phys. Lett.* **2015**, *107*, 252103. [CrossRef]

79. Choi, D.-W.; Chung, K.-B.; Park, J.-S. Low temperature Ga_2O_3 atomic layer deposition using gallium tri-isopropoxide and water. *Thin Solid Films* **2013**, *546*, 31–34. [CrossRef]
80. Mizutani, F.; Higashi, S.; Inoue, M.; Nabatame, T. Atomic layer deposition of high purity Ga_2O_3 films using liquid pentamethyl-cyclopentadienyl gallium and combinations of H_2O and O_2 plasma. *J. Vac. Sci. Technol. A* **2020**, *38*, 022412. [CrossRef]
81. Ilhom, S.; Mohammad, A.; Shukla, D.; Grasso, J.; Willis, B.G.; Okyay, A.K.; Biyikli, N. Low-Temperature As-Grown Crystalline β-Ga_2O_3 Films via Plasma-Enhanced Atomic Layer Deposition. *ACS Appl. Mater. Interfaces* **2021**, *13*, 8538–8551. [CrossRef]
82. Jiao, Y.; Jiang, Q.; Meng, J.; Zhao, J.; Yin, Z.; Gao, H.; Zhang, J.; Deng, J.; Zhang, X. Growth and characteristics of β-Ga_2O_3 thin films on sapphire (0001) by low pressure chemical vapour deposition. *Vacuum* **2021**, *189*, 110253. [CrossRef]
83. Tao, J.; Lu, H.-L.; Gu, Y.; Ma, H.-P.; Li, X.; Chen, J.-X.; Liu, W.-J.; Zhang, H.; Feng, J.-J. Investigation of growth characteristics, compositions, and properties of atomic layer deposited amorphous Zn-doped Ga_2O_3 films. *Appl. Surf. Sci.* **2019**, *476*, 733–740. [CrossRef]
84. Sasaki, K.; Higashiwaki, M.; Kuramata, A.; Masui, T.; Yamakoshi, S. MBE grown Ga_2O_3 and its power device applications. *J. Cryst. Growth* **2013**, *378*, 591–595. [CrossRef]
85. Mazzolini, P.; Falkenstein, A.; Wouters, C.; Schewski, R.; Markurt, T.; Galazka, Z.; Martin, M.; Albrecht, M.; Bierwagen, O. Substrate-orientation dependence of β-Ga_2O_3 (100), (010), (001), and (2¯01) homoepitaxy by indium-mediated metal-exchange catalyzed molecular beam epitaxy (MEXCAT-MBE). *APL Mater.* **2020**, *8*, 011107. [CrossRef]
86. Nepal, N.; Katzer, D.S.; Downey, B.P.; Wheeler, V.D.; Nyakiti, L.O.; Storm, D.F.; Hardy, M.T.; Freitas, J.A.; Jin, E.N.; Vaca, D.; et al. Heteroepitaxial growth of β-Ga_2O_3 films on SiC via molecular beam epitaxy. *J. Vac. Sci. Technol. A Vac. Surf. Films* **2020**, *38*, 063406. [CrossRef]
87. Kamimura, T.; Nakata, Y.; Higashiwaki, M. Effect of $(AlGa)_2O_3$ back barrier on device characteristics of β-Ga_2O_3 metal-oxide-semiconductor field-effect transistors with Si-implanted channel. *Jpn. J. Appl. Phys.* **2021**, *60*, 030906. [CrossRef]
88. Mazzolini, P.; Vogt, P.; Schewski, R.; Wouters, C.; Albrecht, M.; Bierwagen, O. Faceting and metal-exchange catalysis in (010) β-Ga_2O_3 thin films homoepitaxially grown by plasma-assisted molecular beam epitaxy. *APL Mater.* **2019**, *7*, 022511. [CrossRef]
89. Ngo, T.S.; Le, D.D.; Lee, J.; Hong, S.-K.; Ha, J.-S.; Lee, W.-S.; Moon, Y.-B. Investigation of defect structure in homoepitaxial (2¯01) β-Ga_2O_3 layers prepared by plasma-assisted molecular beam epitaxy. *J. Alloys Compd.* **2020**, *834*, 155027. [CrossRef]
90. Ahmadi, E.; Koksaldi, O.S.; Kaun, S.W.; Oshima, Y.; Short, D.B.; Mishra, U.K.; Speck, J.S. Ge doping of β-Ga_2O_3 films grown by plasma-assisted molecular beam epitaxy. *Appl. Phys. Express* **2017**, *10*, 041102. [CrossRef]
91. Cheng, Y.; Xu, Y.; Li, Z.; Zhang, J.; Chen, D.; Feng, Q.; Xu, S.; Zhou, H.; Zhang, J.; Hao, Y.; et al. Heteroepitaxial growth of α-Ga_2O_3 thin films on a-, c- and r-plane sapphire substrates by low-cost mist-CVD method. *J. Alloys Compd.* **2020**, *831*, 154776. [CrossRef]
92. Kaneko, K.; Fujita, S.; Hitora, T. A power device material of corundum-structured α-Ga_2O_3fabricated by MIST EPITAXY®® technique. *Jpn. J. Appl. Phys.* **2018**, *57*, 02CB18. [CrossRef]
93. Ma, T.; Chen, X.; Ren, F.; Zhu, S.; Gu, S.; Zhang, R.; Zheng, Y.; Ye, J. Heteroepitaxial growth of thick α-Ga_2O_3 film on sapphire (0001) by MIST-CVD technique. *J. Semicond.* **2019**, *40*, 012804. [CrossRef]
94. Morimoto, S.; Nishinaka, H.; Yoshimoto, M. Growth and characterization of F-doped α-Ga_2O_3 thin films with low electrical resistivity. *Thin Solid Films* **2019**, *682*, 18–23. [CrossRef]
95. Isomura, N.; Nagaoka, T.; Watanabe, Y.; Kutsuki, K.; Nishinaka, H.; Yoshimoto, M. Determination of Zn-containing sites in β-Ga_2O_3 film grown through mist chemical vapor deposition via X-ray absorption spectroscopy. *Jpn. J. Appl. Phys.* **2020**, *59*, 070909. [CrossRef]
96. Nishinaka, H.; Nagaoka, T.; Kajita, Y.; Yoshimoto, M. Rapid homoepitaxial growth of (010) β-Ga_2O_3 thin films via mist chemical vapor deposition. *Mater. Sci. Semicond. Process.* **2021**, *128*, 105732. [CrossRef]
97. Xu, Y.; Cheng, Y.; Li, Z.; Chen, D.; Xu, S.; Feng, Q.; Zhu, W.; Zhang, Y.; Zhang, J.; Zhang, C.; et al. Ultrahigh-Performance Solar-Blind Photodetectors Based on High Quality Heteroepitaxial Single Crystalline β-Ga_2O_3 Film Grown by Vacuumfree, Low-Cost Mist Chemical Vapor Deposition. *Adv. Mater. Technol.* **2021**, *6*, 2001296. [CrossRef]
98. Lee, S.-D.; Kaneko, K.; Fujita, S. Homoepitaxial growth of beta gallium oxide films by mist chemical vapor deposition. *Jpn. J. Appl. Phys.* **2016**, *55*, 1202B8. [CrossRef]
99. Li, Z.; Jiao, T.; Yu, J.; Hu, D.; Lv, Y.; Li, W.; Dong, X.; Zhang, B.; Zhang, Y.; Feng, Z.; et al. Single crystalline β-Ga_2O_3 homoepitaxial films grown by MOCVD. *Vacuum* **2020**, *178*, 109440. [CrossRef]
100. Feng, Z.; Bhuiyan, A.F.M.A.U.; Karim, R.; Zhao, H. MOCVD homoepitaxy of Si-doped (010) β-Ga_2O_3 thin films with superior transport properties. *Appl. Phys. Lett.* **2019**, *114*, 250601. [CrossRef]
101. Zhang, Y.; Alema, F.; Mauze, A.; Koksaldi, O.S.; Miller, R.; Osinsky, A.; Speck, J.S. MOCVD grown epitaxial β-Ga_2O_3 thin film with an electron mobility of 176 cm2/V s at room temperature. *APL Mater.* **2019**, *7*, 022506. [CrossRef]
102. Alema, F.; Hertog, B.; Osinsky, A.; Mukhopadhyay, P.; Toporkov, M.; Schoenfeld, W.V. Fast growth rate of epitaxial β–Ga_2O_3 by close coupled showerhead MOCVD. *J. Cryst. Growth* **2017**, *475*, 77–82. [CrossRef]
103. Li, Z.; Jiao, T.; Hu, D.; Lv, Y.; Li, W.; Dong, X.; Zhang, Y.; Feng, Z.; Zhang, B. Study on β-Ga_2O_3 Films Grown with Various VI/III Ratios by MOCVD. *Coatings* **2019**, *9*, 281. [CrossRef]
104. Tadjer, M.J.; Alema, F.; Osinsky, A.; Mastro, M.A.; Nepal, N.; Woodward, J.M.; Myers-Ward, R.L.; Glaser, E.R.; Freitas, J.A.; Jacobs, A.G.; et al. Characterization of β-Ga_2O_3 homoepitaxial films and MOSFETs grown by MOCVD at high growth rates. *J. Phys. D Appl. Phys.* **2020**, *54*, 034005. [CrossRef]

105. Alema, F.; Zhang, Y.; Osinsky, A.; Valente, N.; Mauze, A.; Itoh, T.; Speck, J.S. Low temperature electron mobility exceeding 104 cm2/V s in MOCVD grown β-Ga_2O_3. *APL Mater.* **2019**, *7*, 121110. [CrossRef]
106. Chikoidze, E.; Sartel, C.; Madaci, I.; Mohamed, H.; Vilar, C.; Ballesteros, B.; Belarre, F.; del Corro, E.; Castro, P.V.; Sauthier, G.; et al. p-Type Ultrawide-Band-Gap Spinel ZnGa2O4: New Perspectives for Energy Electronics. *Cryst. Growth Des.* **2020**, *20*, 2535–2546. [CrossRef]
107. Chikoidze, E.; Rogers, D.; Teherani, F.; Rubio, C.; Sauthier, G.; Von Bardeleben, H.; Tchelidze, T.; Ton-That, C.; Fellous, A.; Bove, P.; et al. Puzzling robust 2D metallic conductivity in undoped β-Ga_2O_3 thin films. *Mater. Today Phys.* **2019**, *8*, 10–17. [CrossRef]
108. Perez-Tomas, A.; Lira-Cantu, M.; Catalan, G. Above-Bandgap Photovoltages in Antiferroelectrics. *Adv. Mater.* **2016**, *28*, 9644–9647. [CrossRef]
109. Pérez-Tomás, A. Functional Oxides for Photoneuromorphic Engineering: Toward a Solar Brain. *Adv. Mater. Interfaces* **2019**, *6*, 1900471. [CrossRef]
110. Chikoidze, E.; Von Bardeleben, H.J.; Akaiwa, K.; Shigematsu, E.; Kaneko, K.; Fujita, S.; Dumont, Y. Electrical, optical, and magnetic properties of Sn doped α-Ga_2O_3 thin films. *J. Appl. Phys.* **2016**, *120*, 025109. [CrossRef]
111. Guo, D.; Guo, Q.; Chen, Z.; Wu, Z.; Li, P.; Tang, W. Review of Ga_2O_3-based optoelectronic devices. *Mater. Today Phys.* **2019**, *11*, 100157. [CrossRef]
112. Goto, K.; Konishi, K.; Murakami, H.; Kumagai, Y.; Monemar, B.; Higashiwaki, M.; Kuramata, A.; Yamakoshi, S. Halide vapor phase epitaxy of Si doped β-Ga_2O_3 and its electrical properties. *Thin Solid Films* **2018**, *666*, 182–184. [CrossRef]
113. Yan, X.; Esqueda, I.S.; Ma, J.; Tice, J.; Wang, H. High breakdown electric field in β-Ga_2O_3/graphene vertical barristor heterostructure. *Appl. Phys. Lett.* **2018**, *112*, 032101. [CrossRef]
114. Chikoidze, E.; Fellous, A.; Perez-Tomas, A.; Sauthier, G.; Tchelidze, T.; Ton-That, C.; Huynh, T.T.; Phillips, M.; Russell, S.; Jennings, M.; et al. P-type β-gallium oxide: A new perspective for power and optoelectronic devices. *Mater. Today Phys.* **2017**, *3*, 118–126. [CrossRef]
115. Chikoidze, E.; Sartel, C.; Mohamed, H.; Madaci, I.; Tchelidze, T.; Modreanu, M.; Vales-Castro, P.; Rubio, C.; Arnold, C.; Sallet, V.; et al. Enhancing the intrinsic p-type conductivity of the ultra-wide bandgap Ga_2O_3 semiconductor. *J. Mater. Chem. C* **2019**, *7*, 10231–10239. [CrossRef]
116. Kyrtsos, A.; Matsubara, M.; Bellotti, E. On the feasibility of p-type Ga_2O_3. *Appl. Phys. Lett.* **2018**, *112*, 032108. [CrossRef]
117. Varley, J.B.; Janotti, A.; Franchini, C.; Van de Walle, C. Role of self-trapping in luminescence andp-type conductivity of wide band-gap oxides. *Phys. Rev. B* **2012**, *85*, 081109. [CrossRef]
118. Lyons, J.L. A survey of acceptor dopants forβ-Ga_2O_3. *Semicond. Sci. Technol.* **2018**, *33*, 05LT02. [CrossRef]
119. Sun, D.; Gao, Y.; Xue, J.; Zhao, J. Defect stability and electronic structure of doped β-Ga_2O_3: A comprehensive ab initio study. *J. Alloys Compd.* **2019**, *794*, 374–384. [CrossRef]
120. Goyal, A.; Zakutayev, A.; Stevanović, V.; Lany, S. Computational Fermi level engineering and doping-type conversion of Mg:Ga_2O_3 via three-step synthesis process. *J. Appl. Phys.* **2021**, *129*, 245704. [CrossRef]
121. Sabino, F.P.; Cai, X.; Wei, S.-H.; Janotti, A. Bismuth-Doped Ga_2O_3 as Candidate for p-Type Transparent Conducting Material. *arXiv* **2019**, arXiv:190600840.
122. Li, L.; Liao, F.; Hu, X. The possibility of N–P codoping to realize P type β-Ga_2O_3. *Superlattices Microstruct.* **2020**, *141*, 106502. [CrossRef]
123. Ma, J.; Lin, J.; Liu, J.; Li, F.; Liu, Y.; Yang, G. Achieving high conductivity p-type Ga_2O_3 through Al-N and In-N co-doping. *Chem. Phys. Lett.* **2020**, *746*, 137308. [CrossRef]
124. Zhang, L.; Yan, J.; Zhang, Y.; Li, T.; Ding, X. A comparison of electronic structure and optical properties between N-doped β-Ga_2O_3 and N–Zn co-doped β-Ga_2O_3. *Phys. B Condens. Matter* **2012**, *407*, 1227–1231. [CrossRef]
125. Qian, Y.; Guo, D.; Chu, X.; Shi, H.; Zhu, W.; Wang, K.; Huang, X.; Wang, H.; Wang, S.; Li, P.; et al. Mg-doped p-type β-Ga_2O_3 thin film for solar-blind ultraviolet photodetector. *Mater. Lett.* **2017**, *209*, 558–561. [CrossRef]
126. Yue, W.; Yan, J.; Wu, J.; Zhang, L. Structural and optical properties of Zn-doped β-Ga_2O_3films. *J. Semicond.* **2012**, *33*, 073003. [CrossRef]
127. Alema, F.; Hertog, B.; Ledyaev, O.; Volovik, D.; Thoma, G.; Miller, R.; Osinsky, A.; Mukhopadhyay, P.; Bakhshi, S.; Ali, H.; et al. Solar blind photodetector based on epitaxial zinc doped Ga2 O3 thin film. *Phys. Status Solidi* **2017**, *214*, 1600688. [CrossRef]
128. Su, Y.; Guo, D.; Ye, J.; Zhao, H.; Wang, Z.; Wang, S.; Li, P.; Tang, W. Deep Level Acceptors of Zn-Mg Divalent Ions Dopants in b-Ga_2O_3 for the Difficulty to p-Type Conductivity. *J. Alloys Compd.* **2019**, *782*, 299–303. [CrossRef]
129. Feng, Q.; Liu, J.; Yang, Y.; Pan, D.; Xing, Y.; Shi, X.; Xia, X.; Liang, H. Catalytic growth and characterization of single crystalline Zn doped p-type β-Ga_2O_3 nanowires. *J. Alloys Compd.* **2016**, *687*, 964–968. [CrossRef]
130. Islam, M.; Liedke, M.O.; Winarski, D.; Butterling, M.; Wagner, A.; Hosemann, P.; Wang, Y.; Uberuaga, B.; Selim, F.A. Chemical manipulation of hydrogen induced high p-type and n-type conductivity in Ga_2O_3. *Sci. Rep.* **2020**, *10*, 6134. [CrossRef]
131. Wu, Z.; Jiang, Z.; Ma, C.; Ruan, W.; Chen, Y.; Zhang, H.; Zhang, G.; Fang, Z.; Kang, J.; Zhang, T.-Y. Energy-driven multi-step structural phase transition mechanism to achieve high-quality p-type nitrogen-doped β-Ga_2O_3 films. *Mater. Today Phys.* **2021**, *17*, 100356. [CrossRef]
132. Yao, Y.; Davis, R.F.; Porter, L.M. Investigation of different metals as ohmic contacts to β-Ga_2O_3: Comparison and analysis of electrical behavior, morphology, and other physical properties. *J. Electron. Mater.* **2017**, *46*, 2053–2060. [CrossRef]

133. Ji, M.; Taylor, N.R.; Kravchenko, I.; Joshi, P.; Aytug, T.; Cao, L.R.; Paranthaman, M.P. Demonstration of Large-Size Vertical Ga_2O_3 Schottky Barrier Diodes. *IEEE Trans. Power Electron.* **2020**, *36*, 41–44. [CrossRef]
134. Yao, Y.; Gangireddy, R.; Kim, J.; Das, K.K.; Davis, R.F.; Porter, L.M. Electrical behavior of β-Ga_2O_3 Schottky diodes with different Schottky metals. *J. Vac. Sci. Technol. B Nanotechnol. Microelectron. Mater. Process. Meas. Phenom.* **2017**, *35*, 03D113. [CrossRef]
135. Lu, X.; Zhou, L.; Chen, L.; Ouyang, X.; Tang, H.; Liu, B.; Xu, J. X-ray Detection Performance of Vertical Schottky Photodiodes Based on a Bulk β-Ga_2O_3 Substrate Grown by an EFG Method. *ECS J. Solid State Sci. Technol.* **2019**, *8*, Q3046–Q3049. [CrossRef]
136. Higashiwaki, M.; Konishi, K.; Sasaki, K.; Goto, K.; Nomura, K.; Thieu, Q.T.; Togashi, R.; Murakami, H.; Kumagai, Y.; Monemar, B.; et al. Temperature-dependent capacitance–voltage and current–voltage characteristics of Pt/Ga_2O_3 (001) Schottky barrier diodes fabricated on n^-–Ga_2O_3 drift layers grown by halide vapor phase epitaxy. *Appl. Phys. Lett.* **2016**, *108*, 133503. [CrossRef]
137. Harada, T.; Ito, S.; Tsukazaki, A. Electric Dipole Effect in $PdCoO_2$/β-Ga_2O_3 Schottky Diodes for High-Temperature Operation. *Sci. Adv.* **2019**, *5*, eaax5733. [CrossRef]
138. Pérez-Tomás, A.; Mingorance, A.; Tanenbaum, D.; Lira-Cantú, M. Chapter 8—Metal Oxides in Photovoltaics: All-Oxide, Ferroic, and Perovskite Solar Cells. In *The Future of Semiconductor Oxides in Next-Generation Solar Cells*; Lira-Cantu, M., Ed.; Elsevier: Amsterdam, The Netherlands, 2018; pp. 267–356.
139. Carey IV, P.H.; Yang, J.; Ren, R.; Sharma, R.; Law, M.; Pearton, S.J. Comparison of Dual-Stack Dielectric Field Plates on β-Ga_2O_3 Schottky Rectifiers. *ECS J. Solid State Sci. Technol.* **2019**, *8*, 7. [CrossRef]
140. Chen, Y.-T.; Yang, J.; Ren, F.; Chang, C.-W.; Lin, J.; Pearton, S.J.; Tadjer, M.J.; Kuramata, A.; Liao, Y.-T. Implementation of a 900 V Switching Circuit for High Breakdown Voltage β-Ga_2O_3 Schottky Diodes. *ECS J. Solid State Sci. Technol.* **2019**, *8*, 7. [CrossRef]
141. Hu, Z.; Zhou, H.; Feng, Q.; Zhang, J.; Zhang, C.; Dang, K.; Cai, Y.; Feng, Z.; Gao, Y.; Kang, X.; et al. Field-Plated Lateral β-Ga_2O_3 Schottky Barrier Diode with High Reverse Blocking Voltage of More Than 3 kV and High DC Power Figure-of-Merit of 500 MW/cm^2. *IEEE Electron Device Lett.* **2018**, *39*, 1564–1567. [CrossRef]
142. Hu, Z.; Zhou, H.; Dang, K.; Cai, Y.; Feng, Z.; Gao, Y.; Feng, Q.; Zhang, J.; Hao, Y. Lateral β -Ga_2O_3 Schottky Barrier Diode on Sapphire Substrate With Reverse Blocking Voltage of 1.7 kV. *IEEE J. Electron Devices Soc.* **2018**, *6*, 815–820. [CrossRef]
143. Oh, S.; Yang, G.; Kim, J. Electrical Characteristics of Vertical Ni/β-Ga_2O_3 Schottky Barrier Diodes at High Temperatures. *ECS J. Solid State Sci. Technol.* **2016**, *6*, Q3022. [CrossRef]
144. Müller, S.; Thyen, L.; Splith, D.; Reinhardt, A.; Wenckstern, H.V.; Grundmann, M. High-Quality Schottky Barrier Diodes on β-Gallium Oxide Thin Films on Glass Substrate. *ECS J. Solid State Sci. Technol* **2019**, *8*, Q3126. [CrossRef]
145. Tadjer, M.J.; Wheeler, V.D.; Shahin, D.I.; Eddy, C.R., Jr.; Kub, F.J. Thermionic Emission Analysis of TiN and Pt Schottky Contacts to β-Ga_2O_3. *ECS J. Solid State Sci. Technol.* **2017**, *6*, 4. [CrossRef]
146. Du, L.; Xin, Q.; Xu, M.; Liu, Y.; Mu, W.; Yan, S.; Wang, X.; Xin, G.; Jia, Z.; Tao, X.-T.; et al. High-Performance Ga_2O_3 Diode Based on Tin Oxide Schottky Contact. *IEEE Electron Device Lett.* **2019**, *40*, 451–454. [CrossRef]
147. Hu, Z.; Li, J.; Zhao, C.; Feng, Z.; Tian, X.; Zhang, Y.; Zhang, Y.; Ning, J.; Zhou, H.; Zhang, C.; et al. Design and Fabrication of Vertical Metal/TiO2/β-Ga_2O_3 Dielectric Heterojunction Diode With Reverse Blocking Voltage of 1010 V. *IEEE Trans. Electron Devices* **2020**, *67*, 5628–5632. [CrossRef]
148. Fontserè, A.; Pérez-Tomás, A.; Banu, V.; Godignon, P.; Millán, J.; de Vleeschouwer, H.; Parsey, J.M.; Moens, P. A HfO2 based 800V/300 °C Au-Free AlGaN/GaN-on-Si HEMT Technology. In Proceedings of the 2012 24th International Symposium on Power Semiconductor Devices and ICs, Bruges, Belgium, 3–7 June 2012; 2012; pp. 37–40.
149. Zhou, H.; Yan, Q.L.; Zhang, J.C.; Lv, Y.J.; Liu, Z.H.; Zhang, Y.N.; Dang, K.; Dong, P.F.; Feng, Z.Q.; Feng, Q.; et al. High-Performance Vertical β-Ga_2O_3 Schottky Barrier Diode With Implanted Edge Termination. *IEEE Electron Device Lett.* **2019**, *40*, 1788–1791. [CrossRef]
150. Lin, C.-H.; Yuda, Y.; Wong, M.H.; Sato, M.; Takekawa, N.; Konishi, K.; Watahiki, T.; Yamamuka, M.; Murakami, H.; Kumagai, Y.; et al. Vertical Ga_2O_3 Schottky Barrier Diodes With Guard Ring Formed by Nitrogen-Ion Implantation. *IEEE Electron Device Lett.* **2019**, *40*, 1487–1490. [CrossRef]
151. Wang, Y.G.; Lv, Y.J.; Long, S.B.; Zhou, X.Y.; Song, X.B.; Liang, S.L.; Han, T.T.; Tan, X.; Feng, Z.H.; Cai, S.J.; et al. High-Voltage (201) β-Ga_2O_3 Vertical Schottky Barrier Diode With Thermally-Oxidized Termination. *IEEE Electron Device Lett.* **2020**, *41*, 131–134. [CrossRef]
152. Allen, N.; Xiao, M.; Yan, X.; Sasaki, K.; Tadjer, M.J.; Ma, J.; Zhang, R.; Wang, H.; Zhang, Y. Vertical Ga_2O_3 Schottky Barrier Diodes With Small-Angle Beveled Field Plates: A Baliga's Figure-of-Merit of 0.6 GW/cm^2. *IEEE Electron Device Lett.* **2019**, *40*, 1399–1402. [CrossRef]
153. Li, W.S.; Nomoto, K.; Hu, Z.Y.; Jena, D.; Xing, H.L.G. Field-Plated Ga_2O_3 Trench Schottky Barrier Diodes With a BV2/$R_{on,sp}$ of up to 0.95 GW/cm^2. *IEEE Electron Device Lett.* **2020**, *41*, 107–110. [CrossRef]
154. Saitoh, Y.; Sumiyoshi, K.; Okada, M.; Horii, T.; Miyazaki, T.; Shiomi, H.; Ueno, M.; Katayama, K.; Kiyama, M.; Nakamura, T. Extremely Low On-Resistance and High Breakdown Voltage Observed in Vertical GaN Schottky Barrier Diodes with High-Mobility Drift Layers on Low-Dislocation-Density GaN Substrates. *Appl. Phys. Express* **2010**, *3*, 081001. [CrossRef]
155. Xiao, M.; Ma, Y.; Cheng, K.; Liu, K.; Xie, A.; Beam, E.; Cao, Y.; Zhang, Y. 3.3 kV Multi-Channel AlGaN/GaN Schottky Barrier Diodes With P-GaN Termination. *IEEE Electron Device Lett.* **2020**, *41*, 1177–1180. [CrossRef]
156. Kizilyalli, I.C.; Edwards, A.P.; Aktas, O.; Prunty, T.; Bour, D. Vertical Power p-n Diodes Based on Bulk GaN. *IEEE Trans. Electron Devices* **2014**, *62*, 414–422. [CrossRef]

157. Watahiki, T.; Yuda, Y.; Furukawa, A.; Yamamuka, M.; Takiguchi, Y.; Miyajima, S. Heterojunction p-Cu2O/n-Ga_2O_3 diode with high breakdown voltage. *Appl. Phys. Lett.* **2017**, *111*, 222104. [CrossRef]
158. Lu, X.; Zhou, X.; Jiang, H.; Ng, K.W.; Chen, Z.; Pei, Y.; Lau, K.M.; Wang, G. 1-kV Sputtered p-NiO/n-Ga_2O_3 Heterojunction Diodes with an Ultra-Low Leakage Current Below 1uA/cm2. *IEEE Electron Device Lett.* **2020**, *41*, 449–452. [CrossRef]
159. Gong, H.H.; Chen, X.H.; Xu, Y.; Ren, F.-F.; Gu, S.L.; Ye, J.D. A 1.86-kV Double-Layered NiO/β-Ga_2O_3 Vertical p–n Heterojunction Diode. *Appl. Phys. Lett.* **2020**, *117*, 022104. [CrossRef]
160. Pérez-Tomás, A.; Chikoidze, E.; Dumont, Y.; Jennings, M.R.; Russell, S.O.; Vales-Castro, P.; Catalan, G.; Lira-Cantú, M.; Ton–That, C.; Teherani, F.H.; et al. Giant Bulk Photovoltaic Effect in Solar Cell Architectures with Ultra-Wide Bandgap Ga_2O_3 Transparent Conducting Electrodes. *Mater. Today Energy* **2019**, *14*, 100350. [CrossRef]
161. Russell, S.A.O.; Jennings, M.R.; Dai, T.X.; Li, F.; Hamilton, D.P.; Fisher, C.A.; Sharma, Y.K.; Mawby, P.A.; Pérez-Tomás, A. Functional Oxide as an Extreme High-k Dielectric towards 4H-SiC MOSFET Incorporation. *Mater. Sci. Forum* **2017**, *897*, 155–158. [CrossRef]
162. Xia, Z.; Chandrasekar, H.; Moore, W.; Wang, C.; Lee, A.J.; McGlone, J.; Kalarickal, N.K.; Arehart, A.; Ringel, S.; Yang, F.; et al. Metal/$BaTiO_3$/β-Ga_2O_3 dielectric heterojunction diode with 5.7 MV/cm breakdown field. *Appl. Phys. Lett.* **2019**, *115*, 252104. [CrossRef]
163. Razzak, T.; Chandrasekar, H.; Hussain, K.; Lee, C.H.; Mamun, A.; Xue, H.; Xia, Z.; Sohel, S.H.; Rahman, M.W.; Bajaj, S.; et al. $BaTiO_3$/Al0.58Ga0.42N lateral heterojunction diodes with breakdown field exceeding 8 MV/cm. *Appl. Phys. Lett.* **2020**, *116*, 023507. [CrossRef]
164. Kalarickal, N.K.; Feng, Z.; Bhuiyan, A.; Xia, Z.; McGlone, J.F.; Moore, W.; Arehart, A.R.; Ringel, S.A.; Zhao, H.; Rajan, S. Electrostatic engineering using extreme permittivity materials for ultra-wide band gap semiconductor transistors. *arXiv* **2020**, arXiv:2006.02349.
165. Roy, S.; Bhattacharyya, A.; Krishnamoorthy, S. Analytical Modeling and Design of Gallium Oxide Schottky Barrier Diodes Beyond Unipolar Figure of Merit Using High-k Dielectric Superjunction Structures. *arXiv* **2020**, arXiv:2008.00280.
166. Pérez-Tomás, A.; Fontserè, A.; Jennings, M.R.; Gammon, P.M. Modeling the Effect of Thin Gate Insulators (SiO_2, SiN, Al_2O_3 and HfO2) on AlGaN/GaN HEMT Forward Characteristics Grown on Si, Sapphire and SiC. *Mater. Sci. Semicond. Processing* **2013**, *16*, 1336–1345. [CrossRef]
167. Sasaki, K.; Higashiwaki, M.; Kuramata, A.; Masui, T.; Yamakoshi, S. Si-Ion Implantation Doping in β-$Ga_2O_3$and Its Application to Fabrication of Low-Resistance Ohmic Contacts. *Appl. Phys. Express* **2013**, *6*, 086502. [CrossRef]
168. Zhou, H.; Si, M.; Alghamdi, S.; Qiu, G.; Yang, L.; Ye, P.D. High performance depletion/enhancement-mode β-Ga_2O_3 on insulator (GOOI) field-effect transistors with record drain currents of 600/450 mA/mm. *IEEE Electron Device Lett.* **2017**, *38*, 103–106. [CrossRef]
169. Carey, P.H., IV; Yang, J.; Ren, F.; Hays, D.C.; Pearton, S.J.; Jang, S.; Kuramata, A.; Kravchenko, I.I. Ohmic contacts on n-type β-Ga_2O_3 using AZO/Ti/Au. *AIP Adv.* **2017**, *7*, 095313. [CrossRef]
170. Fontserè, A.; Pérez-Tomás, A.; Placidi, M.; Fernández-Martínez, P.; Baron, N.; Chenot, S.; Cordier, Y.; Moreno, J.; Gammon, P.; Jennings, M. Temperature dependence of Al/Ti-based Ohmic contact to GaN devices: HEMT and MOSFET. *Microelectron. Eng.* **2011**, *88*, 3140–3144. [CrossRef]
171. Fontserè, A.; Pérez-Tomás, A.; Placidi, M.; Llobet, J.; Baron, N.; Chenot, S.; Cordier, Y.; Moreno, J.C.; Gammon, P.M.; Jennings, M.R.; et al. Micro and Nano Analysis of 0.2 Ω mm Ti/Al/Ni/Au Ohmic Contact to AlGaN/GaN. *Appl. Phys. Lett.* **2011**, *99*, 213504. [CrossRef]
172. Li, Z.; Liu, Y.; Zhang, A.; Liu, Q.; Shen, C.; Wu, F.; Xu, C.; Chen, M.; Fu, H.; Zhou, C. Quasi-two-dimensional β-Ga_2O_3 field effect transistors with large drain current density and low contact resistance via controlled formation of interfacial oxygen vacancies. *Nano Res.* **2018**, *12*, 143–148. [CrossRef]
173. Chabak, K.D.; McCandless, J.P.; Moser, N.A.; Green, A.J.; Mahalingam, K.; Crespo, A.; Hendricks, N.; Howe, B.M.; Tetlak, S.E.; Leedy, K.; et al. Recessed-gate enhancement-mode-Ga_2O_3 MOSFETs. *IEEE Electron Device Lett.* **2018**, *39*, 67–70. [CrossRef]
174. Hu, Z.; Nomoto, K.; Li, W.; Tanen, N.; Sasaki, K.; Kuramata, A.; Nakamura, T.; Jena, D.; Xing, H.G. Enhancement-Mode Ga_2O_3 Vertical Transistors With Breakdown Voltage >1 kV. *IEEE Electron Device Lett.* **2018**, *39*, 869–872. [CrossRef]
175. Hu, Z.; Nomoto, K.; Li, W.; Jinno, R.; Nakamura, T.; Jena, D.; Xing, H. 1.6 kV Vertical Ga_2O_3 FinFETs With Source-Connected Field Plates and Normally-off Operation. In Proceedings of the 31st International Symposium on Power Semiconductor Devices and ICs (ISPSD), Shanghai, China, 19–23 May 2019; 2019; pp. 483–486.
176. Li, W.; Nomoto, K.; Hu, Z.; Nakamura, T.; Jena, D.; Xing, H.G. Single and multi-fin normally-off Ga_2O_3 vertical transistors with a breakdown voltage over 2.6 kV. Proceedings of IEDM Technical Digest, San Francisco, CA, USA, 1–4 December 2019.
177. Lv, Y.; Zhou, X.; Long, S.; Song, X.; Wang, Y.; Liang, S.; He, Z.; Han, T.; Tan, X.; Feng, Z.; et al. Source-Field-Plated β-Ga_2O_3 MOSFET with Record Power Figure of Merit of 50.4 MW/cm^2. *IEEE Electron Device Lett.* **2019**, *40*, 83–86. [CrossRef]
178. Lv, Y.; Liu, H.; Zhou, X.; Wang, Y.; Song, X.; Cai, Y.; Yan, Q.; Wang, C.; Liang, S.; Zhang, J.; et al. Lateral β-Ga_2O_3 MOSFETs With High Power Figure of Merit of 277 MW/cm^2. *IEEE Electron Device Lett.* **2020**, *41*, 537–540. [CrossRef]
179. Tetzner, K.; Treidel, E.B.; Hilt, O.; Popp, A.; Anooz, S.B.; Wagner, G.; Thies, A.; Ickert, K.; Gargouri, H.; Würfl, J. Lateral 1.8 kV β-Ga_2O_3 MOSFET with 155 MW/cm^2 Power Figure of Merit. *IEEE Electron Device Lett.* **2019**, *40*, 1503–1506. [CrossRef]
180. Sharma, S.; Zeng, K.; Saha, S.; Singisetti, U. Field-plated lateral Ga_2O_3 MOSFETs with polymer passivation and 8.03 kV breakdown voltage. *IEEE Electron Device Lett.* **2020**, *41*, 836–839. [CrossRef]

181. Shibata, D.; Kajitani, R.; Ogawa, M.; Tanaka, K.; Tamura, S.; Hatsuda, T. 1.7 kV/1.0 mΩcm^2 normally-off vertical GaN transistor on GaN substrate with regrown p-GaN/AlGaN/GaN semipolar gate structure. Proceedings of IEDM Technical Digest, San Francisco, CA, USA, 1–4 December 2016.
182. Uemoto, Y.; Shibata, D.; Yanagihara, M.; Ishida, H.; Matsuo, H.; Nagai, S. 8300 V blocking voltage AlGaN/GaN power HFET with thick poly-AlN passivation. Proceedings of IEEE International Electron Devices Meeting, Washington, DC, USA, 10–12 December 2007; pp. 861–864.
183. Zhang, Y.; Neal, A.; Xia, Z.; Joishi, C.; Johnson, J.M.; Zheng, Y.; Bajaj, S.; Brenner, M.; Dorsey, D.; Chabak, K.; et al. Demonstration of high mobility and quantum transport in modulation-doped β-(AlxGa1-x)$_2$O$_3$/Ga$_2$O$_3$ heterostructures. *Appl. Phys. Lett.* **2018**, *112*, 173502. [CrossRef]
184. Song, K.; Zhang, H.; Fu, H.; Yang, C.; Singh, R.; Zhao, Y.; Sun, H.; Long, S. Normally-off AlN/β-Ga$_2$O$_3$ field-effect transistors using polarization-induced doping. *J. Phys. D Appl. Phys.* **2020**, *53*, 345107. [CrossRef]
185. Lyons, J.L. Electronic Properties of Ga$_2$O$_3$ Polymorphs. *ECS J. Solid State Sci. Technol.* **2019**, *8*, Q3226–Q3228. [CrossRef]
186. Roy, R.; Hill, V.G.; Osborn, E.F. Polymorphism of Ga$_2$O$_3$ and the System Ga$_2$O$_3$—H$_2$O. *J. Am. Chem. Soc.* **1952**, *74*, 719–722. [CrossRef]
187. Oshima, T.; Okuno, T.; Arai, N.; Kobayashi, Y.; Fujita, S. β-Al$_{2x}$Ga$_{2-2x}$O$_3$ Thin Film Growth by Molecular Beam Epitaxy. *Jpn. J. Appl. Phys.* **2009**, *48*, 070202. [CrossRef]
188. Jinno, R.; Chang, C.S.; Onuma, T.; Cho, Y.; Ho, S.-T.; Rowe, D.; Cao, M.C.; Lee, K.; Protasenko, V.; Schlom, D.G.; et al. Crystal orientation dictated epitaxy of ultrawide-bandgap 5.4- to 8.6-eV α-(AlGa)$_2$O$_3$ on m-plane sapphire. *Sci. Adv.* **2021**, *7*, eabd5891. [CrossRef] [PubMed]
189. Jinno, R.; Kaneko, K.; Fujita, S. Thermal stability of α-(Al$_x$Ga$_{1-x}$)$_2$O$_3$ films grown on c-plane sapphire substrates with an Al composition up to 90%. *Jpn. J. Appl. Phys.* **2021**, *60*, SBBD13. [CrossRef]
190. Chi, Z.; Tarntair, F.-G.; Frégnaux, M.; Wu, W.-Y.; Sartel, C.; Madaci, I.; Chapon, P.; Sallet, V.; Dumont, Y.; Pérez-Tomás, A.; et al. Bipolar Self-doping in Ultra-wide Bandgap Spinel ZnGa$_2$O$_4$. *Mater. Today Phys.* **2021**, *20*, 100466. [CrossRef]
191. Chen, M.-I.; Singh, A.; Chiang, J.-L.; Horng, R.-H.; Wuu, D.-S. Zinc Gallium Oxide—A Review from Synthesis to Applications. *Nanomaterials* **2020**, *10*, 2208. [CrossRef]
192. Galazka, Z.; Irmscher, K.; Pietsch, M.; Ganschow, S.; Schulz, D.; Klimm, D.; Hanke, I.M.; Schroeder, T.; Bickermann, M. Experimental Hall electron mobility of bulk single crystals of transparent semiconducting oxides. *J. Mater. Res.* **2021**, *36*, 4746–4755. [CrossRef]
193. Galazka, Z.; Klimm, D.; Irmscher, K.; Uecker, R.; Pietsch, M.; Bertram, R.; Naumann, M.; Albrecht, M.; Kwasniewski, A.; Schewski, R.; et al. MgGa$_2$O$_4$ as a new wide bandgap transparent semiconducting oxide: Growth and properties of bulk single crystals. *Phys. Status Solidi* **2015**, *212*, 1455–1460. [CrossRef]

materials

Review

Diamond for Electronics: Materials, Processing and Devices

Daniel Araujo [1,*], Mariko Suzuki [1], Fernando Lloret [2], Gonzalo Alba [1] and Pilar Villar [1]

1 Department of Materials Science and Engineering, University of Cádiz, 11510 Puerto Real, Spain; mariko.suzuki@uca.es (M.S.); gonzalo.alba@uca.es (G.A.); pilar.villar@uca.es (P.V.)
2 Department of Applied Physics, University of Cádiz, 11510 Puerto Real, Spain; fernando.lloret@uca.es
* Correspondence: daniel.araujo@uca.es

Abstract: Progress in power electronic devices is currently accepted through the use of wide bandgap materials (WBG). Among them, diamond is the material with the most promising characteristics in terms of breakdown voltage, on-resistance, thermal conductance, or carrier mobility. However, it is also the one with the greatest difficulties in carrying out the device technology as a result of its very high mechanical hardness and smaller size of substrates. As a result, diamond is still not considered a reference material for power electronic devices despite its superior Baliga's figure of merit with respect to other WBG materials. This review paper will give a brief overview of some scientific and technological aspects related to the current state of the main diamond technology aspects. It will report the recent key issues related to crystal growth, characterization techniques, and, in particular, the importance of surface states aspects, fabrication processes, and device fabrication. Finally, the advantages and disadvantages of diamond devices with respect to other WBG materials are also discussed.

Keywords: diamond; MPCVD growth; power electronics; electron microscopy

Citation: Araujo, D.; Suzuki, M.; Lloret, F.; Alba, G.; Villar, P. Diamond for Electronics: Materials, Processing and Devices. *Materials* **2021**, *14*, 7081. https://doi.org/10.3390/ma14227081

Academic Editor: Fabrizio Roccaforte

Received: 29 August 2021
Accepted: 8 November 2021
Published: 22 November 2021

1. Motivations

In the 2020 and 2030 Climate-Energy Packages, the EU committed to lower greenhouse gas emissions by 20% with respect to 1990 and 55% by 2030 (very recent EU target) and to reach a share of renewables of 20% by 2020 and at least 27% by 2030. Today, there is a great concern about the conflict between energy and the environment. In this context, it is becoming usual to expect that an extraordinary increase of the use of electricity in the energy production, transport, and consumption will help a sustainable future. Enormous energy savings and exciting enhancements in quality of life will be enabled by new power electronics (PE) energy conversion systems. All energy-consuming devices, from pacemakers and home appliances to electric vehicles and industrial waste processing plants, will be affected. All alternative, sustainable and distributed energy (DE) sources, as well as energy storage systems, will be tied to the smart grid (SG) through swift and efficient PEs converters.

For competitive low-carbon renewable energy, transport energy, and smart grid applications, the impact of power electronics is striking where an optimized electrical energy conversion is demanded by the society. Approximately 30% of all electric generated power utilizes power electronics somewhere between the point of generation and its end use. Power electronics is used for more efficient transport, renewable energy production, and distribution including in highly efficient electricity distribution over long distances via high-voltage direct current power lines (HVDC) as well as in the better control of loads in switching power supplies and variable-speed drives for motors that drive fans, pumps, and compressors. By 2030, it is expected that perhaps as much as 80% of all electric power will use power electronics somewhere between generation and consumption. However, with the current state-of-the-art electric equipment, the transformation of the electrical energy occurs with significant losses (in the order of 10% from the source to the point of use) because available semiconductors are not ideal for high power.

The key to the efficient transmission and conversion of low-carbon electrical energy is the improvement of power electronic devices, which must be durable and reliable in high-power environments to eliminate the need for auxiliary systems for its use in transport (airplane, cars, etc.). "Green electronics", i.e., highly efficient electronic devices, are crucially important for our future energy system. A first estimation gives a 75% reduction in losses, representing about a 10 MW energy saving on a 300 MW HVDC converter. The disruptive approach of the use of diamond in electronic devices can contribute knowledge, new approaches, innovative materials, and skills arising from the cross-fertilization with other WBG materials to yield more efficient and cost-competitive energy technologies.

2. Diamond Properties towards Electronic Applications

Silicon is a well-established semiconductor material that has addressed the requirements of energy conversion for more than 50 years. However, it is widely recognised (as shown in research roadmaps on power semiconductor devices [1]) that a real step-improvement in power electronics will be obtained by employing devices based on wide bandgap semiconductor materials. These materials have superior electrical characteristics for power devices when compared to silicon. Many grid applications use multilevel converters, with 3.3 kV (4.5 kV IGBT pressed (or stack) pack is commonly used fo HVDC and now 6.5 kV is also commercially available) silicon power semiconductors that need to be set in series to reach the required voltage level. Higher voltage capability devices are then highly required, 15 kV being a first step. To reach a high level current, paralleling of devices is required. Power electronic devices based on wide bandgap semiconductors are now resulting in substantial improvements in the performance of power electronics systems by offering higher blocking voltages, improved efficiency and reliability (higher performance/cost ratio), easier paralleling, and reduced thermal requirements, thus leading to the realization of more efficient green electronic systems.

Among wide bandgap semiconductors, diamond is considered to be the ultimate semiconductor for applications in high-power electronics due to its exceptional properties. Its dielectric breakdown strength is three times higher than in silicon carbide (SiC) and more than 30 times higher than in silicon (Si). In addition, unlike most other WBG semiconductors, the carrier mobility is very high for both carrier types, and the thermal conductivity is unsurpassed (see Table 1). In the latter, we indicate the commonly reported values of mobilities measured by Hall bars set-ups [2,3] on microwave-plasma-assisted chemical vapor deposition (MPCVD) layers. Note that some authors report much higher values when achieving the measurements with time of flight (TOF) set-ups on very high purity single-crystal diamonds. Values of electron and hole mobilities as high as 4500 and 3800 cm^2/Vs have been reported [4]. Baliga's figure of merit is oriented to static power losses. Another typical way to compare the power semiconductors is to draw the theoretical relationship between unipolar on-resistance versus the breakdown voltage of Schottky barrier diodes (SBDs). This is represented in Figure 1 for different semiconductors [5]. Power device engineering is interested in minimising the on-resistance for a given breakdown voltage capability. Diamond is the best candidate, and even Ga_2O_3, despite its lower carrier mobility, is better than SiC and GaN thanks to its high critical electric field.

Table 1. Diamond's properties are strikingly superior to other semiconductors (see [5]) when considered for use in power electronic devices. Johnson's figure of merit (FoM) is a measure of the ultimate performances of the electronic device at high power and high frequency of a transistor, Keyes' FoM measures the performance limited by heat generation and removal, and Baliga's FoM measures performance limited by losses at high-power and high-frequency operation. The properties/figure where diamond is outstanding for the present project are highlighted in orange.

Property (Unit), (See [5])	Si	SiC-4H	GaN	Ga_2O_3	Diamond
Bandgap (eV)	1.1	3.23	3.42	4.8	5.45
Dielectric constant, ε	11.8	9.7	9	10	5.7
Breakdown field (MV/cm)	0.3	3	2	8	10
Electron mobility (cm^2/Vs)	1500	1000	2000	300	1000
Hole mobility (cm^2/Vs)	480	100	20		2000
Thermal conductivity (W/cmK)	1.5	5	1.5	0.27	22
Johnson's figure of merit (10^{23} $\Omega W/s^2$)	2.3	900	490	1236	2530
Keyes' figure of merit (10^7 W/Ks)	10	53	17	2	218
Baliga's figure of merit (Si = 1)	1	554	188	3214	23,068

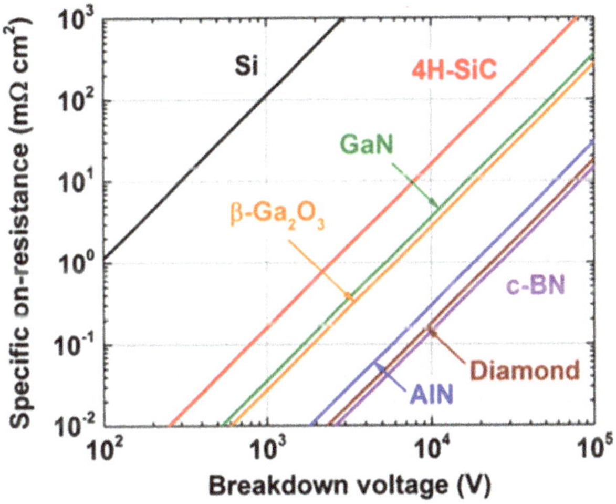

Figure 1. ON-resistance and breakdown voltage of the different semiconductors at room temperature. Note that at high temperature diamond improves its characteristics (Reproduced with permission from Pearton et al. [5] Copyright 2021 ©AIP Publishing).

An important aspect to rise is that Si power semiconductor switches used in 90% of the power applications market are metal-oxide semiconductor (MOS) gate-controlled devices (vertically diffused metal-oxide semiconductor, VDMOS, IGBT). Thyristor-based structures (integrated gate-commutated thyristors IGCT; gate turn-off thyristors, GTO) are still used for high-power high-voltage applications, mainly because of the lack of equivalent performance MOS controlled devices. This is something that diamond could solve considering its very high breakdown field above 10 MV/cm. In general, WBG semiconductors could open the way to novel concepts and applications in the high-voltage field such as electric transport and energy generation and distribution. Diamond, as a material with exceptional properties, could provide solutions to industry by providing diodes and transistors that withstand voltages above 10 kV, but competition with other materials, especially silicon carbide (SiC), and the intrinsic limitations of diamond (hardness, size of the substrate, etc.) require a great deal of effort to improve the performance, especially to reach high currents.

3. Diamond Growth: Substrates, Techniques, and Doping

3.1. Diamond Substrates

Homoepitaxially deposited diamond is carried out over commercially available diamond substrates. Diamond substrates can be synthetized by high-pressure high-temperature (HPHT) or chemical vapor deposition (CVD) techniques. Synthetic and natural diamonds are classified on the basis of their impurity concentrations. All natural and lab-grown diamonds contain some nitrogen (N) impurities [6]. When the N content is high (hundreds or thousands of ppm), so that it can be characterized by infrared (IR) spectroscopy, diamond is classified as Type I. These N atoms can be placed replacing carbon atoms in the lattice forming aggregated (Type Ia) or isolated from each other N atoms (Type Ib). Type Ia is also subdivided into IaA, with nitrogen atoms forming pairs, and IaB, with four nitrogen atoms symmetrically surrounding a vacancy in the diamond structure. On the other hand, when nitrogen content is low enough to not be detected by IR ($<10^{17}$ cm^{-3}), diamond is classified as Type II. Usually, the nitrogen reduction is a consequence of the incorporation of Boron impurities. When both impurities contents are very low, the substrate is named Type IIa, which corresponds to the purest diamond crystals. When the boron concentration increases so it is higher than nitrogen, the substrate is called Type IIb, and it corresponds to a p-type semiconductive diamond crystal.

Above the impurities content, there are two main restrictions related to the available substrate: the crystalline defects density and the substrate size. The HPHT process provides high-purity and low-defect-density substrates ($<10^3$ cm^{-2} in Type IIa). However, the size is restricted to ≤ 1 cm^2 by the technological limitations of the method. Moreover, the prize increases drastically when a high-purity diamond crystal is required. For this reason, Type Ib is the most usual diamond substrate for electronic applications, which means low cost and an acceptable dislocation density of about 10^5 cm^{-2}. Larger substrate sizes can be obtained by CVD growth [7]. The largest single-crystal diamond substrate has been reported to have a diameter of ~3.5-inches based on Ir/YSZ/Si [8]. ® Other groups have reported 2-inch-scale substrates labs also using Ir hetereoepitaxy [9,10]. However, the crystalline quality is still worse than that of HPHT (commonly dislocation density of 10^7 to 10^9 cm^{-2} in heteroepitaxy growth) [11–13]. An alternative method to obtain large diamond substrates is growing on a mosaic configuration by CVD. This technique results in substrates sizes above 5 cm^2 but the bonding boundaries of the mosaic are very strained and defective [14]. Thus, up to now, it has not been possible to obtain larger than 1 cm^2 good-quality diamond substrates.

3.2. MPCVD Growth and Parameters

It was in the 1980s when, for the first time, cheaper synthetic and reproducible grown diamond was carried out using the CVD process [15]. In synthetic diamond growth, the extremely high surface energy of diamond, which reflects the strength of the bonds that must be broken to create a new surface, leads to inefficient wetting of substrates surfaces by growing diamond species. In addition, the complexity of the chemical reactions requires a highly advanced understanding of the technique. The CVD process is quite different from HPHT and the natural diamond formation. As its name implies, chemical vapor deposition involves chemical reaction inside a gas-phase as well as deposition onto a substrate surface. Growth of diamond films by CVD must be conducted under non-equilibrium conditions to avoid the more stable sp^2 graphite phase. Furthermore, during the CVD process, hydrogen radicals (atomic hydrogen) must be present to, among other things, remove non-diamond carbon, including graphite, which is formed on the diamond surface. Methane is commonly used as the carbon source for CVD diamond growth. The process procedure with the whole range of selectable process parameters is shown in Figure 2 [16].

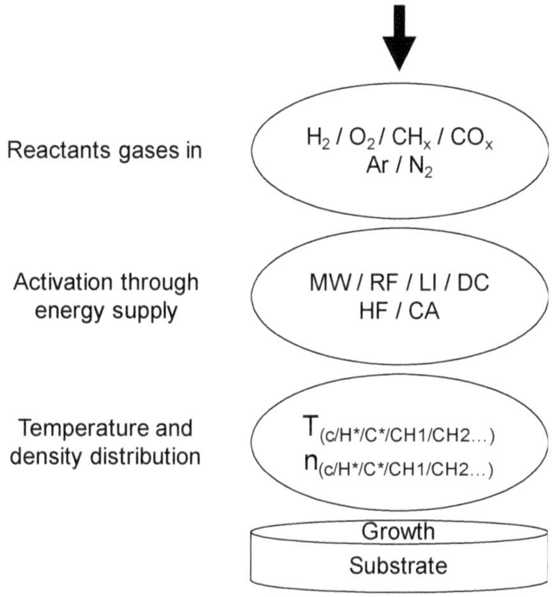

Figure 2. Schematic diagram of the mechanism for diamond CVD growth processes. The acronyms used in the figure correspond to: MW, microwave; RF, radiofrequency; LI, laser induced; HF, hot filament; DC, direct current; CA, chemical activation.

This sketch illustrates the various direct and indirect adjustable parameters. The first group shows the different selectable process gases that can be used for CVD. The second group reflects a selection of energy sources for the activation of the chemical process, followed by ensuring parameters. Below that, there is the substrate with the growing diamond layer. The principal chemical mechanism relevant to the growth of diamond from gaseous hydrogen and hydrocarbon species was presented in 1993 by Butler et al. [17], and the diamond CVD growth processes have been continuously developed since this time.

CVD englobes several growth methods based on the nature of the energy source. Among them, microwave-plasma-assisted CVD (MPCVD) has several advantages for growth of high-quality diamond yielding superior electric and optical properties. The main reason is that a stable plasma can be generated without any electrodes in the vicinity of the diamond growth plasma [18]. In general, growth mechanisms by MPCVD or HPHT have been extensively studied [19–24], and the models developed are well-known and accepted. These models are mainly based on the relative growth velocities of four low index crystal planes: {100}, {110}, {111}, and {113}. The value of the velocity ratios (the so-called growth parameters) describing the global morphology of the crystal during the growth process allows predicting their final shape. This final shape is given by the slowest rate among facets sharing an edge that limits the growth [15,17,18].

However, some critical issues that still exist for the overgrowth of the homoepitaxially MPCVD diamond layers must be resolved before CVD diamond films can be industrialized. Increasing the growth rate of diamonds, Rg, while keeping the crystal quality is one of the most important. This rate is highly dependent on the ratio of the methane flow to the total source gas flow, $C_{me} = CH_4/(H_2 + CH_4)$. When it is increased, the process achieves higher Rg. However, in the same proportion, the crystalline quality of the diamond layer grown tends to become poorer. The increase of carbon atoms in the gas mixture leads to a less effective sp^2 etching process, carried out by the hydrogen. This results in an appreciable increase of the superficial roughness and defects by the formation of secondary-nucleated non-epitaxial crystallites and non-diamond phases. Indeed, most of the reported high-quality diamond (100) films were grown with C_{me} below 1.0%. Consequently, the typical Rg is lower than 1 μm/h [25–29]. In fact, the most appropriate C_{me}s for the growth of atomically flat MPCVD diamond films are 0.05%, which result in Rg of the order of only 0.01 μm/h [27].

Nevertheless, this issue can be, at least, partially overcome using high microwave power. Teraji et al. reported high-quality (free-exciton recombination emissions in cathodoluminescence (CL) spectra at RT) homoepitaxial diamond films at higher Rg by using a MPCVD reactor with 3.8 kW of microwave power in addition to a relatively high C_{me} of 4.0% [30]. It was discovered that a kind of lateral growth was dominant even at such a high C_{me} when the high-power MPCVD was employed [31]. Reaching high Rg MPCVD processes is required for the realization of commercially available diamond-based electronic devices [32,33], but it is not the only issue to face.

3.3. Doping Issues

Another challenge of devices technology faced by novel semiconductors is the need for local doping of n-type and p-type layers, for instance, to build the n+ source and the p-well in a n-channel transistor. Something that seems to be easy in silicon, because of the existence of numerous low energy donors and acceptors atoms, is particularly complex with other semiconductors. The nearest to Si is SiC, even if high temperature implantation is needed to p-type doping. On the other side, it is difficult to p-type dope GaN and Ga_2O_3. Typically, in these materials, it is easier, or rather less difficult, to dope the semiconductor during the growth stage (bulk growth or epilayer growth) than by using local doping techniques such as diffusion or implantation.

Concerning diamond, boron and phosphorus are widely used as p- and n-type dopants of diamond semiconductors, respectively. p-doping is relatively easy during growth using boron atoms; the covalent radii of boron (0.088 nm) and carbon (0.077 nm) are close enough to allow the incorporation of boron as substitutional sites [34,35]. However, it needs an exclusive reactor as boron contaminates the entire gas system, and then the non-intentionally doping level is not controlled but in the range of 10^{15}–10^{16} cm^{-3}. If undoped or n-doping is needed, it should be grown in a different reactor even using different gas lines. On the other hand, phosphorous has a covalent radius of 0.117 nm that makes its incorporation difficult, this being more effective on (111)-oriented diamond substrates. In fact, the first phosphorous-doped {111}-oriented diamond was reported by Koizumi et al. at 1997 [36], whereas the first phosphorous doping of {100}-oriented diamond was not achieved until 2005 [37].

The doping level is also an important aspect due to the difficulty of reaching high doping levels and the dislocations that can be originated by such doping. Concerning the n-type, although first results were obtained more than one decade ago with the consecution of p–n diodes [38], the difficulty of introducing phosophorus in substitutional sites means that only some groups are able to grow such diamond [6,39]. For p-doping, the strain generated by such atoms' incorporation in the diamond lattice introduces dislocations in the grown layers. Either critical thickness [40] resulting from the generated stress or dopant proximity effects [41,42] is the mechanism that can be responsible for the defects generation. Thus, growers should be very vigilant about adequate doping levels with the growth orientation and growth parameters to avoid the introduction of lattice defects.

Impurites such as phosphorous and boron atoms are easy to identify by cathodoluminescence. Doping atoms pine the Fermi level and incorporate either aceptors (by B atoms) or donors (by P atoms) levels in the bandgap. Thus, the doping increases to an impurity band merging with the valence band above the metallic transition. Dean et al. reported the first identification of bound excitons in (natural) diamond in 1965 [43]. The dependence of the isotopic boron-bound exciton of the host-lattice subsequently revealed the change in the diamond band gap based on the purity of its content of ^{12}C or ^{13}C [44–46], while the fine structures of boron-bound boron were observed for the first time [47]. In 1993, the effect of the boron concentration on the relative intensities of bound and free excitons in polycristalline CVD diamond was reported by Kawarada et al. [48] Today, several studies on highly boron-doped (near the metallic transition) diamond have been performed for both single and polycrystal [49]. On the other hand, clear donor characterestics of phosphorus were evidenced in 1997 [36,50]. Since then, many groups have observed the neutral phosphorus-bound exciton [51]. Experimental models have been developed to estimate the content for both impurities in low and high doping ranges [52–54].

3.4. Diamond Surface Roughness Effect

A tentative solution to the doping problems in diamond is surface transfer doping [55–58]. Hydrogen-terminated diamond exhibits p-type surface conductivity after its exposure to air. This process is closely related to the surface of the diamond as it relies on the hydrogen termination of the surface and the contact with a suitable electron-accepting medium. It is well known that surface/interface roughness scattering deals with a negatively impact carrier mobility in other material systems [59–61]. Contrary to what might be expected, the increasing of the surface roughness seems to enhance the hole-mobility in the bidimensional sub-surface of diamond [62]. The authors arributed this phenomenon to an increase in activation sites, which, in turn, led to an increase in carrier density rather than mobility. Other authors observed similar conductances increasing after roughening the diamond surfaces with ICP and RIE plasma etchings. Despite the lack of understanding, the authors considered that it may be linked to removal of surface defects during the plasma process. Indeed, the experimental results in hydrogen-terminated diamond field-effect transistors show that the diamond surface roughness significantly affects the carrier density dependence of the mobility [63]. Modeling FETs behavior showed that mobility drastically decreases with roughness when it is above 1 nm, and atomically flat surfaces are then highly desirable [64]. What is clear is that the roughness of the diamond surface has a great impact on the conductance of the surface. The reasons why this occurs, however, are still an open topic.

3.5. Alternative Growth Geometries

The cubic lattice of diamond is formed by two superimposed face centered cubic (FCC) lattices, with a/4 along each dimension displacement. This structure is anisotropic and results in high dependence with the plane of growth for growth rates, impurities incorporation, and surface passivation. This particularity has been used by authors for growth along unusual orientations for applications such as defects reductions [65,66] or doping optimization [67]. More recently, the anisotropy of diamond has been considered as an advantage for the design of three-dimensional architectures for devices that contributes to overcome the classical issues in diamond technology. Diamond overgrowth over patterned substrates can be predicted based on the growth parameters [68,69]. Thus, it is possible to define a specific lateral surface to be more suitable for the device applications, e.g., an orientation that maximizes the dopant incorporation. In fact, when diamond is homoepitaxially deposited on (100)-oriented patterned substrates using very low methane contents, growth rates along the <100> direction are very low, so lateral facets are maximized [70]. In addition, it is well-known that doping reduces the growth rates [71]. Consequently, it allows the growth of selectively thin doped layers with a high accuracy [72]. Obviously, there is a tendency of planarization during the deposition process, but this planarization can be extended by adapting the growth parameters in order to maximize the lateral sides of the structures. This technology is unequivocally aimed at circumventing the unsolved technological issues of diamond-based device manufacturing, such as the etchings, the doping incorporations, or the defect densities.

4. Structural Characterization Techniques

There are still several open questions concerning diamond that should be answered that are mandatory to manufacture feasible and reproducible devices. First, obtaining substrates of enough quality and low defect density to grow the required diamond structures on them for the device remains a challenge. Then, developing the growing conditions to obtain the desired doping levels (p- or n-type) without defects that could affect the performance of the device is the main concern of the diamond grower community. Issues such as dopant or defects (point defects, dislocations, and planar defects) distributions need to be controlled. Manufacturing the device also requires the capability of obtaining good ohmic and Schottky contacts and fully controlled diamond/diamond, diamond/dielectric, and diamond/metal interfaces, as some examples of technological challenges. Therefore, structural characterization is crucial to achieve all these technological targets. However, from the characterization point of view, some advantages of diamond may become a drawback to carry out some studies: for example, the high mechanical hardness makes difficult not only the sample preparation for transmission electron microscopy (TEM) but also cleaving the sample to make local analysis versus depth or to make laser mirrors. This is the main reason why not many TEM-related results for micro/nanostructural characterization are reported in the bibliography. The commercialization of the FIB-Dual Beam (focused ion beam coupled to a scanning electron microscope, SEM, column) 25 years ago now makes possible the sample preparation for TEM studies [73–78] of single-crystal diamond, although this equipment is still not extensively introduced in laboratories and, moreover, specific diamond TEM sample preparation issues such as amorphization or redeposition during etching have to be fully controlled to be successful in further TEM observations. Therefore, other techniques have been more extensively used for diamond characterization, as Raman or FTIR spectroscopies, X-ray diffraction (XRD), atomic force microscopy (AFM), or even X-ray photoelectron spectrometry (XPS).

This last one provides useful information related to the chemistry, composition, and electronic phenomena of diamond interfaces. The measured kinetic energy of escaping electrons when the material is irradiated by an X-ray beam is mostly dependent on its original core-level energy, which allows its identification. The depth sensitivity of the technique is dependent on the inelastic mean free path, which, in turn, is a function of the kinetic energy of escaping electron and the nature of the material through which it travels. For diamond C 1s electrons excited by an Al-kα source ($h\nu$ = 1486.6 eV), the inelastic mean free path has been experimentally estimated as ~2.4 nm [79], which gives a maximum depth sensitivity of ~10 nm. This very short sensitivity has promoted XPS for diamond surface termination characterization, which is a topic of great importance in further devices manufacture. However, under the mentioned XPS conditions, the intensity of the diamond surface contributions is relatively much lower than that of the bulk contribution, harming the spectra analysis and peak identification. To overcome this issue and obtain even more surface specific information, some authors have opted for two different approaches: the use of synchrotron or other tunable X-ray beam energy systems [80] or the use of angle-resolved XPS (ARXPS) mode [81]. In both cases, the number of studies is very low in comparison to those based on energy-fixed and angle-fixed XPS conditions while being the key for a better comprehension of diamond surface phenomena. Recently, the ARXPS mode has allowed the identification and reinterpretation of a surface downward band bending component on (100)-H-terminated surfaces [82,83]. It is remarkable to keep in mind that the use of adequate and precise models to interpret the material is of maximum importance to correctly analyze XPS peaks obtaining a good fitting with the experimental results. Figure 3 shows, as an example, how the recorded XPS C 1s peak has been decomposed based on a model consisting of a diamond bulk region ($Peak_{bulk}$), a diamond with a downward band bending region ($Peak_{bb}$), and one monolayer of C-H ($Peak_{CH}$) XPS contribution for an H-terminated diamond surface.

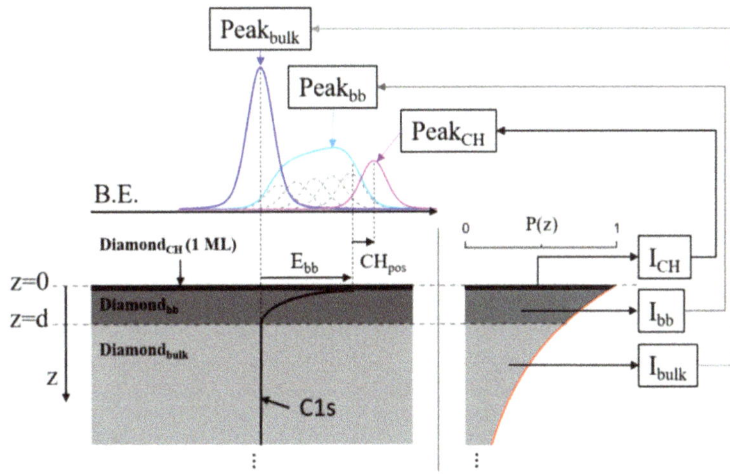

Figure 3. Schematic of the XPS model used for C 1s peak analysis. Sample is divided into three regions: diamond bulk (Diamond$_{bulk}$), diamond with band bending (Diamond$_{bb}$), and diamond surface C-H$_x$ (Diamond$_{CH}$) for the first monolayer of material. The intensities I_{bulk}, I_{bb}, and I_{CH} of the respective XPS peaks Peak$_{bulk}$, Peak$_{bb}$, and Peak$_{CH}$ are obtained from their respective ratios. Peak$_{bulk}$ and Peak$_{CH}$ follow Voigt distributions; Peak$_{bb}$ is defined by its band bending width (d) and band bending (E_{bb}). The latter is set to the Peak$_{bulk}$ position. Adapted from the graphical abstract of Alba et al. [83].

On the other hand, the XPS results on O-terminated surfaces have allowed the attribution of different oxygenated species contributions such as C-O-C bridges, ketones, or hydroxyl [84]. However, due to the short energy distance among some of them and the wide oxygenation methods, there is still no wide agreement on the models of O-terminated surface. In this sense, the ARXPS detection and quantification of a reproducible carbon sp^2 surface contribution could provide a turn to the current vision of (100)-O-terminated reconstructions models [85].

Concerning diamond junctions' interface electronic phenomena, XPS can be used for the estimation of Schottky barrier height (SBH) in metal–diamond junctions [86–88], as well as band-offset in heterojunctions [89–91]. The estimation of the SBH is complementary to the electrical characterization, but its comparison gives an idea of the homogeneity of the contact throughout the contact area. Localized SBH variations could have a critical effect on the I/V performance while remaining negligible in XPS experiments. Thus, it is expected that the XPS method overestimates the SBH in comparison to the electrical characterization. Concerning heterojunction band-offset estimation, XPS is the main experimental method for this purpose. For the estimation of such parameters, the simultaneous XPS detection (within few nanometers) of diamond and the deposited material is required.

Even though the density of dislocations depends on the supplier, the most common ones have still a high density of such defects in addition to the presence of growth sectors where the incorporation of impurities can vary from one to another. The makes it difficult to make reliable electrical characterizations. Optical-related characterization can also deliver informations on such defects, but the wide bandgap makes an excitation above the bandgap energy difficult. In consequence, cathodoluminescence (CL) is particularly well-adapted for the diamond crystal characterization [30–32]. Its indirect bandgap favors the defect-related transitions. Exciton-related transition in the UV range gives important information either on the quality of the diamond crystal or on the dopant present in it. In the optical-range extended defects related to the A-band, point defects are well-known from published studies and charts [92]. During the last two decades, CL has been demonstrated [93] to be an exceptional tool to evaluate the doping level [94] with an accuracy one order of magnitude better than secondary ion mass spectroscopy (SIMS). It also allows determining the type of point defects present in the crystal as H3 that are generated in the mid-gap [95]. Dislocations related to mid-gap levels (A-band) can be related to the presence of interfaces using monochromatic luminescence maps with a sub-micrometric resolution. In addition to CL, carrier densities can be evaluated using optical infrared (IR) spectroscopy (reflection as well as transmission, where peculiar FIB designed geometries can be used, Fourier transformed IR, FTIR). Their relative peak intensities related to the valence band to impurity levels (or band for high doping) permit the deduction of the boron doping level. FTIR and CL both allow the determination of the active boron density (or p-type dopant level), while SIMS gives the total density of boron atoms present in the crystal [96]. Thus, these are complementary techniques. Concerning their spatial resolution, CL is clearly the most efficient one with a resolution below the μm when used on FIB-prepared sample, which allows avoiding the pear-shaped volume of interaction between the high energetic electron beam and the diamond material. The electron directly crosses the FIB lamella (usually in the range of being 0.5 μm thick), and only the e-beam spot size excites the diamond material. Then, carrier transport gives the spatial resolution of the technique.

Other complementary analysis can be carried out by TEM to confirm the optical spectroscopy analysis. In this case, the experimental analysis becomes heavier as the sample preparation requires the fabrication of FIB lamella, similar to those used for CL, but much thinner (less than 100 nm thick). Indeed, diamond is so hard that it is nearly impossible to prepare lamella using traditionally used methods.

The most-used methodology in diamond materials is the lift-off FIB technique, which is summarized in Figure 4. A cross-sectional lamella is obtained for final polishing after trimming a thin diamond wall between previous deep trenches. Protection of the surface with some metal, as Pt, is required, especially if a close-to-surface interface needs to be studied. Final polishing up to some tens of nanometers is mandatory for certain TEM analyses. This final Ga^+ ions etching can induce some amorphization or degradation of the crystalline structure if operation conditions are not well-controlled, which makes it unusable for HREM or HR-STEM studies.

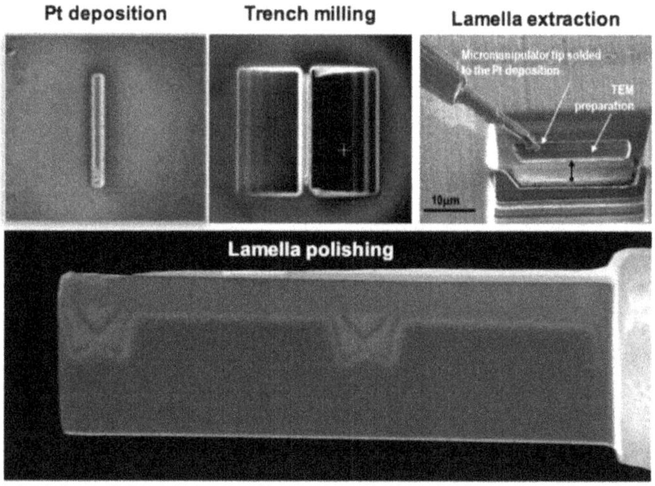

Figure 4. Lift-off methodology for the preparation of TEM lamella in diamond materials. After Pt deposition for surface protection, two parallel trenches are milled at both sides, and, after cutting off and removing using a micromanipulator tip, final polishing is performed.

The TEM analysis also allows assessing spectroscopic analysis as electron energy loss spectroscopy (EELS), where the bonding configurations can be evaluated. It is particularly powerful in the study of the B.C-N system, where the sp, sp^2, or sp^3 carbon hybrids can be evaluated [97]. The recently commercialized TEM microscopes can include a monochromator that leads to zero-loss peaks (ZLP) full width at half maximum (FWHM) in the range of 0, 1 eV. For spectroscopic analysis, this is an important technological improvement, and diamond is then an ideal material, due to its wide bandgap, to make spectroscopical analyses in the low loss range where the influence of the zero-loss peak [98] is now reduced. Damages, defect configuration, and changes in crystal structure attributed to B addition have been studied by EELS on polycrystalline diamond doping [99,100], where the variation of the C hybrids are well-revealed in the 300 eV range peak, which can also be related to the presence of point defects, dislocations, or vacancies. The technique is then complementary to CL, where such defects can also be observed. When HPHT is demonstrated to be able to deliver colorless diamond crystals (removing the usual brown color), the gem community and industry express great interest in understanding the origin of the different possible colors of diamond. Investigations allow attributing the blue to boron doping, the yellow to nitrogen doping, and the red to NV centers, and the brown is then related to the presence of vacancies clusters or nodes of dislocations [101]. However, this last aspect is still an open question.

In addition to carbon bonding configurations, EELS is also sensitive to the boron-binding states in the spectroscopical analysis and also allows the evaluation of the spatial distribution of boron dopant through the mapping facilities of the TEM, either used in the EFTEM mode (energy filtering TEM) or in the STEM (scanning TEM) one. This imaging mode of the TEM can also evidence distribution of B-B entities and point defects. Concerning dopants, due to their different sensitivity and spatial distribution, Raman is then also complementary to CL and EELS, as some vibrational modes of B allow the evaluation of the doping level of the diamond crystal [102].

As diamond has a relatively large bandgap with respect to the FWHM of the ZLP, valence electron energy loss spectroscopy in STEM (VEELS) allows the evaluation of the bandgap and the complex dielectric function with nanometers' resolution. Special attention to the Cherenkov and Plasmon peaks has to be taken, and such experiments are recommended to be carried out at low electron beam energy (<80 keV, typically 60 keV) on relatively thin FIB lamella (<50 nm) [103]. The bandgap and dielectric constant of polycrystalline alumina onto diamond has been then estimated using this methodology [95].

Quantitative compositional information can also be obtained by high-angle annular dark field (HAADF) in the STEM mode [104,105]. Figure 5 shows a comparison between B-dopant determination by HAADF and SIMS. High-angle scattering of electrons is dominated by inelastic scattering, which means that no diffraction effects are produced if the collection angle is high enough so that the scattered intensity depends directly on the square of the atomic scattering factor. This has been demonstrated to be especially useful to quantify boron doping profiles with nanometric resolution in δ-doped layers [106,107] (down to 10^{20} at./cm^{-3} and 5 nm thick). This TEM related technique, however, is suitable to evaluate boron content when the dopant level is high, over the 10^{19} cm^{-3} scale. When boron level is below that, CL on cross section foils constitutes a more suitable methodology [108].

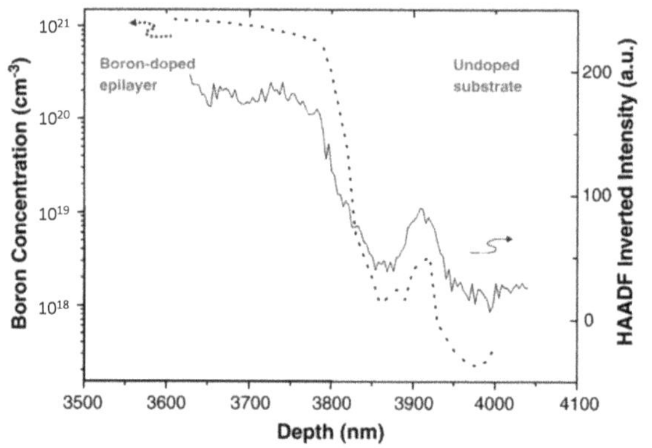

Figure 5. A comparison between HAADF and SIMS is shown for B determination in a boron-doped diamond homoepitaxial layer. The dashed curve corresponds to the SIMS profile and the continuous line to the inverted HAADF profile. Boron can be clearly detected in the 10^{20} cm^{-3} range. Adapted with permission from Araujo et al. [104] Copyright 2021 ©Elsevier.

By far, the control of defects, especially dislocations, in diamond is one of the main goals for diamond researchers, as they can be responsible for future leakage currents in the device. Here again, TEM is a powerful tool to fully analyze the dislocation generation when growing CVD diamond so that some design rules can be offered to diamond community. It is very well-known from III–V semiconductors that a critical thickness on the epilayer is necessary to start dislocation formation as a consequence of an energy balance when plastic relaxation starts [40,109–112]. However, in the case of diamond, proximity effects concerning boron atoms in a closed diamond lattice have been shown to also be effective for that. Therefore, not only a critical thickness is taken into account, but a critical boron content can also be defined [38,39,113], which is of maximum interest to diamond growers. Introduction of other dopants, such as P, also induces defect formation, as can be seen in Figure 6 where a diffraction contrast TEM micrograph is shown. Plenty of dislocations and planar defects are formed in this phosphorous-doped diamond epilayer ([P] = 2.5×10^{20} at./cm^3).

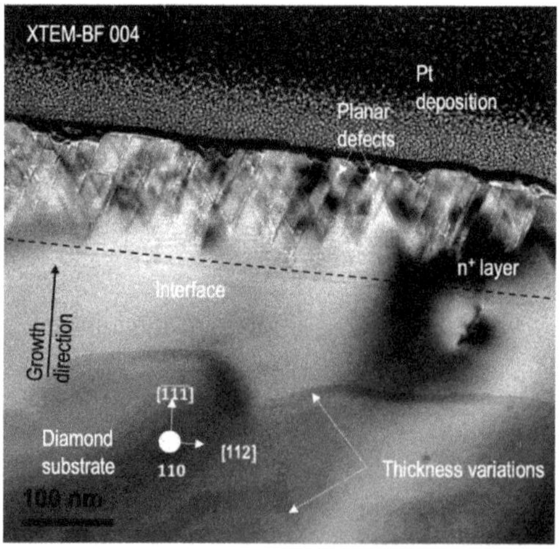

Figure 6. Diffraction contrast TEM micrograph, recorded with the 004 reflection, shows the formation of dislocations and planar defects above the interface for a heavily P-doped diamond epilayer.

5. Diamond Electronic Devices

5.1. Schottky, Ohmic Contacts, and Diamond SBDs

Diamond Schottky barrier diodes (SBDs) have been extensively studied, and high breakdown voltage of >10 kV, high temperature operation, and low on-resistance have been reported [114–118]. Mostly, the p-type layer is used for the drift layer and contact layer, because the p-type layer is easier to control with doping concentration (10^{15}–10^{22} cm^{-3}) [119,120] and it shows higher Hall mobility of ~2000 cm^2 V^{-1} s^{-1} [121] compared to n-type layers.

The quality of metal/diamond interfaces is one of the most important issues to obtain high performances in SBDs. Surface termination has an important role as well as other semiconductors, and it drastically changes the electrical characteristics of diamond surface. Generally, oxygen termination is adapted to perform stable Schottky contacts with higher Schottky barrier height (SBH). An acid mixture (e.g., H_2SO_4 + HNO_3 at 200 °C), oxygen plasma treatment (or ashing), and exposure to ultraviolet (UV) under ozone atmosphere [25,122–126] are widely used to obtain O-terminated surfaces. Metals with a high-temperature melting point, such as Mo, Pt, Ru, and Zr, have resulted in high performances for high-voltage SBDs [127–129], although various metal species have been investigated for Schottky contacts [130–136]. SBH is reported to be 1.2–3.4 eV [28,137,138], depending on the metal species and surface treatments. In contrast, the n-type layer has high resistivity at room temperature due to the large donor activation energy (0.57 eV for P, 1.7 eV for N), and large SBH (4.3–4.5 eV) was found independent of metal species due to strong Fermi level pinning [33,139–142].

For ohmic contacts, titanium (Ti) is the most widely used for both p-type and n-type diamonds. Typically, ohmic electrodes are formed by depositing Ti (Ti/Au or Ti/Pt/Au) and annealing in N_2 or Ar atmosphere. Ohmic characteristic is considered to be improved by a chemical reaction between Ti and diamond, such as carbide formation [128,143]. A low specific contact resistance of 2.8×10^{-7} Ωcm^2 has been reported for p-type diamond (100)/Ti with annealing at 420 °C for 60 min in an Ar atmosphere [144].

Figure 7 shows schematic illustrations of diamond vertical-type SBDs proposed for power devices. These diodes have exhibited a breakdown voltage of 1.8–3.7 kV [115,136,145,146]. A maximum forward current of 10 A (electrode area of 16 mm^2) has been reported for vertical SBD (VSBD) [147]. The electric breakdown field of 7.7 MV/cm has been published [117,129,148]. A reverse leakage current can be explained by thermionic field emission (TFE) + barrier lowering [149–151]. The abrupt leakage current increasing and the breakdown field lowering are suggested to be caused by defects in the diamond derived from substrate, CVD growth, and/or device processing [136,152–154], although effects of crystallographic defects have not yet been clarified. Metal-assisted termination (MAT) has been proposed as a buffer layer for CVD growth to reduce density of threading dislocation and to improve crystal quality and SBD properties [155].

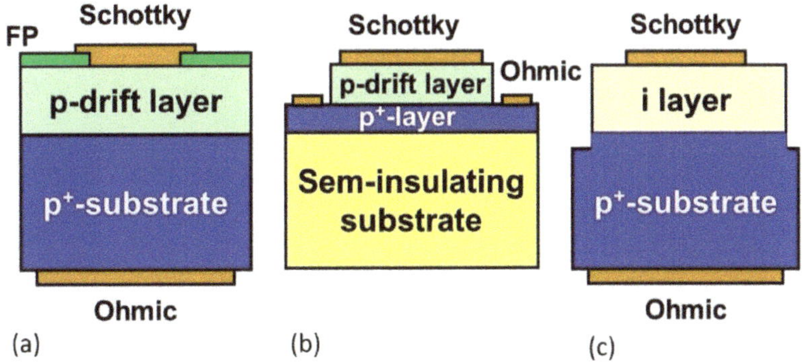

Figure 7. Schematic illustrations of vertical-type diamond Schottky barrier diodes. (**a**) Vertical SBD. (**b**) Pseudo-vertical SBD. (**c**) Metal-intrinsic SBD.

5.2. Diamond pn and PiN Junction Devices

PiN diodes are expected to have high reverse-blocking voltage because a depletion region between the p-type layer and n-type layer (drift layer thickness for punch-through type) can support the electric field. Compared to SBD, it has higher on-voltage due to the bandgap energy and lower switching speed due to the longer recovery time of accumulated carriers in the drift layer. However, in an ultra-high-voltage region such as >10 kV, considerable lowering of specific on-resistance by conduction modulation can be an advantage while Si devices develop very high resistance due to the large thickness of the drift layer.

A high-quality diamond pn junction has been achieved by Koizumi et al. in 2001 [156], followed by a number of reports on optimized diamond pn or pin UV LEDs [157,158]. B-doping and P-doping are widely used for the p-type layer and n-type layer, respectively. Large activation energies of B-acceptor (0.37 eV) and P-donor (0.57 eV) cause high resistivity at room temperature. Nevertheless, recent studies showed that an extremely high doping level (10^{22} cm^{-3}) is possible both for p-type and n-type while keeping high quality of crystal and junctions, which enables high carrier injection current under bipolar regime [159]. In addition, high P-concentration n-type layers on <100>-oriented devices have been achieved by overgrowth on shape-processed (100) diamond [160,161]. Kato et al. have reported on successful diamond bipolar junction transistors with this technique in the n-type layer [162].

Regarding the reverse blocking properties of diamond PiN diodes, a breakdown voltage (*BV*) of 11.5 kV with rectification ratio of 10^7 has been reported [163], as shown in Figure 8. The breakdown was clear and non-destructive as shown. The diode structure was fabricated by MPCVD growth of undoped (intrinsic) and P-doped n-type homoepitaxial layers on an HPHT (100) IIb p^+-type diamond substrate followed by forming Ti/Pt/Au electrodes on both top and bottom surfaces [164]. The thickness of the drift layer (undoped layer) is a 70 μm thick drift layer, and the corresponding breakdown field (F_B) is estimated to be 1.9 MV/cm with assuming a punch-through state by:

$$|BV| = |F_B| - \frac{qN_A d^2}{2\varepsilon_S} \quad (1)$$

where d is the drift layer thickness, q is the electronic charge, N_A is the acceptor concentration of the drift layer, and ε_S is permittivity. Figure 9 shows a comparison of reverse I–V characteristics between diodes with a mesa structure and without a mesa structure for the diamond PiN diodes (36 μm thick drift layer) [163]. It is found that the reverse leakage current is considerably reduced by the mesa structure. This result suggests that the reverse leakage current in the diodes can be passed through the n-type layer or surface. The *BV* and F_B of diamond pin diodes are shown in Figure 10 as a function of the drift layer thickness. The *BV* increased with the increase the drift layer thickness, closely tracking a theoretical calculation result [165], although the value is somewhat smaller than that. The maximum value of F_B, 3.6 MV/cm, was obtained for the PiN diode with 2 μm thick drift layer [163]. This value is higher than that of theoretically predicted values of GaN or SiC. Higher F_B, such as >10 MV/cm, should be realized by proper terminations, device structures, and higher crystal quality.

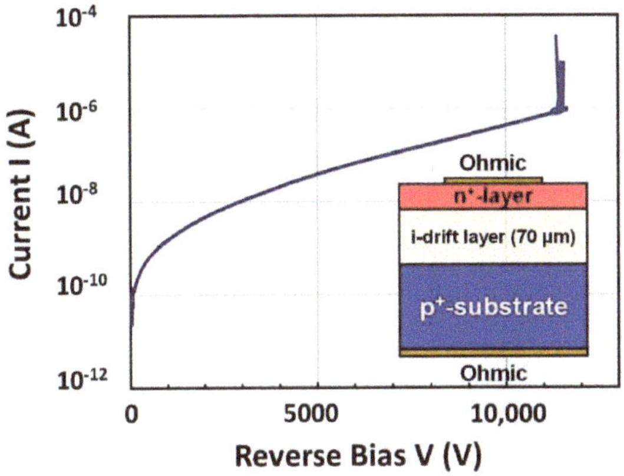

Figure 8. A representative reverse I–V property for diamond PiN diode (drift layer thickness 70 μm).

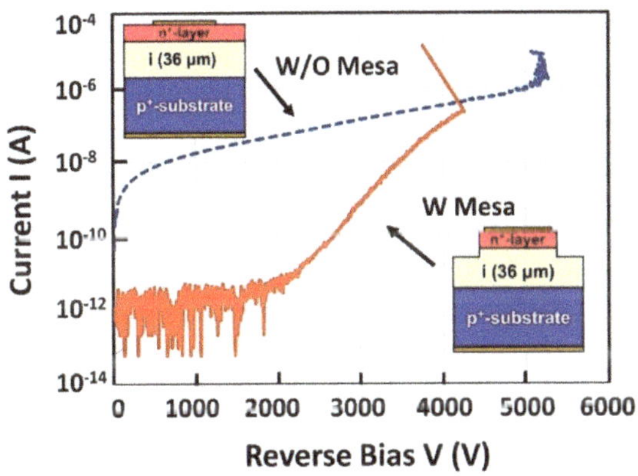

Figure 9. Reverse I–V properties for diamond PiN diode both with mesa structure and without mesa structure (drift layer thickness 36 μm).

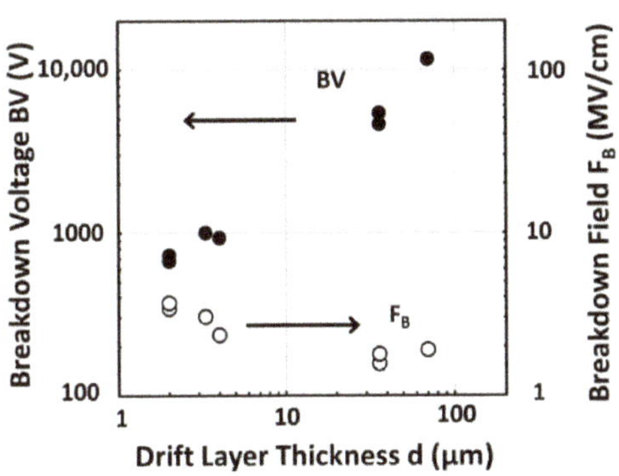

Figure 10. Break down voltage and breakdown field of diamond PiN diodes as a function of the drift layer thickness.

Other diodes with pn junctions have been also proposed for high-power devices, as shown in Figure 11. The Schottky–pn diode (SPND) is tandemly merged SBD with a pn diode (PND) [166] (Figure 11a). This diode is a unipolar device that shows lower on-voltage than that of PND and lower specific on-resistance and higher reverse blocking properties compared to SBD. Forward current density of 60 kA/cm^2 at 6 V (corresponding $R_{on}S$ = 0.03 mΩcm^2) with rectification ratio of 10^{12} and 3.4 MV/cm in a reverse blocking field has been reported for SPND [167]. Schottky PiN diodes have also been demonstrated with a blocking voltage of 500 V [168]. A high-voltage vacuum-power switch with a diamond PiN diode has been also proposed for the ultra-high-voltage region (Figure 11b). This device is utilizing highly efficient electron emission from the diamond PiN diode based on the negative electron affinity (NEA) of diamond. Takeuchi at al. demonstrated a 10 kV vacuum switch with high-power transmission efficiency of 73% using a diamond PiN diode [169].

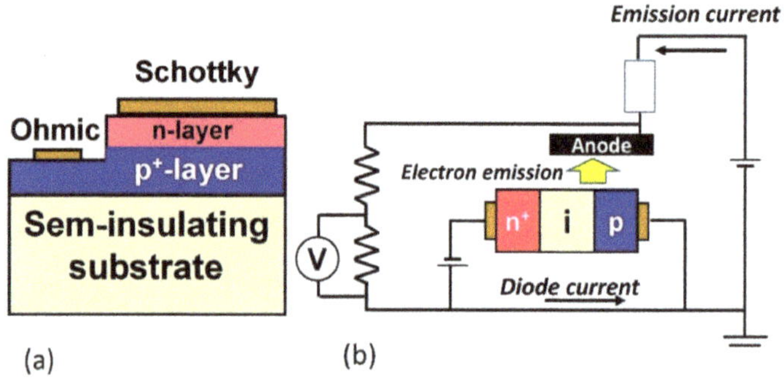

Figure 11. Schematic illustrations of noble diamond diodes. (**a**) Schottky pn diode (SPND). (**b**) Vacuum switch utilizing highly efficient electron emission from diamond PiN diode.

5.3. Diamond FETs

Diamond field effect transistors (FETs) have been also widely studied for high-power and/or high-frequency switching devices, as shown in Figure 12. Diamond metal semiconductor FETs (MESFETs) with a p-type Schottky junction gate (Figure 12a) have exhibited the breakdown voltage of >2 kV [170]. High temperature operation and high radiation tolerance have been reported [171]. Junction FETs (JFETs) are expected to be highly reliable for high-temperature and high-voltage operation because of the pn junctions instead of a gate oxide (Figure 12b). Normally off diamond JFETs with a high current density of 458 A/cm^2 have been achieved [172].

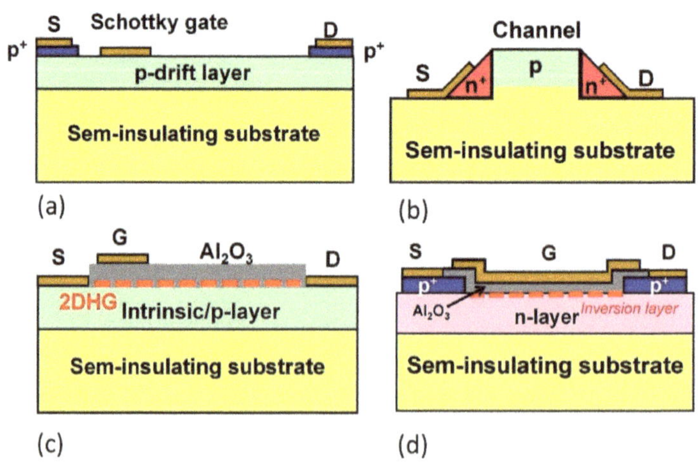

Figure 12. Schematic illustrations of diamond FETs. (**a**) MESFET. (**b**) Lateral pn junction JFET. (**c**) H-FET (C-H 2DHG MOSFET). (**d**) Inversion MOSFET.

Inversion metal-oxide semiconductor FETs (MOSFETs) are one of the most widely used electron devices. High-quality MOS interfaces have been difficult to fabricate on diamond due to the lack of natural oxide layers. Recently, thanks to improvements in the MOS interface by O-H termination with a wet-annealing technique and a higher quality of n-type layer, a diamond inversion-type p-channel MOSFET with normally-off operation has been realized [173] (Figure 12d). The field-effect (inversion channel) mobility has been estimated to be 8 $cm^2\ V^{-1}\ s^{-1}$, and the low mobility can be caused the existence of a high interface state density of 6 $\times$ $10^{12}\ cm^{-2}\ eV^{-1}$. The field-effect mobility was found to be dependent on the interface state density, which increased with the increase in the roughness of the Al_2O_3/n-diamond interface at the channel [174]. The roughness increased with the increase in the phosphorus concentration in the n-layer, and the improved field-effect mobility of 20 $cm^2\ V^{-1}\ s^{-1}$ has been obtained by reducing the interface state density [174]. Moreover, for bulk FETs, recently, deep depletion MOSFET has been proposed and demonstrated [175]. A breakdown field of 4 MV/cm has been obtained for a lateral normally on device consisting of oxygen-terminated diamond [176].

H-terminated FETs have been also extensively investigated. Diamond has the unique property that, near the hydrogen-terminated (C-H) surface, high-density hole accumulation (~$10^{13}\ cm^{-2}$) forms two-dimensional hole gas (2DHG) with very low activation energy and high hole mobility [25,177,178]. Accordingly, hydrogen-terminated diamond FETs with 2DHG show high transistor performances from the point of current density, high transconductance, and high-frequency operation. Both diamond MESFET and MOSFET with 2DHG have exhibited high-frequency operation (GHz) since 2001 [179–182]. These FETs are promising for application to high-power radio-frequency (RF) power amplifiers beyond other wide bandgap semiconductor devices. A cut-off frequency (f_T) of 70 GHz[182] and a maximum oscillation frequency (f_{max}) of 120 GHz [180] have been achieved. These values are comparable to GaN-based HEMTs [183–185]. The maximum drain current density ($I_{D\ max}$) of 1.35 A/mm [186], the blocking voltage of >2 kV for normally off operation devices [187], and the microwave output power (P_{out}) of 3.8 W/mm at 1 GHz [188] and 1.5 W/mm at 3.6 GHz [189] have been also reported. Improvement in sheet resistance and contact resistance can provide further improvement in output power [190]. Recently, vertical (trench gate structure) MOSFETs with side wall 2DHG are also proposed and demonstrated [191,192], which have exhibited maximum drain current density of 710 mA/mm. In addition, the above-mentioned diamond H-terminated 2DHG FETs, which have a p-channel, can be highly promising for complimentary circuits with GaN n-channel FETs in power amplifiers.

In this section, the current status of diamond electronic devices has been reviewed, focusing on power semiconductor devices. Diamond devices have been remarkably improved thanks to the establishment of MPCVD growth techniques including doping control and characterization techniques. In this decade, several diamond devices have exhibited excellent performances beyond other semiconductors based on the material advantage. However, the full potential of remarkable advantage of diamond has not yet been demonstrated. One of the big issues is that inadequate device fabrication techniques are limiting device performances. Etching technique, interfaces in MOS, ion implantation, or selective doping (for edge termination, buried structure, and so on), and passivation materials are key techniques to obtain higher performances. Furthermore, for ultra-high-power electronics applications, bipolar devices should be necessary, such as GTO, thyristor, and IGBT in addition to PiN diodes in which sophisticated device fabrication techniques and doping techniques are crucial. Due to a great deal of effort, techniques of selective growth [193,194], selective doping [63,72,160,195,196], and selective etching [161,197] have made great progress in fabricating device structures. The problems of n-type layers are still not fully solved. However, today, n-type doping level is possible to control from 10^{15}–$10^{20}\ cm^{-3}$, and then significant reduction in resistivity by using hopping conduction has been reported in the pin structure [149], and improvement in the crystal quality of phosphorus-doped (n-type) diamond has been reported [67,174]. As a summary, diamond devices still have many issues for practical use; however, considering the remarkable development in recent years, ultra-high-power diamond devices can be achieved both for HVDC and RF applications in the near future.

6. Summary and Conclusions

This paper gives a brief overview of the state of the art of the diamond technology for power devices. It shows that strong limitations of the material for such applications have now been overcome. However, its high hardness and epitaxial growth out of equilibrium are not straightforwardly resolved and have taken some decades. Example in the characterization method and the growth of p-type and n-type diamond show the high potential of this material. High-quality vertical and lateral growth have been demonstrated, and the example of Schottky and p–n diodes as well as transistors are shown to work under high-power electronic conditions. This shows that the diamond is competitive with other WBG materials for the power electronic niche. Several device architectures have been recently manufactured, even though reliability is still a pending subject for it commercialization. Therefore, the material is close to finding some application as an electronic material, making it possible to take advantage of its outstanding electronic properties.

Funding: The authors thank the Ministerio de Economia y Competitividad (MINECO) of the Spanish Government for funding under Grant Nos. TEC2017-86347-C2-1-R, ESP2017-91820, PID2020-117201RB-C21, and PID2019-110219RB-100 and the Junta de Andalucia (Andalusian Government, Spain) for funding through Nos. P20_00946, FEDER-UCA18- 106470 and FEDER-UCA18-107851 projects.

Institutional Review Board Statement: Not applicable.

Informed Consent Statement: Not applicable.

Data Availability Statement: The data that support the findings of this study are available from the corresponding author upon reasonable request.

Acknowledgments: The authors also thank IMEYMAT and SC-ICYT (University of Cadiz) for the facilities in TEM characterization.

Conflicts of Interest: The authors declare no conflict of interest.

References

1. Ohashi, H. Power devices now and future, strategy of Japan. In Proceedings of the 2012 24th International Symposium on Power Semiconductor Devices and ICs (ISPSD 2012), Bruges, Belgium, 3–7 June 2012.
2. Pernot, J.; Volpe, P.N.; Omnès, F.; Muret, P.; Mortet, V.; Heanen, K.; Teraji, T. Hall hole mobility in boron-doped homoepitaxial diamond. *Phys. Rev. B* **2010**, *81*, 205203. [CrossRef]
3. Pernot, J.; Koisumi, S. Electron mobility in phosphorous doped {111} homoepitaxial diamond. *Appl. Phys. Lett.* **2008**, *93*, 52105. [CrossRef]
4. Isberg, J.; Hammersberg, J.; Johansson, E.; Wilkström, T.; Twitchen, D.; Whitehead, A.J.; Coe, S.E.; Scarsbrook, G.A. High carrier mobility in single-crystal plasma-deposited diamond. *Science* **2002**, *297*, 1670. [CrossRef] [PubMed]
5. Pearton, S.J.; Yang, J.; Cary, P.H.; Ren, F.; Kim, J.; Tadjer, M.J.; Mastro, M.A. A review of Ga_2O_3 materials, processing, and devices. *Appl. Phys. Rev.* **2018**, *5*, 011301. [CrossRef]
6. Mikhail, S.; Howell, D.A. A petrological assessment of diamond as a recorder of the mantle nitrogen cycle. *Am. Mineral.* **2016**, *101*, 780. [CrossRef]
7. Schreck, M.; Asmussen, J.; Shikata, S.; Fujimori, N. Large-area high-quality single crystal diamond. *MRS Bull.* **2014**, *39*, 504–510. [CrossRef]
8. Schreck, M.; Gsell, S.; Brescia, R.; Fischer, M. Ion bombardment induced buried lateral growth: The key mechanism for the synthesis of single crystal diamond wafers. *Sci. Rep.* **2017**, *7*, 44462. [CrossRef] [PubMed]
9. Kim, S.W.; Kawamata, Y.; Takaya, R.; Koyama, K.; Kasu, M. Growth of high-quality one-inch free-standing heteroepitaxial (001) diamond on (110) sapphire substrate. *Appl. Phys. Lett.* **2020**, *117*, 202102. [CrossRef]
10. Levedev, V.; Engels, J.; Kustermann, J.; Weippert, J.; Cimalla, V.; Kirste, L.; Giese, C.; Quellmalz, P.; Graff, A.; Meyer, F.; et al. Growth defects in heteroepitaxial diamond. *J. Appl. Phys.* **2021**, *129*, 165301. [CrossRef]
11. Schreck, M.; Arnault, J.-C. Heteroepitaxy of diamond on Ir/metal-oxide/Si substrates. In *Power Electronics Device Applications of Diamond Semiconductors*; Woodhead: Cambridge, UK, 2018; p. 58.

12. Mehmel, L.; Issaoui, R.; Brinza, O.; Tallaire, A.; Mille, V.; Delchevalrie, J.; Saada, S.; Arnault, J.C.; Benedic, F.; Achard, J. Dislocation density reduction using overgrowth on hole arrays made in heteroepitaxial diamond substrates. *J. Appl. Phys. Lett.* **2021**, *118*, 061901. [CrossRef]
13. Ichikawa, K.; Kurone, K.; Kodama, H.; Suzuki, K.; Sawabe, A. High crystalline quality heteroepitaxial diamond using grid-patterned nucleation and growth on Ir. *Diam. Relat. Mater.* **2019**, *94*, 92. [CrossRef]
14. Yamada, H.; Chayahara, A.; Mokuno, Y.; Kato, Y.; Shikata, S. A 2-in. mosaic wafer made of a single-crystal diamond. *Appl. Phys. Lett.* **2014**, *104*, 102110. [CrossRef]
15. Kamo, M.; Sato, Y.; Matsumoto, S.; Setaka, N. Diamond synthesis from gas phase in microwave plasma. *J. Cryst. Growth* **1983**, *62*, 642. [CrossRef]
16. Schwander, M.; Partes, K. A review of diamond synthesis by CVD processes. *Diam. Relat. Mater.* **2011**, *20*, 1287. [CrossRef]
17. Butler, J.E.; Woodin, R.L. Thin film diamond growth mechanisms. *Phil. Trans. R. Soc. A* **1993**, *342*, 209.
18. Setaka, N. *Synthetic Diamond Emerging CVD Science and Technology*; Spear, K.E., Dismukes, J.P., Eds.; Wiley: New York, NY, USA, 1994.
19. Wild, C.; Kohl, R.; Herres, N.; Müller-Sebert, W.; Koidl, P. Oriented CVD diamond films: Twin formation, structure and morphology. *Diam. Relat. Mater.* **1994**, *3*, 373. [CrossRef]
20. Battaile, C.C.; Srolovitz, D.J.; Butler, J.E. Atomic-scale simulations of chemical vapor deposition on flat and vicinal diamond substrates. *J. Cryst. Growth* **1998**, *194*, 353. [CrossRef]
21. Silva, F.; Achard, J.; Bonnin, X.; Michau, A.; Tallaire, A.; Brinza, O.; Gicquel, A. 3D crystal growth model for understanding the role of plasma pre–treatment on CVD diamond crystal shape. *Phys. Status Solidi A* **2006**, *20*, 3049. [CrossRef]
22. Silva, F.; Bonnin, X.; Achard, J.; Brinza, O.; Michau, A.; Gicquel, A.J. Geometric modeling of homoepitaxial CVD diamond growth: I. The {1 0 0}{1 1 1}{1 1 0}{1 1 3} system. *Cryst. Growth* **2008**, *310*, 187. [CrossRef]
23. Tallaire, A.; Kasu, M.; Ueda, K.; Makimoto, T. Origin of growth defects in CVD diamond epitaxial films. *Diam. Relat. Mater.* **2008**, *17*, 60. [CrossRef]
24. Bogatskiy, A.; Butler, J.E. A geometric model of growth for cubic crystals: Diamond. *Diam. Relat. Mater.* **2015**, *53*, 58. [CrossRef]
25. Hayashi, K.; Yamanaka, S.; Okushi, H.; Kajimura, K. Homoepitaxial diamond films with large terraces. *Appl. Phys. Lett.* **1996**, *68*, 1220. [CrossRef]
26. Gamo, M.N.; Loh, K.P.; Sakaguchi, I. Surface morphology of homoepitaxially grown (111), (001), and (110) diamond studied by low energy electron diffraction and reflection high-energy electron diffraction. *J. Vac. Sci. Technol. A* **1999**, *17*, 2991. [CrossRef]
27. Watanabe, H.; Takeuchi, D.; Yamanaka, S.; Okushi, H.; Kajimura, K. Homoepitaxial diamond film with an atomically flat surface over a large area. *Diam. Relat. Mater.* **1999**, *8*, 1272. [CrossRef]
28. Collins, A.T.; Kamo, M.; Sato, Y. Optical centres related to nitrogen, vacancies and interstitials in polycrystalline diamond films grown by plasma-assisted chemical vapour deposition. *J. Phys. D Appl. Phys.* **1989**, *22*, 1402. [CrossRef]
29. Spitsyn, B.V. Chemical crystallization of diamond from the activated vapor phase. *J. Cryst. Growth* **1990**, *99*, 1162. [CrossRef]
30. Teraji, T.; Ito, T. Homoepitaxial diamond growth by high-power microwave-plasma chemical vapor deposition. *J. Cryst. Growth* **2004**, *271*, 409. [CrossRef]
31. Teraji, T.; Mitani, S.; Ito, T. High rate growth and luminescence properties of high–quality homoepitaxial diamond (100) films. *Phys. Status Solidi A* **2003**, *198*, 395. [CrossRef]
32. Ito, T.; Watanabe, T.; Irie, M.; Nakamura, J.; Teraji, T. Electron emissions from CVD diamond surfaces. *Diam. Relat. Mater.* **2003**, *12*, 434. [CrossRef]
33. Teraji, T.; Yoshizaki, S.; Mitani, S.; Watanabe, T.; Hamada, M.; Ito, T. Highly sensitive UV photodetectors fabricated using high-quality single-crystalline CVD diamond films. *Diam. Relat. Mater.* **2004**, *13*, 858. [CrossRef]
34. Ekimov, E.A.; Sidorov, V.A.; Bauer, E.D.; Mel'nik, N.N.; Curro, N.J.; Thompson, J.D.; Stishov, S.M. Superconductivity in diamond. *Nature* **2004**, *428*, 542. [CrossRef]
35. Sidorov, V.A.; Ekimov, E.A. Superconductivity in diamond. *Diam. Relat. Mater.* **2010**, *19*, 351. [CrossRef]
36. Koizumi, S.; Kamo, M.; Sato, Y.; Ozaki, H.; Inuzuka, T. Growth and characterization of phosphorous doped {111} homoepitaxial diamond thin films. *Appl. Phys. Lett.* **1997**, *71*, 1065. [CrossRef]
37. Makino, T.; Kato, H.; Ri, S.-G.; Cheng, Y.; Okushi, H. Electrical characterization of homoepitaxial diamond p–n+ junction. *Diam. Relat. Mater.* **2005**, *14*, 1995. [CrossRef]
38. Koizumi, S.; Suzuki, M. n-Type doping of diamond. *Phys. Status Solidi A* **2006**, *203*, 3358. [CrossRef]
39. Pinault-Thaury, M.-A.; Temgoua, S.; Gillet, R.; Bensalah, H.; Stenger, I.; Jomard, F.; Issaoui, R.; Barjon, J. Phosphorus-doped (113) CVD diamond: A breakthrough towards bipolar diamond devices. *J. Appl. Phys. Lett.* **2019**, *114*, 112106. [CrossRef]
40. People, R.; Bean, J.C. Calculation of critical layer thickness versus lattice mismatch for Ge_xSi_{1-x}/Si strained–layer heterostructures. *Appl. Phys. Lett.* **1985**, *47*, 322. [CrossRef]
41. Alegre, M.P.; Araujo, D.; Fiori, A.; Piñero, J.C.; Lloret, F.; Villar, M.P.; Achatz, P.; Chicot, G.; Bustarret, E.; Jomard, F. Critical boron-doping levels for generation of dislocations in synthetic diamond. *Appl. Phys. Lett.* **2014**, *105*, 173103. [CrossRef]
42. Lloret, F.; Eon, D.; Bustarret, E.; Fiori, A.; Araujo, D. Boron-doping proximity effects on dislocation generation during non-planar MPCVD homoepitaxial diamond growth. *Nanomaterials* **2018**, *8*, 480. [CrossRef]
43. Dean, P.J.; Lightowlers, E.C.; Wight, D.R. Intrinsic and extrinsic recombination radiation from natural and synthetic aluminum-doped diamond. *Phys. Rev.* **1965**, *140*, A352. [CrossRef]

44. Collins, A.; Lawson, S.; Davies, G.; Kanda, H. Real Time Ellipsometry Characterization and Process Monitoring for Amorphous Carbon Deposition. *Mater. Sci. Forum* **1990**, *65–66*, 19. [CrossRef]
45. Ruf, T.; Cardona, M.; Sternschulte, H.; Wahl, S.; Thonke, K.; Sauer, R.; Pavone, P.; Anthony, T. Cathodoluminescence investigation of isotope effects in diamond. *Solid State Commun.* **1998**, *105*, 311. [CrossRef]
46. Sauer, R.; Sternschulte, H.; Wahl, S.; Thonke, K.; Anthony, T. Cathodoluminescence investigation of isotope effects in diamond. *Phys. Rev. Lett.* **2000**, *84*, 4172. [CrossRef] [PubMed]
47. Sharp, S.; Collins, A.; Davies, G.; Joyce, G. Higher resolution studies of shallow bound exciton luminescence in diamond. *J. Phys. Condens. Matter* **1997**, *9*, L451. [CrossRef]
48. Kawarada, H.; Matsuyama, H.; Yokota, Y.; Sogi, T.; Yamaguchi, A.; Hiraki, A. Excitonic recombination radiation in undoped and boron-doped chemical-vapor-deposited diamonds. *Phys. Rev. B* **1993**, *47*, 3633. [CrossRef] [PubMed]
49. Deneuville, A.; Baron, C.; Wade, M.; Jomard, F.; Chevallier, J. Cathodoluminescence of highly and heavily boron doped (100) homoepitaxial diamond films. *Diam. Relat. Mater.* **2006**, *15*, 597.
50. Hasegawa, M.; Teraji, T.; Koizumi, S. Lattice location of phosphorus in n-type homoepitaxial diamond films grown by chemical-vapor deposition. *Appl. Phys. Lett.* **2001**, *79*, 3068. [CrossRef]
51. Nakazawa, K.; Tanabe, K.; Tachiki, M.; Kawarada, H.; Koizumi, S. Excitonic recombination radiation in phosphorus-doped CVD diamonds. *Phys. Rev. B* **2001**, *64*, 23520. [CrossRef]
52. Barjon, J.; Desfonds, P.; Pinault, M.-A.; Kociniewski, T.; Jomard, F.; Chevallier, J. Determination of the phosphorus content in diamond using cathodoluminescence spectroscopy. *J. Appl. Phys.* **2007**, *101*, 113701. [CrossRef]
53. Omnès, F.; Muret, P.; Volpe, P.N.; Wade, M.; Pernot, J.; Jomard, F. Study of boron doping in MPCVD grown homoepitaxial diamond layers based on cathodoluminescence spectroscopy, secondary ion mass spectroscopy and capacitance–voltage measurements. *Diam. Relat. Mater.* **2011**, *20*, 912–916. [CrossRef]
54. Ghodbane, S.; Omnès, F.; Agnès, C. A cathodoluminescence study of boron doped {111}-homoepitaxial diamond films. *Diam. Relat. Mater.* **2010**, *19*, 273. [CrossRef]
55. Maier, F.; Riedel, M.; Mantel, B.; Ristein, J.; Ley, L. Origin of surface conductivity in diamond. *Phys. Rev. Lett.* **2000**, *85*, 3472. [CrossRef] [PubMed]
56. Chen, W.; Qi, D.; Gao, X.; Wee, A.T.S. Surface transfer doping of semiconductors. *Prog. Surf. Sci.* **2009**, *84*, 279. [CrossRef]
57. Pakes, C.I.; Garrido, J.A.; Kawarada, H. Diamond surface conductivity: Properties, devices, and sensors. *MRS Bull.* **2014**, *39*, 542. [CrossRef]
58. Ristein, J. Diamond surfaces: Familiar and amazing. *Appl. Phys. A Mater. Sci. Process.* **2006**, *82*, 377. [CrossRef]
59. Gökden, S.; Barana, R.; Balkan, N.; Mazzucato, S. The effect of interface roughness scattering on low field mobility of 2D electron gas in GaN/AlGaN heterostructure. *Phys. E* **2004**, *24*, 249. [CrossRef]
60. Gurusinghe, M.N.; Davidsson, S.K.; Andersson, T.G. Two-dimensional electron mobility limitation mechanisms in heterostructures. *Phys. Rev. B* **2005**, *72*, 045316. [CrossRef]
61. Karavolas, V.K.; Smith, M.J.; Fromhold, T.M.; Butcher, P.N.; Mulimani, B.G.; Gallagher, B.L.; Oxley, J.P. The effect of interface roughness scattering and background impurity scattering on the thermopower of a 2DEG in a Si MOSFET. *J. Phys. Condens. Matter* **1990**, *2*, 10401. [CrossRef]
62. Wade, T.; Geis, M.W.; Fedynyshyn, T.H.; Vitale, S.A.; Varghese, J.O.; Lennon, D.M.; Grotjohn, T.A.; Nemanich, R.J.; Hollis, M.A. Effect of surface roughness and H–termination chemistry on diamond's semiconducting surface conductance. *Diam. Relat. Mater.* **2017**, *76*, 79. [CrossRef]
63. Zhang, X.; Matsumoto, T.; Yamasaki, S.; Nebel, C.E.; Inokuma, T.; Tokuda, N.J. Inversion-type p-channel diamond MOSFET issues. *Mater. Res.* 2021. [CrossRef]
64. Sasama, Y.; Kageura, T.; Komatsu, K.; Imura, M.; Watanabe, K.; Taniguchi, T.; Yamaguchi, T. Charge-carrier mobility in hydrogen-terminated diamond field-effect transistors. *J. Appl. Phys.* **2020**, *127*, 185707. [CrossRef]
65. Tallaire, A.; Brinza, O.; Mille, V.; William, L.; Achard, J. Reduction of dislocations in single crystal diamond by lateral growth over a macroscopic hole. *Adv. Mater.* **2017**, *29*, 1604823. [CrossRef] [PubMed]
66. Lloret, F.; Gutierrez, M.; Araujo, D.; Eon, D.; Bustarret, E. MPCVD diamond lateral growth through microterraces to reduce threading dislocations density. *Phys. Status Solidi A* **2017**, *214*, 1700242. [CrossRef]
67. Pinault-Thaury, M.-A.; Stenger, I.; Gillet, R.; Temgoua, S.; Chikoidze, E.; Dumont, Y.; Jomard, F.; Kociniewski, T.; Barjon, J. Attractive electron mobility in (113) n-type phosphorus-doped homoepitaxial diamond. *Carbon* **2021**, *175*, 254e258. [CrossRef]
68. Lloret, F.; Fiori, A.; Araujo, D.; Eon, D.; Villar, M.P.; Bustarret, E. Stratigraphy of a diamond epitaxial three-dimensional overgrowth using doping superlattices. *App. Phys. Lett.* **2016**, *108*, 181901. [CrossRef]
69. Lloret, F.; Araujo, D.; Eon, D.; Bustarret, E. Three-Dimensional Diamond MPCVD Growth over MESA Structures: A Geometric Model for Growth Sector Configuration. *Cryst. Growth Des.* **2018**, *18*, 7628. [CrossRef]
70. Lloret, F.; Araujo, D.; Eon, D.; Villar, M.P.; Gonzalez-Leal, J.M.; Bustarret, E. Influence of methane concentration on MPCVD overgrowth of 100-oriented etched diamond substrates. *Phys. Status Solidi A* **2016**, *213*, 2570. [CrossRef]
71. Zou, Y.; Larsson, K. Effect of boron doping on the CVD growth rate of diamond. *J. Phys. Chem. C* **2016**, *120*, 10658. [CrossRef]
72. Lloret, F.; Eon, D.; Bustarret, E.; Donatini, F.; Araujo, D. Selectively boron doped homoepitaxial diamond growth for power device applications. *Appl. Phys. Lett.* **2021**, *118*, 023504. [CrossRef]

73. Tarutani, M.; Takai, Y.; Shimizu, R.; Ando, T.; Kamo, M.; Bando, Y. Transmission electron microscopy study of interface and internal defect structures of homoepitaxial diamond. *Appl. Phys. Lett.* **1996**, *68*, 2070. [CrossRef]
74. Hickey, D.P.; Kuryliw, E.; Siebein, K.; Jones, K.S.; Chodelka, R.; Elliman, R. Cross-sectional transmission electron microscopy method and studies of implant damage in single crystal diamond. *J. Vac. Sci. Technol. A* **2006**, *24*, 1302. [CrossRef]
75. Jiang, N.; Shinjo, S.; Inaoka, T.; Shintani, Y.; Ito, T.; Makita, H.; Hatta, A.; Hiraki, A. Characterization of N-doped diamond films by transmission electron microscopy. *J. Cryst. Growth* **2001**, *224*, 111. [CrossRef]
76. Yin, L.-W.; Li, M.-S.; Sun, D.-S.; Li, F.-Z.; Hao, Z.-Y. Some aspects of diamond crystal growth at high temperature and high pressure by TEM and SEM. *Mater. Lett.* **2002**, *55*, 397. [CrossRef]
77. Barnes, R.; Bangert, U.; Martineau, P. HR-TEM imaging and image simulation of vacancy clusters in brown diamond. *Phys. Status Solidi* **2006**, *203*, 3081. [CrossRef]
78. Sawada, H.; Ichinose, H.; Watanabe, H.; Takeuchi, D.; Okushi, H. Cross-sectional TEM study of unepitaxial crystallites in a homoepitaxial diamond film. *Diam. Relat. Mater.* **2001**, *10*, 2030. [CrossRef]
79. Kono, S.; Saitou, T.; Kawata, H.; Goto, T.; Kakefuda, Y.; Komeda, T. Characteristic energy band values and electron attenuation length of a chemical-vapor-deposition diamond (0 0 1) 2×1 surface. *Surf. Sci.* **2009**, *603*, 860. [CrossRef]
80. Graupner, R.; Maier, F.; Ristein, J.; Ley, L.; Jung, C. High-resolution surface-sensitive C core-level spectra of clean and hydrogen-terminated diamond (100) and (111) surfaces. *Phys. Rev. B Cond. Matter Mater. Phys.* **1998**, *57*, 12397. [CrossRef]
81. Kono, S.; Sawabe, A.; Kodama, H.; Hayashi, Y.; Kageura, T.; Ogura, M.; Inaba, M.; Kawarada, H.; Teraji, T.; Ri, S.-G.; et al. Carbon 1s X-ray photoelectron spectra of realistic samples of hydrogen-terminated and oxygen-terminated CVD diamond (111) and (001). *Diam. Relat. Mater.* **2019**, *93*, 105. [CrossRef]
82. Kono, S.; Saito, T.; Kang, S.H.; Jung, W.Y.; Kim, B.Y.; Kawata, H.; Goto, T.; Kakefuda, Y.; Yeom, H.W. Band diagram for chemical vapor deposition diamond surface conductive layer: Presence of downward band bending due to shallow acceptors. *Surf. Sci.* **2010**, *604*, 1148. [CrossRef]
83. Alba, G.; Eon, D.; Villar, M.P.; Alcántara, R.; Chicot, G.; Cañas, J.; Letellier, J.; Pernot, J.; Araujo, D. H-Terminated Diamond Surface Band Bending Characterization by Angle-Resolved XPS. *Surfaces* **2020**, *3*, 61–71. [CrossRef]
84. Navas, J.; Araujo, D.; Piñero, J.C.; Sánchez-Coronilla, A.; Blanco, E.; Villar, P.; Alcántara, R.; Montserrat, J.; Florentin, M.; Eon, D.; et al. Oxygen termination of homoepitaxial diamond surface by ozone and chemical methods: An experimental and theoretical perspective. *Appl. Surf. Sci.* **2018**, *433*, 408. [CrossRef]
85. Alba, G.; Villar, M.P.; Alcántara, R.; Navas, J.; Araujo, D. Surface States of (100) O-Terminated Diamond: Towards Other 1×1: O Reconstruction Models. *Nanomaterials* **2020**, *10*, 1193. [CrossRef] [PubMed]
86. Alba, G.; Leinen, D.; Villar, M.P.; Alcántara, R.; Piñero, J.C.; Fiori, A.; Teraji, T.; Araujo, D. Comprehensive nanoscopic analysis of tungsten carbide/Oxygenated-diamond contacts for Schottky barrier diodes. *Appl. Surf. Sci.* **2021**, *537*, 147874. [CrossRef]
87. Li, F.N.; Liu, J.W.; Zhang, J.W.; Wang, X.L.; Wang, W.; Liu, Z.C.; Wang, H.X. Measurement of barrier height of Pd on diamond (100) surface by X-ray photoelectron spectroscopy. *Appl. Surf. Sci.* **2016**, *370*, 496. [CrossRef]
88. Kono, S.; Teraji, T.; Kodama, H.; Ichikawa, K.; Ohnishi, S.; Sawabe, A. Direct determination of the barrier height of Ti-based ohmic contact on p-type diamond (001). *Diam. Relat. Mater.* **2015**, *60*, 117. [CrossRef]
89. Cañas, J.; Alba, G.; Leinen, D.; Lloret, F.; Gutierrez, M.; Eon, D.; Pernot, J.; Gheeraert, E.; Araujo, D. Diamond/γ-alumina band offset determination by XPS. *Appl. Surf. Sci.* **2020**, *535*, 146301. [CrossRef]
90. Liu, J.W.; Liao, M.Y.; Imura, M.; Koide, Y. Spin transfer torque devices utilizing the giant spin Hall effect of tungsten. *Appl. Phys. Lett.* **2012**, *101*, 1.
91. Marechal, E.A.; Aoukar, M.; Vallée, C.; Rivière, C.; Eon, D.; Pernot, J.; Gheeraert, E. Energy-band diagram configuration of Al_2O_3/oxygen-terminated p-diamond metal-oxide-semiconductor. *Appl. Phys. Lett.* **2015**, *107*, 141601. [CrossRef]
92. Zaitsev, A.M. *Optical Properties of Diamond: A Data Handbook*; Springer: Berlin/Heidelberg, Germany, 2001.
93. Tavares, C.; Omnès, F.; Pernot, J.; Bustarret, E. Electronic properties of boron-doped {111}-oriented homoepitaxial diamond layers. *Diam. Relat. Mater.* **2006**, *15*, 582. [CrossRef]
94. Fernández-Lorenzo, C.; Araujo, D.; Martin, J.; Alcántara, R.; Navas, J.; Villar, M.P.; Alegre, M.P.; Volpe, P.N.; Omnés, F.; Bustarret, E. Hydrogen passivation of boron acceptors in as-grown boron-doped CVD diamond epilayers. *Diam. Relat. Mat.* **2010**, *19*, 904–907. [CrossRef]
95. Araújo, D.; Kadri, M.; Wade, M.; Bustarret, E.; Deneuville, A. Excitonic emission and N- and B-incorporation in homoepitaxial CVD-grown diamond investigated by cathodoluminescence. *Phys. Stat. Sol.* **2005**, *2*, 1336. [CrossRef]
96. De la Mata, B.G.; Sanz-Hervás, A.; Dowsett, M.G.; Schwitters, M.; Twitchen, D. Calibration of boron concentration in CVD single crystal diamond combining ultralow energy secondary ions mass spectrometry and high resolution X-ray diffraction. *Diam. Relat. Mater.* **2007**, *16*, 809. [CrossRef]
97. El-Barbary, A.S.; Trasobares, S.; Ewels, C.P.; Stephan, O.; Okotrub, A.V.; Bulusheva, L.G.; Fall, C.J.; Heggie, M.I. Electron spectroscopy of carbon materials: Experiment and theory. *J. Phys. Conf. Ser.* **2006**, *26*, 149. [CrossRef]
98. Cañas, J.; Piñero, J.C.; Lloret, F.; Gutierrez, M.; Pham, T.; Pernot, J.; Araujo, D. Determination of alumina bandgap and dielectric functions of diamond MOS by STEM-VEELS. *Appl. Surf. Sci.* **2018**, *461*, 93. [CrossRef]
99. Hartmann, P.; Bohr, S.; Haubner, R.; Lux, B.; Wurzinger, P.; Griesser, M.; Bergmaier, A.; Dollinger, G.; Sternschulte, H.; Sauer, R. Diamond growth with boron addition. *Int. J. Refract. Met. Hard Mater.* **1998**, *16*, 223. [CrossRef]

100. Waidmann, S.; Bartsch, K.; Endler, I.; Fontaine, F.; Arnold, B.; Knupfer, M.; Leonhardt, A.; Fink, J. Electron energy-loss spectroscopy in transmission of undoped and doped diamond films. *Carbon* **1999**, *37*, 823. [CrossRef]
101. Bangert, U.; Barnes, R. Electron energy loss spectroscopy of defects in diamond. *Phys. Status Solidi* **2007**, *204*, 2201. [CrossRef]
102. Bourgeois, E.; Bustarret, E.; Achatz, P.; Omnès, F.; Blase, X. Impurity dimers in superconducting B-doped diamond: Experiment and first-principles calculations. *Phys. Rev. B* **2006**, *74*, 094509. [CrossRef]
103. Stöger-Pollach, M.; Schattschneider, P. The influence of relativistic energy losses on bandgap determination using valence EELS. *Ultramicroscopy* **2007**, *107*, 1178. [CrossRef] [PubMed]
104. Araujo, D.; Achatz, P.; el Bouayadi, R.; García, A.J.; Alegre, M.P.; Villar, M.P.; Jomard, F.; Bustarret, E. Local boron doping quantification in homoepitaxial diamon structures. *Diam. Relat. Mater.* **2010**, *19*, 972. [CrossRef]
105. Lu, Y.-G.; Turner, S.; Verbeeck, J.; Janssens, S.D.; Wagner, P.; Haenen, K.; Tendeloo, G.V. Direct visualization of boron dopant distribution and coodination in individual chamical vapor deposition nanocrystalline B-doped diamond grains. *Appl. Phys. Lett.* **2012**, *101*, 041907. [CrossRef]
106. Araujo, D.; Alegre, M.P.; Piñero, J.C.; Fiori, A.; Bustarret, E.; Jomard, F. Boron concentration profiling by high angle annular dark field-scanning transmission electron microscopy in homoepitaxial δ-doped diamond layers. *Appl. Phys. Lett.* **2013**, *103*, 042104. [CrossRef]
107. Piñero, J.C.; Lloret, F.; Alegre, M.P.; Villar, M.P.; Fiori, A.; Bustarret, E.; Araújo, D. High resolution boron content profilometry at δ-doping epitaxial diamond interfaces by CTEM. *Appl. Surf. Sci.* **2018**, *461*, 221. [CrossRef]
108. Araujo, D.; Alegre, M.P.; García, A.J.; Villar, M.P.; Bustarret, E.; Achatz, P.; Volpe, P.N.; Omnès, F. Cross sectional evaluation of boron doping and defect distribution in homoespitaxial diamond layers. *Phys. Status Solidi C* **2011**, *8*, 1366.
109. Matthews, J.W.; Blakeslee, A.E. Defects in epitaxial multilayers: I. Misfit dislocations. *J. Cryst. Growth* **1974**, *27*, 118. [CrossRef]
110. Wiesauer, K.; Springholz, G. Critical thickness and strain relaxation in high-misfit heteroepitaxial systems: $PbTe_{1-x}Se_x$ on PbSe (001). *Phys. Rev. B* **2004**, *69*, 245313. [CrossRef]
111. Van der Merwe, J.H. Crystal interfaces. Part II. Finite overgrowths. *J. Appl. Phys.* **1963**, *34*, 123. [CrossRef]
112. Dunstan, J.D.; Young, S.; Dixon, R.H. Geometrical theory of critical thicness and relaxation in strained-layer growth. *J. Appl. Phys.* **1991**, *70*, 3038. [CrossRef]
113. Araujo, D.; Lloret, F.; Alba, G.; Alegre, M.P.; Villar, M.P. Dislocation generation mechanisms in heavily boron-doped diamond epilayers. *Appl. Phys. Lett.* **2021**, *118*, 052108. [CrossRef]
114. Butler, J.E.; Geis, M.W.; Krohn, K.E.; Lawless Jr., J.; Deneault, S.; Lyszczarz, T.M.; Flechtner, D.; Wright, R. Exceptionally high voltage Schottky diamond diodes and low boron doping. *Semicond. Sci. Technol.* **2003**, *18*, 567. [CrossRef]
115. Twitchen, D.J.; Whithead, A.J.; Coe, S.E.; Isberg, J.; Hammerberg, J.; Wikstrom, T.; Johansson, E. High-voltage single-crystal diamond diodes. *IEEE Trans. Electron Devices* **2004**, *5*, 826. [CrossRef]
116. Huang, W.; Chow, T.P.; Yang, J.; Butler, J.E. High-voltage diamond Schottky rectifiers. *Int. J. High Speed Electron. Syst.* **2004**, *14*, 872. [CrossRef]
117. Volpe, P.-N.; Muret, P.; Permot, J.; Omnès, F.; Teraji, T.; Koide, Y.; Jomard, F.; Planson, D.; Brosselard, P.; Dheilly, N.; et al. Extreme dielectric strength in boron doped homoepitaxial diamond. *Appl. Phys. Lett.* **2010**, *97*, 223501. [CrossRef]
118. Umezawa, H.; Kato, Y.; Shikata, S. 1 Ω on-resistance diamond vertical-Schottky barrier diode operated at 250 °C. *Appl. Phys. Express* **2013**, *6*, 011302. [CrossRef]
119. Okushi, H. High quality homoepitaxial CVD diamond for electronic devices. *Diam. Relat. Mater.* **2001**, *10*, 281. [CrossRef]
120. Kawano, A.; Ishiwata, H.; Iriyama, S.; Okada, R.; Yamaguchi, T.; Takano, Y.; Kawarada, H. Superconductor-to-insulator transition in boron-doped diamond films grown using chemical vapor deposition. *Phys. Rev. B* **2010**, *82*, 085318. [CrossRef]
121. Volpe, P.-N.; Pernot, J.; Muret, P.; Omnes, F. High hole mobility in boron doped diamond for power device applications. *Appl. Phys. Lett.* **2009**, *94*, 092102. [CrossRef]
122. Pehrsson, P.E.; Merser, T.W. Oxidation of the hydrogenated diamond (100) surface. *Surf. Sci.* **2000**, *460*, 49. [CrossRef]
123. Ri, S.G.; Nebel, C.E.; Takeuchi, D.; Rezek, B.; Tokuda, N.; Yamasaki, S.; Okushi, H. Surface conductive layers on (111) diamonds after oxygen treatments. *Diam. Relat. Mater.* **2006**, *15*, 692. [CrossRef]
124. Maier, F.; Ristein, J.; Ley, L. Electron affinity of plasma-hydrogenated and chemically oxidized diamond (100) surfaces. *Phys. Rev. B* **2001**, *64*, 165411. [CrossRef]
125. Riedel, M.; Ristein, J.; Ley, L. The impact of ozone on the surface conductivity of single crystall diamond. *Diam. Relat. Mater.* **2004**, *13*, 746. [CrossRef]
126. Sakai, T.; Song, K.; Kanazawa, H.; Nakamura, Y.; Umezawa, H.; Tachiki, M.; Kawarada, H. Ozone-treated channel diamond field-effect transitors. *Diam. Relat. Mater.* **2003**, *12*, 1971. [CrossRef]
127. Umezawa, H.; Ikeda, K.; Kumaresan, R.; Tatsumi, N.; Shikata, S. Increase in reverse operation limit by barrier height control of diamond Schottky barrier diode. *IEEE. Electron Devices Lett.* **2009**, *30*, 960. [CrossRef]
128. Ikeda, K.; Umezawa, H.; Ramanujam, K.; Shikata, S. Thermally stable Schottky barrier diode by Ru/diamond. *Appl. Phys. Express* **2009**, *2*, 011202. [CrossRef]
129. Traoré, A.; Muret, P.; Fiori, A.; Eon, D.; Gheeraert, E.; Pernot, J. Zr/oxidized diamond interface for high power Schottky diodes. *Appl. Phys. Lett.* **2014**, *104*, 052105. [CrossRef]
130. Van der Weide, J.; Nemanich, R.J. Influence of interfacial hydrogen and oxygen on the Schottky barrier height of nickel on (111) and (100) diamond surfaces. *Phys. Rev. B* **1994**, *49*, 13629. [CrossRef] [PubMed]

131. Takeuchi, D.; Yamanaka, S.; Watanabe, H.; Okushi, H. Device grade B-doped homoepitaxial diamond thin films. *Phys. Status Solidi A* **2001**, *186*, 269. [CrossRef]
132. Ri, S.-G.; Takeuchi, D.; Tokuda, N.; Okushi, H.; Yamasaki, S. Fermi level pinning-free interface at metals/homoepitaxial diamond (111) firlms after oxidation treatments. *Appl. Phys. Lett.* **2008**, *92*, 112112. [CrossRef]
133. Liao, M.; Koide, Y.; Alvarez, J. Thermally stable visible-blind diamond photodiode using tungsten carbide Schottky contact. *Appl. Phys. Lett.* **2005**, *87*, 022105.
134. Teraji, T.; Koizumi, S.; Koide, Y.; Ito, T. Electric field breakdown of lateral-type Schottky diodes formed on lightly doped homoepitaxial diamond. *Appl. Surf. Sci.* **2008**, *254*, 6273. [CrossRef]
135. Tsugawa, K.; Noda, H.; Hirose, K.; Kawarada, H. Schottky barrier heights, carrier density, and negative electron affinity of hydrogen-terminated diamond. *Phys. Rev. B* **2010**, *81*, 045303. [CrossRef]
136. Umezawa, H.; Nagase, M.; Kato, Y.; Shikata, S. High temperature application of diamond power device. *Diam. Relat. Mater.* **2012**, *24*, 201. [CrossRef]
137. Matsumoto, T. Schottky and ohmic contacts on diamond. In *Power Electronics Device Applications of Diamond Semiconductors*; Woodhead: Cambridge, UK, 2018; p. 241.
138. Craciun, M.; Saby, C.; Muret, P.; Deneuville, A. A 3.4 eV potential barrier height in Schottky diodes on boron-doped diamond thin films. *Diam. Relat. Mater.* **2004**, *13*, 292. [CrossRef]
139. Kato, H.; Yamasaki, S.; Okushi, H. n-type doping of (001)-oriented single-crystalline diamond by phosphorus. *Appl. Phys. Lett.* **2005**, *86*, 222111. [CrossRef]
140. Suzuki, M.; Yoshida, H.; Sakuma, N.; Ono, T.; Sakai, T.; Koizumi, S. Electrical characterization of phosphorous-doped n-type homoepitaxial diamond layers by Schottky barrier diodes. *Appl. Phys. Lett.* **2004**, *84*, 2349. [CrossRef]
141. Suzuki, M.; Koizumi, S.; Katagiri, M.; Ono, T.; Sakuma, N.; Yoshida, H.; Sakai, T. Electrical characteristics on n-type diamond Schottky diodes and metal/diamond interfaces. *Phys. Status Solidi A* **2006**, *203*, 3128. [CrossRef]
142. Kato, H.; Takeuchi, D.; Tokuda, N.; Umezawa, H.; Yamasaki, S.; Okushi, H. Electrical activity of doped phosphorus atoms in (001) n-type diamond. *Phys. Status Solidi A* **2008**, *205*, 2195. [CrossRef]
143. Jingu, Y.; Hirama, K.; Kawarada, H. Ultrashallow TiC source/drain contacts in fiamond MOSFETs formed by hydrogenation-last approach. *IEEE Trans. Electron Devices* **2010**, *57*, 966. [CrossRef]
144. Ohmagari, S.; Matsumoto, T.; Umezawa, H.; Mokuno, Y. Ohmic contact formation to heavily boron-doped p+ diamond prepared by hot-filament chemical vapor deposition. *MRS Adv.* **2016**, *1*, 3489. [CrossRef]
145. Huang, W.; Chow, T.P.; Yang, J.; Butler, J.E. High-voltage diamond vertical Schottky rectifiers. In Proceedings of the 17th International Symposium on Power Semiconductor Devices & Ics, Santa Barbara, CA, USA, 23–26 May 2005; pp. 319–322.
146. Brezeanu, M.; Rashid, S.J.; Amaratunga, G.A.J.; Rupesinghe, N.L.; Butler, T.; Udrea, F.; Brezeanu, G. On-state behaviour of diamond M-i-P structures. In Proceedings of the 2006 International Semiconductor Conference, Sinaia, Romania, 27–29 September 2006; pp. 311–314.
147. Tarelkin, S.; Bormashov, V.; Buga, S.; Volkov, A.; Teteruk, D.; Kornilov, N.; Kuznetsov, M.; Terentiev, S.; Golovanov, A.; Blank, V. Power diamond vertical Schottky barrier diode with 10 A forward current. *Phys. Status Solidi A* **2015**, *212*, 2621. [CrossRef]
148. Volpe, P.N.; Muret, P.; Pernot, J.; Omnes, F.; Teraji, T.; Jomard, F.; Planson, D.; Brosselard, P.; Dheilly, N.; Vergne, B.; et al. High breakdown voltage Schottky diodes synthesized on p-type CVD diamond layer. *Phys. Status Solidi A* **2010**, *207*, 2088. [CrossRef]
149. Umezawa, H.; Saito, T.; Tokuda, N.; Ogura, M.; Ri, S.G.; Yoshikawa, H.; Shikata, S. Leakage current analysis of diamond Schottky barrier diode. *Appl. Phys. Lett.* **2007**, *90*, 073506. [CrossRef]
150. Eon, D.; Traore, A.; Pernot, J.; Gheeraert, E. Recent progress on diamond Schottky diode. In Proceedings of the 28th International Symposium on Power Semiconductor Devices and Ics (ISPSD), Prague, Czech Republic, 12–16 June 2016; p. 55.
151. Eon, D.; Umezawa, H. Schottky diodes and MESFETS. In *Power Electronics Device Applications of Diamond Semiconductors*; Woodhead: Cambridge, UK, 2018; p. 302.
152. Chicot, G.; Eon, D.; Rouger, N. Optimal drift region for diamond power devices. *Diam. Relat. Mater.* **2016**, *69*, 68. [CrossRef]
153. Umezawa, H.; Gima, H.; Driche, K.; Kato, Y.; Yoshitake, T.; Mokuno, Y.; Gheeraert, E. Defect and field-enhancement characterization through electron-beam-induced current analysis. *Appl. Phys. Lett.* **2017**, *110*, 182103. [CrossRef]
154. Kato, Y.; Teraji, T. Characterization methods of defects in single crystalline diamond for power device application. In *Power Electronics Device Applications of Diamond Semiconductors*; Woodhead: Cambridge, UK, 2018; p. 219.
155. Ohmagari, S.; Yamada, H.; Tsubouchi, N.; Umezawa, H.; Chayahara, A.; Tanaka, S.; Mokuno, Y. Large reduction of threading dislocations in diamond by hot-filament chemical vapor deposition accompanying W incorporations. *Appl. Phys. Lett.* **2018**, *113*, 032108. [CrossRef]
156. Koizumi, S.; Watanabe, K.; Hasegawa, M.; Kanda, H. Ultraviolet Emission from a Diamond pn Junction. *Science* **2001**, *292*, 1899. [CrossRef] [PubMed]
157. Horiuchi, K.; Okajima, Y.; Kawamura, A. Diamond Ultraviolet Light-Emitting Device. In *Advanced Materials 2002: Synthesis and Characterization of Diamond, BN and Others Made from Gas Phase, Proceedings of the 9th International Symposium on Advanced Materials (ISAM 2002), Tsukuba, Japan, 3–7 March 2002*; National Institute for Research in Inorganic Materials: Tsukuba, Japan, 2002.
158. Makino, T.; Kato, H.; Ogura, M.; Watanabe, H.; Ri, S.G.; Yamasaki, S.; Okushi, H. Electrical and optical characterizations of (001)-oriented homoepitaxial diamond p–n junction. *Diam. Relat. Mater.* **2006**, *15*, 513. [CrossRef]

159. Makino, T.; Oyama, K.; Kato, H.; Takeuchi, D.; Ogura, M.; Okushi, H.; Yamasaki, S. Diamond electronic devices fabricated using heavily doped hopping p^+ and n^+ layers. *Jpn. J. Appl. Phys.* **2014**, *53*, 05FA12. [CrossRef]
160. Kato, H.; Makino, T.; Ogura, M.; Tokuda, N.; Okushi, H.; Yamasaki, S. Selective Growth of Buried n^+ Diamond on (001) Phosphorus-Doped n-Type Diamond Film. *Appl. Phys. Express* **2009**, *2*, 055502. [CrossRef]
161. Schreyvogel, C.; Temgoua, S.; Giese, C.; Cimalla, V.; Barjon, J.; Nebel, C.E. Fabrication of n-Type Doped V-Shaped Structures on (100) Diamond. *Phys. Status Solidi A* **2021**, *218*, 2000502. [CrossRef]
162. Kato, H.; Makino, T.; Ogura, M.; Takeuchi, D.; Yamasaki, S. Fabrication of bipolar junction transistor on (001)-oriented diamond by utilizing phosphorus-doped n-type diamond base. *Diam. Relat. Mater.* **2013**, *34*, 41. [CrossRef]
163. Suzuki, M. High voltage diamond pin diodes. *Oyo Buturi* **2016**, *85*, 218.
164. Suzuki, M.; Sakai, T.; Makino, T.; Kato, H.; Takeuchi, D.; Ogura, M.; Okushi, H.; Yamasaki, S. Electrical characterization of diamond PiN diodes for high voltage applications. *Phys. Status Solidi A* **2013**, *210*, 2035. [CrossRef]
165. Hiraiwa, A.; Kawarada, H.J. Blocking characteristics of diamond junctions with a punch-through design. *Appl. Phys.* **2015**, *117*, 124503.
166. Makino, T.; Tanimoto, S.; Kato, H.; Tokuda, N.; Ogura, M.; Takeuchi, D.; Oyama, K.; Ohashi, H.; Okushi, H.; Yamasaki, S. Diamond Schottky-pn diode with high forward current density and fast switching operation. *Phys. Status Solidi A* **2009**, *9*, 2086. [CrossRef]
167. Makino, T.; Kato, H.; Tokuda, N.; Ogura, M.; Takeuchi, D.; Oyama, K.; Tanimoto, S.; Okushi, H.; Yamasaki, S. Diamond Schottky-pn diode without trade-off relationship between on-resistance and blocking voltage. *Phys. Status Solidi A* **2010**, *207*, 2105. [CrossRef]
168. Dutta, M.; Koeck, F.A.M.; Hathwar, R.a.; Goodnick, S.M.; Nemanich, R.J.; Chowdhury, S. Demonstration of Diamond-Based Schottky p-i-n Diode with Blocking Voltage >500 V. *IEEE Electron Device Lett.* **2016**, *37*, 1170. [CrossRef]
169. Takeuchi, D.; Koizumi, S.; Makino, T.; Kato, H.; Ogura, M.; Ohashi, H.; Okushi, H.; Yamasaki, S. Negative electron affinity of diamond and its application to high voltage vacuum power switches. *Phys. Status Solidi A* **2013**, *210*, 1961. [CrossRef]
170. Driche, K.; Umezawa, H.; Makino, T.; Ogura, M.; Okumura, H.; Gheeraert, E. Diamond based metal-semiconductor field effect transistor with over 2 kV breakdown voltages. In Proceedings of the EMRS 2018 Fall Meeting, Warsaw, Poland, 17–21 September 2018.
171. Umezawa, H.; Ohmagari, S.; Mokuno, Y.; Kaneko, J.H. Characterization of X-ray radiation hardness of diamond Schottky barrier diode and metal-semiconductor field-effect transistor. In Proceedings of the 2017 29th International Symposium on Power Semiconductor Devices and IC's (ISPSD), Sapporo, Japan, 28 May–1 June 2017; p. 379.
172. Iwasaki, T.; Kato, H.; Makino, T.; Ogura, M.; Takeuchi, D.; Yamasaki, S.; Hatano, M. High-Temperature Bipolar-Mode Operation of Normally-Off Diamond JFET. *IEEE J. Electron Devices Soc.* **2016**, *5*, 95. [CrossRef]
173. Masumoto, T.; Kato, H.; Oyama, K.; Makino, T.; Ogura, M.; Takeuchi, D.; Inokuma, T.; Tokuda, N.; Yamasaki, S. Inversion channel diamond metal-oxide-semiconductor field-effect transistor with normally off characteristics. *Sci. Rep.* **2016**, *6*, 31585. [CrossRef]
174. Masumoto, T.; Kato, H.; Makino, T.; Ogura, M.; Takeuchi, D.; Yamasaki, S.; Inokuma, T.; Tokuda, N. Inversion channel mobility and interface state density of diamond MOSFET using N-type body with various phosphorus concentrations. *Appl. Phys. Lett.* **2019**, *114*, 242101. [CrossRef]
175. Pham, T.T.; Rouger, N.; Masante, C.; Chicot, G.; Udrea, F.; Eon, D.; Gheeraert, E.; Pernot, J. Deep depletion concept for diamond MOSFET. *Appl. Phys. Lett.* **2017**, *111*, 173503. [CrossRef]
176. Pham, T.T.; Pernot, J.; Masante, C.; Eon, D.; Gheeraert, E.; Chicot, G.; Udrea, F.; Rouger, N. 200 V, 4 MV/cm lateral diamond MOSFET. *IEEE Electron. Device Lett.* **2017**, *38*, 1571. [CrossRef]
177. Kawarada, H. Hydrogen-terminated diamond surfaces and interfaces. *Surf. Sci. Rep.* **1996**, *26*, 205. [CrossRef]
178. Nebel, C.E.; Sauerer, C.; Ertl, F.; Stutzmann, M.; Graeff, C.; Bergonzo, P.; Williams, O.; Jackman, R. Hydrogen-induced transport properties of holes in diamond surface layers. *Appl. Phys. Lett.* **2001**, *79*, 4541. [CrossRef]
179. Taniuchi, H.; Umezawa, H.; Arima, T.; Tachiki, M.; Kawarada, H. High-frequency performance of diamond field-effect transistor. *IEEE Electron Device Lett.* **2001**, *EDL-22*, 390. [CrossRef]
180. Ueda, K.; Kasu, M.; Yamauchi, Y.; Makimoto, T.; Schwitters, M.; Twitchen, D.J.; Scarsbrook, G.A.; Coe, S.E. Diamond FET using high-quality polycrystalline diamond with f_T of 45 GHz and f_{max} of 120 GHz. *IEEE Electron Device Lett.* **2006**, *27*, 570. [CrossRef]
181. Matsudaira, H.; Miyamoto, S.; Ishizaka, H.; Umezawa, H.; Kawarada, H. Over 20-GHz cutoff frequency submicrometer-gate diamond MISFETs. *IEEE Electron Device Lett.* **2004**, *25*, 480. [CrossRef]
182. Yu, X.; Zhou, J.; Qi, C.; Cao, Z.; Kong, Y.; Chen, T. A High Frequency Hydrogen-Terminated Diamond MISFET with f_T/f_{max} of 70/80 GHz. *IEEE Electron Device Lett.* **2018**, *39*, 1373. [CrossRef]
183. Moon, J.-S.; Wong, J.; Chuong, D.; Arkun, E.; Morales, D.V.; Chen, P.; Malek, C.; Fanning, D.; Venkatesan, N.; Fay, P. Power Scaling of Graded-Channel GaN HEMTs with Mini-Field-Plate T-gate and 156 GHz f_T. *IEEE Electron Device Lett.* **2021**, *42*, 796. [CrossRef]
184. Husna, K.H.; Nirmal, D. A review of GaN HEMT broadband power amplifiers. *AEU-Int. J. Electron. Commun.* **2020**, *116*, 153040. [CrossRef]
185. Cui, P.; Chen, H.; Xiao, J.Q.; Zeng, Y. High-Performance HZO/InAlN/GaN MIS-HEMT with f_T/f_{max} of 155/250 GHz. *arXiv* **2021**, arXiv:2101.09760.
186. Hirama, K.; Sato, H.; Harada, Y.; Yamamoto, H.; Kasu, M. Diamond Field-Effect Transistors with 1.3 A/mm Drain Current Density by Al_2O_3 Passivation Layer. *Jpn. J. Appl. Phys.* **2012**, *51*, 0901121.

187. Kitabayashi, Y.; Kudo, T.; Tsuboi, H.; Yamada, T.; Xu, D.; Shibata, M.; Matsumura, D.; Hayashi, Y.; Syamsul, M.; Inaba, M.; et al. Normally-Off C–H Diamond MOSFETs with Partial C–O Channel Achieving 2-kV Breakdown Voltage. *IEEE Electron Device Lett.* **2016**, *37*, 209. [CrossRef]
188. Imanishi, S.; Horikawa, K.; Oi, N.; Okubo, S.; Kageura, T.; Hiraiwa, A.; Kawarada, H. 3.8 W/mm RF Power Density for ALD Al_2O_3-Based Two-Dimensional Hole Gas Diamond MOSFET Operating at Saturation Velocity. *IEEE Electron Device Lett.* **2019**, *40*, 279. [CrossRef]
189. Kudara, K.; Imanishi, S.; Hiraiwa, A.; Komatsuzaki, Y.; Yamaguchi, Y.; Kawamura, Y.; Shinjo, S.; Kawarada, H. High Output Power Density of 2DHG Diamond MOSFETs with Thick ALD-Al_2O_3. *IEEE Trans. Electron Devices* **2021**, *68*, 3942. [CrossRef]
190. Cui, A.; Zhang, J.; Ren, Z.; Zhou, H.; Wang, D.; Wu, Y.; Lei, Y.; Zhang, J.; Hao, Y. Microwave power performance analysis of hydrogen terminated diamond MOSFET. *Diam. Relat. Mater.* **2021**, *118*, 108538. [CrossRef]
191. Iwataki, M.; Oi, N.; Amano, S.; Nishimura, J.; Kageura, T.; Inaba, M.; Hiraiwa, A.; Kawarada, H. Over 12,000 A/cm^2 and 3.2 m Ω cm^2 Miniaturized Vertical-Type Two-Dimensional Hole Gas Diamond MOSFET. *IEEE Electron Device Lett.* **2020**, *41*, 111. [CrossRef]
192. Tsunoda, J.; Iwataki, M.; Oi, N.; Morishita, A.; Hiraiwa, A.; Kawarada, H. (111) vertical-type two-dimensional hole gas diamond MOSFETs with hexagonal trench structures. *Carbon* **2021**, *176*, 349e357. [CrossRef]
193. Tokuda, N. Formation of atomically flat diamond surfaces. In *Power Electronics Device Applications of Diamond Semiconductors*; Woodhead: Cambridge, UK, 2018; p. 256.
194. Tokuda, N. Step-edge growth and doping of diamond. In *Diamonnd for Quantum Applications Part I, Semiconductor and Semimetals*; Academic Press Inc.: Cambridge, MA, USA, 2020; Volume 103, p. 57.
195. Kato, H.; Makino, T.; Ogura, M.; Yamasaki, S.; Jpn, J. Maskless selective growth method for p–n junction applications on (001)-oriented diamond. *Appl. Phys.* **2012**, *51*, 090118.
196. Hoshino, Y.; Kato, H.; Makino, T.; Ogura, M.; Iwasaki, T.; Hatano, M.; Yamasaki, S. Electrical properties of lateral p–n junction diodes fabricated by selective growth of n_+ diamond. *Physica Status Solidi* **2012**, *209*, 1761. [CrossRef]
197. Nagai, N.; Nakamura, Y.; Yamada, T.; Tabakoya, T.; Matsumoto, T.; Inokuma, T.; Nebel, C.E.; Makino, T.; Yamasaki, S.; Tokuda, N. Formation of U-shaped diamond trenches with vertical {111} sidewalls by anisotropic etching of diamond (110) surfaces. *Diam. Relat. Mater.* **2020**, *103*, 107713. [CrossRef]

materials

MDPI

Article

GaAs Nanomembranes in the High Electron Mobility Transistor Technology

Dagmar Gregušová *, Edmund Dobročka, Peter Eliáš, Roman Stoklas, Michal Blaho, Ondrej Pohorelec, Štefan Haščík, Michal Kučera and Róbert Kúdela

Institute of Electrical Engineering, Slovak Academy of Sciences, Dúbravská Cesta 9, 841 04 Bratislava, Slovakia; edmund.dobrocka@savba.sk (E.D.); peter.elias@savba.sk (P.E.); roman.stoklas@savba.sk (R.S.); michal.blaho@savba.sk (M.B.); ondrej.pohorelec@savba.sk (O.P.); stefan.hascik@savba.sk (Š.H.); michal.kucera@savba.sk (M.K.); robert.kudela@savba.sk (R.K.)

* Correspondence: dagmar.gregusova@savba.sk

Abstract: A 100 nm MOCVD-grown HEMT AlGaAs/InGaAs/GaAs heterostructure nanomembrane was released from the growth GaAs substrate by ELO using a 300 nm AlAs layer and transferred to sapphire. The heterostructure contained a strained 10 nm 2DEG $In_{0.23}Ga_{0.77}As$ channel with a sheet electron concentration of 3.4×10^{12} cm^{-2} and Hall mobility of 4590 cm^{2}V^{-1}s^{-1}, which was grown close to the center of the heterostructure to suppress a significant bowing of the nanomembrane both during and after separation from the growth substrate. The as-grown heterostructure and transferred nanomembranes were characterized by HRXRD, PL, SEM, and transport measurements using HEMTs. The InGaAs and AlAs layers were laterally strained: $\sim -1.5\%$ and $\sim -0.15\%$. The HRXRD analysis showed the as-grown heterostructure had very good quality and smooth interfaces, and the nanomembrane had its crystalline structure and quality preserved. The PL measurement showed the nanomembrane peak was shifted by 19 meV towards higher energies with respect to that of the as-grown heterostructure. The HEMTs on the nanomembrane exhibited no degradation of the output characteristics, and the input two-terminal measurement confirmed a slightly decreased leakage current.

Keywords: nanomembrane; hybrid integration; GaAs; InGaAs channel; epitaxial lift-off; HEMT; van der Waals

check for updates

Citation: Gregušová, D.; Dobročka, E.; Eliáš, P.; Stoklas, R.; Blaho, M.; Pohorelec, O.; Haščík, Š.; Kučera, M.; Kúdela, R. GaAs Nanomembranes in the High Electron Mobility Transistor Technology. *Materials* **2021**, *14*, 3461. https://doi.org/10.3390/ma14133461

Academic Editor: Fabrizio Roccaforte

Received: 29 May 2021
Accepted: 16 June 2021
Published: 22 June 2021

1. Introduction

Almost three decades have passed since Boeck and Borghs published their paper on heteroepitaxy versus epitaxial lift-off techniques [1], inspired by the seminal papers of Yablonovitch on epitaxial lift-off (ELO) [2,3]. Since then, a large number of scientific publications have been released regarding thin film epitaxy and film transfer to a variety of host substrates [4]. These publications show that ELO and related techniques are an outgrowth of mature epitaxial growth techniques, such as metal-organic chemical vapor deposition (MOCVD) and molecular beam epitaxy (MBE); they inherently need each other for future microelectronics and nanoelectronics progress.

Although many various two-dimensional materials, such as graphene, MoS_2, WS_2, and WSe_2 [5,6], are intensively studied at present, long well-established thin III-V heterostructures with high-mobility two-dimensional electron gases (2DEG) are exceptionally appropriate for ELO [2,7,8] and for emerging break-through three-dimensional hybrid microelectronic and nanoelectronic technologies [9,10].

In the early years, ELO was mainly used for the separation of relatively thick layers (up to 800 nm and more), and supporting organic layers were very often used in the separation process to increase the sacrificial AlAs etching efficiency and to facilitate the transfer of separated layers [11].

Currently, by contrast, very (ultra) thin, high-quality, single crystalline inorganic semiconductor nanomembranes, with thickness that matches the length scales of important quantum physical processes, are released from their growth substrates and transferred to host substrates. This allows for studies of the basic physics of such nanomembranes [12]. The nanomembranes are released using highly selective etching processes, which utilize sacrificial layers of semiconductor materials, mostly (but not only) AlAs [12,13]. The nanomembranes are attached by van der Waals forces to foreign substrates, such as sapphire, GaN, and plastic flexible substrates [14], and they are used to prepare hybrid equivalents of devices that cannot be produced monolithically [15].

We pursue the hybrid integration of MOCVD-grown III-V- and III-N-based devices, in which GaAs-based heterostructure transistors with excellent dc and high-frequency properties are integrated with devices based on nitride semiconductors on host substrates.

The motivation for our work was to prepare flexible devices based on high electron mobility 2DEG III-V structures. Very thin nanomembranes are extremely flexible, but they can exhibit a low mobility because electrons are increasingly scattered due to interactions at both nanomembrane surfaces. Moreover, the technology of such devices is much more difficult than that of thicker ones. However, 2DEG III-V heterostructures as thin as 100 nm allow for the processing of nanomembranes with mobilities that are comparable with those of the monolithic heterostructures. As such, nanomembranes are very flexible; they can stick to various substrates with van der Waals forces.

2. Materials and Methods

This paper reports on the growth, ELO, and transfer of an AlGaAs/InGaAs/GaAs high electron mobility transistor (HEMT) heterostructure to sapphire. The as-grown heterostructure and transferred nanomembrane heterostructure were characterized by high-resolution X-ray diffraction (HRXRD), photoluminescence (PL), and scanning electron microscopy (SEM), and they were used to produce HEMTs, whose performance was studied. The heterostructure contained a strained 10 nm 2DEG $In_{0.23}Ga_{0.77}As$ channel, which makes the ELO process difficult. This work is inspired by previous studies on hybrid transistor technologies, such as [16–20].

The 2DEG AlGaAs/InGaAs/GaAs heterostructure used in this experiment was designed with the intention to test the limits of heterostructure membrane lift-off and transfer. Therefore, the heterostructure was very thin. Its thickness after separation was nominally $\approx$ 100 nm. In addition, it was appropriate for the preparation of good-quality HEMTs, which are used to evaluate the quality of nanomembrane transfer by looking into the device transport properties.

The 2DEG channel of the heterostructure was based on a strained $In_{1\text{-}X}Ga_XAs$ layer, whose thickness and composition were optimized by calculation and experimentation to prevent the generation of misfit dislocations, which is necessary to achieve high electron mobilities and sheet concentrations. Similarly, an appropriate doping level was found to fill the 2DEG channel with electrons and inhibit the formation of a parallel conducting channel with a low electron mobility in the delta doped layer. The strained $In_{1\text{-}X}Ga_XAs$ layer was grown close to the center of the heterostructure to avoid a significant bowing of the heterostructure nanomembrane during and after separation from the growth substrate.

The heterostructure was grown using low-pressure MOVPE in an Aixtron AIX 200 reactor (Aixtron, SE Dornkaulstr. 2, 52134 Herzogenrath, Germeny) at 700 °C. Hydrogen was used as the carrier gas. The precursors were trimethylgallium, trimethylindium, trimethylaluminum, arsine, and diluted silane.

The heterostructure consisted of a GaAs (001) substrate, 300 nm AlAs sacrificial layer, GaAs buffer layer, 10 nm 2DEG $In_{0.23}Ga_{0.77}As$ channel, 4 nm $Al_{0.3}Ga_{0.7}As$ spacer, delta-doped Si layer, 30 nm $Al_{0.3}Ga_{0.7}As$ top layer, and 5 nm GaAs cap layer. The delta-doped Si layer was prepared during a 30 s growth interruption under an overpressure of arsine. The growth rates of AlAs, GaAs, $In_{0.23}Ga_{0.77}As$, and $Al_{0.3}Ga_{0.7}As$ were 0.29, 0.17, 0.2, and 0.23 nm/s, respectively. The heterostructure was designed to have a nominal sheet electron concentration of 2.10^{12} cm^{-2}. The as-grown heterostructure exhibited a sheet electron concentration of 3.4×10^{12} cm^{-2} and Hall mobility of 4590 $cm^2V^{-1}s^{-1}$.

3. Results and Discussion

3.1. Separation

A III-V device structure can be separated from its growth substrate by a technique called epitaxial lift-off. It involves the selective lateral wet etching of a thin sacrificial layer. Once the layer is completely etched away, the III-V structure becomes separated from the underlying layers and growth substrate [21,22]. We tailored this technique to detach our HEMT structures from GaAs substrate. The rate of the heterostructure release was studied with dependence on the composition of an etchant based on hydrofluoric acid (HF). The thickness of the AlAs interlayer was the parameter (three values), and the etching was carried out at room temperature. The results are presented in Figure 1.

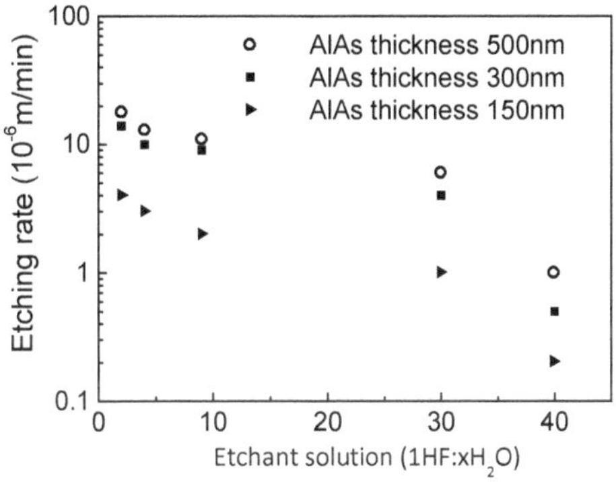

Figure 1. Dependence of the release etching rate of HEMT heterostructure nanomembranes on the composition of an HF-based solution with the thickness of the AlAs sacrificial layer as the parameter.

Thin AlAs layers (<100 nm) were not conducive to the HEMT heterostructure release because the sacrificial etching self-terminated at short distances from the edges of lithographically defined nanomembrane areas (a similar outcome was published previously) [23]. The self-termination of the etching was attributed to a very low wettability of the channel formed as the result of the AlAs etching, as the separation between the underlying GaAs buffer layer and the HEMT heterostructure was too narrow. The situation where the etching of the sacrificial layer (thickness of AlAs is 20 nm) was stopped at a short distance is documented in Figure 2. The wettability problem was alleviated by adding a small amount of a hydrophilic agent to the etching solution.

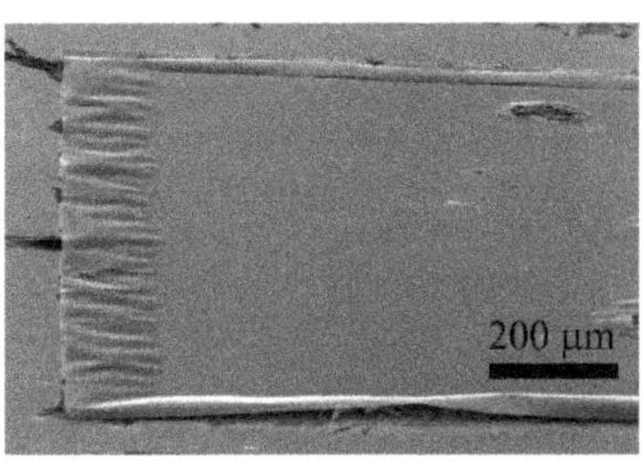

Figure 2. SEM image of a sample where the sacrificial etching of a 20 nm thick AlAs separation layer was self-terminated.

Nanomembranes of the HEMT heterostructure that were processed and studied in this experiment were 1.5×1.5 cm^2 in size. Their release was achieved by the sacrificial etching of a 300 nm thick AlAs layer. To assure a successful transfer of the nanomembranes, it was necessary to have all surfaces of the nanomembranes as smooth and clean as possible. This would guarantee that the heterostructure adhered well to the host substrate by means of van der Waals forces, especially the surface that was exposed by the sacrificial etching at the original interface between the GsAs layer; the heterostructure had to be clean from any remnants of the AlAs sacrificial etching. To achieve this, the AlAs layer had to have very sharp interfaces with the overlying HEMT heterostructure and underlying GaAs buffer layer.

To look at the bottom surface of the nanomembranes, some were flipped over and attached upside down to sapphire host substrate, as is shown in the SEM micrographs of Figure 3. The white arrows in the main picture and the inset point to the edge of an HEMT nanomembrane where the heterostructure was delineated. The main SEM image shows the back side of a nanomembrane surface, which is that of the GaAs layer, after the sacrificial AlAs layer was completely etched away. This surface was originally the interface between the GaAs layer and the AlAl sacrificial layer, and it should ideally be completely flat and smooth. However, the SEM image revealed that it was not fully smooth but contained a shallow wavy structure with prolonged shallow depressions. The relatively large shallow depressions may have originated due to etching at crystal defects. The shallow wavy structure may have stemmed from the intermixing of Al and Ga atoms as the epitaxial growth was switched from AlAs to GaAs. As a result, a very narrow inhomogeneous AlGaAs layer was formed. As the HF-based etchant etched the AlAs layer away during the release process, it reached this inhomogeneous intermixed layer, etched it off (although more slowly compared with the AlAs layer), and left the shallow surface imprint behind. We believe that it did not significantly hamper the bonding of the nanomembranes to the sapphire substrate.

Figure 3. SEM images of an HEMT nanomembrane that was flipped over to expose the surface of its GaAs back side. It contained a shallow wavy structure with prolonged shallow depressions. The arrows mark an edge of the nanomembrane with the HEMT heterostructure delineated.

Figure 4a shows an SEM image of a GaAs nanomembrane attached to the host sapphire substrate. The nanomembrane evidently adhered well to the substrate. Figure 4b shows a detail of the transferred heterostructure nanomembrane with sharp interfaces between the layers of the heterostructure.

Once the nanomembranes became separated from the growth substrate, the etching solution was carefully diluted with water. The membranes were then taken out of the solution and transferred to sapphire substrate. They were dried and thoroughly cleaned in preparation for HEMT processing. Van der Waals forces between the HEMT nanomembranes and sapphire were strong enough for the subsequent processing steps.

Figure 4. (**a**) SEM image of a GaAs nanomembrane attached to the host sapphire substrate by van der Waals forces. (**b**) SEM close-up of the transferred heterostructure nanomembrane with sharp interfaces between the layers.

3.2. XRD Analysis

Successful production of final devices required the growth of high-quality heterostructure layers with smooth interfaces. To verify the structural quality of our layers, HRXRD measurements were performed on the HEMT heterostructure (including the AlAs sacrificial layer) grown on GaAs substrate. To assess the effect of release and transfer procedures on the quality of the heterostructure, the HRXRD measurement was repeated after sticking the structure to a host sapphire substrate. The X-ray analysis was performed in high-resolution mode with a Bruker D8 DISCOVER diffractometer(Bruker AXS Advanced X-ray Solutions GmbH, Östliche Rheinbrückenstraße 49, 76187 Karlsruhe, Germany) equipped with a rotating Cu anode that was operated at 12 kW (40 kV/300 mA). A parabolic Goebel mirror and a Bartels monochromator were inserted into the primary beam. Standard angular $2\theta/\omega$ scans were recorded to determine the composition and thickness of the layers. Linear scans in reciprocal space were performed to evaluate the degree of relaxation of the layers.

The structures were analyzed by measuring high-resolution $2\theta/\omega$ curves of the 004 diffraction. The thickness and composition of the layers were determined by simulation of the theoretical curves using LEPTOS 3.04 software (provided by Bruker Company). Results of the X-ray measurements are exemplified in Figure 5, which compares the measured and simulated $2\theta/\omega$ curves. The measurements show that the diffraction maxima corresponding to the InGaAs and AlAs layers are clearly distinguished, along with the GaAs substrate maximum. The thickness fringes are also well resolved, indicating a high quality of the analyzed heterostructure. To find out whether the layers of the heterostructure were relaxed, linear scans (not shown here) across the asymmetric 224 diffraction were performed in the perpendicular (l scans) and parallel (h scans) directions with respect to the sample surface, respectively. It was found that the l scan of the 224 diffraction perfectly coincided with the l scan of the corresponding symmetric 004 diffraction. The h scans across the 224 diffraction maxima were measured at the values of l coordinates corresponding to the InGaAs and AlAs layers. The maxima of both curves were found to be precisely at $h = 2.000$. These results clearly indicated that the layers did not undergo any relaxation, and the in-plane lattice parameters of the layers were therefore equal to the bulk value of the GaAs lattice parameter. Both InGaAs and AlAs layers were laterally strained, and the estimated values of the strain were $\sim -1.5\%$ and $\sim -0.15\%$, respectively. These values were used as the input parameters in the simulation process of the theoretical $2\theta/\omega$ curves.

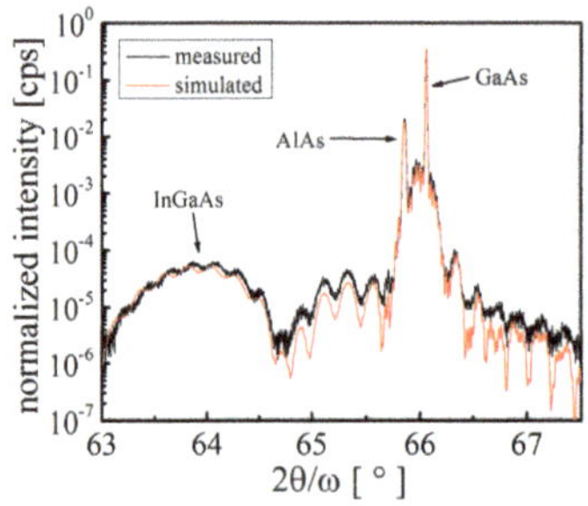

Figure 5. Comparison of the measured and simulated $2\theta/\omega$ curves.

The $2\theta/\omega$ curve of the 004 diffraction of the heterostructure fixed to the host sapphire substrate is shown in Figure 6. Only the maxima of the GaAs buffer layer and of the ~10 nm thick InGaAs layer were visible, the AlAs separation layer was removed at the releasing step, and the corresponding maximum was missing. The most pronounced feature of the diffraction curve is the overall decrease in intensity by about two orders of magnitude with respect to the initial heterostructure on GaAs substrate (Figure 6). This can be partially ascribed to the size of the sticked heterostructure. The dimension of the measured sample in this experiment was only ~1 mm^2. The second effect influencing intensity is the possible loss of the planarity of the heterostructure that cannot be avoided during fine manipulation of the sample. The broadening of the GaAs diffraction maximum and disappearance of the thickness fringes seem to support this reasoning. Generally, X-ray diffraction is extremely sensitive to any distortion of the diffracting area. Hence, a slight bowing of the sample, which has no effect on the functionality of an electronic device, can seriously disturb the interference of X-rays and can result in the observed changes in the diffraction curve.

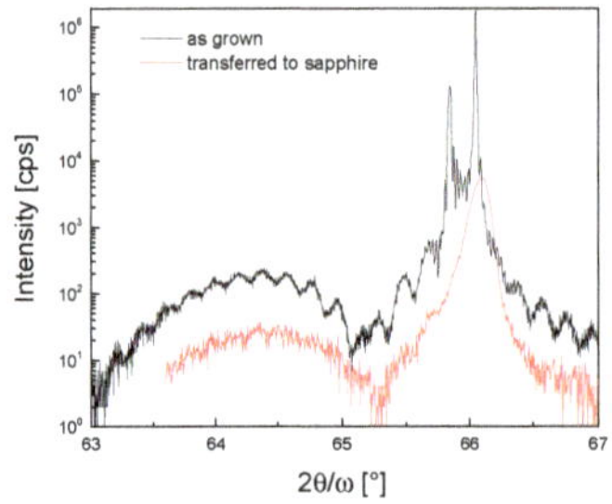

Figure 6. X-ray diffraction of the as-grown AlGaAs/InGaAs/GaAs heterostructure (with the AlAs layer) and of the same heterostructure after being released and fixed to sapphire.

The X-ray measurements revealed that the heterostructures prepared on GaAs substrates were of very good quality with smooth interfaces. The release and transfer procedures did not damage the crystalline structure of the heterostructures and the main features of their diffraction curves were preserved.

3.3. PL Analysis

The quality of the release and transfer of the heterostructure were also investigated using PL measurement at room temperature. A 488 nm line of argon ion laser served for the sample pumping. PL radiation from the sample was filtered via a quarter-meter monochromator (Monochromator DIGIKROM 240, CVI Laser Corporation, Albuquerque, NM, USA.) and detected by a liquid-nitrogen-cooled InGaAs photodiode. The detector signal was amplified and recorded by a standard lock-in technique. At first, we measured the PL signal of the as-grown heterostructure on the parent substrate. The measurement was subsequently repeated when the heterostructure was released and transferred to sapphire. Figure 7 compares both PL signals.

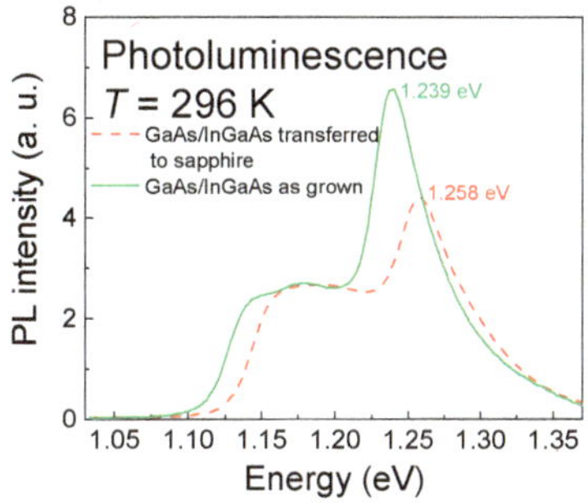

Figure 7. PL spectra of the as-grown heterostructure and released heterostructure nanomembrane.

The figure shows that the PL peaks of the as-grown heterostructure shifted after the heterostructure nanomembrane was released and fixed to the sapphire substrate. Each PL spectrum exhibits two resonances that consist of an asymmetric broadened feature at the lower-energy side and a distinct feature at the higher-energy side. According to [24], the broadened feature at lower energies, designated as 11H, originates in optical recombination from the first conduction subband to the first heavy-hole valence subband. The distinctive feature at higher energies, labeled as 21H, results from the recombination of electrons in the second electron subband and optically excited holes. We observed a shift of 19 meV between the measured peaks. The energy shift can be explained with a re-distribution of charge in the nanomembrane. A part of the charge occupies deep levels at the bottom surface of the nanomembrane. This surface was originally the interface between the bottom GaAs layer and the sacrificial AlAs layer; it was exposed as the nanomembrane was released from the growth substrate. This leads to a change in the electric field intensity in the quantum well of the nanomembrane; hence, the PL transition energy is also changed. As the quantum well (QW) lost some of its charge, the conduction and valence band energies of InGaAs were decreased near the bottom InGaAs/GaAs interface. A lower electron concentration in the QW of the nanomembrane corresponds with the electrical characteristics of the transistors described below.

3.4. HEMT Processing

The HEMTs were processed by standard photolithographic techniques using AZ 5214E positive tone photoresist. At first, MESA etching was performed in an Oxford PlasmaLab apparatus (OxfordInstruments GmbH, Borsigstrasse15a, Wiesbaden, D 65205 Germany) using SiCl4. The etching was stopped on the sapphire substrate. Ohmic contact metallic layers based on Ni/90 nm AuGe /Ni were deposited through a photoresist mask. The NiGe layer was evaporated to achieve a eutectic alloy of 88% Ni and 12% Ge. The layers were lifted off and annealed at 450 °C in N2 for 30 s. The ohmic contacts exhibited a contact resistance of 0.3 Ωmm. The gate electrode (prepared using similar lithographic steps) was composed of non-alloyed 15nmTi/30nmPt/50nmAu layers.

HEMTs were simultaneously processed on the growth GaAs substrate and host sapphire substrate for comparison. Basic dc transistor properties were measured. The output characteristics of both types of HEMT suggest that no degradation in the HEMT properties occurred. The input two-terminal measurement showed that the HEMTs processed on the heterostructure nanomembrane transferred to sapphire exhibited slightly decreased leakage current. The results are in Figure 8.

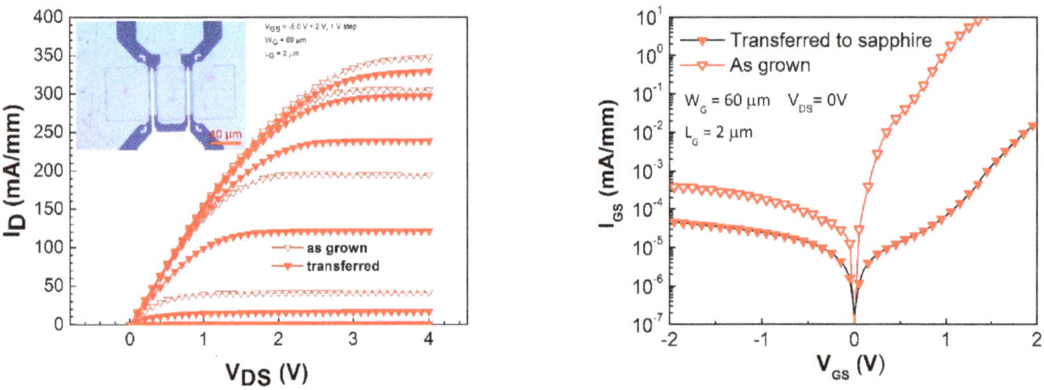

(**a**) (**b**)

Figure 8. (**a**) Output characteristics of transistors processed on an HEMT heterostructure nanomembrane transferred to sapphire compared with those of transistors prepared on the as-grown heterostructure on GaAs substrate. The inset shows an image of an HEMT on the host substrate. (**b**) Comparison of the gate leakage current in two terminal measurement characteristics of HEMTs on sapphire substrate with those of HEMTs on GaAs substrate.

4. Conclusions

A 100 nm 2DEG AlGaAs/InGaAs/GaAs HEMT heterostructure was grown by MOCVD on GaAs substrate, released by HF: H_2O epitaxial lift-off through a 300 nm AlAs layer, transferred, and attached to sapphire by van der Waals forces.

The heterostructure contained a strained 10 nm 2DEG $In_{0.23}Ga_{0.77}As$ channel with a sheet electron concentration of 3.4×10^{12} cm^{-2} and Hall mobility of 4590 cm^{2}V^{-1}s^{-1}. This channel layer was grown close to the center of the heterostructure to suppress a significant bowing of the heterostructure nanomembrane during and after separation from the growth substrate.

The nanomembrane release rate was studied with respect to the thickness of AlAs (varied between 20 to 500 nm) and to the composition of HF: xH_2O (for x varied between 2 to 40) at RT. Thin AlAs layers (<100 nm) did not facilitate the release of the nanomembrane as the etching of AlAs self-terminated.

The as-grown heterostructure and nanomembranes attached to sapphire were characterized by HRXRD, PL, and SEM. The InGaAs and AlAs layers were laterally strained: $\sim -1.5\%$ and $\sim -0.15\%$. HRXRD showed that the as-grown heterostructure had very good quality and smooth interfaces, and the transferred nanomembrane had its crystalline structure and quality preserved. PL showed that the nanomembrane peak was shifted by 19 meV towards higher energies with respect to that of the as-grown heterostructure.

The 2DEG channel transport properties were measured using HEMTs processed on the as-grown heterostructure and on 1.5×1.5 cm^2 nanomembranes attached to sapphire. The nanomembrane HEMTs showed no degradation of the output characteristics and their input two-terminal measurement confirmed a slight decrease in leakage current.

This work demonstrated that the properties of the HEMTs were not adversely affected by the transfer of the HEMT heterostructure to sapphire. This is very promising for our follow-up work aimed at the hybrid integration of III-V- and III-N-based devices, in which transistors with excellent dc and high-frequency properties will be attached to host substrates [25].

Author Contributions: D.G.: conceptualization, original draft writing, investigation, methodology; E.D.: investigation, data curation; P.E.: investigation, data curation, writing, editing review; R.S.: investigation, data curation; M.B.: investigation, data curation; O.P.: investigation, data curation;

Š.H.: investigation, data curation; M.K.: investigation, data curation; R.K.: investigation, supervision. All authors have read and agreed to the published version of the manuscript.

Funding: This work was supported by the Slovak Grant Agency APVV-15-0243, and by the VEGA Grant No, 2/0068/21.

Institutional Review Board Statement: Not applicable.

Informed Consent Statement: Not applicable.

Data Availability Statement: Not applicable.

Conflicts of Interest: The authors declare no conflict of interest.

References

1. De Boeck, J.; Borghs, G. III–V on Si: Heteroepitaxy versus lift-off techniques. *J. Cryst. Growth* **1993**, *127*, 85–92. [CrossRef]
2. Yablonovitch, E.; Hwang, D.M.; Gmitter, T.J.; Florez, L.T.; Harbison, J.P. Van dar Waals bonding of GaAs epitaxial liftoff films onto arbitrary substrates. *Appl. Phys. Lett.* **1990**, *56*, 2419–2421. [CrossRef]
3. Yablonovitch, E.; Gmitter, T.; Harbison, J.P.; Bhat, R. Extreme selectivity in the lift-off of epitaxial GaAs films. *Appl. Phys. Lett.* **1987**, *51*, 2222–2224. [CrossRef]
4. Wu, F.-L.; Ou, S.-L.; Kao, Y.-C.; Chen, C.-L.; Tseng, M.-C.; Lu, F.-C.; Lin, M.-T.; Horng, R.-H. Thin-film vertical-type AlGaInP LEDs fabricated by epitaxial lift-off process via the patterned design of Cu substrate. *Opt. Express* **2015**, *23*, 18156–18165. [CrossRef]
5. Akinwande, D.; Huyghebaert, C.; Wang, C.-H.; Serna, M.I.; Goossens, S.; Li, L.-J.; Wong, H.-S.P.; Koppens, F.H.L. Graphene and two-dimensional materials for silicon technology. *Nature* **2019**, *573*, 507–518. [CrossRef]
6. Geim, A.K.; Grigorieva, I.V. Van der Waals heterostructures. *Nature* **2013**, *499*, 419–425. [CrossRef] [PubMed]
7. Maeda, J.-I.; Sasaki, Y.; Dietz, N.; Shibahara, K.; Yokoyama, S.; Miyazaki, S.; Hirose, M. High-Rate GaAs Epitaxial Lift-Off Technique for Optoelectronic Integrated Circuits. *Jpn. J. Appl. Phys.* **1997**, *36*, 1554–1557. [CrossRef]
8. Yoon, J.; Jo, S.; Chun, I.S.; Jung, I.; Kim, H.-S.; Meitl, M.; Menard, E.; Li, X.; Coleman, J.J.; Paik, U.; et al. GaAs photovoltaics and optoelectronics using releasable multilayer epitaxial assemblies. *Nature* **2010**, *465*, 329–333. [CrossRef] [PubMed]
9. Rogers, J.A.; Someya, T.; Huang, Y. Materials and Mechanics for Stretchable Electronics. *Science* **2010**, *327*, 1603–1607. [CrossRef]
10. Rogers, J.A.; Lagally, M.G.; Nuzzo, R.G. Synthesis, assembly and applications of semiconductor nanomembranes. *Nature* **2011**, *477*, 45–53. [CrossRef]
11. Demeester, P.; Pollentier, I.; De Dobbelaere, P.; Brys, C.; Van Daele, P.; Demeester, P.; Pollentier, I.; De Dobbelaere, P.; Brys, C.; Van Daele, P. Epitaxial lift-off and its applications. *Semicond. Sci. Technol.* **1993**, *8*, 1124–1135. [CrossRef]
12. Fang, H.; Bechtel, H.A.; Plis, E.; Martin, M.C.; Krishna, S.; Yablonovitch, E.; Javey, A. Quantum of optical absorption in two-dimensional semiconductors. *Proc. Natl. Acad. Sci. USA* **2013**, *110*, 11688–11691. [CrossRef]
13. Balocchi, A.; Curran, A.; Graham, T.C.M.; Bradford, C.; Prior, K.A.; Warburton, R.J. Epitaxial lift off of ZnSe-based heterostructures using a II–VI release layer. *Appl. Phys. Lett.* **2005**, *86*, 011915. [CrossRef]
14. Cheng, C.-W.; Shiu, K.-T.; Li, N.; Han, S.-J.; Shi, L.; Sadana, D.K. Epitaxial lift-off process for gallium arsenide substrate reuse and flexible electronics. *Nat. Commun.* **2013**, *4*, 1–7. [CrossRef] [PubMed]
15. Mikulics, M.; Zhang, J.; Serafini, J.; Adam, R.; Grützmacher, D.; Sobolewski, R. Subpicosecond electron-hole recombination time and terahertz-bandwidth photoresponse in freestanding GaAs epitaxial mesoscopic structures. *Appl. Phys. Lett.* **2012**, *101*, 031111. [CrossRef]
16. Shah, D.; Chan, W.; Caneau, C.; Gmitter, T.; Song, J.-I.; Hong, B.; Micelli, P.; De Rosa, F. Epitaxial Lift-off GaAs HEMT's. *IEEE Trans. Electron Devices* **1995**, *42*, 1877–1881. [CrossRef]
17. Ho, C.-L.; Wu, M.-C.; Ho, W.-J.; Liaw, J.-W. GaAs MESFET with low current capability grafted onto quartz substrate. *IEE Proc. Cireuils Devices Syst.* **1999**, *146*, 135–138. [CrossRef]
18. Morf, T.; Biber, C.; Bächtold, W. Effects of epitaxial lift-off on the dc, rf and thermal properties of MESFETs on varrious host materials. *IEEE Trans. Electron Devices* **1998**, *46*, 1407–1413. [CrossRef]
19. Jeong, Y.; Shindo, M.; Akabori, M.; Suzuki, T.-K. Epitaxial Lift-Off of InGaAs/InAlAs Metamorphic High Electron Mobility Heterostructures and Their van der Waals Bonding on AlN Ceramic Substrates. *Appl. Phys. Express* **2008**, *1*, 021201. [CrossRef]
20. Rammohan, K.; Rich, D.H.; MacDougal, M.H.; Dapkus, P.D. Thermal processing of strained InGaAs/GaAs quantum well heterostructures bonded to Si via an epitaxial lift-off technique. *Appl. Phys. Lett.* **1997**, *70*, 1599–1601. [CrossRef]
21. Voncken, M.M.A.J.; Schermer, J.J.; Van Niftrik, A.T.J.; Bauhuls, C.J.; Mulder, P.; Larsen, P.K.; Peters, T.P.J.; De Bruin, B.; Klassen, A. Etching AlAs with HF for epitaxial lift-off application. *J. Electrochem. Soc.* **2004**, *151*, G347–G352. [CrossRef]
22. Wang, J.; Youtsey, C.; McCarthy, R.; Reddy, R.; Allen, N.; Guido, L.; Xie, J.; Beam, E.; Fay, P. Thin-film GaN Schottky diodes formed by epitaxial lift-off. *Appl. Phys. Lett.* **2017**, *110*, 173503. [CrossRef]
23. Wu, F.-L.; Ou, S.-L.; Horng, R.-H.; Kao, Y.-C. Improvement in separation rate of epitaxial lift-off by hydrophilic solvent for GaAs solar cell application. *Sol. Energy Mater. Sol. Cells* **2014**, *122*, 233–240. [CrossRef]

24. Lin, D.Y.; Liang, S.H.; Huang, Y.S.; Tiong, K.K.; Pollak, F.H.; Evans, K.R. Room-temperature photoreflectance and photoluminescence characterization of the AlGaAs/InGaAs/GaAs pseudomorphic high electron mobility transistor structure with varied quantum well compositional profiles. *J. Appl. Phys.* **1999**, *85*, 8235–8241. [CrossRef]
25. Gucmann, F.; Gregušová, D.; Stoklas, R.; Dérer, J.; Kudela, R.; Frohlich, K.; Kordoš, P. InGaAs/GaAs metal-oxide-semiconductor heterostructure field-effect transistors with oxygen-plasma oxide and Al2O3 double -layer insulator. *Appl. Phys. Lett.* **2014**, *105*, 18350. [CrossRef]

materials

MDPI

Article

Induced Superconducting Transition in Ultra-Thin Iron-Selenide Films by a Mg-Coating Process

Zhiqiang Cao [1], Longqing Chen [2,*], Zhenxiang Cheng [3] and Wenbin Qiu [2,*]

1 Instrumental Analysis Center, University of Shanghai for Science and Technology, Shanghai 200093, China; caozhiqiang@usst.edu.cn
2 Key Laboratory of Radiation Physics and Technology of Ministry of Education, Institute of Nuclear Science and Technology, Sichuan University, Chengdu 610064, China
3 Institute for Superconducting and Electronic Materials, Australian Institute for Innovative Materials, University of Wollongong, Squires Way, North Wollongong, NSW 2500, Australia; cheng@uow.edu.au
* Correspondence: chenlongqing@scu.edu.cn (L.C.); wbqiu@scu.edu.cn (W.Q.)

Abstract: Binary Iron selenide (FeSe) thin films have been widely studied for years to unveil the high temperature superconductivity in iron-based superconductors. However, the origin of superconducting transition in this unconventional system is still under debate and worth deep investigations. In the present work, the transition from insulator to superconductor was achieved in non-superconducting FeSe ultrathin films (~8 nm) grown on calcium fluoride substrates via a simple in-situ Mg-coating by a pulsed laser deposition technique. The Mg-coated FeSe film with an optimized amount of Mg exhibited a superconducting critical temperature as 9.7 K and an upper critical field as 30.9 T. Through systematic characterizations on phase identification, carrier transport behavior and high-resolution microstructural features, the revival of superconductivity in FeSe ultrathin films is mostly attributed to the highly crystallized FeSe and extra electron doping received from external Mg-coating process. Although the top few FeSe layers are incorporated with Mg, most FeSe layers are intact and protected by a stable magnesium oxide layer. This work provides a new strategy to induce superconductivity in FeSe films with non-superconducting behavior, which might contribute to a more comprehensive understanding of iron-based superconductivity and the benefit to downstream applications such as magnetic resonance imaging, high-field magnets and electrical cables.

Keywords: thin film; iron-based superconductor; pulsed laser deposition; transmission electron microscopy

Citation: Cao, Z.; Chen, L.; Cheng, Z.; Qiu, W. Induced Superconducting Transition in Ultra-Thin Iron-Selenide Films by a Mg-Coating Process. *Materials* **2021**, *14*, 6383. https://doi.org/10.3390/ma14216383

Academic Editor: Fabrizio Roccaforte

Received: 18 September 2021
Accepted: 22 October 2021
Published: 25 October 2021

1. Introduction

Among the family of electrical materials, high-temperature superconducting material always attracts considerable attention not only considering its huge potential in high-efficiency electric transport and high-field magnets, but also due to the probable complement in condensed matter physics [1]. Iron-based superconductors [2–5] are considered one of the promising candidates that might unveil the mechanism of high-temperature superconductivity (HTS) since dramatic enhancement in critical temperature (T_c) has been repeatedly achieved in binary iron selenide (FeSe) composite through a variety of strategies, including elemental substitution [6–8]/intercalation [9,10], pressurization [11,12], liquid-gate [13], and the substrate-induced heavy electron doping into FeSe unit cells [14–16]. Apparently, the HTS achieved in FeSe system reflects the exotic and unique properties in the unconventional Fe-based superconductors and is worth further investigation.

Although doping processes has been confirmed to dramatically affect the iron-based superconductivity, a comprehensive understanding with regards to the doping dependence of FeSe superconductivity is still being expected. Various approaches have been developed to investigate the doping effect onto FeSe superconductors and boosted T_c was realized in many ways. It is well accepted that the limited electron dopants transferred from STO substrate is not sufficient to drive the HTS in the second and beyond unit-cells [14,17–19].

A challenging issue has been raised which concerns how to supply more charge carriers. Miyata et al. [20] firstly reported the revival of HTS (~48 K) in multi-layer FeSe thin films by an in-situ post-deposition of K element. Extra electrons were introduced to FeSe thin films and the highest level of electron-doping in FeSe thin films was achieved. Shiogai et al. [21] proposed a novel method involving electrochemical etching and electric double-layer transistor to realize the accurate thickness controlling as well as the electron-doping into ultrathin FeSe thin films. Lei et al. [13] published the success of introducing massive electron carriers into FeSe thin flakes by a liquid-gating technique. A T_c^{onset} of 48 K was induced in bulk FeSe and a Lifshitz transition was observed, indicating the dramatic change occurring in the Fermi energy as a result of the applied gate voltage. The deposition effect of Li element onto FeSe thin films was studied by Phan et al. [22] An enhanced T_c of 43 K was reported in a heavily-doped multilayer FeSe film. However, the lattice relaxation found in Li-doped FeSe films is in sharp contrast to the case of K-doping. Apart from the common scenarios of electron-doping, Sun et al. [23] transformed the superconducting phase of FeSe single-crystal from low-T_c to high-T_c by increasing pressure. They pointed out that the pressurization effect brings about inter-band spin fluctuation and reconstructs the Fermi surface of FeSe to a hole-dominated condition which is similar to the high-T_c FeAs superconductors and opposite to the heavily electron-doped FeSe thin films.

Inspired by the works of alkali metal (K, Li) deposition, we previously designed a similar post-deposition process for FeSe superconducting films [24] by coating typical alkaline-earth metal, Mg, which is much less active than K or Li so that a better chemical stability of coated FeSe composite film could be expected. A batch of pulsed laser deposition (PLD) prepared FeSe thin films with practical thickness (60 nm) were post-treated by an in-situ Mg-coating process and received an enhancement of T_c from 10.7 K to 13.4 K owing to the mild electron-doping effect [24]. Recently, a vanish of superconducting transition was found in ultrathin (~8 nm) PLD-prepared FeSe films grown on CaF_2 substrates based on the investigation on thickness effect [25].

Therefore, herein, we propose a possible alternative to revive superconductivity in non-superconducting FeSe ultrathin films by introducing the external Mg-coating process. The pristine FeSe films with an average thickness as 8 nm has been certified to exhibit semiconducting behavior in ρ-T curve without superconducting transition. External electron-doping from the simple Mg-coating process and corresponding enhancement in superconducting performance of FeSe thin films are anticipiated if Mg-coating is proved to be a generic process to induce the superconductivity in FeSe.

2. Materials and Methods

Ultrathin FeSe films (~8 nm) were firstly grown on CaF_2 (100) single crystal substrate (5 mm $\times$ 5 mm) via PLD (Nd: YAG, λ = 355 nm, 10 Hz, output ~2 W) with an average deposition rate as 1.7 nm/min. Right after the deposition of the FeSe layer, in-situ Mg-coating was performed at the same substrate temperature and the amount of Mg-coating was organized by controlling the deposition time as 3, 6.5, 10, 15, and 20 min. The chamber was evacuated to a vacuum state better than 5×10^{-6} Torr and the substrate temperature was fixed at 300 °C. The Mg-coated FeSe films were then cooled down to room temperature. All the samples are denoted in the order of Mg-amount as #UFM0 (pristine ultrathin FeSe), #UFM1 (3 min-Mg), #UFM2 (6.5 min-Mg), #UFM3 (10 min-Mg), #UFM4 (15 min-Mg), and #UFM5 (20 min-Mg), respectively.

Cu Kα X-ray diffraction (XRD, GBC MMA) θ-2θ scans were employed to identify phase structure. The electrical transport measurements including electrical conductivity and Hall measurement were carried out in a physical properties measurement system (PPMS 9 T, Quantum Design, San Diego, CA, USA). In order to investigate microstructural features and elemental information, an aberration-corrected scanning transmission electron microscope (STEM, ARM-200F, JEOL, Akishima, Japan) and an energy dispersive X-ray spectroscopy (EDS, Centurio SDD, JEOL, Akishima, Japan) were utilized. Further high-resolution chemical characterizations were conducted by electron energy loss spectroscopy

(EELS) equipped on STEM at a spectral resolution of 0.05 eV. The electron-transparent lamellae for STEM observation was prepared using an in-situ lift-out technique in a focused-ion-beam (FIB, FEI Helios 600 NanoLab, Hillsboro, OR, USA) system.

3. Results and Discussions

Typical XRD θ-2θ results for the pristine FeSe ultrathin film (#UFM0) and Mg-coated FeSe samples with a gradient of Mg amount (#UFM1, #UFM3, #UFM5) are illustrated in Figure 1a, ranging from 12° to 75°. Similar to our previous work [24], highly (00l) oriented FeSe texture based on PbO structure is observed in #UFM0. Once Mg was introduced, the diffraction peak indexed as FeSe (101) plane emerged in all Mg-coated samples, meanwhile the suppression of FeSe (00l) peaks indicated the reduction in the degree of orientation along *c*-axis upon Mg coating. As none of other phase was detected in all samples apart from those of FeSe and CaF_2 substrate, we deduce that the mechanism in Mg-coated FeSe film might be different from the case of Mg-doped bulks [26] in which MgSe formed. From an enlarged interval near FeSe (001) peak (Figure 1b), a tiny peak shift toward lower angle in #UFM3 compared with pristine #UFM0 is noticed, implying the possible elongation of *c*-axis parameter after Mg-coating process. The phenomenon can be explained based on the interaction between Mg atoms and Fe-vacancies. During the Mg-coating process, Mg enters into FeSe lattice and occupies Fe-vacancies [24,27] in as-grown FeSe thin films, resulting in the elongation of FeSe lattice parameter. The filling effect of Fe-vacancies by Mg and the corresponding lattice elongation might leave impact on the superconducting behavior of Mg-coated FeSe films in consideration of their sensitivity to the subtle variation in lattice parameters [3]. Besides, residual stress is also possible to leave influence on superconducting performance with the compressed stress and tensile stress being beneficial and detrimental to the superconductivity of FeSe films in most cases [8,24,28–30]. Reflected by XRD results, the peak shift of FeSe almost saturates in #UFM3, in which the stress reaches to a maximum level and the highest T_c is obtained.

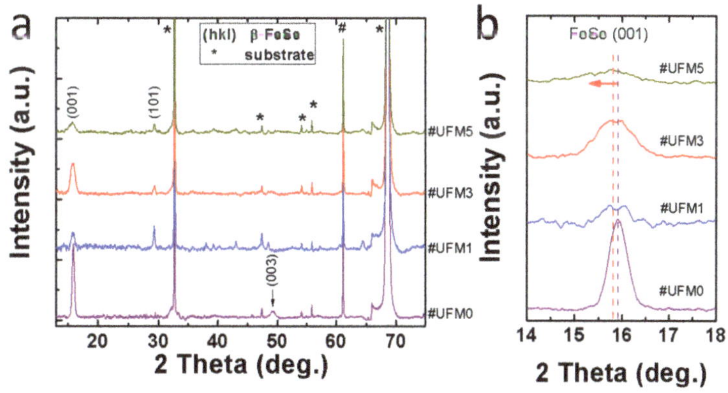

Figure 1. X-ray diffraction (XRD) θ-2θ patterns of #UFM0, #UFM1, #UFM3 and #UFM5. (hkl) signs represent the diffraction peaks of β-FeSe, while asterisk signs are originated from CaF_2 substrate. (**a**) 2θ ranges from 12° to 75°; (**b**) A magnified interval near β-FeSe (001) peak from 14° to 18°. The peak position of #UFM0 and #UFM3 are marked by dashed lines, indicating a clear peak shift of FeSe (001) toward lower 2θ angle in Mg-coated #UFM3.

Figure 2a illustrates the temperature dependence (2–20 K) of normalized electrical resistivity ρ/ρ^{300K} for all six thin film samples in a logarithmic scale. The full range (up to room temperature) results are given in the inset of Figure 2a, demonstrating the metallic ρ-T behavior before T_c for the Mg-coated samples. Intriguingly, distinct superconducting transition is observed in all Mg-coated FeSe thin films, even for #UFM1 with a tiny amount of Mg addition. It is in sharp contrast to the pristine #UFM0 with a semiconducting behavior that is consistent with the case of other ultrathin Fe-chalcogenide thin films prepared by PLD and magnetron sputtering methods [21,25,31]. A similar phenomena of T_c revival in 8 nm non-superconducting FeSe thin films has been already observed in one of previous work in which a FeTe coating layer was introduced [32]. Here, for the case of T_c triggered in Mg-coated FeSe ultra-thin films, we prefer to deduce a possibility of a universal origin, which is attributed to a varied electronic orbital structure and a possibly lower electron occupancy [27] in Fe orbital near Mg/FeSe interfacial region [32]. The temperature dependence of resistivity under external field up to 6 T (parallel to *c*-axis) is shown in Figure 2b with the results of #UFM3 given as an example. The upper critical fields (H_{c2}) of #UFM3 and #UFM5 are calculated depending on linear extrapolated T_c^{mid} (the temperature at which resistivity drops by a half of the value of 14 K), with the H_{c2}-T_c^{mid} results plotted in Figure 2c. The estimated H_{c2} value for #UFM3 is about 30.9 T, exhibiting its good potential for the superconducting applications under high-field environments. The detailed specifications including T_c^{onset}, T_c^{zero}, and ΔT_c are provided in Figure 2d. With increasing the amount of Mg-coating, T_c^{onset} raises from 5.2 K in #UFM1 to 9.7 K in #UFM3. The narrowest transition in #UFM3 (~1.9 K) suggests the good crystallinity of this sample, which can also be deduced based on the largest FeSe (001) diffraction peak intensity compared with that of other Mg-coated samples (as shown in Figure 1). Further increment in Mg addition gives rise to the degradation in T_c^{onset} so that a dome-shaped dependence of T_c^{onset} is obtained depending on the amount of Mg-coating. The transformation of resistivity behavior of FeSe layer from semiconducting to metallic should be related to the effect of extra electron-doping from external Mg-coating, which is believed to be one of the key factors that induce the superconductivity in ultrathin FeSe thin films.

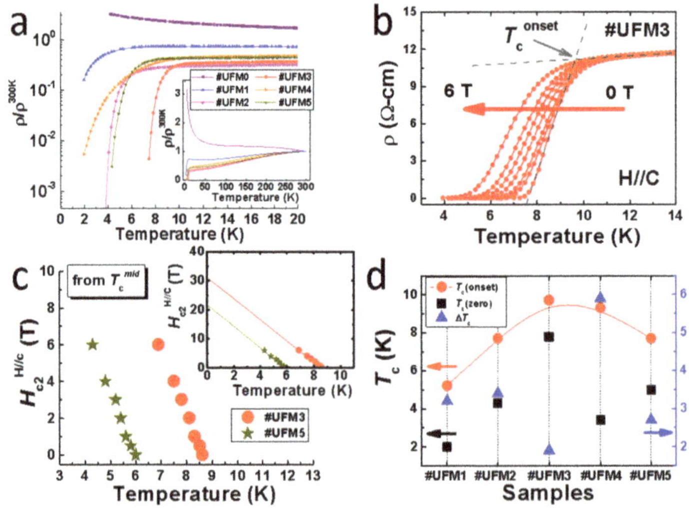

Figure 2. The results of resistivity behaviours of all samples in this work. (**a**) The temperature dependence of normalized resistivity ρ/ρ^{300K} up to 20 K. The *y*-axis is in logarithmic scale. Inset: temperature range up to 300 K. The *y*-axis is in linear scale; (**b**) ρ-T measurements for #UFM3 under external fields up to 6 T (parallel to c-axis); (**c**) Plots of H_{c2} as a function of T_c^{mid} for #UFM3 and #UFM5. Inset: the linear extrapolations to T = 0 K; (**d**) The evolution of T_c^{onset}, T_c^{zero}, and ΔT_c with different amount of Mg-coating. The left *y*-axis refers to the T_c^{onset} and T_c^{zero}, and the right *y*-axis stands for the ΔT_c.

It is well known that charge carrier concentration is one of the most crucial factors that determine the superconductivity in FeSe superconductors [17,20,33,34]. Considering that Mg element belongs to the group of alkaline-earth-metal in the periodic table, abundant electron carriers are supposed to be provided via Mg doping. Here, Hall measurements were performed to investigate the transport behavior of charge carriers. Figure 3a illustrates the temperature dependences of Hall coefficient (R_H), which is defined as $R_H = \rho_{xy}/B$, where ρ_{xy} stands for the Hall transverse resistivity and B is designated field under fixed temperatures ranging from 20 K to 300 K. From room temperature to 150 K, R_H of all samples was almost temperature independent. Below 100 K, positive R_H values were obtained in all the samples, indicating a hole-dominated situation of charge carriers. The correlation between R_H and carrier concentration (n) is deduced by

$$R_H = \frac{E_y}{j_x B} = \frac{V_H t}{IB} = -\frac{1}{ne} \quad (1)$$

(E_y—induced electric field, j—the current density of the carrier electrons, B—magnetic field, V_H—Hall voltage, t—the thickness of the plate, I—the current across the plate length, e—elementary charge). As n is inversely proportional to R_H, the increasing R_H in positive side at lower temperature region represents that n decreases with lowering temperature. The ultrathin #UFM0 without cap layer led to the fluctuation in R_H at low temperature region. Large absolute value of R_H was obtained in #UFM5 with the highest amount of Mg addition, suggesting a severe reduction in carrier concentration. It could be the result of the Mg oxide layer formed on the film surface due to the excessive Mg-coating. The transport property of #UFM5 was hence altered dramatically, which is consistent with the decaying superconducting performance in this sample (shown in Figure 2a). Thus, Hall's measurements offer strong evidence for the degradation of the superconductivity in #UFM5 by revealing the collapse in carrier concentration.

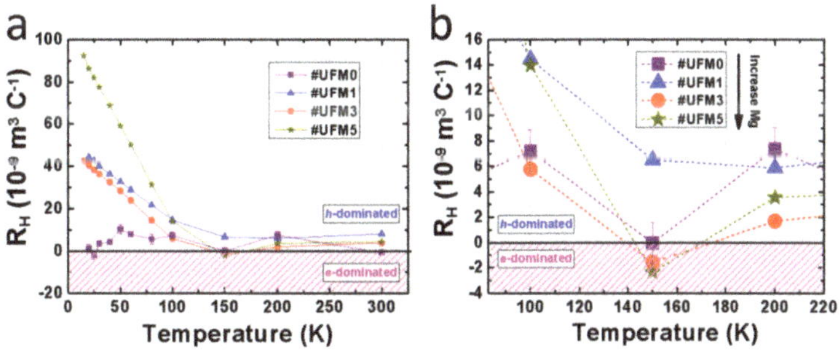

Figure 3. Hall coefficient R_H as a function of temperature for #UFM0, #UFM1, #UFM3 and #UFM5. (**a**) R_H-T plots in a temperature range from 20 K to 300 K; (**b**) A magnified area from 90 K to 220 K, showing the phenomenon of sign reversal in #UFM3 and #UFM5.

The sign reversal of R_H usually suggests a dramatic change occurring to the transport property of charge carriers. An enlarged view of R_H-T ranging from 90 K to 220 K is displayed in Figure 3b. We noted that the sign of R_H of #UFM3 and #UFM5 switched twice during the cooling process near 150 K, while that of the other two samples stayed on the positive side. Similar phenomenon of R_H sign reversal has been reported by Sun et al. [35] in their FeSe and $FeSe_{0.86}S_{0.14}$ single crystals. Obviously, more contribution to the transport performance is provided by electron-type carriers in #UFM3 and #UFM5 near 150 K, evidencing the effective electron-doping into ultrathin FeSe thin films by Mg-coating. However, the external electron-doping was not high enough to completely overturn the hole-dominated condition in pristine FeSe or Mg-coated FeSe composite films. It explains why the T_c in Mg-coated FeSe is not comparable to the case of K-coated multilayer FeSe [20,36,37] or liquid-gating [13] treated FeSe thin flakes which exhibited electron-domination even at low-temperature region. Further increment in Mg addition results in no trace of more electron contribution, reflecting the limited capacity of electron-doping by external Mg-coating. Therefore, we consider the sign reversal of R_H as a signature of higher T_c value in this work induced by electron-doping.

Figure 4 shows the STEM analyses on #UFM3 sample. The cross-sectional region covering the entire FeSe layer is clearly displayed in Figure 4a. About 15 FeSe unit layers (~8 nm in all) are clearly illustrated with the top and bottom layers being adjacent to Mg-Coating and CaSe [25] interlayer, respectively. The FeSe unit layers are fully intact without any obvious disorder or defect, indicating the good quality of the pristine ultrathin FeSe layer prepared on CaF_2 substrate and the absence of large-scale interaction between FeSe layer and Mg-coating. The corresponding fast Fourier transform (FFT) pattern is given in the inset of Figure 4a, demonstrating a typical $P4/nmm$ space group from $[0\bar{1}0]$ zone axis. A linear profile based on contrast fluctuation was conducted throughout the entire FeSe layer with the results of position dependent intensity shown in Figure 4b. Considering of the layered structure of FeSe and the zone axis of $[0\bar{1}0]$, the widths in the position axis between two neighboring valleys indicated the lattice parameter of each FeSe layer along *c*-axis and were individually measured. We noticed that three parts could be divided from the width distribution: the part near Mg-coating, the part of core FeSe layers, and the part near CaSe interlayer. For the case of the bottom FeSe layer with a *c*-axis lattice parameter as 5.75 nm, the expanded value derives from the CaSe interlayer [25] with a lattice parameter (c = 5.92 Å) much larger than that of FeSe. The top few FeSe layers possess the *c*-axis lattice parameters that gradually increase toward Mg-coating region (5.30 nm $\rightarrow$ 5.64 nm $\rightarrow$ 5.86 nm) owing to the filling effect of Fe-vacancies, which is consistent with XRD results.

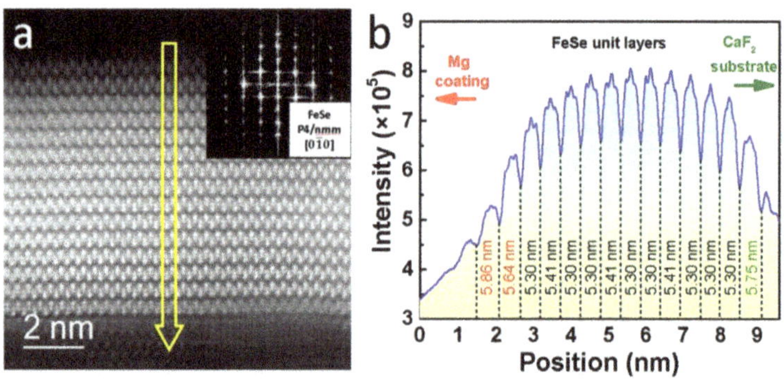

Figure 4. Scanning transmission electron microscopy (STEM) analyses for #UFM3 sample focusing on the cross-sectional region covering the entire FeSe layer. (**a**) Dark-field image. The inset provides a fast Fourier transform (FFT) pattern for FeSe core layers (zone axis $[0\bar{1}0]$); (**b**) A linear profile based on the contrast fluctuation of the arrow indicated in (**a**). The left/right side refers to the region of Mg-coating/CaF_2 substrate. The width values represent the *c*-axis lattice parameters of each FeSe unit layer.

In order to further investigate the properties of Mg-coating and its interaction with FeSe films, detailed STEM characterizations were performed with the results given in Figure 5. The EDS linear-scanning results for Mg, O, Fe, Se and Ca elements are illustrated in Figure 5a. According to the diverse appearance of the net-count curves, the distribution of different elements can be estimated. Apparently, most of the Mg dwelled above FeSe region together with a big amount of O. The inter-diffusion of Mg into FeSe layer degraded rapidly and almost no Mg element was detected deep inside core FeSe layers. Figure 5b is an EELS contour containing the position-dependent energy loss spectra for Mg-*K* and Se-*K* edges. Darker contrast indicates higher intensity of the characteristic edges. EELS results are very sensitive in distinguishing different elements because that the mode of orbital excitation in every element is unique [32,38,39]. In addition to the Mg coating on the top of FeSe layer which was verified by EDS-mappings, a region with the presence of both Mg-*K* and Se-*K* edges was distinguished near the Mg/FeSe interface. Therefore, two conclusions can be made based on the STEM-EDS and EELS results: (1) Most of the Mg dwelled above FeSe layers and almost no Mg element diffused into core FeSe layers. It excluded the possibility of Mg-intercalation in this work. (2) Mg diffused into the top few layers of FeSe film, which was probably responsible for the evolution of the electrical transport and superconducting properties. A bright-field STEM image captured from the zone axis of $[\bar{1}01]$ is displayed in Figure 5c. The FeSe layers dwell between Mg-coating and CaF_2 regions. Above FeSe layer, the region of Mg-coating contains both polycrystalline lattices and amorphous phases. FFT was conducted onto a 10 nm × 10 nm area in Mg-coating region (square in Figure 5c) with the results shown in Figure 5d. The corresponding patterns is well matched with the cubic magnesium oxide (MgO, a = 4.2170 Å, space group: $Fm\bar{3}m$). Thus, we deduce that the excess Mg-coating on the surface of FeSe layer finally got oxidized and transformed into polycrystalline MgO. The high stability of MgO is considered to offer protection to the beneath ultrathin FeSe films.

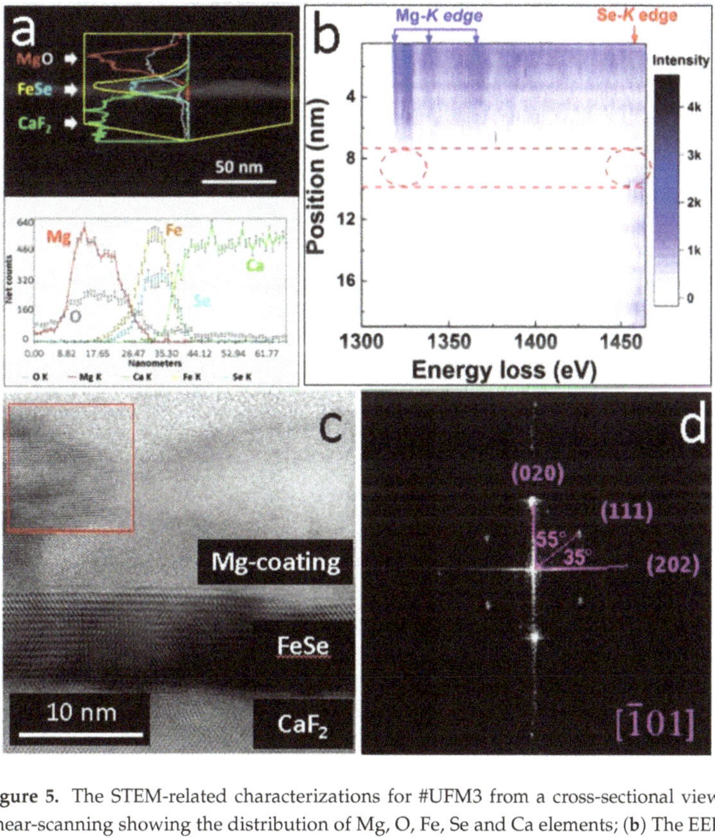

Figure 5. The STEM-related characterizations for #UFM3 from a cross-sectional view. (**a**) EDS linear-scanning showing the distribution of Mg, O, Fe, Se and Ca elements; (**b**) The EELS contour image illustrating the evolution of Mg-*K* and Se-*K* edges. The region in which Mg and Se coexist is highlighted between dashed lines; (**c**) A bright-field STEM image captured from the zone axis of $[\bar{1}01]$; (**d**) The FFT pattern of the polycrystalline region in the square in (**c**).

4. Conclusions

In summary, we carried out a simple Mg-coating process onto non-superconducting FeSe ultrathin (~8 nm) films and obtained a positive result of superconducting transition (T_c = 9.7 K) as well as evidence of extra electron-doping from external Mg-coating. Once Mg-coating was introduced into highly crystallized FeSe (001) grown on CaF_2 substrate, an

FeSe (101) orientation emerged and lattice was slightly elongated along *c*-axis. According to the Hall measurements, a double sign reversal of R_H near 150 K was detected in heavily Mg-coated FeSe films and verified the effective electron-doping which induced the higher T_c values in these samples. High-resolution STEM, EDS, and EELS characterizations revealed that most FeSe layers were intact and defectless while only top few FeSe layers were diffused by Mg, resulting in the expansion in lattice parameter and variation in electrical transport properties. The excessive Mg-coating above FeSe region converted to MgO and provided extra protection to the beneath FeSe ultrathin films. This work is the first attempt to achieve superconducting transition non-superconducting FeSe ultrathin films via external Mg-coating process. Based on the results, a unified mechanism is deduced that the introduction of the elements from the family of alkali/alkaline-earth metals will cause a generic consequence of electron-doping, which is one of the prerequisites that might trigger HTS in FeSe system. The clarification might also profit the downstream applications of iron-based superconducting thin films, including magnetic resonance imaging, high-field magnets, electrical cables, etc.

Author Contributions: Conceptualization, Z.C. (Zhiqiang Cao) and Z.C. (Zhenxiang Cheng); methodology, W.Q.; validation, Z.C. (Zhiqiang Cao) and W.Q.; investigation, Z.C. (Zhenxiang Cheng) and W.Q.; resources, Z.C. (Zhenxiang Cheng); data curation, L.C.; writing—original draft preparation, Z.C. (Zhiqiang Cao) and W.Q.; writing—review and editing, L.C. and Z.C. (Zhenxiang Cheng); supervision, Z.C. (Zhenxiang Cheng); funding acquisition, L.C., Z.C. (Zhenxiang Cheng) and W.Q. All authors have read and agreed to the published version of the manuscript.

Funding: This work was supported by the National Natural Science Foundation of China (No. 12005152, 12005151). Z.C. (Zhenxiang Cheng) acknowledges support through an ARC Discovery Project (DP170104116).

Conflicts of Interest: The authors declare no conflict of interest.

References

1. Zhou, X.; Lee, W.-S.; Imada, M.; Trivedi, N.; Phillips, P.; Kee, H.-Y.; Törmä, P.; Eremets, M. High-temperature superconductivity. *Nat. Rev. Phys.* **2021**, *3*, 462–465. [CrossRef]
2. Kamihara, Y.; Watanabe, T.; Hirano, M.; Hosono, H. Iron-based Layered Superconductor La[$O_{1-x}F_x$]FeAs (x = 0.05 − 0.12) with T_c = 26 K. *J. Am. Chem. Soc.* **2008**, *130*, 3296–3297. [CrossRef] [PubMed]
3. Haindl, S. Thin Film Growth of Fe-Based Superconductors. In *Iron-Based Superconducting Thin Films*; Springer International Publishing: Cham, Switzerland, 2021; pp. 27–148.
4. Cai, Y.; Huang, J.; Miao, T.; Wu, D.; Gao, Q.; Li, C.; Xu, Y.; Jia, J.; Wang, Q.; Huang, Y.; et al. Genuine electronic structure and superconducting gap structure in ($Ba_{0.6}K_{0.4}$)Fe_2As_2 superconductor. *Sci. Bull.* **2021**, *66*, 1839–1848. [CrossRef]
5. Mukasa, K.; Matsuura, K.; Qiu, M.; Saito, M.; Sugimura, Y.; Ishida, K.; Otani, M.; Onishi, Y.; Mizukami, Y.; Hashimoto, K.; et al. High-pressure phase diagrams of $FeSe_{1-x}Te_x$: Correlation between suppressed nematicity and enhanced superconductivity. *Nat. Commun.* **2021**, *12*, 381. [CrossRef]
6. Yeh, K.-W.; Huang, T.-W.; Huang, Y.-L.; Chen, T.-K.; Hsu, F.-C.; Wu, P.M.; Lee, Y.-C.; Chu, Y.-Y.; Chen, C.-L.; Luo, J.-Y.; et al. Tellurium Substitution Effect on Superconductivity of the α-phase Iron Selenide. *EPL* **2008**, *84*, 37002. [CrossRef]
7. Moore, S.A.; Curtis, J.L.; Di Giorgio, C.; Lechner, E.; Abdel-Hafiez, M.; Volkova, O.S.; Vasiliev, A.N.; Chareev, D.A.; Karapetrov, G.; Iavarone, M. Evolution of the superconducting properties in $FeSe_{1-x}S_x$. *Phys. Rev. B* **2015**, *92*, 235113. [CrossRef]
8. Ma, Q.; Lan, F.; Qiu, W.; Ma, Z.; Li, H.; Liu, Y. The fabrication of high-quality superconducting $FeSe_{1-x}S_x$ films via pulsed laser deposition. *Supercond. Sci. Technol.* **2020**, *33*, 035001. [CrossRef]
9. Guo, J.; Jin, S.; Wang, G.; Wang, S.; Zhu, K.; Zhou, T.; He, M.; Chen, X. Superconductivity in the iron selenide $K_xFe_2Se_2$ ($0 \leq x \leq 1.0$). *Phys. Rev. B* **2010**, *82*, 180520. [CrossRef]
10. Wang, A.F.; Ying, J.J.; Yan, Y.J.; Liu, R.H.; Luo, X.G.; Li, Z.Y.; Wang, X.F.; Zhang, M.; Ye, G.J.; Cheng, P.; et al. Superconductivity at 32 K in single-crystalline $Rb_xFe_{2-y}Se_2$. *Phys. Rev. B* **2011**, *83*, 060512. [CrossRef]
11. Mizuguchi, Y.; Tomioka, F.; Tsuda, S.; Yamaguchi, T.; Takano, Y. Superconductivity at 27K in Tetragonal FeSe Under High Pressure. *Appl. Phys. Lett.* **2008**, *93*, 152505. [CrossRef]
12. Bendele, M.; Ichsanow, A.; Pashkevich, Y.; Keller, L.; Strässle, T.; Gusev, A.; Pomjakushina, E.; Conder, K.; Khasanov, R.; Keller, H. Coexistence of superconductivity and magnetism in $FeSe_{1-x}$ under pressure. *Phys. Rev. B* **2012**, *85*, 064517. [CrossRef]
13. Lei, B.; Cui, J.H.; Xiang, Z.J.; Shang, C.; Wang, N.Z.; Ye, G.J.; Luo, X.G.; Wu, T.; Sun, Z.; Chen, X.H. Evolution of High-Temperature Superconductivity from a Low-T_c Phase Tuned by Carrier Concentration in FeSe Thin Flakes. *Phys. Rev. Lett.* **2016**, *116*, 077002. [CrossRef]

14. Wang, Q.; Li, Z.; Zhang, W.; Zhang, Z.; Zhang, J.; Li, W.; Ding, H.; Ou, Y.; Deng, P.; Chang, K.; et al. Interface-Induced High-Temperature Superconductivity in Single Unit-Cell FeSe Films on $SrTiO_3$. *Chin. Phys. Lett.* **2012**, *29*, 037402. [CrossRef]
15. Wang, Z.F.; Zhang, H.; Liu, D.; Liu, C.; Tang, C.; Song, C.; Zhong, Y.; Peng, J.; Li, F.; Nie, C.; et al. Topological Edge States in a High-temperature Superconductor FeSe/$SrTiO_3$(001) Film. *Nat. Mater.* **2016**, *15*, 968–973. [CrossRef] [PubMed]
16. Ge, J.; Liu, Z.; Liu, C.; Gao, C.; Qian, D.; Xue, Q.-K.; Liu, Y.; Jia, J. Superconductivity Above 100 K in Single-layer FeSe Films on Doped $SrTiO_3$. *Nat. Mater.* **2015**, *14*, 285–289. [CrossRef]
17. He, S.; He, J.; Zhang, W.; Zhao, L.; Liu, D.; Liu, X.; Mou, D.; Ou, Y.-B.; Wang, Q.-Y.; Li, Z.; et al. Phase diagram and electronic indication of high-temperature superconductivity at 65 K in single-layer FeSe films. *Nat. Mater.* **2013**, *12*, 605–610. [CrossRef] [PubMed]
18. Shen, B.; Feng, Z.-P.; Huang, J.-W.; Hu, Y.; Gao, Q.; Li, C.; Yu, X.; Liu, G.-D.; Yu, L.; Zhao, L.; et al. Electronic structure and nematic phase transition in superconducting multiple-layer FeSe films grown by pulsed laser deposition method. *Chin. Phys. B* **2017**, *26*, 077402. [CrossRef]
19. Tan, S.; Zhang, Y.; Xia, M.; Ye, Z.; Chen, F.; Xie, X.; Peng, R.; Xu, D.; Fan, Q.; Xu, H.; et al. Interface-induced Superconductivity and Strain-dependent Spin Density Waves in FeSe/$SrTiO_3$ Thin Films. *Nat. Mater.* **2013**, *12*, 634–640. [CrossRef] [PubMed]
20. Miyata, Y.; Nakayama, K.; Sugawara, K.; Sato, T.; Takahashi, T. High-temperature Superconductivity in Potassium-coated Multilayer FeSe Thin Films. *Nat. Mater.* **2015**, *14*, 775–779. [CrossRef]
21. Shiogai, J.; Ito, Y.; Mitsuhashi, T.; Nojima, T.; Tsukazaki, A. Electric-field-induced superconductivity in electrochemically etched ultrathin FeSe films on $SrTiO_3$ and MgO. *Nat. Phys.* **2016**, *12*, 42–46. [CrossRef]
22. Phan, G.N.; Nakayama, K.; Kanayama, S.; Kuno, M.; Sugawara, K.; Sato, T.; Takahashi, T. High-Temperature Superconductivity and Lattice Relaxation in Lithium-Deposited FeSe on $SrTiO_3$. *J. Phys. Soc. Jpn.* **2017**, *86*, 033706. [CrossRef]
23. Sun, J.P.; Ye, G.Z.; Shahi, P.; Yan, J.Q.; Matsuura, K.; Kontani, H.; Zhang, G.M.; Zhou, Q.; Sales, B.C.; Shibauchi, T.; et al. High-T_c Superconductivity in FeSe at High Pressure: Dominant Hole Carriers and Enhanced Spin Fluctuations. *Phys. Rev. Lett.* **2017**, *118*, 147004. [CrossRef] [PubMed]
24. Qiu, W.; Ma, Z.; Liu, Y.; Hossain, M.S.A.; Wang, X.; Cai, C.; Dou, S.X. Tuning Superconductivity in FeSe Thin Films via Magnesium Doping. *ACS Appl. Mater. Interfaces* **2016**, *8*, 7891–7896. [CrossRef]
25. Qiu, W.; Ma, Z.; Patel, D.; Sang, L.; Cai, C.; Shahriar Al Hossain, M.; Cheng, Z.; Wang, X.; Dou, S.X. The Interface Structure of FeSe Thin Film on CaF_2 Substrate and Its Influence on the Superconducting Performance. *ACS Appl. Mater. Interfaces* **2017**, *9*, 37446–37453. [CrossRef] [PubMed]
26. Lan, F.; Ma, Z.; Liu, Y.; Chen, N.; Cai, Q.; Li, H.; Barua, S.; Patel, D.; Hossain, M.S.A.; Kim, J.H.; et al. The formation of nano-layered grains and their enhanced superconducting transition temperature in Mg-doped $FeSe_{0.9}$ bulks. *Sci. Rep.* **2014**, *4*, 6481. [CrossRef]
27. Chen, T.K.; Chang, C.C.; Chang, H.H.; Fang, A.H.; Wang, C.H.; Chao, W.H.; Tseng, C.M.; Lee, Y.C.; Wu, Y.R.; Wen, M.H.; et al. Fe-vacancy order and superconductivity in tetragonal beta-Fe_{1-x}Se. *Proc. Natl. Acad. Sci. USA* **2014**, *111*, 63–68. [CrossRef]
28. Nabeshima, F.; Kawai, M.; Ishikawa, T.; Shikama, N.; Maeda, A. Systematic study on transport properties of FeSe thin films with various degrees of strain. *Jpn. J. Appl. Phys.* **2018**, *57*, 120314. [CrossRef]
29. Phan, G.N.; Nakayama, K.; Sugawara, K.; Sato, T.; Urata, T.; Tanabe, Y.; Tanigaki, K.; Nabeshima, F.; Imai, Y.; Maeda, A.; et al. Effects of strain on the electronic structure, superconductivity, and nematicity in FeSe studied by angle-resolved photoemission spectroscopy. *Phys. Rev. B* **2017**, *95*, 224507. [CrossRef]
30. Maeda, A.; Nabeshima, F.; Takahashi, H.; Okada, T.; Imai, Y.; Tsukada, I.; Hanawa, M.; Komiya, S.; Ichinose, A. Synthesis, characterization, Hall effect and THz conductivity of epitaxial thin films of Fe chalcogenide superconductors. *Appl. Surf. Sci.* **2014**, *312*, 43–49. [CrossRef]
31. Schneider, R.; Zaitsev, A.G.; Fuchs, D.; Löhneysen, H.V. Superconductor-insulator quantum phase transition in disordered FeSe thin films. *Phys. Rev. Lett.* **2012**, *108*, 257003. [CrossRef] [PubMed]
32. Qiu, W.; Ma, Q.; Ma, Z.; Tang, J.; Sang, L.; Cai, C.; Hossain, M.S.A.; Cheng, Z.; Wang, X.; Liu, Y.; et al. Enhanced superconductivity induced by several-unit-cells diffusion in an FeTe/FeSe bilayer heterostructure. *Phys. Rev. B* **2019**, *99*, 064502. [CrossRef]
33. Wang, Q.Y.; Zhang, W.H.; Zhang, Z.C.; Sun, Y.; Xing, Y.; Wang, Y.Y.; Wang, L.L.; Ma, X.C.; Xue, Q.K.; Wang, J. Thickness dependence of superconductivity and superconductor–insulator transition in ultrathin FeSe films on $SrTiO_3$ (001) substrate. *2D Mater.* **2015**, *2*, 044012. [CrossRef]
34. Zhang, W.; Li, Z.; Li, F.; Zhang, H.; Peng, J.; Tang, C.; Wang, Q.; He, K.; Chen, X.; Wang, L.; et al. Interface Charge Doping Effects on Superconductivity of Single-unit-cell FeSe Films on $SrTiO_3$ Substrates. *Phys. Rev. B* **2014**, *89*, 060506. [CrossRef]
35. Sun, Y.; Pyon, S.; Tamegai, T. Electron carriers with possible Dirac-cone-like dispersion in $FeSe_{1-x}S_x$ (x = 0 and 0.14) single crystals triggered by structural transition. *Phys. Rev. B* **2016**, *93*, 104502. [CrossRef]
36. Song, C.; Zhang, H.; Zhong, Y.; Hu, X.; Ji, S.; Wang, L.; He, K.; Ma, X.; Xue, Q.-K. Observation of Double-Dome Superconductivity in Potassium-Doped FeSe Thin Films. *Phys. Rev. Lett.* **2016**, *116*, 157001. [CrossRef]
37. Tang, C.; Liu, C.; Zhou, G.; Li, F.; Ding, H.; Li, Z.; Zhang, D.; Li, Z.; Song, C.; Ji, S.; et al. Interface-enhanced Electron-phonon Coupling and High-temperature Superconductivity in Potassium-coated Ultrathin FeSe Films on $SrTiO_3$. *Phys. Rev. B* **2016**, *93*, 020507. [CrossRef]
38. Egerton, R.F. *Electron Energy-Loss Spectroscopy in the Electron Microscope*, 3rd ed.; Springer: Boston, MA, USA, 2011; pp. 433–491.
39. Taftø, J.; Krivanek, O.L. Site-Specific Valence Determination by Electron Energy-Loss Spectroscopy. *Phys. Rev. Lett.* **1982**, *48*, 560–563. [CrossRef]

materials

MDPI

Review

Intrinsic Point Defects in Silica for Fiber Optics Applications

Giuseppe Mattia Lo Piccolo [1,2,*], Marco Cannas [1,*] and Simonpietro Agnello [1,3]

[1] Dipartimento di Fisica e Chimica, Università Degli Studi di Palermo, Via Archirafi 36, 90123 Palermo, Italy; simonpietro.agnello@unipa.it
[2] Dipartimento di Fisica e Astronomia, Università Degli Studi di Catania, Via Santa Sofia 64, 95123 Catania, Italy
[3] ATeN Center, Università Degli Studi di Palermo, Viale delle Scienze Ed. 18, 90128 Palermo, Italy
* Correspondence: giuseppemattia.lopiccolo@unipa.it (G.M.L.P.); marco.cannas@unipa.it (M.C.)

Abstract: Due to its unique properties, amorphous silicon dioxide (a-SiO_2) or silica is a key material in many technological fields, such as high-power laser systems, telecommunications, and fiber optics. In recent years, major efforts have been made in the development of highly transparent glasses, able to resist ionizing and non-ionizing radiation. However the widespread application of many silica-based technologies, particularly silica optical fibers, is still limited by the radiation-induced formation of point defects, which decrease their durability and transmission efficiency. Although this aspect has been widely investigated, the optical properties of certain defects and the correlation between their formation dynamics and the structure of the pristine glass remains an open issue. For this reason, it is of paramount importance to gain a deeper understanding of the structure–reactivity relationship in a-SiO_2 for the prediction of the optical properties of a glass based on its manufacturing parameters, and the realization of more efficient devices. To this end, we here report on the state of the most important intrinsic point defects in pure silica, with a particular emphasis on their main spectroscopic features, their atomic structure, and the effects of their presence on the transmission properties of optical fibers.

Keywords: silica point defects; optical fibers; radiation effects

Citation: Lo Piccolo, G.M.; Cannas, M.; Agnello, S. Intrinsic Point Defects in Silica for Fiber Optics Applications. *Materials* **2021**, *14*, 7682. https://doi.org/10.3390/ma14247682

Academic Editor: Alexander N. Obraztsov

Received: 24 October 2021
Accepted: 8 December 2021
Published: 13 December 2021

1. Introduction

The integration of silica glasses in optical and electronic devices are, at present, limited by the effects of high-energy radiation on the transmission and reflection properties of the material. As reported in a large number of studies [1–11], the irradiation of silica with either photons or energetic particles (neutrons, electrons, protons, heavy ions) activates a wide range of damage processes that result in the formation of point defects. These localized irregularities of the network are characterized by one or more energy levels lying in the band gap of the dielectric and are responsible for the shift in the optical absorption edge of the glass to lower energies. The phenomenon is known as photodarkening (or solarization) and is even more pronounced in silica optical fibers (OFs) which, due to their strained glass structure, exhibit a higher sensitivity to radiation as compared to the respective preforms [12–22].

In this review, we analyze the main intrinsic point defects of silica absorbing in the ultraviolet (UV) and visible (Vis) spectral domains and investigate their origin and formation pathways. In the discussion, we adopted a new structure–reactivity approach to understand the difference in the likelihood of defect formation observed in bulk silica and optical fiber silica. In the last part of the review, we explore the consequences that defects have on the transmission properties of optical fibers and how their formation can be prevented. This aspect is particularly important for the development of applications such as fiber laser systems, daylighting systems, and new-generation photovoltaic devices, in which a high transparency in the UV–Vis range is required to guarantee an optimal performance.

2. The Geometrical Properties of Silica

SiO_2 is the chemical formula of a group of minerals constituting about 95% of the Earth's rocks and soils [23,24]. The most stable polymorph is α-quartz, which has a trigonal crystal structure and exists in both a right- and left-handed form. As shown in Figure 1, when α-quartz is heated to 573 °C at atmospheric pressure, it transforms into β-quartz and assumes a hexagonal crystal structure [25,26]. By further increasing the temperature, β-quartz turns into HP-tridymite (870 °C) and then into β-cristobalite (1470 °C). At 1713 °C, solid β-cristobalite melts into fused silica, which is a homogeneous and isotropic material characterized by a highly randomized structure. Since phase transitions are reversible, a slow cooling of the melt leads back to the crystalline form. However, if the temperature drop is rapid, the system freezes in a metastable state and eventually relaxes to silica. This non-crystalline form of SiO_2 is characterized by a disordered network lacking long-range order and by the presence of low-density regions.

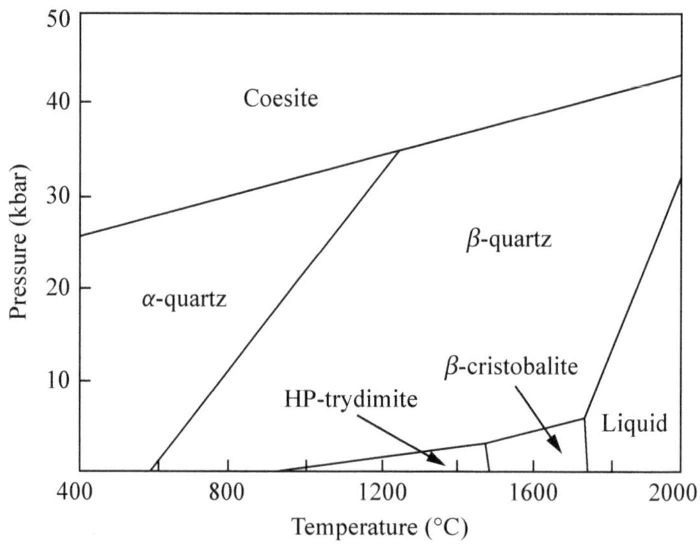

Figure 1. Phase diagram of SiO_2 showing its main crystalline forms as well as the liquid (melted) phase.

The building block of all SiO_2 polymorphs is the $[SiO_4]^0$ tetrahedron (Figure 2), that is, the three-dimensional unit formed by a silicon atom and four oxygen atoms bonded to it. Each O atom bridges two Si atoms and serves as a connection between adjacent tetrahedral units. The Si–O bond has a mixed covalent/ionic character and is described by a wavefunction with both $\sigma[\mathrm{Si}(sp^3)\text{–}\mathrm{O}(2p)]$ and $\pi[\mathrm{Si}(3d)\text{–}\mathrm{O}(2p)]$ contributions [27,28]. This determines the shortening of the Si–O bond length (1.605 Å), the increase in the Si–O–Si angle, and the variability in the intertetrahedral angles [29]. On the other hand, the O–Si–O bond angle (ϕ) is 109.47°, indicating that the $[SiO_4]^0$ unit has a perfect tetrahedral geometry [30,31].

The relative position of two corner-sharing tetrahedra is determined by the Si–O–Si bond angle (θ) and the Si–O–Si–O and O–Si–O–Si torsion angles. Many experimental works have attempted to estimate θ, but the results greatly vary from study to study. The reason for this uncertainty is that the angle is correlated with the Si–O and Si–Si distances and its value cannot directly be determined from experimental data. In fact, the interference function obtained by X-ray and neutron diffraction techniques only provides the distribution of interatomic distances and the estimation of bond and torsion angles requires the mathematical modeling of the total correlation function [32,33]. Similarly, in solid-state nuclear magnetic resonance (NMR) spectra, the dependence of the ^{29}Si chemical shift and the ^{17}O quadrupolar coupling on θ requires an accurate correlation function in order to extract the bond angle distribution [34–36]. In either case, the structural modeling is generally underconstrained and the reliability of the results depends on the goodness of the starting assumptions [37]. Considering the data reported in the literature and the outcomes of their own simulations, Yuan and Cormack proposed a distribution of Si–O–Si angles with a mean value of 147° and a standard deviation of 10–13° [38]. This result was confirmed by Carpentier et al., who derived a mean value of $147° \pm 11°$ by combining molecular dynamics with first-principles calculations of NMR parameters [33]. More recently, Malfait et al. analyzed a large set of data and located the most probable Si–O–Si angle at 149° with a full width at half maximum (FWHM) of 16° [39]. However, this value only corresponds to the peak of the mean (i.e., averaged over the four corners of each SiO_4 tetrahedron) Si–O–Si angle distribution derived from different NMR spectra. To calculate the individual Si–O–Si angle distribution, one should first know the statistical distribution of θ at the corners of the tetrahedra. Since a certain degree of correlation among these angles is inevitable, the individual Si–O–Si bond angles are not statistically independent and their distribution cannot be derived from experimental measurements.

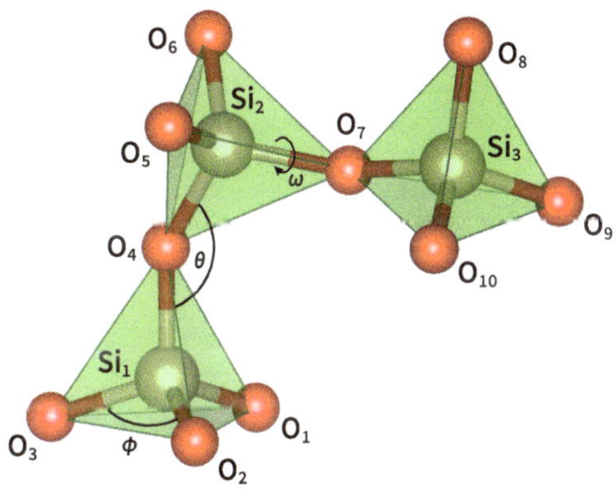

Figure 2. Parameters defining the topology of silica. The angle ϕ corresponds to the O–Si–O bond angle, θ corresponds to the Si–O–Si bond angle, while ω is the dihedral angle between the planes formed by atoms O–Si–O and Si–O–Si.

Since the publication of Wright's paper [40], the three-dimensional organization of silica tetrahedra has been discussed in terms of two distinct torsion angles. Using the notation given in Figure 2, the angle ω_1 is defined as the dihedral angle between O_1 (or, equivalently, O_2 or O_3), Si_1, O_4, Si_2, while the angle ω_2 is defined by the atoms Si_1, O_4, Si_2, O_5 (or, equivalently, O_6 or O_7). However, from a geometrical perspective, the existence of two independent torsion angles is only justified if we define a conventional order to report the structure of silica (as for the primary structure of proteins). Since it is not possible to assign a univocal direction of rotation to silica rings, the two torsion angles are equivalent and carry the same information. For this reason it is sufficient to consider only one of the two torsion angles which we will call ω. From the analysis reported in the literature [38,41], it is found that ω varies as a function of the Si–O–Si bond angle and that, for θ in the 140–160° range, it has three maxima of around 60°, 180°, and 300°. These values correspond to the three staggered conformations of the $O_3Si–O–SiO_3$ moiety, as viewed along the Si–Si axis, and are such that next-nearest-neighbor oxygen atoms are at a maximum distance.

3. The Topology of Silica

The analysis of the network topology of silica was first addressed by King, who proposed the shortest-path criterion to identify closed paths of alternating Si–O bonds [42].

According to this model, a ring is defined as the shortest closed path connecting two oxygen atoms bonded to the same Si atom. If the path is formed by n bonds (or, equivalently, by $2n$ atoms) the ring is called an n-membered ring (n-MR) [43,44]. By applying King's algorithm to α-quartz and β-cristobalite, both polymorphs are composed of regular six-membered rings arranged into different three-dimensional structures [45]. The former network is made up of interconnected hexagonal and trigonal helices of SiO_4 tetrahedra, whereas the latter consists of stacked layers of tetrahedra alternately pointing up and down [46]. In the case of amorphous silica, the flexibility of the Si–O–Si and Si–O–Si–O angles determines a high degree of structural disorder and this, in turn, leads to a quasi-random orientation of the tetrahedral units and a much broader ring-size distribution [47].

Many theoretical works have been dedicated to the statistical analysis of computer-modeled a-SiO_2 structures. In Figure 3, we report the ring-size distribution of three models obtained by different computational methods. The first dataset was calculated by Rino et al. from a structural model obtained using classical molecular dynamics [31]. The second set comes from first-principle molecular dynamics calculations [48], while the last one is from a model simulated using a combination of classical and first-principles molecular dynamics [49]. The ring statistics of the three models shows that five- and six-membered rings are always the most frequent, followed by seven-membered rings. Three- and four-membered rings, however, have different populations and their concentration depends on the simulation conditions and thermal history of the sample. The prevalence of 6-MRs can be explained by considering the volume–temperature diagram in Figure 4. In fact, when melted silica is cooled at a very slow rate, the atoms in the melt have enough time to rearrange and form crystalline β-cristobalite. Conversely, if the cooling is fast (quenching), the liquid cannot equilibrate with the forming solid and the result is the formation of amorphous silica [50]. Despite their different properties, the two polymorphs ideally derive from the same silica melt and their short- and medium-range topology shows a prevalence of six-membered rings.

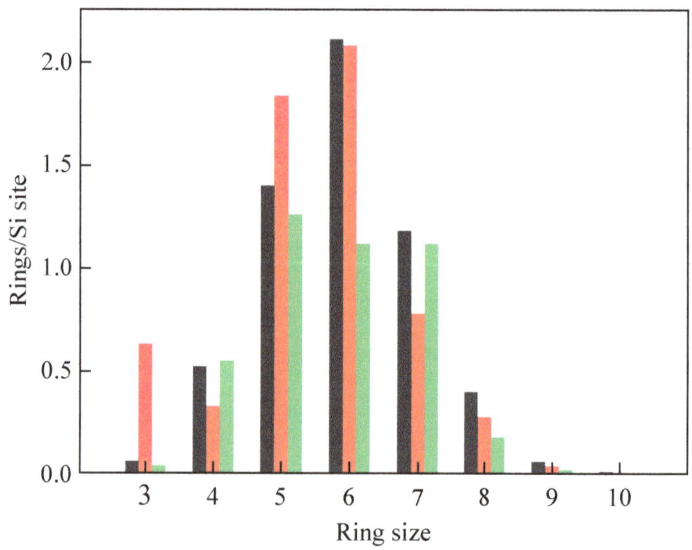

Figure 3. Shortest-path ring statistics of the silica models reported by Rino et al. (grey) [31], Pasquarello and Car (red) [48], and Giacomazzi et al. (green) [49].

The presence of small-sized rings in vitreous silica is strongly correlated with the kinetics of the relaxation process. In fact, when the temperature of the melt is lowered avoiding crystallization, the liquid enters the supercooled phase and its structure continuously rearranges to follow the system temperature. As a consequence, the first-order thermodynamic properties of the liquid (e.g., volume and enthalpy) decrease without any abrupt changes and the viscosity increases accordingly. When the viscosity becomes too high, the atomic motion slows down to the extent that the atoms cannot rearrange themselves into the volume characteristic of that temperature and pressure. At this point, the enthalpy begins to deviate from the equilibrium line and starts to follow a curve of gradually decreasing slope. Eventually, the structure of the system becomes fixed and the supercooled liquid solidifies into silica [51]. The range of temperatures over which the transition occurs is called the glass transformation range and the temperature at which the structure of the supercooled liquid is *frozen-in* in the solid state is the fictive temperature (T_f). Since the departure of the enthalpy from the equilibrium curve is determined by the viscosity of the liquid, i.e., by kinetic factors, a slower cooling rate allows the enthalpy to follow the equilibrium curve to a lower temperature. In this case, the fictive temperature shifts to lower values and the resulting glass has a different atomic arrangement than a more rapidly cooled one [52]. In particular, a slow quenching gives rise to a glass containing mostly medium-sized rings (6- and 7-MRs), whereas an increase in the quenching rate leads to a glass with a greater population of small- and large-sized rings [53].

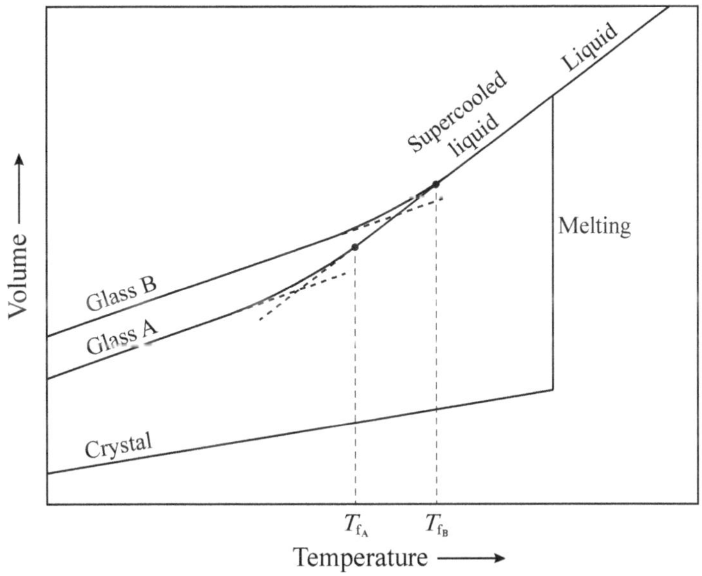

Figure 4. Volume–temperature diagram representing the solidification of melted silica into crystalline or glassy SiO_2. The supercooled liquid can be cooled at different quenching rates to produce glasses with varying fictive temperatures.

4. Generation of Point Defects

During the formation of a glass, part of the disorder characterizing the supercooled liquid is frozen-in in the solid state and the excess energy is stored as strained Si–O–Si bond angles [32,54]. The presence of distorted angles determines the formation of local high-energy structures like three- and four-membered rings. For regular planar three-membered rings, θ takes the value of 130.5° whereas, for regular planar four-membered rings, its value is 160.5°. In both cases, θ is far from its optimal value (144–155°) and, for this reason, the rings tend to release the excess energy by breaking a strained bond and turn into a bigger ring [55,56].

While in bulk silica the concentration of three- and four-membered rings is relatively low, silica optical fibers exhibit a higher concentration of small rings due to the residual stress generated during the manufacturing process [57]. In fact, the rapid cooling of optical fibers results in a T_f higher than that measured in bulk samples [58]. This is mainly due to the influence of the drawing speed on the quenching dynamics. A higher drawing speed means a faster cooling rate which, in turn, results in a higher fictive temperature. In addition, the residual stress of the tensile load applied during the drawing process has a strong impact on increasing the concentration of defect precursors in non-irradiated samples [15,16]. This explains the higher sensitivity of OFs to radiation and the higher concentration of room-temperature stable point defects [59,60]. In the remaining part of this section, we will describe the three main intrinsic defects induced by radiation in silica-based materials.

4.1. E' Centers

The best-known defect in crystalline SiO_2 and silica is the E' center, that is, an unpaired electron in a Si dangling bond. In their experimental work, Hosono et al. demonstrated that strained bonds contained in three- and four-membered rings absorb F_2-excimer laser light (7.9 eV, 157 nm) and jump to the lowest electronic spin triplet state (T_1), causing the opening of the ring and the consequent formation of a pair of point defects [53]. The photolytic reaction can be written as

$$\equiv\text{Si}-\text{O}-\text{Si}\equiv \longrightarrow [\equiv\text{Si}^{-}\text{---}{}^{+}\text{O}-\text{Si}\equiv]^{*} \longrightarrow \equiv\text{Si}^{\bullet} \quad {}^{\bullet}\text{O}-\text{Si}\equiv, \tag{1}$$

where the three dashes represent three separated Si–O bonds, the asterisk indicates an electronically excited state, and the dot represents an unpaired electron. The first step of the reaction involves the transition of an electron from the valence to the conduction band and the formation of a self-trapped exciton (STE) consisting of an excited electron (e^-) located on a silicon atom and a hole (h^+) trapped at one or more neighbouring oxygen atoms [61,62]. This structure is kept metastable by the spontaneous creation of a localised distortion in the SiO_2 network, which lowers the total energy of the system and traps the e^-–h^+ pair at the distortion site [63]. Using self-consistent quantum chemical calculations, Shluger proposed a model in which the exciton self-trapping is accompanied by the weakening of a Si–O bond and the displacement (0.3 Å) of the oxygen atom towards an interstitial position [64]. Other calculations showed that, after electronic excitation, the system relaxes to the nearest energy minimum on the excited-state energy surface by breaking a Si–O bond and moving the Si atom into a planar sp^2 configuration [62,63]. In either case, the relaxation of the STE to the ground state (and the possible rehybridization of the silicon orbitals to an sp^3 configuration) leads to the formation of a silicon and an oxygen dangling bond known, respectively, as E'_α center and non-bridging oxygen hole center (NBOHC) [65].

A second pathway for E' center formation was proved by Tsai and Griscom while studying the effect of highly-focused ArF excimer laser light (6.4 eV, 193 nm) on silica specimens [66]. As in the previous case, the first step of the proposed mechanism consists in the promotion of an electron to the conduction band and the formation of a self-trapped exciton. Here, however, its de-excitation proceeds through the displacement of the oxygen atom and the formation of an interstitial O^0 atom and a ≡Si−Si≡ oxygen vacancy (Frenkel defect pair) [67].

$$\equiv\text{Si}-\text{O}-\text{Si}\equiv \longrightarrow [\equiv\text{Si}^{-}\text{---}{}^{+}\text{O}-\text{Si}\equiv]^{*} \longrightarrow \equiv\text{Si}-\text{Si}\equiv + \text{O}^0 \tag{2}$$

The process involves a non-radiative decay of the electronic excited state, and its efficiency is higher in densified silica where the mean Si–O–Si angle is lower and the mean Si–O bond length is higher [68,69]. By further exciting the oxygen vacancy, the system evolves towards the formation of a Si dangling bond and a nearly planar $\equiv Si^+$ unit [70,71]. To distinguish this variant of E' center from that obtained via the mechanism (1), we label it an E'_γ center.

$$\equiv\text{Si}-\text{Si}\equiv \longrightarrow \equiv\text{Si}^{\bullet} \quad {}^{+}\text{Si}\equiv + e^- \tag{3}$$

Although the major channel for the formation of E' centers has not yet been identified, it appears that the bond-dissociation mechanism (1) prevails at higher irradiation energies whereas the Frenkel-type mechanism (2) and (3) predominates at lower energies. This view is supported by experimental results showing that the concentration of E'_α and NBOHC defects linearly increase with the pulse energy of F_2 lasers (7.9 eV) [53], while that of E'_γ centers and O^0 quadratically increases with the pulse energy of KrF (5.0 eV) and ArF lasers (6.4 eV) [72,73]. As a consequence, it has been suggested that the rupture of the Si–O bond described by Equation (1) is assisted by the absorption of a single photon [74], whereas the cleavage of the O atom and the successive formation of the E'_γ center is induced by a two-photon absorption process [66]. In the latter case, two alternative pathways have also been proposed. If the absorption is not simultaneous, the first photon is responsible for the formation of the STE while the second serves to ionise the oxygen vacancy and produce the Si dangling bond [75]. These two stpdf correspond to reactions (2) and (3) and each of them is activated by the absorption of one photon. Contrarily, if the two photons are absorbed at the same time, the reaction goes through a biexciton process in which one of the two STEs decays as reported in Equation (2), while the other supplies a hole to foster the reaction [76]:

$$\equiv\text{Si}-\text{Si}\equiv + h^+ \longrightarrow \equiv\text{Si}^\bullet \quad {}^+\text{Si}\equiv. \tag{4}$$

Two additional variants of E' defects have seldom been observed in irradiated silica. E'_s is a hemi-center typically observed on silica surfaces or interfaces, comprising only a threefold coordinated Si atom with the unpaired electron. The E'_β center, on the other hand, features a $\equiv\text{Si}^\bullet$ moiety coupled with a hydrogen-saturated oxygen vacancy [77,78]. This defect is generally found in hydrogen-rich silica, where the concentration of silanol groups (SiOH) is higher. In fact, the irradiation of these glasses with F_2 lasers or ionizing beams leads to the rupture of the O–H bond and the formation of a NBOHC and a neutral hydrogen atom (H^0) as follows:

$$\equiv\text{SiOH} \longrightarrow \equiv\text{SiO}^\bullet + \text{H}^0. \tag{5}$$

The free hydrogen may then diffuse in the silica matrix and react with a pre-existing $\equiv\text{Si}-\text{Si}\equiv$ oxygen vacancy to generate an E'_β center [79]:

$$\equiv\text{Si}-\text{Si}\equiv + \text{H}^0 \longrightarrow \equiv\text{Si}^\bullet \quad \text{H}-\text{Si}\equiv. \tag{6}$$

Since all the E' variants have very similar electronic structures, their optical absorption (OA) spectrum is characterized by the same band, peaking at about 5.8 eV (214 nm) with a full width at half maximum (FWHM) of 0.8 eV and an oscillator strength $f = 0.14 \pm 0.1$ [80,81]. An experimental OA spectrum obtained after γ-ray irradiation of a wet silica sample is reported in Figure 5. The electronic transition corresponding to this band is still debated. One hypothesis is that the OA originates from the charge transfer from valence band states (i.e., a $2p$ orbital of one of the three O atoms bonded to the E' center) to the empty state of the Si dangling bond [82]. This is supported by density functional theory calculations showing that the lower part of the absorption spectrum of a-SiO_2 corresponds to the superposition of the $O(2p) \longrightarrow Si(sp^3)$ transition with that from the occupied $Si(sp^3)$ state to the diffuse states in the lower part of the conduction band [83].

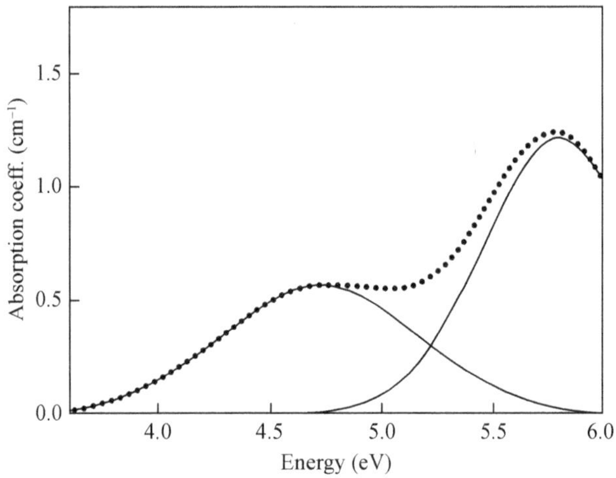

Figure 5. Optical absorption spectrum of a synthetic silica sample showing a band assigned to the E' center (5.8 eV) and another band assigned to NBOHC (4.8 eV). Adapted from Cannas et al. [84].

4.2. Non-Bridging Oxygen Hole Centers

The non-bridging oxygen hole center (NBOHC) is the simplest oxygen-related intrinsic defect in silica. It corresponds to a Si atom bonded to an O atom having a dangling bond, i.e., an unpaired electron in a 2*p*-like non-bonding orbital [65]. As mentioned above, there are two major pathways through which NBOHCs can be created. The first is called the intrinsic mechanisms and consists of the photolysis of a strained Si–O–Si bond, as given in reaction (1). The second is the extrinsic mechanism, which corresponds to the homolytic dehydrogenation of a silanol group bond, shown in reaction (5). The predominance of one mechanism over the other is determined by the presence of pre-existing defects, their concentration, and the energy of the excitation beam [85].

The electronic structure of NBOHCs was fully described by Suzuki et al. using *ab initio* cluster calculations [86]. The proposed energy level diagram is illustrated in Figure 6. Apart from the bonding and antibonding σ orbitals, the diagram shows a series of O non-bonding orbitals whose degeneracy is lifted by the interaction of the defect with the surrounding atoms of the amorphous network. In particular, the interaction splits the non-bonding orbitals of the bridging oxygens into two sets of multiply-degenerate $n_{p_x}(\mathrm{O_B})$ and $n_{p_z}(\mathrm{O_B})$ orbitals, and the lone-pair orbitals of the non-bridging oxygen into two $n_{p_x}(\mathrm{O_{NB}})$ and $n_{p_y}(\mathrm{O_{NB}})$ levels. The highest occupied molecular orbital (HOMO) is the singly-occupied $n_{p_y}(\mathrm{O_{NB}})$ orbital whereas the lowest unoccupied molecular orbital (LUMO) coincides with the antibonding $\sigma^*_{p_z}(\mathrm{Si\text{-}O_{NB}})$ orbital. The promotion of an electron from the ground-state levels to the HOMO gives rise to three OA bands corresponding to three distinct electronic transitions [87,88]:

(i) An asymmetric Pekarian-shaped band peaked at 1.97 eV (FWHM $= 0.17$ eV, $f \approx 1.5 \times 10^{-4}$ eV) attributed to the $\sigma_{p_z}(\mathrm{Si\text{-}O_{NB}}) \longrightarrow n_{p_y}(\mathrm{O_{NB}})$ transition from the bonding σ orbital to the half-filled orbital of the non-bridging O atom (the HOMO of the cluster).
(ii) A band centered at 4.8 eV (FWHM $= 1.07$ eV, $f \approx 0.05$) originating from the $n_{p_y}(\mathrm{O_B}) \longrightarrow n_{p_y}(\mathrm{O_{NB}})$ transition between the $\mathrm{O_B}$ lone-pair orbital perpendicular to the Si–O–Si plane and the HOMO.
(iii) A band at 6.8 eV (FWHM ≈ 1.8 eV, $f = 0.05$) related to the $n_{p_x}(\mathrm{O_B}) \longrightarrow n_{p_y}(\mathrm{O_{NB}})$ transition from the $\mathrm{O_B}$ lone-pair orbital lying in the Si–O–Si plane to the HOMO.

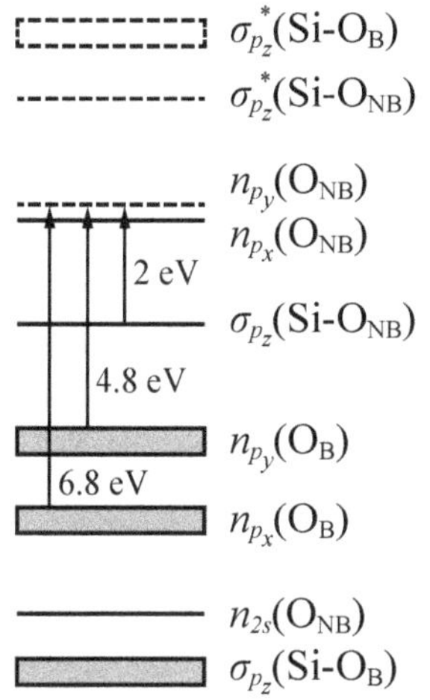

Figure 6. Energy level diagrams of a non-bridging oxygen hole center. Vertical arrows correspond to optical transitions between bondind and non-bonding orbitals, while grey boxes represent multiply degenerated levels. Adapted from Suzuki et al. [86].

The decay of the excited state created by the above transitions gives rise to a photoluminescence (PL) band at 1.91 eV with an FWHM = 0.17 eV and a lifetime of ~14 µs [80,89]. As was generally expected, the excitation from the σ orbital to the HOMO should imply a weakening of the Si−O$^•$ bond and a lengthening of the mean bond distance. Instead, experimental results indicate that the frequencies of the Si−O$^•$ symmetric stretching mode in the ground (890 cm^{-1}) and the excited state (860 cm^{-1}) are almost the same and the Stokes shift between the excitation and emission bands is as small as 0.06 eV [90–92]. This anomalous behaviour is caused by the interaction of the doubly occupied n_{p_y}(O$_{NB}$) orbital in the excited state with a symmetry-adapted combination of the three empty $\sigma^*_{p_y}$(Si-O$_B$) orbitals [86]. This so-called "negative hyperconjugation" is responsible for the partial delocalization of the electron density from the oxygen to the silicon atom and for the resulting stabilization of the σ_{p_z}(Si-O$_{NB}$) MO in electronically excited NBOHCs. In this way, the small electron–phonon coupling typical of NBOHC and the almost-equal Si−O$^•$ bond length in the ground and excited state can be explained.

4.3. *Oxygen-Deficient Centers*

Oxygen-deficient centers (ODCs) are the basic type of neutral oxygen monovacancies in non-stoichiometric silica and generally correspond to a Si–Si dimer configuration [93]. They are naturally present in unirradiated silica but their concentration considerably rises when a glass is irradiated with UV, X-ray, or γ-ray beams [80]. The major formation pathway of ODCs is given in Equation (2), where an energetic photon causes the release of an interstitial oxygen atom from the silica network to form a Si–Si bond.

The first spectroscopic studies on ODCs were performed in the mid-1950s by Garino-Canina [94], Mitchell and Paige [95], and Cohen [96], among others. Their results led to the identification of two optical absorption bands called "E-band" (7.6 eV) and "B$_2$-band" (5.0 eV), which were tentatively assigned to interstitial oxygen atoms and divalent silicon atoms, respectively. In addition, three photoluminescence bands (called α, β, and γ) were observed at approximately 4.3 eV, 3.1 eV, and 2.7 eV, and associated with substitutional Ge atoms at oxygen-vacancy sites. In 1983, O'Reilly and Robertson [97] calculated the electronic structure of the main defects in SiO$_2$ and suggested two different variants for the oxygen-deficient center. The so-called ODC(I) was proposed to be a relaxed ≡Si−Si≡ oxygen vacancy, while ODC(II) was identified with an unrelaxed ≡Si−Si≡ bond of length 3.06 Å. They also calculated the energy levels for both structures and demonstrated that the 7.6 eV

OA band could be associated with the $\sigma \longrightarrow \sigma^*$ transition of the relaxed Si–Si bond. This picture was partially confirmed by Hosono et al. [98] and Imagawa et al. [99] who found that, by heating unirradiated SiO_2 glasses in either hydrogen or oxygen gas flow, the intensity of the E-band decreased in accordance with the assumption that H_2 and O_2 neutralize pre-existing ODCs, as shown by

$$\equiv\text{Si}-\text{Si}\equiv + \text{H}_2 \longrightarrow \equiv\text{Si}-\text{H} \quad \text{H}-\text{Si}\equiv \tag{7}$$

and

$$\equiv\text{Si}-\text{Si}\equiv + \frac{1}{2}\text{O}_2 \longrightarrow \equiv\text{Si}-\text{O}-\text{Si}\equiv. \tag{8}$$

These reactions provided the definitive proof of ODC(I)'s structure and optical activity, but did not clarify the nature of the other defect variant. As experimental evidence built up, two alternative structural models were put forward to explain the spectroscopic behavior of ODCs(II). One is the neutral oxygen vacancy (V_O^0) model originally proposed by O'Reilly and Robertson [97] and subsequently adopted by Imai et al. [75,100] to interpret their findings. While irradiating dehydrated oxygen-deficient silica with ArF and KrF excimer laser, they observed a non-linear decrease in both the B_2-band and the PL α-band, accompanied by a dose-dependent generation of E' centers. The intensity of the E-band, however, did not show any change during the experiments and remained at the original intensity level. The analysis of the concentration of E' with irradiation time revealed that the growth curve can be decomposed into an initial, saturating part due to ODC(II) and a larger, linear-growth component attributable to ODC(I). The formation mechanism of E' centers was thus proposed to proceed via the direct photoionization of ODCs or hole-trapping, as given in reactions (3) and (4). The different response of the 7.6 eV and 5.0 eV bands to laser irradiation can be explained by assuming that the formation efficiency of E' from ODC(II) is much higher than that from ODC(I), probably because of the similarity between the unrelaxed oxygen vacancy and the $\equiv$Si$^+$ atom accompanying the E'_γ center.

The second model, called the twofold-coordinated silicon (Si_2^0) model, was proposed by Skuja to interpret the origin of the OA band at 5.0 eV and the PL bands at 4.3 eV and 2.7 eV [101]. In the Si_2^0 notation, the "2" stands for the coordination number of the Si atom and the "0" for its net electric charge. The structure proposed for the ODC(II) was that of a divalent silicon bonded to two bridging O atoms and with a lone pair in an sp^2 orbital. The transition that gave rise to the B_2-band was identified with the $S_0 \longrightarrow S_1$ excitation of the twofold Si atom, while those associated with the α- and γ-bands were the transitions $S_1 \longrightarrow S_0$ and $T_1 \longrightarrow S_0$ at the same defect site. This model was supported by the studies of Tsai and Griscom on the structure of a hydrogenated variant of the E' center, called an H(I) center [102]. They demonstrated that the electron paramagnetic resonance (EPR) features of this defect were compatible with an sp^3 silicon atom bonded to two oxygens and one H atom, and resumed the pathway described by Radtsig [103] as a formation mechanism:

$$\equiv\text{Si}-\text{O}^\bullet + \text{H}_2 \longrightarrow \equiv\text{Si}-\text{OH} + \text{H}^0 \tag{9}$$

$$=\text{Si}\colon + \text{H}^0 \longrightarrow =\text{Si}^\bullet-\text{H}. \tag{10}$$

The first step corresponds to the annealing of an NBOHC to give a silanol group and a hydrogen atom. The second step is the reaction of latter with an ODC(II) to convert it into an H(I) center. Both reactions have been confirmed in a number of experimental and theoretical studies and represent the cornerstone of the Si_2^0 model [104–107].

In 1989, the ODC(II) issue got even more complicated when Tohmon et al. reported the existence of two accidentally overlapping bands making up the B_2-band [108]. The first contribution (called $B_{2\alpha}$) was centered at 5.02 eV (FWHM = 0.35 eV) and was related to two emission bands at 4.42 and 2.7 eV. The other band ($B_{2\beta}$) was peaked at 5.15 eV (FWHM = 0.48 eV) and was linked to emission at 4.24 eV and 3.16 eV. By analyzing the decay lifetime of the photoluminescence, the authors proposed that the $B_{2\alpha}$-band corresponds to the singlet-to-triplet ($S_0 \longrightarrow T_1$) transition of the relaxed Si–Si bond while the corresponding luminescence is due to the inverse transition $T_1 \longrightarrow S_0$. The origin of the $B_{2\beta}$-band was not discussed by the authors but it was subsequently attributed by Kohketsu et al. [109] to the =Si: center, together with the corresponding 4.24 and 3.16 eV PL bands. These assignments were largely criticized mainly because the $B_{2\alpha}$-band lifetime ($\tau \approx 100\,\mu s$) did not match the value of 10 ns reported in many other studies [110–112]. Moreover, Anedda et al. observed that the 4.4 eV emission band was due to an intrinsic defect and that it shifted to 4.2 eV when the defect was perturbed by an unidentified impurity [113]. The two PL bands were thus called α_I and α_E, where I stands for intrinsic and E for extrinsic. On the basis of these observations, Skuja re-elaborated the Si_2^0 model including the possibility for Ge and Sn to form ODC-like defects which could contribute to the optical properties of low-purity and Ge-doped SiO_2 glasses. According to this new T_2^0 model (with T standing for Si, Ge, or Sn), the divalent =Si: center is responsible for the OA band peaked at 5.02 eV ($S_0 \longrightarrow S_1$ transition) [112,114], as well as for the PL α_I ($S_1 \longrightarrow S_0$) and β ($T_1 \longrightarrow S_0$) bands [101,115]. Similarly, the isoelectronic =Ge: defect gives rise to the 5.15 eV OA band and the two PL bands α_E and γ (transitions as before). This model has gained a wide acceptance in recent years and is backed up by both theoretical [116–119] and experimental [120–122] investigations.

5. Photodarkening in Optical Fibers

When silica optical fibers are exposed to ionizing radiation, some of the strained chemical bonds present in the core are broken by the incoming light to give rise to point defects. The microscopical damages are primarily manifested as a degradation of the optical signal-to-noise ratio and a decrease of the optical power along the waveguide. As discussed above, the glass found in OFs is more sensitive than bulk silica due to its higher content of small-sized rings. This is determined by the particular conditions encountered during the manufacturing of the fibers (i.e., higher quenching rate, drawing speed, and applied strain) and is reflected in the greater propensity of OFs to undergo photodarkening. Furthermore, the glass stoichiometry, the content of hydroxyl groups (OH), and the concentration of impurities also impact the radiation sensitivity of a optical fibers.

Radiation-induced losses in silica OFs have been shown to be caused primarily due to Si-related defects or to absorbing species related to chlorine impurities. The most important contribution is given by E' centers originating either from oxygen-deficient centers, extrinsic Si–H bonds, or strained Si–O bonds [15–17,123]. When E' centers come from pre-existing defects (i.e., ODCs), the growth curve observed under irradiation shows a saturating profile due to the exhaustion of precursors. Conversely, when they are generated from the photo-assisted dissociation of strained bonds, the growth kinetics of E' centers is linearly correlated with that of NBOHCs and both have the same dependence on the dose of radiation received by the material [123]. In many case, the decomposition with Gaussian absorption bands of the radiation-induced attenuation spectra measured for irradiated OFs reveals the presence of ODC(I)s and ODC(II)s at lower concentrations than E' and NBOHCs [124]. At greater energies, the addition of two absorption bands centered at 3.26 eV (FWHM = 1.2 eV) and 3.78 eV (FWHM = 0.57 eV) is often necessary in order to improve the quality of the fit. These two bands have been associated with chlorine species (namely, Cl^0 and Cl_2) deriving from the detachment of Cl atoms from the silicon tetrachloride used to make optical fibers [124,125]. Since all these defects absorb light in the 180–700 nm spectral window, the radiation-induced attenuation is usually higher in the UV–Vis domain and lower in the near-infrared range (700 to 2000 nm). This determined that optical fibers have historically been used in telecommunication and sensing systems to transmit signals in the infrared domain [126–129]. However, the recent technological advances in fields like UV photolithography [130,131], laser surgery [132–134], and fiber-based photovoltaics [135–141] have given a new impulse towards the development of waveguides able to withstand the damaging effects of high-energy photons and maintain a high transmittance of UV and optical signals.

Radiation-hardened optical fibers are specialty waveguides whose composition has been optimized to strongly reduce their radiation sensitivity. Hydrogen loading of pure silica-core OFs was shown to considerably increase the resistance to ionizing radiation and, at the same time, decrease the intrinsic attenuation level at 700 nm [142–145]. In fact, the presence of molecular hydrogen in the silica matrix can help the recovery of optical fibers by annealing photo-induced E' centers and NBOHCs as follows [145]:

$$\equiv\text{Si}^{\bullet} + \text{H}_2 \longrightarrow \equiv\text{Si}-\text{H} + \text{H}^0 \tag{11}$$

$$\equiv\text{Si}-\text{O}^{\bullet} + \text{H}_2 \longrightarrow \equiv\text{Si}-\text{OH} + \text{H}^0. \tag{12}$$

However, it has been shown that most of the hydrogen quickly diffuses out of the core material and only a "stable" amount of H_2 remains in the silica matrix [146]. This means that the anti-radiation efficiency of these fibers depends on the time interval between treatment and irradiation, and that their use should be limited to applications requiring only short-term stability [147–149]. To increase the useful life of H_2-loaded fibers, hermetic coatings made of carbon or metal have been developed to prevent the diffusion of hydrogen out of the fiber [150,151]. This technique was shown to improve the resistance to prolonged irradiation but the strict conditions required to deposit the coating material on the surface of silica still prevent its use in common devices [19]. More recently, Hartung et al. designed a so-called anti-resonant hollow-core fiber (AR-HCF), which is able to guide light in three transmission bands in the UV region with minimum attenuation (1 to 10 $\text{dB} \cdot \text{m}^{-1}$) [152]. Unlike refractive-index guiding fibers, this particular type of waveguide features a microstructured cladding enveloping an air/vacuum core, which guides light via the anti-resonant reflection optical waveguiding mechanism [153,154]. Due to their low modal overlap factor, AR-HCFs usually exhibit high laser damage threshold and high radiation resistance, as well as low modal dispersion, low material absorption, and low optical nonlinearity. All these characteristics make anti-resonant hollow-core fibers good candidates for challenging applications such as high-power UV laser, nonlinear and ultrafast optics, plasma physics, and surgery [155–157]. Nonetheless, their relatively high cost and low availability in the market still favor the use of more economic silica-based OFs for common applications.

6. Summary and Future Directions

In this review, we investigated the structural and optical properties of the most common UV–Vis absorbing point defects found in pure silica. We have focused our attention on the relationship between pristine glass structure and the photo-induced generation of E' centers, NBOHCs, and ODCs. By analyzing the short-range properties of a-SiO_2, we observed that the spatial arrangement of the silicon and oxygen atoms can be described in terms of the Si–O bond length (1.605 Å), the O–Si–O (ϕ) and Si–O–Si (θ) bond angles, and a single Si–O–Si–O dihedral angle (ω). The amplitude of ϕ is fixed to 109.47° by the rigid tetrahedral geometry of the $[SiO_4]^0$ unit, whereas the value of θ follows an asymmetric distribution that peaked at approximately 147° with an FWHM of $\sim$16°. The torsion angle ω was found to be strongly correlated with the Si–O–Si angle and, for $140° \leq \theta \leq 160°$, it takes the value 60°, 180°, and 300°. Traditionally, the arrangement of adjacent SiO_4 tetrahedra has been described in terms of two distinct torsional angles defined by the sequences of consecutive atoms O–Si–O–Si and Si–O–Si–O. However, since these sequences are not ordered in silica, the two angles are equivalent and the three-dimensional properties of the network can be described by using just one of them.

Due to the flexibility of the angles θ and ω, the tetrahedral SiO_4 units arrange themselves to form closed structures called rings. The most abundant rings are composed of six tetrahedra, followed by those constituting five and seven units. Rings of three and four units are characterized by high steric and angular hindrance and, for this reason, they have a high free-energy content. Although their concentration in natural silica is relatively low, glasses obtained by melt quenching (such as those of optical fibers) show an increased concentration of small-sized rings due to the freezing-in of unrelaxed local structures typical of the supercooled liquid state. This feature has important consequences for the optical properties of the glass and its resistance to ionizing radiation. In fact, as we have pointed out in this review, the atoms contained in three- and four-membered rings have a higher propensity to interact with the incoming radiation and release excess energy by breaking a Si–O bond and forming a larger ring.

The most straightforward example of the photo-assisted cleavage of Si–O bonds is the formation of E' centers. The first step of the reaction is the absorption of a photon by a strained bond and the formation of a metastable self-trapped exciton (STE). Depending on the irradiation energy and the structural properties of the glass, the STE can decay to form either an E'_{α}–NBOHC pair or an ODC plus a free O^0 atom. By further exciting the ODC, the Si–Si bond is ionized and the system evolves towards the formation of an E'_{γ} center. Two additional variants of E' have also been observed on the surface of silica samples (E'_s) and in irradiated wet silica (E'_{β}). Despite their different chemical make-up, all E' centers share the same electronic structure and are characterized by a similar OA band, peaking at 5.8 eV. The formation reaction of E'_{α} shown in Equation (1) and that of E'_{β} shown in Equation (5) are also considered the two major pathways for the formation of NBOHCs in silica. Due to the complex electronic structure of this defect, its absorption spectrum features three distinct OA bands centered at 1.97 eV, 4.8 eV, and 6.8 eV. The transitions assigned to these bands originate from different initial levels but have the same oxygen non-bonding orbital as a final level. When the excited state created by these transitions decays, the emitted photons give rise to a PL band peaking at 1.91 eV. The first step of the E'_{γ} formation mechanism Equation (2) is also the major channel for the photo-induced generation of the color center called ODC(I). The structure of this defect has been largely debated but there is an almost universal consensus at present, recognizing it as a $\equiv$Si$-$Si$\equiv$ dimer configuration with an OA band at 7.6 eV. A second variant of oxygen-deficiency center known as ODC(II) has been observed in a large number of experimental works, but its structural features remain controversial. A first model elaborated in the 1980s treated this defect as an unrelaxed oxygen vacancy consisting of an elongated Si–Si bond. However, this model failed to justify a number of successive experimental evidences and, for this reason, it was partially discarded. The newest and most widely adopted model is that proposed by Skuja, which relates the ODC(II) with a twofold coordinated silicon. The optical activity of this defect was correctly correlated with an OA band centered at 5.02 eV and with a strong emission at 4.42 eV and a very weak emission at 2.7 eV. Moreover, the model accounted for the existence of Ge- and Sn-based oxygen-deficient centers, contributing to the optical spectra of impure silica glasses. In this case, the 5.15 eV OA band as well as the 2.7 and 4.24 eV PL bands are considered to originate from electronic transitions at =Ge**:** centers isoelectronic to ODC(II).

As all the analyzed point defects have absorption bands in the 180–700 nm portion of the spectrum, it is clear that transmitting UV–Vis light in optical fibers in a harsh environment is a very challenging task. To mitigate the photodarkening effects of energetic photons, a large plethora of specialty fibers have been developed in the last two decades. The first and most widespread solution is represented by H_2-loaded optical fibers, that is, waveguides that have been infused with molecular hydrogen. The presence of the gas in the silica matrix act as a buffer towards the formation of certain point defects by annealing them and keeping the attenuation levels low. To prevent the out-flowing of hydrogen from the core, H_2-loaded fibers have also been hermetically coated with carbon or metals with a specific high-temperature, high-pressure process. In 2014, a new type of optical fibers, called anti-resonant hollow-core fibers, have been developed for the transmission of high-energy pulses in the UV–Vis domain. These waveguides are not silica-based and allow for the transmission of signals in the core via the anti-resonant effect with a very low attenuation. Nonetheless, their low availability in the market and the high production costs still prevent their diffusion in common devices. For this reason, it is important to further optimize silica optical fibers by controlling the fabrication process parameters to manipulate the nature and concentration of the point defects that are responsible for their degradation.

Funding: This research was funded by the National Operational Programme for Research and Innovation 2014–2020, grant number DOT1308583.

Institutional Review Board Statement: Not applicable.

Informed Consent Statement: Not applicable.

Data Availability Statement: Not applicable.

Conflicts of Interest: The authors declare no conflict of interest.

References

1. Griscom, D.L. Nature of Defects And Defect Generation in Optical Glasses. In *Radiation Effects on Optical Materials*; Levy, P.W., Ed.; SPIE: Bellingham, WA, USA, 1985. [CrossRef]
2. Griscom, D.L. Point Defects in Amorphous SiO_2: What Have We Learned from 30 Years of Experimentation? *MRS Proc.* **1985**, *61*, 213–221. [CrossRef]
3. Griscom, D.L. A Minireview of the Natures of Radiation-Induced Point Defects in Pure and Doped Silica Glasses and Their Visible/Near-IR Absorption Bands, with Emphasis on Self-Trapped Holes and How They Can Be Controlled. *Phys. Res. Int.* **2013**, *2013*, 379041. [CrossRef]
4. Imai, H.; Hirashima, H. Intrinsic- and extrinsic-defect formation in silica glasses by radiation. *J. Non Cryst. Solids* **1994**, *179*, 202–213. [CrossRef]
5. Devine, R. Macroscopic and microscopic effects of radiation in amorphous SiO_2. *Nucl. Instruments Methods Phys. Res. Sect. B* **1994**, *91*, 378–390. [CrossRef]
6. Weeks, R.A. The many varieties of E' centers: A review. *J. Non Cryst. Solids* **1994**, *179*, 1–9. [CrossRef]
7. Boizot, B.; Agnello, S.; Reynard, B.; Boscaino, R.; Petite, G. Raman spectroscopy study of β-irradiated silica glass. *J. Non Cryst. Solids* **2003**, *325*, 22–28. [CrossRef]
8. Kajihara, K.; Hirano, M.; Skuja, L.; Hosono, H. Intrinsic defect formation in amorphous SiO_2 by electronic excitation: Bond dissociation versus Frenkel mechanisms. *Phys. Rev. B* **2008**, *78*, 094201. [CrossRef]
9. Alessi, A.; Agnello, S.; Sporea, D.; Oproiu, C.; Brichard, B.; Gelardi, F. Formation of optically active oxygen deficient centers in Ge-doped SiO_2 by γ- and β-ray irradiation. *J. Non Cryst. Solids* **2010**, *356*, 275–280. [CrossRef]
10. Beaudier, A.; Wagner, F.R.; Natoli, J.Y. Using NBOHC fluorescence to predict multi-pulse laser-induced damage in fused silica. *Opt. Commun.* **2017**, *402*, 535–539. [CrossRef]
11. Ollier, N.; Girard, S.; Peuget, S. Radiation Effects in Glass. In *Encyclopedia of Glass Science, Technology, History, and Culture*; John Wiley & Sons, Ltd.: Hoboken, NJ, USA, 2021; pp. 405–414. [CrossRef]
12. Kajihara, K.; Hirano, M.; Skuja, L.; Hosono, H. Frenkel defect process in amorphous silica. In *Damage to VUV, EUV, and X-ray Optics III*; Juha, L., Bajt, S., London, R.A., Eds.; SPIE: Bellingham, WA, USA, 2011. [CrossRef]
13. Tsai, H.S.; Chao, D.S.; Wu, Y.H.; He, Y.T.; Chueh, Y.L.; Liang, J.H. Spectroscopic investigation of gamma radiation-induced coloration in silicate glass for nuclear applications. *J. Nucl. Mater.* **2014**, *453*, 233–238. [CrossRef]
14. Brichard, B.; Agnello, S.; Nuccio, L.; Dusseau, L. Comparison Between Point Defect Generation by γ-rays in Bulk and Fibre Samples of High Purity Amorphous SiO_2. *IEEE Trans. Nucl. Sci.* **2008**, *55*, 2121–2125. [CrossRef]
15. Girard, S.; Ouerdane, Y.; Origlio, G.; Marcandella, C.; Boukenter, A.; Richard, N.; Baggio, J.; Paillet, P.; Cannas, M.; Bisutti, J.; et al. Radiation Effects on Silica-Based Preforms and Optical Fibers—I: Experimental Study With Canonical Samples. *IEEE Trans. Nucl. Sci.* **2008**, *55*, 3473–3482. [CrossRef]
16. Girard, S.; Richard, N.; Ouerdane, Y.; Origlio, G.; Boukenter, A.; Martin-Samos, L.; Paillet, P.; Meunier, J.P.; Baggio, J.; Cannas, M.; et al. Radiation Effects on Silica-Based Preforms and Optical Fibers—II: Coupling Ab initio Simulations and Experiments. *IEEE Trans. Nucl. Sci.* **2008**, *55*, 3508–3514. [CrossRef]
17. Girard, S.; Kuhnhenn, J.; Gusarov, A.; Brichard, B.; Uffelen, M.V.; Ouerdane, Y.; Boukenter, A.; Marcandella, C. Radiation Effects on Silica-Based Optical Fibers: Recent Advances and Future Challenges. *IEEE Trans. Nucl. Sci.* **2013**, *60*, 2015–2036. [CrossRef]
18. Khalilov, V.K.; Klein, K.F.; Amosov, A.V. Influence of fiber drawing on optical UV attenuation of all-silica fibers with undoped core. In *Surgical-Assist Systems*; Bogner, M.S., Charles, S.T., Grundfest, W.S., Harrington, J.A., Katzir, A., Lome, L.S., Vannier, M.W., Hanwehr, R.V., Eds.; SPIE: Bellingham, WA, USA, 1998. [CrossRef]
19. Girard, S.; Alessi, A.; Richard, N.; Martin-Samos, L.; De Michele, V.; Giacomazzi, L.; Agnello, S.; Di Francesca, D.; Morana, A.; Winkler, B.; et al. Overview of radiation induced point defects in silica-based optical fibers. *Rev. Phys.* **2019**, *4*, 100032. [CrossRef]
20. De Michele, V.; Marcandella, C.; Morana, A.; Campanella, C.; Vidalot, J.; Paillet, P.; Gaillardin, M.; Marin, E.; Ouerdane, Y.; Boukenter, A.; et al. Pulsed X-Ray Radiation Response of Ultra-Low Loss Pure-Silica-Core Optical Fiber. *Phys. Status Solidi* **2021**. [CrossRef]

21. Campanella, C.; De Michele, V.; Morana, A.; Mélin, G.; Robin, T.; Marin, E.; Ouerdane, Y.; Boukenter, A.; Girard, S. Radiation Effects on Pure-Silica Multimode Optical Fibers in the Visible and Near-Infrared Domains: Influence of OH Groups. *Appl. Sci.* **2021**, *11*, 2991. [CrossRef]
22. De Michele, V.; Sciortino, A.; Messina, F.; Cannas, M.; Boukenter, A.; Marin, E.; Girard, S.; Ouerdane, Y. NBOHCs' photocycle revealed in synthetic silica by transient absorption measurements. In Proceedings of the The 22nd International Conference on Ultrafast Phenomena 2020, Online, 16–19 November 2020. [CrossRef]
23. Ringdalen, E. Changes in Quartz During Heating and the Possible Effects on Si Production. *JOM* **2014**, *67*, 484–492. [CrossRef]
24. Drees, L.R.; Wilding, L.P.; Smeck, N.E.; Senkayi, A.L. Silica in Soils: Quartz and Disordered Silica Polymorphs. In *Minerals in Soil Environments*; Dixon, J.B., Weed, S.B., Eds.; Soil Science Society of America: Madison, WI, USA, 2018; pp. 913–974. [CrossRef]
25. Raman, C.V.; Nedungadi, T.M.K. The α–β Transformation of Quartz. *Nature* **1940**, *145*, 147. [CrossRef]
26. Flörke, O.W.; Graetsch, H.A.; Brunk, F.; Benda, L.; Paschen, S.; Bergna, H.E.; Roberts, W.O.; Welsh, W.A.; Libanati, C.; Ettlinger, M. Silica. In *Ullmann's Encyclopedia of Industrial Chemistry*; Wiley: Hoboken, NJ, USA, 2000. [CrossRef]
27. Newton, M.D.; Gibbs, G.V. Ab initio calculated geometries and charge distributions for H_4SiO_4 and $H_6Si_2O_7$ compared with experimental values for silicates and siloxanes. *Phys. Chem. Miner.* **1980**, *6*, 221–246. [CrossRef]
28. Noritake, F.; Kawamura, K. The Nature of Si-O-Si Bonding via Molecular Orbital Calculations. *J. Comput. Chem.* **2015**, *14*, 124–130. [CrossRef]
29. Pauling, L. The nature of silicon–oxygen bonds. *Am. Mineral.* **1980**, *65*, 321–323.
30. Awazu, K.; Kawazoe, H. Strained Si–O–Si bonds in amorphous SiO_2 materials: A family member of active centers in radio, photo, and chemical responses. *J. Appl. Phys.* **2003**, *94*, 6243–6262. [CrossRef]
31. Rino, J.P.; Ebbsjö, I.; Kalia, R.K.; Nakano, A.; Vashishta, P. Structure of rings in vitreous SiO_2. *Phys. Rev. B* **1993**, *47*, 3053–3062. [CrossRef]
32. Wright, A.C. Neutron scattering from vitreous silica. V. The structure of vitreous silica: What have we learned from 60 years of diffraction studies? *J. Non Cryst. Solids* **1994**, *179*, 84–115. [CrossRef]
33. Charpentier, T.; Kroll, P.; Mauri, F. First-Principles Nuclear Magnetic Resonance Structural Analysis of Vitreous Silica. *J. Phys. Chem. C* **2009**, *113*, 7917–7929. [CrossRef]
34. Sternberg, U. The bond angle dependence of the asymmetry parameter of the oxygen-17 electric field gradient tensor. *Solid State Nucl. Magn. Reson.* **1993**, *2*, 181–190. [CrossRef]
35. Clark, T.M.; Grandinetti, P.J. Dependence of bridging oxygen ^{17}O quadrupolar coupling parameters on Si–O distance and Si–O–Si angle. *J. Phys. Condens. Matter* **2003**, *15*, S2387–S2395. [CrossRef]
36. Smith, J.V.; Blackwell, C.S. Nuclear magnetic resonance of silica polymorphs. *Nature* **1983**, *303*, 223–225. [CrossRef]
37. Clark, T.M.; Grandinetti, P.J.; Florian, P.; Stebbins, J.F. Correlated structural distributions in silica glass. *Phys. Rev. B* **2004**, *70*, 064202. [CrossRef]
38. Yuan, X.; Cormack, A. Si–O–Si bond angle and torsion angle distribution in vitreous silica and sodium silicate glasses. *J. Non Cryst. Solids* **2003**, *319*, 31–43. [CrossRef]
39. Malfait, W.J.; Halter, W.E.; Verel, R. ^{29}Si NMR spectroscopy of silica glass: T_1 relaxation and constraints on the Si–O–Si bond angle distribution. *Chem. Geol.* **2008**, *256*, 269–277. [CrossRef]
40. Wright, A.C. Diffraction studies of glass structure. *J. Non Cryst. Solids* **1990**, *123*, 129–148. [CrossRef]
41. Ginhoven, R.M.V.; Jónsson, H.; Corrales, L.R. Silica glass structure generation for ab initio calculations using small samples of amorphous silica. *Phys. Rev. B* **2005**, *71*. [CrossRef]
42. King, S.V. Ring Configurations in a Random Network Model of Vitreous Silica. *Nature* **1967**, *213*, 1112–1113. [CrossRef]
43. Guttman, L. Ring structure of the crystalline and amorphous forms of silicon dioxide. *J. Non Cryst. Solids* **1990**, *116*, 145–147. [CrossRef]
44. Franzblau, D.S. Computation of ring statistics for network models of solids. *Phys. Rev. B* **1991**, *44*, 4925–4930. [CrossRef] [PubMed]
45. Görlich, E. The structure of SiO_2 — Current views. *Ceram. Int.* **1982**, *8*, 3–16. [CrossRef]
46. Papike, J.J. Chemistry of the rock-forming silicates: Multiple-chain, sheet, and framework structures. *Rev. Geophys.* **1988**, *26*, 407–444. [CrossRef]
47. Sarnthein, J.; Pasquarello, A.; Car, R. Structural and Electronic Properties of Liquid and Amorphous SiO_2: An *Ab Initio* Molecular Dynamics Study. *Phys. Rev. Lett.* **1995**, *74*, 4682–4685. [CrossRef]
48. Pasquarello, A.; Car, R. Identification of Raman Defect Lines as Signatures of Ring Structures in Vitreous Silica. *Phys. Rev. Lett.* **1998**, *80*, 5145–5147. [CrossRef]
49. Giacomazzi, L.; Umari, P.; Pasquarello, A. Medium-range structure of vitreous SiO_2 obtained through first-principles investigation of vibrational spectra. *Phys. Rev. B* **2009**, *79*. [CrossRef]
50. Conradt, R. Thermodynamics and Kinetics of Glass. In *Springer Handbook of Glass*; Springer International Publishing: Berlin/Heidelberg, Germany, 2019; pp. 51–77. [CrossRef]
51. Karmakar, B. Fundamentals of Glass and Glass Nanocomposites. In *Glass Nanocomposites*; Elsevier: Amsterdam, The Netherlands, 2016; pp. 3–53. [CrossRef]
52. Lopes, M. *Introduction to Glass Science and Technology*; The Royal Society of Chemistry: London, UK, 2005. [CrossRef]

53. Hosono, H.; Ikuta, Y.; Kinoshita, T.; Kajihara, K.; Hirano, M. Physical Disorder and Optical Properties in the Vacuum Ultraviolet Region of amorphous SiO_2. *Phys. Rev. Lett.* **2001**, *87*. [CrossRef]
54. Jones, F.; Huff, N. The structure and properties of glass fibres. In *Handbook of Textile Fibre Structure*; Elsevier: Amsterdam, The Netherlands, 2009; pp. 307–352. [CrossRef]
55. Galeener, F. Planar rings in glasses. *Solid State Commun.* **1982**, *44*, 1037–1040. [CrossRef]
56. Barrio, R.A.; Galeener, F.L.; Martínez, E.; Elliott, R.J. Regular ring dynamics in AX_2 tetrahedral glasses. *Phys. Rev. B* **1993**, *48*, 15672–15689. [CrossRef] [PubMed]
57. Hibino, Y.; Hanawa, F.; Horiguchi, M. Drawing-induced residual stress effects on optical characteristics in pure-silica-core single-mode fibers. *J. Appl. Phys.* **1989**, *65*, 30–34. [CrossRef]
58. Lancry, M.; Régnier, E.; Poumellec, B. Fictive temperature in silica-based glasses and its application to optical fiber manufacturing. *Prog. Mater. Sci.* **2012**, *57*, 63–94. [CrossRef]
59. Friebele, E.J.; Sigel, G.H.; Griscom, D.L. Drawing-induced defect centers in a fused silica core fiber. *Appl. Phys. Lett.* **1976**, *28*, 516–518. [CrossRef]
60. Alessi, A.; Girard, S.; Cannas, M.; Agnello, S.; Boukenter, A.; Ouerdane, Y. Influence of Drawing Conditions on the Properties and Radiation Sensitivities of Pure-Silica-Core Optical Fibers. *J. Light. Technol.* **2012**, *30*, 1726–1732. [CrossRef]
61. Song, K.S.; Williams, R.T. Silicon Dioxide. In *Self-Trapped Excitons*; Song, K.S., Williams, R.T., Eds.; Springer: Berlin/Heidelberg, Germany, 1993. [CrossRef]
62. Ginhoven, R.M.V.; Jónsson, H.; Corrales, L.R. Characterization of exciton self-trapping in amorphous silica. *J. Non Cryst. Solids* **2006**, *352*, 2589–2595. [CrossRef]
63. Ismail-Beigi, S.; Louie, S.G. Self-Trapped Excitons in Silicon Dioxide: Mechanism and Properties. *Phys. Rev. Lett.* **2005**, *95*. [CrossRef]
64. Shluger, A.L. The model of a triplet self-trapped exciton in crystalline SiO_2. *J. Phys. C Solid State Phys.* **1988**, *21*, L431–L434. [CrossRef]
65. Stapelbroek, M.; Griscom, D.; Friebele, E.; Sigel, G. Oxygen-associated trapped-hole centers in high-purity fused silicas. *J. Non Cryst. Solids* **1979**, *32*, 313–326. [CrossRef]
66. Tsai, T.E.; Griscom, D.L. Experimental evidence for excitonic mechanism of defect generation in high-purity silica. *Phys. Rev. Lett.* **1991**, *67*, 2517–2520. [CrossRef]
67. Hosono, H.; Kawazoe, H.; Matsunami, N. Experimental Evidence for Frenkel Defect Formation in Amorphous SiO_2 by Electronic Excitation. *Phys. Rev. Lett.* **1998**, *80*, 317–320. [CrossRef]
68. Itoh, C.; Suzuki, T.; Itoh, N. Luminescence and defect formation in undensified and densified amorphous SiO_2. *Phys. Rev. B* **1990**, *41*, 3794–3799. [CrossRef] [PubMed]
69. Devine, R.A.B. Radiation-sensitivity enhancement and annealing variation in densified, amorphous SiO_2. *Phys. Rev. B* **1987**, *35*, 9783–9789. [CrossRef]
70. Griscom, D.L. Trapped-electron centers in pure and doped glassy silica: A review and synthesis. *J. Non Cryst. Solids* **2011**, *357*, 1945–1962. [CrossRef]
71. Agnello, S.; Boscaino, R.; Buscarino, G.; Cannas, M.; Gelardi, F.M. Structural relaxation of E'_γ centers in amorphous silica. *Phys. Rev. B* **2002**, *66*, 113201. [CrossRef]
72. Devine, R.A.B. Defect Creation and Two-Photon Absorption in Amorphous SiO_2. *Phys. Rev. Lett.* **1989**, *62*, 340–340. [CrossRef]
73. Glinka, Y.; Lin, S.H.; Chen, Y.T. Two-photon-excited luminescence and defect formation in SiO_2 nanoparticles induced by 6.4-eV ArF laser light. *Phys. Rev. B* **2000**, *62*, 4733–4743. [CrossRef]
74. Kajihara, K.; Ikuta, Y.; Hirano, M.; Hosono, H. Power dependence of defect formation in SiO_2 glass by F_2 laser irradiation. *Appl. Phys. Lett.* **2002**, *81*, 3164–3166. [CrossRef]
75. Imai, H.; Arai, K.; Imagawa, H.; Hosono, H.; Abe, Y. Two types of oxygen-deficient centers in synthetic silica glass. *Phys. Rev. B* **1988**, *38*, 12772–12775. [CrossRef] [PubMed]
76. Kajihara, K.; Ikuta, Y.; Oto, M.; Hirano, M.; Skuja, L.; Hosono, H. UV–VUV laser induced phenomena in SiO_2 glass. *Nucl. Instruments Methods Phys. Res. Sect. B* **2004**, *218*, 323–331. [CrossRef]
77. Griscom, D.L. Characterization of three E'-center variants in X- and γ-irradiated high purity a-SiO_2. *Nucl. Instruments Methods Phys. Res. Sect.* **1984**, *1*, 481–488. [CrossRef]
78. Griscom, D.L. Optical Properties and Structure of Defects in Silica Glass. *J. Ceram. Soc.* **1991**, *99*, 923–942. [CrossRef]
79. Kajihara, K.; Skuja, L.; Hirano, M.; Hosono, H. In situobservation of the formation, diffusion, and reactions of hydrogenous species in F_2-laser-irradiated SiO_2 glass using a pump-and-probe technique. *Phys. Rev. B* **2006**, *74*. [CrossRef]
80. Skuja, L. Optical properties of defects in silica. In *Defects in SiO_2 and Related Dielectrics: Science and Technology*; Pacchioni, G., Skuja, L., Griscom, D.L., Eds.; Springer: Berlin/Heidelberg, Germany, 2000; pp. 73–116. [CrossRef]
81. Buscarino, G.; Boscaino, R.; Agnello, S.; Gelardi, F.M. Optical absorption and electron paramagnetic resonance of the E'_α center in amorphous silicon dioxide. *Phys. Rev. B* **2008**, *77*, 155214. [CrossRef]
82. Agnello, S.; Buscarino, G.; Gelardi, F.M.; Boscaino, R. Optical absorption band at 5.8 eV associated with the E'_γ centers in amorphous silicon dioxide: Optical absorption and EPR measurements. *Phys. Rev. B* **2008**, *77*, 195206. [CrossRef]
83. Giordano, L.; Sushko, P.V.; Pacchioni, G.; Shluger, A.L. Optical and EPR properties of point defects at a crystalline silica surface: Ab initio embedded-cluster calculations. *Phys. Rev. B* **2007**, *75*, 024109. [CrossRef]

84. Cannas, M.; Leone, M. Photoluminescence at 1.9 eV in synthetic wet silica. *J. Non Cryst. Solids* **2001**, *280*, 183–187. [CrossRef]
85. Nishikawa, H.; Nakamura, R.; Ohki, Y.; Hama, Y. Correlation of preexisting diamagnetic defect centers with induced paramagnetic defect centers by ultraviolet or vacuum-ultraviolet photons in high-purity silica glasses. *Phys. Rev. B* **1993**, *48*, 15584–15594. [CrossRef]
86. Suzuki, T.; Skuja, L.; Kajihara, K.; Hirano, M.; Kamiya, T.; Hosono, H. Electronic Structure of Oxygen Dangling Bond in Glassy SiO_2: The Role of Hyperconjugation. *Phys. Rev. Lett.* **2003**, *90*, 186404. [CrossRef]
87. Skuja, L.; Kajihara, K.; Hirano, M.; Hosono, H. Visible to vacuum-UV range optical absorption of oxygen dangling bonds in amorphous SiO_2. *Phys. Rev. B* **2011**, *84*, 205206. [CrossRef]
88. Hosono, H.; Kajihara, K.; Suzuki, T.; Ikuta, Y.; Skuja, L.; Hirano, M. Vacuum ultraviolet optical absorption band of non-bridging oxygen hole centers in SiO_2 glass. *Solid State Commun.* **2002**, *122*, 117–120. [CrossRef]
89. Sakurai, Y. Low temperature dependence of photoluminescence band near 2.0 eV in silica glass. *J. Appl. Phys.* **2000**, *87*, 755–759. [CrossRef]
90. Skuja, L. The origin of the intrinsic 1.9 eV luminescence band in glassy SiO_2. *J. Non Cryst. Solids* **1994**, *179*, 51–69. [CrossRef]
91. Vaccaro, L.; Cannas, M.; Boscaino, R. Phonon coupling of non-bridging oxygen hole center with the silica environment: Temperature dependence of the 1.9 eV emission spectra. *J. Lumin.* **2008**, *128*, 1132–1136. [CrossRef]
92. Skuja, L.; Suzuki, T.; Tanimura, K. Site-selective laser-spectroscopy studies of the intrinsic 1.9-eV luminescence center in glassy SiO_2. *Phys. Rev. B* **1995**, *52*, 15208–15216. [CrossRef]
93. Amossov, A.; Rybaltovsky, A. Oxygen-deficient centers in silica glasses: a review of their properties and structure. *J. Non Cryst. Solids* **1994**, *179*, 75–83. [CrossRef]
94. Garino-Canina, V. La luminescence de la silice vitreuse en fonction de la temperature. *Comptes Rendus de l'Academie des Sci.* **1954**, *238*, 1577.
95. Mitchell, E.W.J.; Paige, E.G.S. CXI. The optical effects of radiation induced atomic damage in quartz. *Philos. Mag.* **1956**, *1*, 1085–1115. [CrossRef]
96. Cohen, A.J. Neutron Specific Color Center in Fused Silica and an Impurity Band of Identical Wavelength. *Phys. Rev.* **1957**, *105*, 1151–1155. [CrossRef]
97. O'Reilly, E.P.; Robertson, J. Theory of defects in vitreous silicon dioxide. *Phys. Rev. B* **1983**, *27*, 3780–3795. [CrossRef]
98. Hosono, H.; Abe, Y.; Imagawa, H.; Imai, H.; Arai, K. Experimental evidence for the Si–Si bond model of the 7.6-eV band in SiO_2 glass. *Phys. Rev. B* **1991**, *44*, 12043–12045. [CrossRef]
99. Imagawa, H.; Arai, T.; Hosono, H.; Imai, H.; Arai, K. Reaction kinetics of oxygen-deficient centers with diffusing oxygen molecules in silica glass. *J. Non Cryst. Solids* **1994**, *179*, 70–74. [CrossRef]
100. Imai, H.; Arai, K.; Hosono, H.; Abe, Y.; Arai, T.; Imagawa, H. Dependence of defects induced by excimer laser on intrinsic structural defects in synthetic silica glasses. *Phys. Rev. B* **1991**, *44*, 4812–4818. [CrossRef] [PubMed]
101. Skuja, L.; Streletsky, A.; Pakovich, A. A new intrinsic defect in amorphous SiO_2: Twofold coordinated silicon. *Solid State Commun.* **1984**, *50*, 1069–1072. [CrossRef]
102. Tsai, T.E.; Griscom, D.L. On the structures of hydrogen-associated defect centers in irradiated high-purity *a*-SiO_2:OH. *J. Non Cryst. Solids* **1987**, *91*, 170–179. [CrossRef]
103. Radtsig, V. Paramagnetic centers on fresh surfaces of quartz. Interactions with molecules of H2 and D2. *Kinet. Katal.* **1979**, *20*, 456–464.
104. Amosov, A.; Malyshkin, S. Role of oxygen vacancy type defects in formation of radiation-induced color centers in quartz glasses. *Fizika i Khimiya Stekla* **1984**, *10*, 305–309.
105. Nishikawa, H.; Nakamura, R.; Ohki, Y.; Hama, Y. Enhanced photogeneration of E' centers from neutral oxygen vacancies in the presence of hydrogen in high-purity silica glass. *Phys. Rev. B* **1993**, *48*, 2968–2973. [CrossRef]
106. Radzig, V.A.; Bagratashvili, V.N.; Tsypina, S.I.; Chernov, P.V.; Rybaltovskii, A.O. Photoinduced Reactions of Oxygen-Deficient Centers with Molecular Hydrogen in Silica Glasses. *J. Phys. Chem.* **1995**, *99*, 6640–6647. [CrossRef]
107. Edwards, A.H.; Germann, G. Interaction of hydrogenated molecules with intrinsic defects in *a*-SiO_2. *Nucl. Instruments Methods Phys. Res. Sect. B* **1988**, *32*, 238–247. [CrossRef]
108. Tohmon, R.; Mizuno, H.; Ohki, Y.; Sasagane, K.; Nagasawa, K.; Hama, Y. Correlation of the 5.0- and 7.6-eV absorption bands in SiO_2 with oxygen vacancy. *Phys. Rev. B* **1989**, *39*, 1337–1345. [CrossRef]
109. Kohketsu, M.; Awazu, K.; Kawazoe, H.; Yamane, M. Photoluminescence Centers in VAD SiO_2 Glasses Sintered under Reducing or Oxidizing Atmospheres. *Jpn. J. Appl. Phys.* **1989**, *28*, 615–621. [CrossRef]
110. Skuja, L.N.; Silin, A.R.; Mareš, J. Decay time and polarization properties of luminescence centers in vitreous silica. *Phys. Status Solidi* **1978**, *50*, K149–K152. [CrossRef]
111. Stathis, J.H.; Kastner, M.A. Time-resolved photoluminescence in amorphous silicon dioxide. *Phys. Rev. B* **1987**, *35*, 2972–2979. [CrossRef] [PubMed]
112. Skuja, L. Isoelectronic series of twofold coordinated Si, Ge, and Sn atoms in glassy SiO_2: A luminescence study. *J. Non Cryst. Solids* **1992**, *149*, 77–95. [CrossRef]
113. Anedda, A.; Congiu, F.; Raga, F.; Corazza, A.; Martini, M.; Spinolo, G.; Vedda, A. Time resolved photoluminescence of a centers in neutron irradiated SiO_2. *Nucl. Instruments Methods Phys. Res. Sect. B* **1994**, *91*, 405–409. [CrossRef]

114. Skuja, L. Direct singlet-to-triplet optical absorption and luminescence excitation band of the twofold-coordinated silicon center in oxygen-deficient glassy SiO_2. *J. Non Cryst. Solids* **1994**, *167*, 229–238. [CrossRef]
115. Skuja, L.N.; Trukhin, A.N.; Plaudis, A.E. Luminescence in Germanium-Doped Glassy SiO_2. *Phys. Status Solidi* **1984**, *84*, K153–K157. [CrossRef]
116. Zhang, B.L.; Raghavachari, K. Photoabsorption and photoluminescence of divalent defects in silicate and germanosilicate glasses: First-principles calculations. *Phys. Rev. B* **1997**, *55*, R15993–R15996. [CrossRef]
117. Pacchioni, G.; Ferrario, R. Optical transitions and EPR properties of two-coordinated Si, Ge, Sn and related H(I), H(II), and H(III) centers in pure and doped silica from ab initio calculations. *Phys. Rev. B* **1998**, *58*, 6090–6096. [CrossRef]
118. Pacchioni, G. Ab initio theory of point defects in SiO_2. In *Defects in SiO_2 and Related Dielectrics: Science and Technology*; Pacchioni, G., Skuja, L., Griscom, D.L., Eds.; Springer: Berlin/Heidelberg, Germany, 2000; pp. 161–195. [CrossRef]
119. Sokolov, V.O.; Sulimov, V.B. Theory of Twofold Coordinated Silicon and Germanium Atoms in Solid Silicon Dioxide. *Phys. Status Solidi* **1994**, *186*, 185–198. [CrossRef]
120. Agnello, S.; Boscaino, R.; Cannas, M.; Gelardi, F.M.; Leone, M. γ-ray-induced bleaching in silica: Conversion from optical to paramagnetic defects. *Phys. Rev. B* **2000**, *61*, 1946–1951. [CrossRef]
121. Gelardi, F.; Agnello, S. Gamma rays induced conversion of native defects in natural silica. In *Defects in SiO_2 and Related Dielectrics: Science and Technology*; Pacchioni, G., Skuja, L., Griscom, D.L., Eds.; Springer: Berlin/Heidelberg, Germany, 2000; pp. 285–305. [CrossRef]
122. Agnello, S.; Boscaino, R.; Mattina, F.L.; Grandi, S.; Magistris, A. Hydrogen-Related Paramagnetic Centers in Ge-Doped Sol-Gel Silica Induced by γ-Ray Irradiation. *J. Sol-Gel Sci. Technol.* **2006**, *37*, 63–68. [CrossRef]
123. Morana, A.; Girard, S.; Cannas, M.; Marin, E.; Marcandella, C.; Paillet, P.; Périsse, J.; Macé, J.R.; Boscaino, R.; Nacir, B.; et al. Influence of neutron and gamma-ray irradiations on rad-hard optical fiber. *Opt. Mater. Express* **2015**, *5*, 898–911. [CrossRef]
124. Girard, S.; Marcandella, C.; Origlio, G.; Ouerdane, Y.; Boukenter, A.; Meunier, J.P. Radiation-induced defects in fluorine-doped silica-based optical fibers: Influence of a pre-loading with H2. *J. Non Cryst. Solids* **2009**, *355*, 1089–1091. [CrossRef]
125. Lo Piccolo, G.M.; Morana, A.; Alessi, A.; Boukenter, A.; Girard, S.; Ouerdane, Y.; Gelardi, F.M.; Agnello, S.; Cannas, M. Ultraviolet–visible light-induced solarisation in silica-based optical fibres for indoor solar applications. *J. Non Cryst. Solids* **2021**, *552*, 120458. [CrossRef]
126. Agrawal, G. *Fiber-Optic Communication Systems*; Wiley: Hoboken, NJ, USA, 2021.
127. Hartog, A.H. *An Introduction to Distributed Optical Fibre Sensors*; CRC Press: Boca Raton, FL, USA, 2017. [CrossRef]
128. Bao, X.; Chen, L. Recent Progress in Distributed Fiber Optic Sensors. *Sensors* **2012**, *12*, 8601–8639. [CrossRef] [PubMed]
129. Lu, P.; Lalam, N.; Badar, M.; Liu, B.; Chorpening, B.T.; Buric, M.P.; Ohodnicki, P.R. Distributed optical fiber sensing: Review and perspective. *Appl. Phys. Rev.* **2019**, *6*, 041302. [CrossRef]
130. Gilliam, W.; Fleming, B.T.; Vorobiev, D.; Winter, B.; Wadsworth, W.; Birks, T. Far-ultraviolet optical fibers for instrumentation in the sub-200 nm regime. In *UV/Optical/IR Space Telescopes and Instruments: Innovative Technologies and Concepts X*; Breckinridge, J.B., Stahl, H.P., Barto, A.A., Eds.; SPIE: Bellingham, WA, USA, 2021. [CrossRef]
131. Waldo, W. Techniques and Tools for Optical Lithography. In *Handbook of VLSI Microlithography*; Elsevier: Amsterdam, The Netherlands, 2001; pp. 472–643. [CrossRef]
132. Makarov, N.S.; Ramasamy, K.; Jackson, A.; Velarde, A.; Castaneda, C.; Archuleta, N.; Hebert, D.; Bergren, M.R.; McDaniel, H. Fiber-Coupled Luminescent Concentrators for Medical Diagnostics, Agriculture, and Telecommunications. *ACS Nano* **2019**, *13*, 9112–9121. [CrossRef]
133. Wang, J.; Dong, J. Optical Waveguides and Integrated Optical Devices for Medical Diagnosis, Health Monitoring and Light Therapies. *Sensors* **2020**, *20*, 3981. [CrossRef] [PubMed]
134. Artyushenko, V.G. Fiber solutions from 355 nm to 16 μm in advanced diagnostics and laser surgery (Conference Presentation). In *Optical Fibers and Sensors for Medical Diagnostics and Treatment Applications XX*; Gannot, I., Ed.; SPIE: Bellingham, WA, USA, 2020. [CrossRef]
135. Ullah, I.; Whang, A. Development of Optical Fiber-Based Daylighting System and Its Comparison. *Energies* **2015**, *8*, 7185–7201. [CrossRef]
136. Antón, I.; Silva, D.; Sala, G.; Bett, A.; Siefer, G.; Luque-Heredia, I.; Trebst, T. The PV-FIBRE concentrator: a system for indoor operation of 1000x MJ solar cells. *Prog. Photovoltaics Res. Appl.* **2007**, *15*, 431–447. [CrossRef]
137. Núñez, R.; Antón, I.; Sala, G. Hybrid lighting-CPV, a new efficient concept mixing illumination with CPV. *Opt. Express* **2013**, *21*, 4864–4874. [CrossRef]
138. Núñez, R.; Antón, I.; Sala, G. Proof-of-concept of a building-integrated hybrid concentrator photovoltaics-lighting system. *Light. Res. Technol.* **2017**, *50*, 1082–1090. [CrossRef]
139. Aslian, A.; Asli, B.H.S.; Tan, C.J.; Adikan, F.R.M.; Toloei, A. Design and Analysis of an Optical Coupler for Concentrated Solar Light Using Optical Fibers in Residential Buildings. *Int. J. Photoenergy* **2016**, *2016*, 3176052. [CrossRef]
140. Vu, N.H.; Shin, S. Flat Optical Fiber Daylighting System with Lateral Displacement Sun-Tracking Mechanism for Indoor Lighting. *Energies* **2017**, *10*, 1679. [CrossRef]
141. Vu, N.H.; Pham, T.T.; Shin, S. Modified optical fiber daylighting system with sunlight transportation in free space. *Opt. Express* **2016**, *24*, A1528–A1545. [CrossRef] [PubMed]

142. Henschel, H.; Kohn, O.; Weinand, U. Radiation hardening of pure silica optical fibres by high pressure hydrogen treatment. In Proceedings of the 2001 6th European Conference on Radiation and Its Effects on Components and Systems (Cat. No.01TH8605), Grenoble, France, 10–14 September 2001. [CrossRef]
143. Karlitschek, P.; Hillrichs, G.; Klein, K.F. Influence of hydrogen on the colour center formation in optical fibers induced by pulsed UV-laser radiation. Part 1: all silica fibers with high-OH undoped core. *Opt. Commun.* **1998**, *155*, 376–385. [CrossRef]
144. Karlitschek, P.; Hillrichs, G.; Klein, K.F. Influence of hydrogen on the colour center formation in optical fibers induced by pulsed UV-laser radiation. *Opt. Commun.* **1998**, *155*, 386–397. [CrossRef]
145. Oto, M.; Kikugawa, S.; Sarukura, N.; Hirano, M.; Hosono, H. Optical fiber for deep ultraviolet light. *IEEE Photonics Technol. Lett.* **2001**, *13*, 978–980. [CrossRef]
146. Brichard, B.; Fernandez, A.; Ooms, H.; Borgermans, P.; Berghmans, F. Dependence of the POR and NBOHC defects as function of the dose in hydrogen-treated and untreated KU1 glass fibers. *IEEE Trans. Nucl. Sci.* **2003**, *50*, 2024–2029. [CrossRef]
147. Girard, S.; Vivona, M.; Laurent, A.; Cadier, B.; Marcandella, C.; Robin, T.; Pinsard, E.; Boukenter, A.; Ouerdane, Y. Radiation hardening techniques for Er/Yb doped optical fibers and amplifiers for space application. *Opt. Express* **2012**, *20*, 8457–8465. [CrossRef] [PubMed]
148. Zotov, K.V.; Likhachev, M.E.; Tomashuk, A.L.; Bubnov, M.M.; Yashkov, M.V.; Gur'yanov, A.N. Radiation-resistant erbium-doped silica fibre. *Quantum Electron.* **2007**, *37*, 946–949. [CrossRef]
149. Zotov, K.; Likhachev, M.; Tomashuk, A.; Bubnov, M.; Yashkov, M.; Guryanov, A.; Klyamkin, S. Radiation-resistant erbium-doped fiber for spacecraft applications. In Proceedings of the 2007 9th European Conference on Radiation and Its Effects on Components and Systems, Deauville, France, 10–14 September 2007. [CrossRef]
150. Lemaire, P.J.; Lindholm, E.A. Hermetic Optical Fibers. In *Specialty Optical Fibers Handbook*; Elsevier: Amsterdam, The Netherlands, 2007; pp. 453–490. [CrossRef]
151. Bogatyrev, V.A.; Semjonov, S. Metal-Coated Fibers. In *Specialty Optical Fibers Handbook*; Elsevier: Amsterdam, The Netherlands, 2007; pp. 491–512. [CrossRef]
152. Hartung, A.; Kobelke, J.; Schwuchow, A.; Wondraczek, K.; Bierlich, J.; Popp, J.; Frosch, T.; Schmidt, M.A. Double antiresonant hollow core fiber–guidance in the deep ultraviolet by modified tunneling leaky modes. *Opt. Express* **2014**, *22*, 19131–19140. [CrossRef] [PubMed]
153. White, T.P.; McPhedran, R.C.; de Sterke, C.M.; Litchinitser, N.M.; Eggleton, B.J. Resonance and scattering in microstructured optical fibers. *Opt. Lett.* **2002**, *27*, 1977. [CrossRef] [PubMed]
154. Litchinitser, N.M.; Abeeluck, A.K.; Headley, C.; Eggleton, B.J. Antiresonant reflecting photonic crystal optical waveguides. *Opt. Lett.* **2002**, *27*, 1592–1594. [CrossRef] [PubMed]
155. Debord, B.; Amrani, F.; Vincetti, L.; Gérôme, F.; Benabid, F. Hollow-Core Fiber Technology: The Rising of "Gas Photonics". *Fibers* **2019**, *7*, 16. [CrossRef]
156. Ni, W.; Yang, C.; Luo, Y.; Xia, R.; Lu, P.; Hu, D.J.J.; Danto, S.; Shum, P.P.; Wei, L. Recent Advancement of Anti-Resonant Hollow-Core Fibers for Sensing Applications. *Photonics* **2021**, *8*, 128. [CrossRef]
157. Yu, F.; Knight, J.C. Low loss anti-resonant hollow-core fibers and applications. In Proceedings of the 2017 19th International Conference on Transparent Optical Networks (ICTON), Girona, Spain, 2–6 July 2017. [CrossRef]

materials

MDPI

Review

Towards Perfect Absorption of Single Layer CVD Graphene in an Optical Resonant Cavity: Challenges and Experimental Achievements

Abedin Nematpour [1], Maria Luisa Grilli [1,*], Laura Lancellotti [2] and Nicola Lisi [1]

1 Energy Technologies and Renewable Sources Department, Italian National Agency for New Technologies, Energy and Sustainable Economic Development (ENEA), Casaccia Research Centre, Via Anguillarese 301, 00123 Roma, Italy; abedinnemat@grad.kashanu.ac.ir (A.N.); nicola.lisi@enea.it (N.L.)

2 Energy Technologies and Renewable Sources Department, Italian National Agency for New Technologies, Energy and Sustainable Economic Development (ENEA), Portici Research Centre, P.le E. Fermi 1, 80055 Portici, Italy; laura.lancellotti@enea.it

* Correspondence: marialuisa.grilli@enea.it

Abstract: Graphene is emerging as a promising material for the integration in the most common Si platform, capable to convey some of its unique properties to fabricate novel photonic and optoelectronic devices. For many real functions and devices however, graphene absorption is too low and must be enhanced. Among strategies, the use of an optical resonant cavity was recently proposed, and graphene absorption enhancement was demonstrated, both, by theoretical and experimental studies. This paper summarizes our recent progress in graphene absorption enhancement by means of Si/SiO_2-based Fabry–Perot filters fabricated by radiofrequency sputtering. Simulations and experimental achievements carried out during more than two years of investigations are reported here, detailing the technical expedients that were necessary to increase the single layer CVD graphene absorption first to 39% and then up to 84%. Graphene absorption increased when an asymmetric Fabry–Perot filter was applied rather than a symmetric one, and a further absorption increase was obtained when graphene was embedded in a reflective rather than a transmissive Fabry–Perot filter. Moreover, the effect of the incident angle of the electromagnetic radiation and of the polarization of the light was investigated in the case of the optimized reflective Fabry–Perot filter. Experimental challenges and precautions to avoid evaporation or sputtering induced damage on the graphene layers are described as well, disclosing some experimental procedures that may help other researchers to embed graphene inside PVD grown materials with minimal alterations.

Keywords: graphene absorption; Fabry–Perot filter; radio frequency sputtering; CVD graphene

check for updates

Citation: Nematpour, A.; Grilli, M.L.; Lancellotti, L.; Lisi, N. Towards Perfect Absorption of Single Layer CVD Graphene in an Optical Resonant Cavity: Challenges and Experimental Achievements. *Materials* **2022**, *15*, 352. https://doi.org/10.3390/ma15010352

Academic Editor: Fabrizio Roccaforte

Received: 28 November 2021
Accepted: 28 December 2021
Published: 4 January 2022

1. Introduction

Graphene-based absorbers are receiving a considerable interest due to their potential applications in photovoltaics [1–3], as wave modulators [4–8], biological sensors [9–11], photodetectors [12–15], etc. Achieving graphene-based perfect absorbers is quite challenging because single layer graphene has got a weak and spectrally broad absorption of 2.3% over a wideband wavelength range from Vis to Far-infrared [16–21].

In the THz and IR regions, however, high quality graphene may show strong interaction with light thanks to generation of surface plasmon polaritons (SPPs), making graphene a promising alternative to typical plasmonic materials [22–25]. This is thanks to the capability of graphene to support plasmon modes with extremely tight confinement, long lifetime, and low losses at IR and THz frequencies [26–29].

In addition, tunable graphene electromagnetic response may be achieved by chemical and/or electrical doping or by using a magnetic field [30–32].

In the Vis and NIR regions, on the contrary, absence of SSPs makes it necessary to couple graphene with resonant structures such as metamaterials, photonic crystals and plasmonic materials [33–39].

Many examples of perfect graphene-based absorbers have been reported in the literature basing on the critical coupling concept [40–42].

Thongrattanasiri et al. have simulated 100% absorption in doped graphene nanodisks by exploiting the critical coupling conditions [43].

Piper et al. [44] have demonstrated numerically perfect absorption in unpatterned monolayer graphene in the Vis and NIR range by means of critical coupling with guided resonances of a photonic crystal slab with a back reflector.

Liu et al. [45] have demonstrated experimentally close to total absorption (85%) in monolayer graphene based on critical coupling with guided resonances in Fano resonance photonic NIR filters, by using a structure with a back reflector.

Xiao et al. [46] have reported about a general theoretical method for tailoring the absorption bandwidth of graphene via critical coupling in the NIR range by using a simple two-port resonant structure composed of a graphene-covered two-dimensional photonic crystal slab.

Jin et al. [47] have reviewed recent advances of graphene-based architectures for perfect absorption from Vis to THz band, both narrowband and broadband, and have discussed also about criticalities in the practical implementation of the simulated structures.

Another strategy to enhance graphene absorption is to exploit the electric field enhancement by resonant cavities [48–50]. Table 1 reports the graphene absorption values obtained by many authors by exploiting the electric field enhancement inside Fabry–Perot resonant cavities.

Table 1. Summary of simulated and experimental absorption of graphene inside Fabry–Perot structures in the Vis-THz region, as reported in the literature.

Authors	Wavelength	Absorption	Simulation	Experiment	Ref.
Nulli et al.	Wavelength regardless	0.5	Yes	No	[48]
Ferreira et al.	Wavelength regardless	~1	Yes	No	[49]
Xu et al.	Telecommunication wavelengths	0.5	Yes	No	[51]
Shuhan Chen	1600 nm	0.889	Yes	No	[52]
Yu et al.	550 nm	0.995	Yes	No	[53]
Wei et al.	1537 and 1579 nm	0.995	Yes	No	[54]
Bian et al.	2.5146 THz	0.997	Yes	No	[55]
Vasić et al.	Near-infrared to Terahertz	~1	Yes	No	[56]
Deng et al.	THz range	~1	Yes	No	[57]
Zheng et al.	600 nm	~1	Yes	No	[58]
Bian et al.	THz range	~1	Yes	No	[59]
Doukas et al.	1550 nm	~1	Yes	No	[60]
Zand et al.	1200, 1550 and 1900 nm	~1	Yes	No	[61]
Chen et al.	THz range	~1	Yes	No	[62]
Furchi et al.	855 nm	0.6	Yes	Yes	[63]

Nulli et al. reported theoretically that the optical absorption of single-layer graphene can be enhanced up to 50% by increasing the electric field on the surface of undoped graphene by placing graphene inside symmetric Fabry–Perot structures [48]. Moreover, Ferreira et al. have calculated the same absorption result for single layer graphene by utilizing one Fabry–Perot cavity and they have simulated also that graphene optical absorption can increase up to 100% using two symmetric Fabry–Perot cavities [50].

Zand et al. have used a genetic optimization algorithm coupled to a transfer matrix code to design one-dimensional aperiodic multilayer microstructures embedding single layer graphene, where near-total absorptions at selected wavelengths is obtained by existence of critical coupling [61]. They simulated that, compared with asymmetric-Fabry–Perot-based designs, aperiodic structures may provide higher efficiency for the spatial selective localization of the resonant modes.

Furchi et al. [63] have demonstrated experimentally in 2012 a graphene-based photodetector. Absorption enhancement up to 60% at 855 nm was demonstrated by inserting exfoliated μm-sized graphene flakes inside a dielectric multilayer stack grown by combining plasma-enhanced chemical vapor deposition and molecular beam epitaxy.

Many of the theoretical studies found in the literature simulating perfect graphene absorption consider for single layer graphene a very high mobility μ of about 10.000 $cm^2 \cdot V^{-1}/s^{-1}$, which is generally much higher than the real one (non-isolated from the environment, CVD graphene), leading sometimes to too overoptimistic predictions [64,65].

Apart from the work from Furchi et al., the enhancement of graphene absorption by using an optical resonant cavity has been rarely explored experimentally due to the many experimental challenges related to embedding graphene inside metals or dielectric stacks [63].

The graphene absorption enhancement inside an optical resonant cavity is generally a narrowband and is, therefore, more suitable for application in photodetectors, sensors, or absorption filters [63,66,67]

In the present work, we summarize our recent studies [67–69] on CVD graphene absorption inside three different Fabry–Perot (FP) filters fabricated by radiofrequency sputtering. Experimental challenges related to graphene-based Fabry–Perot filters fabrication are described in detail with the aim to provide a useful recipe which can help researchers to embed graphene inside materials grown by conventional physical vapor deposition (PVD) techniques, without altering its properties.

In the optimized structure, a high absorption of 84% at 3150 nm was measured [67] in case of large area (1 inch) single layer CVD graphene, which is the highest value of absorption so far experimentally achieved for single layer graphene inside a Fabry–Perot optical cavity.

2. Materials and Methods

2.1. Fabry–Perot Filter: Simulation

A Fabry–Perot optical filter consists of a cavity separated by two flat and parallel high reflecting mirrors where light experiences multiple reflections. Simulation of FP filters was carried out by using the Wave Optics Module of Comsol Multiphysics software. Our Fabry–Perot filters consisted of a sequence of alternate quarter wave thick high refractive index Si (H) and low refractive index SiO_2 (L) layers, for use in the Near to Mid IR. Three different Fabry–Perot filters were simulated: (i) a symmetric FP filter with identical top and bottom mirrors (FP1), (ii) an asymmetric FP filter where the bottom mirror had a higher reflectance due to a larger number of layers with respect to the top mirror (FP2), and (iii) an asymmetric reflective FP filter obtained by further increasing the bottom mirror layers (FP3). FP1 and FP2 filters worked in transmissive mode, while FP3 worked in reflective mode, due to the high reflectance of the bottom mirror (99.7%). The structure of FP1, FP2, and FP3 filters was the following: (i) air/HLH LL HLH/sub, (ii) air/HLHL HH LHLHLH/sub, and (iii) air/HLH LL HLHLHLHLH/sub, respectively. For FP1 and FP3 the cavity was (LL), i.e., constituted by a half wavelength thick SiO_2 layer, for FP2 the cavity (HH) was made by a half wavelength thick Si layer. This last FP structure, a bit different from the other two, was chosen because of its particular distribution of the electric field characterized by two maxima. A single layer graphene (SLG) was embedded inside the FP structure and located at the position where the electric field was maximum to enhance its absorption. SLG was covered by a 30 nm MgF_2 layer, meant to protect graphene during the later sputtering deposition. Figure 1 shows the three FP filters embedding SLG, and Figure 2 shows the simulated reflectance of the bottom mirrors in the three cases, R1, R2, and R3, R1 also has the reflectance of the top mirror, being identical for the three filters.

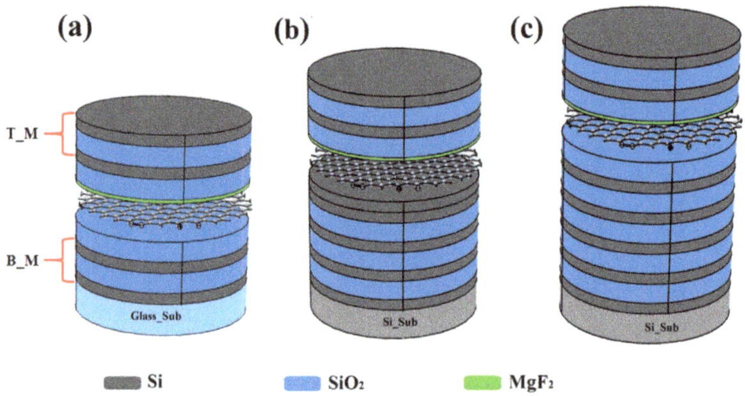

Figure 1. Scheme of the Fabry–Perot filters embedding SLG: (**a**) symmetric FP (FP1), (**b**) asymmetric FP (FP2) and (**c**) asymmetric reflective FP (FP3). T_M and B_M stand for top and bottom mirrors.

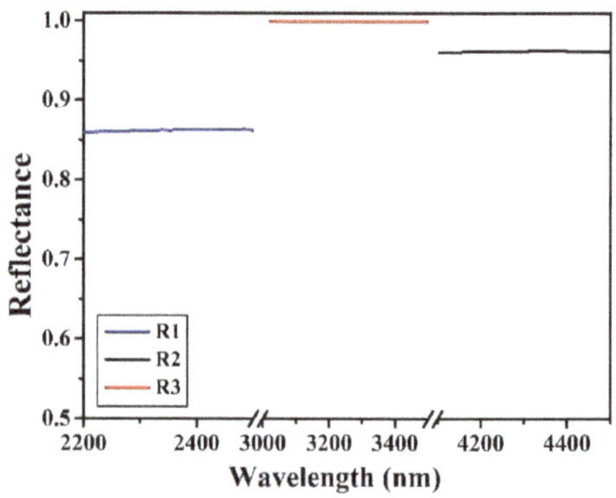

Figure 2. Simulated reflectance curves of the bottom mirrors shown in Figure 1. R1, R2, and R3 are the reflectance of bottom mirrors of FP1, FP2, and FP3, respectively.

Figure 3a shows the simulated transmittance (T), reflectance (R), and absorption (A) of the FP1, FP2, and FP3 filters, centered at $\lambda = 2315$, 4342, and 3150 nm, respectively. Figure 3b shows the electric field distribution inside the filters. The simulation has been carried in case of a TE polarized light at normal incident angle.

Table 2 reports the simulated and experimental absorption values of SLG embedded in the three FP filters and the value of electric field (E_G) where SLG is positioned. The central wavelength of the Fabry–Perot filters is different for the three cases because it results from different experiments, however, central wavelength position in the considered wavelength range, only causes small variations of T, R, and A due to the wavelength dependence of Si and SiO_2 optical constants. As we can observe from Table 2, the highest graphene absorption is obtained in the structure where the electric field is maximum.

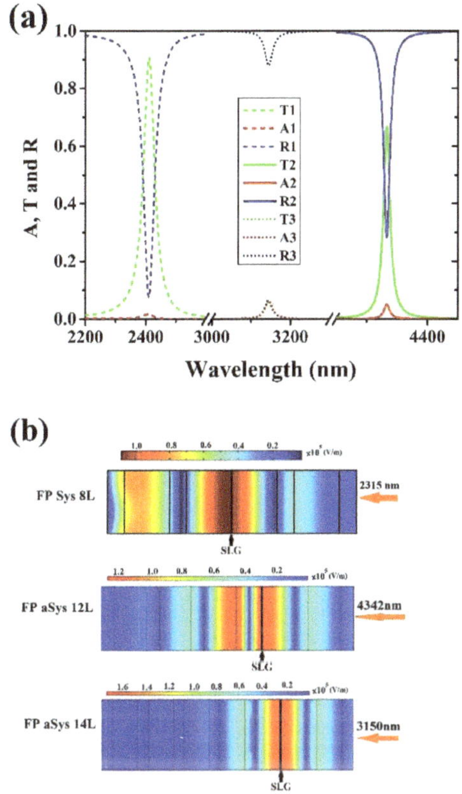

Figure 3. (**a**) Simulated transmittance (T), reflectance (R), and absorption (A) of Fabry–Perot filters FP1, FP2, and FP3; (**b**) electric field distribution inside the Fabry–Perot filters. SLG position is indicated in the three filters by a black line positioned at electric field maximum (red color).

Table 2. Graphene-based Fabry–Perot characteristics. Simulated and experimental absorption and electric field value at SLG position (maximum of electric field).

	Wavelength (nm)	A_SIM	A_EXP	E_G (V/m)
FP1	2315	40	39	106,604
FP2	4342	53	50	125,941
FP3	3150	87	84	172,663

2.2. Fabry–Perot Filter: Fabrication

The Fabry–Perot filters consisted of alternate layers of Si and SiO_2 with optical thickness equal to a quarter wavelength ($\lambda/4$, where λ is the central wavelength of the FP filter). Si and SiO_2 layers were fabricated by radio frequency sputtering in MRC (Material Research Corporation) systems, starting from a 99.999% purity Si and/or SiO_2 targets, as described elsewhere [67–69]. For FP1 and FP2, two different sputtering conditions were chosen: in case of all the layers except for the first 30 nm of the SiO_2 layer covering graphene, a radiofrequency sputtering of 200 W and a sputtering pressure of 1 Pa were used. For the first 30 nm of SiO_2 layer deposited onto graphene, a milder sputtering condition was used: i.e., sputtering power of 50 W and sputtering pressure of 3 Pa. For FP3 filter, grown in a different system, the same sputtering conditions (sputtering power of 200 W and sputtering pressure of 0.53 Pa) were used for all SiO_2 layers, due to the fact that such system geometry allowed a milder sputtering process.

MgF_2 layer was carefully evaporated in a Balzer BAE 250 evaporation system, and the evaporation rate was controlled by a quartz crystal microbalance.

2.3. CVD Graphene: Fabrication and Transfer

Single layer graphene was grown by CVD on Cu foils previously submitted to a pre-oxidation treatment at 250 °C for 90 min. CVD growth conditions were the following: substrate temperature 1070 °C, 10 mbar pressure by using 0.025 SCCM ethanol vapor with 20 SCCM Ar and 10 SCCM H_2. After the growth, graphene was removed from the back of the Cu substrate by means of an oxygen plasma cleaner (100 W, 4 min). For graphene transfer, a procedure reported elsewhere was applied [70]. Summarizing: a few drops of cyclododecane (20% solution in dichloromethane) were spin coated on top of graphene, and the Cu foils were left floating on an ammonium persulfate (PSA) water solution (120 g/L) for 3 h at 4 °C, until complete copper etching occurred. Afterword, the sample was rinsed in distilled water and scooped directly with the bottom multilayer, then submitted to a heating treatment @70 °C for a few hours and to ethyl acetate vapor cleaning to remove cyclododecane residues. The same transfer procedure was applied for transferring SLG onto other kinds of substrates (Si or quartz) used for graphene characterization.

2.4. Embedding Graphene Inside the Fabry–Perot Filters

Figure 4 shows the experimental steps used for the insertion of SLG inside the FP filters. In our experiments, to maximize absorption, graphene was always positioned inside the multilayer where the electric field was maximum [48,57]. In the symmetric FP1 filter, graphene was positioned in the middle of the cavity (LL), i.e., sandwiched between the two SiO_2 layers which separate the top and bottom mirrors. In the asymmetric FP2 filter, two maxima of electric field occurred, and graphene was positioned onto the upper Si layer of the cavity (HH). In the asymmetric reflective FP3 filter, graphene was positioned in the middle of the cavity (LL), i.e., sandwiched between the two SiO_2 layers.

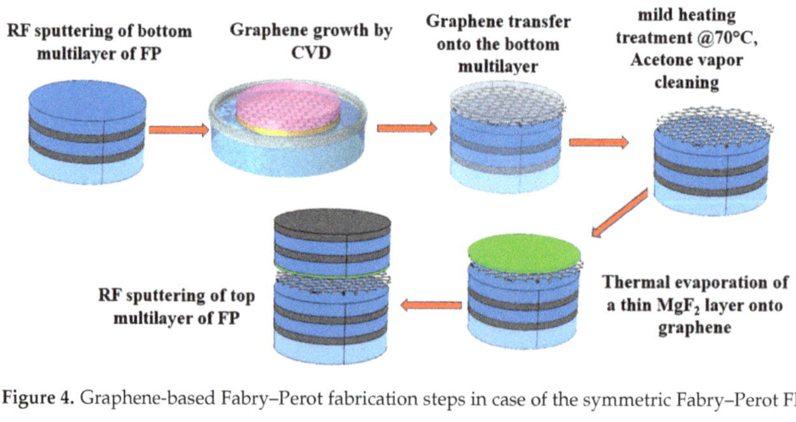

Figure 4. Graphene-based Fabry–Perot fabrication steps in case of the symmetric Fabry–Perot FP1.

The fabrication procedure of the multilayer structure was the following. First, the bottom multilayer structure and two CVD graphene layers were fabricated separately in the sputtering and CVD chambers, respectively. For each sputtering deposition run, two substrates (quartz or Si) were loaded in the chamber, one for the reference and the other for the graphene-based FP filters. Similarly, two graphene layers were prepared in the same run, one for insertion into the FP filter and the other for characterizations. Then, the graphene layer was transferred on the top of the bottom multilayer by the mild transfer process described above. The multilayer structure topped by graphene and the reference multilayer (without graphene) were transferred into the evaporation chamber for the growth of 30 nm MgF_2 protective layer. After the evaporation of the MgF_2 layer, the two samples were transferred back into the sputtering chamber for the deposition of the top multilayer structure.

2.5. Materials Characterizations

Transmittance and reflectance measurements in the NIR were carried out with a PerkinElmer 900 spectrophotometer, while MIR measurements were performed with a high resolution FTIR Perkin Elmer instrument. The reflectance measurements were obtained by using calibrated standards, i.e., an Ocean Optics Al mirror certified in the range 250–2500 nm in the first case and an Al infrared reflectance standard with a SiO overcoat certified by the National Physics Laboratory (NPL) in the wavenumber range of 4000–200 cm^{-1}. In case of the FTIR measurements, a Fixed-Angle Specular Reflectance Accessory from Perkin Elmer was used.

The Raman measurements were carried with a Renishaw inVia Reflex Raman spectrometer using a 514.5 nm excitation source.

3. Results and Discussion

3.1. Graphene Absorption Simulation

Figure 5a shows the simulated absorption of graphene inside the three FP filters while Figure 5b shows the FPs' absorption without the single layer graphene. The wavelength-dependent absorption A(λ) was inferred from the wavelength-dependent transmittance T(λ) and reflectance R(λ) (A(λ) = 1 − R(λ) − T(λ)) [71,72], by using the Wave Optics Module of Comsol Multiphysics software. Graphene optical properties were simulated by a wavelength-dependent complex refractive index $n_g(\omega) = \sqrt{\varepsilon_g(\omega)}$ [73,74], where the single-layer permittivity graphene was [75–77]:

$$\varepsilon_g(\omega) = \varepsilon_b + \frac{i\sigma(\omega)}{t_G \omega \varepsilon_0}$$

being ε_b (2.5) [78,79], t_G (0.35 nm) [80,81], and $\sigma(\omega)$ the intrinsic contribution to the graphene relative permittivity, the thickness of single layer graphene and the graphene optical conductivity, respectively.

The graphene optical conductivity was considered as the sum of interband transitions and Drude-like intraband conductivity [82–85]:

$$\begin{aligned} \sigma(\omega) &= \tfrac{\sigma_0}{2}[\tanh(\tfrac{\hbar\omega+2E_f}{4k_BT}) + \tanh(\tfrac{\hbar\omega-2E_f}{4k_BT})] \\ &-\tfrac{i\sigma_0}{2\pi}\log(\tfrac{(\hbar\omega+2E_f)^2}{(\hbar\omega-2E_f)^2+(2k_BT)^2}) \\ &+i\tfrac{4\sigma_0}{\pi}\tfrac{E_f}{\hbar\omega+i\hbar\gamma} \end{aligned}$$

with $\sigma_0 = (\frac{e^2}{4\hbar})$ [82,86], E_f the Fermi energy, and γ the intraband scattering rate. In the simulation, $\hbar\gamma$ = 40 meV [83] and E_f = 190 meV corresponding to slightly doped graphene as reported elsewhere [83,87].

Single layer permittivity was simulated also as a function of the wavelength and Fermi level, as shown in Figure 6.

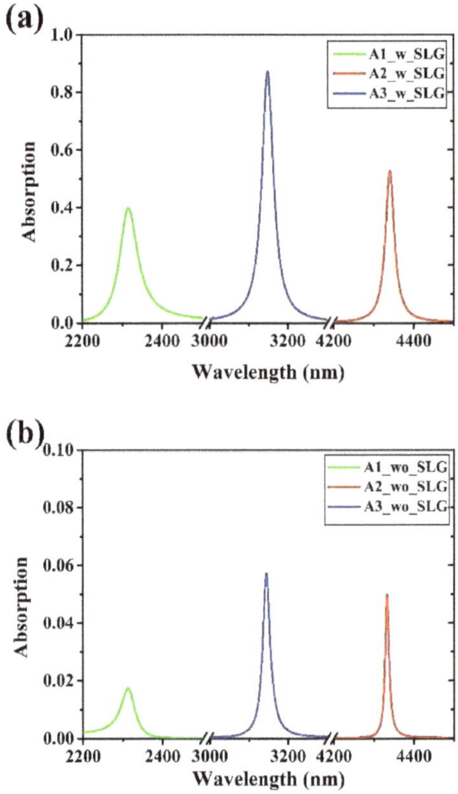

Figure 5. (**a**) Simulated absorption curves of SLG inside the Fabry–Perot filters in case of symmetric FP1 (A1), asymmetric FP2 (A2), and asymmetric reflective FP3 (A3). (**b**) Simulated absorption of FP filters without graphene insertion.

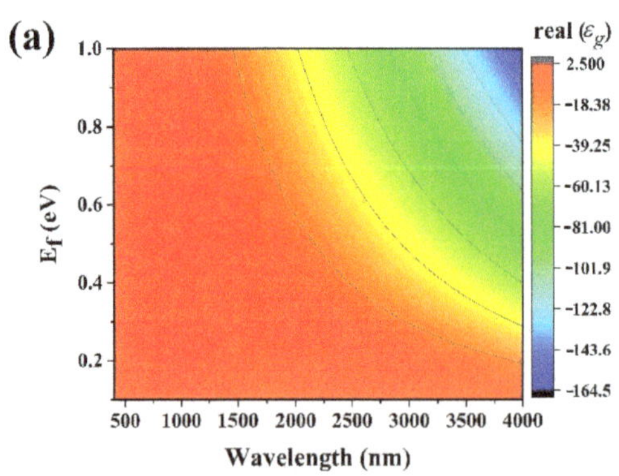

Figure 6. *Cont.*

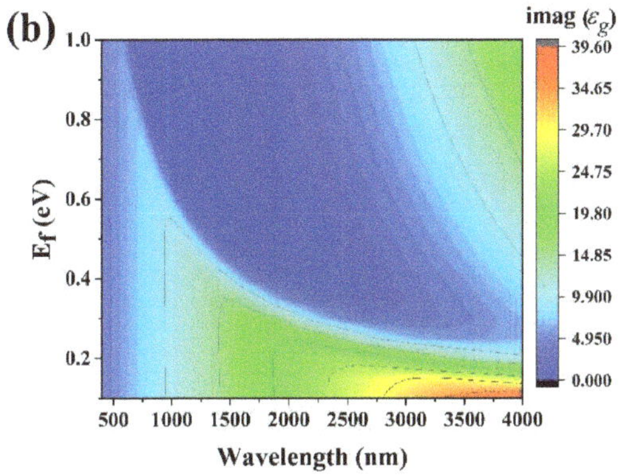

Figure 6. Simulated SLG permittivity as a function of wavelength and Fermi level. (**a**) Real and (**b**) imaginary parts.

3.2. Graphene Quality Assessment

The quality of graphene was checked after each experimental step, i.e., after MgF_2 evaporation, after SiO_2 sputtering on MgF_2/graphene layers and after the whole filter fabrication process.

Initially, for the growth of the first 30 nm of SiO_2 layer directly onto SLG, a mild sputtering condition was applied by increasing the deposition pressure and decreasing the radiofrequency power. Notwithstanding, graphene damage could not be avoided, as demonstrated by Raman analysis (not shown here), and a further graphene protection was necessary. Therefore, to avoid the sputtering of the SiO_2 layer directly onto graphene, a thin protective MgF_2 layer was evaporated prior to the sputtering of SiO_2. We have chosen MgF_2 due to its refractive index and extinction coefficient values, very close to those of SiO_2. This is necessary since MgF_2 becomes part of L (the low index layer) of the multilayer stack. Other fluoride materials could have served the purpose. Furthermore, the evaporation of the MgF_2 layer required some care since we observed that if evaporation occurred at a high rate, graphene defects band D increased. Figure 7a compares the Raman spectra of pristine graphene and graphene after evaporation of MgF_2 at high (2 Å/s) and low (0.1 Å/s) evaporation rates. Only the latter case left graphene unchanged. The effect of deposition rate of SiO_2 was preliminarily evaluated in experiments with ML graphene with unoptimized MgF_2 deposition (curve of Figure 7b). The effect of the optimized, low power and high pressure SiO_2 deposition on SL graphene can be seen in the red and blue curves of Figure 7c. Here the comparison of the red (Raman from the back of the peeled stack) and blue curves (Raman from the top through the SiO_2), suggests that some extent of the modifications of the D and 2D bands by SiO_2 is not due to lattice damage but to optical and electronic "proximity" influences on graphene by the stack.

Despite the increase of the defectiveness D band, the optical properties of graphene were preserved, as shown by the comparison of UV-Vis absorbance curves of pristine graphene transferred on a quartz substrate and graphene after MgF_2 and SiO_2 deposition (Figure 7d). Moreover, two-point probe measurements of SLG resistance prior and after the MgF_2 and SiO_2 sputtering confirmed that graphene resistance was almost unaffected by these processes. SLG sheet resistance increased, in fact, of less than 10% with respect to the initial value of about 800 Ω/sq.

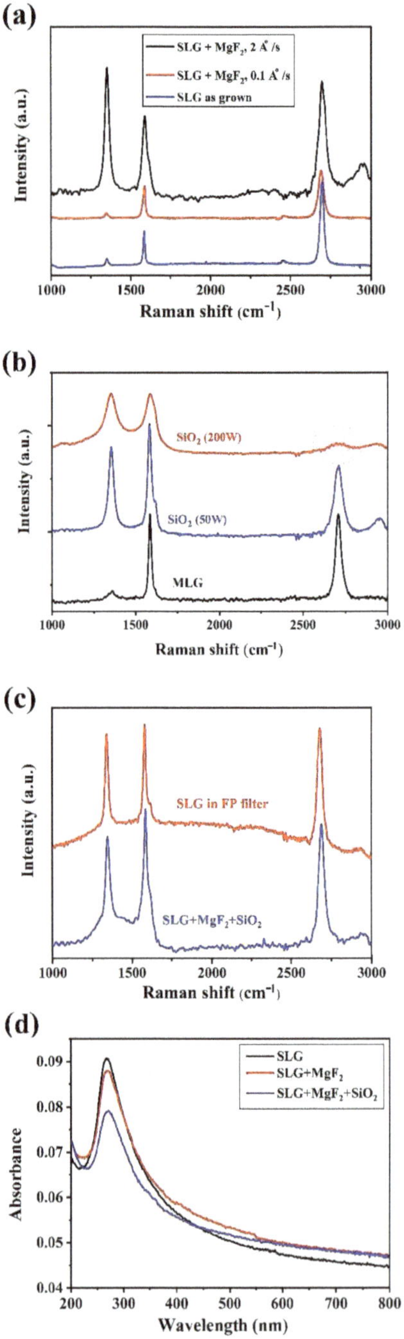

Figure 7. Raman spectra of: (**a**) SLG showing the effect of MgF_2 evaporation rate, (**b**) MLG showing the effect of SiO_2 sputtering conditions, and (**c**) SLG showing the effect of insertion inside the Fabry–Perot filter. (**d**) Comparison of absorbance of SLG pristine (as transferred on quartz substrate) and after MgF_2 evaporation and SiO_2 sputtering.

The results obtained after several attempts to not damage graphene can be summarized as following: (i) a thin MgF_2 layer was essential to protect graphene, (ii) evaporation of the thin MgF_2 layer should occur with a very low rate, (iii) the thin MgF_2 layer alone was not enough to avoid damage if sputtering process is too energetic, (iv) combination of evaporated MgF_2 and mild sputtering condition during SiO_2 growth produces only an increase in the graphene defective band (D band) but preserves the optical and electrical properties of pristine graphene, and (v) we could not quantify the damage of graphene inside the stack due to possible proximity effects by the embedding layers on D and 2D bands.

3.3. Graphene Absorption Inside the Fabry Perot Filters: Experimental

Figure 8a–c shows the comparison between the simulated and measured total absorption curves of graphene inside the three Fabry–Perot filters. The total absorption values (Table 2) include, also, the contribution of SLG absorption (2.3%) and of the FP materials (Figure 5b). A very good agreement between the simulated and measured absorption values was obtained, and only some broadening of the experimental curves was found in all cases. As it can be noticed, the experimental graphene absorption trend is to increase when passing from a symmetric structure to an asymmetric one and it increases further by increasing the reflectance of the bottom mirror. Even though a comparison of the absolute absorption values is not possible due to the different FPs' central wavelengths, the absorption trend strongly suggests that inside a Fabry–Perot cavity a high graphene absorption may be obtained only in asymmetric reflective structures. In our experiments, for example graphene absorption values increased from 39 to 84% when passing from a symmetric to an asymmetric reflective structure. The highest absorption value obtained in the asymmetric reflective FP3 filter is due to the fact that the incident light inside the cavity is reflected a greater number of times which allows to graphene a higher number of multiple absorptions. The number of reflections is in fact related to the Fabry–Perot finesse which is defined by [68,88–90]:

$$\Gamma_s = \frac{\pi\sqrt{R}}{1-R} \text{ and } \Gamma_{as} = \frac{\pi\sqrt[4]{R_{Top}R_{Bottom}}}{1-\sqrt{R_{Top}R_{Bottom}}}$$

Above, the left-hand side formula represents the finesse of the symmetric FP filter (FP1), where R is the reflectance of the top and bottom mirrors, while on the right-hand is the finesse of the asymmetric filters (FP2 and FP3). In our FP filters, the finesse increased from 21 to 31.5 when passing from the symmetric structure of Figure 1a (FP1) to the asymmetric reflective structure of Figure 1c (FP3).

Absorption of unpatterned and undoped single layer graphene up to 84% has been never reported experimentally in a Fabry–Perot filter. Comparable or even higher absorption values have instead been obtained by other authors by using graphene coupled to non-planar resonant plasmonic structures [45,91,92].

Enhancing the light-matter interaction in single-layer graphene by the use of optical microcavity requires no stringent constrains on the graphene electrical properties, such as the sheet resistance, and may find application in a variety of graphene-based devices, such as optical absorption modulators, light emitters, and optical attenuators, and provides new routes to graphene photonics for applications in spectroscopy, communications, sensing, and security.

Results, validated here at NIR and MIR wavelengths, may be applied in principle at different wavelengths ranging from Vis to THz by properly choosing the materials of the layers and of the substrate. Moreover, as obtained by our previous simulations, modification of the filter amplitude and bandwidth may be achieved by: (i) using multilayer graphene instead of single layer graphene [68], (ii) using two or more SLGs positioned in the filter where the electric field has its maxima [69], and (iii) increasing the number of cavities of the multilayer structure [68].

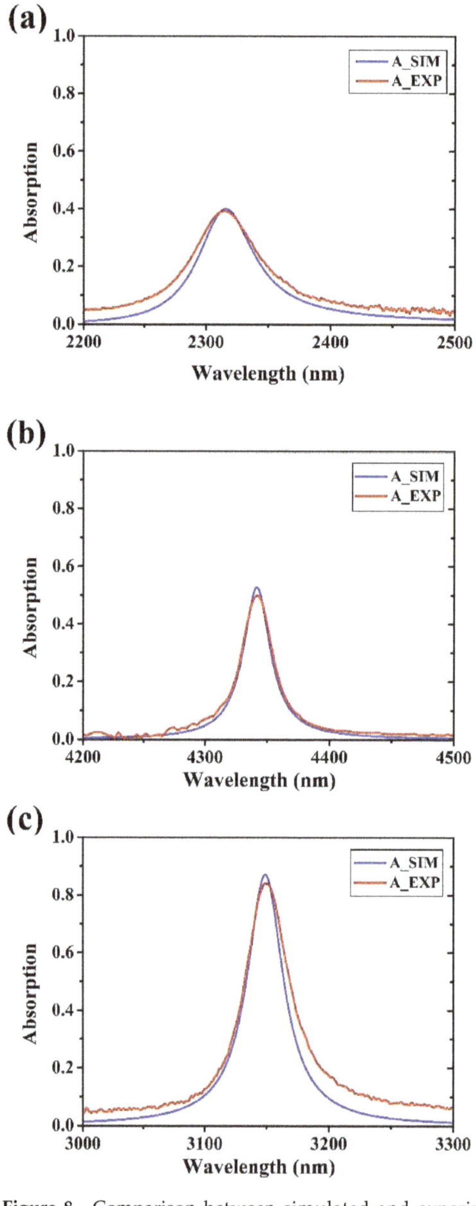

Figure 8. Comparison between simulated and experimental absorption curves of single layer graphene inside the Fabry–Perot filters: (**a**) symmetric FP1, (**b**) asymmetric FP2, and (**c**) asymmetric reflective FP3.

Figure 9a shows, as an example, the absorption of a single layer graphene, of a double-layers graphene (DLG) and of a five-layers graphene (MLG) embedded in the symmetric FP1. As it can be noticed, absorption bandwidth increases by increasing the number of layers, while absorption amplitude increases in DLG and decreases in MLG. The latter behavior can be attributed to the higher intrinsic absorption of MLG which reduces the reflections of the light inside the optical cavity, and to the fact that MLG perturbs largely the optical quality of the Fabry–Perot cavity [68].

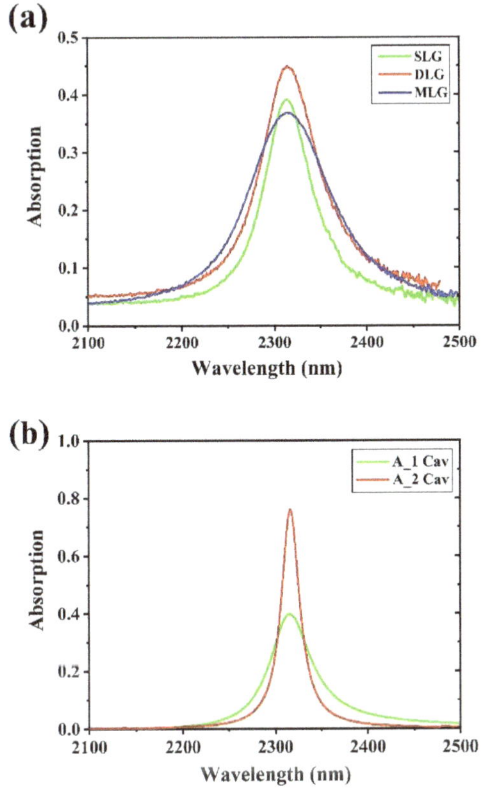

Figure 9. Absorption of graphene layers inside the FP1 as a function of number of graphene layers (**a**) and absorption of a single layer graphene in a single Fabry–Perot cavity and in a dual Fabry–Perot cavity (**b**).

Figure 9b shows the effect of the number of Fabry–Perot cavities on the absorption of a single layer graphene [68], showing a higher graphene absorption inside a dual cavity Fabry–Perot filter.

Results are in accordance with what has been reported by other authors. In [46], Xiao et al. have numerically modeled, in a structure different from ours, the bandwidth absorption as a function of the number of graphene layers finding a broader absorption band by increasing the number of graphene layers from 1 to 7. In [50], Ferreira et al. have numerically simulated that absorption of single layer graphene increases by using a dual cavity Fabry–Perot instead of a single one.

Graphene absorption discussed so far was obtained in the case of a TE polarization. For the optimized FP structure, i.e., the FP3 filter, the simulation was carried out, also, in case of TM polarization. Figure 10 shows the comparison between the simulated absorption of SLG for TE and TM polarizations as a function of the incident angle. As it can be noticed, in case of TE polarization the absorption maximum is reached for incident radiation, while for TM polarization it occurs at a higher incident angle of 60°.

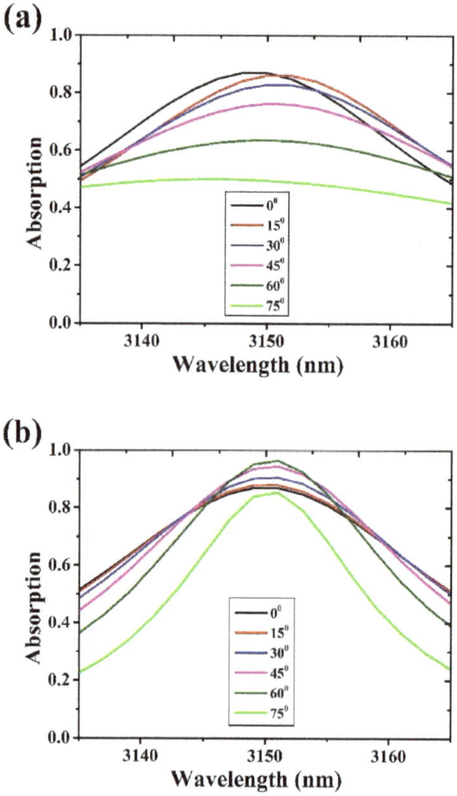

Figure 10. SLG optical absorption as a function of wavelength for different incident angles at (**a**) TE and (**b**) TM polarization.

4. Conclusions

The increase of single layer graphene absorption obtained by exploiting the electric field enhancement inside a resonant optical cavity was modeled and experimentally demonstrated in three different Fabry–Perot filters with central wavelengths varying in the NIR-MIR spectral ranges. The Fabry–Perot filters were fabricated by radiofrequency sputtering, and consisted of alternate quarter wave thick Si and SiO_2 layers. Results demonstrated that graphene absorption greatly increases when graphene is embedded inside an asymmetric Fabry–Perot structure, reaching its maximum in case of a reflective Fabry–Perot filter. SLG properties were preserved during the sputtering process by applying a thin, slowly evaporated MgF_2 layer, allowing the effective embedding of graphene inside thick monolithic structures fabricated by conventional PVD techniques. Such a high graphene absorption discloses exciting potentiality for exploitation of 2D materials in new optoelectronic devices for application in the NIR-MIR spectral range.

Author Contributions: Conceptualization, M.L.G. and A.N.; supervision of the study, M.L.G.; simulation, A.N.; investigation, A.N., M.L.G., N.L. and L.L.; data curation, M.L.G. and A.N.; writing—original draft preparation, M.L.G.; writing—review and editing, M.L.G., A.N. and N.L. All authors have read and agreed to the published version of the manuscript.

Funding: This research received no external funding.

Data Availability Statement: Data are available on request from the corresponding author.

Acknowledgments: Authors acknowledges the support from the International Centre for Theoretical Physics, ICTP TRIL fellowship number 3356. Authors thank Angelo Gentili for his excellent technical maintenance of the sputtering and evaporation systems.

Conflicts of Interest: The authors declare no conflict of interest.

References

1. Nematpour, A.; Nikoufard, M. Plasmonic thin film InP/graphene-based Schottky-junction solar cell using nanorods. *J. Adv. Res.* **2018**, *10*, 15–20. [CrossRef] [PubMed]
2. Das, S.; Pandey, D.; Thomas, J.; Roy, T. The role of graphene and other 2D materials in solar photovoltaics. *Adv. Mater.* **2019**, *31*, 1802722. [CrossRef] [PubMed]
3. O'keeffe, P.; Catone, D.; Paladini, A.; Toschi, F.; Turchini, S.; Avaldi, L.; Martelli, F.; Agresti, A.; Pescetelli, S.; Del Rio Castillo, A.; et al. Graphene-induced improvements of perovskite solar cell stability: Effects on hot-carriers. *Nano Lett.* **2019**, *19*, 684–691. [CrossRef] [PubMed]
4. Agarwal, H.; Terrés, B.; Orsini, L.; Montanaro, A.; Sorianello, V.; Pantouvaki, M.; Watanabe, K.; Taniguchi, T.; Van Thourhout, D.; Romagnoli, M.; et al. 2D-3D integration of hexagonal boron nitride and a high-κ dielectric for ultrafast graphene-based electro-absorption modulators. *Nat. Commun.* **2021**, *12*, 1070. [CrossRef]
5. Phare, C.T.; Lee, Y.-H.D.; Cardenas, J.; Lipson, M. Graphene electro-optic modulator with 30 GHz bandwidth. *Nat. Photonics* **2015**, *9*, 511–514. [CrossRef]
6. Amin, R.; Ma, Z.; Maiti, R.; Khan, S.; Khurgin, J.B.; Dalir, H.; Sorger, V.J. Attojoule-efficient graphene optical modulators. *Appl. Opt.* **2018**, *57*, D130–D140. [CrossRef]
7. Guo, C.; Zhang, J.; Xu, W.; Liu, K.; Yuan, X.; Qin, S.; Zhu, Z. Graphene-based perfect absorption structures in the visible to terahertz band and their optoelectronics applications. *Nanomaterials* **2018**, *8*, 1033. [CrossRef]
8. Kakenov, N.; Ergoktas, M.S.; Balci, O.; Kocabas, C. Graphene based terahertz phase modulators. *2D Mater.* **2018**, *5*, 035018. [CrossRef]
9. Goldsmith, B.R.; Locascio, L.; Gao, Y.; Lerner, M.; Walker, A.; Lerner, J.; Kyaw, J.; Shue, A.; Afsahi, S.; Pan, D.; et al. Digital biosensing by foundry-fabricated graphene sensors. *Sci. Rep.* **2019**, *9*, 434. [CrossRef]
10. Hajian, R.; Balderston, S.; Tran, T.; DeBoer, T.; Etienne, J.; Sandhu, M.; Wauford, N.A.; Chung, J.-Y.; Nokes, J.; Athaiya, M.; et al. Detection of unamplified target genes via CRISPR–Cas9 immobilized on a graphene field-effect transistor. *Nat. Biomed. Eng.* **2019**, *3*, 427–437. [CrossRef]
11. Tajik, S.; Dourandish, Z.; Zhang, K.; Beitollahi, H.; Van Le, Q.; Jang, H.W.; Shokouhimehr, M. Carbon and graphene quantum dots: A review on syntheses, characterization, biological and sensing applications for neurotransmitter determination. *RSC Adv.* **2020**, *10*, 15406–15429. [CrossRef]
12. De Arquer, F.P.G.; Armin, A.; Meredith, P.; Sargent, E.H. Solution-processed semiconductors for next-generation photodetectors. *Nat. Rev. Mater.* **2017**, *2*, 16100. [CrossRef]
13. Polat, E.O.; Mercier, G.; Nikitskiy, I.; Puma, E.; Galan, T.; Gupta, S.; Montagut, M.; Piqueras, J.J.; Bouwens, M.; Durduran, T.; et al. Flexible graphene photodetectors for wearable fitness monitoring. *Sci. Adv.* **2019**, *5*, eaaw7846. [CrossRef]
14. Muench, J.E.; Ruocco, A.; Giambra, M.A.; Miseikis, V.; Zhang, D.; Wang, J.; Watson, H.F.; Park, G.C.; Akhavan, S.; Sorianello, V.; et al. Waveguide-integrated, plasmonic enhanced graphene photodetectors. *Nano Lett.* **2019**, *19*, 7632–7644. [CrossRef]
15. Grotevent, M.J.; Hail, C.U.; Yakunin, S.; Dirin, D.N.; Thodkar, K.; Borin Barin, G.; Guyot-Sionnest, P.; Calame, M.; Poulikakos, D.; Kovalenko, M.V.; et al. Nanoprinted quantum dot–graphene photodetectors. *Adv. Opt. Mater.* **2019**, *7*, 1900019. [CrossRef]
16. Withers, F.; Bointon, T.H.; Craciun, M.F.; Russo, S. All-graphene photodetectors. *ACS Nano* **2013**, *7*, 5052–5057. [CrossRef]
17. Huang, H.; Ke, S.; Wang, B.; Long, H.; Wang, K.; Lu, P. Numerical study on plasmonic absorption enhancement by a rippled graphene sheet. *J. Lightwave Technol.* **2017**, *35*, 320–324. [CrossRef]
18. Sang, T.; Gao, J.; Yin, X.; Qi, H.; Wang, L.; Jiao, H. Angle-insensitive broadband absorption enhancement of graphene using a multi-grooved metasurface. *Nanoscale Res. Lett.* **2019**, *14*, 105. [CrossRef]
19. Ye, L.; Zeng, F.; Zhang, Y.; Liu, Q.H. Composite graphene-metal microstructures for enhanced multiband absorption covering the entire terahertz range. *Carbon* **2019**, *148*, 317–325. [CrossRef]
20. Du, Z.; Chen, X.; Zhang, Y.; Que, X.; Liu, P.; Zhang, X.; Ma, H.-L.; Zhai, M. One-pot hydrothermal preparation of Fe3O4 decorated graphene for microwave absorption. *Materials* **2020**, *13*, 3065. [CrossRef]
21. Berry, V. Impermeability of graphene and its applications. *Carbon* **2013**, *62*, 1–10. [CrossRef]
22. Yin, J.; Wang, H.; Peng, H.; Tan, Z.; Liao, L.; Lin, L.; Sun, X.; Koh, A.L.; Chen, Y.; Peng, H. Selectively enhanced photocurrent generation in twisted bilayer graphene with van Hove singularity. *Nat. Commun.* **2016**, *7*, 10699. [CrossRef] [PubMed]
23. Su, Y.; Lin, Q.; Zhai, X.; Luo, X.; Wang, L.-L. Controlling terahertz surface plasmon polaritons in Dirac semimetal sheets. *Opt. Mater. Express* **2018**, *8*, 884–892. [CrossRef]
24. Farhat, M.; Guenneau, S.; Bağcı, H. Exciting graphene surface plasmon polaritons through light and sound interplay. *Phys. Rev. Lett.* **2013**, *111*, 237404. [CrossRef]

25. Hajian, H.; Serebryannikov, A.E.; Ghobadi, A.; Demirag, Y.; Butun, B.; Vandenbosch, G.A.; Ozbay, E. Tailoring far-infrared surface plasmon polaritons of a single-layer graphene using plasmon-phonon hybridization in graphene-LiF heterostructures. *Sci. Rep.* **2018**, *8*, 13209. [CrossRef]
26. Liu, S.; Zhang, C.; Hu, M.; Chen, X.; Zhang, P.; Gong, S.; Zhao, T.; Zhong, R. Coherent and tunable terahertz radiation from graphene surface plasmon polaritons excited by an electron beam. *Appl. Phys. Lett.* **2014**, *104*, 201104. [CrossRef]
27. Nong, J.; Tang, L.; Lan, G.; Luo, P.; Guo, C.; Yi, J.; Wei, W. Wideband tunable perfect absorption of graphene plasmons via attenuated total reflection in Otto prism configuration. *Nanophotonics* **2020**, *9*, 645–655. [CrossRef]
28. Bandurin, D.A.; Svintsov, D.; Gayduchenko, I.; Xu, S.G.; Principi, A.; Moskotin, M.; Tretyakov, I.; Yagodkin, D.; Zhukov, S.; Taniguchi, T.; et al. Resonant terahertz detection using graphene plasmons. *Nat. Commun.* **2018**, *9*, 5392. [CrossRef]
29. Castilla, S.; Terrés, B.; Autore, M.; Viti, L.; Li, J.; Nikitin, A.Y.; Vangelidis, I.; Watanabe, K.; Taniguchi, T.; Lidorikis, E.; et al. Fast and sensitive terahertz detection using an antenna-integrated graphene pn junction. *Nano Lett.* **2019**, *19*, 2765–2773. [CrossRef]
30. Chen, P.-Y.; Argyropoulos, C.; Farhat, M.; Gomez-Diaz, J.S. Flatland plasmonics and nanophotonics based on graphene and beyond. *Nanophotonics* **2017**, *6*, 1239–1262. [CrossRef]
31. Lee, S.H.; Choi, M.; Kim, T.-T.; Lee, S.; Liu, M.; Yin, X.; Choi, H.K.; Lee, S.S.; Choi, C.-G.; Choi, S.-Y. Switching terahertz waves with gate-controlled active graphene metamaterials. *Nat. Mater.* **2012**, *11*, 936–941. [CrossRef]
32. Fang, Z.; Wang, Y.; Liu, Z.; Schlather, A.; Ajayan, P.M.; Koppens, F.H.; Nordlander, P.; Halas, N.J. Plasmon-induced doping of graphene. *ACS Nano* **2012**, *6*, 10222–10228. [CrossRef]
33. Ogawa, S.; Kimata, M. Metal-insulator-metal-based plasmonic metamaterial absorbers at visible and infrared wavelengths: A review. *Materials* **2018**, *11*, 458. [CrossRef]
34. Lin, K.-T.; Lin, H.; Yang, T.; Jia, B. Structured graphene metamaterial selective absorbers for high efficiency and omnidirectional solar thermal energy conversion. *Nat. Commun.* **2020**, *11*, 1389. [CrossRef]
35. Li, Q.; Lu, J.; Gupta, P.; Qiu, M. Engineering optical absorption in graphene and other 2D materials: Advances and applications. *Adv. Opt. Mater.* **2019**, *7*, 1900595. [CrossRef]
36. Xiong, L.; Forsythe, C.; Jung, M.; McLeod, A.; Sunku, S.; Shao, Y.; Ni, G.; Sternbach, A.; Liu, S.; Edgar, J.; et al. Photonic crystal for graphene plasmons. *Nat. Commun.* **2019**, *10*, 4780. [CrossRef]
37. Rao, W.-F.; Zhang, M.-T.; Zheng, G.-G. Voltage-controlled enhancement of optical absorption in a graphene monolayer with a one-dimensional photonic crystal. *Appl. Phys. B* **2017**, *123*, 232. [CrossRef]
38. Hu, J.; Liu, W.; Xie, W.; Zhang, W.; Yao, E.; Zhang, Y.; Zhan, Q. Strong coupling of optical interface modes in a 1D topological photonic crystal heterostructure/Ag hybrid system. *Opt. Lett.* **2019**, *44*, 5642–5645. [CrossRef]
39. Lin, Y.; Chou, S.; Hsueh, W.-J. Tunable light absorption of graphene using topological interface states. *Opt. Lett.* **2020**, *45*, 4369–4372. [CrossRef]
40. Yariv, A. Universal relations for coupling of optical power between microresonators and dielectric waveguides. *Electron. Lett.* **2000**, *36*, 321–322. [CrossRef]
41. Wang, R.; Ren, X.-G.; Yan, Z.; Jiang, L.-J.; Wei, E.; Shan, G.-C. Graphene based functional devices: A short review. *Front. Phys.* **2019**, *14*, 13603. [CrossRef]
42. Yildirim, D.U.; Ghobadi, A.; Ozbay, E. Near-absolute polarization insensitivity in graphene based ultra-narrowband perfect visible light absorber. *Sci. Rep.* **2018**, *8*, 15210. [CrossRef]
43. Thongrattanasiri, S.; Koppens, F.H.; De Abajo, F.J.G. Complete optical absorption in periodically patterned graphene. *Phys. Rev. Lett.* **2012**, *108*, 047401. [CrossRef]
44. Piper, J.R.; Fan, S. Total absorption in a graphene monolayer in the optical regime by critical coupling with a photonic crystal guided resonance. *ACS Photonics* **2014**, *1*, 347–353. [CrossRef]
45. Liu, Y.; Chadha, A.; Zhao, D.; Piper, J.R.; Jia, Y.; Shuai, Y.; Menon, L.; Yang, H.; Ma, Z.; Fan, S. Approaching total absorption at near infrared in a large area monolayer graphene by critical coupling. *Appl. Phys. Lett.* **2014**, *105*, 181105. [CrossRef]
46. Xiao, S.; Liu, T.; Wang, X.; Liu, X.; Zhou, C. Tailoring the absorption bandwidth of graphene at critical coupling. *Phys. Rev. B* **2020**, *102*, 085410. [CrossRef]
47. Jin, H.; Guo, C.; Liu, X.; Liu, J.; Vasileff, A.; Jiao, Y.; Zheng, Y.; Qiao, S.-Z. Emerging two-dimensional nanomaterials for electrocatalysis. *Chem. Rev.* **2018**, *118*, 6337–6408. [CrossRef]
48. Nulli, S.A.; Ukhtary, M.S.; Saito, R. Significant enhancement of light absorption in undoped graphene using dielectric multilayer system. *Appl. Phys. Lett.* **2018**, *112*, 073101. [CrossRef]
49. Ferreira, A.; Peres, N.; Ribeiro, R.; Stauber, T. Graphene-based photodetector with two cavities. *Phys. Rev. B* **2012**, *85*, 115438. [CrossRef]
50. Zhang, K.; Liu, Y.; Xia, F.; Li, S.; Kong, W. Tuning of the polariton modes induced by longitudinal strong coupling in the graphene hybridized DBR cavity. *Opt. Lett.* **2020**, *45*, 3669–3672. [CrossRef]
51. Xu, W.; Zhu, Z.; Liu, K.; Zhang, J.; Yuan, X.; Lu, Q.; Qin, S. Enhancement of near-infrared light graphene interaction by nanobeam resonator. *IEEE Photonics Technol. Lett.* **2015**, *27*, 2023–2026. [CrossRef]
52. Chen, S. Dual-mode near-infrared light tunable absorber based on graphene by optical Tamm state and microcavity. *Opt. Commun.* **2021**, *492*, 126962. [CrossRef]
53. Yu, Y.; Zhou, J.; Cai, Q.; Chu, Z.; Deng, J.; Lu, W.; Li, Z.; Chen, X. Dynamically tunable ultra-narrowband perfect absorbers for the visible-to-infrared range based on a microcavity integrated graphene pair. *Opt. Lett.* **2021**, *46*, 2236–2239. [CrossRef]

54. Wei, T.; Su, Z.; Wang, Y. Graphene-based dual-band near-perfect absorption in Rabi splitting between topological edge and Fabry–Perot cavity modes. *J. Opt.* **2021**, *23*, 125003. [CrossRef]
55. Bian, L.-A.; Deng, Z.; Hong, Y.; Qiu, Y.; Liu, Z.; Xiao, P.; Li, G. Double-mode absorption in double-defect photonic crystal with one graphene multilayer. *Opt. Quantum Electron.* **2020**, *52*, 145. [CrossRef]
56. Vasić, B.; Gajić, R. Tunable Fabry–Perot resonators with embedded graphene from terahertz to near-infrared frequencies. *Opt. Lett.* **2014**, *39*, 6253–6256. [CrossRef] [PubMed]
57. Deng, X.-H.; Liu, J.-T.; Yuan, J.; Wang, T.-B.; Liu, N.-H. Tunable THz absorption in graphene-based heterostructures. *Opt. Express* **2014**, *22*, 30177–30183. [CrossRef]
58. Zheng, J.; Barton, R.A.; Englund, D. Broadband coherent absorption in chirped-planar-dielectric cavities for 2D-material-based photovoltaics and photodetectors. *Acs Photonics* **2014**, *1*, 768–774. [CrossRef]
59. Bian, L.-A.; Yang, L.; Liu, P.; Chen, Y.; Liu, H.; Zhou, Q. Controllable perfect absorption in a double-cavity photonic crystal with one graphene monolayer. *J. Phys. D Appl. Phys.* **2017**, *51*, 025106. [CrossRef]
60. Doukas, S.; Chatzilari, A.; Dagkli, A.; Papagiannopoulos, A.; Lidorikis, E. Deep and fast free-space electro-absorption modulation in a mobility-independent graphene-loaded Bragg resonator. *Appl. Phys. Lett.* **2018**, *113*, 011102. [CrossRef]
61. Zand, I.; Dalir, H.; Chen, R.T.; Dowling, J.P. Multispectral selective near-perfect light absorption by graphene monolayer using aperiodic multilayer microstructures. *Appl. Phys. Express* **2018**, *11*, 035101. [CrossRef]
62. Chen, Y.; Bian, L.-a.; Xie, Y.; Li, G.; Lin, M.; Liu, P.; Zha, S.; Yi, B. Controllable tri-mode perfect absorption in hybrid resonance cavity with single graphene monolayer. *Opt. Commun.* **2019**, *445*, 84–89. [CrossRef]
63. Furchi, M.; Urich, A.; Pospischil, A.; Lilley, G.; Unterrainer, K.; Detz, H.; Klang, P.; Andrews, A.M.; Schrenk, W.; Strasser, G. Microcavity-integrated graphene photodetector. *Nano Lett.* **2012**, *12*, 2773–2777. [CrossRef]
64. Wang, X.-C.; Tretyakov, S.A. Graphene-based Perfect Absorbers: Systematic Design and High Tunability. In Proceedings of the 2018 IEEE International Symposium on Antennas and Propagation & USNC/URSI National Radio Science Meeting, Boston, MA, USA, 8–13 July 2018; pp. 919–920.
65. Fan, Y.; Guo, C.; Zhu, Z.; Xu, W.; Wu, F.; Yuan, X.; Qin, S. Monolayer-graphene-based perfect absorption structures in the near infrared. *Opt. Express* **2017**, *25*, 13079–13086. [CrossRef] [PubMed]
66. Engel, M.; Steiner, M.; Lombardo, A.; Ferrari, A.C.; Löhneysen, H.V.; Avouris, P.; Krupke, R. Light–matter interaction in a microcavity-controlled graphene transistor. *Nat. Commun.* **2012**, *3*, 906. [CrossRef] [PubMed]
67. Nematpour, A.; Lisi, N.; Lancellotti, L.; Chierchia, R.; Grilli, M.L. Experimental Mid-Infrared Absorption (84%) of Single Layer Graphene in a Reflective Asymmetric Fabry–Perot Filter: Implications for Photodetectors. *ACS Appl. Nano Mater.* **2021**, *4*, 1495–1502. [CrossRef]
68. Nematpour, A.; Lisi, N.; Piegari, A.; Lancellotti, L.; Hu, C.; Grilli, M.L. Experimental near infrared absorption enhancement of graphene layers in an optical resonant cavity. *Nanotechnology* **2019**, *30*, 445201. [CrossRef]
69. Nematpour, A.; Lisi, N.; Chierchia, R.; Grilli, M.L. Experimental demonstration of mid-IR absorption enhancement in single layer CVD graphene. *Opt. Lett.* **2020**, *45*, 3861–3864. [CrossRef]
70. Lisi, N.; Dikonimos, T.; Buonocore, F.; Pittori, M.; Mazzaro, R.; Rizzoli, R.; Marras, S.; Capasso, A. Contamination-free graphene by chemical vapor deposition in quartz furnaces. *Sci. Rep.* **2017**, *7*, 9927. [CrossRef]
71. Hu, S.; Yang, S.; Liu, Z.; Quan, B.; Li, J.; Gu, C. Broadband and polarization-insensitive absorption based on a set of multisized Fabry–Perot-like resonators. *J. Phys. Chem. C* **2019**, *123*, 13856–13862. [CrossRef]
72. Wang, X.; Wang, J.; Hu, Z.-D.; Sang, T.; Feng, Y. Perfect absorption of modified-molybdenum-disulfide-based Tamm plasmonic structures. *Appl. Phys. Express* **2018**, *11*, 062601. [CrossRef]
73. Li, X.; Tao, L.; Chen, Z.; Fang, H.; Li, X.; Wang, X.; Xu, J.-B.; Zhu, H. Graphene and related two-dimensional materials: Structure-property relationships for electronics and optoelectronics. *Appl. Phys. Rev.* **2017**, *4*, 021306. [CrossRef]
74. Vorobev, A.S.; Bianco, G.V.; Bruno, G.; D'Orazio, A.; O'Faolain, L.; Grande, M. Tuning of Graphene-Based Optical Devices Operating in the Near-Infrared. *Appl. Sci.* **2021**, *11*, 8367. [CrossRef]
75. Xia, S.-X.; Zhai, X.; Wang, L.-L.; Sun, B.; Liu, J.-Q.; Wen, S.-C. Dynamically tunable plasmonically induced transparency in sinusoidally curved and planar graphene layers. *Opt. Express* **2016**, *24*, 17886–17899. [CrossRef] [PubMed]
76. Xia, S.-X.; Zhai, X.; Wang, L.-L.; Wen, S.-C. Plasmonically induced transparency in double-layered graphene nanoribbons. *Photonics Res.* **2018**, *6*, 692–702. [CrossRef]
77. Chen, F.; Cheng, Y.; Luo, H. A broadband tunable terahertz metamaterial absorber based on single-layer complementary gammadion-shaped graphene. *Materials* **2020**, *13*, 860. [CrossRef]
78. Wang, X.; Chen, C.; Pan, L.; Wang, J. A graphene-based Fabry-Pérot spectrometer in mid-infrared region. *Sci. Rep.* **2016**, *6*, 32616. [CrossRef]
79. Sui, G.; Wu, J.; Zhang, Y.; Yin, C.; Gao, X. Microcavity-integrated graphene waveguide: A reconfigurable electro-optical attenuator and switch. *Sci. Rep.* **2018**, *8*, 12445. [CrossRef]
80. Cao, K.; Feng, S.; Han, Y.; Gao, L.; Ly, T.H.; Xu, Z.; Lu, Y. Elastic straining of free-standing monolayer graphene. *Nat. Commun.* **2020**, *11*, 284. [CrossRef]
81. Guo, M.; Qian, Y.; Qi, H.; Bi, K.; Chen, Y. Experimental measurements on the thermal conductivity of strained monolayer graphene. *Carbon* **2020**, *157*, 185–190. [CrossRef]

82. Tasolamprou, A.C.; Koulouklidis, A.D.; Daskalaki, C.; Mavidis, C.P.; Kenanakis, G.; Deligeorgis, G.; Viskadourakis, Z.; Kuzhir, P.; Tzortzakis, S.; Kafesaki, M. Experimental demonstration of ultrafast THz modulation in a graphene-based thin film absorber through negative photoinduced conductivity. *ACS Photonics* **2019**, *6*, 720–727. [CrossRef]
83. Chang, Y.-C.; Liu, C.-H.; Liu, C.-H.; Zhang, S.; Marder, S.R.; Narimanov, E.E.; Zhong, Z.; Norris, T.B. Realization of mid-infrared graphene hyperbolic metamaterials. *Nat. Commun.* **2016**, *7*, 10568. [CrossRef] [PubMed]
84. Sato, S.A.; Hirori, H.; Sanari, Y.; Kanemitsu, Y.; Rubio, A. High-order harmonic generation in graphene: Nonlinear coupling of intraband and interband transitions. *Phys. Rev. B* **2021**, *103*, L041408. [CrossRef]
85. Jiang, H.; Choudhury, S.; Kudyshev, Z.A.; Wang, D.; Prokopeva, L.J.; Xiao, P.; Jiang, Y.; Kildishev, A.V. Enhancing sensitivity to ambient refractive index with tunable few-layer graphene/hBN nanoribbons. *Photonics Res.* **2019**, *7*, 815–822. [CrossRef]
86. Dashti, M.; Carey, J.D. Graphene microstrip patch ultrawide band antennas for THz communications. *Adv. Funct. Mater.* **2018**, *28*, 1705925. [CrossRef]
87. Chang, Y.-C.; Liu, C.-H.; Liu, C.-H.; Zhong, Z.; Norris, T.B. Extracting the complex optical conductivity of mono-and bilayer graphene by ellipsometry. *Appl. Phys. Lett.* **2014**, *104*, 261909. [CrossRef]
88. Xiao, Z.; Huang, H.; Lu, X.-X. Resonant tunneling dynamics and the related tunneling time. *Int. J. Mod. Phys. B* **2015**, *29*, 1550052. [CrossRef]
89. Muller, A.; Flagg, E.B.; Lawall, J.R.; Solomon, G.S. Ultrahigh-finesse, low-mode-volume Fabry–Perot microcavity. *Opt. Lett.* **2010**, *35*, 2293–2295. [CrossRef]
90. Suter, M.; Dietiker, P. Calculation of the finesse of an ideal Fabry–Perot resonator. *Appl. Opt.* **2014**, *53*, 7004–7010. [CrossRef]
91. Kim, S.; Jang, M.S.; Brar, V.W.; Mauser, K.W.; Kim, L.; Atwater, H.A. Electronically tunable perfect absorption in graphene. *Nano Lett.* **2018**, *18*, 971–979. [CrossRef]
92. Guo, C.C.; Zhu, Z.H.; Yuan, X.D.; Ye, W.M.; Liu, K.; Zhang, J.F.; Xu, W.; Qin, S.Q. Experimental demonstration of total absorption over 99% in the near infrared for monolayer-graphene-based subwavelength structures. *Adv. Opt. Mater.* **2016**, *4*, 1955–1960. [CrossRef]

materials

Review

A Review on Chemical Vapour Deposition of Two-Dimensional MoS_2 Flakes

Luca Seravalli * and **Matteo Bosi** *

IMEM-CNR, Parco Area delle Scienze 37A, 43124 Parma, Italy
* Correspondence: authors: luca.seravalli@imem.cnr.it (L.S.); matteo.bosi@imem.cnr.it (M.B.)

Abstract: Two-dimensional (2D) materials such as graphene, transition metal dichalcogenides, and boron nitride have recently emerged as promising candidates for novel applications in sensing and for new electronic and photonic devices. Their exceptional mechanical, electronic, optical, and transport properties show peculiar differences from those of their bulk counterparts and may allow for future radical innovation breakthroughs in different applications. Control and reproducibility of synthesis are two essential, key factors required to drive the development of 2D materials, because their industrial application is directly linked to the development of a high-throughput and reliable technique to obtain 2D layers of different materials on large area substrates. Among various methods, chemical vapour deposition is considered an excellent candidate for this goal thanks to its simplicity, widespread use, and compatibility with other processes used to deposit other semiconductors. In this review, we explore the chemical vapour deposition of MoS_2, considered one of the most promising and successful transition metal dichalcogenides. We summarize the basics of the synthesis procedure, discussing in depth: (i) the different substrates used for its deposition, (ii) precursors (solid, liquid, gaseous) available, and (iii) different types of promoters that favour the growth of two-dimensional layers. We also present a comprehensive analysis of the status of the research on the growth mechanisms of the flakes.

Keywords: Chemical Vapour Deposition; 2D materials; MoS_2

Citation: Seravalli, L.; Bosi, M. A Review on Chemical Vapour Deposition of Two-Dimensional MoS_2 Flakes. *Materials* **2021**, *14*, 7590. https://doi.org/10.3390/ma14247590

Academic Editor: Abbes Tahraoui

Received: 28 October 2021
Accepted: 7 December 2021
Published: 10 December 2021

1. Introduction

The information technology revolution is driven by the continuous improvement of electronic devices and their constant scaling down. However, this process is rapidly approaching its limit. In order to continue the innovation process, introduction of new concepts and new materials is required. Some solutions have been identified in (i) the use of strain engineering and high-k gate dielectrics, (ii) silicon–germanium (SiGe) alloys and germanium, because of their higher electron mobility and lower need for power, and (iii) nanostructures and 2D materials, for the possibility of introducing brand-new device designs and concepts.

For the latter, graphene and similar 2D materials recently emerged as promising candidates because they exhibit interesting mechanical, electronic, optical, and transport properties with peculiar differences from those of their bulk counterparts. Graphene, despite its exceptional physical properties, has the major drawback of lacking an electronic bandgap, a limit that poses problems in realizing logic circuits and transistors. Bandgap engineering in graphene, although possible, has several drawbacks, such as increased complexity of the process and degradation of material quality, so it is not usually considered a viable choice for certain applications.

Transition metal dichalcogenides (TMD), silicon, germanium, and boron nitride can be thinned down to monolayers similarly to graphene, exhibiting weak interplane interaction and strong in-plane bonds. MoS_2 is one of the most promising materials of this family because of the relative easiness of its synthesis and its interesting physical properties.

It exhibits a hexagonal lattice structure consisting of a layer of transition metal atoms embedded in two chalcogen layers, with strong covalent bonds in each 2D plane and a weaker interaction between different planes. The properties of MoS_2, similarly to those of other 2D materials, depend mainly on the vertical thickness of MoS_2 rather than its lateral size. The most important difference with respect to graphene is that 2D MoS_2 shows a direct bandgap of about 1.8 eV, making it an ideal candidate for applications in electronics, photonics, photovoltaics, energy storage, and catalysis [1].

Several proof-of-concept MoS_2 devices have been already demonstrated. FETs with mobility up to 320 $cm^2\ V^{-1}\ s^{-1}$ and on/off ratio of 10^8 at room temperature were realized [2,3], as well as inverters with gain up to 16 [4]. Several kinds of phototransistors, photosensors [5], and gas and biological sensors [6] were demonstrated, and the mechanical properties of MoS_2 were exploited in the realization of strain sensors [7]. Moreover, the light emission properties of this material are also attracting considerable interest [8–11].

The future development of research is trending towards the integration of different building blocks and different 2D materials into a single device, aiming for the fabrication of high-performance CMOS-based circuits, sensors, and efficient photocatalytic systems [12,13].

The interest in developing new devices from 2D structures is also driven by several novel properties observed in these materials, such as valley polarization, in which conduction and valence bands of MoS_2 monolayers present two inequivalent valleys at points K and K_0, giving a new degree of freedom to carriers. This would permit radical new device concepts based on valleytronics and on spin–valley coupling, permitting control of information using circular polarized light [14].

Despite the results obtained so far on MoS_2 devices and transistors in particular, it was observed that their performance was still far from the best theoretical predictions. The transport properties of MoS_2 are limited by defects such as vacancies, antisites, charge traps, grain boundaries, and Coulomb impurities at the flake interface [15,16]. For example, the maximum predicted electron mobility of MoS_2 at room temperature is about 410 $cm^2\ V^{-1}\ s^{-1}$, leaving room for improvement. Understanding these limits and how to overcome them is a major issue on the MoS_2 roadmap.

Reliable and controlled doping in MoS_2 flakes is also a fundamental requirement to achieve precise control of their electrical and optical properties. Usually, substitutional doping is the dominating process as compared to interstitial doping. On this topic, fundamental contributions were the very recent reviews of Lin et al. [17] and Rai et al. [18], who considered the issues of Schottky barriers and contact resistance at the interface between MoS_2 and a metal and the problems related to the doping of a wide range of 2D TMD materials, respectively. N-type doping of MoS_2 could be obtained by adding ReO_3 to the solid precursors [19] or by using ZnS as a sulfur precursor [20], while chloride permitted decreasing the resistivity of MoS_2 flakes from some KΩ to about 0.5 KΩ, allowing for the realization of Schottky barriers with enhanced properties [21]. p-doping could be obtained by adding Nb_2O_5 to NaCl promoter when using solid precursors [22].

Another topic of particular interest is the possibility of including elements such as manganese (Mn) and iron (Fe), aiming to add new functionalities related to magnetism and spin to MoS_2. First reports indicated the possibility of incorporating Mn in MoS_2, but with some issues relative to the substrate used, as Mn was detected only in structures deposited on graphene and not on those deposited on SiO_2/Si [23].

Despite their peculiar properties and the achievements so far obtained, the controllable synthesis of large-area flakes and the reproducibility of the process are still challenging issues; these are two essential, key factors required to sustain the future development of 2D materials. Mechanical exfoliation has so far granted a very high material quality, leading to the realization of devices and to an in-depth study of their properties. However, this method is very impractical and poses many limits on large-scale production. Industrial application of TMD is directly linked to the development of a high-throughput and reliable technique to obtain 2D layers of different materials on large area substrates, with a simple, reproducible, and scalable method. Chemical vapour deposition (CVD) is a good candidate for this task, since it has already been widely adopted for the mass production of many kinds of devices based on III-V and III-N materials and has thus the potential to permit the integration of 2D materials with "standard" semiconductors or other compounds. Many excellent reviews on MoS_2 are already available, giving a broad scenario of all the work done on this material [24,25].

Although MoS_2 flakes with large size (>500 μm) have been obtained with relevant success from several groups [26–28], in our opinion, research is still needed on many points that are usually overlooked: (i) the effect of different parameters (such as carriers' flows or sulfur partial pressure), (ii) the possibility of engineering layers by surface treatments—this is crucial for the realization of heterostructures and possibly hybrid nanostructures, (iii) the growth reproducibility of these structures. It is a common experience to not be able to obtain the same 2D layers in different growth runs, and this is still a consistent problem for the technology's transfer towards industrial applications. A reliable growth protocol that can be transferred to different reactors with minimal setup time is a strong requirement to allow 2D MoS_2 to make the jump to mass production.

The preparation of this material is still challenging, and there is a lot of room for improving both the process and the material quality. Despite its relative simplicity, CVD remains one of the most widely used techniques to synthesize TMD flakes. Recently, new protocols and precursors have been introduced to optimize the deposition, to obtain better quality material, and to improve the reproducibility of the process.

The aim of this work was to collect and critically present the knowledge acquired so far in this field, providing the reader with a comprehensive review of the state of the art of the CVD growth processes used to prepare MoS_2.

2. CVD Setup for Growth of MoS_2

The basic concept of CVD is very simple, and it is schematically presented in Figure 1a: the precursors—either only S or both S and Mo compounds—are delivered in gaseous form to a substrate (placed on a susceptor of graphite or similar material) kept at high temperature, where the chemical reactions needed for the deposition of MoS_2 occur. Most of the results reported in literature have been obtained using a standard horizontal reactor, such as that sketched in Figure 1. This configuration, although widely used, has the drawback of a nonuniform precursor gradient along the flow direction, resulting in a nonuniform deposition and in intrinsic difficulty in optimizing the growth parameter, since small changes in substrate position or in precursor supply may cause nonreproducible results. Development of vertical CVD reactors may help in solving these issues, since with this configuration, temperature and precursor supply may be more easily controlled and homogenized [29].

In the standard horizontal configuration, a quartz tube about 1 m long with a diameter of 2–5 cm is used. The heating is provided by a resistance placed around the tube. The reaction for MoS_2 deposition requires a temperature in the 600–800 °C range, but in some cases, the supply of sulfur requires a second heating zone in the 100–200 °C range. In this case, to obtain better reproducibility and control, it is more convenient to separately and independently heat this additional zone (T2 in Figure 1a) in which the sulfur precursor is placed.

It is also possible to deposit the molybdenum precursors directly on the substrate, either in solid form (Figure 1b) or as a liquid solution (Figure 1c). Growth promoters, to obtain better control of the process, can be added either directly on the growth substrate or on a different substrate (Figure 1d). As a last method, the use of gaseous precursors (Figure 1e) avoids the need for placing the precursors inside the tube before the beginning of the process.

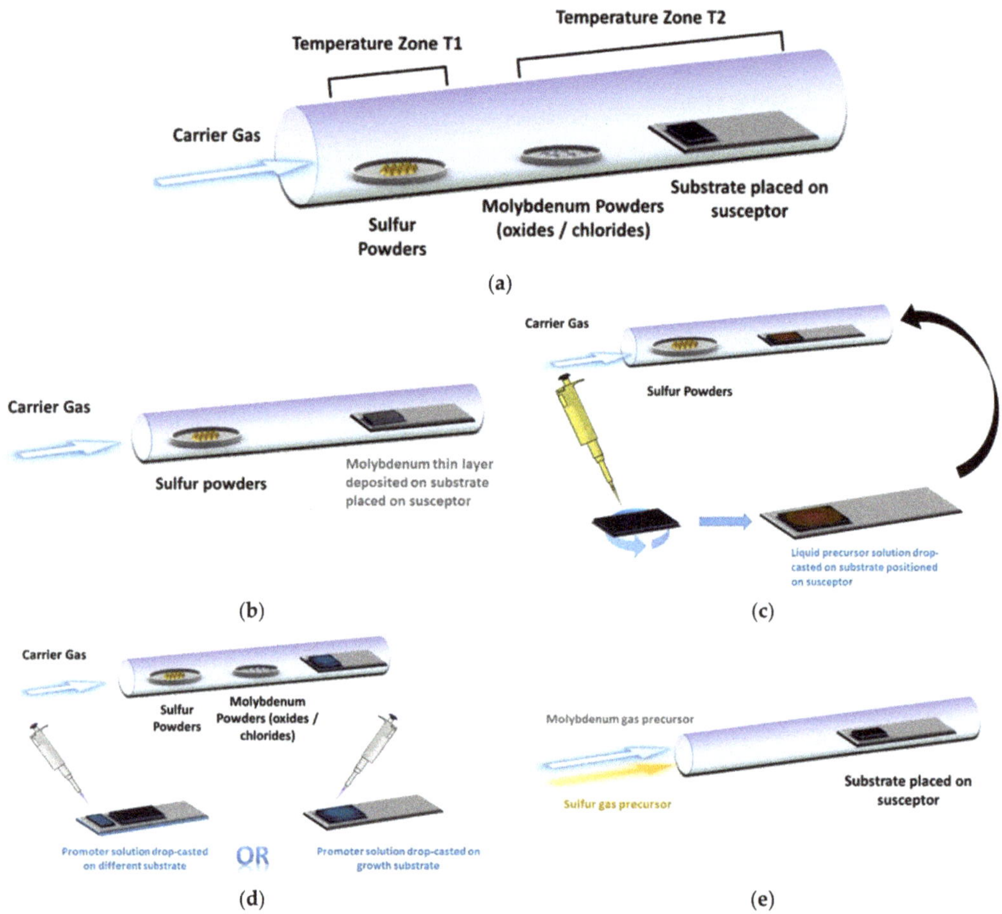

Figure 1. Schematic view of the CVD tube in different configurations for MoS_2 flake growth: (**a**) with solid precursors separated by substrate; (**b**) with solid molybdenum deposited on growth substrate; (**c**) with liquid molybdenum precursors; (**d**) with solid precursors and drop-casted promoters (either on growth substrate or a different substrate); (**e**) with gaseous precursors.

A standard routine to minimize the contamination from external air and O_2 in the reaction tube is to purge the system with inert gas several times before starting the growth, eventually evacuating the tube with a rotary pump and then flushing with purified Ar or N_2 up to ambient pressure, repeating this procedure several times.

The actual CVD process depends on the chemical status of the used precursors (solid, liquid, gaseous), but it is nevertheless possible to divide it into the following steps:

The precursors are brought into gaseous form and diluted in an inert carrier gas; e.g., powders are evaporated/sublimated, or a controlled amount of gas is measured by means of a mass flow controller.

The reactive species are transported by the carrier gas to the substrate. Chemical reactions may also occur in this step, e.g., reducing reactions.

The precursors diffuse towards the substrate surface.

The precursors are adsorbed at the surface, where adatom adsorption and migration occurs. The MoS_2 flakes synthesis occurs in this step. By-products re-evaporate or desorb into the gas streams and are carried away into the exhaust.

The heating ramp times are usually in the range of 10–30 min from room temperature to growth temperature. The heating of sulfur powders typically starts when the substrate is already at the target temperature; this permits avoiding injecting S into the system before a constant substrate temperature is reached, limiting prereactions. The MoS_2 growth proceeds for about 10–20 min, and then the system naturally cool downs to ambient temperature. If powders are used as reagents, heating of the substrate and of the MoO_3 powders occurs simultaneously, since they are usually placed very near the substrate. This means that Mo starts to evaporate before the substrate reaches the target temperature.

Flake synthesis is dependent on different parameters such as the growth temperature and the distance between the substrate and the powders. Flow rate, chamber pressure, precursor supply, powder dimension, and purity also affect the final outcome of the growth. Despite the simplicity of the process, if excellent control is not achieved, the results may not be completely reproducible, and large-scale deposition can be very hard to obtain.

While it has been shown that different carrier gases can strongly influence the growth of graphene [30], considerably less attention has been devoted to this topic for the growth of MoS_2. Considering that S is a strong reducing agent, the use of a purified inert gas (N_2 or Ar) is usually sufficient for the sulfurization reaction to occur, and the carrier gases can be considered as inert in the reactions at the basis of 2D nucleation and growth. The use of H_2 is thus not common in the MoS_2 CVD process. Nevertheless, it was argued that H_2 can have beneficial effects, as it can inhibit the thermally induced etching effect and promote the desulfurization reaction [31]. By carefully adjusting the amount of hydrogen in a H–Ar mixed carrier gas, authors were able to obtain layers with high crystallinity and a nearly perfect S/Mo atomic ratio. It was also suggested that, in the case of MoS_2 growth using $Mo(CO)_6$ with $(C_2H_5)_2S$ as precursors, the presence of H_2 is necessary for removing carbonaceous species generated during the MOCVD growth, permitting increasing the average grain size from hundreds of nanometres to more than 10 μm [32].

Carrier gas flow rate was reported to affect the morphology of flakes, as well as their size and density: lowering the gas carrier flow during the heating stage from 35 to 15 sccm permitted increasing the density and size of the flakes [33]. Moreover, the shape of MoS_2 flakes changed from zigzag to triangular as the gas carrier flow was decreased, indicating higher material quality. This behaviour was directly related to the quantity of S in the gas phase. As the carrier flow increased, the S transport grew more efficient, and S reacted with MoO_3 powders, suppressing their evaporation, lowering MoO_3 partial pressure, and promoting zigzag edge termination.

In order to have better control over S and MoO_3 reactions during the heating stage of the process, a procedure of "flow reversal" was proposed [34]. During the heating of the furnace, Ar was introduced from the side of the Mo powders towards the side of the S powders (Figure 2). In this way, S reached neither the substrate nor the Mo powders, and any unintentional reaction was prevented; growth and nucleation of MoS_2 far from the steady growth regime was inhibited. Once the setpoint temperatures of both substrate and S powders were reached, the Ar flow direction was switched back to the standard direction, so that it entered from the S powder side and delivered S vapours to the Mo powders and the substrate. Using this two-stage growth, the mean side length of MoS_2 flakes was increased up to 250–300 μm.

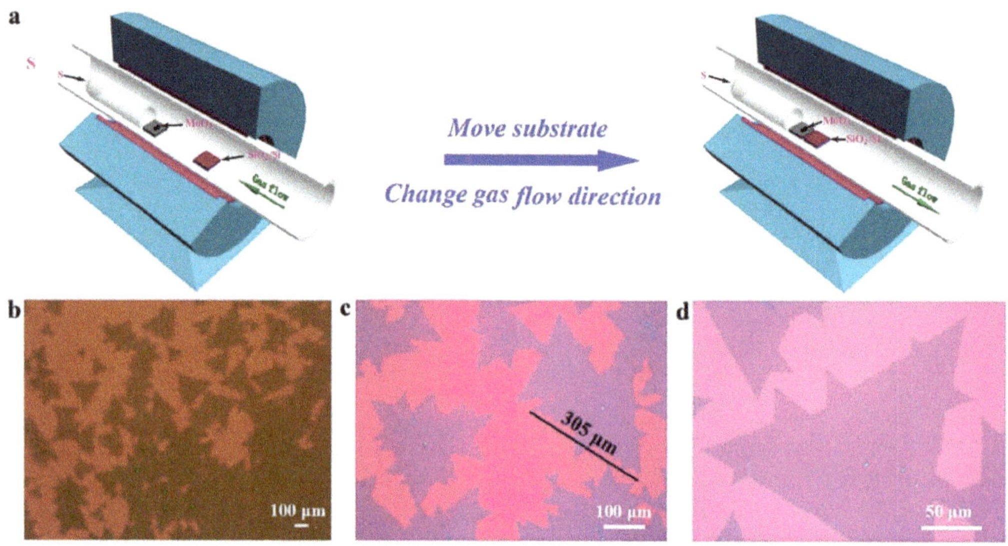

Figure 2. (**a**) Flow reversal method to improve MoS_2 deposition; (**b**–**d**) typical optical images of MoS_2 flakes obtained with the modified procedure. Adapted from [34].

Another custom setup to obtain better control of the precursor supply and to avoid cross-contamination of MoO_3 and S powders consisted of placing S powders in a small tube nested inside another, larger tube so that S did not contaminate the MoO_3 powders before entering the reaction zone [35].

3. Substrates

A typical characteristic of 2D materials is the very weak interlayer Van der Waals forces, in contrast with the stronger in-layer chemical bonds, which are ionic or covalent. This is the reason why bulk crystals of TMD materials, and MoS_2 in particular, can be easily cleaved and exfoliated to produce few-layer flakes. As a consequence, the dangling bonds of the 2D plane are relatively inactive, and few interactions occur out-of-plane. For this reason, one could argue that in the 2D synthesis, the substrate's role should be less important with respect to standard epitaxy, in which lattice and thermal matching conditions between the substrate and the epilayer are of paramount importance, determining whether defects are generated at the interface. The weak interactions between the substrate and the 2D layer reduce the importance of lattice mismatch, but the choice of the substrate remains an important parameter, not to be overlooked. The reason should be sought in the processes still occurring at the substrate surface that involve the reactive species, such as adsorption, diffusion, and nucleation. It was indeed observed that the substrate played an important role in influencing both the growth process and the properties of 2D layers, as evidenced in the excellent review on this topic in [36].

SiO_2/Si substrate is the most commonly used for CVD synthesis of MoS_2 because of: (i) the ease of observing and characterizing flakes on it through optical microscopy (from this point of view, the thickness of the oxide layer is of paramount importance [37]); (ii) its high melting point; (iii) its compatibility with silicon processing. On the other hand, for further characterization and/or applications, it is necessary to transfer the 2D material from the SiO_2/Si substrate on other substrates or TEM grids. This implies a delicate chemical process composed of various steps: (i) coating of the material with poly(methyl methacrylate) (PMMA) film on the MoS_2 layers by spin coating, (ii) etching of the SiO_2 layer to separate it from the Si substrate in a solution (KOH, HF, NaOH), (iii) transfer of the floating PMMA/MoS_2 film from the surface of the solution to the substrate of interest, and (iv) removal of the PMMA coating by rinsing in acetone and isopropanol [38].

The possibility of growing MoS_2 on rigid metals has been investigated, in particular for the case of gold, as it was demonstrated that excellent nanostructures can be obtained on Au foils [39]. If one considers surfaces of Au(111), very large nanoflakes can be obtained [40] thanks to the lattice matching between Au(111) and MoS_2 and the fact that the step edge of Au(111) favours the unidirectional nucleation of the film. The advantage of this substrate was confirmed recently in [41], where a unidirectional growth ratio over 99% of MoS_2 flakes was achieved on Au(111) films sputtered on c-sapphire. The growth on Ag(111) was also demonstrated, opening up interesting perspectives on the realization of metal contacts on this 2D material [42].

Liquid metals are interesting, as their uniform and smooth surface can inhibit the inhomogeneous nucleation of 2D materials. Although convincing results have been published on other 2D materials, the growth of MoS_2 on these substrates has been reported only for the MoS_2/h-BN heterojunction grown on liquid metal Ni–Ga alloy [43].

MoS_2 has also been grown with considerable success on c-sapphire, as it has a stable hexagon single crystal structure, matching the lattice symmetry of 2D material [27,44,45]. Sapphire surface can be prepared with a thermal treatment at 1100 °C before MoS_2 growth to obtain a clean and atomically flat surface. Scaling up to a Metal–Organic Vapour Phase Epitaxy (MOVPE) process for 6-inch wafers was obtained, and the possibility of transferring the obtained flakes to a secondary carrier substrate opens up the possibility to reuse the sapphire substrate, making this choice an interesting alternative to the common SiO_2/Si approach [46]. Very recently, the growth of monolayer flakes as large as 50 μm on a C-plane sapphire was demonstrated by designing the miscut orientation towards the A axis. This resulted in a break in the degeneracy of nucleation energy for the antiparallel MoS_2 domains [47].

The growth of a 6-inch uniform monolayer of MoS_2 on solid soda-lime glass was achieved because of the advantageous effect of sodium in enhancing the formation of flakes that were homogenously distributed in glass [48]. Na acts as a growth promoter; its effects are discussed in depth in a subsequent section

Recently, the possibility of growing MoS_2 on flexible substrates such as polyimide was reported [49]. The issues related to the stability of the substrate were solved thanks to a CVD growth protocol allowing for low temperatures (<300 °C), and the feasibility of developing flexible gas sensors taking advantage of the properties of MoS_2 was demonstrated.

In an interesting work, MoS_2 monolayers were reproducibly deposited on a variety of substrates, amorphous (SiO_2/Si and fused quartz), crystalline (bare Si and sapphire), and layered-flexible (mica), using the same growth conditions with solid precursors and mixed NaCl [50].

CVD growth of MoS_2 on gallium nitride (GaN) was also reported, a very relevant step towards the realization of heterostructures for many applications [51]. This possibility is of particular interest because it would eliminate the delicate step of detachment and transfer of 2D flakes, allowing realizing MoS_2 heterojunctions directly on a nitride template.

Following the ideas widely explored with III-N and III-V semiconductors, in which AlGaAs–InGaAs or AlGaN–InGaN heterostructures are used to exploit the possibility of bandgap engineering to design advanced devices, an emerging topic of research is the realization of so-called 2D Van der Waals heterostructures. By vertically stacking layers of different 2D materials coupled by very weak forces, it would be possible to avoid the presence of defects due to lattice mismatch observed in conventional 3D heterostructures [52]. A considerable effort has thus been devoted to exploring the synthesis of MoS_2 and other 2D compounds onto graphene and other 2D materials. First results highlighted how it was possible to obtain high-quality 2D flakes by adding some gases to gas carriers (N or Ar) for the treatment of graphene surfaces, such as H_2 [53] or ozone [54]. This allows for the preparation of high-quality nanoflakes of MoS_2 because of the reduction in the oxidation of graphene during the growth process.

In [55,56], few-layer MoS_2 was grown on graphene oxide films or flakes. These results suggested that carbon-based materials can significantly promote the growth rate and yield of MoS_2. In [57], the differences in using graphene, sapphire, and SiO_2/Si substrates were investigated, highlighting the role of surface diffusion mechanisms in determining different properties of MoS_2 flakes. The authors remarked that the growth on graphene was very stable, resulting in the realization of a strain-free 2D layer.

A topic of paramount importance for the development of nanodevices is the growth on patterned substrates to increase the spatial control of MoS_2 nanoflakes and to obtain a controlled nucleation only on certain substrate zones. The first pioneering work on this issue was the one by Najmaei et al. [58], who observed a catalytic effect of the edges of the substrate. Based on this effect, authors proposed and demonstrated a method to control the growth of 2D flakes using lithography to pattern the substrate. Later, it was demonstrated that, by using patterned seeds of molybdenum, it was possible to obtain flakes of MoS_2 at predetermined locations with a good spatial resolution [59]. Another approach to control the nucleation site of MoS_2 flakes was to use droplets of $(NH_4)_2MoS_4$ in dimethyl-formamide suspended from the tip of a micropipette that is dragged across a sapphire single-crystalline substrate by a controlled substrate movement. This resulted in MoS_2 films with alternating mono- and few-layer regions that had distinct optical properties [60].

Other metal seeds proven as very valuable for the spatial control of the nucleation of flakes include Pt/Ti and Au [61–63]. More recently, the synthesis of patterned MoS_2 nanoflakes using an industrial inkjet fed with an aqueous solution of ammonium molybdate tetrahydrate as liquid precursor was demonstrated (Figure 3). The possibility of obtaining large-area patterned 2D films in centimetre size with good controllability of the thickness and good reproducibility opens up interesting possibilities for the development of complex devices based on this material [64].

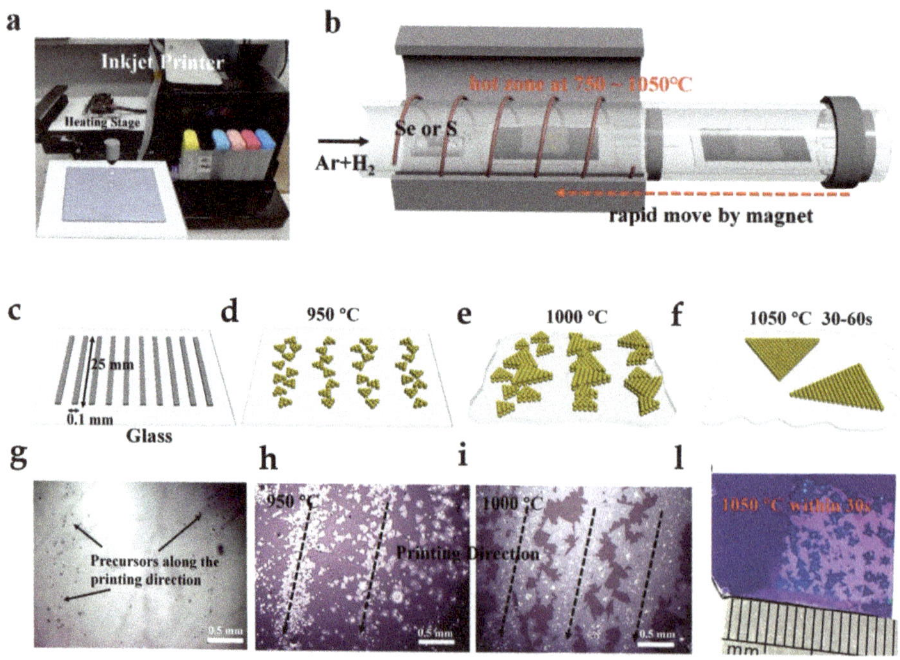

Figure 3. Synthesis of MoS_2 flakes from inkjet-printed aqueous precursors: (**a**) a photo of the customized inkjet printer; (**b**) schematic of the synthesis process with the fast annealing process; (**c**–**f**) growth of flakes at different growth temperatures; (**g**–**l**) corresponding optical images. Adapted from [64]. Copyright 2021 John Wiley & Sons.

4. Precursors for MoS_2 Growth

The literature has reported the use of several kinds of Mo and S precursors for the CVD growth of MoS_2. It is possible to divide these precursors into three main categories according to their physical status: solid, liquid, and gaseous. To optimize the synthesis process and to obtain good control over the number of layers, it is important to understand the limits, advantages, and chemical and physical properties of each precursor. Moreover, for some precursors, several safety issues should be considered. In the following paragraphs, a brief description of the available precursors for each category is given.

5. Solid Precursors

Mo and S powders are still the precursors of choice for many groups to deposit mono- and few-layer MoS_2, although the reproducibility of the process remains an issue. This approach is very simple, since the handling of powders is easy, has few hazard problems, and nevertheless permits obtaining high quality flakes. The drawback is that the flake dimension is somewhat limited to some tens of microns; scalability to large areas and enlargement of the single flake are quite difficult, and reproducibility remains an issue.

5.1. Molybdenum

The most commonly used powders for molybdenum are MoO_3, available in high purity (99.98%). The vapour pressure of MoO_3 was reported in [65]. A quantity of 10–200 mg of MoO_3 powders is placed in a quartz crucible near the substrate in the high temperature zone of the reactor. There are several approaches in positioning the substrate; some groups place the substrate facing down directly above the powders, while others place it facing up immediately after the crucible. It should be noted that some impurities might be present in these solid precursors. Robertson et al. [66] showed that Cr and V atoms were found in CVD grown MoS_2, most probably from elements present in solid precursors, causing some charge trapping in the grown 2D material. The issue of unintentional doping is very relevant in 2D materials; in [67], the effect of unintentional carbon doping during CVD was discovered and thoroughly discussed.

Precursors such as molybdenum hexacarbonyl ($Mo(CO)_6$) and molybdenum (V) chloride ($MoCl_5$) are mostly used for ALD and MOVPE. $Mo(CO)_6$ is used with H_2O or ozone to deposit thin MoO_3 films through ALD, to be sulfurized in a subsequent step. $Mo(CO)_6$ has a vapour pressure of about 0.10–0.15 mbar at room temperature and is very volatile (melting point of 150 °C, boiling point of 156 °C) [68].

Molybdenum (V) chloride ($MoCl_5$) powders were also used to grow MoS_2 with S powders [69,70]. However, $MoCl_5$ is air sensitive and toxic, so its use poses more hazards than that of MoO_3 powders.

Deposition of a thin Mo layer prior to growth could also be included in the solid precursor category. Electron-beam evaporation or sputtering are used to obtain a very thin and controlled Mo layer on the substrate, which is converted to MoS_2 because of the exposure of the surface to S vapours. The final MoS_2 layers are dependent on the thickness of the initial Mo layer, so it is important to have very precise control over the metal thickness and good homogeneity. MoO_3 deposited by evaporation is an alternative approach to metallic Mo [71], since the oxide has a lower evaporation temperature and is easier to sulfurize.

The use of a Mo metal foil was suggested to obtain large flakes on 6-inch substrates [48]. The foil was placed 10 mm above a soda–lime glass substrate and using a mixture of Ar and O_2 as carrier, flakes with later size higher than 400 μmm were obtained. Reactive MoO_2 was produced thanks to the oxidation of Mo foil by the O_2 carrier, and S powders were used as an S precursor. The dimensions of triangular MoS_2 flakes were adjusted by controlling the distance between the Mo foil precursor and the substrate.

5.2. Sulfur

S powders are nontoxic and melt at about 115 °C, evaporating to gaseous sulfur. S vapour pressure is reported in [72] but the amount of precursors present in the vapour phase depends also on the quantity of powder placed in the crucible and on the their dimension: a good practice would be to replace the S in the crucible after each growth in order to increase reproducibility and have a better control of the deposited layers.

S powder with a weight in the range of 0.1–1 g is usually placed upstream in a quartz crucible about 10–15 cm away from the substrate. It is heated at temperatures of about 100–200 °C, and the vapour is delivered to the substrate by the carrier gas. In these conditions, the S vapour pressure is in the range of $10^{-2} - 1$ torr. Since the evaporation of S is quite fast, a common technique to ensure repeatability is to independently heat the S crucible only after the substrate has reached the growth temperature setpoint. Moreover, in order to obtain a continuous supply of precursor during the whole process, the use of two different crucibles placed in different reactor position was suggested [73]. As already discussed above, changing the direction of the carrier gas to deliver the S vapours away from the substrate during the heating process may deliver more consistent results [34].

6. Liquid Precursors

The necessity to overcome the reproducibility and size limitations imposed by the use of powders fostered the research of alternative precursors. The idea of sulfurizing a thin and controlled Mo layer deposited on the substrate (originally obtained by sputtering or e-beam) was at the basis of the method of spinning liquid precursors. This approach, well known from the deposition of resists and protective layers, consists of dropping a viscous liquid with the Mo precursor onto the substrate and then using a fast-rotating spinner to spread and homogenize the liquid onto the surface. Once the centripetal and viscous forces are well balanced, the resulting layer has very good thickness reproducibility. Then, the substrate is placed in the reactor chamber and sulfurized using standard S powders.

6.1. Molybdenum

The literature has reported the use of two liquid precursors, sodium molybdate dihydrate (SMD—$Na_2MoO_4{\cdot}2H_2O$) and ammonium molybdate tetrahydrate (AMT—$(NH_4)_6Mo_7O_{24}$). It is important to remember that SMD is incompatible with alkali metals, most common metals, and oxidizing agents. These precursors are usually diluted in water (0.1–0.2 g in 10–50 mL of H_2O) and mixed with a promoter such as NaOH (0.1 g in 50 mL of H_2O) and a density gradient medium such as iodixanol (Opti-Prep) to facilitate the spinning and substrate adhesion process. Standard proportions for these three components are in the range 1:(1–8):0.5, and they are then spun onto the substate for 30–60 s at 2000–4000 revolutions per minute. An hydrophilic surface is required to obtain a uniform coating molybdenum precursor, so the substrate is usually treated with O_2 plasma and/or sodium hydroxide before the drop-casting of the solution [74].

SMD was also used in an alternative approach by embedding it into two thin pieces of glass that were fused together after heating. Thanks to the high temperature, the SMD melted and diffused through the molten glass to its surface with a dissolution–precipitation process. When the metal source reached the upper surface, it reacted with sulfur to grow MoS_2 on the molten glass surface. Through this original method, highly uniform and monolayer MoS_2 flakes on centimetre-scale glass substrates were obtained thanks to a uniform distribution of the metal precursor [75].

The liquid-phase precursor approach was used to obtain controlled doping of TMD and related heterostructures. Mixing of Mo liquid precursors along with selected dopants (Fe, Re, V) was reported to produce controllable doping in MoS_2 with excellent uniformity. More complex structures, such as V-doped in-plane $W_xMo_{1-x}S_2/Mo_xW_{1-x}S_2$, were also obtained [76]. Liquid, metal-organic-like precursors are still not common for MoS_2 synthesis. Efforts are being directed towards developing suitable reagents to be used in MOCVD systems, in order to enhance process reproducibility and controllability. A notable effort was directed towards bis(tertbutylimido)bis(dimethylamido)molybdenum $(Nt Bu)_2(NMe2)_2Mo$, which was stored in a container kept at 45 °C and used as molybdenum precursor with H_2S the as sulfurizing agent. A rapid and scalable process using a CVD approach on 2-inch sapphire substrate was developed to synthesize, layer-by-layer, atomically thin MoS_2 films with good spatial homogeneity and wafer-scale uniformity [77].

6.2. Sulfur

Dimethyl disulfide (CH_3SSCH_3, DMDS) is an organic, flammable liquid with a garlic-like odour that can be used as a S precursor. It decomposes into methanethiol (CH_3SH), ethene ($CH_2{=}CH_2$), carbon disulfide (CS_2), and H_2S. Reaction by-products of this precursor pose some safety hazards (see the following section) [78].

Dodecane-1-thiol (or dodecyl mercaptan, $C_{12}H_{25}SH$) is a sulfur-containing liquid thiol that can be bubbled and delivered into the reactor by using a carrier gas in order to obtain a precise and uniform concentration of S during the process [79].

7. Gaseous Precursors

Sulfur

Despite its toxicity, hydrogen sulfide (H_2S) has been used by some groups as a precursor for S. It decomposes very quickly at high temperature, and for this reason, it should be more efficient than S powder for TMD synthesis [80]. Its partial pressure during a typical TMD growth is in the range 10^{-1}–10^{-2} [81,82]. H_2S may also act as a reducing agent for metal oxides. It should be remembered that H_2S is also corrosive, flammable, and explosive, with severe consequences in the event of exposure to eyes and skin and inhalation, and the design of a reactor with a H_2S line should be carefully planned [83]. Despite these drawbacks due to safety reasons, a gas permit can be used to obtain a very controllable precursor supply, which is reflected in the reproducibility of the process.

8. Safety aspects of Mo and S Precursors

Ingestion of Mo or S related compounds is considered a very unlikely event in a research laboratory if standard precautions for the manipulation of substances are taken into consideration (e.g., wearing gloves, washing hands, etc.). However, it should be taken into consideration that a low level of Mo or S vapours could be present in the ambient because of the volatility of these substances at room temperature. Use of a laboratory hood should be always taken into consideration. Exposure to excess levels of Mo has been associated with adverse health outcomes such as respiratory effects following inhalation exposure to molybdenum trioxide, decreases in body weight, kidney damage, decreases in sperm count, and anaemia following oral exposure. A review of human and laboratory animal health effects for Mo can be found in [84]. Oral exposure to sulfur may cause problems in the gastrointestinal tract, central nervous system, kidneys, liver, heart, lungs, blood, and salivary glands. Sulfur has induced skin damage and affected various other tissues (https://www.bibra-information.co.uk/downloads/toxicity-profile-for-sulphur-1990/ Accesed on 12 September 2021). Handling of H_2S exhaust could include a bubbler through an aqueous solution of ethylene glycol to dissolve H_2S before releasing it to the atmosphere.

9. The Use of Growth Promoters in MoS_2 Synthesis

The growth of 2D layers is driven by the chemical potentials and surface energies of the material and substrate. Depending on the balance between adhesion to the surface of atoms and adatom cohesive force, 2D layer growth might be favoured with respect to island nucleation. If the surface adhesion is insufficient to allow layer growth, the use of additional molecules (that are not involved in the chemical processes underlying the growth) is an attractive method to obtain 2D layers. These elements are generally known as "promoters", and their use has been proven to be very beneficial for many aspects of the growth of MoS_2.

The inclusion of other elements that may favour the nucleation and lateral growth of 2D layers was recognized as crucial in the very early steps of the research on the CVD growth of this material. In particular, the pioneering work of Ling et al. [85] analysed many possible options for promoters, giving an ample view of the possibilities; this was a fundamental paper to have a wide insight on many possible substances to be used as promoters. We limit our discussion to the most successful and useful promoters reported so far, to allow the reader a general view of the state of the art on this topic (Table 1). Many approaches have been proposed in these years, and we consider it useful to divide growth promoters into two main categories: inorganic solid state and organic. A section on the effect of surface treatments is also included, as this is another important element in determining the quality of the grown 2D layers.

From the point of view of understanding the mechanisms underlying the effects of promoters, theoretical work is currently underway, with different proposals to be investigated. For example, in [86], the effect of alkali metal compounds as promoters was studied, and the authors provided an explanation based on eutectic intermediates containing alkali metal molybdates and molybdenum oxides. Because of their low melting point, their mobility was enhanced in comparison with that of other molybdenum compounds, reducing the nucleation of new nanoislands and favouring lateral growth of existing ones.

Table 1. Growth promoters for the synthesis of MoS_2.

Seeding Promoter	Method	Substrate/Surface	Growth Temperature (°C)	Reference
Organic				
PTAS	dispersed on different substrate	SiO_2/Si	>600	[85]
PTAS	dispersed as solution on growth substrate	SiO_2/Si, quartz, sapphire, TiO_2	650–680	[87–89]
PTAS	solution drop-casted on oxone-treated surface	graphene	-	[54]
PTAS	solution drop-casted on surface at 90 °C	SiO_2/Si	750	[90]
PTARG	solution drop-casted on surface at 90 °C	SiO_2/Si	750	[90]
F_{16}CuPc	prior thermal evaporation on the growth substrate	SiO_2/Si	650	[85]
PTCDA	solution drop-casted on surface	SiO_2/Si, sapphire	650	[91]
PTCDA	solution dispersed on different substrate downstream	SiO_2/Si	750	[92]
rGO	hydrazine solution drop-casted on surface	SiO_2/Si, sapphire	650	[91]
CuPc	solution drop-casted on surface	SiO_2/Si	680	[85,88]
CV (crystal violet)	solution drop-casted on surface	SiO_2/Si	680	[88,93]
H2TPP (porphyrin)	thermal evaporation of thin film	SiO_2/Si coated with carbon nanotubes	-	[94]
p-THPP	fibres dipped in promoter solution	graphene oxide fibres	650	[95]
Zn(II)-THPP	metalation of pTHPP promoter	graphene oxide fibres	650	[95]
Inorganic				
NaOH	added in liquid precursor solution	SiO_2/Si	780	[96]
NaCl	on substrate facing the growth substrate	SiO_2/Si	750	[97]
NaCl	mixed with solid Mo precursor	SiO_2/Si, sapphire, Si, fused quartz, mica	650	[50]

Table 1. *Cont.*

Seeding Promoter	Method	Substrate/Surface	Growth Temperature (°C)	Reference
Alkali metal halides (NaCl, KI)	placed upstream to growth substrate	sapphire	800	[98]
IIa metal chlorides (CaCl2, SrCl2)	spin-coating of solution on substrate	SiO_2/Si, sapphire	850	[35]
Gold	EBL-patterned arrays of Au nanoparticles	SiO_2/Si	650	[62]
Gold	drop-casting on colloidal Au nanoparticles	SiO_2/Si	785	[63]

9.1. Inorganic Solid State Promoters

Among the promoters in this category, sodium was the first and most studied, as its beneficial effects on the lateral growth of flakes were recognized very early. The presence of Na on the substrate acts as a promoter for MoS_2 flake nucleation. Moreover, the presence of Na facilitates the peeling of flakes, because it has a low energy barrier between MoS_2 and the substrate during the growth process. This permits covering the flakes with a PMMA layer and then detaching the PMMA with the flakes with a simple immersion in water [48]. Na has been used in solution as sodium hydroxide (NaOH), particularly when liquid precursors were used, as it can be easily added to the solution to be spun on the substrate [96].

Another substance containing sodium that has shown usefulness in growing 2D materials is sodium chloride (NaCl). In particular, its use allowed obtaining high-quality MoS_2–WS_2 in-plane heterostructures [97]. NaCl's role is related to the enhanced formation of micrometre-sized particles at lower temperature and to the weakening of interlayer adhesion, allowing for the lateral growth of materials. Different processes have been proposed for the role of NaCl, such as: (i) the formation of low-melting-point intermediates, which facilitates a sufficient supply of metal precursors; (ii) the decrease in the energy barrier for atom bonding at the edges of 2D flakes; and (iii) the generation of low-melting point-eutectic intermediates, increasing the surface mobility of precursors [35]. The beneficial effect of mixing NaCl into the MoO_3 solid precursors was studied in [50], where 2D flakes were deposited on a variety of substrates. The authors indicated that the inclusion of NaCl allowed for a reduction in the growth temperature from 900 to 650 °C, making it a cost-effective growth promoter for high-quality, large-area monolayer flakes. In [99], the use of this salt was instrumental in obtaining large-area 2D layers of a large series of metal chalcogenides, including MoS_2. The authors stated that NaCl helped in reducing the melting points of metal precursors and in increasing the chemical reaction rate to grow 2D TMCs.

In [98], alkali metal halides such as KI and NaCl were considered, highlighting their beneficial effect to the growth of MoS_2 nanoflakes, which was due to the suppression of the nucleation of new MoS_2 domains during growth and during subsequent enhancement of lateral growth of existing domains. In [100], the role of salts in promoting the 2D growth was interpreted as due to: (i) reaction between alkali metal salts with transition metal precursors forming nonvolatile liquid alkali metal molybdates that reduced the nucleation density and promoted lateral growth; (ii) catalytic effects that increased the surface reaction rate; (iii) the formation of highly active volatile metal oxychlorides possessing relatively low evaporation temperatures; and (iv) reduction in the activation energy on the specific surfaces of nonlayered materials hindering perpendicular growth.

Recently, the use of some group IIA metal chlorides ($CaCl_2$ and $SrCl_2$) as promoters was proposed [35], as it resulted in a relevant increase in the size of flakes, as shown in Figure 4. As in other cases, this promoter was spin-coated on the surface of the substrate as an aqueous solution, while the Mo and S precursors were in solid state. In this comprehensive work, different metal elements (Li, Na, Sr, Mg, Ca) were considered and an acid–base model was developed, allowing the authors to attribute the promoting effect to acid–base interactions between precursors and substrates influencing the adsorption of atoms for the formation of the 2D material.

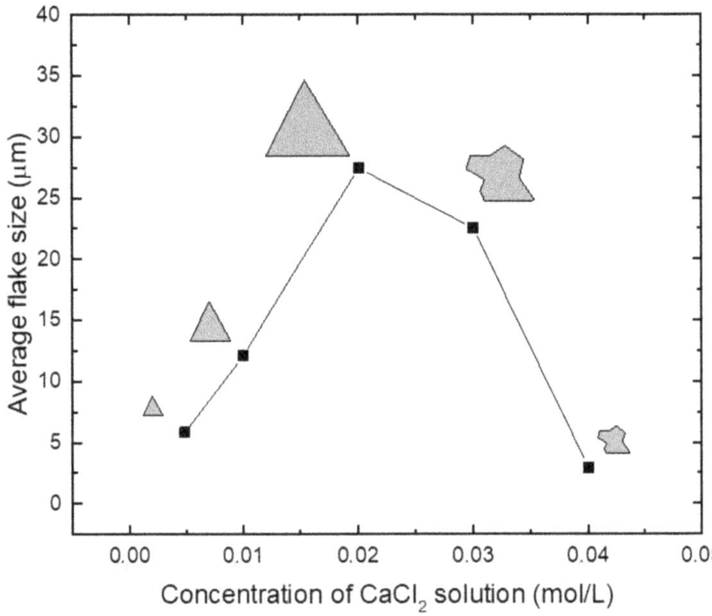

Figure 4. Dependence of the average size of flakes on $CaCl_2$ solution concentration. Symbols indicate schematically the morphology of flakes.

Beneficial effects are not limited to these substances, since it has been reported that metals in solid state can also favourably influence the 2D growth of MoS_2, as discussed above when the use of patterned substrates was introduced. More recently, it was reported that an increase in size of MoS_2 flakes occurred after drop-casting a solution with colloidal gold nanoparticles [63]. This effect was explained by gold nanoparticles acting as catalytic seeding points for the initial synthesis of the MoS_2 monolayer, similarly to what happens in 1D nanowires [101]. This is a topic of very relevant interest, as the integration of metal nanoparticles (such as gold or silver) with 2D nanostructures is very actively researched because of their effectiveness in enhancing and controlling the light-emission properties of these novel materials, allowing for in-depth studies of light–matter interaction in 2D systems [102–104].

9.2. Organic Promoters

Among the most-used organic promoters, perylene-3,4,9,10-tetracarboxylic acid tetrapotassium salt (PTAS) and perylene-3,4,9,10-tetracarboxylic dianhydride (PTCDA) have provided the most interesting results, thanks to the property of aromatic molecules of increasing the wettability of the growth surface and of lowering the free energy for nucleation. The first report on the usefulness of treating surfaces with such substances in obtaining large-area, high-quality TMD materials dates to 2012, when Y.H. Lee et al. used PTAS, PTCDA, and reduced graphene oxide (rGO) [91].

PTAS can be obtained by alkaline hydrolysis of PTCDA by diluting it in ethanol and adding KOH aqueous solution while refluxing. Ethyl ether can be added dropwise to the solution until a solid product begins to separate out. The precipitate can then be filtered and dissolved in deionized water, and the obtained aqueous solution must be filtered again to remove any insoluble residuals. Adding ethanol to the resulting aqueous filtrate allows precipitating the PTAS final product [105].

In a later study, Ling et al. [85] showed that PTAS permitted the growth of good-quality MoS_2 monolayers at 650 °C, while the use of PTCDA allowed the use of lower temperatures while maintaining good quality in the nanostructures. As illustrated in Figure 5, when the promoter is deposited on a different substrate with respect of the growth one, the diffusion of the seeding molecules causes a concentration gradient from the promoter substrate over to the growth region.

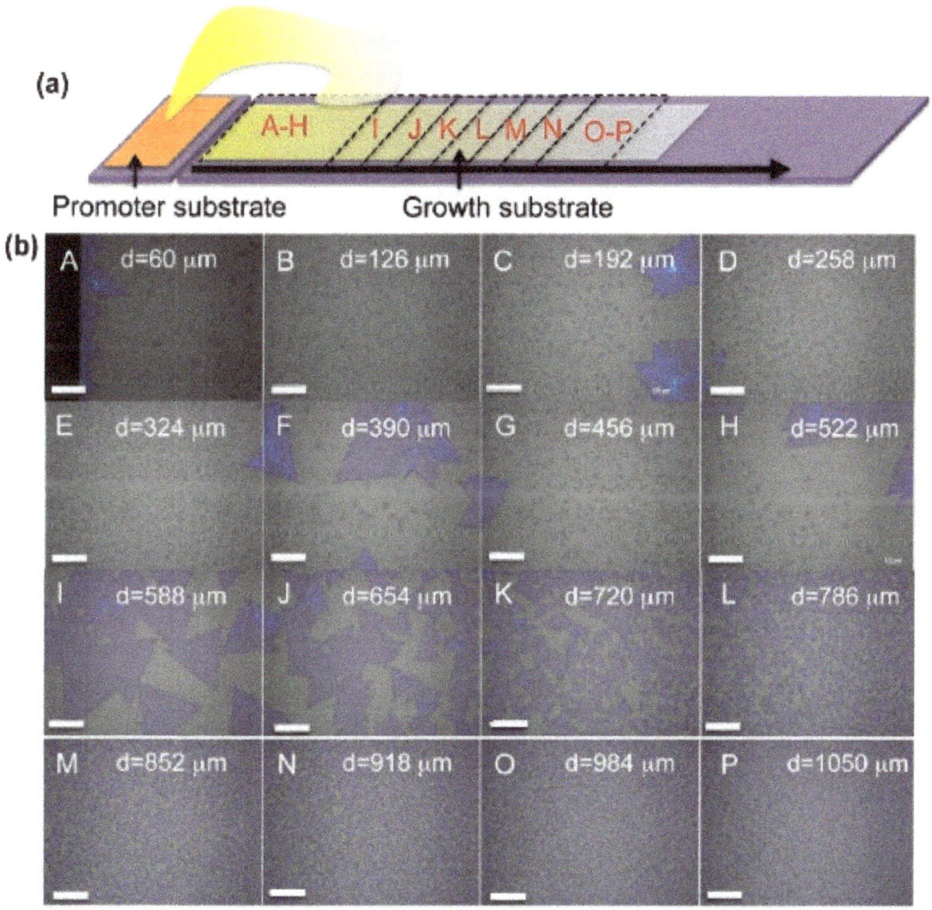

Figure 5. (**a**) Schematic of substrates used in PTAS-promoted growth of MoS_2 flakes. (**b**) Optical images of structures in different regions of the growth substrate, as identified above. Scale bars: 20 μm. Adapted with permission from [85]. Copyright 2014 American Chemical Society.

The use of PTAS was recently demonstrated as very effective in achieving lateral heterostructures wherein MoS_2 and other 2D materials can be parallel stitched [106]. Ref. [89] provided the possible growth mechanisms underlying the growth of nanoflakes, showing that this 2D material could be obtained on different surfaces such as Si particles, TiO_2 aggregates, and quartz.

In [92], the effect of gradient of PTCDA on the growth dynamics was clarified, highlighting that a low amount of this organic promoter led to reduced lateral growth and boosted vertical growth, while a high amount enhanced lateral growth and suppressed vertical growth. Figure 6 illustrates the proposed growth dynamics due to the two opposite flows of precursors and PCTDA. This configuration allowed studying the effect of the gradient of this organic promoter on the MoS_2 structure morphology and flake sizes along the length of the growth substrate.

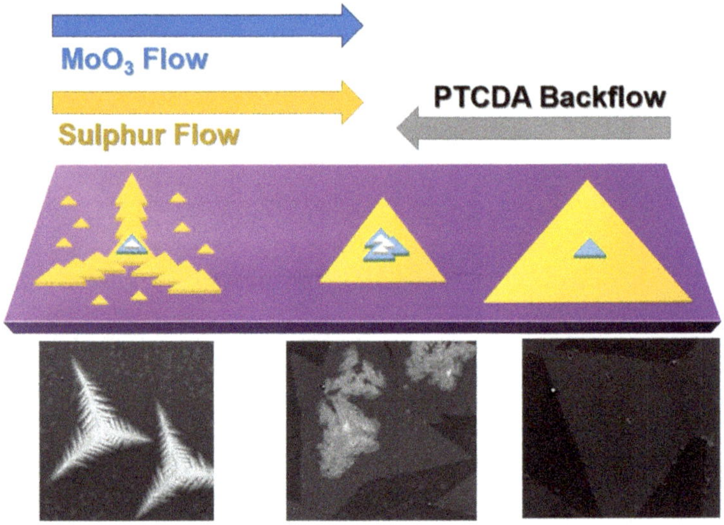

Figure 6. Illustration of the growth dynamics of MoS_2 due to the flow gradient of PTCDA organic promoter, with SEM images from different areas of the substrate. Reprinted from [92].

As was shown recently by Martella et al. [91], the use of organic promoters can also be useful for tailoring the properties of 2D structures. Using two different perylene-based molecules (PTAS and perylene-3,4,9,10-tetracarboxylic dianhydride—PTARG) as seeding promoters resulted in a change in the local electronic polarizability of MoS_2 monolayers due to extra charges trapped in the MoS_2 monolayers.

Copper phthalocyanine (CuPc) and copper(II)-hexadecafluoro-29H,31H-phthalocyanine (F_{16}CuPc) have also given interesting results [85,88], with performance comparable to PTAS and PTCDA in facilitating the growth of large-area, high-quality, and uniform monolayer and few-layer flakes. F_{16}CuPc has the attractive property of high stability at high temperature, making this a good choice for high-temperature growth processes.

Another interesting organic compound is crystal violet (CV), thanks to the possibility of engineering its configuration by simply varying the polarity of solvent of the solution. Ko et al. [93] studied the effects of different CV conditions on the growth of MoS_2 nanoflakes both from an experimental and a theoretical perspective. Their results gave interesting insight on the growth of monolayer MoS_2 from aromatic seeding promoters.

Thin films of 5, 10, 15, 20-tetraphenylporphyrin (H2TPP) were used as a promoter layer for the realization of hybrid structures composed of carbon nanotubes and MoS_2 nanosheets for the realization of flexible chemical sensors [94] and for the growth of vertical MoS_2 nanosheets [107]. The use of this promoter also allows for the simultaneous doping of the material by performing a metalation of the H2TPP films, obtaining different metalloporphyrins such as Al(III)-tetraphenyl porphyrin (Al(III)TPP) or Zn(II) meso-tetra(4-hydroxyphenyl) porphyrin (Zn(II)THPP) [108]. Porphyrin-based organic molecules such as 5,10,15,20-tetrakis(4-hydroxyphenyl)-21H,23H-porphyrin (p-THPP) were used also for the growth of MoS_2 on graphene oxide fibres. In [95], these fibres were dipped in the promoter solution. The use of Zn(II)THPP to achieve zinc-doping of the flakes was also demonstrated.

9.3. Surface Treatments

In their pioneering work, van der Zande et al. [26] showed that a careful treatment of SiO_2/Si substrates (2 h in H_2SO_4/H_2O_2 (3:1) followed by 5 min of O_2 plasma) and a minimal exposure of the precursors to air during storage was sufficient to obtain large flakes up to 100 µm. From these experimental observations, it is evident that the state of the surface on which the growth is carried on is of paramount relevance for the quality of flakes. From this point of view, many approaches to optimize this surface have been studied in the past. For example, the treatment of SiO_2/Si substrates with oxygen plasma resulted in layer-controlled and large-area CVD MoS_2 films [109].

Another interesting approach was studied recently that relied on the treatment of the SiO_2 substrate by a piranha etching solution prior to deposition. The authors stated that this treatment reduced the surface free energy of the substrate, making the use of promoters unnecessary [110]. An interesting method to treat surfaces to ease the transfer of flakes was proposed by Shinde et al. [111]. By etching an SiO_2 surface with hydrofluoric acid, the hydrophilicity is increased, allowing water to penetrate at the interface between polymer-capped MoS_2 and the substrate, resulting in direct transfer of flakes over a large area.

10. Growth Mechanisms

The synthesis of MoS_2 by CVD is the result of several steps, which can be summarized as follows: (i) transport of precursors to the substrate by a carrier gas; (ii) diffusion through the boundary layer from the gas phase to the substrate surface; (iii) adsorption of molecules to the substrate surface; (iv) diffusion of adsorbed molecules on the surface; (v) heterogeneous reactions on the surface resulting in TMD growth; (vi) desorption of by-products. In order to control the process and obtain a monolayer or few-layer material as an outcome, a high degree of control of the heterogeneous reactions should be pursued [35]. The exact evolution and progress of the synthesis and its mechanisms depend, of course, on the chosen precursors. The "holy grail" of MoS_2 CVD would be to develop a precise, controllable, and reproducible process to uniformly deposit large-area monodimensional flakes over a large area without grain boundaries and with a low density of defects. Understanding the growth mechanisms for the different available precursors and considering the possible presence of growth promoters could help to recognize the strengths and weaknesses of these precursors.

Considering the experience gathered on graphene, but at the same keeping in mind that the synthesis of MoS_2 proceeds through completely different mechanisms, in order to promote the growth of large 2D flakes, it is important to decrease the density of the nuclei and to increase the lateral growth rate while at the same time suppressing 3D nucleation [112]. Nucleation of MoO_{3-x} species at the beginning of growth is the first step to allow the formation of MoS_2 nanoparticles and large flakes in the later stage of the synthesis [113]. It was suggested that this growth mode is essentially controlled by the deposition rate, which should be kept below a certain threshold to avoid the formation of thicker islands. Kang et al. [32] observed that this occurred when a low partial pressure of Mo vapour was used. With a higher Mo partial pressure, the layer-by-layer growth mode was abandoned in favour of the nucleation of a mixture of monolayer, multilayer, and no-growth regions. To obtain uniform monolayer growth, potentially on a large substrate, a precise and constant Mo partial pressure for the entire duration of growth is necessary. This requirement is not straightforward when Mo powders are used as precursors, since they deplete and are consumed as the growth proceeds, but is more easily obtained using metal-organic-like precursors such as $Mo(CO)_6$.

The competition between the precursor mass flow towards the substrate and the reactions occurring at the surface determines the ultimate evolution of the growth. Zhou et al. [114] presented a general growth model for the deposition of a wide variety of TMDs, combining 12 transition metals and 3 chalcogens to synthesize up to 47 different 2D materials. They suggested that the mass flow supply determines the amount of metal precursors involved in the formation of the nucleus and the growth of domains, while the growth rate, essentially determined by the temperature, dominates the grain size of the layer. The interplay between growth rate and mass flow determines the formation of continuous monolayers, small flakes, atomic clusters, or large 2D single crystals. The two contradictory statements presented in [32,114] may be interpreted as a sign that the deposition of MoS_2 with powders is very dependent on the used growth system and setup, and finding a unifying explanation of the experimental observation is not straightforward.

The main reason for the lateral enlargement of MoS_2 flakes is the high chemical reactivity of the atoms at the edge of the flakes and in the presence of preferred nucleation sites, with high chemical reactivity, compared to the absence of dangling bonds in the vertical direction. A low and constant Mo precursor partial pressure may help in growing uniform monolayer thin films over a large area [32]. However, the growth of vertically standing MoS_2 nanosheets was also reported [115]. They were supposed to be caused by the reduction in the elastic strain energy that forms during the horizontal growth of MoS_2 nanosheets. The transition from 2D to 3D structure was achieved by controlling the quantity and distribution of the precursor concentration, placing the substrate in different orientations and positions with respect to the source powders [107]
when using Mo powders, the proposed reactions for MoS_2 synthesis are:

$$2\ MoO_3 + S \rightarrow 2\ MoO_{3-x} + SO_2$$

$$2\ MoO_{3-x} + (7 - x)\ S \rightarrow 2\ MoS_2 + (3 - x)\ SO_2$$

MoO_3 is first reduced to suboxides (MoO_{3-x}) by S vapours, further reduced to oxisulfides, and then transported to the substrate, where a further reaction with S vapour to form MoS_2 occurs [116]. Since a MoO_3 suboxide is involved in the growth reaction, the use of MoO_2 instead of MoO_3 was proposed in order to obtain better control of the flakes [117]. In this case, the synthesis reaction would proceed by the mechanism:

$$MoO_2 + 3\ S \rightarrow MoS_2 + O_2$$

This reaction avoids intermediate chemistry and permits obtaining higher-quality flakes on different substrates such as SiO_2, Si, quartz, and SiN. The quantity of starting Mo powders greatly influences the reaction process. In [118], a comprehensive summary of typical powder weights, along with growth temperature and setup, was presented.

In order to have a detailed outlook on the growth process of MoS_2, Cain et al. [119] deposited MoS_2 using S and MoO_3 powders on ultrathin SiO_2 membranes to allow investigation of the nucleation mechanism with aberration-corrected STEM and elemental EDS mapping. They confirmed the presence of nucleation centres (10–30 nm) only at the centres of triangular flakes, representing the early stages of TMD growth. The second or third layer of the flakes also started at this site. Moreover, the nuclei showed a nanoscale core–shell structure, similar to that of inorganic TMD fullerenes, with inhomogeneous distribution of Mo and S. The growth process was essentially defined by the concentration of the chalcogen vapour: if the atmosphere was weakly reducing (sulfur poor), oxi-chalcogenide particles with an orthorhombic crystal structure were formed. In a moderately reducing sulfur-rich atmosphere, MoS_2 formation was favoured. These experimental observations confirm the importance of promoting the formation of nucleation sites.

A convenient and empirical method to tune growth parameters towards the optimal growth conditions is to observe the flake shape at the optical microscope. Yang et al. [120] studied the morphological evolution of MoS_2 flakes for various Mo–S ratios, using MoO_3 and S powders as reagents at different temperatures. Obtaining a continuous film with large triangular MoS_2 flakes should be possible at higher temperature and with a high S content (Figure 7).

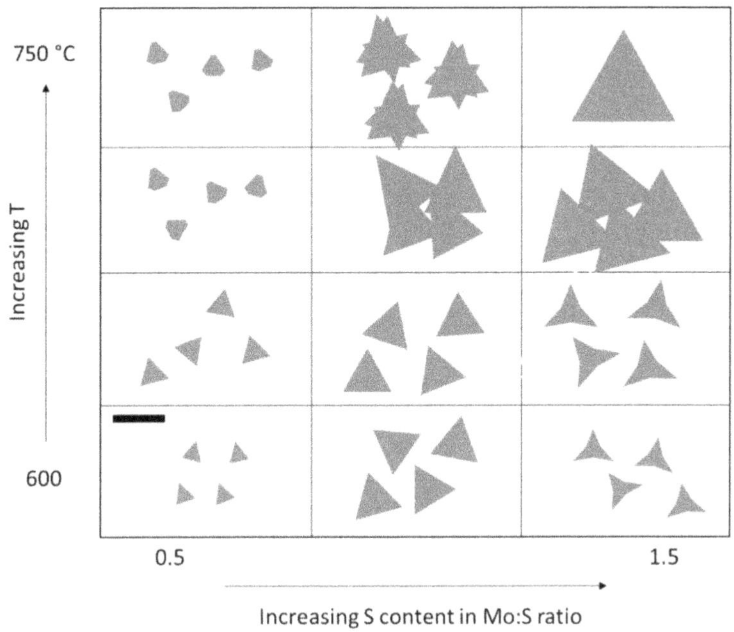

Figure 7. Schematic illustration of the evolution of the morphology of MoS_2 flakes for changing Mo–S ratios and temperatures. Marker represents a flake size of 100 μm. Derived from data from [120].

Obtaining large flakes, by preventing the cohesion of smaller flakes, would also limit the presence of grain boundaries at flake coalescence regions, which may limit electrical transport in the layer [16]. Moreover, it was observed that the MoS_2 bandgap could be tuned by controlling the distance from a grain boundary [121], so it is important to achieve good control over the flakes' size over the substrate area in order to avoid undesired modulations of the bandgap. Nevertheless, it was demonstrated that MoS_2 grain boundaries could exhibit a memristor behaviour [122], which may open new applications for this material.

An important source of carrier scattering in devices is defects and charged particles at the interface (Coulomb scattering). These defects are usually introduced in postprocessing fabrication steps, and it was suggested that proper passivation with a suitable dielectric could help mitigating the issue, increasing the performance of the device [123].

In general, point defects in MoS_2 act as important electron scattering centres and may be categorized as vacancies or antisite defects [124,125]. Sulfur vacancies behave as deep donors and induce midgap defect states, making the material n-type. Furthermore, vacancies and antisites induce a modification in the electrical properties of MoS_2, resulting in n-type doping and introducing localized states in the band gap, lowering the carrier mobility. They can cause Fermi level pinning and high contact resistance in electrical devices. Inclusion of other elements related to promoters (Na, Au) can also cause unintentional doping effects. These states also reduce the emission efficiency for photonic devices based on this material.

On the other hand, control of defect density and nature can be an effective tool to engineer material properties. It was shown that the electrical properties of MoS_2 could be changed by exposing the samples to oxygen plasma, going from the semiconducting to the insulating regime [50]. A recent method to reduce sulfur vacancies was introduced by Durairaj et al. [126]; in this study, a SiO_2/Si wafer was placed next to the growth substrate, providing oxygen and effectively enabling oxygen passivation of sulfur-vacancy defects in monolayer flakes. The authors observed a threefold increase in the PL efficiency due to the elimination of defect-related bound exciton emission.

In Figure 8, the reaction pathways that lead to the formation of MoS_2 through intermediate species are summarized [118,127].

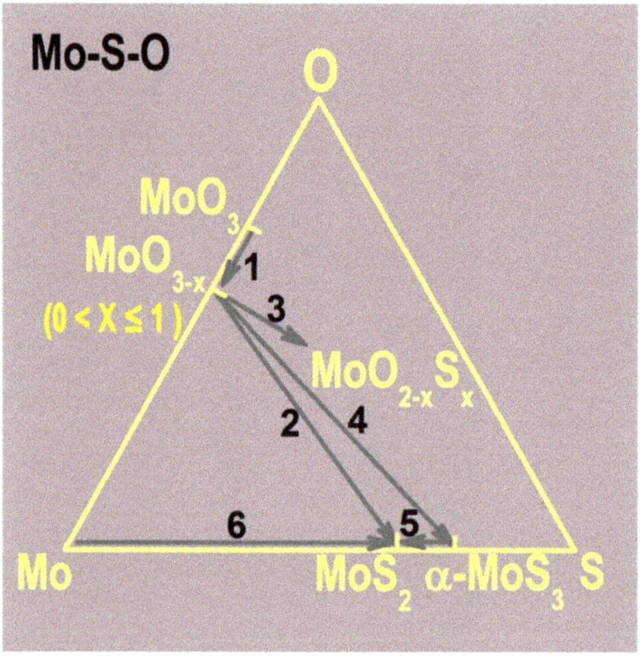

Figure 8. Ternary phase diagram for Mo-S-O, showing the possible reactions to produce MoS_2. Adapted from [127]; reprinted with permission from AAAS.

If $MoCl_5$ is used instead of MoO_3 powders, the proposed reactions are different from those reported above. Since the ratio between S and $MoCl_5$ is very high (>1000), the conversion of $MoCl_5$ to MoS_2 is supposed to be completed in the gas phase, with a negligible concentration of intermediate suboxides. By precisely controlling the amounts of $MoCl_5$ and S powders, their flow in the growth environment, and the total pressure in the growth chamber, Yu et al. [69] were able to control the number of MoS_2 layers, with good uniformity over an area of several cm^2. The key role was suggested to be the self-limiting nature of the layer-by-layer process, which could be controlled by tuning the flow of the precursors as well as the total pressure of the system: increasing the amount of precursors (from 1 mg of $MoCl_5$ to 25 mg) and the total pressure (from 2 to 750 torr) resulted in the deposition of thicker films.

One could argue that the use of powder Mo precursors, because of their time-dependent evaporation and the necessity of precise and continuous flow control, would lead to worse and less reproducible results with respect to the sulfurization of a thin Mo layer predeposited on the substrate [118]. In the latter case, the resulting MoS_2 layer would depend only on the starting conditions, which in principle could be precisely optimized by thermal evaporation or sputtering of the thin Mo or MoO_x layer. However, starting with a Mo film deposited on the substrate, the sulfurization process should occur at very high temperature to be efficient. This results in undesired surface evaporation of the starting material and leads to inhomogeneities, since the balance between the surface mobility and surface evaporation leads to limited grain sizes and/or limited yields of monolayer films. Moreover, it is still difficult to obtain an optimal starting film density and uniformity. However, one of the advantages of this approach is the ease of saturating the growth chamber with sulfur vapours (either from powders or from H_2S), so this is not the process-limiting factor.

To promote bilayer growth, the interactions between the adatoms and the surface should be larger than the adatom–adatom interactions. If these conditions are not met, the deposition evolves towards a 3D growth. It was suggested that by increasing the intermolecular acid–base interactions between admolecules and the substrate surface, it should be possible to increase the adsorption. This can be done by selecting precursors and substrates with complementary acid/base properties [35]. A convenient way to increase the surface forces is to use growth promoters or seed molecules to change the surface energy, reduce the free energy barrier, and lower the nucleation energy necessary to obtain monolayer MoS_2. It was argued that the best promoters are aromatic or graphene-like molecules that enhance the wettability of the substrate, lowering the free energy for the nucleation. In this way, better control of nucleation can be achieved [91]. Thus, an optimal concentration of the seeding promoter is essential for MoS_2 monolayer deposition [128]. One important factor to be considered is the polarity of the growth promoter molecule, as pointed out by Ko et al. [93]. With density function theory calculations coupled to experimental evidence, they suggested that the polar part of the crystal violet molecule used as a promoter provides a preferred site for S adsorption, initiating the MoS_2 nucleation. The orientation of the molecule during the spinning on the substrate, mediated by appropriate solvents, can effectively promote monolayer growth instead of multilayer MoS_2 deposition.

It has been shown that the use of alkali-based promoters allows for larger flakes [98], but their excess might result in loss of MoS_2 crystallinity, with the creation of additional defects [128]. Extensive work is currently underway on controlling defects in MoS_2 flakes acting on growth parameters, not only to reduce their deleterious effects on devices, but also to take advantage of some of their features. For example, Na cations coming from promoters can cure interface defects to achieve low intrinsic defect levels and enhance electrical properties [129]. The use of liquid precursors spun in a controllable way on the substrate could overcome the problems due to the limited supply control of powder precursors. Spinning SMD and/or AMT with a density gradient medium and a growth promoter such as NaOH or NaCl could permit obtaining a reproducible layer that could be sulfurized in a controllable way. Kang et al. [130] presented a comprehensive study on the role of the different growth parameters in the synthesis of MoS_2 flakes using liquid precursor, including the roles of temperature, S pressure, gas carrier flow, composition of the initial spun solution, and time. By an optimization of the process, MoS_2 flakes with dimensions up to 100 μm were obtained in a reproducible way. AFM analysis of the MoS_2 surface revealed the presence of precursor and promoter residues, particularly at the interfaces between different flakes. AMT is converted to MoO_3 at about 300 °C. Using Na as a promoter, if further reacts with NaOH to produce sodium molybdate (Na_6MoO_4), which finally converts to MoS_2 thanks to the presence of S vapour. Using SiO_2 as a substrate, in this case, permits the reaction between Na_2MoO_4 and the oxide substrate to form sodium silicon oxide, promoting the lateral growth of the MoS_2 layer [97].

11. Conclusions and Outlook

The rising interest in 2D MoS_2 comes from its peculiar physical, chemical, and mechanical properties, including, among other features, peculiar light adsorption characteristics and promising electrical transport properties.

However, the synthesis of reproducible, uniform, and single-layer MoS_2 flakes on large-area wafers is still a challenge, and open issues remain to be solved. Increasing control of the deposition process is mandatory to fine-tune the electrical and optical properties of MoS_2. At the same time, obtaining flakes with larger dimensions, higher density, and a low concentration of defects is necessary to pave the way for the industrial application of this material. In this review, we explore the CVD synthesis of mono- and few-layer MoS_2, giving an overview of the growth method, different substrates for deposition, and the used precursors and growth promoters. The topic of the doping of this material is also considered, and different growth mechanisms involving different types of precursors are overviewed.

From the critical analysis of the results obtained so far, it emerged that homogenous precursor feed, and in particular the achievement of fine control of Molybdenum supply during the CVD process, are necessary conditions for the optimization of the process. Because of their ease of use, the most used precursors for MoS_2 synthesis are still Mo and S powders, but this approach presents several issues, such as poor control of the precursor supply causing low reproducibility. The use of growth promoters has allowed overcoming these problems, but the process is far from being optimized for an industrial scale-up. A more recent approach involves the use of liquid precursor spun on the substrate to obtain a controllable thin layer containing Mo and Na as promoter. This method permitted overcoming the problems faced by the other techniques to increase the reproducibility and yield of the process, resulting in very large flakes on very wide areas.

The vast majority of the CVD reactors used for MoS_2 deposition have a horizontal geometry, which is known to have intrinsic flow inhomogeneities due to precursor depletion. This may be among the causes of the generally low reproducibility often observed in standard setups. Switching to a vertical gas distribution may help to solve these issues and obtain a better run-to-run consistency.

It is well known that CVD protocols are not easily transferrable from one laboratory to another, as even minimal changes in reactor parameters and growth conditions strongly affect reproducibility of the growth of 2D materials. It seems that many processes and recipes presented in literature worked only in a specific growth setup and configuration. There is a need to identify universal conditions and reaction schemes that will permit better control of the deposition of 2D MoS_2 flakes. For this reason, further work on the theoretical modelling of growth is urgently needed to help identify the most crucial parameters. More experimental efforts focusing on the effects of other parameters somewhat overlooked (sulfur partial pressure, carrier gas flows, reactor geometry, and size) could also be useful to gain more knowledge.

From the point of view of reproducibility, the use of liquid precursors seems to have some advantages, as it allows a more precise and reproducible control of the amount of Mo source materials. Similarly, the use of sulfur powders is a crucial element, as the quantity of vapourized sulfur also depends on the mass of powder put in the crucible. The use of sulfur gas has shown some advantages, despite its safety drawbacks. Also, the precise control of gas flows is a fundamental parameter to achieve reproducibility of results.

In our opinion, high-quality, large flakes can be obtained in a controllable and reproducible way using a combination of these three fundamental elements: (i) liquid precursors, (ii) controlled surface treatments (etching and/or plasma), and (iii) organic promoters (Na-based ones being the most effective). However, the optimal combination of these elements has still not been clearly identified, and exact protocols and recipes are still very cumbersome; more research work is needed to identify the optimal conditions.

From the work presented in the literature, it seems that the uniformity of large flakes is strongly dependent on the state of the growing surface. Reduced graphene and sapphire seem to have an advantage, but by careful etching and cleaning of the substrate (by methods such as plasma cleaning), very good results can be obtained also on more common and less expensive systems such as SiO_2/Si. These excellent results are fostering a large increase in studies and activities on the growth of this material, considering different substrates on which to grow it, various elements for doping, and the development of reliable models to gain insight on the growth mechanisms. Variations on the basic CVD concepts to achieve better control of precursor supply through various hardware modifications of the apparatus have been proposed with encouraging results.

From this perspective, the knowledge on the CVD growth of MoS_2 seems to be mature enough to envision the design and realization of complex heterostructures based on MoS_2, in which not only the materials but their dimensionality are different (so-called hybrid heterostructures).

However, we think that at the moment, there is a bottleneck on the realization of complex heterostructures, as the transfer of flakes from one substrate to another is a very complex and delicate step that could represent a second major hurdle for the transfer of the technology for this material to the development of commercial devices. Therefore, it would be desirable to achieve large-area flakes on other substrates, such as metals, semiconductors, and possibly flexible substrates, on which devices can be directly fabricated.

Moreover, we are keen to consider that the next major breakthrough for MoS_2 is related to the possibility of realizing patterned deposition of this 2D material, aiming for a method to obtain designed microcircuits. From this point of view, the possibility of "printing" MoS_2 layers seems to us a more appealing approach than delicate and cumbersome transfer and manipulation of large flakes.

Author Contributions: Writing—review and editing, L.S. and M.B. All authors have read and agreed to the published version of the manuscript.

Funding: This research received no external funding.

Institutional Review Board Statement: Not applicable.

Informed Consent Statement: Not applicable.

Data Availability Statement: No new data were created or analyzed in this study.

Conflicts of Interest: The authors declare no conflict of interest.

References

1. Gupta, D.; Chauhan, V.; Kumar, R. A comprehensive review on synthesis and applications of molybdenum disulfide (MoS_2) material: Past and recent developments. *Inorg. Chem. Commun.* **2020**, *121*, 108200. [CrossRef]
2. Zhang, S.; Le, S.T.; Richter, C.A.; Hacker, C.A. Improved contacts to p-type MoS_2 transistors by charge-transfer doping and contact engineering. *Appl. Phys. Lett.* **2019**, *115*, 73106. [CrossRef] [PubMed]
3. Abuzaid, H.; Williams, N.X.; Franklin, A.D. How good are 2D transistors? An application-specific benchmarking study. *Appl. Phys. Lett.* **2021**, *118*, 30501. [CrossRef]
4. Wang, H.; Yu, L.; Lee, Y.-H.; Shi, Y.; Hsu, A.; Chin, M.L.; Li, L.-J.; Dubey, M.; Kong, J.; Palacios, T. Integrated Circuits Based on Bilayer MoS_2 Transistors. *Nano Lett.* **2012**, *12*, 4674–4680. [CrossRef] [PubMed]
5. Nalwa, H.S. A review of molybdenum disulfide (MoS_2) based photodetectors: From ultra-broadband, self-powered to flexible devices. *RSC Adv.* **2020**, *10*, 30529–30602. [CrossRef]
6. Pumera, M.; Loo, A.H. Layered transition-metal dichalcogenides (MoS_2 and WS2) for sensing and biosensing. *TrAC Trends Anal. Chem.* **2014**, *61*, 49–53. [CrossRef]
7. Zhu, M.; Du, X.; Liu, S.; Li, J.; Wang, Z.; Ono, T. A review of strain sensors based on two-dimensional molybdenum disulfide. *J. Mater. Chem. C* **2021**, *9*, 9083–9101. [CrossRef]
8. Butun, S.; Tongay, S.; Aydin, K. Enhanced Light Emission from Large-Area Monolayer MoS_2 Using Plasmonic Nanodisc Arrays. *Nano Lett.* **2015**, *15*, 2700–2704. [CrossRef]
9. Ponomarev, E.; Gutiérrez-Lezama, I.; Ubrig, N.; Morpurgo, A.F. Ambipolar Light-Emitting Transistors on Chemical Vapor Deposited Monolayer MoS_2. *Nano Lett.* **2015**, *15*, 8289–8294. [CrossRef]
10. Golovynskyi, S.; Irfan, I.; Bosi, M.; Seravalli, L.; Datsenko, O.I.; Golovynska, I.; Li, B.; Lin, D.; Qu, J. Exciton and trion in few-layer MoS_2: Thickness- and temperature-dependent photoluminescence. *Appl. Surf. Sci.* **2020**, *515*, 146033. [CrossRef]
11. Golovynskyi, S.; Bosi, M.; Seravalli, L.; Li, B. MoS_2 two-dimensional quantum dots with weak lateral quantum confinement: Intense exciton and trion photoluminescence. *Surfaces Interfaces* **2021**, *23*, 100909. [CrossRef]
12. Cho, A.-J.; Park, K.C.; Kwon, J.-Y. A high-performance complementary inverter based on transition metal dichalcogenide field-effect transistors. *Nanoscale Res. Lett.* **2015**, *10*, 1–6. [CrossRef] [PubMed]
13. Swain, G.; Sultana, S.; Parida, K. A review on vertical and lateral heterostructures of semiconducting 2D-MoS_2 with other 2D materials: A feasible perspective for energy conversion. *Nanoscale* **2021**, *13*, 9908–9944. [CrossRef] [PubMed]
14. Xiao, D.; Liu, G.-B.; Feng, W.; Xu, X.; Yao, W. Coupled Spin and Valley Physics in Monolayers of MoS_2 and Other Group-VI Dichalcogenides. *Phys. Rev. Lett.* **2012**, *108*, 196802. [CrossRef]
15. Yu, Z.; Ong, Z.-Y.; Li, S.-L.; Xu, J.; Zhang, G.; Zhang, Y.-W.; Shi, Y.; Wang, X. Analyzing the Carrier Mobility in Transition-Metal Dichalcogenide MoS_2 Field-Effect Transistors. *Adv. Funct. Mater.* **2017**, *27*. [CrossRef]
16. Ly, T.H.; Perello, D.J.; Zhao, J.; Deng, Q.; Kim, H.; Han, G.H.; Chae, S.H.; Jeong, H.Y.; Lee, Y.H. Misorientation-angle-dependent electrical transport across molybdenum disulfide grain boundaries. *Nat. Commun.* **2016**, *7*, 10426. [CrossRef]
17. Lin, Y.; Torsi, R.; Geohegan, D.B.; Robinson, J.A.; Xiao, K. Controllable Thin-Film Approaches for Doping and Alloying Transition Metal Dichalcogenides Monolayers. *Adv. Sci.* **2021**, *8*, 2004249. [CrossRef]

18. Rai, A.; Movva, H.C.P.; Roy, A.; Taneja, D.; Chowdhury, S.; Banerjee, S.K. Progress in Contact, Doping and Mobility Engineering of MoS_2: An Atomically Thin 2D Semiconductor. *Crystals* **2018**, *8*, 316. [CrossRef]
19. Zhang, K.; Bersch, B.M.; Joshi, J.; Addou, R.; Cormier, C.R.; Zhang, C.; Xu, K.; Briggs, N.C.; Wang, K.; Subramanian, S.; et al. Tuning the Electronic and Photonic Properties of Monolayer MoS_2 via In Situ Rhenium Substitutional Doping. *Adv. Funct. Mater.* **2018**, *28*, 1706950. [CrossRef]
20. Xu, E.Z.; Liu, H.M.; Park, K.; Li, Z.; Losovyj, Y.; Starr, M.; Werbianskyj, M.; Fertig, H.A.; Zhang, S.X. p-Type transition-metal doping of large-area MoS_2 thin films grown by chemical vapor deposition. *Nanoscale* **2017**, *9*, 3576–3584. [CrossRef]
21. Yang, L.; Majumdar, K.; Liu, H.; Du, Y.; Wu, H.; Hatzistergos, M.; Hung, P.Y.; Tieckelmann, R.; Tsai, W.; Hobbs, C.; et al. Chloride Molecular Doping Technique on 2D Materials: WS2 and MoS. *Nano Lett.* **2014**, *14*, 6275–6280. [CrossRef]
22. Li, M.; Yao, J.; Wu, X.; Zhang, S.; Xing, B.; Niu, X.; Yan, X.; Yu, Y.; Liu, Y.; Wang, Y. P-type Doping in Large-Area Monolayer MoS_2 by Chemical Vapor Deposition. *ACS Appl. Mater. Interfaces* **2020**, *12*, 6276–6282. [CrossRef] [PubMed]
23. Zhang, K.; Feng, S.; Wang, J.; Azcatl, A.; Lu, N.; Addou, R.; Wang, N.; Zhou, C.; Lerach, J.; Bojan, V.; et al. Manganese Doping of Monolayer MoS_2: The Substrate Is Critical. *Nano Lett.* **2015**, *15*, 6586–6591. [CrossRef] [PubMed]
24. Li, X.; Zhu, H. Two-dimensional MoS_2: Properties, preparation, and applications. *J. Materiomics* **2015**, *1*, 33–44. [CrossRef]
25. Mouloua, D.; Kotbi, A.; Deokar, G.; Kaja, K.; El Marssi, M.; EL Khakani, M.; Jouiad, M. Recent Progress in the Synthesis of MoS_2 Thin Films for Sensing, Photovoltaic and Plasmonic Applications: A Review. *Materials* **2021**, *14*, 3283. [CrossRef]
26. Van der Zande, A.; Huang, P.; Chenet, D.A.; Berkelbach, T.C.; You, Y.; Lee, G.-H.; Heinz, T.F.; Reichman, D.R.; Muller, D.; Hone, J.C. Grains and grain boundaries in highly crystalline monolayer molybdenum disulphide. *Nat. Mater.* **2013**, *12*, 554–561. [CrossRef] [PubMed]
27. Dumcenco, D.; Ovchinnikov, D.; Marinov, K.; Lazić, P.; Gibertini, M.; Marzari, N.; Sanchez, O.L.; Kung, Y.-C.; Krasnozhon, D.; Chen, M.-W.; et al. Large-Area Epitaxial Monolayer MoS_2. *ACS Nano* **2015**, *9*, 4611–4620. [CrossRef] [PubMed]
28. Yang, P.; Yang, H.; Wu, Z.; Liao, F.; Guo, X.; Deng, J.; Xu, Q.; Wang, H.; Sun, J.; Chen, F.; et al. Large-Area Monolayer MoS_2 Nanosheets on GaN Substrates for Light-Emitting Diodes and Valley-Spin Electronic Devices. *ACS Appl. Nano Mater.* **2021**, *4*, 12127–12136. [CrossRef]
29. Tang, L.; Li, T.; Luo, Y.; Feng, S.; Cai, Z.; Zhang, H.; Liu, B.; Cheng, H.-M. Vertical Chemical Vapor Deposition Growth of Highly Uniform 2D Transition Metal Dichalcogenides. *ACS Nano* **2020**, *14*, 4646–4653. [CrossRef]
30. Zhang, X.; Ning, J.; Li, X.; Wang, B.; Hao, L.; Liang, M.; Jin, M.; Zhi, L. Hydrogen-induced effects on the CVD growth of high-quality graphene structures. *Nanoscale* **2013**, *5*, 8363–8366. [CrossRef]
31. Li, X.; Li, X.; Zang, X.; Zhu, M.; He, Y.; Wang, K.; Xie, D.; Zhu, H. Role of hydrogen in the chemical vapor deposition growth of MoS_2 atomic layers. *Nanoscale* **2015**, *7*, 8398–8404. [CrossRef] [PubMed]
32. Kang, K.; Xie, S.; Huang, L.; Han, Y.; Huang, P.; Mak, K.F.; Kim, C.-J.; Muller, D.; Park, J. High-mobility three-atom-thick semiconducting films with wafer-scale homogeneity. *Nature* **2015**, *520*, 656–660. [CrossRef] [PubMed]
33. Liu, H.; Zhu, Y.; Meng, Q.; Lu, X.; Kong, S.; Huang, Z.; Jiang, P.; Bao, X. Role of the carrier gas flow rate in monolayer MoS_2 growth by modified chemical vapor deposition. *Nano Res.* **2017**, *10*, 643–651. [CrossRef]
34. Chen, J.; Tang, W.; Tian, B.; Liu, B.; Zhao, X.; Liu, Y.; Ren, T.; Liu, W.; Geng, D.; Jeong, H.Y.; et al. Chemical Vapor Deposition of High-Quality Large-Sized MoS_2 Crystals on Silicon Dioxide Substrates. *Adv. Sci.* **2016**, *3*, 1500033. [CrossRef]
35. Li, S.; Wang, S.; Xu, T.; Zhang, H.; Tang, Y.; Liu, S.; Jiang, T.; Zhou, S.; Cheng, H. Growth mechanism and atomic structure of group-IIA compound-promoted CVD-synthesized monolayer transition metal dichalcogenides. *Nanoscale* **2021**, *13*, 13030–13041. [CrossRef] [PubMed]
36. Qin, B.; Ma, H.; Hossain, M.; Zhong, M.; Xia, Q.; Li, B.; Duan, X. Substrates in the Synthesis of Two-Dimensional Materials via Chemical Vapor Deposition. *Chem. Mater.* **2020**, *32*, 10321–10347. [CrossRef]
37. Zhang, H.; Wan, Y.; Ma, Y.; Wang, W.; Wang, Y.; Dai, L. Interference effect on optical signals of monolayer MoS_2. *Appl. Phys. Lett.* **2015**, *107*, 101904. [CrossRef]
38. Salvatore, G.A.; Münzenrieder, N.; Barraud, C.; Petti, L.; Zysset, C.; Büthe, L.; Ensslin, K.; Tröster, G. Fabrication and Transfer of Flexible Few-Layers MoS_2 Thin Film Transistors to Any Arbitrary Substrate. *ACS Nano* **2013**, *7*, 8809–8815. [CrossRef]
39. Shi, J.; Ma, D.; Han, G.-F.; Zhang, Y.; Ji, Q.; Gao, T.; Sun, J.; Song, X.; Li, C.; Zhang, Y.; et al. Controllable Growth and Transfer of Monolayer MoS_2 on Au Foils and Its Potential Application in Hydrogen Evolution Reaction. *ACS Nano* **2014**, *8*, 10196–10204. [CrossRef] [PubMed]
40. Yang, P.; Zhang, S.; Pan, S.; Tang, B.; Liang, Y.; Zhao, X.; Zhang, Z.; Shi, J.; Huan, Y.; Shi, Y.; et al. Epitaxial Growth of Centimeter-Scale Single-Crystal MoS_2 Monolayer on Au(111). *ACS Nano* **2020**, *14*, 5036–5045. [CrossRef]
41. Li, J.; Wang, S.; Jiang, Q.; Qian, H.; Hu, S.; Kang, H.; Chen, C.; Zhan, X.; Yu, A.; Zhao, S.; et al. Single-Crystal MoS_2 Monolayer Wafer Grown on Au (111) Film Substrates. *Small* **2021**, *17*, 2100743. [CrossRef] [PubMed]
42. Amaral, G.M.D.; Tonon, I.D.C.; Román, R.J.P.; Plath, H.D.O.; Taniguchi, T.M.; de Lima, L.H.; Zagonel, L.F.; Landers, R.; de Siervo, A. Epitaxial growth, electronic hybridization and stability under oxidation of monolayer MoS_2 on Ag(1 1 1). *Appl. Surf. Sci.* **2021**, *538*, 148138. [CrossRef]
43. Fu, L.; Sun, Y.; Wu, N.; Mendes, R.G.; Chen, L.; Xu, Z.; Zhang, T.; Rümmeli, M.H.; Rellinghaus, B.; Pohl, D.; et al. Direct Growth of MoS_2/h-BN Heterostructures via a Sulfide-Resistant Alloy. *ACS Nano* **2016**, *10*, 2063–2070. [CrossRef] [PubMed]
44. Ma, Z.; Wang, S.; Deng, Q.; Hou, Z.; Zhou, X.; Li, X.; Cui, F.; Si, H.; Zhai, T.; Xu, H. Epitaxial Growth of Rectangle Shape MoS_2 with Highly Aligned Orientation on Twofold Symmetry a-Plane Sapphire. *Small* **2020**, *16*, e2000596. [CrossRef] [PubMed]

45. Aljarb, A.; Cao, Z.; Tang, H.-L.; Huang, J.-K.; Li, M.; Hu, W.; Cavallo, L.; Li, L.-J. Substrate Lattice-Guided Seed Formation Controls the Orientation of 2D Transition-Metal Dichalcogenides. *ACS Nano* **2017**, *11*, 9215–9222. [CrossRef]
46. Cun, H.; Macha, M.; Kim, H.; Liu, K.; Zhao, Y.; Lagrange, T.; Kis, A.; Radenovic, A. Wafer-scale MOCVD growth of monolayer MoS_2 on sapphire and SiO_2. *Nano Res.* **2019**, *12*, 2646–2652. [CrossRef]
47. Li, T.; Guo, W.; Ma, L.; Li, W.; Yu, Z.; Han, Z.; Gao, S.; Liu, L.; Fan, D.; Wang, Z.; et al. Epitaxial growth of wafer-scale molybdenum disulfide semiconductor single crystals on sapphire. *Nat. Nanotechnol.* **2021**, *16*, 1201–1207. [CrossRef] [PubMed]
48. Yang, P.; Zou, X.; Zhang, Z.; Zhongfan, L.; Shi, J.; Chen, S.; Shulin, C.; Zhao, L.; Jiang, S.; Zhou, X.; et al. Batch production of 6-inch uniform monolayer molybdenum disulfide catalyzed by sodium in glass. *Nat. Commun.* **2018**, *9*, 1–10. [CrossRef] [PubMed]
49. Zhao, Y.; Song, J.-G.; Ryu, G.H.; Ko, K.Y.; Woo, W.J.; Kim, Y.; Kim, D.; Lim, J.H.; Lee, S.; Lee, Z.; et al. Low-temperature synthesis of 2D MoS_2 on a plastic substrate for a flexible gas sensor. *Nanoscale* **2018**, *10*, 9338–9345. [CrossRef]
50. Singh, A.; Moun, M.; Sharma, M.; Barman, A.; Kapoor, A.K.; Singh, R. NaCl-assisted substrate dependent 2D planar nucleated growth of MoS_2. *Appl. Surf. Sci.* **2021**, *538*, 148201. [CrossRef]
51. Yan, P.; Tian, Q.; Yang, G.; Weng, Y.; Zhang, Y.; Wang, J.; Xie, F.; Lu, N. Epitaxial growth and interfacial property of monolayer MoS_2 on gallium nitride. *RSC Adv.* **2018**, *8*, 33193–33197. [CrossRef]
52. Li, J.; Chen, X.; Zhang, D.W.; Zhou, P. Van der Waals Heterostructure Based Field Effect Transistor Application. *Crystals* **2017**, *8*, 8. [CrossRef]
53. Chen, T.; Zhou, Y.; Sheng, Y.; Wang, X.; Zhou, S.; Warner, J.H. Hydrogen-Assisted Growth of Large-Area Continuous Films of MoS_2 on Monolayer Graphene. *ACS Appl. Mater. Interfaces* **2018**, *10*, 7304–7314. [CrossRef]
54. Nguyen, V.T.; Kim, Y.C.; Ahn, Y.H.; Lee, S.; Park, J.-Y. Large-area growth of high-quality graphene/MoS_2 vertical heterostructures by chemical vapor deposition with nucleation control. *Carbon* **2020**, *168*, 580–587. [CrossRef]
55. Shi, Y.; Li, H.; Wong, J.I.; Zhang, X.; Wang, Y.; Song, H.; Yang, H.Y. MoS_2 Surface Structure Tailoring via Carbonaceous Promoter. *Sci. Rep.* **2015**, *5*, 10378. [CrossRef] [PubMed]
56. Fan, H.; Wang, J.; Li, X.; You, H.; Li, X.; Pei, C.; Huang, X.; Li, H. Direct CVD growth of MoS_2 on chemically and thermally reduced graphene oxide nanosheets for improved photoresponse. *Apll. Mater.* **2021**, *9*, 51105. [CrossRef]
57. Sitek, J.; Plocharski, J.; Pasternak, I.; Gertych, A.; McAleese, C.; Conran, B.R.; Zdrojek, M.; Strupinski, W. Substrate-induced variances in morphological and structural properties of MoS_2 grown by chemical vapor deposition on epitaxial graphene and SiO_2. *ACS Appl. Mater. Interfaces* **2020**, *12*, 45101–45110. [CrossRef] [PubMed]
58. Najmaei, S.; Liu, Z.; Zhou, W.; Zou, X.; Shi, G.; Lei, S.; Yakobson, B.I.; Idrobo, J.C.; Ajayan, P.M.; Lou, J. Vapour phase growth and grain boundary structure of molybdenum disulphide atomic layers. *Nat. Mater.* **2013**, *12*, 754–759. [CrossRef] [PubMed]
59. Han, G.H.; Kybert, N.J.; Naylor, C.H.; Lee, B.S.; Ping, J.; Park, J.H.; Kang, J.; Lee, S.Y.; Lee, Y.H.; Agarwal, R.; et al. Seeded growth of highly crystalline molybdenum disulphide monolayers at controlled locations. *Nat. Commun.* **2015**, *6*, 6128. [CrossRef] [PubMed]
60. Pareek, D.; Roach, K.G.; Gonzalez, M.A.; Büsing, L.; Parisi, J.; Gütay, L.; Schäfer, S. Micro patterned deposition of MoS_2 ultrathin-films by a controlled droplet dragging approach. *Sci. Rep.* **2021**, *11*, 13993. [CrossRef]
61. Wang, Z.; Huang, Q.; Chen, P.; Guo, S.; Liu, X.; Liang, X.; Wang, L. Metal Induced Growth of Transition Metal Dichalcogenides at Controlled Locations. *Sci. Rep.* **2016**, *6*, 38394. [CrossRef] [PubMed]
62. Li, Y.; Hao, S.; Distefano, J.G.; Murthy, A.A.; Hanson, E.D.; Xu, Y.; Wolverton, C.; Chen, X.; Dravid, V.P. Site-Specific Positioning and Patterning of MoS_2 Monolayers: The Role of Au Seeding. *ACS Nano* **2018**, *12*, 8970–8976. [CrossRef]
63. Seravalli, L.; Bosi, M.; Fiorenza, P.; Panasci, S.E.; Orsi, D.; Rotunno, E.; Cristofolini, L.; Rossi, F.; Giannazzo, F.; Fabbri, F. Gold nanoparticle assisted synthesis of MoS_2 monolayers by chemical vapor deposition. *Nanoscale Adv.* **2021**, *3*, 4826–4833. [CrossRef]
64. Wan, X.; Miao, X.; Yao, J.; Wang, S.; Shao, F.; Xiao, S.; Zhan, R.; Chen, K.; Zeng, X.; Gu, X.; et al. In Situ Ultrafast and Patterned Growth of Transition Metal Dichalcogenides from Inkjet-Printed Aqueous Precursors. *Adv. Mater.* **2021**, *33*, 2100260. [CrossRef] [PubMed]
65. Stull, D.R. Vapor Pressure of Pure Substances. Organic and Inorganic Compounds. *Ind. Eng. Chem.* **1947**, *39*, 517–540. [CrossRef]
66. Robertson, A.W.; Lin, Y.-C.; Wang, S.; Sawada, H.; Allen, C.S.; Chen, Q.; Lee, S.; Lee, G.-D.; Lee, J.; Han, S.; et al. Atomic Structure and Spectroscopy of Single Metal (Cr, V) Substitutional Dopants in Monolayer MoS_2. *ACS Nano* **2016**, *10*, 10227–10236. [CrossRef] [PubMed]
67. Liang, T.; Habib, M.R.; Xiao, H.; Xie, S.; Kong, Y.; Yu, C.; Iwai, H.; Fujita, D.; Hanagata, N.; Chen, H.; et al. Intrinsically Substitutional Carbon Doping in CVD-Grown Monolayer MoS_2 and the Band Structure Modulation. *ACS Appl. Electron. Mater.* **2020**, *2*, 1055–1064. [CrossRef]
68. Diskus, M.; Nilsen, O.; Fjellvåg, H. Growth of thin films of molybdenum oxide by atomic layer deposition. *J. Mater. Chem.* **2011**, *21*, 705–710. [CrossRef]
69. Yu, Y.; Li, C.; Liu, Y.; Su, L.; Zhang, Y.; Cao, L. Controlled Scalable Synthesis of Uniform, High-Quality Monolayer and Few-layer MoS_2 Films. *Sci. Rep.* **2013**, *3*, 1866. [CrossRef] [PubMed]
70. Mann, J.; Sun, D.; Ma, Q.; Chen, J.-R.; Preciado, E.; Ohta, T.; Diaconescu, B.; Yamaguchi, K.; Tran, T.; Wurch, M.; et al. Facile growth of monolayer MoS_2 film areas on SiO_2. *Eur. Phys. J. B* **2013**, *86*, 1–4. [CrossRef]
71. Lin, Y.-C.; Zhang, W.; Huang, J.-K.; Liu, K.-K.; Lee, Y.-H.; Liang, C.-T.; Chu, C.-W.; Li, L.-J. Wafer-scale MoS_2 thin layers prepared by MoO_3 sulfurization. *Nanoscale* **2012**, *4*, 6637–6641. [CrossRef]
72. Meyer, B. Elemental sulfur. *Chem. Rev.* **1976**, *76*, 367–388. [CrossRef]

73. Liu, H.; Antwi, K.K.A.; Chua, S.; Chi, D. Vapor-phase growth and characterization of Mo1−xWxS2($0 \leq x \leq 1$) atomic layers on 2-inch sapphire substrates. *Nanoscale* **2013**, *6*, 624–629. [CrossRef]
74. Boandoh, S.; Choi, S.H.; Park, J.-H.; Park, S.Y.; Bang, S.; Jeong, M.S.; Lee, J.S.; Kim, H.J.; Yang, W.; Choi, J.-Y.; et al. A Novel and Facile Route to Synthesize Atomic-Layered MoS_2 Film for Large-Area Electronics. *Small* **2017**, *13*. [CrossRef] [PubMed]
75. Cai, Z.; Lai, Y.; Zhao, S.; Zhang, R.; Tan, J.; Feng, S.; Zou, J.; Tang, L.; Lin, J.; Liu, B.; et al. Dissolution-precipitation growth of uniform and clean two dimensional transition metal dichalcogenides. *Natl. Sci. Rev.* **2020**, *8*. [CrossRef] [PubMed]
76. Zhang, T.; Fujisawa, K.; Zhang, F.; Liu, M.; Lucking, M.; Gontijo, R.N.; Lei, Y.; Liu, H.; Crust, K.; Granzier-Nakajima, T.; et al. Universal In Situ Substitutional Doping of Transition Metal Dichalcogenides by Liquid-Phase Precursor-Assisted Synthesis. *ACS Nano* **2020**, *14*, 4326–4335. [CrossRef] [PubMed]
77. Shinde, N.B.; Francis, B.; Rao, M.S.R.; Ryu, B.D.; Chandramohan, S.; Eswaran, S.K. Rapid wafer-scale fabrication with layer-by-layer thickness control of atomically thin MoS_2 films using gas-phase chemical vapor deposition. *APL Mater.* **2019**, *7*, 81113. [CrossRef]
78. Jin, Z.; Shin, S.; Kwon, D.H.; Han, S.-J.; Min, Y.-S. Novel chemical route for atomic layer deposition of MoS_2 thin film on SiO_2/Si substrate. *Nanoscale* **2014**, *6*, 14453–14458. [CrossRef]
79. Feng, S.; Tan, J.; Zhao, S.; Zhang, S.; Khan, U.; Tang, L.; Zou, X.; Lin, J.; Cheng, H.; Liu, B. Synthesis of Ultrahigh-Quality Monolayer Molybdenum Disulfide through In Situ Defect Healing with Thiol Molecules. *Small* **2020**, *16*, e2003357. [CrossRef]
80. Robinson, Z.; Schmucker, S.W.; McCreary, K.M.; Cobas, E.D. Chemical Vapor Deposition of Two-Dimensional Crystals. In *Handbook of Crystal Growth—Thin Films and Epitaxy*, 2nd ed.; Elsevier: Amsterdam, The Netherlands, 2015; pp. 785–833. [CrossRef]
81. Kumar, V.K.; Dhar, S.; Choudhury, T.H.; Shivashankar, S.A.; Raghavan, S. A predictive approach to CVD of crystalline layers of TMDs: The case of MoS_2. *Nanoscale* **2015**, *7*, 7802–7810. [CrossRef]
82. Song, J.-G.; Park, J.; Lee, W.; Choi, T.; Jung, H.; Lee, C.W.; Hwang, S.-H.; Myoung, J.M.; Jung, J.-H.; Kim, S.-H.; et al. Layer-Controlled, Wafer-Scale, and Conformal Synthesis of Tungsten Disulfide Nanosheets Using Atomic Layer Deposition. *ACS Nano* **2013**, *7*, 11333–11340. [CrossRef] [PubMed]
83. Dasgupta, N.P.; Mack, J.F.; Langston, M.C.; Bousetta, A.; Prinz, F.B. Design of an atomic layer deposition reactor for hydrogen sulfide compatibility. *Rev. Sci. Instrum.* **2010**, *81*, 44102. [CrossRef] [PubMed]
84. US Department of Health and Human Sevices. *Toxicological Profile for Molybdenum*; US Department of Health and Human Sevices: Washington, DC, USA, 2020.
85. Ling, X.; Lee, Y.-H.; Lin, Y.; Fang, W.; Yu, L.; Dresselhaus, M.S.; Kong, J. Role of the Seeding Promoter in MoS_2 Growth by Chemical Vapor Deposition. *Nano Lett.* **2014**, *14*, 464–472. [CrossRef]
86. Wang, P.; Lei, J.; Qu, J.; Cao, S.; Jiang, H.; He, M.; Shi, H.; Sun, X.; Gao, B.; Liu, W. Mechanism of Alkali Metal Compound-Promoted Growth of Monolayer MoS_2: Eutectic Intermediates. *Chem. Mater.* **2019**, *31*, 873–880. [CrossRef]
87. Zhu, W.; Low, T.; Lee, Y.-H.; Wang, H.; Farmer, D.B.; Kong, J.; Xia, F.; Avouris, P. Electronic transport and device prospects of monolayer molybdenum disulphide grown by chemical vapour deposition. *Nat. Commun.* **2014**, *5*, 3087. [CrossRef] [PubMed]
88. Yang, P.; Yang, A.-G.; Chen, L.; Chen, J.; Zhang, Y.; Wang, H.; Hu, L.; Zhang, R.-J.; Liu, R.; Qu, X.-P.; et al. Influence of seeding promoters on the properties of CVD grown monolayer molybdenum disulfide. *Nano Res.* **2019**, *12*, 823–827. [CrossRef]
89. Lee, Y.-H.; Yu, L.; Wang, H.; Fang, W.; Ling, X.; Shi, Y.; Lin, C.-T.; Huang, J.-K.; Chang, M.-T.; Chang, C.-S.; et al. Synthesis and Transfer of Single-Layer Transition Metal Disulfides on Diverse Surfaces. *Nano Lett.* **2013**, *13*, 1852–1857. [CrossRef]
90. Martella, C.; Kozma, E.; Tummala, P.P.; Ricci, S.; Patel, K.A.; Andicsovà-Eckstein, A.; Bertini, F.; Scavia, G.; Sordan, R.; Nobili, L.G.; et al. Changing the Electronic Polarizability of Monolayer MoS_2 by Perylene-Based Seeding Promoters. *Adv. Mater. Interfaces* **2020**, *7*, 2000791. [CrossRef]
91. Lee, Y.-H.; Zhang, X.-Q.; Zhang, W.; Chang, M.-T.; Lin, C.-T.; Chang, K.-D.; Yu, Y.-C.; Wang, J.T.-W.; Chang, C.-S.; Li, L.-J.; et al. Synthesis of Large-Area MoS_2 Atomic Layers with Chemical Vapor Deposition. *Adv. Mater.* **2012**, *24*, 2320–2325. [CrossRef]
92. Rotunno, E.; Bosi, M.; Seravalli, L.; Salviati, G.; Fabbri, F. Influence of organic promoter gradient on the MoS_2 growth dynamics. *Nanoscale Adv.* **2020**, *2*, 2352–2362. [CrossRef]
93. Ko, H.; Kim, H.S.; Ramzan, M.S.; Byeon, S.; Choi, S.H.; Kim, K.K.; Kim, Y.-H.; Kim, S.M. Atomistic mechanisms of seeding promoter-controlled growth of molybdenum disulphide. *2D Mater.* **2020**, *7*, 15013. [CrossRef]
94. Kim, S.; Han, J.; Kang, M.-A.; Song, W.; Myung, S.; Kim, S.-W.; Lee, S.S.; Lim, J.; An, K.-S. Flexible chemical sensors based on hybrid layer consisting of molybdenum disulphide nanosheets and carbon nanotubes. *Carbon* **2018**, *129*, 607–612. [CrossRef]
95. Kim, S.J.; Kang, M.-A.; Jeon, I.-S.; Ji, S.; Song, W.; Myung, S.; Lee, S.S.; Lim, J.; An, K.-S. Fabrication of high-performance flexible photodetectors based on Zn-doped MoS_2 /graphene hybrid fibers. *J. Mater. Chem. C* **2017**, *5*, 12354–12359. [CrossRef]
96. Kim, H.; Han, G.H.; Yun, S.J.; Zhao, J.; Keum, D.H.; Jeong, H.Y.; Ly, T.H.; Jin, Y.; Park, J.-H.; Moon, B.H.; et al. Role of alkali metal promoter in enhancing lateral growth of monolayer transition metal dichalcogenides. *Nanotechnology* **2017**, *28*, 36LT01. [CrossRef] [PubMed]
97. Wang, Z.; Xie, Y.; Wang, H.; Wu, R.; Nan, T.; Zhan, Y.; Sun, J.; Jiang, T.; Zhao, Y.; Lei, Y.; et al. NaCl-assisted one-step growth of MoS_2 –WS2in-plane heterostructures. *Nanotechnology* **2017**, *28*, 325602. [CrossRef] [PubMed]
98. Kim, H.; Ovchinnikov, D.; Deiana, D.; Unuchek, D.; Kis, A. Suppressing Nucleation in Metal–Organic Chemical Vapor Deposition of MoS_2 Monolayers by Alkali Metal Halides. *Nano Lett.* **2017**, *17*, 5056–5063. [CrossRef]
99. Zhou, J.; Lin, J.; Huang, X.; Zhou, Y.; Chen, Y.; Xia, J.; Wang, H.; Xie, Y.; Yu, H.; Lei, J.; et al. A library of atomically thin metal chalcogenides. *Nature* **2018**, *556*, 355–359. [CrossRef] [PubMed]

100. Xie, C.; Yang, P.; Huan, Y.; Cui, F.; Zhang, Y. Roles of salts in the chemical vapor deposition synthesis of two-dimensional transition metal chalcogenides. *Dalton Trans.* **2020**, *49*, 10319–10327. [CrossRef] [PubMed]
101. Seravalli, L.; Bosi, M.; Beretta, S.; Rossi, F.; Bersani, D.; Musayeva, N.; Ferrari, C. Extra-long and taper-free germanium nanowires: Use of an alternative Ge precursor for longer nanostructures. *Nanotechnology* **2019**, *30*, 415603. [CrossRef] [PubMed]
102. Li, Y.; Cain, J.D.; Hanson, E.D.; Murthy, A.; Hao, S.; Shi, F.; Li, Q.; Wolverton, C.; Chen, X.; Dravid, V.P. Au@MoS_2 Core–Shell Heterostructures with Strong Light–Matter Interactions. *Nano Lett.* **2016**, *16*, 7696–7702. [CrossRef] [PubMed]
103. Irfan, I.; Golovynskyi, S.; Bosi, M.; Seravalli, L.; Yeshchenko, O.A.; Xue, B.; Dong, D.; Lin, Y.; Qiu, R.; Li, B.; et al. Enhancement of Raman Scattering and Exciton/Trion Photoluminescence of Monolayer and Few-Layer MoS_2 by Ag Nanoprisms and Nanoparticles: Shape and Size Effects. *J. Phys. Chem. C* **2021**, *125*, 4119–4132. [CrossRef]
104. Shi, Y.; Huang, J.-K.; Jin, L.; Hsu, Y.-T.; Yu, S.F.; Li, L.-J.; Yang, H.Y. Selective Decoration of Au Nanoparticles on Monolayer MoS_2 Single Crystals. *Sci. Rep.* **2013**, *3*, 1839. [CrossRef] [PubMed]
105. Wang, W.; Bando, Y.; Zhi, C.; Fu, W.; Wang, E.; Golberg, D. Aqueous Noncovalent Functionalization and Controlled Near-Surface Carbon Doping of Multiwalled Boron Nitride Nanotubes. *J. Am. Chem. Soc.* **2008**, *130*, 8144–8145. [CrossRef]
106. Ling, X.; Lin, Y.; Ma, Q.; Wang, Z.; Song, Y.; Yu, L.; Huang, S.; Fang, W.; Zhang, X.; Hsu, A.L.; et al. Parallel Stitching of 2D Materials. *Adv. Mater.* **2016**, *28*, 2322–2329. [CrossRef] [PubMed]
107. Kang, M.-A.; Kim, S.K.; Han, J.K.; Chang, S.-J.; Park, C.-Y.; Myung, S.; Song, W.; Lee, S.S.; Lim, J.; An, K.-S.; et al. Large scale growth of vertically standing MoS_2 flakes on 2D nanosheet using organic promoter. *2D Mater.* **2017**, *4*, 025042. [CrossRef]
108. Kim, S.J.; Kang, M.-A.; Kim, S.H.; Lee, Y.; Song, W.; Myung, S.; Lee, S.S.; Lim, J.; An, K.-S. Large-scale Growth and Simultaneous Doping of Molybdenum Disulfide Nanosheets. *Sci. Rep.* **2016**, *6*, 24054. [CrossRef] [PubMed]
109. Jeon, J.; Jang, S.K.; Jeon, S.M.; Yoo, G.; Jang, Y.H.; Park, J.-H.; Lee, S. Layer-controlled CVD growth of large-area two-dimensional MoS_2 films. *Nanoscale* **2015**, *7*, 1688–1695. [CrossRef]
110. Gnanasekar, P.; Periyanagounder, D.; Nallathambi, A.; Subramani, S.; Palanisamy, M.; Kulandaivel, J. Promoter-free synthesis of monolayer MoS_2 by chemical vapour deposition. *CrystEngComm* **2018**, *20*, 4249–4257. [CrossRef]
111. Shinde, S.M.; Das, T.; Hoang, A.T.; Sharma, B.K.; Chen, X.; Ahn, J.-H. Surface-Functionalization-Mediated Direct Transfer of Molybdenum Disulfide for Large-Area Flexible Devices. *Adv. Funct. Mater.* **2018**, *28*, 1706231. [CrossRef]
112. Zhou, H.; Yu, W.J.; Liu, L.; Cheng, R.; Chen, Y.; Huang, X.; Liu, Y.; Wang, Y.; Huang, Y.; Duan, X. Chemical vapour deposition growth of large single crystals of monolayer and bilayer graphene. *Nat. Commun.* **2013**, *4*, 2096. [CrossRef] [PubMed]
113. Liang, T.; Xie, S.; Huang, Z.; Fu, W.; Cai, Y.; Yang, X.; Chen, H.; Ma, X.; Iwai, H.; Fujita, D.; et al. Elucidation of Zero-Dimensional to Two-Dimensional Growth Transition in MoS_2 Chemical Vapor Deposition Synthesis. *Adv. Mater. Interfaces* **2017**, *4*, 1600687. [CrossRef]
114. Zou, X.; Ji, L.; Lu, X.; Zhou, Z. Facile electrosynthesis of silicon carbide nanowires from silica/carbon precursors in molten salt. *Sci. Rep.* **2017**, *7*, 1–9. [CrossRef]
115. Kong, D.; Wang, H.; Cha, J.J.; Pasta, M.; Koski, K.J.; Yao, J.; Cui, Y. Synthesis of MoS_2 and MoSe2 Films with Vertically Aligned Layers. *Nano Lett.* **2013**, *13*, 1341–1347. [CrossRef]
116. Zhang, Y.; Zhang, Y.; Ji, Q.; Ju, J.; Yuan, H.; Shi, J.; Gao, T.; Ma, D.; Liu, M.; Chen, Y.; et al. Controlled Growth of High-Quality Monolayer WS2 Layers on Sapphire and Imaging Its Grain Boundary. *ACS Nano* **2013**, *7*, 8963–8971. [CrossRef]
117. Bilgin, I.; Liu, F.; Vargas, A.; Winchester, A.; Man, M.K.L.; Upmanyu, M.; Dani, K.; Gupta, G.; Talapatra, S.; Mohite, A.; et al. Chemical Vapor Deposition Synthesized Atomically Thin Molybdenum Disulfide with Optoelectronic-Grade Crystalline Quality. *ACS Nano* **2015**, *9*, 8822–8832. [CrossRef] [PubMed]
118. Liu, H.F.; Wong, S.L.; Chi, D.Z. CVD Growth of MoS_2 -based Two-dimensional Materials. *Chem. Vap. Depos.* **2015**, *21*, 241–259. [CrossRef]
119. Cain, J.D.; Shi, F.; Wu, J.; Dravid, V.P. Growth Mechanism of Transition Metal Dichalcogenide Monolayers: The Role of Self-Seeding Fullerene Nuclei. *ACS Nano* **2016**, *10*, 5440–5445. [CrossRef] [PubMed]
120. Yang, S.Y.; Shim, G.W.; Seo, S.-B.; Choi, S.-Y. Effective shape-controlled growth of monolayer MoS_2 flakes by powder-based chemical vapor deposition. *Nano Res.* **2017**, *10*, 255–262. [CrossRef]
121. Huang, Y.L.; Chen, Y.; Zhang, W.; Quek, S.Y.; Chen, C.-H.; Li, L.-J.; Hsu, W.-T.; Chang, W.-H.; Zheng, Y.; Chen, W.; et al. Bandgap tunability at single-layer molybdenum disulphide grain boundaries. *Nat. Commun.* **2015**, *6*, 6298. [CrossRef] [PubMed]
122. Sangwan, V.K.; Jariwala, D.; Kim, I.S.; Chen, K.-S.; Marks, T.J.; Lauhon, L.; Hersam, M.C. Gate-tunable memristive phenomena mediated by grain boundaries in single-layer MoS_2. *Nat. Nanotechnol.* **2015**, *10*, 403–406. [CrossRef] [PubMed]
123. Radisavljevic, B.; Kis, A. Mobility engineering and a metal–insulator transition in monolayer MoS_2. *Nat. Mater.* **2013**, *12*, 815–820. [CrossRef] [PubMed]
124. Zhou, W.; Zou, X.; Najmaei, S.; Liu, Z.; Shi, Y.; Kong, J.; Lou, J.; Ajayan, P.M.; Yakobson, B.I.; Idrobo, J.C. Intrinsic Structural Defects in Monolayer Molybdenum Disulfide. *Nano Lett.* **2013**, *13*, 2615–2622. [CrossRef] [PubMed]
125. Hong, J.; Hu, Z.; Probert, M.; Li, K.; Lv, D.; Yang, X.; Gu, L.; Mao, N.; Feng, Q.; Xie, L.; et al. Exploring atomic defects in molybdenum disulphide monolayers. *Nat. Commun.* **2015**, *6*, 6293. [CrossRef] [PubMed]
126. Durairaj, S.; Ponnusamy, K.; Shinde, N.B.; Eswaran, S.K.; Asokan, V.; Park, J.B.; Chandramohan, S. Oxygen-Driven Growth Regulation and Defect Passivation in Chemical Vapor Deposited MoS_2 Monolayers. *Cryst. Growth Des.* **2021**, *21*, 6793–6801. [CrossRef]

127. Feldman, Y.; Wasserman, E.; Srolovitz, D.J.; Tenne, R. High-Rate, Gas-Phase Growth of MoS_2 Nested Inorganic Fullerenes and Nanotubes. *Science* **1995**, *267*, 222–225. [CrossRef]
128. Zhang, K.; Bersch, B.M.; Zhang, F.; Briggs, N.C.; Subramanian, S.; Xu, K.; Chubarov, M.; Wang, K.; Lerach, J.O.; Redwing, J.M.; et al. Considerations for Utilizing Sodium Chloride in Epitaxial Molybdenum Disulfide. *ACS Appl. Mater. Interfaces* **2018**, *10*, 40831–40837. [CrossRef]
129. Han, S.W.; Yun, W.S.; Woo, W.J.; Kim, H.; Park, J.; Hwang, Y.H.; Nguyen, T.K.; Le, C.T.; Kim, Y.S.; Kang, M.; et al. Interface Defect Engineering of a Large-Scale CVD-Grown MoS_2 Monolayer via Residual Sodium at the SiO 2 /Si Substrate. *Adv. Mater. Interfaces* **2021**, *8*, 2100428. [CrossRef]
130. Kang, S.K.; Lee, H.S. Study on Growth Parameters for Monolayer MoS_2 Synthesized by CVD Using Solution-based Metal Precursors. *Appl. Sci. Converg. Technol.* **2019**, *28*, 159–163. [CrossRef]

materials

MDPI

Review

Recent Advances in High-Throughput Nanomaterial Manufacturing for Hybrid Flexible Bioelectronics

Nathan Zavanelli [1], Jihoon Kim [1] and Woon-Hong Yeo [1,2,3,*]

[1] George W. Woodruff School of Mechanical Engineering, Center for Human-Centric Interfaces and Engineering at the Institute for Electronics and Nanotechnology, Georgia Institute of Technology, Atlanta, GA 30332, USA; nzavanelli@gatech.edu (N.Z.); louisjihoonkim@gatech.edu (J.K.)

[2] Wallace H. Coulter Department of Biomedical Engineering, Georgia Institute of Technology, Atlanta, GA 30332, USA

[3] Parker H. Petit Institute for Bioengineering and Biosciences, Neural Engineering Center, Institute for Materials, Institute for Robotics and Intelligent Machines, Georgia Institute of Technology, Atlanta, GA 30332, USA

* Correspondence: whyeo@gatech.edu; Tel.: +1-404-385-5710

Abstract: Hybrid flexible bioelectronic systems refer to integrated soft biosensing platforms with tremendous clinical impact. In this new paradigm, electrical systems can stretch and deform with the skin while previously hidden physiological signals can be continuously recorded. However, hybrid flexible bioelectronics will not receive wide clinical adoption until these systems can be manufactured at industrial scales cost-effectively. Therefore, new manufacturing approaches must be discovered and studied under the same innovative spirit that led to the adoption of novel materials and soft structures. Recent works have taken mature manufacturing approaches from the graphics industry, such as gravure, flexography, screen, and inkjet printing, and applied them to fully printed bioelectronics. These applications require the cohesive study of many disparate parts. For instance, nanomaterials with optimal properties for each specific application must be dispersed in printable inks with rheology suited to each printing method. This review summarizes recent advances in printing technologies, key nanomaterials, and applications of the manufactured hybrid bioelectronics. We also discuss the existing challenges of the available nanomanufacturing methods and the areas that need immediate technological improvements.

Keywords: nanomanufacturing; high-throughput method; material printing; flexible bioelectronics

check for updates

Citation: Zavanelli, N.; Kim, J.; Yeo, W.-H. Recent Advances in High-Throughput Nanomaterial Manufacturing for Hybrid Flexible Bioelectronics. *Materials* **2021**, *14*, 2973. https://doi.org/10.3390/ma14112973

Academic Editor: Fabrizio Roccaforte

Received: 29 April 2021
Accepted: 27 May 2021
Published: 31 May 2021

1. Introduction

There is a fundamental mismatch between biological systems, which are soft and deformable, and traditional electronics, which are rigid and impermeable to sweat and liquids [1–5]. This incongruity places a significant constraint on the development of bioelectronics systems [4,5]. Traditional systems cannot conform well to the human body, making most wearable devices susceptible to large noise during motion, uncomfortable and obtrusive to wear, and limited to very specific regions on the body, such as the wrist, chest, and finger, that allow for easy attachment of rigid systems [1,3]. Furthermore, rigid implantable electronics and surgical instruments cannot easily integrate with the soft systems for which they are targeted [2]. As a result, current applications of wearable electronics, biosensors, and implantable healthcare are highly limited, leaving millions of people with serious undiagnosed diseases and allowing the steady progression towards heart attack and stroke to remain undetected [2,5]. In contrast, recent advances in hybrid electronics have yielded new classes of electronic devices and sensors that integrate well with the human body [6–9]. These systems can achieve high stretchability or flexibility through two paradigms: first, metal depositions on the order of 10 μm or less can easily bend in accordance with Euler–Bernoulli theory because of their minimal height, and fractal geometric patterns can be introduced to allow for stretching with minimal local strain as the patterns unfold [10]. Second, conductive nanomaterial–polymer matrices can be made intrinsically stretchable by maintaining conductive pathways as the polymer undergoes strain [11]. Both novel systems can stretch and deform when needed, allowing for seamless integration with the skin [8]. As a result, previously inaccessible physiological signals and implantable healthcare targets can be realized, promising a transformation in modern medicine [7]. Likewise, traditional biosensing methods, such as florescent microarrays, lateral flow immunoassays, DNA microarrays, enzyme-linked immunosorbent assays, and polymerase chain reaction-based methods require expensive reagents and laboratory equipment, and are limited by slow signal processing methods; however, fully printed sensors allow for real time, continuous biomarker quantification in a simple, affordable, and mass producible package [12,13]. Printed biomolecule sensors promise a transformative way to continuously assess crucial biomarkers, making their development critical in the future of healthcare development [12,14–16]. These systems were initially fabricated with traditional micro- and nano-electromechanical systems (MEMS/NEMS) approaches, but such techniques are poorly suited for the industrial scales necessary to commercialize hybrid electronics. For hybrid bioelectronics to achieve their potential, the same innovative spirit that led to the adoption of new materials must be applied to the study of new manufacturing approaches.

Fully printed electronics methods have gained significant interest in recent years because they are cheaper, more efficient, and more scalable than traditional MEMS/NEMS processes and capable of direct printing on many flexible and stretchable substrates, such as polyethylene terephthalate (PET), polyimide (PI), polydimethylsiloxane (PDMS), thermoplastic polyurethane (TPU) and paper [17–19]. However, these approaches are also limited by product throughputs and yields, making it very difficult to manufacture hybrid devices in a manner that would allow for their mass adoption [17]. For instance, aerosol jet printing has demonstrated very high print resolutions and control over material heights and microarchitectures, but the additive deposition of numerous nano thickness layers makes it by necessity a low-speed manufacturing option [20]. Inkjet printing is another attractive printing method, but it is limited by the requirement that particles are small and well dispersed enough not to clog the inkjet nozzle, and traditional inkjet printing is also too low throughput for industrial scales [17,21]. However, roll-to-roll inkjet printing has been demonstrated, making it a potential target for high-speed device manufacturing [22]. Electrohydrodynamic printing is an exciting new method that can overcome several of the key challenges in inkjet printing by pulling ionized inks directly to the substrate, but it is likewise limited by constrictive stand-off height requirements and low manufacturing yields [21]. In contrast, contact printing methods, such as screen, gravure, flexographic, slot die, and doctor blade printing are easily integrated with high throughput roll to roll systems, making them an excellent option for industrial scale hybrid bioelectronics manufacturing. A summary of each of the high throughput nanomaterial fabrication methods discussed in this review is provided in Table 1 [17,23–32].

There are four crucial design challenges that must be met to make high throughput manufacturing of hybrid bioelectronics a reality, as depicted in Figure 1 [16,33–39], and all four are highly interdependent on each other. First, conductive nanomaterials must be identified and produced based on the final application requirements. Second, these nanomaterials must be dispersed in a fully printable ink with rheology and viscosity well suited to the specific printing method. Third, innovative manufacturing techniques must be explored that can achieve scales suited for mass production. Finally, the printed inks must be sintered and cured after printing in a way that does not limit manufacturing speed.

In this review, we will summarize recent attempts to develop high-throughput manufacturing of nanomaterials for hybrid bioelectronics, with a specific focus on new printing methods, nanomaterial selection, synthesis, dispersion and printing, post-print processing and demonstrated applications. We will begin with a discussion on several of the key manufacturing methods under investigation, emphasizing how deposition physics leads to key constraints on ink design, print resolutions, deposition heights, device yield, and print speed and novel approaches to push the field beyond its traditional limitations. Next, we will summarize the key nanomaterials that are used in printed bioelectronics. For each material, we will summarize the key mechanical, chemical, and electrical properties that determine the material's functionality, recent advances, and challenges in ink formulations, demonstrated bioelectronics devices fabricated with high throughput methods, and novel high throughput sintering methods. Finally, we will comment on the state-of-the-art in the field, assess the key limitations yet to be solved, and look forward to the future development of high-throughput nanomaterial manufacturing for soft bioelectronics.

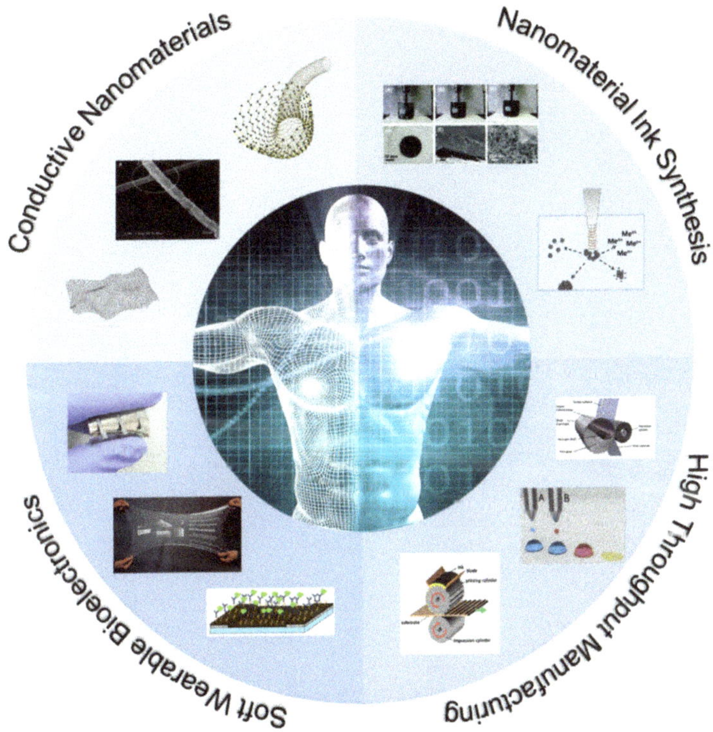

Figure 1. Overview of high-throughput nanomaterial fabrication methods and applications in soft bioelectronics. (Figures are adapted or reprinted, clockwise from the top: (1) *RSC Adv.* (2013), 3, 24812, Copyright 2013, RSC, (2) *J. Nanoparticle Res.* (2016), 18, 285, Copyright 2106, Springer, (3,7) *Ind. Eng. Chem. Res.* (2019), 58, 43, 19909–19916, Copyright 2019, ACS, (4) *Adv. Mater. Interfaces* (2018), 5, 1701561. Copyright 2018, Wiley, (5) *Adv. Mater.* (2019), 31, 1806702. Copyright 2020, Wiley, (6) Creative Commons License CC BY from *J. Electrochem. Soc.* (2018), 165 B3084, (8) *ACS Appl. Mater. Interfaces* (2020), 12, 4146797–46803. Copyright 2020, ACS, (10) Creative Commons license CC BY from *Sci. Rep.* (2017), 7(1)).

Table 1. Summary of nanomaterial fabrication methods with key advantages and limitations.

Printing Method	Typical Min Resolution [μm]	Printing Speed [m/min]	Ink Viscosities [mPa.s]	Ink Surface Tension [mN/m]	Pros	Cons
Gravure	30	1–20	100–500	20–40	High speed [40] Good reliability [27] Long production runs [27]	High startup costs [27] Expensive prototyping [27]
Flexography	30	1–20	50–500	10–30	High Speed [41] Easier to prototype than gravure [29]	High startup costs [29] Lower durability than gravure [41]
Screen	50	0.1–15	500–10,000	35–50	Inexpensive to prototype [42] Balance between speed, reliability, and cost [35] Simple process optimization [43]	Limited resolution [24] Strict ink rheology requirements [44]
Inkjet	30	0.01–15	1–20	10–30	No additional cost to prototype [45] Excellent resolution and pattern control [46]	Complicated to integrate with roll–to–roll systems Nozzle clogging [47] Coffee ring effect [48]
Slot die	40	1–50	2–500	–	Efficient and precise coating of homogeneous films [30]	Not suited for complex patterning [49]

2. Printing Fundamentals

2.1. Gravure Printing

Gravure printing is a mature manufacturing method that has been employed for high throughput image printing since the 19th century [27]. Gravure printing is achieved in four phases, as shown in Figure 2a,b [27,34]. First, ink is poured on a rotating gravure roll and fills the recessed cells in the roll [40]. Second, a doctor blade removes the excess ink from the roll, leaving only an ink thickness corresponding to the depth of the cell [40]. Third, the ink is brought in contact with a substrate, which is itself being rolled at the same speed as the gravure roll, and the ink is pulled from the roll onto the substrate as a result of adhesive forces between the ink and the substrate and the ink's surface tension [27,40]. Finally, the print will stabilize on the substrate and spread based on the theoretical contact angle that the liquid–gas interface makes to the substrate, which is determined by Young's equation. $cos\theta_c = \frac{\gamma_{sg}-\gamma_{sl}}{\gamma_{lg}}$, where γ_{sg}, γ_{sl}, γ_{lg} are the surface energies for solid–liquid, liquid-gas, and solid-gas, respectively [27,50,51]. Although this is true for all direct printing methods, it is particularly important for gravure printing because gravure patterns consist of individual cells which must spread into each other to form a cohesive print [52]. In addition, print resolution, quality, and speed are primarily limited by the complex fluid dynamics occurring when excess ink is removed by the doctor blade [51]. Printing faults during this phase are broadly characterized by two processes: lubrication residue and ink drag out [27,53]. First, the doctor blade will always leave a small residual ink layer on the roll, and this layer's thickness must be substantially reduced to prevent electrical shorts and erroneous material depositions [27]. The doctor blade's efficiency depends heavily on the relative magnitudes of viscous forces and surface tension. This relationship is captured in the capillary number $C_a = \frac{viscous\ forces}{surface\ tension} = \frac{\mu U}{\sigma}$, where μ is the ink viscosity, U is the print speed, and σ is the ink surface tension [51].

At high capillary numbers, the residual thickness is often unacceptable [27]. Therefore, reducing print speed and ink viscosity is essential in limiting lubrication residue. Second, the doctor blade may pull ink out of the cells as it passes and deposits the ink on the roll in a process termed drag out [53]. This process has been analytically and empirically shown to depend heavily on capillary flow, which is limited at high capillary numbers [51,52]. In this condition, the print velocity is too high compared to the capillary flow characteristic velocity for drag out to occur [27]. Therefore, high capillary numbers prevent drag out and low capillary numbers limit lubrication residue. In practice, achieving a capillary number of Ca $\approx$ 1 is necessary for high-quality gravure printing, although the ideal capillary number also depends on pattern geometry, orientation, substrate wetting, and print thickness [54]. Because these interactions are often complicated to determine analytically a priori, especially with viscoelastic inks, many researchers optimize their process with statistical design of experiments techniques, such as analysis of variance and Taguchi methods [51]. Figure 2c depicts results from one such experiment to determine the optimal ink viscosity and cell spacing for a process with a fixed speed (3 m/min) [53]. Three graphene inks were formulated with various viscosities (i), and single dots were printed (ii–iv), with the low viscosity ink (iv) producing an unlevel print with an extended residue tail [53].

In addition to optimizing the ink rheology and substrate wetting, the cell pattern is crucial in achieving high-resolution prints [55]. Printing continuous lines, which is referred to as Intaglio printing, is avoided because the drag-out effect is amplified with long prints oriented in the printing direction [56]. In order to produce high resolution, level prints, minimizing cell dimensions is critical [27,52,53]. As shown in Figure 2c, previously discussed 2.5 Pa·S ink was also printed with cell spacings of 50 μm (v), 25 μm (vi) and 5 μm (vii), and the line uniformity increased significantly with a decrease in cell spacing [53]. Further increasing resolution and quality in gravure printing is complicated, however, because traditional print head fabrication methods, such as electromechanical and laser engraving, are unable to produce cells with dimensions < 10 μm, and they are also likely to produce additional roughness on the gravure roll near the cell as a result of the engraving process [52,53]. Therefore, recent works have employed silicon microfabrication techniques to design very high-resolution gravure rolls [53]. For instance, Secor et al. used photolithography to design a silicon-based gravure roll capable of producing high resolution trace <30 μm with conductivities >10,000 S/m [53]. To further reduce trace widths, Lee et al. experimented with various cell depth profiles under the hypothesis that curved cell walls would reduce drag out [52]. As shown in Figure 2d, a curved gradient pattern was able to reduce print width by 65%, yielding a final pattern of <10 μm [52]. These recent developments in high resolution and high throughput gravure manufacturing, combined with the novel advances in printable, conductive nanomaterial inks discussed in Section 3, make them well suited to numerous bioelectronics applications, including multilayer circuit fabrication and sensor manufacturing, such as the sweat sensor demonstrated in Figure 2e [40]. However, gravure printing also presents very high startup costs, incurs high costs to prototype, places rigid requirements on ink rheology, and often requires substrate surface modifications in order to achieve optimal printing [17,27,57].

2.2. Flexographic Printing

Flexography is another high throughput, a roll-to-roll fabrication method for printed electronics the origins of which can be traced to late 19th-century image printing [29]. Flexographic printing consists of five subprocesses, as depicted in Figure 3a. First, ink is pulled from a reservoir by the fountain roller. Second, the ink is transferred to an intermediate anilox roller containing millions of miniature engraved cells. Third, a blade removes excess ink from the anilox roller. Fourth, ink is transferred from the anilox to a flexible photopolymer plate containing a mirror engraved pattern. Finally, the substrate is rolled between the flexographic plate and an impression cylinder, yielding an ink deposition on the substrate. Because the printing roller is made of a flexible polymer wrapped around a metal cylinder, the prototyping and startup costs are significantly lower in flexography than gravure printing. However, plate deformation is a significant limitation to be overcome in high-resolution flexographic printing [41]. Another key difference between gravure printing and flexography is the presence of an anilox roll, which allows for a wider range of ink rheology to be printed, but whose geometry, pressure, and speed must be carefully optimized [41]. These challenges exist in addition to those faced by gravure printing, which is one explanation for the greater adoption of gravure for printed electronics. However, recent works have significantly improved flexographic printing capabilities, making flexography an exciting and fast-developing approach with a significantly lower barrier to entry than gravure printing. For instance, the surface energy of the flexography roll relative to the anilox and substrate can be modified to improve ink transfer in each phase, the print speed and pressure can be optimized for the specific transfer chemistry and pattern geometries, and the geometries themselves can be improved [41,58,59]. However, these innovations still result in resolutions >50 μm because of fundamental material limitations in the photopolymer flexographic roll. As a result, Kim et al. developed a microstructured, nanoporous carbon nanotube (CNT) stamp to replace the traditional roll with carefully controlled porosity, mechanics, and surface chemistry [60]. As shown in Figure 3b, the CNT nanopillars leave a precise open area in which the ink can reside. During printing, the stamp is brought into conformal contact with the substrate due to the mechanical flexibility of the CNTs, and a highly controlled deposition is produced as the stamp is removed, as shown in Figure 3c. This mechanism overcomes many of the key challenges in flexographic transfer by storing the ink in the stamp pores, then transferring directly to the substrate, and yield high-quality prints with a variety of nanomaterial inks of <20 μm were demonstrated [60,61]. In light of these innovations, flexography is now considered an exciting new field in printed electronics with great opportunities for further improvement.

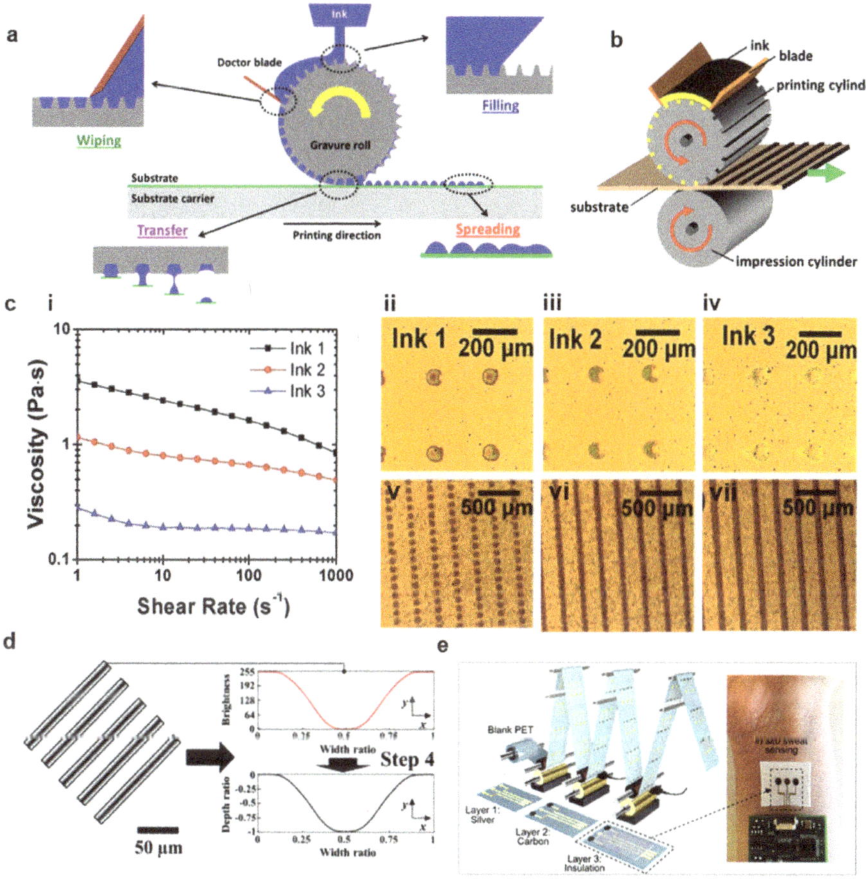

Figure 2. Gravure printing of hybrid bioelectronics. (**a**) Overview of the gravure printing process (reprinted with permission from *Flex. Print. Electron* (2016), 1 023003. Copyright 2016, IOP). (**b**) Illustration of gravure printing linear traces against an impression roll. (reprinted with permission from *Adv. Mater.* (2019), 31, 1806702. Copyright 2020, Wiley). (**c**) Optimizing graphene inks for gravure printing. (**i**) Characterization of viscosity for the three different ink formulations. (**ii–iv**) Images of printed dots for each ink using a gravure cell of 50 µm. (**v–vii**) Images showing line formation as the cell spacing is reduced, corresponding to 50, 25, and 5 µm spacing for a cell size of 50 µm. (reproduced with permission from *Adv. Mater.* (2014), 26: 4533–4538. Copyright 2014, Wiley). (**d**) Optimized cell patterns achieved with gradation engraving to achieve high-resolution gravure printing. (reprinted with permission from *Precis. Eng.* (2021), 69: 1–7. Copyright 2021, Elsevier). (**e**) Illustration of roll-to-roll printed sweat sensors. (reprinted with permission from *ACS Nano* (2018), 12(7): 6978–6987, Copyright 2018, ACS).

2.3. Screen Printing

Screen printing is an ancient printing method that has been employed in garment processing for centuries [19]. Today, it is a mature industrial process used in textiles, graphics, printed circuit silkscreens, in-mold electronics, capacitive touch sensors, printed heaters, and chemical sensors [47]. Significantly, screen printing is highly suitable for roll-to-roll manufacturing and high throughput processing. Unlike gravure printing and flexography, Screen printing involves the active transfer of ink from a mesh to a target substrate mediated by pressure and shear applied by a blade termed the squeegee [62,63]. Printing occurs in six distinct phases, as shown in Figure 4a [43,62]. First (I), ink enters the mesh after the application of gentle pressure such that it occupies the enter open mesh area, but does not run out from the bottom of the mesh [63]. Second (II), the mesh is brought into contact with the substrate as a result of applied pressure and the highly pseudoplastic ink becomes highly thin with applied pressure [64]. Third (III), the ink adheres to both the mesh and the substrate [43]. Fourth (IV), the mesh is pulled upwards as the squeegee progresses down the print, causing the ink to rise [43]. Fifth (V), the ink begins to form filaments underneath the mesh wires as the mesh is continually raised [43]. Finally (VI), the filaments break and the print levels, resulting in a deposition thickness that depends on the mesh open area and the ink adhesion to both the mesh and the substrate [43]. In traditional screen-printing applications, the substrate is placed on a flat plate below the mesh, as shown in Figure 4b [34]. In roll-to-roll screen-printing, the mesh is folded into a cylinder with the squeegee blade inside the cylinder [35]. The substrate is then rolled against the mesh and the impression cylinder, which causes an applied pressure against the squeegee and shear proportional to the print velocity. This process is depicted in Figure 4c, and an example roll-to-roll machine is shown in Figure 4d [35]. The same six steps described previously also apply to roll-to-roll screen printing, but some non-idealities in the mesh liftoff are caused by the curved substrate, especially when the radius of curvature is small [35,65].

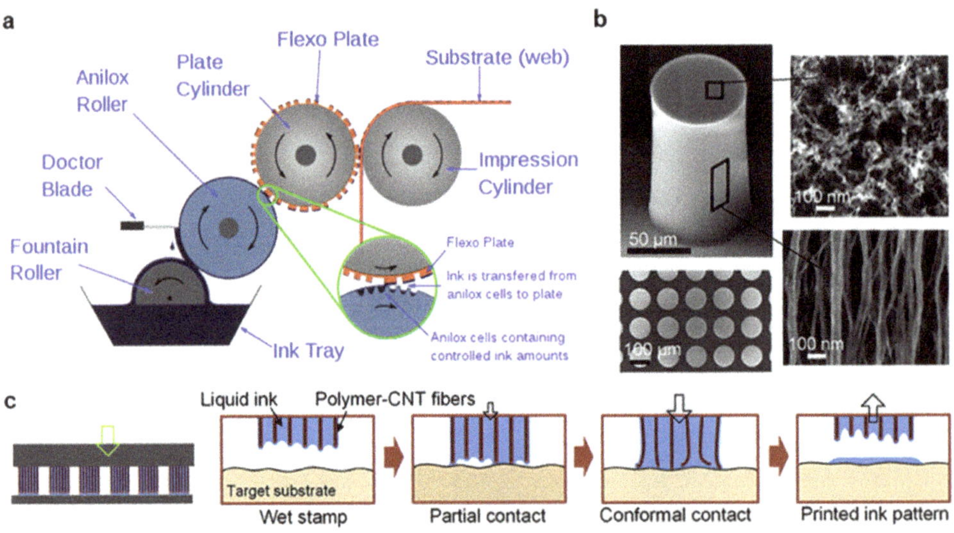

Figure 3. Advanced flexographic printing techniques. (**a**) Diagram of a standard flexographic printer to illustrate the key operating principles. (reproduced under creative commons license CC BY-SA 4.0). (**b**) Scanning electron microscope images of a CNT array (100 µm pillar diameter, 150 µm height) used for high-resolution flexography, and close-up top and side surfaces of a micropillar (reproduced with permission from *Langmuir* (2019), 35, 24, 7659–7671. Copyright 2019, ACS). (**c**) Simplified schematic of ink transfer from carbon nanotubes (CNTs) micropillar stamp loaded with ink. (reproduced with permission from *Langmuir* (2019), 35, 24, 7659–7671. Copyright 2019, ACS).

In screen printing, the mesh height from the substrate and mesh geometry are crucial parameters, but the squeegee speed and pressure are not highly correlated with print quality [66]. This is because the sheer and compressive forces applied are typically large enough to elicit a strong sheer thinking response in ink and prevent ink hydroplaning before the squeegee [24]. Instead, optimizing ink rheology for this complicated fluid dynamics is crucial in screen printing [64]. Ink viscosities are typically high (10–30 PaS) and highly pseudoplastic so that they can avoid running through the mesh preprinting, flow easily during applied shear, and rapidly coalesce post print into a steep deposition without slumping on the substrate [11,67]. Furthermore, the mesh area that is not to be printed is blocked by an ultraviolet (UV) cured emulsion mask, and limiting this emulsion's roughness is important in creating a high resolution and even print [66]. Like gravure and flexographic printing, the minimal resolution achievable in screen printing is limited fundamentally by the mesh quality, even if many inks with suboptimal rheology cannot approach this limit [63]. Specifically, screen printing meshes are limited by lithography resolution in emulsion etching, emulsion smoothness, and mesh geometries [11]. Creating a finer mesh with more weaves per unit area improves print resolution, but the reduction in mesh open area leads to a thinner print deposition [11]. Mesh counts generally reach their minimization limits beyond 140 threads per centimeter, and screens with around these mesh counts and optimized emulsions are capable of printing resolutions of around 70 μm [62]. However, Hyun et al. recently demonstrated a screen-printing stencil derived from a thin silicon wafer (90 μm thickness) with photolithographically defined openings to produce high quality depositions of graphene and AgNP inks with widths of 40 μm, and the silicon stencil fabrication and graphene printing is illustrated in Figure 4e [11]. In summary, screen printing is attractive for high throughput printed bioelectronics because it is a mature industrial process with significantly lower startup and prototyping costs than gravure and flexographic printing, and new innovations in mesh or stencil design open new opportunities for increased print resolutions.

2.4. Roll to Roll Inkjet Printing

Inkjet printing is an extensively developed technology that is widely employed in conventional printing applications, and it is exceptionally well suited to rapid, low-cost prototyping [25,31,68]. In inkjet printing, pressurized ink is forced through a nozzle, forming droplets that fall onto the substrate and collapse due to their momentum and substrate wettability [46]. Inkjet printing is achieved through two approaches, although drop on demand (DOD) printing is greatly preferred over continuous inkjet printing (CIJ) for bioelectronics because it allows for higher placement accuracy and higher resolutions [17]. In DOD printing, the ink is forced through the nozzle through either a contractile force applied from a piezoelectric actuator or a thermal disturbance that produces a shockwave capable of ejecting the ink [25]. In contrast, a CIJ printer charges ink droplets and continually passes them through an electric field formed between two deflection plates, allowing one to control the ink depositions [25]. Both inkjet printing processes are illustrated in Figure 5a. DOD printing is highly attractive because it allows for excellent control over deposition thickness, high resolution down to 40 μm, very inexpensive prototyping, and minimal startup costs, but it is also limited by clogging in the minuscule nozzle head, the uneven flow of material to the edge of the print in a process termed the coffee ring effect, and lower throughputs than the previously described methods [11,22,68].

Each of these limitations has been thoroughly studied using traditional graphics inks, and the challenge in bioelectronics fabrication is to apply these lessons to nanomaterial-based inks [25,69,70]. The fluid mechanics during printing is characterized primarily by three dimensionless quantities, the Weber number (We), Reynolds number (Re), and Ohnesorge number (Oh):

$$\mathrm{We} = \frac{\zeta\rho v^2}{\gamma}$$
$$\mathrm{Re} = \frac{\zeta\rho v}{\eta}$$
$$\mathrm{Oh} = \frac{\sqrt{\mathrm{We}}}{\mathrm{Re}} = \frac{\eta}{\sqrt{\zeta\rho\gamma}}$$

where η, ρ, and γ are the ink viscosity, density, and surface tension, respectively, v is the print velocity, and ζ is characteristic printing length, which is in most cases simply the diameter of the print head nozzle [25,26,70]. In almost all inkjet applications, Oh must be between 1 and 1/10 to achieve a quality print, as illustrated in Figure 5b [28]. At high Oh values, the ink viscosity will prevent stable drop formation [28]. When Oh is too low, the ink forms many uncontrolled drops instead of a single, well-defined drop, which results in an unusable print [28,69]. In addition, the particle size cannot be $>\zeta/50$ in order to avoid immediate nozzle clogging [71]. As we will discuss in Section 3, these requirements greatly complicate the printing of Ag nanowires (AgNWs) and CNTs, which are usually much longer than $\zeta/50$, and carbon-based nanomaterials, which are difficult to disperse with both low viscosity and high material loadings [47,72,73]. Another crucial challenge in inkjet printing is the accumulation of the deposited material along the edge of the print, commonly termed the coffee ring effect [48]. This occurs when the edge of a droplet on a substrate is fixed in place and capillary flow induced by evaporation of the drop causes material to flow from the interior towards this fixed edge [48]. This process is combatted by Marangoni flow within the drop, but many surfactants and even added water tend to have very weak Marangoni flows [74]. There are numerous methods employed to combat coffee ring formation, including careful control of the surfactant mediated interactions between particles and the liquid–gas interface [72,74], mixing high and low boiling point solvents [75], heating the substrate [76], depinning the contact line (which reduces print definition) [77], alternating voltage electrowetting [78], and dual drop inkjet printing [33]. In an example of the first method, Anyfantakis et al. mixed surfactants and colloids with opposite charges and observed that particles that absorbed the surfactants become hydrophobic, giving them a greater affinity to the liquid–gas interface [72]. These particles on the drop surface prevented capillary flow from collapsing the structure, leading to a uniform deposition, as shown in Figure 5c [72]. In the later method, two main approaches are employed. First, the Langmuir–Blodgett concept is applied to the picolitre depositions by first depositing a supporting layer, then adding a functional ink on top containing colloidal nanoparticles that assemble as the solvent dissolves to produce a highly uniform layer, as illustrated in Figure 5d,e [33]. Second, antisolvent crystallization can be used to form highly uniform semiconducting films at the liquid–air interface in a mixed droplet [79]. This occurs after printing an antisolvent layer, then a semiconductor solution. The undissolved nuclei form a cohesive film on the drop surface, preventing the drop from collapsing as the solvent evaporates [79].

Finally, significant commercial interest in inkjet printing has led to many efforts to improve manufacturing throughput, and numerous inkjet printers can achieve speeds far beyond those achieved in home-use graphics printers [73]. However, there are still key tradeoffs between print resolution, deposition uniformity, and throughput [26,73]. The greatest improvements in throughput generally come through roll-to-roll processing, stringent quality control on component manufacturing, and precise temperature control, all of which have been thoroughly investigated by private companies [73]. Even in the most advanced systems, nozzle clogging is still a crucial issue with nanomaterial inks that limits manufacturing throughput, and continuous cleaning of the systems is therefore necessary [73]. Inkjet printing is highly attractive for printing bioelectronics because complex systems can be very rapidly prototyped during development, then easily scaled to mass production, but there are also very strict requirements on nanomaterial ink properties, lower demonstrated throughputs than alternative methods, and key challenges relating to nozzle clogging that complicate high throughput fabrication.

2.5. Slot Die and Blade Coating

Slot die and blade coating, which are sometimes referred to as bar coating or knife coating, are high throughput methods to deposit homogenous films for applications that do not require complex patterns to be formed [34,49]. In blade coating, ink is placed before the blade, and deposition is left as the blade swipes across the substrate [34]. The thickness of the resultant deposition depends largely on the blade height relative to the substrate, the print velocity, ink viscosity, and ink-substrate wetting contact angle [34]. In slot coating, ink is continually pumped from a slot inside a print head, which can be masked to print unidirectional lines [49]. In addition, the print head can be displaced perpendicular to the print direction to yield curved lines [80]. The print quality and film thickness in slot die coating is determined by the meniscus forming between the print head and the substrate, and this meniscus can be controlled by the same parameters mentioned for the doctor blade in addition to the pumping rate and temperature control of the ink [30]. Slot die and blade coating are mature manufacturing processes for depositing homogeneous films, which are desired in pressure, chemical, and electrophysiological sensors for soft bioelectronics; however, these methods are not well suited to more complicated printing applications that require sophisticated patterning.

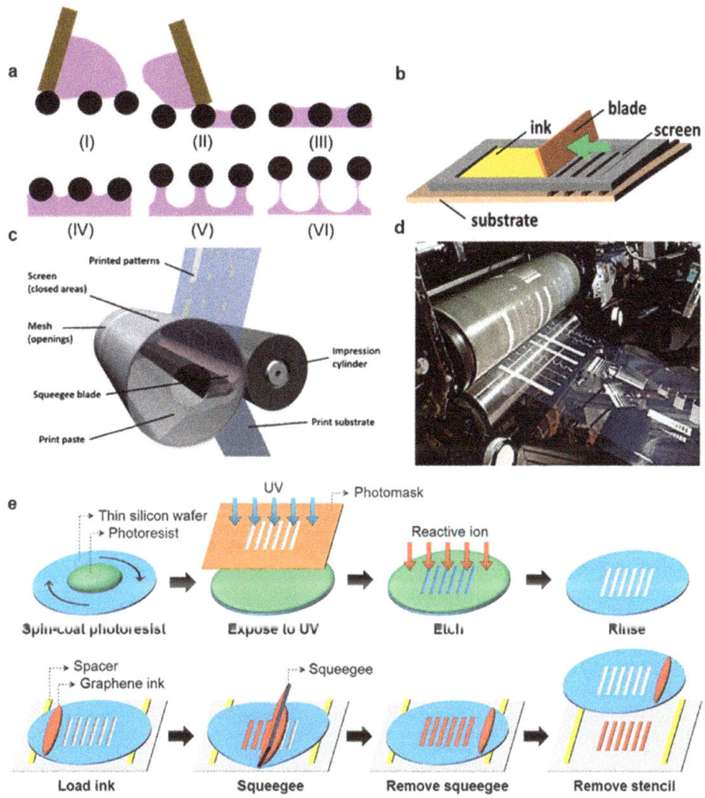

Figure 4. High-throughput screen-printing approaches. (**a**) Illustration of the six stages of screen printing, as proposed by Messerschmitt et al. and investigated Abbott et al. (**i**) Ink flooded into the mesh. (**ii**) Squeegee pressure brings the mesh in contact with the substrate. (**iii**) Ink adheres to both the substrate and mesh. (**iv–vi**) As the mesh is raised off the substrate, the ink first (**iv**) forms a continuous structure, then (**v**) forms filaments, which then (**vi**) collapse and level to form a deposition. (reprinted with permission from *ACS Omega* (2021), 6, 14, 9344–9351. Copyright 2021, ACS). (**b**) Illustration of a sheet-to-sheet screen-printer. (reprinted with permission from *Adv. Mater.* (2019), 31, 1806702. Copyright 2020, Wiley). (**c**) Illustration of a roll-to-roll screen printer, demonstrating the key operating principles. (reprinted with permission from Ind. *Eng. Chem. Res.* (2019), 58, 43, 19909–19916, Copyright 2020, ACS). (**d**) Image of a roll-to-roll screen-printer used in nanomaterial printing. (reprinted with permission from *Ind. Eng. Chem. Res.* (2019), 58, 43, 19909–19916, Copyright 2020, ACS). (**e**) Fabrication of a thin silicon screen printing stencil for high-resolution printing and printing process implanting this stencil. (reprinted with permission from *Adv. Mater.* (2014), 27: 109–115. Copyright 2014, Wiley).

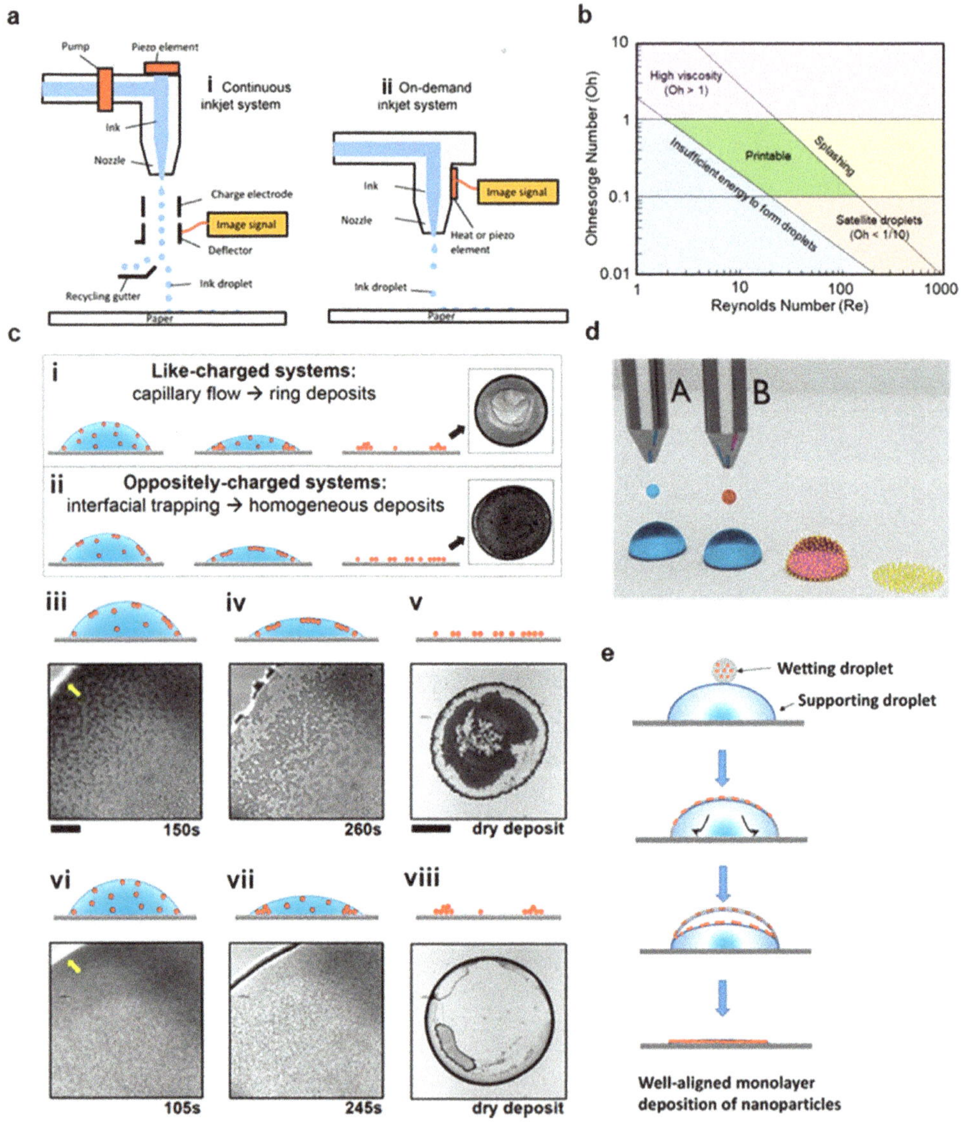

Figure 5. Roll-to-roll inkjet printing for hybrid bioelectronics. (**a**) Illustration of CIJ (**left**) and DOD (**right**) inkjet printing techniques. (reprinted with permission from *Micromachines* (2017), 8(6), 194. Copyright 2017, MDPI). (**b**) Reynolds number and Ohnesorge numbers that yield a high-quality inkjet deposition. Weber numbers can be calculated based on the ratio $\mathrm{Oh} = \sqrt{\mathrm{We}}/\mathrm{Re}$. (**c**) Example images and illustrations of coffee ring formation due to capillary flow in evaporating droplets (**i**,**vi–viii**). This is compared to uniform depositions produced with added DTAB to promote particle trapping at the liquid–gas interface, which created particle skins that lead to homogenous disk like patterns upon drying (**ii–v**). (reprinted with permission from *Langmuir* (2015), 31, 14, 4113–4120, Copyright 2015, ACS). (**d**) Illustration of the dual drop inkjet printing process, where the blue ink is the supporting droplet, the red ink is the wetting droplet, and the gold represents the nanoparticles to be deposited. (reprinted with permission from *Adv. Mater. Interfaces* (2018), 5, 1701561. Copyright 2018, Wiley). (**e**) Illustration of the dual drop process used to deposit a uniform nanoparticle monolayer. (reprinted with permission from *Adv. Mater. Interfaces* (2018), 5, 1701561. Copyright 2018, Wiley).

3. Conductive Nanomaterial Printing

3.1. Fundamentals

Printed nanomaterial applications typically follow the same four-step process: first, nanomaterials are produced either through top-down methods, where the nanomaterial is broken off from bulk material, or bottom-down approaches, where the particles are synthesized from atomic precursors [81–85]. Second, these nanomaterials are dispersed in printable inks with viscosities and rheology that are optimized for the printing method of choice [11,26,86]. Third, the inks are printed on a substrate and create a deposition based on the fluid mechanics during printing and free energy effects at the liquid–gas and liquid–solid interfaces [11,44,63,69]. Finally, the solvent is evaporated, and, in some cases, the nanomaterials are sintered to yield a conductive structure [50,87,88]. When choosing a nanomaterial for a specific bioelectronics' application, the material's electrical and mechanical properties, the tendency to agglomerate, required loading to produce a conductive network, and particle aspect ratio are crucial considerations. For instance, graphene nanoplatelets and carbon nanotubes (CNTs) have excellent conductivities, are easily functionalized, and have high durability, but they are difficult to disperse in printable inks because of strong intermolecular forces [67,83,89]. In most nanomaterial inks, the solvent is highly polar, and the nanomaterial is nonpolar [90,91]. An amphiphilic dispersion agent, such as polyvinyl pyrrolidone (PVP) and sodium dodecyl sulfate (SDS), is introduced [64]. The nonpolar region binds to the nanomaterial surface, leaving a polar tail that allows the material complex to be dissolved in the solvent and creates interparticle repulsive forces that prevent agglomeration. In silver nanomaterials, PVP is highly attractive because the nitrogen and oxygen atoms enable effective absorption into the surfaces of Ag seeds or particles, whereas SDS is effectively absorbed into CNT surfaces in the presence of ultrasonication energy [64,91,92]. However, SDS is not biocompatible, and it must be effectively removed or reduced in concentration either before or after printing if the CNTs will be skin-contacting [91]. On the other hand, PVP is biocompatible, making it more attractive for many bioelectronics applications [93]. In the following sections, we will summarize recent developments in the synthesis, dispersion, high throughput printing, and sintering for each nanomaterial and demonstrated the soft electronics devices created with these methods.

3.2. Metal Nanoparticles (NPs)

3.2.1. Material Properties, Synthesis, and Ink Formation

AgNPs and CuNPs are low aspect ratio particles, typically with spherical geometries and radii from 10–100 nm for printing applications, that are often formed through wet chemistry from ionic precursors [85]. Example images of printed Ag and Au nanoparticles with spherical geometries are shown in Figure 6a [94]. In wet chemistry NP synthesis, a metal ion precursor, such as $AgNO_3$ and $Cu(NO_3)_2$, is reacted with a reducing agent, such as ethylene glycol (EG) or ascorbic acid in solution with a capping agent, such as PVP and SDS [85]. In addition to wet chemical synthesis, NPs may also be formed through physical methods, such as evaporation condensation [95] and laser ablation [96], additional chemical methods, such as microemulsion [97], UV or other photonic source initiated photoreduction [98], electrochemical synthesis [84], irradiation [99], microwave-assisted synthesis [100], and biosynthesis techniques, either through bacteria, fungi, algae or plants [100]. Spherical metal NPs tend to agglomerate strongly because of their large surface areas, strong interparticle attractions, and particle symmetry regardless of orientation. As a result, the NP surface must be functionally modified to aid in dispersion. Furthermore, their low aspect ratios require high material loadings in order to form conductive networks, but loadings over 60% complicate the design of inks for printing methods requiring low viscosities or which tend to clog, such as inkjet printing. On the other hand, the excellent material symmetry, high material loading, and surface pre-melting allow for very effective, low-temperature sintering at around 200 °C into uniform conductive films. An example of a printed AgNP film is provided in Figure 6b, and the resultant AgNP network after the solvent is dissolved is readily seen [101].

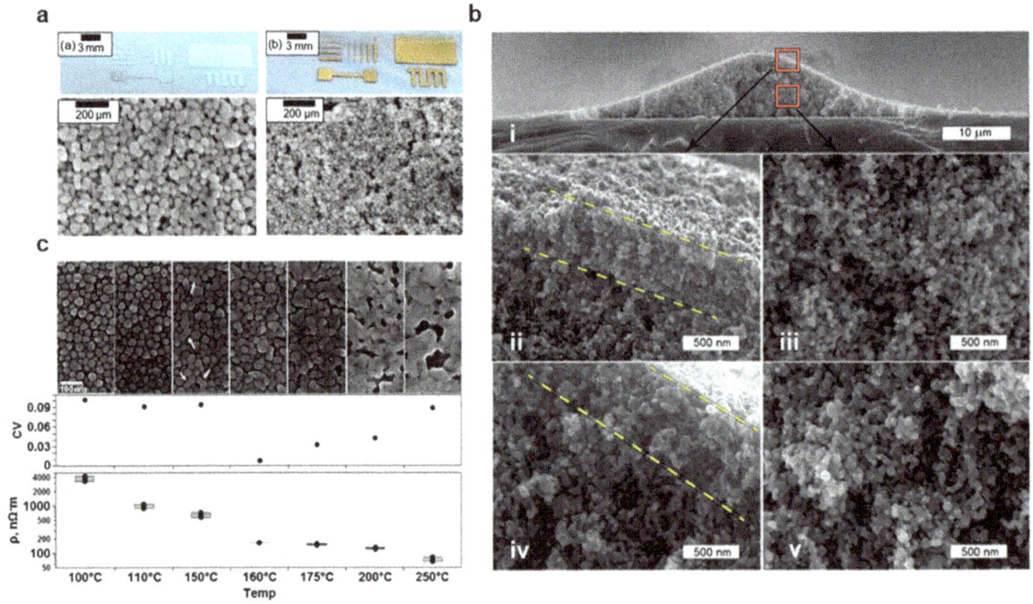

Figure 6. Metal nanoparticles for printed electronics. (**a**) photos and SEM images of inkjet-printed films with Ag (left) and Au (right) NPs. (reprinted under Creative Commons license CC BY 4.0 from *Adv. Radio Sci.* (2019), 17:119–127.) (**b**) SEM images of an AgNP deposition cross-section, showing the overall structure (**i**), surface (**ii** and **iv**) and interior (**iii** and **v**) after rapid laser sintering. (reprinted with permission from *Appl. Sci.* (2020), 10(1), 246. Copyright 2019, MDPI). (**c**) SEM images of AgNP films sintered at various temperatures, with corresponding graphs depicting the coefficient of variance and resistivity (reprinted with permission from *Materials* (2011), 4(6), 963–979, Copyright 2011, MDPI).

NPs inks are typically synthesized with 40–88% material loadings and dispersed with high concentrations of dispersants, such as 1:1 PVP mixtures. Because this drastically reduces the ink viscosity, PVP concentrations must be limited for screen printing. For instance, Wang et al. preheated and magnetically stirred a 0.3 M solution of PVP and ethylene glycol (EG) to increase the ability of PVP to bind to the AgNP surface, allowing them to disperse the NPs with a 1:2 PVP/$AgNO_3$ ratio [102]. After mixing 60 mL of 0.3 M PVP-EG solution and 40 mL 0.29 M $AgNO_3$-EG, the solution was mixed with N,N-dimethylformamide, hydroxyethyl cellulose, and ethylene glycol (EG) to yield a 45 wt.% ink with viscosity and rheology optimized for screen printing. When printed on PI, and sintered at 220 °C, the inks demonstrated a remarkably low resistivity of 8.3×10^{-6} Ω·cm, which is only five times greater than the bulk silver resistivity [102]. For gravure printing, Shiokawa et al. created an organic protection layer on AgNPs to improve dispersibility and printability. $AgNO_3$ (22 wt.%) was mixed with oxalic acid dihydrate (9 wt.%), n-Hextlamine, N,N-dimethyl-1,3diaminopropane and oleic acid, and AgNPs were synthesized through thermal decomposition of an oxalate-bridged silver alkylamine complex [103]. The resultant powder was then dispersed in tetralin, tetradecane, and dodecane with 80 wt.%, and it was determined that the tetralin solution had the highest printability [103]. The ink was then gravure printed on a glass slide with widths of 20 µm with 4.4 µΩ cm [103]. For flexographic printing, Benson et al. developed an AuNP ink that was used to create biocompatible sites on a PI substrate for the enzyme attachment in glucose sensing [104]. AuNPs were synthesized by reducing $HAuCL_4$ (0.2 g) with $NaBH_4$

(0.05 g) in the presence of PVP (0.15 g) in 30 mL DI. The solution was centrifuged to yield an AuNP pellet, which was subsequently redispersed in 70% IPA and 30% dionized water (DI) via ultrasonication. Electrodes were fabricated by flexographic printing of a carbon layer, then the AuNP layer, with a printing force of 125 N, anilox force of 125 N, and speed of 0.6 m/s. After functionalization with glucose oxidase, the electrodes demonstrated a high sensitivity of 5.52 $\mu A\ mM^{-1}\ cm^{-2}$ with a detection limit of 26 μM [104]. In addition, NPs are highly attractive for inkjet printing because of their low aspect ratios, which can avoid nozzle clogging, and they have thus been carefully studied [26]. For instance, Fernandes et al. designed an experiment to assess the printability and conductivity of AgNP inks with a variety of solvents and additives [47]. Silver nanoparticles were synthesized by reduction of 100 mL 0.006 M $AgNO_3$ and 0.008 M PVP in DI water by 8 mL of 0.529 M sodium borohydride ($NaBH_4$), centrifuged at 1500 RPM for 1 h, then dispersed in a range of ethanal based solutions with viscosities ranging from 3.7–7.4 mPa.s and material loadings from 8–16 wt.% [47]. It was determined that the EG, ethanol, ethanolamine, and hyperdispersant (Solsperse 20000) ink with 5.25 mPa.s viscosity resulted in the greatest printability due to the addition of humectants (i.e., ethylene glycol and ethanolamine) combined with low resistivity ($1.6 \times 10^{-4}\ \Omega$.cm) [47] Finally, AgNPs are not well suited to skin contact because of poor biocompatibility, so they either must be well insulated or replaced with AuNPs for such applications [26].

3.2.2. Post Print Processing

After printing, NP depositions must be cured to remove the solvent, and many, but not all, inks are also sintered to form conductive sheets [94,100]. Sintering is not typically employed for printing on TPU, PET and paper because of low substrate melting points, in printing stretchable interconnects, where the unconnected particles form effective conductive networks with strain, and for biosensor applications where increased surface area is preferred (e.g., glucose sensors) [104]. Sintering, however, is highly advantageous for forming conductive sheets with low resistances and high yield stress [47,105,106]. The SEM images in Figure 6c clearly show the formation of a more uniform metal sheet with increased temperature in thermal sintering, and this is reflected in the decreased resistivity [107]. Although thermal and chemical sintering are easily employed in sheet-to-sheet processes, alternative methods are needed for roll-to-roll integration [105]. One approach with significant promise is photonic sintering, where energy is provided by an ultrafast pulsed laser source with a wavelength tuned to match the ink's absorption spectrum [105,106]. For instance, Hösel et al. demonstrated a single exposure system integrated into roll-to-roll flexography printing with speeds of 2.5 m/min [106]. In addition, electric, plasma, and microwave sintering are well suited for roll-to-roll processes [108,109]. Allen et al. demonstrated effective electric sintering with a directly applied voltage, but the method has not been explored for roll-to-roll processes, likely because of the need to create direct and secure contact between the pattern and electrode [108]. In contrast, indirect methods, such as microwave sintering, can be well integrated into roll-to-roll processes, but their throughput is greatly limited compared to photonic and electrical methods [105]. For instance, Fujii et al. demonstrated effective sintering in 1.5 min, compared to milliseconds in other methods [110]. Finally, plasma sintering is a promising sintering method, but it is limited for thick or multilayer depositions by a slow depth penetration, which is an issue for high throughput applications [105].

3.3. *Metal Nanowires (NWs)*

3.3.1. Material Properties, Synthesis, and Ink Formation

Unlike NPs, NWs are differentiated by their large aspect ratios, with lengths often 1000 times greater than their widths [3,11,111]. As a result of these aspect ratios, NWs can from conductive networks with very minimal loading, exhibit minimal bending stiffness and exceptional yield strength approaching the theoretical value of E (Young's modulus)/10, high optical transmittance, and electrical conductivities that are dominated by quantum effects [17,45,112]. When the NW widths become too small, conductivity is greatly diminished by edge effects from atoms at the material surface and scattering, setting a practical limit on widths for printed inks [112]. Compared to NPs, NW inks are significantly easier to synthesize because NWs in random orientations are much more resistant to agglomeration [11]. Unlike sintered NP sheets, these NW networks can stretch and deform when embedded in a polymer matrix [45]. NWs can also be made biocompatible because of the inability of small Ag particles to migrate into the skin [111]. In addition, NWs may be laser welded for the rapid formation of highly conductive sheets. Nanowires are traditionally synthesized through the polyol method for printed inks, but the template method is also widely employed [11,17,113]. In polyol synthesis, the solution temperature, PVP molar ratio to $AgNO_3$, stirring rate, the introduction of platinum seeds or other nucleation agents, and the addition of chloride or bromide ions can all be used to control the material dimensions and AgNW quality [93,113,114]. Figure 7a shows SEM images of AgNWs synthesized in various PVP solutions along with quantitative measurements of average NW diameter and length, and this experiment demonstrates that PVP solutions must be carefully optimized for Polyol synthesis [93]. During this synthesis method, AgNWs are typically produced from Ag seeds reduced from $AgNO_3$, and these seeds are capped by the presence of PVP [93,114]. Although the exact mechanism by which AgNWs are synthesized in the polyol process is not fully known, it is likely that the differential affinity of PVP to the <100> plane than <111> plane in silver leads to unidirectional growth [115]. Despite a rigorous theoretical model, empirical findings allow for precise control of material aspect ratios and purities [93].

AgNW inks typically contain much lower material loadings than AgNP inks, simplifying ink design. As a result, greater resolutions are often achievable. For instance, Liang et al. experimented with different material loadings in AgNW inks for high-resolution screen printing [11]. AgNWs with aspect ratios of 500 were mixed with (hydroxypropyl)methyl cellulose (HMC), Zonyl FC-300, and defoamer MO-2170 in a distilled water solution and sonicated [11]. HMC is a viscoelastic polymer with hydroxy groups that bind strongly to AgNWs to aid in dispersion and that serves as an emulsifier and thickening agent [11]. Zonyl FC-300 was used to decrease the surface tension of the ink and promote substrate wettability for high-resolution printing, and defoamer MO-2170 was necessary to prevent foaming during mechanical agitation. It was determined that a 6.6 wt.% AgNW ink had the greatest pseudo-plasticity and lowest viscoelasticity (i.e., the ink had the highest difference in viscosity during low and high shear and recovered viscosity the quickest after applied shear was removed), which allowed for screen printing of highly conductive (4.67×10^4 S/cm) 50 μm width traces [11]. Likewise, Huang et al. investigated various material loadings for gravure printing AgNW inks, arriving at an optimal value of 5.0 wt.% that yielded 50 μm width traces and 5.34×10^4 S/cm conductivity [50]. AgNWs were synthesized in the presence of PVP (50 mL 0.09 M in EG) and NaCl (150 μL 0.1 M in EG) and centrifuged with acetone and ethanol to remove the solvent and surfactant. The AgNWs were then dispersed in a Poly(ethylene oxide) solution. At 1.5 mm/s, the 5.0 wt.% ink demonstrated a capillary number of 1.09 and viscosity of 20.9 Pa.s, making it suitable for gravure printing [50]. Optical (left) and SEM (right) of the resultant prints are shown in Figure 7b [50]. It is also possible to pattern a nanowire precursor on a substrate and grow the nanowires in situ, and this process has been demonstrated using flexography for ZnO NW functionalization of electrochemical biosensors [116]. In this work, commercially available carbon and AgCl inks (Gwent, PontyPool, UK) were flexographically printed on flexible PI to form a conductive electrode, and the ZnO precursor ink (1.1 g of zinc acetate in 10 mL DI and 40 mL IPA) was printed with an anilox volume of 12 cm^3/m^2, anilox force of 125 N, printing force of 150 N and printing speed of 0.2 m/s. The ZnO wires were hydrothermally synthesized in situ in an aqueous solution of 10 mM hexamethylenetetramine to yield a flexible glucose sensor with a sensitivity of 1.2 ± 0.2 μA mM^{-1} cm^{-2} with a linear response to the addition of glucose over a concentration range of 0.1 mM to 3.6 mM [116]. Finally, inkjet printing AgNWs have been demonstrated, but the printing process must be carefully controlled to prevent nozzle clogging. In a sheet-to-sheet process, Al-Milaji et al. created an AgNW ink for inkjet printing by synthesizing AgNWs with an average of diameter of 100 nm and length of 14.5 μm in a polyol process, then dispersing the resultant precipitate in ethanol [45]. The resultant ink was printed on an uncured liquid PDMS layer spin coated on PET, and the AgNW ink was absorbed into the PDMS to create a stretchable interconnect. The connectors demonstrated high reliability during strain and bending, but initial resistances were high (0.68 kΩ over 25 mm) [45]. In contrast, Finn et al. sonicated commercially purchased NWs to reduce particle length, dispersed in IPA, and optimized inkjet parameters to yield sheet resistances of 8 Ω/sq and conductivities of 105 S/m in traces with widths of 1–10 mm and thickness of 0.5–2 μm after curing at 110 °C [117]. In order to reduce clogging, a Dimatix printer with 16 nozzles of diameter 21.5 μm spaced 254 μm apart was used to create 10-pL droplets at 5 kHz with a spacing of 20 μm and 50% overlap [117].

3.3.2. Post Print Processing

NW networks are often cured without sintering because the particles naturally contact when randomly dispersed, but welding NWs can significantly reduce wire-to-wire resistances when significant PVP coatings are present. For instance, Lee et al. used thermal sintering at 200 °C for 20 min to reduce resistance in a printed AgNW trace from 1000 to 100 Ω/sq [118] and Li et al. photonically welded NWs to reduce sheet resistances from 53 to 7.1 Ω/sq [23]. An example SEM image of laser-welded AgNWs is provided in Figure 7c [119]. Finally, NWs can be welded by NPs embedded in a matrix film. In one demonstration, Triambulo created a highly conductive (5.0–7.3 × 10^5 S/m) AgNW-AgNP matrix film on a flexible PET substrate with similar optical transmittance (>90%) compared to a pure AgNW film, and SEM images of the resultant network are provided in Figure 7d [120]. Despite the advantages of welding NWs for improving conductivity, the ability to process NWs at room temperature for many ink formulations is a key advantage in roll-to-roll integration, especially when attempting to limit start-up costs [11].

3.4. *Graphene*

3.4.1. Material Properties, Synthesis, and Ink Formation

Graphene is a zero-gap semiconductor with exceptional conductivity, biocompatibility, and high mechanical strength that is easily functionalized with numerous materials for sensing applications [81]. As a result, graphene is highly attractive for printed bioelectronics [53,67,86]. However, graphene is very difficult to print because of its low dispersibility in printable inks [86]. Typically, graphene is formed through exfoliation from graphite, either through ultrasonic or mechanical methods, but numerous additional mechanisms have been explored, including electrochemical synthesis, chemical vapor deposition, laser processing and sodium ethoxide pyrolysis [81]. An example SEM image of graphene sheets for screen printing is provided in Figure 8a [121]. Once synthesized, dispersing graphene in printable inks is a key challenge. Although graphene oxide (GO) is easily dispersed, it must be reduced after printing, limiting throughput, and creating numerous defects that detract from the material's electrical and sensing capabilities.

To create a screen printable ink, He et al. dispersed 5 g of graphene nanoplatelets (GNPs) in 50 mL EG and 0.5 g PVP and printed traces on the PI with conductivities of 8.81 × 10^4 S/m [67]. Graphene's natural tendency to agglomerate was used to increase the ink viscosity to a range reasonable for screen printing (1 Pa·s) [67]. For gravure printing, Secor et al. noted that stabilizing graphene with ethyl cellulose (EC) greatly aids in dispersion [53]. Graphene was exfoliated from graphite in ethanol with EC, excess graphite was removed by centrifuging. The resultant graphene–EC precipitate was then redispersed in ethanol and terpineol, and the specific quantities of each substance were altered to yield inks with various viscosities [53]. As described in Section 2.1, the optimized ink was able to be printed with trace widths of <30 μm with conductivities >10,000 S/m [53]. Although initial feasibility studies have investigated flexographic graphene printing, no successful use in soft bioelectronics has been reported to date [122]. Likewise, many inkjet applications select to use GO instead of graphene, but graphene printing has been successfully reported. For instance, Li et al. exfoliated graphene from graphite in dimethylformamide (DMF), and the toxic DMF is distilled out in a terpineol solution [86]. Graphene in this state is normally stable for only hours, but in this work, EC was added to protect the graphene from agglomeration. After the solvent exchange, the graphene/toluene dispersion was mixed in ethanol in a volume ratio of 3:1 to yield a printable viscosity and rheology. Figure 8b,c depicts this optimized ink (b) directly after printing and (c) after curing, demonstrating a mostly uniform deposition with a minimal coffee ring effect [86]. Finally, the resultant ink was printed in 80 μm traces on both plastic, and silicon substrates and printed supercapacitors were able to achieve a specific capacitance of 0.59 mF cm^{-2} [86].

3.4.2. Laser Synthesis of Graphene

Laser printing is an emerging technology whereby a thin film of material is selectively removed from a carrier substrate via a laser beam and irradiated to a receiver substrate [123]. This approach allows for integration with roll-to-roll laser printers, printing without harsh chemicals, high spatial resolution, and control of edge plane functionalization, which makes laser printing of great interest for bioelectronics applications [123,124]. For instance, Rahimi et al. demonstrated a high throughput process by which graphene can be irradiated onto a PDMS substrate to yield strain sensors sensitive up to 100% strain with a gauge factor of up to 20,000 [124]. It was reported that laser power and speed greatly affected print quality and conductivity, and the authors optimized the process to 0.5–1.9 m/s and 4.5–8.25 W for printable traces [124]. Laser printed graphene is also of great utility in a number of biosensor applications. For instance, Ortiz-Gómez et al. ablated a PI film with a 12 W CO_2 laser operating at 2.4 W and 0.15 m/s to create a graphene heater for a microfluidic device that used fluorescent silicon nanodots to detect total carbohydrates [125]. In addition, GO can be reduced by laser excitation through the conversion of sp^3 carbon to sp^2 and the removal of oxygen functional groups, and the photothermal and photochemical processes involved in the reduction of GO can be well controlled by altering the laser wavelength [126]. For instance, Zahed et al. used a CO_2 laser to reduce GO for an electrocardiography (ECG) sensor with comparable signal quality to commercial Ag/AgCl electrodes (12.9 dB vs. 13.3 dB) [127].

3.4.3. Post Print Processing

Because graphene ink stability is predicated heavily on the addition of strong solvents and polymer stabilizers, these chemicals must be evaporated, dissolved, decomposed, or otherwise removed in order to yield optimal conductivity and material properties [128]. Post-print processing is highly dependent on the choice of chemical additives. For instance, graphene dispersed in high concentrations of EC must be treated at 300–400 C, whereas EG-PVP mixtures can be cured at 120 °C [82]. Furthermore, several inks that do not evaporate solvents can be treated at room temperature, although these inks will exhibit lower conductivities as a result [129,130]. While thermal curing beyond 120 °C is not suited for flexible electronics on many substrates, such as PET and TPU, novel laser treatment approaches are able to efficiently treat printed patterns without damaging the underlying substrate [131]. For instance, Jabari et al. reported a laser treatment method to cure printed graphene with similar conductivities to traditional thermal curing [131]. Likewise, Secor et al. demonstrated an intense pulsed light annealing for inkjet-printed graphene that is suited to a variety of substrates and can result in fewer impurities than thermal alternatives [132]. Finally, GO is much easier to disperse than graphene, but it must be reduced after printing with harsh chemicals and high temperatures, which often results in defects and poor conductivity [128]. As a result, GO is not as attractive as graphene for bioelectronics applications.

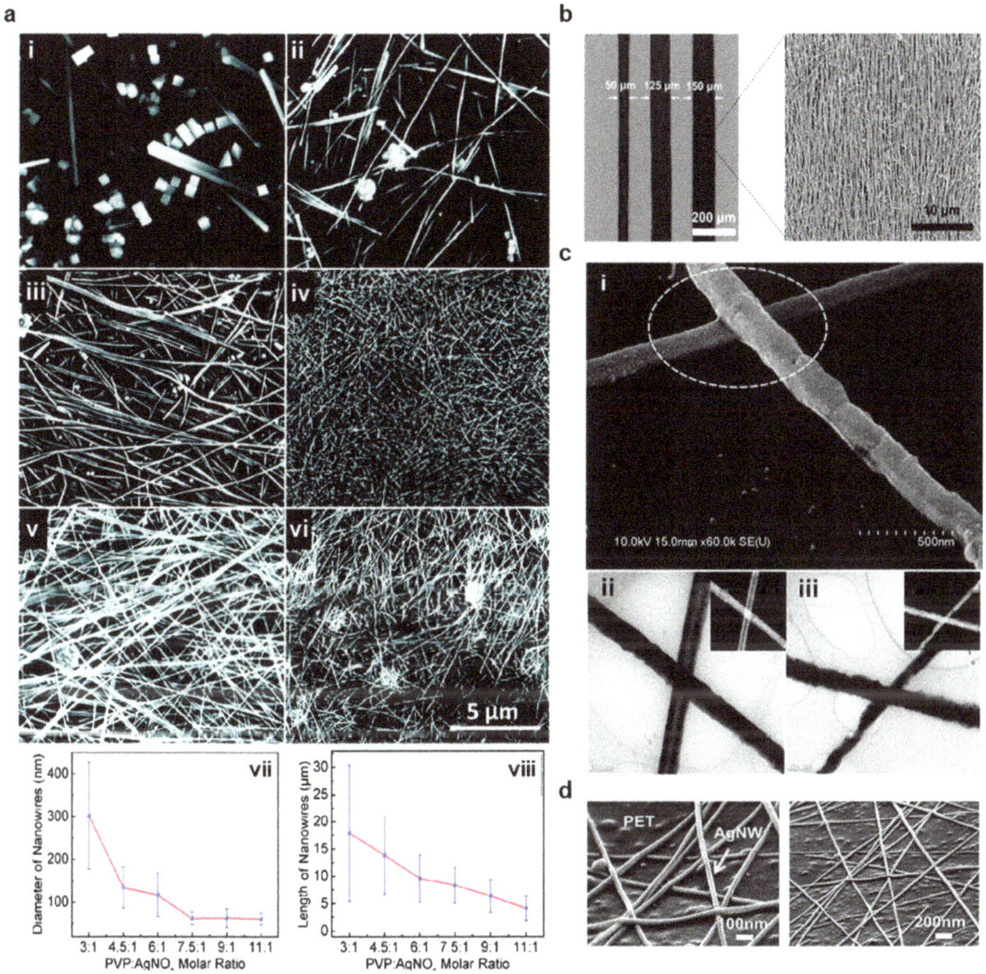

Figure 7. Metal nanowire synthesis, printing, and welding. (**a**) SEM images of Ag nanowires synthesized at different PVP:AgNO3 molar ratios of (**i**) 3:1, (**ii**) 4.5:1, (**iii**) 6:1, (**iv**) 7.5:1, (**v**) 9:1, and (**vi**) 11:1. All scales are the same. Changes in (**vii**) nanowire diameter and (**viii**) length with PVP:AgNO3 molar ratio are also shown. (reprinted with permission from *Cryst. Growth Des.* (**2011**), 11, 11, 4963–4969, Copyright 2011, ACS). (**b**) Images of gravure printed AgNW traces with various thicknesses and an SEM image demonstrating the aligned AgNW network. (reprinted with permission from *Sci. Rep.* (2018), 8, 15167, Copyright 2018, Nature Publishing Group). (**c**) SEM image of CuNWs nanowelded with laser irradiation (**i**) and TEM images of CuNWs before (**ii**) and after (**iii**) laser irradiation. (reprinted under Creative Commons license CC BY from *Sci. Rep.* (2017), 7(1)). (**d**) SEM images of sparse, randomly oriented AgNWs on a PET film (reprinted courtesy of *Org. Electron.* (2014) 15(11), 2685–2695, Copyright 2014, Elsevier).

3.4.4. Graphene Functionalization for Biosensor Applications

Organo-functionalized graphene has played a crucial role in the development of novel biosensors, and the ability to print such sensors in roll-to-roll methods would have a transformative effect on healthcare [12]. Although each of the four materials covered in this review has been successfully explored biosensors, those based on graphene have generated the most recent interest because of the great degree of freedom in material functionalization [15]. Graphene functionalization occurs either covalently or non-covalently.

In covalent functionalization, graphene is oxidized to GO, and covalent bonds are formed to organic functional groups on a sensing material [128]. For instance, a carboxylic group on GO can covalently bond to glucose oxidase to form a glucose sensor [12,16,128]. Examples of covalent sensing systems on functionalized GO are illustrated in Figure 8c [13]. Non-covalent bonding occurs when functional groups are attracted to graphene through Van der Waals and electrostatic forces, but this bonding is typically nonstable for long durations [12,133]. Instead, target biomolecules may be directly absorbed into the graphene, allowing the graphene to serve as a sensor through non-covalent functionalization [133]. The most common form of glucose-based biosensors is electrochemical sensors. Graphene functionalized with biological receptors is employed as a working electrode to detect analytes through electrochemical oxidation or reduction of analytes [12]. For instance, Kinnamon et al. screen-printed GO on a textile substrate and bound 1-Pyrenebutyric acid-*N*-hydrosuccinimide ester (PANHS) as a crosslinker to bind to an influenza A-specific antibody [16]. The textile sensor demonstrated high stability with washing (~4.6% variability) and accurate sensing over a range of virus expression of 10 ng/mL to 10 μg/mL with a limit of detection of 10 ng/mL. The sensor also exhibited very good specificity, and the sensing range is well suited to the average human viral expression of 50 ng/mL [16]. In addition, graphene field effect transistor (FET) biosensors may be used to control the flow of current as a function of charge accumulating on a functionalized graphene gate of channel, as shown in Figure 8d [12]. For instance, Xiang et al. used inkjet printing to deposit a graphene channel for a fully printed FET on the PI with low resistivity (110 Ω/sq) that was subsequently functionalized in cystamine solution (Figure 8e) [133]. Norovirus antibodies were then bonded, and bovine serum albumin was introduced to prevent non-specific binding of other biomolecules. It was determined that the voltage gain from source to drain with an applied 10 GHz wave generates a linear response from 0.07 to 3.70 dB when the concentration of Norovirus protein increases from 0.1 to 100 μg/mL [133].

3.5. Carbon Nanotubes

3.5.1. Material Properties, Synthesis, and Ink Formation

CNTs offer very attractive elasticity, biocompatibility, surface area, aspect ratios, strength, and conductivity, making them of great interest for electronics applications, but very strong van der Waals interactions greatly complicate particle dispersion [83]. CNTs consist of rolled graphene sheets that consist of either one tube (single-walled CNT, or SWCNT) or multiple tubes (multi-walled CNT, or MWCNT) held together with Van der Waals attractions [90]. The direction in which CNTs are rolled greatly affects their observed properties, and illustrations of several common orientations are provided in Figure 9a [134]. "Armchair" CNTs are highly preferred for interconnects or conductive planes because their identical chiral indices create highly uniform conductivity [92], but zigzag or chiral CNT orientations are widely employed for their semiconducting effects, and they are also of great interest for printed transistor fabrication [91]. CNTs are typically synthesized through three processes: chemical vapor deposition (CVD), arc discharge, and laser ablation, although CVD is the most widely employed. In CVD, metal NPs of the CNT diameter are introduced in the presence of a carbon-based gas, such as CO_2, to form CNTs, and this process is illustrated in Figure 9b [82,91]. In order to remove the NPs and other impurities, the CNT powder is typically sonicated or treated with acid, and CNT purity is crucial in achieving optimal material properties [83,135]. The final product is CNTs like those shown in the AFM images in Figure 9c [135].

Once the CNTs have been synthesized, dispersing them in printable ink is a key challenge [92]. In designing a screen printable CNT ink, Menon et al. dispersed CNTs in an ethanol SDS solution optimized to 7.5 wt.%, then added various PVP loadings and assessed printability [135]. It was determined that PVP weights equal to half that of the CNTs were most suited for screen printing [135]. In addition, Shi et al. demonstrated that sonication is crucial in SDS facilitated CNT dispersion because the sonication forcibly breaks apart CNT clusters, exposing the CNT surface to SDS [87]. Figure 9d depicts TEM image results of one such experiment, where dispersion clearly improves after sonicating for 6 instead of 4 h [87]. Gravure printed semiconducting SWCNTs have been thoroughly studied for thin-film transistor applications, and Sun et al. recently demonstrated a thin film transistor active-matrix (TFT-AM) electrophoretic sheet on PET that could be used as a wearable display [136]. Metallic CNTs were removed from a mixed semiconducting-metal powder with poly(9,9-didodecylfluorene) (PFDD) in a PFDD/CNT ratio of 1.25/1.00 to yield a semiconducting purity of 99.9%. The PFDD was then exchanged with a polythiophene derivative (P3ME4MT) in toluene and dispersed in 1-octanol to produce a printable viscosity and suitable capillary number. After gravure printing at 6 m/min at 30 μm depth and 150 μm cell opening for 10 PPI resolution and 10 μm depth and 35 μm opening for 40 PPI, TFT-AMs with average mobility of 0.23 ± 0.12 cm^2 V^{-1} s^{-1}, the average on-off ratio of 104.1, and threshold voltage variation of ±13% was demonstrated [136]. Images of the printed TFT-AM are provided in Figure 9d [136]. Finally, inkjet printing of CNTs has been demonstrated for numerous biosensor, conductor, and semiconductor applications [137]. For instance, Okimoto et al. improved on previous CNT semiconductor performances by optimizing the CNT density in a novel SWCNT inkjet printing ink [138]. The SWCNTs were prepared by laser vaporizing carbon rods doped with Co/Ni in an argon environment and purified with H_2O_2, HCl, and NaOH. The SWCNTs were dispersed in DMF in a mixture of 0.04 μg/mL, sonicated, centrifuged, and filtered through poly(tetrafluoroethylene) membrane filters. The ink was printable with a 30 μm nozzle at 500 Hz, and the fabricated CNT TFT yielded mobility of 1.6 to 4.2 cm^2 V^{-1} s^{-1} and an on/off ratio of 4–5 digits [138]. Because CNT inks are both highly desirable for commercial sensing and TFT applications and the large challenges in designing printable inks, numerous commercial inks are now available, and many reported works in the literature are using these inks for inkjet and gravure printing (Figure 9e) [91,92]. In summary, CNT ink printing is highly attractive for many essential bioelectronics' applications, and new continued investigations into high throughput fabrication methods are essential in translating these novel discoveries to industrial and clinical use.

3.5.2. Post Print Processing

Post print processing for CNT inks varies greatly based on the specific dispersants and polymers employed in the ink synthesis [136,139]. Generally, the processing is complex, which creates an incentive to remove as much of the polymer residue and dispersant before printing as possible [135]. For instance, sonication, centrifugation, washing and filtering before printing are typically essential measures to create an environment in which post-print processing is feasible [87]. In addition, careful selection of polymers and dispersants and effective processing can be employed to yield simple and effective processing [92,140]. Although several complicated polymer removal strategies have been studied, such as metal–chelation-assisted polymer removal (McAPR) and yttrium oxide coating, washing and annealing are still the most preferred because of simplicity, cost, and scalability [91]. For example, Yu et al. recently removed polycarbazole (PCz) from a CNT print via THF washing [140]. Although some PCz remained in ink, this method is an effective and simple mechanism for biocompatible post-print CNT processing [140]. Another common washing solvent is toluene which is often used with elevated temperatures to improve solubility. Annealing is also highly effective, but high-temperature restraints (above 300 °C) make it not suited for many flexible substrates, such as PET and TPU [83]. In a modified annealing process, one may exchange the polymer for a different material, and this process is both effective and suited to lower temperatures [139]. For instance, Sun et al. exchanged PFDD for P_3ME_4MT, as discussed previously, to create a high mobility electrophoretic deposition [136]. Overall, post-print processing for CNT inks is an area of high research interest, and novel advancements are greatly needed to implement CNT imprinting in high throughput fabrication processes fully.

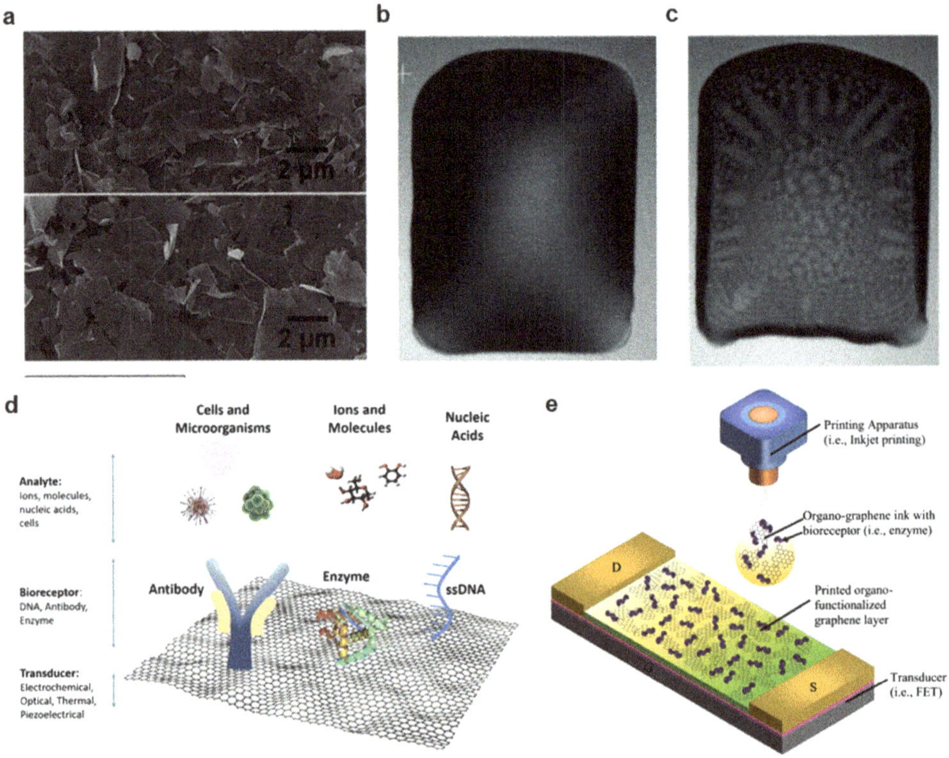

Figure 8. Graphene printing and functionalization for bioelectronics sensors. (**a**) Atomic force microscopy (AFM) images of screen-printed graphene (reprinted with permission from *J. Colloid Interface* Sci. (2021), 582(A), 15. Copyright 2021, Elsevier). (**b**,**c**) SEM images of an inkjet-printed graphene deposition (**b**) before and (**c**) after curing, with a minor coffee ring effect. (reprinted with permission from *Adv. Mater.* (2013), 25(29), 3985–3992. Copyright 2013, Wiley). (**d**) Examples of covalently bonded bioreceptors on a functionalized GO deposition. (reprinted with permission from *J. Nanobiotechnol.* (2018), 16,75. Copyright 2018, Springer Nature). (**e**) Graphical depiction of a FET biosensor with organo-functionalization. (reprinted with permission from *Biosens. Bioelectron.* (2017). 87, 7–17. Copyright 2017, Elsevier).

3.6. Novel 2D Nanomaterials

One of the critical advantages of nanomaterial printing is the opportunity to tune material properties to address various application needs finely. The development of novel 2D materials is essential in the high throughput printing of advanced biosensors and bioelectronics [141]. For instance, two-dimensional transition metal dichalcogenides (TMDs), such as WSe_2, WS_2, $MoSe_2$, and MoS_2, are direct bandgap monolayers with high flexibility that can be used alone or in combination with graphene to create various flexible sensors [141–144]. TMDs have several exceptional material properties that make them highly suited for many electronics applications. They contain no inversion center, which allows the k-valley index to be manifest as a new degree of freedom charge carrier [144]. Strong spin-orbit coupling leads to spin-orbit splitting, making them well suited to spin transport electronics applications, commonly termed spintronics [145]. Printable TMD inks can be synthesized from bulked cellular samples through liquid-phase exfoliation (LPE). This has been demonstrated for applications such as screen-printed oxygen sensing electrodes [42] and wearable heterostructure photodetectors [143]. Despite recent advances in LPE processes by optimizing dispersion agent concentrations, polymers, stabilizers, and binders, TMDs can be challenging to disperse in printable inks. Many LPE processes still rely on toxic and hazardous materials that do not demonstrate biocompatibility for wearable applications [146]. Lee et al. developed a zwitterion-assisted LPE process to synthesize TMDs in water to address this concern, allowing for the development of highly-biocompatible TMD inks [146].

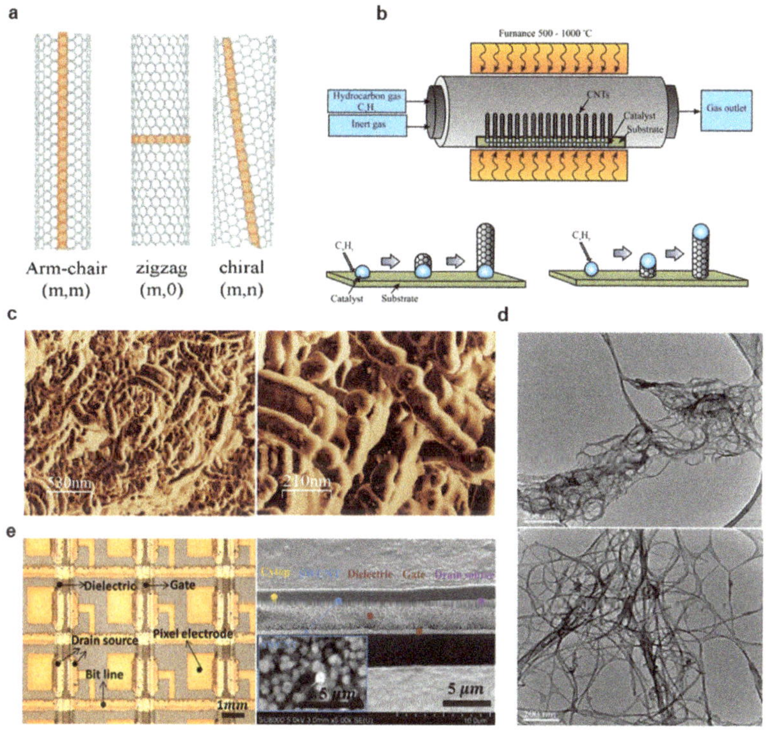

Figure 9. Carbon nanotubes for high throughput bioelectronics printing. (**a**) Armchair, zigzag, and chiral CNT geometries, each of which exhibits unique material properties. (reprinted with permission from *Physica E Low Dimens. Syst. Nanostruct.* (2014), 59:186–191. Copyright 2014, Elsevier) (**b**) Schematic representation of the CVD process for CNT synthesis, with illustrations of the base growth (bottom left) and tip growth (bottom right) CNT synthesis methods. (reprinted with permission from *Chem. Biol. Technol. Agric.* (2016), 3(17). Copyright 2016, Springer Nature) (**c**) AFM images of printed MWCNTs at different magnifications. (reprinted with permission from *RSC Adv.* (2017), 7, 44076–44081. Copyright 2017, RSC). (**d**) TEM images of SWCNTs in an SDS solution after sonication for 4 h (top) and 6 h (bottom). (reprinted with permission from *J. Surf. Eng. Mater. Adv. Technol.* (2013), 3, 6–12. (**e**) Image of pixels in a roll-to-roll gravure printed TFT-active matrix with 10 PPI resolution (left) and cross-sectional FIB-SEM of printed SWCNTs on the printed dielectric (right). (reprinted with permission from *Adv. Electron. Mater.* (2020), 6, 1901431, Copyright 2020, Wiley).

Additionally, hexagonal boron nitride (h-BN) is a high bandgap, biocompatible, nanomaterial isostructural to graphene that is highly suited for nanophotonics. It is a natural hyperbolic material in the mid-IR range [147], attractive for use as a substrate for graphene transistors because of its atomic-scale smoothness [32], advantageous for electrochemical sensing [148], of great interest as a capacitive dielectric [130], and potentially suited for the in-situ formation of 1D conducting channels [57]. Printable h-BN monolayers may be synthesized through top-down approaches, such as mechanical and chemical exfoliation, or bottom-down approaches, such as PVD and CVD [148]. Because h-BN has strong in-plane covalent bonds and weak inter-plane van der Waals forces compared to graphene, h-BN is an attractive 2D material for printable inks. Although h-BN has long been of interest, its potential for high-throughput fabrication via screen and inkjet printing has just recently been appreciated [148]. For instance, h-BN is now well understood for capacitive, dielectric, and transistor substrate applications. Still, new investigations into printable optic devices and electrochemical sensors will be needed to unlock this material's full potential. In one recent work, Desai et al. optimized h-BN nanoplatelet geometries synthesized through exfoliation and deposition thicknesses to yield a printed photo-capacitor with excellent thermal stability ranging from 6–350 K [130]. Additionally, Angizi et al. used edge functionalized h-BN dispersed in ethanol for screen printed Vitamin C detection in a flexible biosensor [149].

4. Applications for Bioelectronics

4.1. Electrical Interconnections

Conductive interconnections are the backbone of all fully integrated electronic devices [10]. These traces form the basis of circuits, and they must exhibit high conductivity and reliability. Furthermore, many applications require the circuit to stretch and deform [8]. For instance, a skin-mounted electrophysiology sensor is highly degraded by motion artifacts, and a stretchable circuit can greatly reduce these artifacts [4,17]. Interconnections for these soft, flexible, and stretchable devices have followed a three-stage development process: first came the development of stretchable interconnections based on fractal geometries fabricated with traditional MEMS processes [10,150]. Second, recent works have sought to fully print these systems on non-conventional substrates, such as TPU and PET, that are not compatible with MEMS fabrication [150]. Finally, these printed methods are being scaled with high throughput methods to make them suitable for commercial scales.

Several key challenges must be overcome in this third stage of interconnection fabrication. Crucially, they must be printed with high resolutions, speed, conductivity, and reliability on a variety of non-traditional substrates, and many applications require high resistive stability with local strain [9,10,17]. Although local strain can be alleviated with optimized geometries, printed interconnects are often embedded in a polymer matrix, which allows them to form conductive networks that remain conductive with strain [11,35,64]. These interconnections are typically stretchable up to 10% for wearable applications, although in some cases, stretchability up to 100% has been demonstrated [151]. However, the addition of polymer matrices often limits interconnect conductivity, which can often approach the limit set by bulk metals when sintered on temperature-stable substrates. As mentioned previously, these thin films are made flexible despite their high modulus through thin deposition heights, and they can stretch as a system without high local strain through optimized geometries [35]. However, these geometries may require spatial resolutions approaching the limits of fully printed technologies [152].

A summary of recently reported interconnections fabricated with high throughput processing is provided in Table 2, showing the different substrates, materials, fabrication methods, and curing approaches employed in state-of-the-art processes. In addition, resistances and resolutions are compared for each system, indicating which methods are preferred for each specific use case and application. When high conductivity is required, NPs and NWs inks with high material loadings are preferred, and sintering is often required in the case of NPs [17,153,154]. However, Scheideler et al. and Ohsawa et al. were able to achieve high conductivities on polyethylene naphthalate (PEN) substrates using NWs and NPs, respectively, without sintering. In addition to conductivity, the inks should not be significantly higher modulus when cured than the substrate, or advanced geometries are needed to alleviate local strain [35]. With optimized ink compositions and judicious trace patterning, very high reliability during bending, washing, and other wearable use can be demonstrated [82]. Finally, the same geometries that are effective in strain relief for high modulus MEMS interconnects are not always the ideal choice for stretchable interconnects because of the complex mechanics introduced when the substrate itself stretches and deforms from Poisson effects and the substrate-ink modulus mismatch [35]. Therefore, Huttunen et al. performed an experiment to assess different trace geometries of AgNP inks on a PDMS substrate, determining that triangular patterns maintained conductivity with higher applied strain [35]. An example of several patterns printed on a stretchable PDMS sheet is provided in Figure 10a [35]. In summary, recent developments in high throughput interconnect printing allow one to produce patterns with conductivity and stretchability optimized to many bioelectronic applications. Still, more thorough testing and process optimization are required for these systems to become commercially adopted.

Table 2. High-throughput nanomaterial interconnection fabrication.

Reference	Material	Method	Substrate	Curing	Printing Speed [m/min]	Sheet Resistance [Ω/sq]	Resolution [μm]
[53]	Graphene	Gravure	PI	Room temperature	0.3	6.25	30
[22]	AgNP	Inkjet	PEN	Laser Sintering	10	2.5	50
[155]	AgNP	Flexography	PET	130 °C for 5 min	5	45	150
[35]	AgNP	Screen	PDMS	140 °C for 8 min	2	2.5	125
[50]	AgNW	Gravure	PET	150 °C for 5 min	1.5 mm/s	~20	50–150
[153]	AgNW	Screen printing	PET	Flash Light Sintering	0.2	9.6	20
[154]	AgNW	Gravure	PEN	170 °C for 10 min	1	9.3	Film was tested
[88]	AgNP	Gravure	PEN	100 °C for 1 min	Not reported, roll-to-roll	4.9	40 mm × 80 mm

4.2. Biosensors

Biomolecule sensing devices, such as glucose-sensing patches, promise a transformative way to continuously assess crucial biomarkers, making their development critical in the future of healthcare development [12,14–16]. Traditional biosensing methods, as discussed in the introduction, are highly limited because of exceptional costs and the inability to record results continuously in real-time, whereas printed biosensors are well suited to long-term, continuous monitoring in an affordable and wearable package [12,13]. The clinical implication of such technology is clear, and the development of high throughput biosensor fabrication, such as the slot die process that is shown in Figure 10b [57] It is of very high importance [13]. Generally, biosensors consist of a receptor, e.g., an antibody, and transducer, e.g., a nanomaterial sheet capable of transmitting an electrical signal from the receptor to a circuit element [13,16]. To bind the receptor to the transducer, the material must be functionalized, as discussed in Section 3.4. Although many nanomaterials may be effectively functionalized, graphene is one of the most favorable materials because of the ease in which one can attach a variety of organic and inorganic functional groups [13]. For instance, graphene can be oxidized, then functionalized with 1-ethyl-3-(3-dimethylaminopropyl) carbodiimide hydrochloride (EDC)/N-hydroxysuccinimide (NHS) (EDC/NHS) to facilitate antibody binding [156]. The target molecule detection can be achieved through several methods, although electrochemistry is the most employed [12,13]. An example of a potentiometric electrochemistry analysis is provided in Figure 10c [12]. In these systems, functionalized working (where the reaction occurs) and reference (where the current is provided) electrodes are implemented, and the transducer is able to record changes in current, resistance, or potentially caused during the binding reaction [13]. Another printed biosensor method is based on FET technology, where the binding of a target molecule is used to modulate the flow of current through a channel [13].

Table 3 summarizes recent demonstrations of high throughput biosensor fabrication, with an emphasis on device performance [16,40,157–160]. Although each work incorporated high throughput fabrication methods, many did not specifically state key process parameters required to translate this technology, such as print speed, roll pressure, ink viscosity, and in some cases, curing [12]. Instead, these works focused primarily on device efficacy, likely because there are significant unanswered questions in biosensor printing relating to material choice and functionalization. However, Cagnani et al. were able to achieve an exceptionally high 30 m/min printing speed using a slot die coating on PET for dopamine detection [151]. Bariya et al. developed a high throughput gravure printing method for wearable sweat sensor fabrication capable of 6 m/min printing [40]. The majority of additional works focused primarily on sensor stability, which itself is highly dependent on material functionalization, purity, and receptor choice. For instance, Narakathu et al. used gravure printing to fabricate AuNP electrodes for the detection of a variety of chemicals, such as mercury sulfide (HgS), lead sulfide (PbS), D-proline, and sarcosine, demonstrating high sensitivity down to pico-molar concentrations [158]. In another experiment, Favero et al. demonstrated that graphene and MWCNT functionalized electrodes can be improved with the ingrafting of AuNPs to increase conductivity, noticing a >10% increase in electroactive area. A corresponding increase in R correlations indicates linearity after the addition of AuNPs [157]. Although biosensor printing is an active area of research with significant hurdles to overcome, the study of high throughput fabrication methods for the sensors that have been well tested, such as glucose and sweat monitors, is highly needed to scale these methods into clinical practice.

4.3. Additional Applications

There is a great diversity of potential bioelectronics applications, and this review will focus on those that have gained attention for high throughput fabrication. However, many other applications, such as implantable cerebrovascular and arterial stents, brain–machine interfaces, and fully printed wearable devices, are of tremendous interest [10,150]. Other systems have been successfully fabricated with high throughput methods, as summarized in Table 4 [37,129,162–166]. One area of critical interest is wearable electrophysiology monitoring. Traditional electrocardiogram (ECG), electromyogram (EMG), electrooculogram (EOG), and electroencephalogram (EEG) electrodes are based on a hydrogel that can cause irritations in long term use, especially in neonates and those with sensitive skin, and they are highly prone to motion artifacts [8,17]. In contrast, printed dry electrodes can conform to the patient's skin and interface without any damaging gels, making them excellently suited to continuous monitoring, even during patient motion [9,10]. In two reported works, Tan et al. and Chalihawi et al. used carbon black and AgNP and MWCNTs, respectively, to fabricate dry electrodes. Tan et al. used doctor blade coating on a TPU substrate to produce high-performing electrodes for textile integration that can endure over 50 washing cycles [165]. In addition, Chalihawi et al. screen printed AgNP interconnections and an electrode pad, then used doctor blade coating to deposit a functional MWCNT sensor, which was shown to achieve similar ECG signals when compared to a gel-based Ag/AgCl sensor [166]. Images of the fabricated electrodes with (i) Ag layer and (ii) MWCNT layer are provided in Figure 10d, and the ECG performance is shown in (iii) [166]. Although it was not assessed, it would be of great interest to determine these electrode's performance during patient motion. Another interesting area of research is the development of capacitive touch sensors, which have been widely reported in the literature using traditional MEMS fabrication. Lee et al. created such a touch sensor with an air gap instead of PDMS dielectric, and noted that the increased dielectric constant of air allowed for highly improved sensitivity ($\Delta C/C_0$ (%) of 0.118%) and high linear sensing range from 0–20 KPa [37]. One area of high interest is in printing on TPU substrates, and this was the focus of a recent investigation by Jansson et al. using screen-printed AgNP inks. In this experiment, various dimensions were cut in a roll-to-roll laser process and filled with AgNP inks, and ink was filled from both the cutting side and the opposite side [162]. It was determined that the via diameter had a minor impact on conductivity and reliability, but the match between via diameter and screen opening, optimization of printing thickness and side from which the via is printing were of high importance [162]. In addition, Alsuradi et al. demonstrated a very high control of capacitive and inductive behavior in screen-printed traces based on geometries adapted from integrated microwave circuits, then optimized for thicker depositions common in screen printing [129]. As a result, inductances and capacitances could be reliably controlled to within 5% error, which is considered acceptable for many commercial passive components [129]. Finally, polymer materials like poly(3,4-ethylenedioxythiophene) polystyrene sulfonate (PEDOT:PSS) are outside the scope of this review, but it is worth noting that polymer inks can be printed with high throughput methods, such as screen printing, to manufacture bioelectronics systems. For instance, Khan et al. demonstrated a fully printed PEDOT:PSS photoplethysmography array based on screen printing for use in patients recovering from skin graft surgery. The device is shown along with sensitivities to oxygenated and deoxygenated hemoglobin in Figure 10e [163]. In summary, many additional bioelectronics applications could be scaled with high throughput fabrication methods. It is an open challenge to the reader to apply these techniques to their area of expertise.

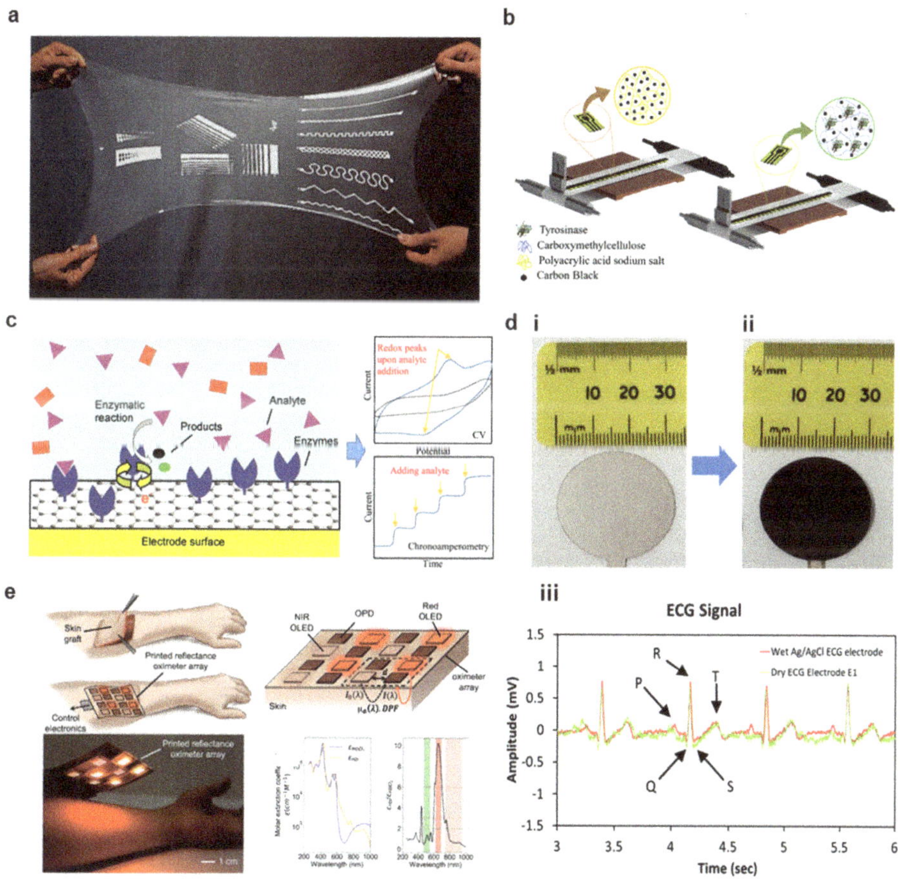

Figure 10. Bioelectronics applications (**a**) Image of screen-printed conductive traces. (reprinted with permission from *Ind. Eng. Chem. Res.* (2019), 58, 43, 19909–19916, Copyright 2020, ACS). (**b**) Illustration of roll-to-roll slot-die coated electrochemical sensors. (reprinted with permission from *Biosens. Bioelectron.* (2020), 165, 112428. Copyright 2020 Elsevier). (**c**) Schematic of a graphene-based enzymatic biosensor. (reprinted with permission from *Biosens. Bioelectron.* (2017), 87, 7–17. Copyright 2017 Elsevier). (**d**) Screen-printed ECG electrodes with (**i**) Ag and (**ii**) MWCNT layers. (**iii**) Example ECG signals are shown compared to commercial Ag/AgCl gel electrodes. (reprinted with permission under Creative Commons license CC BY-NC-ND 4.0 from *Sens Biosensing Res* (2018), 20, 9–15.) (**e**) Overview and operation of a screen-printed reflectance oximeter array (ROA). (**Top left**) placement of the ROA after skin graft surgery. (**Top right**) illustration of ROA pixel array. (**Bottom left**) Image of the printed ROA. (**Bottom right**) Molar extinction coefficients for oxygenated and deoxygenated hemoglobin as a function of wavelength. (reprinted with permission from *PNAS*, (2018) 115 (47) E11015–E11024, Copyright 2018, PNAS).

Table 3. High-throughput nanomaterial-based biosensor fabrication.

Reference	Material	Method	Substrate	Curing	Printing Speed [m/min]	Application	Reported Efficacy
[161]	Carbon	Slot Die	PET	60 °C for 2 min	30	Dopamine detection	Sensitivity of 0.32 μA L/μmol with limit of detection (LOD) of 0.09 μmol/L
[160]	AgNP	Ink-jet	PET	No Post-print treatment	-	Antibiotic detection in milk	100–10,000 μg/mL with LOD of 10 μg/mL
[40]	AgNP	Gravure	Paper	120 °C for 2 min	6	Sweat sensing	Error of 1.4% over a range of ~4–100 [Na^+] (mM)
[159]	AgNP	Gravure	PET	Not reported	Not reported	IgG sensing	2–5% sensitivity to IgG over 10 pM-10 μM concentrations
[158]	AgNP	Gravure	PET	Not reported	Not reported	Sacrosine sensing	Resistance changed from 299 Ω to 325 Ω with varying concentration from 1 pM to 100 mM
[16]	Graphene	Screen printing	Textile	Not reported	Not reported	Influenza sensing	Stabile sensing over 10 ng/mL to 10 μg/mL with a limit of detection of 10 ng/mL.
[157]	AuNPs and MWCNTs	Screen printing	Glass	Not reported	Not reported	Laccase sensing	Linear range of 1–100 μM with a LOD = 0.5 μM and a sensitivity of 0.051 $\mu A \cdot \mu M^{-1}$

Table 4. Additional bioelectronics applications fabricated with high throughput methods.

Reference	Material	Method	Substrate	Curing	Printing Speed [m/min]	Application	Reported Efficacy
[165]	Carbon Black	Doctor Blade Coating	TPU	Belt fed convection oven at 80 °C	Not reported	ECG monitoring	Quality signals after 50 wash cycles.
[166]	AgNP and MWCNT	Screen print AgNP then Doctor blade MWCNTs	PET	120 °C for 20 min	Not reported	ECG monitoring	Equivalent ECG signals to gel Ag/AgCl comparison
[37]	AgNP	Slot Die	PET	150 °C for 10 min.	0.5	Cantilever touch sensor	$\Delta C/C_0$ (%) of 0.118% for the range 0–20 kPa
[162]	AgNP	Screen printing	TPU	140 °C for 2 min.	2	Via Filling	Low (10–40Ω resistance over the range 0–100 stretching cycles with <10% strain
[164]	AgNP	Screen printing	TPU	Not reported	Not reported	Oxygen sensing	0.1 ppm of O_2 sensitivity with 40 PPM LOD
[129]	AgNP	Screen printing	hydrocolloid dressings	Not reported	Not reported	Passive elements (e.g., capacitors)	Control of impedance and capacitance within 5% error
[163]	PEDOT:PSS	Blade Coating	PEN	120 °C for 10 min.	0.6	Photoplethysmography (PPG) array for SpO_2 monitoring	Mean error of 1.1% compared to a commercial device

5. Conclusions and Future Outlook

Recently, various high-throughput nanomaterial fabrication methods have been demonstrated for hybrid bioelectronics. But, there remain substantial challenges to be overcome. Recent attempts to optimize printing parameters for gravure, flexography, screen, inkjet, and slot die printing have opened new possibilities for highly scalable soft electronics and critically needed hybrid biosensors. In addition, novel approaches to high throughput printing, such as flexography aided by CNT stamps, set the leading edge in print resolution, homogeneity, and quality [60]. Significant progress within the last decade on ink rheology optimization and material-interface studies allows for the high-resolution patterning of many functional nanomaterials. And these materials are being extensively studied for a diverse set of bioelectronics applications. This field, however, remains in its infancy, and there are several critical challenges to be overcome. First, electronic circuits require more than interconnects, and the further study of printed vias and material adhesion in multiple layer prints is of high importance. Second, printing resolutions remain low for many methods; thus, new approaches to increasing resolution must be investigated. Third, inkjet printing offers substantial advantages in prototyping and manufacturing costs. Still, recent works have not implemented commercially tested roll-to-roll inkjet printing to the degree necessary to make inkjet printing a desirable method for high throughput nanomaterial fabrication. Finally, such as gravure roll geometries and materials, printing parameters should be reimagined for nanomaterial applications instead of simply relying on processes optimized for inks without large-volume loadings or dispersion challenges. Overall, recent progress in the field of nanomaterial printing offers great hope that a new class of hybrid flexible bioelectronics can provide the affordable, long term usable devices necessary to help the millions of people suffering from undiagnosed diseases. However, new investigations into novel printing approaches and further optimization of current methods are greatly needed before this vision is made a reality.

Author Contributions: N.Z. and W.-H.Y. conceived and designed the materials in this paper. All authors conducted reviews of materials and printing technologies, designed figures, and reviewed all sections. All authors wrote the paper together. All authors have read and agreed to the published version of the manuscript.

Funding: We acknowledge the support from the IEN Center for Human-Centric Interfaces and Engineering at Georgia Tech.

Institutional Review Board Statement: Not applicable.

Informed Consent Statement: Not applicable.

Data Availability Statement: No new data were created or analyzed in this study.

Conflicts of Interest: The authors declare no conflict of interest.

References

1. Niu, Y.; Liu, H.; He, R.; Li, Z.; Ren, H.; Gao, B.; Guo, H.; Genin, G.M.; Xu, F. The New Generation of Soft and Wearable Electronics for Health Monitoring in Varying Environment: From Normal to Extreme Conditions. *Mater. Today* **2020**, *41*, 219–242. [CrossRef]
2. Song, Y.; Min, J.; Gao, W. Wearable and Implantable Electronics: Moving toward Precision Therapy. *ACS Nano* **2019**, *13*, 12280–12286. [CrossRef]
3. Zhang, S.; Li, S.; Xia, Z.; Cai, K. A Review of Electronic Skin: Soft Electronics and Sensors for Human Health. *J. Mater. Chem. B* **2020**, *8*, 852–862. [CrossRef]
4. Sana, F.; Isselbacher, E.M.; Singh, J.P.; Heist, E.K.; Pathik, B.; Armoundas, A.A. Wearable Devices for Ambulatory Cardiac Monitoring: JACC State-of-the-Art Review. *J. Am. Coll. Cardiol.* **2020**, *75*, 1582–1592. [CrossRef]
5. Mahajan, A.; Pottie, G.; Kaiser, W. Transformation in Healthcare by Wearable Devices for Diagnostics and Guidance of Treatment. *ACM Trans. Comput. Healthc.* **2020**, *1*, 1–12. [CrossRef]
6. Chung, H.U.; Rwei, A.Y.; Hourlier-Fargette, A.; Xu, S.; Lee, K.; Dunne, E.C.; Xie, Z.; Liu, C.; Carlini, A.; Kim, D.H.; et al. Skin-Interfaced Biosensors for Advanced Wireless Physiological Monitoring in Neonatal and Pediatric Intensive-Care Units. *Nat. Med.* **2020**, *26*, 418–429. [CrossRef]

7. Lee, K.; Ni, X.; Lee, J.Y.; Arafa, H.; Pe, D.J.; Xu, S.; Avila, R.; Irie, M.; Lee, J.H.; Easterlin, R.L.; et al. Mechano-Acoustic Sensing of Physiological Processes and Body Motions via a Soft Wireless Device Placed at the Suprasternal Notch. *Nat. Biomed. Eng.* **2020**, *4*, 148–158. [CrossRef] [PubMed]
8. Kim, H.; Kim, Y.S.; Mahmood, M.; Kwon, S.; Zavanelli, N.; Kim, H.S.; Rim, Y.S.; Epps, F.; Yeo, W.H. Fully Integrated, Stretchable, Wireless Skin-Conformal Bioelectronics for Continuous Stress Monitoring in Daily Life. *Adv. Sci.* **2020**, *7*, 2000810. [CrossRef] [PubMed]
9. Gao, W.; Emaminejad, S.; Nyein, H.Y.Y.; Challa, S.; Chen, K.; Peck, A.; Fahad, H.M.; Ota, H.; Shiraki, H.; Kiriya, D.; et al. Fully Integrated Wearable Sensor Arrays for Multiplexed In Situ Perspiration Analysis. *Nature* **2016**, *529*, 509–514. [CrossRef]
10. Rogers, J.A.; Someya, T.; Huang, Y. Materials and Mechanics for Stretchable Electronics. *Science* **2010**, *327*, 1603–1607. [CrossRef]
11. Liang, J.; Tong, K.; Pei, Q. A Water-Based Silver-Nanowire Screen-Print Ink for the Fabrication of Stretchable Conductors and Wearable Thin-Film Transistors. *Adv. Mater.* **2016**, *28*, 5986–5996. [CrossRef] [PubMed]
12. Wisitsoraat, A.; Mensing, J.P.; Karuwan, C.; Sriprachuabwong, C.; Jaruwongrungsee, K.; Phokharatkul, D.; Daniels, T.M.; Liewhiran, C.; Tuantranont, A. Printed Organo-Functionalized Graphene for Biosensing Applications. *Biosens. Bioelectron.* **2017**, *87*, 7–17. [CrossRef] [PubMed]
13. Pena-Bahamonde, J.; Nguyen, H.N.; Fanourakis, S.K.; Rodrigues, D.F. Recent Advances in Graphene-Based Biosensor Technology with Applications in Life Sciences. *J. Nanobiotechnol.* **2018**, *16*, 75. [CrossRef]
14. Beitollahi, H.; Mohammadi, S.Z.; Safaei, M.; Tajik, S. Applications of Electrochemical Sensors and Biosensors Based on Modified Screen-Printed Electrodes: A Review. *Anal. Methods* **2020**, *12*, 1547–1560. [CrossRef]
15. Morales-Narvaez, E.; Baptista-Pires, L.; Zamora-Galvez, A.; Merkoci, A. Graphene-Based Biosensors: Going Simple. *Adv. Mater.* **2017**, *29*, 1604905. [CrossRef]
16. Kinnamon, D.S.; Krishnan, S.; Brosler, S.; Sun, E.; Prasad, S. Screen Printed Graphene Oxide Textile Biosensor for Applications in Inexpensive and Wearable Point-of-Exposure Detection of Influenza for At-Risk Populations. *J. Electrochem. Soc.* **2018**, *165*, B3084–B3090. [CrossRef]
17. Huang, Q.; Zhu, Y. Printing Conductive Nanomaterials for Flexible and Stretchable Electronics: A Review of Materials, Processes, and Applications. *Adv. Mater. Technol.* **2019**, *4*, 1800546. [CrossRef]
18. Khan, Y.; Thielens, A.; Muin, S.; Ting, J.; Baumbauer, C.; Arias, A.C. A New Frontier of Printed Electronics: Flexible Hybrid Electronics. *Adv. Mater.* **2020**, *32*, e1905279. [CrossRef] [PubMed]
19. Zavanelli, N.; Yeo, W.-H. Advances in Screen Printing of Conductive Nanomaterials for Stretchable Electronics. *ACS Omega* **2021**, *6*, 9344–9351. [CrossRef] [PubMed]
20. Secor, E.B. Principles of Aerosol Jet Printing. *Flex. Print. Electron.* **2018**, *3*, 035002. [CrossRef]
21. Abbel, R.; Pit, T.; Rubingh, E.; van Lammeren, T.; Cauchois, R.; Everaars, M.; Valeton, J.; van de Geijn, S.; Groen, P. Industrial-scale Inkjet Printed Electronics Manufacturing—Production Up-Scaling from Concept Tools to A Roll-To-Roll Pilot Line. *Transl. Mater. Res.* **2014**, *1*, 015002. [CrossRef]
22. Ren, P.; Liu, Y.; Song, R.; O'Connor, B.; Dong, J.; Zhu, Y. Achieving High-Resolution Electrohydrodynamic Printing of Nanowires on Elastomeric Substrates through Surface Modification. *ACS Appl. Electron. Mater.* **2020**, *3*, 192–202. [CrossRef]
23. Li, W.; Yang, S.; Shamim, A. Screen Printing of Silver Nanowires: Balancing Conductivity with Transparency while Maintaining Flexibility and Stretchability. *NPJ Flex. Electron.* **2019**, *3*, 1–8. [CrossRef]
24. Riemer, D. The Theoretical Fundamentals of the ScreenPrinting Process. *Microelectron. Int.* **1989**, *6*, 8–17. [CrossRef]
25. Singh, M.; Haverinen, H.M.; Dhagat, P.; Jabbour, G.E. Inkjet Printing-Process and Its Applications. *Adv. Mater.* **2010**, *22*, 673–685. [CrossRef]
26. Nayak, L.; Mohanty, S.; Nayak, S.K.; Ramadoss, A. A Review on Inkjet Printing of Nanoparticle Inks for Flexible Electronics. *J. Mater. Chem. C* **2019**, *7*, 8771–8795. [CrossRef]
27. Grau, G.; Cen, J.; Kang, H.; Kitsomboonloha, R.; Scheideler, W.J.; Subramanian, V. Gravure-Printed Electronics: Recent Progress in Tooling Development, Understanding of Printing Physics, and Realization of Printed Devices. *Flex. Print. Electron.* **2016**, *1*, 023002. [CrossRef]
28. Krainer, S.; Smit, C.; Hirn, U. The Effect of Viscosity and Surface Tension on Inkjet Printed Picoliter Dots. *RSC Adv.* **2019**, *9*, 31708–31719. [CrossRef]
29. Izdebska, J. Flexographic Printing. In *Printing on Polymers Fundamentals and Applications*; William Andrew: Norwich, NY, USA, 2016; Chapter 11, pp. 179–197.
30. Seong, J.; Park, J.; Lee, J.; Ahn, B.; Yeom, J.H.; Kim, J.; Hassinen, T.; Rhee, S.; Ko, S.; Lee, D.; et al. Practical Design Guidelines for the Development of High-Precision Roll-to-Roll Slot-Die Coating Equipment and the Process. *IEEE Trans. Compon. Packag. Manuf. Technol.* **2016**, *6*, 1677–1686. [CrossRef]
31. Wu, W. Inorganic Nanomaterials for Printed Electronics: A review. *Nanoscale* **2017**, *9*, 7342–7372. [CrossRef]
32. Chung, S.; Cho, K.; Lee, T. Recent Progress in Inkjet-Printed Thin-Film Transistors. *Adv. Sci.* **2019**, *6*, 1801445. [CrossRef] [PubMed]
33. Al-Milaji, K.N.; Secondo, R.R.; Ng, T.N.; Kinsey, N.; Zhao, H. Interfacial Self-Assembly of Colloidal Nanoparticles in Dual-Droplet Inkjet Printing. *Adv. Mater. Interfaces* **2018**, *5*, 1701561. [CrossRef]
34. Howard, I.A.; Abzieher, T.; Hossain, I.M.; Eggers, H.; Schackmar, F.; Ternes, S.; Richards, B.S.; Lemmer, U.; Paetzold, U.W. Coated and Printed Perovskites for Photovoltaic Applications. *Adv. Mater.* **2019**, *31*, e1806702. [CrossRef]

35. Huttunen, O.-H.; Happonen, T.; Hiitola-Keinänen, J.; Korhonen, P.; Ollila, J.; Hiltunen, J. Roll-To-Roll Screen-Printed Silver Conductors on a Polydimethyl Siloxane Substrate for Stretchable Electronics. *Ind. Eng. Chem. Res.* **2019**, *58*, 19909–19916. [CrossRef]
36. Kharissova, O.V.; Kharisov, B.I.; de Casas Ortiz, E.G. Dispersion of Carbon Nanotubes in Water and Non-Aqueous Solvents. *RSC Adv.* **2013**, *3*, 24812–24852. [CrossRef]
37. Lee, S.H.; Lee, S. Fabrication and Characterization of Roll-to-Roll-Coated Cantilever-Structured Touch Sensors. *ACS Appl. Mater. Interfaces* **2020**, *12*, 46797–46803. [CrossRef]
38. Pradhan, S.; Hedberg, J.; Blomberg, E.; Wold, S.; Odnevall Wallinder, I. Effect of Sonication on Particle Dispersion, Administered Dose and Metal Release of Non-Functionalized, Non-Inert Metal Nanoparticles. *J. Nanopart. Res.* **2016**, *18*, 285. [CrossRef]
39. Rickey, K.M.; Nian, Q.; Zhang, G.; Chen, L.; Suslov, S.; Bhat, S.V.; Wu, Y.; Cheng, G.J.; Ruan, X. Welding of Semiconductor Nanowires by Coupling Laser-Induced Peening and Localized Heating. *Sci. Rep.* **2015**, *5*, 16052. [CrossRef]
40. Bariya, M.; Shahpar, Z.; Park, H.; Sun, J.; Jung, Y.; Gao, W.; Nyein, H.Y.Y.; Liaw, T.S.; Tai, L.C.; Ngo, Q.P.; et al. Roll-to-Roll Gravure Printed Electrochemical Sensors for Wearable and Medical Devices. *ACS Nano* **2018**, *12*, 6978–6987. [CrossRef] [PubMed]
41. Zhong, Z.W.; Ee, J.H.; Chen, S.H.; Shan, X.C. Parametric Investigation of Flexographic Printing Processes for R2R Printed Electronics. *Mater. Manuf. Process.* **2020**, *35*, 564–571. [CrossRef]
42. Rowley-Neale, S.J.; Smith, G.C.; Banks, C.E. Mass-Producible 2D-MoS2-Impregnated Screen-Printed Electrodes that Demonstrate Efficient Electrocatalysis toward the Oxygen Reduction Reaction. *ACS Appl. Mater. Interfaces* **2017**, *9*, 22539–22548. [CrossRef]
43. Potts, S.J.; Phillips, C.; Jewell, E.; Clifford, B.; Lau, Y.C.; Claypole, T. High-Speed Imaging the Effect of Snap-Off Distance and Squeegee Speed on the Ink Transfer Mechanism of Screen-Printed Carbon Pastes. *J. Coat. Technol. Res.* **2020**, *17*, 447–459. [CrossRef]
44. Reinhardt, K.; Hofmann, N.; Eberstein, M. The Importance of Shear Thinning, Thixotropic and Viscoelastic Properties of Thick Film Pastes to Predict Effects on Printing Performance. In Proceedings of the 2017 21st European Microelectronics and Packaging Conference (EMPC) & Exhibition, Warsaw, Poland, 10–13 September 2017; IEEE: Piscataway, NJ, USA, 2018; Volume 2018, pp. 1–7.
45. Al-Milaji, K.N.; Huang, Q.; Li, Z.; Ng, T.N.; Zhao, H. Direct Embedment and Alignment of Silver Nanowires by Inkjet Printing for Stretchable Conductors. *ACS Appl. Electron. Mater.* **2020**, *2*, 3289–3298. [CrossRef]
46. Gao, M.; Li, L.; Song, Y. Inkjet Printing Wearable Electronic Devices. *J. Mater. Chem. C* **2017**, *5*, 2971–2993. [CrossRef]
47. Fernandes, I.J.; Aroche, A.F.; Schuck, A.; Lamberty, P.; Peter, C.R.; Hasenkamp, W.; Rocha, T. Silver Nanoparticle Conductive Inks: Synthesis, Characterization, and Fabrication of Inkjet-Printed Flexible Electrodes. *Sci. Rep.* **2020**, *10*, 8878. [CrossRef] [PubMed]
48. Deegan, R.B.; Dupont, O.; Huber, T.; Nagel, G.; Witten, S. Capillary Flow as the Cause of Ring Stains from Dried Liquid Drops. *Nature* **1997**, *389*, 827–829. [CrossRef]
49. Lee, H.L.; Lee, D.; Hwang, J.; Nam, D.; Byeon, C.; Ko, S.H.; Lee, S. Silver Nanoparticle Piezoresistive Sensors Fabricated by Roll-to-Roll Slot-Die Coating and Laser Direct Writing. *Opt. Express* **2014**, *22*, 8919–8927. [CrossRef] [PubMed]
50. Huang, Q.; Zhu, Y. Gravure Printing of Water-based Silver Nanowire ink on Plastic Substrate for Flexible Electronics. *Sci. Rep.* **2018**, *8*, 15167. [CrossRef]
51. Nguyen, H.A.D.; Lee, C.; Shin, K.-H.; Lee, D. An Investigation of the Ink-Transfer Mechanism During the Printing Phase of High-Resolution Roll-to-Roll Gravure Printing. *IEEE Trans. Compon. Packag. Manuf. Technol.* **2015**, *5*, 1516–1524. [CrossRef]
52. Lee, J.; Kim, J.; Noh, Y.; Jeong, H.; Lee, D. Advanced Technique for Achieving 10-Mm-Width Fine Lines in Roll-to-Roll Continuous Gravure Printing. *Precis. Eng.* **2021**, *69*, 1–7. [CrossRef]
53. Secor, E.B.; Lim, S.; Zhang, H.; Frisbie, C.D.; Francis, L.F.; Hersam, M.C. Gravure Printing of Graphene for Large-Area Flexible Electronics. *Adv. Mater.* **2014**, *26*, 4533–4538. [CrossRef]
54. Scheideler, W.J.; Jang, J.; Ul Karim, M.A.; Kitsomboonloha, R.; Zeumault, A.; Subramanian, V. Gravure-Printed Sol-Gels on Flexible Glass: A Scalable Route to Additively Patterned Transparent Conductors. *ACS Appl. Mater. Interfaces* **2015**, *7*, 12679–12687. [CrossRef]
55. Kim, K.; Kim, J.; Kim, B.; Ko, S. Fabrication of Microfluidic Structure Based Biosensor Using Roll-to-Roll Gravure Printing. *Int. J. Precis. Eng. Manuf. Green Technol.* **2018**, *5*, 369–374. [CrossRef]
56. Sung, D.; de la Fuente Vornbrock, A.; Subramanian, V. Scaling and Optimization of Gravure-Printed Silver Nanoparticle Lines for Printed Electronics. *IEEE Trans. Compon. Packag. Technol.* **2010**, *33*, 105–114. [CrossRef]
57. Park, H.J.; Cha, J.; Choi, M.; Kim, J.H.; Tay, R.Y.; Teo, E.H.T.; Park, N.; Hong, S.; Lee, Z. One-dimensional hexagonal boron nitride conducting channel. *Sci. Adv.* **2020**, *6*, eaay4958. [CrossRef] [PubMed]
58. Mogg, B.T.; Claypole, T.; Deganello, D.; Phillips, C. Flexographic printing of ultra-thin semiconductor polymer layers. *Transl. Mater. Res.* **2016**, *3*. [CrossRef]
59. Tomašegović, T.; Beynon, D.; Claypole, T.; Poljaček, S.M. Modifying the qualitative properties of print by surface treatment of flexographic printing plate. *J. Print Media Technol. Res.* **2017**, *2*, 57–67.
60. Kim, S.; Sojoudi, H.; Zhao, H.; Mariappan, D.; McKinley, G.H.; Gleason, K.K.; Hart, A.J. Ultrathin High-Resolution Flexographic Printing Using Nanoporous Stamps. *Sci. Adv.* **2016**, *2*, e1601660. [CrossRef] [PubMed]
61. Mariappan, D.D.; Kim, S.; Boutilier, M.S.H.; Zhao, J.; Zhao, H.; Beroz, J.; Muecke, U.; Sojoudi, H.; Gleason, K.; Brun, P.T.; et al. Dynamics of Liquid Transfer from Nanoporous Stamps in High-Resolution Flexographic Printing. *Langmuir* **2019**, *35*, 7659–7671. [CrossRef]
62. Messerschmitt, E. Rheological Considerations for Screen Printing Inks. *Screen Print* **1982**, *72*, 62–65.

63. Kapur, N.; Abbott, S.J.; Dolden, E.D.; Gaskell, P.H. Predicting the Behavior of Screen Printing. *IEEE Trans. Compon. Packag. Manuf. Technol.* **2013**, *3*, 508–515. [CrossRef]
64. Xu, C.; Willenbacher, N. How Rheological Properties Affect Fine-Line Screen Printing of Pastes: A Combined Rheological and High-Speed Video Imaging Study. *J. Coat. Technol.* **2018**, *15*, 1401–1412. [CrossRef]
65. Sliz, R.; Huttunen, O.-H.; Jansson, E.; Kemppainen, J.; Schroderus, J.; Kurkinen, M.; Fabritius, T. Reliability of R2R-Printed, Flexible Electrodes for E-Clothing Applications. *NPJ Flex. Electron.* **2020**, *4*, 1–9. [CrossRef]
66. Philip, B.; Jewell, E.; Greenwood, P.; Weirman, C. Material and Process Optimization Screen Printing Carbon Graphite Pastes for Mass Production of Heating Elements. *J. Manuf. Process.* **2016**, *22*, 185–191. [CrossRef]
67. He, P.; Cao, J.; Ding, H.; Liu, C.; Neilson, J.; Li, Z.; Kinloch, I.A.; Derby, B. Screen-Printing of a Highly Conductive Graphene Ink for Flexible Printed Electronics. *ACS Appl. Mater. Interfaces* **2019**, *11*, 32225–32234. [CrossRef] [PubMed]
68. Lau, G.K.; Shrestha, M. Ink-Jet Printing of Micro-Electro-Mechanical Systems (MEMS). *Micromachines* **2017**, *8*, 194. [CrossRef]
69. Bartolo, D.; Boudaoud, A.; Narcy, G.; Bonn, D. Dynamics of non-Newtonian droplets. *Phys. Rev. Lett.* **2007**, *99*, 174502. [CrossRef]
70. Wijshoff, H. Drop dynamics in the inkjet printing process. *Curr. Opin. Colloid Interface Sci.* **2018**, *36*, 20–27. [CrossRef]
71. Cai, L.; Zhang, S.; Zhang, Y.; Li, J.; Miao, J.; Wang, Q.; Yu, Z.; Wang, C. Direct Printing for Additive Patterning of Silver Nanowires for Stretchable Sensor and Display Applications. *Adv. Mater. Technol.* **2018**, *3*, 1700232. [CrossRef]
72. Anyfantakis, M.; Geng, Z.; Morel, M.; Rudiuk, S.; Baigl, D. Modulation of the coffee-ring effect in particle/surfactant mixtures: The importance of particle-interface interactions. *Langmuir* **2015**, *31*, 4113–4120. [CrossRef]
73. Zapka, W. *Handbook of Industrial Inkjet Printing: A Full System Approach*; John Wiley & Sons: Hoboken, NJ, USA, 2017; pp. 1–22.
74. Hu, H.L.; Larson, R.G. Marangoni Effect Reverses Coffee-Ring Depositions. *J. Phys. Chem. B* **2006**, *110*, 7090–7094. [CrossRef]
75. de Gans, B.J.; Schubert, U.S. Inkjet Printing of Well-Defined Polymer Dots and Arrays. *Langmuir* **2004**, *20*, 7789–7793. [CrossRef]
76. Patil, N.D.; Bange, P.G.; Bhardwaj, R.; Sharma, A. Effects of Substrate Heating and Wettability on Evaporation Dynamics and Deposition Patterns for a Sessile Water Droplet Containing Colloidal Particles. *Langmuir* **2016**, *32*, 11958–11972. [CrossRef] [PubMed]
77. McBride, S.A.; Dash, S.; Varanasi, K.K. Evaporative Crystallization in Drops on Superhydrophobic and Liquid-Impregnated Surfaces. *Langmuir* **2018**, *34*, 12350–12358. [CrossRef]
78. Eral, H.B.; Augustine, D.M.; Duits, M.H.; Mugele, F. Suppressing the coffee stain effect: How to control colloidal self-assembly in evaporating drops using electrowetting. *Soft Matter.* **2011**, *7*, 7090–7094. [CrossRef]
79. Minemawari, H.; Yamada, T.; Matsui, H.; Tsutsumi, J.; Haas, S.; Chiba, R.; Kumai, R.; Hasegawa, T. Inkjet Printing of Single-Crystal Films. *Nature* **2011**, *475*, 364–367. [CrossRef] [PubMed]
80. Vak, D.; Hwang, K.; Faulks, A.; Jung, Y.-S.; Clark, N.; Kim, D.-Y.; Wilson, G.J.; Watkins, S.E. 3D Printer Based Slot-Die Coater as a Lab-to-Fab Translation Tool for Solution-Processed Solar Cells. *Adv. Energy Mater.* **2015**, *5*, 1401539. [CrossRef]
81. Whitener, K.E.; Sheehan, P.E. Graphene Synthesis. *Diam. Relat. Mater.* **2014**, *46*, 25–34. [CrossRef]
82. Zaytseva, O.; Neumann, G. Carbon Nanomaterials: Production, Impact on Plant Development, Agricultural and Environmental Applications. *Chem. Biol. Technol. Agric.* **2016**, *3*, 1–26. [CrossRef]
83. Chen, K.; Gao, W.; Emaminejad, S.; Kiriya, D.; Ota, H.; Nyein, H.Y.; Takei, K.; Javey, A. Printed Carbon Nanotube Electronics and Sensor Systems. *Adv. Mater.* **2016**, *28*, 4397–4414. [CrossRef]
84. Manoj, D.; Saravanan, R.; Santhanalakshmi, J.; Agarwal, S.; Gupta, V.K.; Boukherroub, R. Towards Green Synthesis of Monodisperse Cu Nanoparticles: An Efficient and High Sensitive Electrochemical Nitrite Sensor. *Sens. Actuators B Chem.* **2018**, *266*, 873–882. [CrossRef]
85. Iravani, S.; Korbekandi, H.; Mirmohammadi, S.V.; Zolfaghari, B. Synthesis of silver nanoparticles: Chemical, physical and biological methods. *Res. Pharm. Sci.* **2014**, *9*, 385–406.
86. Li, J.; Ye, F.; Vaziri, S.; Muhammed, M.; Lemme, M.C.; Ostling, M. Efficient inkjet printing of graphene. *Adv. Mater.* **2013**, *25*, 3985–3992. [CrossRef] [PubMed]
87. Shi, Y.; Ren, L.; Li, D.; Gao, H.; Yang, B. Optimization Conditions for Single-Walled Carbon Nanotubes Dispersion. *J. Surf. Eng. Mater. Adv. Technol.* **2013**, *3*, 6–12. [CrossRef]
88. Ohsawa, M.; Hashimoto, N. Bending reliability of flexible transparent electrode of gravure offset printed invisible silver-grid laminated with conductive polymer. *Microelectron. Reliab.* **2019**, *98*, 124–130. [CrossRef]
89. Yeom, C.; Chen, K.; Kiriya, D.; Yu, Z.; Cho, G.; Javey, A. Large-area compliant tactile sensors using printed carbon nanotube active-matrix backplanes. *Adv. Mater.* **2015**, *27*, 1561–1566. [CrossRef] [PubMed]
90. Kholghi Eshkalak, S.; Chinnappan, A.; Jayathilaka, W.A.D.M.; Khatibzadeh, M.; Kowsari, E.; Ramakrishna, S. A Review on Inkjet Printing of CNT Composites for Smart Applications. *Appl. Mater. Today* **2017**, *9*, 372–386. [CrossRef]
91. Gupta, N.; Gupta, S.M.; Sharma, S.K. Carbon Nanotubes: Synthesis, Properties and Engineering Applications. *Carbon Lett.* **2019**, *29*, 419–447. [CrossRef]
92. Tortorich, R.; Choi, J.-W. Inkjet Printing of Carbon Nanotubes. *Nanomaterials* **2013**, *3*, 453–468. [CrossRef] [PubMed]
93. Coskun, S.; Aksoy, B.; Unalan, H.E. Polyol Synthesis of Silver Nanowires: An Extensive Parametric Study. *Cryst. Growth Des.* **2011**, *11*, 4963–4969. [CrossRef]
94. Cruz, B.; Albrecht, A.; Eschlwech, P.; Biebl, E. Inkjet Printing of Metal Nanoparticles for Green UHF RFID Tags. *Adv. Radio Sci.* **2019**, *17*, 119–127. [CrossRef]

95. Jung, J.H.; Cheol Oh, H.; Soo Noh, H.; Ji, J.H.; Soo Kim, S. Metal Nanoparticle Generation Using a Small Ceramic Heater With a Local Heating Area. *J. Aerosol Sci.* **2006**, *37*, 1662–1670. [CrossRef]
96. Shih, C.Y.; Shugaev, M.V.; Wu, C.; Zhigilei, L. The effect of pulse duration on nanoparticle generation in pulsed laser ablation in liquids: Insights from large-scale atomistic simulations. *Phys. Chem. Chem. Phys* **2020**, *22*, 7077–7099. [CrossRef]
97. Richard, B.; Lemyre, J.L.; Ritcey, A.M. Nanoparticle Size Control in Microemulsion Synthesis. *Langmuir* **2017**, *33*, 4748–4757. [CrossRef]
98. du Toit, H.; Macdonald, T.J.; Huang, H.; Parkin, I.P.; Gavriilidis, A. Continuous Flow Synthesis of Citrate Capped Gold Nanoparticles Using UV Induced Nucleation. *RSC Adv.* **2017**, *7*, 9632–9638. [CrossRef]
99. Fuentes-Garcia, J.A.; Santoyo-Salzar, J.; Rangel-Cortes, E.; Goya, G.F.; Cardozo-Mata, V.; Pescador-Rojas, J.A. Effect of Ultrasonic Irradiation Power on Sonochemical Synthesis of Gold Nanoparticles. *Ultrason. Sonochem.* **2021**, *70*, 105274. [CrossRef] [PubMed]
100. Phuruangrat, A.; Siri, S.; Wadbua, P.; Thongtem, S.; Thongtem, T. Microwave-Assisted Synthesis, Photocatalysis and Antibacterial Activity of Ag Nanoparticles Supported on Zno Flowers. *J. Phys. Chem. Solids* **2019**, *126*, 170–177. [CrossRef]
101. Khabarov, K.; Kornyushin, D.; Masnaviev, B.; Tuzhilin, D.; Saprykin, D.; Efimov, A.; Ivanov, V. The Influence of Laser Sintering Modes on the Conductivity and Microstructure of Silver Nanoparticle Arrays Formed by Dry Aerosol Printing. *Appl. Sci.* **2019**, *10*, 246. [CrossRef]
102. Wang, Z.; Liang, X.; Zhao, T.; Hu, Y.; Zhu, P.; Sun, R. Facile Synthesis of Monodisperse Silver Nanoparticles for Screen Printing Conductive Inks. *J. Mater. Sci. Mater. Electron.* **2017**, *28*, 16939–16947. [CrossRef]
103. Shiokawa, D.; Izumi, K.; Sugano, R.; Sekine, T.; Minami, T.; Kumaki, D.; Tokito, S. Development of a Silver Nanoparticle Ink for Fine Line Patterning Using Gravure Offset Printing. *Jpn. J. Appl. Phys.* **2017**, *56*, 05EA04. [CrossRef]
104. Benson, J.; Fung, C.M.; Lloyd, J.S.; Deganello, D.; Smith, N.A.; Teng, K.S. Direct Patterning of Gold Nanoparticles Using Flexographic Printing for Biosensing Applications. *Nanoscale Res. Lett.* **2015**, *10*, 127. [CrossRef] [PubMed]
105. Perelaer, J.; Schubert, U.S. Novel Approaches for Low Temperature Sintering of Inkjet-Printed Inorganic Nanoparticles for Roll-to-Roll (R2R) Applications. *J. Mater. Res.* **2013**, *28*, 564–573. [CrossRef]
106. Roberson, D.A.; Wicker, R.B.; Murr, L.E.; Church, K.; MacDonald, E. Microstructural and Process Characterization of Conductive Traces Printed from Ag Particulate Inks. *Materials* **2011**, *4*, 963–979. [CrossRef] [PubMed]
107. Hösel, M.; Krebs, F.C. Large-Scale Roll-to-Roll Photonic Sintering of Flexo Printed Silver Nanoparticle Electrodes. *J. Mater. Chem.* **2012**, *22*, 15683–15688. [CrossRef]
108. Allen, M.L.; Aronniemi, M.; Mattila, T.; Alastalo, A.; Ojanpera, K.; Suhonen, M.; Seppa, H. Electrical Sintering of Nanoparticle Structures. *Nanotechnology* **2008**, *19*, 175201. [CrossRef]
109. Allen, M.; Alastalo, A.; Suhonen, M.; Mattila, T.; Leppaniemi, J.; Seppa, H. Contactless Electrical Sintering of Silver Nanoparticles on Flexible Substrates. *IEEE Trans. Microw. Theory Tech.* **2011**, *59*, 1419–1429. [CrossRef]
110. Fujii, S.; Kawamura, S.; Mochizuki, D.; Maitani, M.M.; Suzuki, E.; Wada, Y. Microwave Sintering of Ag-Nanoparticle Thin Films on a Polyimide Substrate. *AIP Adv.* **2015**, *5*, 127226. [CrossRef]
111. Chen, S.; Guan, Y.; Li, Y.; Yan, X.; Ni, H.; Li, L. A Water-Based Silver Nanowire Ink for Large-Scale Flexible Transparent Conductive Films and Touch Screens. *J. Mater. Chem. C* **2017**, *5*, 2404–2414. [CrossRef]
112. Tilke, A.T.; Simmel, F.C.; Lorenz, H.; Blick, R.H.; Kotthaus, J.P. Quantum interference in a one-dimensional silicon nanowire. *Phys. Rev. B* **2003**, *68*, 075311. [CrossRef]
113. Zhang, P.; Wyman, I.; Hu, J.; Lin, S.; Zhong, Z.; Tu, Y.; Huang, Z.; Wei, Y. Silver nanowires: Synthesis technologies, growth mechanism and multifunctional applications. *Mater. Sci. Eng. B* **2017**, *223*, 1–23. [CrossRef]
114. Hemmati, S.; Harris, M.T.; Barkey, D.P. Polyol Silver Nanowire Synthesis and the Outlook for a Green Process. *J. Nanomater.* **2020**, *2020*, 1–25. [CrossRef]
115. Sun, Y.; Mayers, B.; Herricks, T.; Xia, X. Polyol Synthesis of Uniform Silver Nanowires: A Plausible Growth Mechanism and the Supporting Evidence. *Nano Lett.* **2003**, *3*, 955–960. [CrossRef]
116. Fung, C.M.; Lloyd, J.S.; Samavat, S.; Deganello, D.; Teng, K.S. Facile Fabrication of Electrochemical Zno Nanowire Glucose Biosensor Using Roll to Roll Printing Technique. *Sens. Actuators B Chem.* **2017**, *247*, 807–813. [CrossRef]
117. Finn, D.J.; Lotya, M.; Coleman, J.N. Inkjet Printing of Silver Nanowire Networks. *ACS Appl. Mater. Interfaces* **2015**, *7*, 9254–9261. [CrossRef]
118. Lee, J.-Y.; Connor, S.T.; Cui, Y.; Peumans, P. Solution-Processed Metal Nanowire Mesh Transparent Electrodes. *Nano Lett.* **2008**, *8*, 689–692. [CrossRef] [PubMed]
119. Tran, N.H.; Duong, T.H.; Kim, H.C. A Fast Fabrication of Copper Nanowire Transparent Conductive Electrodes by Using Pulsed Laser Irradiation. *Sci. Rep.* **2017**, *7*, 15093. [CrossRef]
120. Triambulo, R.E.; Cheong, H.-G.; Park, J.-W. All-Solution-Processed Foldable Transparent Electrodes of Ag Nanowire Mesh and Metal Matrix Films for Flexible Electronics. *Org. Electron.* **2014**, *15*, 2685–2695. [CrossRef]
121. Liu, L.; Shen, Z.; Zhang, X.; Ma, H. Highly Conductive Graphene/Carbon Black Screen Printing Inks for Flexible Electronics. *J. Colloid Interface Sci.* **2021**, *582*, 12–21. [CrossRef] [PubMed]
122. Hassan, S.M.; Yusof, M.S.; Maksud, M.I.; Nodin, M.N.; Rejab, N.A. Feasibility Study of Roll to Roll Printing of Graphene. *Appl. Mech. Mater.* **2015**, *799–800*, 402–406. [CrossRef]
123. Wan, Z.; Nguyen, N.-T.; Gao, Y.; Li, Q. Laser Induced Graphene for Biosensors. *Sustain. Mater. Technol.* **2020**, *25*, e00205. [CrossRef]

124. Rahimi, R.; Ochoa, M.; Yu, W.; Ziaie, B. Highly Stretchable and Sensitive Unidirectional Strain Sensor via Laser Carbonization. *ACS Appl. Mater. Interfaces* **2015**, *7*, 4463–4470. [CrossRef] [PubMed]
125. Ortiz-Gómez, I.; Toral-López, V.; Romero, F.J.; de Orbe-Payá, I.; García, A.; Rodríguez, N.; Capitán-Vallvey, L.F.; Morales, D.P.; Salinas-Castillo, A. In Situ Synthesis of Fluorescent Silicon Nanodots for Determination of Total Carbohydrates in a Paper Microfluidic Device Combined with Laser Prepared Graphene Heater. *Sens. Actuators B Chem.* **2021**, *332*, 129506. [CrossRef]
126. Wan, Z.; Streed, E.W.; Lobino, M.; Wang, S.; Sang, R.T.; Cole, I.S.; Thiel, D.V.; Li, Q. Laser-Reduced Graphene: Synthesis, Properties, and Applications. *Adv. Mater. Technol.* **2018**, *3*, 1700315. [CrossRef]
127. Zahed, M.A.; Das, P.S.; Maharjan, P.; Barman, S.C.; Sharifuzzaman, M.; Yoon, S.H.; Park, J.Y. Flexible and Robust Dry Electrodes Based on Electroconductive Polymer Spray-Coated 3D Porous Graphene for Long-Term Electrocardiogram Signal Monitoring System. *Carbon* **2020**, *165*, 26–36. [CrossRef]
128. Tölle, F.J.; Fabritius, M.; Mülhaupt, R. Emulsifier-Free Graphene Dispersions with High Graphene Content for Printed Electronics and Freestanding Graphene Films. *Adv. Funct. Mater.* **2012**, *22*, 1136–1144. [CrossRef]
129. Alsuradi, H.; Yoo, J. Screen Printed Passives and Interconnects on Bio-Degradable Medical Hydrocolloid Dressing for Wearable Sensors. *Sci. Rep.* **2019**, *9*, 17467. [CrossRef] [PubMed]
130. Desai, J.A.; Bandyopadhyay, A.; Min, M.; Saenz, G.; Kaul, A.B. A Photo-Capacitive Sensor Operational from 6 K to 350 K with A Solution Printable, Thermally-Robust Hexagonal Boron Nitride (H-BN) Dielectric and Conductive Graphene Electrodes. *Appl. Mater. Today* **2020**, *20*, 100660. [CrossRef]
131. Jabari, E.; Toyserkani, E. Laser Heat Treatment of Aerosol-Jet Additive Manufactured Graphene Patterns. *J. Phys. D Appl. Phys.* **2015**, *48*, 375503. [CrossRef]
132. Secor, E.B.; Ahn, B.Y.; Gao, T.Z.; Lewis, J.A.; Hersam, M.C. Rapid and Versatile Photonic Annealing of Graphene Inks for Flexible Printed Electronics. *Adv. Mater.* **2015**, *27*, 6683–6688. [CrossRef]
133. Xiang, L.; Wang, Z.; Liu, Z.; Weigum, S.E.; Yu, Q.; Chen, M.Y. Inkjet Printed Flexible Biosensor Based on Graphene Field Effect Transistor. *IEEE Sens. J.* **2016**, *16*, 8359–8364. [CrossRef]
134. Boumia, L.; Zidour, M.; Benzair, A.; Tounsi, A. A Timoshenko Beam Model for Vibration Analysis of Chiral Single-Walled Carbon Nanotubes. *Phys. E Low. Dimens. Syst. Nanostruct.* **2014**, *59*, 186–191. [CrossRef]
135. Menon, H.; Aiswarya, R.; Surendran, K.P. Screen Printable MWCNT Inks for Printed Electronics. *RSC Adv.* **2017**, *7*, 44076–44081. [CrossRef]
136. Sun, J.; Sapkota, A.; Park, H.; Wesley, P.; Jung, Y.; Maskey, B.B.; Kim, Y.; Majima, Y.; Ding, J.; Ouyang, J.; et al. Fully R2R-Printed Carbon-Nanotube-Based Limitless Length of Flexible Active-Matrix for Electrophoretic Display Application. *Adv. Electron. Mater.* **2020**, *6*, 1901431. [CrossRef]
137. Jung, M.; Kim, J.; Noh, J.; Lim, N.; Lim, C.; Lee, G.; Kim, J.; Kang, H.; Jung, K.; Leonard, A.D.; et al. All-Printed and Roll-to-Roll-PrinTable 13.56-MHz-Operated 1-bit RF Tag on Plastic Foils. *IEEE Trans. Electron. Devices* **2010**, *57*, 571–580. [CrossRef]
138. Okimoto, H.; Takenobu, T.; Yanagi, K.; Miyata, Y.; Shimotani, H.; Kataura, H.; Iwasa, Y. Tunable Carbon Nanotube Thin-Film Transistors Produced Exclusively via Inkjet Printing. *Adv. Mater.* **2010**, *22*, 3981–3986. [CrossRef]
139. Lu, S.; Franklin, A.D. Printed Carbon Nanotube Thin-Film Transistors: Progress on Printable Materials and the Path to Applications. *Nanoscale* **2020**, *12*, 23371–23390. [CrossRef] [PubMed]
140. Yu, X.; Liu, D.; Kang, L.; Yang, Y.; Zhang, X.; Lv, Q.; Qiu, S.; Jin, H.; Song, Q.; Zhang, J.; et al. Recycling Strategy for Fabricating Low-Cost and High-Performance Carbon Nanotube TFT Devices. *ACS Appl. Mater. Interfaces* **2017**, *9*, 15719–15726. [CrossRef]
141. Briggs, N.; Subramanian, S.; Lin, Z.; Li, X.; Zhang, X.; Zhang, K.; Xiao, K.; Geohegan, D.; Wallace, R.; Chen, L.-Q.; et al. A Roadmap for Electronic Grade 2D Materials. *2D Mater.* **2019**, *6*, 022001. [CrossRef]
142. Carey, T.; Cacovich, S.; Divitini, G.; Ren, J.; Mansouri, A.; Kim, J.M.; Wang, C.; Ducati, C.; Sordan, R.; Torrisi, F. Fully Inkjet-Printed Two-Dimensional Material Field-Effect Heterojunctions for Wearable and Textile Electronics. *Nat. Commun.* **2017**, *8*, 1202. [CrossRef]
143. Hossain, R.F.; Deaguero, I.G.; Boland, T.; Kaul, A.B. Biocompatible, Large-Format, Inkjet Printed Heterostructure Mos2-Graphene Photodetectors on Conformable Substrates. *NPJ 2D Mater. Appl.* **2017**, *1*, 1–10. [CrossRef]
144. Splendiani, A.; Sun, L.; Zhang, Y.; Li, T.; Kim, J.; Chim, C.Y.; Galli, G.; Wang, F. Emerging Photoluminescence in Monolayer Mos2. *Nano. Lett.* **2010**, *10*, 1271–1275. [CrossRef]
145. Feng, Y.P.; Shen, L.; Yang, M.; Wang, A.; Zeng, M.; Wu, Q.; Chintalapati, S.; Chang, C.-R. Prospects of Spintronics Based On 2D Materials. *Wiley Interdiscip. Rev. Comput. Mol. Sci.* **2017**, *7*, e1313. [CrossRef]
146. Lee, H.; Koo, M.; Park, C.; Patel, M.; Han, H.; Park, T.H.; Kumar, P.; Koh, W.-G.; Park, C. Zwitterion-Assisted Transition Metal Dichalcogenide Nanosheets for Scalable and Biocompatible Inkjet Printing. *Nano Res.* **2020**, *13*, 2726–2734. [CrossRef]
147. Caldwell, J.D.; Aharonovich, I.; Cassabois, G.; Edgar, J.H.; Gil, B.; Basov, D.N. Photonics with Hexagonal Boron Nitride. *Nat. Rev. Mater.* **2019**, *4*, 552–567. [CrossRef]
148. Angizi, S.; Khalaj, M.; Alem, S.A.A.; Pakdel, A.; Willander, M.; Hatamie, A.; Simchi, A. Review—Towards the Two-Dimensional Hexagonal Boron Nitride (2D h-BN) Electrochemical Sensing Platforms. *J. Electrochem. Soc.* **2020**, *167*, 126513. [CrossRef]
149. Angizi, S.; Hatamie, A.; Ghanbari, H.; Simchi, A. Mechanochemical Green Synthesis of Exfoliated Edge-Functionalized Boron Nitride Quantum Dots: Application to Vitamin C Sensing through Hybridization with Gold Electrodes. *ACS Appl. Mater. Interfaces* **2018**, *10*, 28819–28827. [CrossRef] [PubMed]

150. Kwon, Y.T.; Kim, Y.S.; Kwon, S.; Mahmood, M.; Lim, H.R.; Park, S.W.; Kang, S.O.; Choi, J.J.; Herbert, R.; Jang, Y.C.; et al. All-Printed Nanomembrane Wireless Bioelectronics Using a Biocompatible Solderable Graphene for Multimodal Human-Machine Interfaces. *Nat. Commun.* **2020**, *11*, 3450. [CrossRef] [PubMed]
151. Yuan, W.; Wu, X.; Gu, W.; Lin, J.; Cui, Z. Printed Stretchable Circuit on Soft Elastic Substrate for Wearable Application. *J. Semicond.* **2018**, *39*, 015002. [CrossRef]
152. Hyun, W.J.; Secor, E.B.; Hersam, M.C.; Frisbie, C.D.; Francis, L.F. High-resolution patterning of graphene by screen printing with a silicon stencil for highly flexible printed electronics. *Adv. Mater.* **2015**, *27*, 109–115. [CrossRef] [PubMed]
153. Li, W.; Yarali, E.; Bakytbekov, A.; Anthopoulos, T.D.; Shamim, A. Highly Transparent and Conductive Electrodes Enabled by Scalable Printing-and-Sintering of Silver Nanowires. *Nanotechnology* **2020**, *31*, 395201. [CrossRef] [PubMed]
154. Scheideler, W.J.; Smith, J.; Deckman, I.; Chung, S.; Arias, A.; Subramaniam, V. A Robust, Gravure-Printed, Silver Nanowire/Metal Oxide Hybrid Electrode for High-Throughput Patterned Transparent Conductors. *J. Mater. Chem. C* **2016**, *4*, 3248–3255. [CrossRef]
155. Deganello, D.; Cherry, J.A.; Gethin, D.T.; Claypole, T.C. Impact of Metered Ink Volume on Reel-to-Reel Flexographic Printed Conductive Networks for Enhanced Thin Film Conductivity. *Thin Solid Film.* **2012**, *520*, 2233–2237. [CrossRef]
156. Nair, M.; Johal, R.K.; Hamaia, S.W.; Best, S.M.; Cameron, R.E. Tunable Bioactivity and Mechanics of Collagen-Based Tissue Engineering Constructs: A Comparison of EDC-NHS, Genipin and TG2 Crosslinkers. *Biomaterials* **2020**, *254*, 120109. [CrossRef]
157. Favero, G.; Fusco, G.; Mazzei, F.; Tasca, F.; Antiochia, R. Electrochemical Characterization of Graphene and MWCNT Screen-Printed Electrodes Modified with AuNPs for Laccase Biosensor Development. *Nanomaterials* **2015**, *5*, 1995–2006. [CrossRef]
158. Narakathu, B.B.; Reddy, S.; Atashbar, M.Z.; Rebrosova, E.; Rebros, M.; Joyce, M.K. A Novel Gravure Printed Impedance Based Flexible Electrochemical Sensor. In Proceedings of the SENSORS 2011 IEEE, Limerick, Ireland, 28–31 October 2011; pp. 577–580. [CrossRef]
159. Reddy, A.S.G.; Narakathu, B.B.; Atashbar, M.Z.; Rebros, M.; Rebrosova, E.; Joyce, M.K. Gravure Printed Electrochemical Biosensor. *Procedia Eng.* **2011**, *25*, 956–959. [CrossRef]
160. Rosati, G.; Ravarotto, M.; Scaramuzza, M.; De Toni, A.; Paccagnella, A. Silver Nanoparticles Inkjet-Printed Flexible Biosensor for Rapid Label-Free Antibiotic Detection in Milk. *Sens. Actuators B Chem.* **2019**, *280*, 280–289. [CrossRef]
161. Cagnani, G.R.; Ibanez-Redin, G.; Tirich, B.; Goncalves, D.; Balogh, D.T.; Oliveira, O.N., Jr. Fully-Printed Electrochemical Sensors Made with Flexible Screen-Printed Electrodes Modified by Roll-to-Roll Slot-Die Coating. *Biosens. Bioelectron.* **2020**, *165*, 112428. [CrossRef]
162. Jansson, E.; Korhonen, A.; Hietala, M.; Kololuoma, T. Development of a Full Roll-to-Roll Manufacturing Process of Through-Substrate Vias with Stretchable Substrates Enabling Double-Sided Wearable Electronics. *Int. J. Adv. Manuf. Technol.* **2020**, *111*, 3017–3027. [CrossRef]
163. Khan, Y.; Han, D.; Pierre, A.; Ting, J.; Wang, X.; Lochner, C.M.; Bovo, G.; Yaacobi-Gross, N.; Newsome, C.; Wilson, R.; et al. A Flexible Organic Reflectance Oximeter Array. *Proc. Natl. Acad. Sci. USA* **2018**, *115*, E11015–E11024. [CrossRef] [PubMed]
164. Martinez-Olmos, A.; Fernandez-Salmeron, J.; Lopez-Ruiz, N.; Rivadeneyra Torres, A.; Capitan-Vallvey, L.F.; Palma, A.J. Screen Printed Flexible Radiofrequency Identification Tag for Oxygen Monitoring. *Anal. Chem.* **2013**, *85*, 11098–11105. [CrossRef]
165. Tan, R.L.S.; Tey, J.N.; Kerk, W.T.; Lok, B.K. Roll-to-Roll Manufacturing of Printed Biosensor Electrodes for ECG Monitoring. In Proceedings of the 2020 IEEE 22nd Electronics Packaging Technology Conference (EPTC), Singapore, 2–4 December 2020; pp. 213–215.
166. Chlaihawi, A.A.; Narakathu, B.B.; Emamian, S.; Bazuin, B.J.; Atashbar, M.Z. Development of Printed and Flexible Dry ECG Electrodes. *Sens. Bio. Sens. Res.* **2018**, *20*, 9–15. [CrossRef]

materials

MDPI

Review

Recent Advances in Materials and Flexible Sensors for Arrhythmia Detection

Matthew Guess [1,2], Nathan Zavanelli [1,2] and Woon-Hong Yeo [1,2,3,4,*]

1 George W. Woodruff School of Mechanical Engineering, Georgia Institute of Technology, Atlanta, GA 30332, USA; mguess7@gatech.edu (M.G.); nzavanelli@gatech.edu (N.Z.)
2 Center for Human-Centric Interfaces and Engineering, Institute for Electronics and Nanotechnology, Georgia Institute of Technology, Atlanta, GA 30332, USA
3 Wallace H. Coulter Department of Biomedical Engineering, Georgia Institute of Technology, Atlanta, GA 30332, USA
4 Parker H. Petit Institute for Bioengineering and Biosciences, Neural Engineering Center, Institute for Materials, Institute for Robotics and Intelligent Machines, Georgia Institute of Technology, Atlanta, GA 30332, USA
* Correspondence: whyeo@gatech.edu; Tel.: +1-404-385-5710

Abstract: Arrhythmias are one of the leading causes of death in the United States, and their early detection is essential for patient wellness. However, traditional arrhythmia diagnosis by expert evaluation from intermittent clinical examinations is time-consuming and often lacks quantitative data. Modern wearable sensors and machine learning algorithms have attempted to alleviate this problem by providing continuous monitoring and real-time arrhythmia detection. However, current devices are still largely limited by the fundamental mismatch between skin and sensor, giving way to motion artifacts. Additionally, the desirable qualities of flexibility, robustness, breathability, adhesiveness, stretchability, and durability cannot all be met at once. Flexible sensors have improved upon the current clinical arrhythmia detection methods by following the topography of skin and reducing the natural interface mismatch between cardiac monitoring sensors and human skin. Flexible bioelectric, optoelectronic, ultrasonic, and mechanoelectrical sensors have been demonstrated to provide essential information about heart-rate variability, which is crucial in detecting and classifying arrhythmias. In this review, we analyze the current trends in flexible wearable sensors for cardiac monitoring and the efficacy of these devices for arrhythmia detection.

Keywords: arrhythmia detection; cardiovascular monitoring; soft biosensors; wearable sensors; flexible electronics

check for updates

Citation: Guess, M.; Zavanelli, N.; Yeo, W.-H. Recent Advances in Materials and Flexible Sensors for Arrhythmia Detection. *Materials* **2022**, 15, 724. https://doi.org/10.3390/ma15030724

Academic Editor: Fabrizio Roccaforte

Received: 1 December 2021
Accepted: 16 January 2022
Published: 18 January 2022

1. Introduction

Arrhythmia is the presence of abnormal cardiac rhythms. In 2018, more than 500,000 American deaths included arrhythmia as a contributing factor, demonstrating its deleterious impact on patient health [1]. Furthermore, the lifetime risk of atrial fibrillation in the United States is estimated to be one in three among Caucasians and one in five among African Americans [2]. Arrhythmias occur when the electrical pulses of the heart are not functioning properly, causing the heart to beat either too fast, too slow, or skip beats. Impulse-production arrhythmias can be grouped into six categories: premature beats, non-sinus rhythm, fibrillation, tachycardias, bradycardias, and flutter. Premature beats are abnormally timed beats that occur before the sinus rhythm and are caused by the heart being unable to fill with the appropriate amount of blood [3]. Atrial fibrillation, the most common arrhythmia, occurs when the electrical pulses between the upper chambers of the heart, the atria, do not sync with the pulses in the lower chambers of the heart, the ventricles. Ventricular fibrillation, on the other hand, occurs when there is a mismatch between the right and left atria, which makes the heart unable to pump blood to the body [3]. Tachycardias occur when the heart is beating too fast, generally more than 100 beats per minute, and bradycardias occur when the heart is beating too slow, generally less than 40 beats per minute. In addition, impulse-conduction arrhythmia types include atrioventricular block, bundle branch block, Wolff—Parkinson—White syndrome, and escape beats [4]. An atrioventricular block occurs when the impulses between the atria and ventricles become blocked due to a failure in the heart's conduction system. Bundle branch block occurs as a result of blockages in the pathways in the heart. Wolff—Parkinson—White syndrome occurs when additional electrical pathways are made between the atria and ventricles, resulting in a rapid heartbeat [5].

Despite the clear need for early arrythmia detection to avoid these serious complications, existing detection mechanisms have proven insufficient. Arrhythmias have traditionally been diagnosed by medical professionals based on qualitative data, a patient's medical history, and clinical examinations. Electrocardiography (ECG) has proven instrumental in identifying arrhythmias. The importance of continuous monitoring for specific arrhythmias has been increasingly identified, as both asymptomatic arrhythmias and paroxysmal diseases remain difficult to detect through intermittent clinical ECG recordings [6,7]. The 12-lead Holter monitor has long been the clinical standard for detection and diagnosis of heart-rate diseases using long-term monitoring of ECG [8]. Though these devices are widely used, they are prone to poor patient compliance because of their bulkiness and reliance on wired leads [9]. In addition, these devices experience signal deterioration over time due to the drying of the conductive gels [10]. The advent of miniaturized, one-lead devices has offered an alternative to multi-lead ECG devices. However, they are susceptible to motion artifacts that disrupt data collection [11].

Flexible devices have emerged as alternatives to these rigid devices, eliminating motion artifacts by increasing sensor-to-skin adhesion. Recently, new areas of research have been developed to make these heart-rate monitoring devices cheaper and faster to manufacture, expanding accessibility for these previously costly devices. Table 1 shows the currently available flexible devices for arrhythmia monitoring. Additionally, new alternatives to ECG can provide information that ECG alone cannot. For instance, photoplethysmography, ultrasound, seismocardiography, and ballistocardiography can characterize the heart's electromechanics. Figure 1 shows these soft-sensor types, along with the desirable qualities of the devices. New arrhythmia-detection methodologies also offer more accurate, automatic information to patients for a low cost. For example, deep neural networks, which are capable of learning important features and patterns without extensive preprocessing or feature engineering, are becoming extremely accurate in predicting types of arrhythmia [12].

In this review, we summarize the types of sensors used to detec arrhythmias, with an emphasis on non-implantable devices and recent advances in the flexibility of previously rigid sensor types. Different sensing methods can offer low-cost alternatives to traditional sensing or provide information that is unobtainable with other sensors. We explore the materials needed to fabricate these flexible devices and discuss the mechanical, chemical, and electrical properties of these materials, as well as the effect of these properties on the detection of arrhythmias. Next, arrhythmia detection methodologies of various arrhythmia-detection devices are explored, along with their limitations. Finally, we comment on the current cutting-edge research in the field, the current problems and possible solutions, and the future development of wearable sensors for arrhythmia detection.

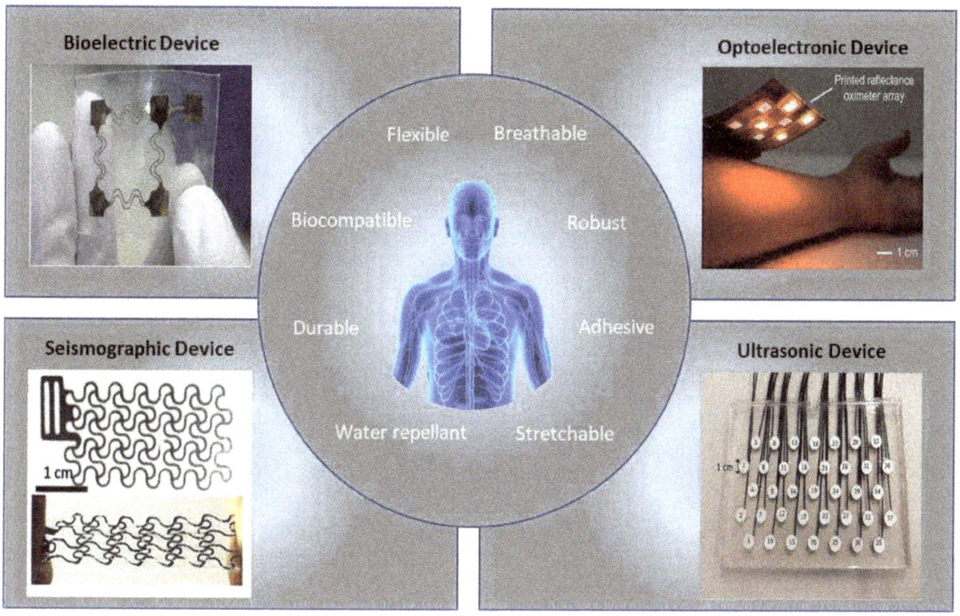

Figure 1. Examples of flexible sensors and functions for accurate arrhythmia detection [13–16]. (Figures are adapted or reprinted, clockwise, from the top–left: (1) *Sensors Actuators A Phys.* 2018, *272*, 92–101, Copyright 2019, Elsevier; (2) *Proc. Natl. Acad. Sci.* 2018, *115*, E11015–E11024, Copyright 2018, National Academy of Sciences; (3) Creative Common License by MDPI; (4) Creative Common License by Wiley.

Table 1. Comparison of flexible devices for heart-rate monitoring.

Reference	Measured Signal	Sensor Location	Substrate Material	Sensor Material	Flexibility
[14]	PPG	Wrist	PEN	PEDOT:PSS	Flexible
[17]	ECG	Arm	unknown	Ag/AgCl	Flexible
[18]	ECG	Chest	PDMS	Carbon black-PDMS nanocomposite	Stretchable
[19]	PPG	Finger	PI	Sb2Se3	Rigid
[20]	ECG	Wrist	Polythiophene	Polyvinyl alcohol/cellulose /PEDOT:PSS	Flexible
[21]	ECG	Forearm	unknown	PEDOT:PSS/WPU/D-sorbitol	Flexible
[22]	Ultrasound	Neck	PI	1–3 Piezoelectric composite	Stretchable
[16]	SCG, ECG	Chest	Tegaderm	PVDF	Stretchable

2. Bioelectric Signals

2.1. Mechanics of ECG

ECG is the practice of measuring the heart's electrical activity using pairs of electrodes on the skin. Clinically, this is achieved with 12 leads using 10 electrodes to measure cardiac signals from many angles [23]. A healthy ECG cycle typically consists of five different waves: P, Q, R, S, and T. The P-wave is the first positive wave and corresponds to atrial depolarization. The QRS complex, which consists of a negative Q-wave, a large positive R-wave, and a negative S-wave, represents ventricular depolarization and ventricular contraction. The T-wave represents ventricular repolarization and ventricular diastole [24]. Distortion of these waves can indicate abnormalities in the heart rhythm [25]. The Pan-Tompkins algorithm is commonly used to find the QRS complex in a regular ECG rhythm [26]. Although 12-lead ECGs provide the best clinical method for diagnosing arrhythmias, the switch to 3-lead ECGs or single-lead ECGs offers many advantages [23]. Cardiovascular patches that use adhesives have become increasingly popular due to their unobtrusiveness. For example, the FDA-approved Zio patch (iRhythm Technologies, Inc., San Francisco, CA, USA) has shown clinical suitability for detecting arrhythmias when compared with a 12-lead Holter monitor over 14 days [27].

2.2. Materials for Flexible ECG Devices

Many ECG electrodes currently used in clinics are made up of three parts: (1) a conductive metal, traditionally Ag/AgCl electrodes; (2) a conductive gel; and (3) an adhesive patch. Conductive gels reduce the impedance from the electrode to the skin. However, they dry up over time, which causes signal quality to deteriorate during long-term monitoring. Therefore, many materials have been explored as alternatives to traditional Ag/AgCl that are conformal to the skin. Although they generally have higher impedances than wet electrodes, flexible dry electrodes are quickly gaining in popularity. Electrodes can be made from any conductive materials. Since metals have high Young's moduli, ultra-thin metal films can be arranged in serpentine or fractal geometries to provide flexibility or stretchability, as shown in Figure 2a [28–31]. Chlaihawi et al. reported an electrode that was screen printed with Ag flake ink [32]. The electrode with the largest area reported a 0.95 correlation coefficient with traditional wet Ag/AgCl electrodes. In addition, the study showed the feasibility of using the high-throughput process of screen printing for the development of flexible dry electrodes.

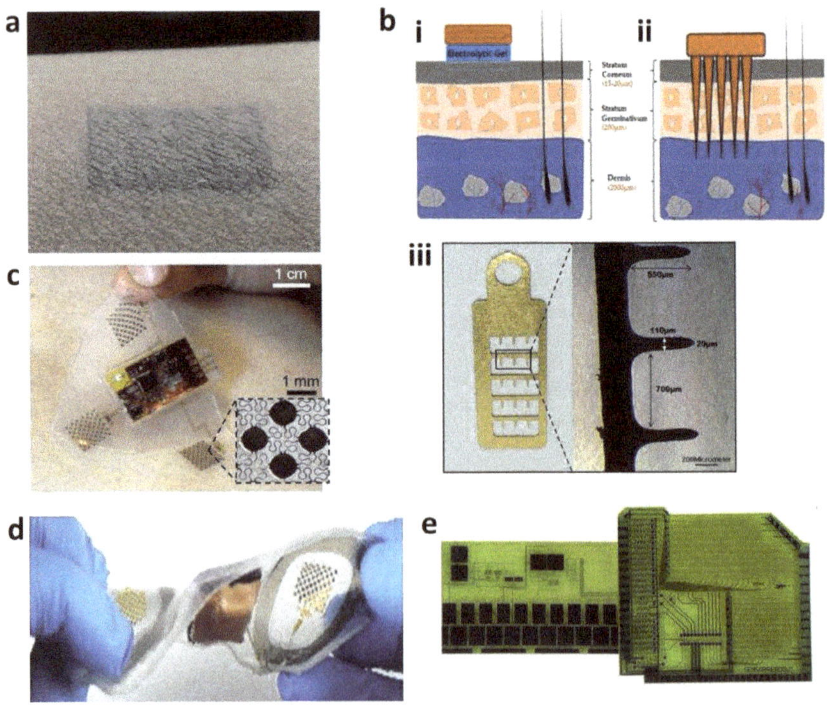

Figure 2. Electrocardiography. (**a**) Photo of a skin-conformal electrode. (**b**) Microneedle array-based ECG. Illustration of (**i**) traditional Ag/AgCl electrode and (**ii**) microneedle array electrode. (**iii**) Photo of a microneedle array electrode. (**c**) Photo of a stretchable hybrid-electronics device. (**d**) Photo of a soft strain-isolated bioelectric device. (**e**) Foil micrograph of a flexible ECG patch.

The stretchability of these thin-film metals can be increased by introducing conductive polymers. Polyethylene terephthalate (PET) and polydimethylsiloxane (PDMS) are commonly chosen as polymers due to their biocompatibility, wide availability, and low Young's modulus. To make these polymers conductive, materials such as activated carbon or metal micro/nanoparticles are added to form networks of conductivity. For example, Jung et al. showed a carbon nanotube (CNT)/PDMS composite-based dry electrode to combat motion and sweat artifacts [33]. The performance of the electrode was able to be tuned by adjusting the CNT concentration. This electrode showed no signal degradation over a seven-day period of continuous monitoring, providing similar motion-artifact reduction as wet electrodes and more motion-artifact reduction than other dry electrodes. Zhang et al. showed an electrode that had stretchability up to 500% by combining Ag nanowires (NWs) with polymers [34]. This sensor improved upon the CNT/PDMS structure, which is subject to weak connection of conductive materials when stretched. This polymer/Ag NW sensor also increased the durability over 1000 cycles and exhibited good fatigue resistance.

In recent years, the conductive polymer poly (3,4-ethylene dioxthiophene): polystyrene trans acid (PEDOT:PSS) has been the most common polymer for textile-based electrodes due to its high sensitivity to biological molecules and high response time [35]. Wang et al. showed that PEDOT:PSS could be used to achieve even higher flexibility and lower skin impedance by combining it with a flexible cellulose/polyvinyl alcohol substrate [20]. This process provides new ideas for low-cost manufacturing of environmentally friendly ECG devices.

In addition to material advances, recent studies have decreased the impedance of surface electrodes by changing the form factor. For example, semi-invasive strategies, such as microneedle-based approaches, have been demonstrated to reduce motion artifacts [36,37]. Satti et al. reported a microneedle array electrode (MNE), as shown in Figure 2b, that showed no mechanical failure under compression forces of 16 N and showed that while the signal quality of wet Ag and AgCl electrodes decreased after 3 days and 1 week, respectively, the MNE showed no signs of signal quality deterioration [36].

Despite the many developments in electrode technology, there is still a lack of all-in-one integration. To address these sensor-only systems, innovations in packaging have become important as well. For example, the general impedance challenges of sweat buildup on electrodes have been addressed by integrating hydrophilic poly(urethane-acrylate) into Ag electrodes to increase conductivity during sweating and increasing the breathability of the substrate, as shown in Figure 2c,d [37–41]. All-in-one systems featuring wireless charging, wireless data communication, and onboard data analysis have been developed, as in Figure 2e [42–45].

3. Optoelectronic Signals

3.1. Mechanics of PPG

Photoplethysmography (PPG) is emerging as an alternative to ECG for cardiovascular monitoring due to its small size and ability to capture many different physiological parameters. The LED operates at red and near-infrared (NIR) frequencies, and the light intensity reaching the photodiode changes depending on the volumetric changes in the veins and arteries [46]. PPG sensors consist of two basic components: a light-emitting diode (LED) and a photodiode. The PPG sensor can function in two main modes: (1) transmission, where the LED and photodiode are placed on opposite sides of the medium; or (2) reflection, where the LED and photodiode are placed on the same side of the medium [47]. Due to the many factors affecting blood flow, including cardiac, neural, and respiratory factors, it is possible to look at many physiological parameters. PPG is currently used to measure several different aspects of heart health, including blood oxygen saturation (SpO2), blood pressure, heart rate, and respiratory rate [48–52].

Because PPG sensors only require two simple components, an LED and photodiode, they are commonly implemented in already-existing devices, such as watches and smartphones, at low cost. However, the fundamental mismatch between the shape and rigidity of these devices with the skin makes them prone to heavy motion-artifact noise. Thus, flexible materials are essential for increasing the accuracy of optoelectronic sensors.

3.2. Materials for Flexible PPG Devices

Both flexible LEDs and flexible diodes have been developed to reduce the effects of motion artifacts, which have been difficult to remove with filtering alone [53]. The most common photodiodes currently in use are silicon photodiodes, as they are widely available and flexible. Kim et al. used flexible (PIN) silicon diodes in combination with near-field communication (NFC) to deliver power, eliminating the need for a battery [54]. The photodiodes were paired with red and infrared LEDs, and the signals were amplified to coils and sent to a smartphone using the NFC platform. Li et al. offered an improvement on the conventional optoelectronic architecture by designing an epidermal silicon-based device by using a specific strain-isolation design, nanodiamond thinning, and hybrid transfer printing [55]. Through the thinning process, the thickness of the LEDs and PD was reduced to 20 μm. Mechanical deformation was addressed by adding a flexible island in a sandwich structure, with PI and PDMS helping the device to show stable operation, even under a strain of 35%. This device promised the possibility for functional optoelectronic devices to be directly mounted on the skin. Gallium arsenide (GaAs) is a frequently used III-V semiconductor material that can be used as an alternative to Si-based materials based on its excellent charge-carrier mobility and high stability. Hong et al. demonstrated a GaAs-based flexible photodetector array that was hetero-epitaxially grown on a Si wafer [56]. This innovative manufacturing method showed promising results that could lower the cost of inorganic photodiodes, which are normally expensive. This platform shows promising possibilities for large-scale creation of flexible photodiodes.

Organic materials for PPG sensors have become increasingly attractive due to their low fabrication cost and environmentally friendly footprint. For example, Yokota et al. developed a flexible pulse oximeter consisting of a polymer LED (PLED) and an organic photodiode (OPD) (Figure 3a,b) [57]. The device addressed a large barrier in organic optoelectronics, which is the ability to form a high-quality passivation layer on an ultraflexible substrate by making the passivation layers very thin using a low-temperature process. The PLED was constructed using light-emitting polymers and indium tin oxide electrodes. The OPD was constructed with a poly(3-hexylthiophene) (P3HT):(6,6)-phenyl-C61-butyric acid methyl ester (PCBM) active layer, which was manufactured on a 1 μm Parylene substrate, which was used as the passive layer. The lightweight device, which is only 3 μm thick, showed robustness, even under repeated 60% compression. Khan et al. showed a flexible oximeter array in which the active materials for the organic LED (OLED) and the OPD were fabricated on polymer substrates and placed in a grid consisting of photodiodes, red LEDs, and near-infrared LEDS [14]. The device offers a solution to the fundamental problem of only being able to measure PPG signals at a single location by using a reflectance-based array and is therefore capable of measuring blood oxygenation, even in the absence of pulsatile arterial blood signal. The device was able to measure SpO2 with a mean error of 1.1%.

Exciting new optoelectronic research on quantum and nano-based materials is emerging thanks to the ability to tune the performance due to the size of the particles. As shown in Figure 3c, Polat et al. introduced a photodiode made with graphene sensitized with semiconducting quantum dots [58]. Quantum-dot-based graphene photodetectors have high responsivity due to their built-in photoconductive gain. Therefore, the readout electronics can be placed far from the sensor, preserving the form factor of the active sensing area. In addition, the detector's transparency can be changed by changing the thickness of the quantum-dot layer, which alters the responsivity. This transparent device used ambient light for low power consumption and communicated wirelessly using near-field communication circuitry, as represented by a correlation coefficient of $\rho = 0.98$, with a state-of-the-art clinical PPG sensor. Kim et al. demonstrated a spirally wrapped CNT-based microelectrode, which is shown in Figure 3d [59]. A CNT-based solution was printed on an agarose hydrogel substrate, where it was then spirally wrapped around a microfiber surface, such as nylon. The CNT electrodes demonstrated a current ratio of ~10^5 and a maximum field-effect mobility of 0.68 $cm^2 V^{-1} s^{-1}$, which is comparable to similar flat devices.

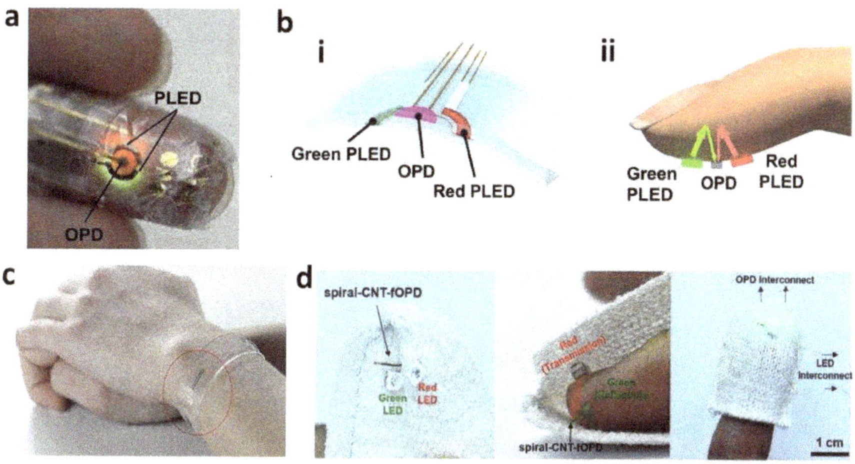

Figure 3. Photoplethysmography. (**a**) Photo of an ultra-flexible organic optical sensor. (**b**) Illustration of polymer LED (**i**) and organic photodiode (**ii**) pulse oximeters. (**c**) Photo of a graphene-based flexible sensor in a heart-rate monitoring bracelet. (**d**) Photos of CNT-based microelectrodes for a fiber optoelectronic device.

For arrhythmia detection, a high signal-to-noise ratio is essential. Thus, the specific design and its optimization are crucial. Pribadi et al. optimized a flexible OLED-OPD patch using an optical simulation [60]. The group optimized the AC/DC ratio of a square-type and cross-type patch. Their results showed that the square-type OPD was the best patch due to the wide area of the OPD, with an OLED drive current between 0.1 and 0.4 mA. The flexibility of the resulting design was 130°, and the heart-rate measurement accuracy was 95%. Khan et al. optimized the geometry of OLEDs and OPDs by designing three geometries: a rectangular geometry, a bracket geometry, and a circular geometry [61]. Both the bracket geometry and the circular geometry showed clear improvement over the rectangular design, where the bracket geometry showed a 39.7% improvement in the red PPG-signal magnitude and an 18.2% improvement in the NIR-channel magnitude, while the circular geometry showed a 48.6% improvement in the red-channel magnitude and a 9.2% improvement in the NI-channel magnitude. These results show promising form-factor and geometry optimization that could increase sensor accuracy and reduce power consumption in future wearable devices.

4. Other Signals

4.1. Ultrasonic Signals

Doppler ultrasound has been used to track changes in arterial diameter, which can be used to track heart rate. However, commercially available ultrasound monitors are handheld and rigid and therefore not suitable for continuous monitoring. The active layer for ultrasonic transducers is most commonly lead zirconate titanate (PZT) or composites of PZT, as it exhibits high piezoelectric properties and high electromechanical properties [62]. A conformal ultrasonic device was suggested by Wang et al. that could withstand strains of up to 60% [22]. The device utilized a piezoelectric pillar that was hybridized with soft, stretchable components. Liu et al. built on this concept by arranging stretchable ultrasound sensors into a two-dimensional array based on row and column electrodes [63]. The design consisted of PZT blocks, serving as the piezoelectric islands, connected with polyimide (PI) serpentine hinges, allowing for stretchability between the otherwise rigid blocks. Lee et al. showed that a calcium-modified silk could be used as an interface layer between sensor and skin for ultrasound transducer arrays and that it has a similar acoustic impedance to human skin [64]. Hamelmann et al. also introduced an ultrasound array based on PZTs (Figure 4a) [15]. Other materials for stretchable ultrasound substrates include polyethylene naphthalate (PEN), PDMS, acrylic, and PET [60,64,65].

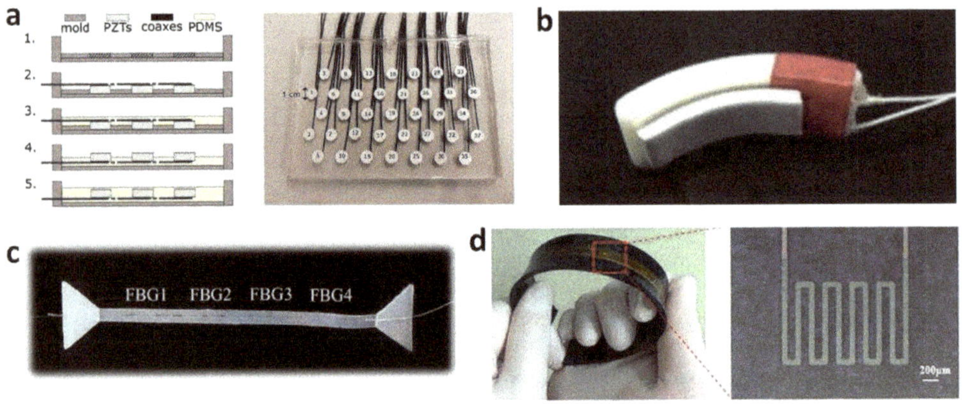

Figure 4. (**a**) Fabrication processes (left) and photo of a fabricated ultrasonic transducer (right). (**b**) Photo of a finger-worn SCG sensor. (**c**) Photo of a skin-like SCG sensor with fibers. (**d**) Flexible strain sensor for heartbeat monitoring.

4.2. Mechanoelectric Signals

Both ballistocardiography (BCG) and seismocardiography (SCG) have been used to measure cardiac activity based on the heart's displacement, velocity, and accelerations. BCG measures entire body movement due to cardiac ejection, whereas SCG is a local chest measurement that registers cardiac-induced vibrations. Both are measured in terms of acceleration [65]. Since SCG is typically measured on the body, while BCG is measured using non-contact sensors, SCG has been more commonly used in wearable platforms. While the relationship between SCG and cardiac events is still being studied, the literature has estimated the correlation between certain SCG waves and event timing. For example, a low-frequency acceleration wave can be seen at the start of atrial systole [66]. However, the rigid mismatch between current accelerometers and human skin can introduce whole-body inertia measurements into the SCG signal. Skin-compatible SCG sensors have attempted to address this problem. Ha et al. showed a stretchable e-tattoo SCG based on polyvinylidene fluoride (PVDF) [16]. The sensor also showed a great correlation between the systolic time interval and blood-pressure measurements, and it can simultaneously record ECG signals. Other studies have used PVDF for the SCG sensor while incorporating elements essential for long-term monitoring, like wireless charging and communication (Figure 4b,c) [61,67–69]. Wearable phonocardiography (PCG) sensors have also been used to assess the heart. Accelerometer-based mechano-acoustic sensors function similarly to SCG and can pick up auditory frequencies that the human ear cannot hear with a traditional stethoscope. Flexible, wearable stethoscopes based on accelerometry have been developed with flexible substrates and electrodes [67,68]. Kwak et al. also demonstrated a strain-gauge-based heart-rate sensor that can detect not only the timing of the heart pulsation but also the amplitude and shape of the pulse (Figure 4d) [69].

4.3. Electrochemical Signals

Metabolic factors, such as glucose level, have been linked to increased risk for arrhythmias. However, monitoring of metabolite levels has traditionally been performed with invasive devices. Recently, wearable glucose-detection devices have allowed for repeated measurement of glucose levels without the burden of an implantable device [70]. Bandodkar et al. reported a tattoo-based noninvasive glucose-monitoring system based on Ag/AgCl ink electrodes and a reagent layer [71]. In vitro characterization proved the system's ability to detect micromolar levels of glucose, and on-body evaluation proved its ability to detect a rise in glucose after a meal. Sempionatto et al. improved this by creating a stretchable patch for both hemodynamic and metabolic monitoring by combining PZT ultrasound transducers and printed polymer composites [72]. Blood pressure could be monitored through the sensor array, while chemical sensing was realized through sweat detection. This study showed the potential of the integration of acoustic and electrochemical sensing.

5. Arrhythmia Detection

Computer-aided ECG and PPG analysis has vastly improved the detection and classification of arrhythmias. The most current process for classification of arrhythmias consists of: (1) pre-processing, where baseline wander and unwanted noises and frequencies are filtered out; (2) feature extraction, where the most important features of a wave are identified, and (3) classification, where the most important features are input into a model to predict the class of arrhythmia of a given signal [73]. For filtering ECG signals, the P- and T-waves are typically found between 0.5 Hz and 10 Hz, while the QRS complex is found between 4 Hz and 20 Hz. Discrete wavelet transforms (DWTs) can be used in combination with low- and high-pass filters to remove unwanted frequencies [74,75]. For PPG signals, wavelet decomposition has also been investigated. However, filtering of motion artifacts for PPG signals has only been proven for weak noise, and very noisy PPG signals need to be discarded [76].

Current arrhythmia-detection models are often based on machine learning, as their accuracy is easily increased with large training datasets. These machine learning methods require feature extraction in either the time domain, frequency domain, time frequency domain, or nonlinear domain. They are often based on physiological incidents. For example, dimensionality-reduction techniques, such as principal component analysis (PCA), have been used in the time domain, and Fourier transforms have been used in the frequency domain [77,78]. Machine learning models for both ECG and PPG signals have included support vector machine (SVM), multilayer perceptron (MLP), and decision tree (DT) models [78–81]. The cutting-edge machine learning area is deep learning. Deep learning methods, such as convolutional neural network (CNN), deep belief network (DBN), recurrent neural network (RNN), long short-term memory (LSTM), and gated recurrent unit (GRU), have been applied to arrhythmia detection [82–86]. Table 2 shows these machine learning methods for arrhythmia detection and classification.

Kaisti et al. were able to perfectly distinguish between 13 sinus-rhythm subjects and seven subjects with atrial fibrillation using a k-means clustering-based approach [87]. The input to the algorithm was time-frequency data derived from a soft, band-based MEMS pressure-sensor array, showing the feasibility of combining flexible devices and current arrhythmia detection algorithms to provide high detection accuracy. Improving on this concept, Dong et al. used an arrhythmia-detection system consisting DWT and SVM algorithms, with a novel acetylene carbon black/PDMS ECG recording patch as the input, which achieved a high online classification accuracy of 98.7% [88].

Table 2. Comparison of arrhythmia-detection methodologies using wearable devices.

Reference	Device	Target Signal	Arrhythmia Type	Detection Methodology	Accuracy
[12]	iRhythm Zio monitor	ECG	10 types	Deep neural network	ROC = 0.97 F_1 = 0.837
[89]	Apple watch	PPG, ACC	Atrial fibrillation	Deep neural network	Sens = 0.98 Spec = 0.90
[90]	2-lead Holter monitor	ECG	Atrial fibrillation, atrial flutter, AV junctional rhythm	Hybrid CNN-LSTM	Sens = 0.9787 Spec = 0.9929
[91]	Fingertip pulse oximeter	PPG	Atrial fibrillation	CNN, RNN	AOC = 0.998 AOC = 0.996
[73]	MIT-BIH arrhythmia database	ECG	Ventricular fibrillation	CNN	Acc = 0.9318
[92]	Point-of-care ultrasound	Ultrasound images	Atrial fibrillation	Semi-supervised deep learning network	Acc = 0.79

6. Substrate Materials and Skin Interfaces

Due to their flexibility, ease of manufacturing, and low cost, substrates are often made from polymers or fabrics. Elastomers, such as PET, PI, PEN, polyetherimide (PEI), and parylene, are common materials for thin-film-based substrates. Their weak intramolecular forces enable greater elongation and therefore greater stretchability. Substrates such as these offer many opportunities for breakthroughs. For example, Wonryung et al. reported an active, ultra-flexible, multielectrode array that using a 1.2 μm-thick parylene substrate [93]. The 2.6 μm sensor can be used for long-term ECG of dynamically moving hearts due to a 15% strain. Likewise, Shahandashti et al. showed dry stretchable electrodes based on PDMS, which has great biocompatibility, stretchability, and chemical inertness [13]. The substrate showed similar contact impedance to standard wet Ag/AgCl electrodes, though pressure was applied between the electrode and skin using a transparent tape. To improve the adhesion between substrate and skin, Zhang et al. explored a blended film of PEDOT:PSS, waterborne polyurethane, and D-sorbitol prepared by solution processing. These films exhibited low electrode-skin electrical impedances in the frequency range of 1 Hz–10 MHz and adhesion forces above 0.4 N/cm.

7. Wearable Devices

Long-term, real-time, continuous monitoring is essential in arrhythmia detection. Thus, advances in form factors and device's comfort are critical for better wearable devices for successful cardiovascular monitoring. Hardware platforms, such as watches and smartphones, commonly include rigid sensors for measuring ECG and PPG. However, these devices suffer from motion artifacts, data loss, and low accuracy. Recently, form-factor innovations in lightweight wearable devices have increased user comfort and compliance. Breakthroughs in epidermal electronics, for example, have minimized the bulkiness of wearable devices. Wang et al. reported low-cost electronic sensors based on epidermal electronics that minimized motion and sweat artifacts [94]. The sensors reported up to 45% stretchability, adhering to the skin using only van der Waals forces. The 13-μm thick sensor is tape-free and disposable, allowing for ease of both patient compliance and comfort. The durability of wearable devices is also essential in arrhythmia detection. For example, the conductive gels in the traditional Ag/AgCl electrodes degrade over time, resulting in a signal decrease. In addition, the buildup of moisture, dead skin, and material degradation can impact the durability of wearable devices. Xu et al. attempted to mitigate some of these problems through a washable and screen-printed graphene electrode on textiles [95]. The ECG sensor showed negligible change over nine washing cycles and 2000 bending cycles.

8. Limitations

Although flexible devices have demonstrated tremendous potential, they are rarely commercialized or used clinically. Low throughput or expensive manufacturing methods make many of these devices difficult to implement widely. For example, non-traditional substrates like PET or TPU cannot be manufactured with large-scale conventional cleanroom fabrication [96,97]. High-throughput manufacturing processes, such as screen printing, have been demonstrated, yet optimal parameters have not yet been discovered. In addition, the resolution of these processes remains low. The elimination of motion artifacts also presents significant challenges. For example, dry electrodes still suffer high impedance with the skin. In addition, powering these devices for long-term monitoring remains a challenge, especially for battery-less devices, and low power consumption is an essential trait for continuous monitoring devices. The addition of wireless transmission also increases the power consumption needed. Bluetooth, for example, consumes up to 5 mW of power, which is more than many thin-film batteries can provide. New machine learning methods can help detect the presence of arrhythmia in heart rates. However, distinguishing between types of arrhythmias and between other classes of heart disease has proven difficult. Machine learning is also limited by the quantity and quality of the training data. In addition, many low-training-data machine learning models are prone to overfitting data and are therefore unable to generalize the testing data. Machine learning is also computationally intensive, making real-time data classification difficult. Finally, the "black-box" nature of machine learning is inherently complex for doctors and clinicians to interpret.

9. Conclusions

This review provides many examples to discuss recent advances in arrhythmia detection using flexible and wearable systems that utilize advanced soft materials, flexible designs, and integrated sensors. We summarize the current bioelectric, optoelectric, mechanoelectric, and ultrasonic sensing methods for monitoring various physiological signals related to arrhythmia. These methods are used to detect and classify arrhythmia accurately. We believe that the future of arrhythmia detection lies in further advancements in flexible wearable sensors and automated classification tools using machine learning algorithms. For applications of portable wearable devices in clinical diagnosis, there are areas to improve in terms of materials and sensor performance, such as sensing materials, sensor-to-skin contact quality, impedance control, power-consumption management, miniaturization, wireless data transmissibility, and detection and classification.

Author Contributions: Conceptualization, M.G. and W.-H.Y.; writing—original draft preparation, M.G., N.Z. and W.-H.Y.; writing—review and editing, M.G., N.Z. and W.-H.Y.; supervision, W.-H.Y.; project administration, W.-H.Y.; funding acquisition, W.-H.Y. All authors have read and agreed to the published version of the manuscript.

Funding: We acknowledge the support from the IEN Center for Human-Centric Interfaces and Engineering at Georgia Tech. This study was partially supported by the Institute of Information & communications Technology, Planning & Evaluation (IITP) grant funded by the Korean government (MSIT) (2021-0-01517).

Institutional Review Board Statement: Not applicable.

Informed Consent Statement: Not applicable.

Data Availability Statement: No new data were created or analyzed in this study.

Conflicts of Interest: The authors declare no conflict of interest.

References

1. Virani, S.S.; Alonso, A.; Aparicio, H.J.; Benjamin, E.J.; Bittencourt, M.S.; Callaway, C.W.; Carson, A.P.; Chamberlain, A.M.; Cheng, S.; Delling, F.N.; et al. Heart disease and stroke statistics—2021 update. *Circulation* **2021**, *143*, e254–e743. [CrossRef]
2. Mou, L.; Norby, F.L.; Chen, L.Y.; O'Neal, W.T.; Lewis, T.T.; Loehr, L.R.; Soliman, E.Z.; Alonso, A. Lifetime risk of atrial fibrillation by race and socioeconomic status. *Circ. Arrhythm. Electrophysiol.* **2018**, *11*, e006350. [CrossRef] [PubMed]

3. Richards, K.J.C.; Cohen, A.T. Cardiac arrhythmias in the critically ill. *Anaesth. Intensive Care Med.* **2006**, *7*, 289–293. [CrossRef]
4. Kléber, A.G.; Rudy, Y. Basic mechanisms of cardiac impulse propagation and associated arrhythmias. *Physiol. Rev.* **2004**, *84*, 431–488. [CrossRef] [PubMed]
5. Sardana, H.K.; Kanwade, R.; Tewary, S. Arrhythmia detection and classification using ECG and PPG techniques: A review. *Phys. Eng. Sci. Med.* **2021**, *44*, 1027–1048. [CrossRef]
6. Rho, R.W.; Page, R.L. Asymptomatic atrial fibrillation. *Prog. Cardiovasc. Dis.* **2005**, *48*, 79–87. [CrossRef] [PubMed]
7. Williams, S.E.; O'Neill, M.; Kotadia, I.D. Supraventricular tachycardia: An overview of diagnosis and management. *Clin. Med.* **2020**, *20*, 43. [CrossRef]
8. DiMarco, J.P.; Philbrick, J.T. Use of ambulatory electrocardiographic (Holter) monitoring. *Ann. Intern. Med.* **1990**, *113*, 53–68. [CrossRef]
9. Fung, E.; Järvelin, M.-R.; Doshi, R.N.; Shinbane, J.S.; Carlson, S.K.; Grazette, L.P.; Chang, P.M.; Sangha, R.S.; Huikuri, H.V.; Peters, N.S. Electrocardiographic patch devices and contemporary wireless cardiac monitoring. *Front. Physiol.* **2015**, *6*, 149. [CrossRef]
10. Chen, Y.H.; Op de Beeck, M.; Vanderheyden, L.; Carrette, E.; Mihajlović, V.; Vanstreels, K.; Grundlehner, B.; Gadeyne, S.; Boon, P.; van Hoof, C. Soft, comfortable polymer dry electrodes for high quality ECG and EEG recording. *Sensors* **2014**, *14*, 23758. [CrossRef]
11. Ackermans, P.A.J.; Solosko, T.A.; Spencer, E.C.; Gehman, S.E.; Nammi, K.; Engel, J.; Russell, J.K. A user-friendly integrated monitor-adhesive patch for long-term ambulatory electrocardiogram monitoring. *J. Electrocardiol.* **2012**, *45*, 148–153. [CrossRef]
12. Hannun, A.Y.; Rajpurkar, P.; Haghpanahi, M.; Tison, G.H.; Bourn, C.; Turakhia, M.P.; Ng, A.Y. Cardiologist-level arrhythmia detection and classification in ambulatory electrocardiograms using a deep neural network. *Nat. Med.* **2019**, *25*, 65–69. [CrossRef]
13. Fayyaz Shahandashti, P.; Pourkheyrollah, H.; Jahanshahi, A.; Ghafoorifard, H. Highly conformable stretchable dry electrodes based on inexpensive flex substrate for long-term biopotential (EMG/ECG) monitoring. *Sens. Actuators A Phys.* **2019**, *295*, 678–686. [CrossRef]
14. Khan, Y.; Han, D.; Pierre, A.; Ting, J.; Wang, X.; Lochner, C.M.; Bovo, G.; Yaacobi-Gross, N.; Newsome, C.; Wilson, R.; et al. A flexible organic reflectance oximeter array. *Proc. Natl. Acad. Sci. USA* **2018**, *115*, E11015–E11024. [CrossRef]
15. Hamelmann, P.; Mischi, M.; Kolen, A.F.; van Laar, J.O.E.H.; Vullings, R.; Bergmans, J.W.M. Fetal heart-rate monitoring Implemented by dynamic adaptation of transmission power of a flexible ultrasound transducer array. *Sensors* **2019**, *19*, 1195. [CrossRef]
16. Ha, T.; Tran, J.; Liu, S.; Jang, H.; Jeong, H.; Mitbander, R.; Huh, H.; Qiu, Y.; Duong, J.; Wang, R.L.; et al. A chest-laminated ultrathin and stretchable e-tattoo for the measurement of electrocardiogram, seismocardiogram, and cardiac time intervals. *Adv. Sci.* **2019**, *6*, 1900290. [CrossRef]
17. Villegas, A.; McEneaney, D.; Escalona, O. Arm-ECG wireless sensor system for wearable long-term surveillance of heart arrhythmias. *Electronics* **2019**, *8*, 1300. [CrossRef]
18. Li, Y.; Luo, Y.; Nayak, S.; Liu, Z.; Chichvarina, O.; Zamburg, E.; Zhang, X.; Liu, Y.; Heng, C.H.; Thean, A.V.Y. A stretchable-hybrid low-power monolithic ECG patch with microfluidic liquid-metal interconnects and stretchable carbon-black nanocomposite electrodes for wearable heart monitoring. *Adv. Electron. Mater.* **2019**, *5*, 1800463. [CrossRef]
19. Chen, E.; Jiang, J.; Su, R.; Gao, M.; Zhu, S.; Zhou, J.; Huo, Y. A new smart wristband equipped with an artificial intelligence algorithm to detect atrial fibrillation. *Heart Rhythm* **2020**, *17*, 847–853. [CrossRef] [PubMed]
20. Wang, Y.; Zhong, X.; Wang, W.; Yu, D. Flexible cellulose/polyvinyl alcohol/PEDOT:PSS electrodes for ECG monitoring. *Cellulose* **2021**, *28*, 4913–4926. [CrossRef]
21. Lai, D.; Bu, Y.; Su, Y.; Zhang, X.; Ma, C.S. A flexible multilayered dry electrode and assembly to single-lead ECG patch to monitor atrial fibrillation in a real-life scenario. *IEEE Sens. J.* **2020**, *20*, 12295–12306. [CrossRef]
22. Wang, C.; Li, X.; Hu, H.; Zhang, L.; Huang, Z.; Lin, M.; Zhang, Z.; Yin, Z.; Huang, B.; Gong, H.; et al. Monitoring of the central blood pressure waveform via a conformal ultrasonic device. *Nat. Biomed. Eng.* **2018**, *2*, 687–695. [CrossRef] [PubMed]
23. Hammad, M.; Maher, A.; Wang, K.; Jiang, F.; Amrani, M. Detection of abnormal heart conditions based on characteristics of ECG signals. *Measurement* **2018**, *125*, 634–644. [CrossRef]
24. Chen, S.; Qi, J.; Fan, S.; Qiao, Z.; Yeo, J.C.; Lim, C.T. Flexible wearable sensors for cardiovascular health monitoring. *Adv. Healthc. Mater.* **2021**, *10*, 2100116. [CrossRef]
25. Maršánová, L.; Němcová, A.; Smíšek, R.; Vítek, M.; Smital, L. Advanced P wave detection in ecg signals during pathology: Evaluation in different arrhythmia contexts. *Sci. Rep.* **2019**, *9*, 19053. [CrossRef]
26. Pan, J.; Tompkins, W.J. A Real-Time QRS Detection Algorithm. *IEEE Trans. Biomed. Eng.* **1985**, *BME-32*, 230–236. [CrossRef]
27. Barrett, P.M.; Komatireddy, R.; Haaser, S.; Topol, S.; Sheard, J.; Encinas, J.; Fought, A.J.; Topol, E.J. Comparison of 24-hour holter monitoring with 14-day novel adhesive patch electrocardiographic monitoring. *Am. J. Med.* **2014**, *127*, 95.e11–95.e17. [CrossRef] [PubMed]
28. Dong, W.; Cheng, X.; Xiong, T.; Wang, X. Stretchable bio-potential electrode with self-similar serpentine structure for continuous, long-term, stable ECG recordings. *Biomed. Microdevices* **2019**, *21*, 6. [CrossRef] [PubMed]
29. Fan, J.A.; Yeo, W.H.; Su, Y.; Hattori, Y.; Lee, W.; Jung, S.Y.; Zhang, Y.; Liu, Z.; Cheng, H.; Falgout, L.; et al. Fractal design concepts for stretchable electronics. *Nat. Commun.* **2014**, *5*, 3266. [CrossRef]
30. Morikawa, Y.; Yamagiwa, S.; Sawahata, H.; Numano, R.; Koida, K.; Ishida, M.; Kawano, T.; Morikawa, Y.; Yamagiwa, S.; Sawahata, H.; et al. Ultrastretchable kirigami bioprobes. *Adv. Healthc. Mater.* **2018**, *7*, 1701100. [CrossRef]

31. Xu, B.; Akhtar, A.; Liu, Y.; Chen, H.; Yeo, W.-H.; Park, S.I.I.; Boyce, B.; Kim, H.; Yu, J.; Lai, H.-Y.; et al. An epidermal stimulation and sensing platform for sensorimotor prosthetic control, management of lower back exertion, and electrical muscle activation. *Adv. Mater.* **2016**, *28*, 4462–4471. [CrossRef]
32. Chlaihawi, A.A.; Narakathu, B.B.; Emamian, S.; Bazuin, B.J.; Atashbar, M.Z. Development of printed and flexible dry ECG electrodes. *Sens. Bio-Sens. Res.* **2018**, *20*, 9–15. [CrossRef]
33. Jung, H.; Moon, J.; Baek, D.; Lee, J.; Choi, Y.; Hong, J.; Lee, S. CNT/PDMS composite flexible dry electrodesfor long-term ECG monitoring. *IEEE Trans. Biomed. Eng.* **2012**, *59*, 1472–1479. [CrossRef] [PubMed]
34. Zhang, Y.; He, P.; Luo, M.; Xu, X.; Dai, G.; Yang, J. Highly stretchable polymer/silver nanowires composite sensor for human health monitoring. *Nano Res.* **2020**, *13*, 919–926. [CrossRef]
35. Pani, D.; Dessì, A.; Saenz-Cogollo, J.F.; Barabino, G.; Fraboni, B.; Bonfiglio, A. Fully textile, PEDOT:PSS based electrodes for wearable ECG monitoring systems. *IEEE Trans. Biomed. Eng.* **2016**, *63*, 540–549. [CrossRef] [PubMed]
36. Satti, A.T.; Park, J.; Park, J.; Kim, H.; Cho, S. Fabrication of parylene-coated microneedle array electrode for wearable ECG device. *Sensors* **2020**, *20*, 5183. [CrossRef] [PubMed]
37. Hou, Y.; Li, Z.; Wang, Z.; Yu, H. Miura-ori structured flexible microneedle array electrode for biosignal recording. *Microsyst. Nanoeng.* **2021**, *7*, 53. [CrossRef] [PubMed]
38. Lv, J.; Thangavel, G.; Li, Y.; Xiong, J.; Gao, D.; Ciou, J.; Tan, M.W.M.; Aziz, I.; Chen, S.; Chen, J.; et al. Printable elastomeric electrodes with sweat-enhanced conductivity for wearables. *Sci. Adv.* **2021**, *7*, eabg8433. [CrossRef] [PubMed]
39. Fang, Y.; Li, Y.; Li, Y.; Ding, M.; Xie, J.; Hu, B. Solution-processed submicron free-standing, conformal, transparent, breathable epidermal electrodes. *ACS Appl. Mater. Interfaces* **2020**, *12*, 23689–23696. [CrossRef]
40. Haddad, P.A.; Servati, A.; Soltanian, S.; Ko, F.; Servati, P. Breathable dry silver/silver chloride electronic textile electrodes for electrodermal activity monitoring. *Biosensors* **2018**, *8*, 79. [CrossRef]
41. Rodeheaver, N.; Herbert, R.; Kim, Y.-S.; Mahmood, M.; Kim, H.; Jeong, J.-W.; Yeo, W.-H. Strain-isolating materials and interfacial physics for soft wearable bioelectronics and wireless, motion artifact-controlled health monitoring. *Adv. Funct. Mater.* **2021**, *31*, 2104070. [CrossRef]
42. Liu, C.; Kim, J.-T.; Kwak, S.S.; Hourlier-Fargette, A.; Avila, R.; Vogl, J.; Tzavelis, A.; Chung, H.U.; Lee, J.Y.; Kim, D.H.; et al. Wireless, skin-interfaced devices for pediatric critical care: Application to continuous, noninvasive blood pressure monitoring. *Adv. Healthc. Mater.* **2021**, *10*, 2100383. [CrossRef]
43. Chung, H.U.; Rwei, A.Y.; Hourlier-Fargette, A.; Xu, S.; Lee, K.; Dunne, E.C.; Xie, Z.; Liu, C.; Carlini, A.; Kim, D.H.; et al. Skin-interfaced biosensors for advanced wireless physiological monitoring in neonatal and pediatric intensive-care units. *Nat. Med.* **2020**, *26*, 418–429. [CrossRef]
44. Kim, Y.-S.; Mahmood, M.; Lee, Y.; Kim, N.K.; Kwon, S.; Herbert, R.; Kim, D.; Cho, H.C.; Yeo, W.-H. All-in-one, wireless, stretchable hybrid electronics for smart, connected, and ambulatory physiological monitoring. *Adv. Sci.* **2019**, *6*, 1900939. [CrossRef] [PubMed]
45. Zulqarnain, M.; Stanzione, S.; Rathinavel, G.; Smout, S.; Willegems, M.; Myny, K.; Cantatore, E. A flexible ECG patch compatible with NFC RF communication. *npj Flex. Electron.* **2020**, *4*, 13. [CrossRef]
46. Elgendi, M.; Fletcher, R.; Liang, Y.; Howard, N.; Lovell, N.H.; Abbott, D.; Lim, K.; Ward, R. The use of photoplethysmography for assessing hypertension. *npj Digit. Med.* **2019**, *2*, 60. [CrossRef] [PubMed]
47. Tamura, T.; Maeda, Y.; Sekine, M.; Yoshida, M. Wearable photoplethysmographic sensors—Past and present. *Electronics* **2014**, *3*, 282–302. [CrossRef]
48. Harfiya, L.N.; Chang, C.-C.; Li, Y.-H. Continuous blood pressure estimation using exclusively photopletysmography by LSTM-based signal-to-signal translation. *Sensors* **2021**, *21*, 2952. [CrossRef]
49. Lee, I.; Park, N.; Lee, H.; Hwang, C.; Kim, J.H.; Park, S. Systematic review on human skin-compatible wearable photoplethysmography sensors. *Appl. Sci.* **2021**, *11*, 2313. [CrossRef]
50. Askarian, B.; Jung, K.; Chong, J.W. Monitoring of heart rate from photoplethysmographic signals using a samsung galaxy note8 in underwater environments. *Sensors* **2019**, *19*, 2846. [CrossRef]
51. Mohan, P.M.; Nisha, A.A.; Nagarajan, V.; Jothi, E.S.J. Measurement of arterial oxygen saturation (SpO_2) using PPG optical sensor. In Proceedings of the 2016 International Conference on Communication and Signal Processing (ICCSP), Melmaruvathur, India, 6–8 April 2016; pp. 1136–1140.
52. L'Her, E.; N'Guyen, Q.-T.; Pateau, V.; Bodenes, L.; Lellouche, F. Photoplethysmographic determination of the respiratory rate in acutely ill patients: Validation of a new algorithm and implementation into a biomedical device. *Ann. Intensive Care* **2019**, *9*, 11. [CrossRef]
53. Chong, J.W.; Cho, C.H.; Tabei, F.; Le-Anh, D.; Esa, N.; Mcmanus, D.D.; Chon, K.H. Motion and noise artifact-resilient atrial fibrillation detection using a smartphone. *IEEE J. Emerg. Sel. Top. Circuits Syst.* **2018**, *8*, 230–239. [CrossRef]
54. Kim, J.; Salvatore, G.A.; Araki, H.; Chiarelli, A.M.; Xie, Z.; Banks, A.; Sheng, X.; Liu, Y.; Lee, J.W.; Jang, K.-I.; et al. Battery-free, stretchable optoelectronic systems for wireless optical characterization of the skin. *Sci. Adv.* **2022**, *2*, e1600418. [CrossRef]
55. Li, H.; Xu, Y.; Li, X.; Chen, Y.; Jiang, Y.; Zhang, C.; Lu, B.; Wang, J.; Ma, Y.; Chen, Y.; et al. Epidermal inorganic optoelectronics for blood oxygen measurement. *Adv. Healthc. Mater.* **2017**, *6*, 1601013. [CrossRef] [PubMed]

56. Hong, N.; Chu, R.J.; Kang, S.S.; Ryu, G.; Han, J.-H.; Yu, K.J.; Jung, D.; Choi, W.J. Flexible GaAs photodetector arrays hetero-epitaxially grown on GaP/Si for a low-cost III-V wearable photonics platform. *Opt. Express* **2020**, *28*, 36559–36567. [CrossRef] [PubMed]
57. Yokota, T.; Zalar, P.; Kaltenbrunner, M.; Jinno, H.; Matsuhisa, N.; Kitanosako, H.; Tachibana, Y.; Yukita, W.; Koizumi, M.; Someya, T. Ultraflexible organic photonic skin. *Sci. Adv.* **2016**, *2*, e1501856. [CrossRef] [PubMed]
58. Polat, E.O.; Mercier, G.; Nikitskiy, I.; Puma, E.; Galan, T.; Gupta, S.; Montagut, M.; Piqueras, J.J.; Bouwens, M.; Durduran, T.; et al. Flexible graphene photodetectors for wearable fitness monitoring. *Sci. Adv.* **2021**, *5*, eaaw7846. [CrossRef]
59. Kim, H.; Kang, T.-H.; Ahn, J.; Han, H.; Park, S.; Kim, S.J.; Park, M.-C.; Paik, S.; Hwang, D.K.; Yi, H.; et al. Spirally wrapped Carbon nanotube microelectrodes for fiber optoelectronic devices beyond geometrical limitations toward smart wearable e-textile applications. *ACS Nano* **2020**, *14*, 17213–17223. [CrossRef] [PubMed]
60. Pribadi, E.F.; Pandey, R.K.; Chao, P.C.P. Optimizing a novel PPG sensor patch via optical simulations towards accurate heart rates. *Microsyst. Technol.* **2020**, *26*, 3409–3420. [CrossRef]
61. Khan, Y.; Han, D.; Ting, J.; Ahmed, M.; Nagisetty, R.; Arias, A.C. Organic multi-channel optoelectronic sensors for wearable health monitoring. *IEEE Access* **2019**, *7*, 128114–128124. [CrossRef]
62. La, T.-G.; Le, L.H. Flexible and wearable ultrasound device for medical applications: A review on materials, structural designs, and current challenges. *Adv. Mater. Technol.* **2021**, 2100798. [CrossRef]
63. Liu, S.; Zhang, S.; Zhang, Y.; Geng, X.; Zhang, J.; Zhang, H. A novel flexible pressure sensor array for depth information of radial artery. *Sens. Actuators A Phys.* **2018**, *272*, 92–101. [CrossRef]
64. Lee, S.-M.; Lee, T.; Kim, H.; Jo, Y.; Kim, M.-G.; Kim, S.; Bae, H.-M.; Lee, H.J. Calcium-modified silk patch as a next-generation ultrasound coupling medium. *ACS Appl. Mater. Interfaces* **2021**, *13*, 55827–55839. [CrossRef]
65. Taebi, A.; Solar, B.E.; Bomar, A.J.; Sandler, R.H.; Mansy, H.A. Recent advances in seismocardiography. *Vibration* **2019**, *2*, 64–86. [CrossRef]
66. Salernod, M.; Zanetti, J. Seismocardiography: A new technique for recording cardiac vibrations. Concept, method, and initial observations. *J. Cardiovasc. Technol. (N. Y.)* **1990**, *9*, 111–118.
67. Liu, Y.; Norton, J.J.; Qazi, R.; Zou, Z.; Ammann, K.R.; Liu, H.; Yan, L.; Tran, P.L.; Jang, K.I.; Lee, J.W.; et al. Epidermal mechano-acoustic sensing electronics for cardiovascular diagnostics and human-machine interfaces. *Sci. Adv.* **2021**, *2*, e1601185. [CrossRef]
68. Lee, K.; Ni, X.; Lee, J.Y.; Arafa, H.; Pe, D.J.; Xu, S.; Avila, R.; Irie, M.; Lee, J.H.; Easterlin, R.L.; et al. Mechano-acoustic sensing of physiological processes and body motions via a soft wireless device placed at the suprasternal notch. *Nat. Biomed. Eng.* **2020**, *4*, 148–158. [CrossRef] [PubMed]
69. Kwak, Y.H.; Kim, W.; Park, K.B.; Kim, K.; Seo, S. Flexible heartbeat sensor for wearable device. *Biosens. Bioelectron.* **2017**, *94*, 250–255. [CrossRef]
70. Lee, H.; Hong, Y.J.; Baik, S.; Hyeon, T.; Kim, D.-H. Enzyme-based glucose sensor: From invasive to wearable device. *Adv. Healthc. Mater.* **2018**, *7*, 1701150. [CrossRef] [PubMed]
71. Bandodkar, A.J.; Jia, W.; Yardımcı, C.; Wang, X.; Ramirez, J.; Wang, J. Tattoo-based noninvasive glucose monitoring: A proof-of-concept study. *Anal. Chem.* **2015**, *87*, 394–398. [CrossRef] [PubMed]
72. Sempionatto, J.R.; Lin, M.; Yin, L.; De la Paz, E.; Pei, K.; Sonsa-Ard, T.; de Loyola Silva, A.N.; Khorshed, A.A.; Zhang, F.; Tostado, N.; et al. An epidermal patch for the simultaneous monitoring of haemodynamic and metabolic biomarkers. *Nat. Biomed. Eng.* **2021**, *5*, 737–748. [CrossRef] [PubMed]
73. Hagiwara, Y.; Fujita, H.; Oh, S.L.; Tan, J.H.; Tan, R.S.; Ciaccio, E.J.; Acharya, U.R. Computer-aided diagnosis of atrial fibrillation based on ECG Signals: A review. *Inf. Sci.* **2018**, *467*, 99–114. [CrossRef]
74. Martis, R.J.; Acharya, U.R.; Adeli, H. Current methods in electrocardiogram characterization. *Comput. Biol. Med.* **2014**, *48*, 133–149. [CrossRef] [PubMed]
75. Addison, P.S. Wavelet transforms and the ECG: A review. *Physiol. Meas.* **2005**, *26*, R155–R199. [CrossRef]
76. Raghuram, M.; Madhav, K.V.; Krishna, E.H.; Komalla, N.R.; Sivani, K.; Reddy, K.A. Dual-tree complex wavelet transform for motion-artifact reduction of PPG signals. In Proceedings of the 2012 IEEE International Symposium on Medical Measurements and Applications Proceedings, Budapest, Hungary, 18–19 May 2012; pp. 1–4.
77. Gupta, V.; Mittal, M. Arrhythmia detection in ECG signal using fractional wavelet transform with principal component analysis. *J. Inst. Eng. Ser. B* **2020**, *101*, 451–461. [CrossRef]
78. Liu, S.; Shao, J.; Kong, T.; Malekian, R. ECG arrhythmia classification using high order spectrum and 2D graph fourier transform. *Appl. Sci.* **2020**, *10*, 4741. [CrossRef]
79. Shan, S.; Tang, S.; Huang, P.; Lin, Y.; Huang, W.; Lai, D.; Wu, A.A. Reliable PPG-based algorithm in atrial fibrillation detection. In Proceedings of the 2016 IEEE Biomedical Circuits and Systems Conference (BioCAS), Shanghai, China, 17–19 October 2016; pp. 340–343.
80. Kanawade, R.; Tewary, S.; Sardana, H.K. Photoplethysmography based arrhythmia detection and classification. In Proceedings of the 2019 6th International Conference on Signal Processing and Integrated Networks (SPIN), Noida, India, 7–8 March 2019; pp. 944–948.

81. Homaeinezhad, M.R.; Atyabi, S.A.; Tavakkoli, E.; Toosi, H.N.; Ghaffari, A.; Ebrahimpour, R. ECG arrhythmia recognition via a neuro-SVM–KNN hybrid classifier with virtual QRS image-based geometrical features. *Expert Syst. Appl.* **2012**, *39*, 2047–2058. [CrossRef]
82. Savalia, S.; Emamian, V. Cardiac arrhythmia classification by multi-layer perceptron and convolution neural networks. *Bioengineering* **2018**, *5*, 35. [CrossRef]
83. Singh, S.; Pandey, S.K.; Pawar, U.; Janghel, R.R. Classification of ECG arrhythmia using recurrent neural networks. *Procedia Comput. Sci.* **2018**, *132*, 1290–1297. [CrossRef]
84. Ebrahimi, Z.; Loni, M.; Daneshtalab, M.; Gharehbaghi, A. A review on deep learning methods for ECG arrhythmia classification. *Expert Syst. Appl. X* **2020**, *7*, 100033. [CrossRef]
85. Liu, M.; Kim, Y. Classification of heart diseases based on ecg signals using long short-term memory. In Proceedings of the 2018 40th Annual International Conference of the IEEE Engineering in Medicine and Biology Society (EMBC), Honolulu, HI, USA, 18–21 July 2018; pp. 2707–2710.
86. Sayantan, G.; Kien, P.T.; Kadambari, K.V. Classification of ECG beats using deep belief network and active learning. *Med. Biol. Eng. Comput.* **2018**, *56*, 1887–1898. [CrossRef]
87. Kaisti, M.; Panula, T.; Leppänen, J.; Punkkinen, R.; Jafari Tadi, M.; Vasankari, T.; Jaakkola, S.; Kiviniemi, T.; Airaksinen, J.; Kostiainen, P.; et al. Clinical assessment of a non-invasive wearable MEMS pressure sensor array for monitoring of arterial pulse waveform, heart rate and detection of atrial fibrillation. *npj Digit. Med.* **2019**, *2*, 39. [CrossRef] [PubMed]
88. Dong, L.; Zhang, B.; Xin, Y.; Liu, Y.; Xing, R.; Yang, Y.; Wang, X.; Que, W.; Liu, W.; Geng, L. Fully integrated flexible long-term electrocardiogram recording patch with gel-less adhesive electrodes for arrhythmia detection. *Sens. Actuators A Phys.* **2021**, *332*, 113063. [CrossRef]
89. Tison, G.; Sanchez, J.M.; Ballinger, B.; Singh, A.; Olgin, J.E.; Pletcher, M.J.; Vittinghoff, E.; Lee, E.S.; Fan, S.M.; Gladstone, R.A.; et al. Passive detection of atrial fibrillation using a commercially available smartwatch. *JAMA Cardiol.* **2018**, *3*, 409–416. [CrossRef]
90. Petmezas, G.; Haris, K.; Stefanopoulos, L.; Kilintzis, V.; Tzavelis, A.; Rogers, J.A.; Katsaggelos, A.K.; Maglaveras, N. Automated atrial fibrillation detection using a hybrid CNN-LSTM network on imbalanced ECG datasets. *Biomed. Signal Process. Control* **2021**, *63*, 102194. [CrossRef]
91. Kwon, S.; Hong, J.; Choi, E.-K.; Lee, E.; Hostallero, D.E.; Kang, W.J.; Lee, B.; Jeong, E.-R.; Koo, B.-K.; Oh, S.; et al. Deep learning approaches to detect atrial fibrillation using photoplethysmographic signals: Algorithms development study. *JMIR mHealth uHealth* **2019**, *7*, e12770. [CrossRef] [PubMed]
92. Dezaki, F.T.; Ginsberg, T.; Luong, C.; Vaseli, H.; Rohling, R.; Gin, K.; Abolmaesumi, P.; Tsang, T. Echo-rhythm net: Semi-supervised learning for automatic detection of atrial fibrillation in echocardiography. In Proceedings of the 2021 IEEE 18th International Symposium on Biomedical Imaging (ISBI), Nice, France, 13–16 April 2021; pp. 110–113.
93. Lee, W.; Kobayashi, S.; Nagase, M.; Jimbo, Y.; Saito, I.; Inoue, Y.; Yambe, T.; Sekino, M.; Malliaras, G.G.; Yokota, T.; et al. Nonthrombogenic, stretchable, active multielectrode array for electroanatomical mapping. *Sci. Adv.* **2022**, *4*, eaau2426. [CrossRef] [PubMed]
94. Wang, Y.; Qiu, Y.; Ameri, S.K.; Jang, H.; Dai, Z.; Huang, Y.; Lu, N. Low-cost, μm-thick, tape-free electronic tattoo sensors with minimized motion and sweat artifacts. *npj Flex. Electron.* **2018**, *2*, 6. [CrossRef]
95. Xu, X.; Luo, M.; He, P.; Yang, J. Washable and flexible screen printed graphene electrode on textiles for wearable healthcare monitoring. *J. Phys. D Appl. Phys.* **2020**, *53*, 125402. [CrossRef]
96. Kwon, Y.-T.; Kim, Y.-S.; Kwon, S.; Mahmood, M.; Lim, H.-R.; Park, S.-W.; Kang, S.-O.; Choi, J.J.; Herbert, R.; Jang, Y.C.; et al. All-printed nanomembrane wireless bioelectronics using a biocompatible solderable graphene for multimodal human-machine interfaces. *Nat. Commun.* **2020**, *11*, 3450. [CrossRef]
97. Zavanelli, N.; Kim, J.; Yeo, W.-H. Recent advances in high-throughput nanomaterial manufacturing for hybrid flexible bioelectronics. *Materials* **2021**, *14*, 2973. [CrossRef] [PubMed]

MDPI
St. Alban-Anlage 66
4052 Basel
Switzerland
Tel. +41 61 683 77 34
Fax +41 61 302 89 18
www.mdpi.com

Materials Editorial Office
E-mail: materials@mdpi.com
www.mdpi.com/journal/materials

www.ingramcontent.com/pod-product-compliance
Ingram Content Group UK Ltd.
Pitfield, Milton Keynes, MK11 3LW, UK
UKHW060121010826
14089UKWH00006B/100

9 783036 532271